Endothelial Dysfunction: From Pathophysiology to Novel Therapeutic Approaches

Endothelial Dysfunction: From Pathophysiology to Novel Therapeutic Approaches

Editor

Byeong Hwa Jeon

MDPI • Basel • Beijing • Wuhan • Barcelona • Belgrade • Manchester • Tokyo • Cluj • Tianjin

Editor
Byeong Hwa Jeon
Chungnam National University
Korea

Editorial Office
MDPI
St. Alban-Anlage 66
4052 Basel, Switzerland

This is a reprint of articles from the Special Issue published online in the open access journal *Biomedicines* (ISSN 2227-9059) (available at: https://www.mdpi.com/journal/biomedicines/special_issues/endothelial_dysfunction_pathophysiology_therapeutic).

For citation purposes, cite each article independently as indicated on the article page online and as indicated below:

LastName, A.A.; LastName, B.B.; LastName, C.C. Article Title. *Journal Name* **Year**, *Volume Number*, Page Range.

ISBN 978-3-0365-3036-9 (Hbk)
ISBN 978-3-0365-3037-6 (PDF)

Cover image courtesy of Byeong Hwa Jeon.

© 2022 by the authors. Articles in this book are Open Access and distributed under the Creative Commons Attribution (CC BY) license, which allows users to download, copy and build upon published articles, as long as the author and publisher are properly credited, which ensures maximum dissemination and a wider impact of our publications.

The book as a whole is distributed by MDPI under the terms and conditions of the Creative Commons license CC BY-NC-ND.

Contents

About the Editor ix

Preface to "Endothelial Dysfunction: From Pathophysiology to Novel Therapeutic Approaches" xi

Byeong Hwa Jeon
Endothelial Dysfunction: From Pathophysiology to Novel Therapeutic Approaches
Reprinted from: *Biomedicines* **2021**, *9*, 1571, doi:10.3390/biomedicines9111571 1

Prabhatchandra Dube, Armelle DeRiso, Mitra Patel, Dhanushya Battepati, Bella Khatib-Shahidi, Himani Sharma, Rajesh Gupta, Deepak Malhotra, Lance Dworkin, Steven Haller and David Kennedy
Vascular Calcification in Chronic Kidney Disease: Diversity in the Vessel Wall
Reprinted from: *Biomedicines* **2021**, *9*, 404, doi:10.3390/biomedicines9040404 5

Esteban Colombo, Antonio Signore, Stefano Aicardi, Angelina Zekiy, Anatoliy Utyuzh, Stefano Benedicenti and Andrea Amaroli
Experimental and Clinical Applications of Red and Near-Infrared Photobiomodulation on Endothelial Dysfunction: A Review
Reprinted from: *Biomedicines* **2021**, *9*, 274, doi:10.3390/biomedicines9030274 27

Inês V. da Silva, Courtney A. Whalen, Floyd J. Mattie, Cristina Florindo, Neil K. Huang, Sandra G. Heil, Thomas Neuberger, A. Catharine Ross, Graça Soveral and Rita Castro
An Atherogenic Diet Disturbs *Aquaporin 5* Expression in Liver and Adipocyte Tissues of Apolipoprotein E-Deficient Mice: New Insights into an Old Model of Experimental Atherosclerosis
Reprinted from: *Biomedicines* **2021**, *9*, 150, doi:10.3390/biomedicines9020150 51

Wa Du, Lu Ren, Milton H. Hamblin and Yanbo Fan
Endothelial Cell Glucose Metabolism and Angiogenesis
Reprinted from: *Biomedicines* **2021**, *9*, 147, doi:10.3390/biomedicines9020147 67

Kondababu Kurakula, Valérie F. E. D. Smolders, Olga Tura-Ceide, J. Wouter Jukema, Paul H. A. Quax and Marie-José Goumans
Endothelial Dysfunction in Pulmonary Hypertension: Cause or Consequence?
Reprinted from: *Biomedicines* **2021**, *9*, 57, doi:10.3390/biomedicines9010057 85

Laura Toma, Camelia Sorina Stancu and Anca Volumnia Sima
Endothelial Dysfunction in Diabetes Is Aggravated by Glycated Lipoproteins; Novel Molecular Therapies
Reprinted from: *Biomedicines* **2021**, *9*, 18, doi:10.3390/biomedicines9010018 109

Teresa Salvatore, Pia Clara Pafundi, Raffaele Galiero, Luca Rinaldi, Alfredo Caturano, Erica Vetrano, Concetta Aprea, Gaetana Albanese, Anna Di Martino, Carmen Ricozzi, Simona Imbriani and Ferdinando Carlo Sasso
Can Metformin Exert as an Active Drug on Endothelial Dysfunction in Diabetic Subjects?
Reprinted from: *Biomedicines* **2021**, *9*, 3, doi:10.3390/biomedicines9010003 137

Eunsik Yun, Yunjin Kook, Kyung Hyun Yoo, Keun Il Kim, Myeong-Sok Lee, Jongmin Kim and Aram Lee
Endothelial to Mesenchymal Transition in Pulmonary Vascular Diseases
Reprinted from: *Biomedicines* **2020**, *8*, 639, doi:10.3390/biomedicines8120639 163

Giovanna Casili, Marika Lanza, Sarah Adriana Scuderi, Salvatore Messina, Irene Paterniti, Michela Campolo and Emanuela Esposito
The Inhibition of Prolyl Oligopeptidase as New Target to Counteract Chronic Venous Insufficiency: Findings in a Mouse Model
Reprinted from: *Biomedicines* 2020, *8*, 604, doi:10.3390/biomedicines8120604 181

Liang-Yin Ke, Shi Hui Law, Vineet Kumar Mishra, Farzana Parveen, Hua-Chen Chan, Ye-Hsu Lu and Chih-Sheng Chu
Molecular and Cellular Mechanisms of Electronegative Lipoproteins in
Cardiovascular Diseases
Reprinted from: *Biomedicines* 2020, *8*, 550, doi:10.3390/biomedicines8120550 201

Yu Ran Lee, Hee Kyoung Joo, Eun-Ok Lee, Myoung Soo Park, Hyun Sil Cho, Sungmin Kim, Hao Jin, Jin-Ok Jeong, Cuk-Seong Kim and Byeong Hwa Jeon
Plasma APE1/Ref-1 Correlates with Atherosclerotic Inflammation in ApoE$^{-/-}$ Mice
Reprinted from: *Biomedicines* 2020, *8*, 366, doi:10.3390/biomedicines8090366 223

Jessica Gambardella, Wafiq Khondkar, Marco Bruno Morelli, Xujun Wang, Gaetano Santulli and Valentina Trimarco
Arginine and Endothelial Function
Reprinted from: *Biomedicines* 2020, *8*, 277, doi:10.3390/biomedicines8080277 239

Chih-Sheng Chu, Shi Hui Law, David Lenzen, Yong-Hong Tan, Shih-Feng Weng, Etsuro Ito, Jung-Chou Wu, Chu-Huang Chen, Hua-Chen Chan and Liang-Yin Ke
Clinical Significance of Electronegative Low-Density Lipoprotein Cholesterol
in Atherothrombosis
Reprinted from: *Biomedicines* 2020, *8*, 254, doi:10.3390/biomedicines8080254 265

Jang Mi Han, Ye Seul Choi, Dipesh Dhakal, Jae Kyung Sohng and Hye Jin Jung
Novel Nargenicin A1 Analog Inhibits Angiogenesis by Downregulating the Endothelial VEGF/ VEGFR2 Signaling and Tumoral HIF-1α/VEGF Pathway
Reprinted from: *Biomedicines* 2020, *8*, 252, doi:10.3390/biomedicines8080252 281

Maria Perticone, Raffaele Maio, Benedetto Caroleo, Angela Sciacqua, Edoardo Suraci, Simona Gigliotti, Francesco Martino, Francesco Andreozzi, Giorgio Sesti and
Francesco Perticone
Serum γ-Glutamyltransferase Concentration Predicts Endothelial Dysfunction in Naïve Hypertensive Patients
Reprinted from: *Biomedicines* 2020, *8*, 207, doi:10.3390/biomedicines8070207 293

Yusuke Takeda, Keiichiro Matoba, Kensuke Sekiguchi, Yosuke Nagai, Tamotsu Yokota, Kazunori Utsunomiya and Rimei Nishimura
Endothelial Dysfunction in Diabetes
Reprinted from: *Biomedicines* 2020, *8*, 182, doi:10.3390/biomedicines8070182 303

Wei Li, Heegeun Park, Erling Guo, Wooyeon Jo, Kyu Min Sim and Sang Ki Lee
Aerobic Exercise Training Inhibits Neointimal Formation via Reduction of PCSK9 and LOX-1 in Atherosclerosis
Reprinted from: *Biomedicines* 2020, *8*, 92, doi:10.3390/biomedicines8040092 319

Yu Ran Lee, Hee Kyoung Joo and Byeong Hwa Jeon
The Biological Role of Apurinic/Apyrimidinic Endonuclease1/Redox Factor-1 as a Therapeutic Target for Vascular Inflammation and as a Serologic Biomarker
Reprinted from: *Biomedicines* 2020, *8*, 57, doi:10.3390/biomedicines8030057 331

Ginevra Nannelli, Marina Ziche, Sandra Donnini and Lucia Morbidelli
Endothelial Aldehyde Dehydrogenase 2 as a Target to Maintain Vascular Wellness and Function in Ageing
Reprinted from: *Biomedicines* **2020**, *8*, 4, doi:10.3390/biomedicines8010004 . . . . **343**

About the Editor

Byeong Hwa Jeon is a vascular physiologist in Daejeon, Korea, and is affiliated with Chungnam National University. He received his medical degree from Chungnam National University, Korea, and completed his post-doctoral training at Johns Hopkins University, USA. He is interested in the redox regulation of vascular endothelial cells in the field of vascular pathophysiology, and is working as a director at a research institute of biomedical sciences.

Preface to "Endothelial Dysfunction: From Pathophysiology to Novel Therapeutic Approaches"

This Special Issue, "Endothelial Dysfunction: From Pathophysiology to Novel Therapeutic Approaches", focuses on the pathophysiology of endothelial dysfunction, new biomarkers for endothelial dysfunction related to cardiovascular disorders or tumors, and novel therapeutic approaches for endothelial dysfunctions. It includes 13 review articles and 6 research papers, in which several novel biomarkers or target proteins associated with endothelial dysfunction are described. We would like to thank the many scientists who participated in peer reviews as well as the Managing Editors of this Special Issue of *Biomedicines*.

Byeong Hwa Jeon
Editor

Editorial

Endothelial Dysfunction: From Pathophysiology to Novel Therapeutic Approaches

Byeong Hwa Jeon [1,2]

1. Research Institute for Medical Sciences, College of Medicine, Chungnam National University, 266 Munhwa-ro, Jung-gu, Daejeon 35015, Korea; bhjeon@cnu.ac.kr
2. Department of Physiology, College of Medicine, Chungnam National University, 266 Munhwa-ro, Jung-gu, Daejeon 35015, Korea

The vascular endothelium is an active tissue that plays a crucial role in the maintenance of vascular homeostasis. Vascular endothelial cells in adults are composed of about 10-60 trillion cells, weighing about 1 kg and having a body surface area of 1~7 m² [1]. Vascular endothelial cells are found in all tissues and are active tissues that deliver nutrients and secrete various modulators. Endothelial cells also play important roles in regulating blood flow, making them a target tissue for blood circulation [2]. The chronic exposure of risk factors, such as hypertension, hypercholesterolemia, or oxidative stress, induces endothelial dysfunctions and results in a loss of endothelial integrity, smooth muscle cell proliferation, and macrophage recruitment. In 1980, Furchgott and Zawadzki [3] discovered a substance, an endothelial cell-derived relaxing factor, and identified it as nitric oxide. This discovery was a great turning point in the development of drugs and therapeutics in the field of vascular medicine. The pathophysiology of endothelial dysfunction is complex, and multifactorial factors are involved, such as oxidative stress or chronic inflammation. The primary prevention of cardiovascular risk factors and endothelial dysfunctions, as well as the early detection or molecular imaging techniques for endothelial dysfunction, help to prevent the development of cardiovascular disorders. Novel therapeutic approaches or drug delivery systems for endothelial dysfunctions have had a promising beneficial effect at preclinical or clinical levels, by affecting the progression of atherosclerotic changes, tumor angiogenesis, and host–immune reactions near tumor environments.

This Special Issue, entitled "Endothelial Dysfunction: From Pathophysiology to Novel Therapeutic Approaches", is focused on the pathophysiology of endothelial dysfunction, new biomarkers for endothelial dysfunction related to cardiovascular disorders or tumors, and novel therapeutic approaches for endothelial dysfunctions. This Special Issue includes 13 review articles and 6 research papers, in which several novel biomarkers or target proteins associated with endothelial dysfunction are described. New concepts, such as new biomarkers, therapeutic targets, and treatment technologies for endothelial dysfunction are, included.

In this short editorial, I would like to highlight the paper that introduced a new concept related to endothelial cell dysfunction. Apurinic/apyrimidinic endonuclease 1/redox factor-1 (APE1/Ref-1) is an essential multifunctional protein. In 2013, the secretion of APE1/Ref-1 into the cultured medium in response to hyperacetylation was first reported [4]. Lee et al. described the usefulness of APE1/Ref-1, a novel biomarker for vascular inflammation, suggesting its potential as a serologic biomarker for cardiovascular disease [5]. APE1/Ref-1 expression is upregulated in aortic endothelial cells/macrophages of atherosclerotic mice, suggesting that plasma APE1/Ref-1 levels could predict atherosclerotic inflammation [6]. The concept of "electronegative LDL" was first proposed in 1979 [7]. By using fast-protein liquid chromatography, low-density lipoproteins (LDLs) can be divided into five subfractions (L1~L5). Among the LDL subfractions, the L5 LDL showed, in a novel concept, that it can be used as a clinical biomarker in chronic vascular thrombotic

Citation: Jeon, B.H. Endothelial Dysfunction: From Pathophysiology to Novel Therapeutic Approaches. *Biomedicines* **2021**, *9*, 1571. https://doi.org/10.3390/biomedicines9111571

Received: 24 October 2021
Accepted: 27 October 2021
Published: 29 October 2021

Publisher's Note: MDPI stays neutral with regard to jurisdictional claims in published maps and institutional affiliations.

Copyright: © 2021 by the author. Licensee MDPI, Basel, Switzerland. This article is an open access article distributed under the terms and conditions of the Creative Commons Attribution (CC BY) license (https://creativecommons.org/licenses/by/4.0/).

disease, including cardiometabolic disorders, acute ischemic events, and autoimmune diseases [8,9]. Chu et al. summarized that electronegative low-density lipoprotein cholesterol is a promising biomarker. A reference value of L5 LDL in serum was also presented so that this guideline for the treatment strategy could be used clinically [8]. In diabetes, vascular endothelial cell damage and endothelial cell dysfunction can be induced by changes in the activity of vascular endothelial cells and perivascular macrophages [10]. In particular, the transition from M2 (anti-inflammatory function) to M1 (inflammatory function) contributes to endothelial dysfunction and insulin resistance. Takeda et al. [11] described the mechanism of action of drugs that promote various endothelial cell functions. Sodium–glucose cotransporter 2 (SGLT2) inhibitors, glucagon-like peptide-1 (GLP-1), and dipeptidyl peptidase-4 (DDP-4) inhibitors, which inhibit M1 transition or promote the M2 macrophage, may provide good strategies to suppress endothelial dysfunction and promote the browning of white adipose tissue.

Nannelli G et al. focused on the role of the detoxifying enzyme aldehyde dehydrogenase 2 (ALDH2) in the maintenance of endothelial function [12]. ALDH2 in mitochondria is primarily involved in the detoxification of acetaldehyde. The impairment of ALDH2 is associated with oxidative stress, aging, and endothelial dysfunction [12]. The development of therapeutic target drugs that increase the expression of ALDH2 will contribute to the development of therapeutic agents for cardiovascular diseases. In diabetes, the diverse role of glycation products needs to be investigated. Hemoglobin A1c (HbA1c) is being used as a blood biomarker, showing the chronic status of diabetes. Toma et al. summarized the role of glycated lipoprotein on endothelial cell dysfunction in diabetes and its interaction with receptors for advanced glycation end products [13]. In diabetes mellitus, the appearance of advanced glycation end products (AGE) in plasma proteins is an important etiology of endothelial dysfunction. Concepts for the glycosylation of lipoprotein, including glycated LDL or glycated HDL, would be contributed to endothelial dysfunction and/or atherosclerosis [13].

There is a new technique for treating endothelial cell dysfunction. Red and near-infrared photobiomodulation is a technology that uses light of various wavelengths to inhibit inflammation, angiogenesis, and promote blood vessel function. Although such long-wavelength light treatment technology requires extensive randomized clinical trials, it has been partially used in clinical practice [14]. Regular exercise contributes to the prevention and treatment of arteriosclerosis, diabetes, and hyperlipidemia. Regular exercise protects vascular endothelial cells and inhibits neointimal formation [15]. Proprotein convertase subtilisin/Kexin type 9 (PCSK9) is a target protein that induces arteriosclerosis, and PCSK9 antibody therapy has been developed and used in clinical practice [16]. As an interesting study, regular exercise significantly inhibits PCSK9 expression in rodent animal experiments [17].

Finally, scientific efforts for endothelium dysfunction will contribute to the development of therapeutic and preventive substances that inhibit endothelial damage in cardiovascular diseases. With the worldwide spread of COVID-19, many researchers are working to develop specific inhibitors to inhibit vascular endothelial dysfunction. We would like to thank the many scientists who participated in peer reviews, as well as the managing editors of this Special Issue of *Biomedicines*.

Funding: This work was supported by grants from the Basic Science Research Program through the National Research Foundation of Korea (NRF) funded by the Ministry of Education (NRF-2014R1A6A1029617).

Conflicts of Interest: The author declares no conflict of interest.

References

1. Augustin, H.G.; Kozian, D.H.; Johnson, R.C. Differentiation of endothelial cells: Analysis of the constitutive and activated endothelial cell phenotypes. *Bioessays* **1994**, *16*, 901–906. [CrossRef] [PubMed]
2. Cines, D.B.; Pollak, E.S.; Buck, C.A.; Loscalzo, J.; Zimmerman, G.A.; McEver, R.P.; Pober, J.S.; Wick, T.M.; Konkle, B.A.; Schwartz, B.S.; et al. Endothelial cells in physiology and in the pathophysiology of vascular disorders. *Blood* **1998**, *91*, 3527–3561. [PubMed]

3. Furchgott, R.F.; Zawadzki, J.V. The obligatory role of endothelial cells in the relaxation of arterial smooth muscle by acetylcholine. *Nature* **1980**, *288*, 373–376. [CrossRef] [PubMed]
4. Choi, S.; Lee, Y.R.; Park, M.S.; Joo, H.K.; Cho, E.J.; Kim, H.S.; Kim, C.S.; Park, J.B.; Irani, K.; Jeon, B.H. Histone deacetylases inhibitor trichostatin A modulates the extracellular release of APE1/Ref-1. *Biochem. Biophys. Res. Commun.* **2013**, *435*, 403–407. [CrossRef] [PubMed]
5. Lee, Y.R.; Joo, H.K.; Jeon, B.H. The Biological Role of Apurinic/Apyrimidinic Endonuclease1/Redox Factor-1 as a Therapeutic Target for Vascular Inflammation and as a Serologic Biomarker. *Biomedicines* **2020**, *8*, 57. [CrossRef] [PubMed]
6. Lee, Y.R.; Joo, H.K.; Lee, E.-O.; Park, M.S.; Cho, H.S.; Kim, S.; Jin, H.; Jeong, J.-O.; Kim, C.-S.; Jeon, B.H. Plasma APE1/Ref-1 Correlates with Atherosclerotic Inflammation in ApoE−/− Mice. *Biomedicines* **2020**, *8*, 366. [CrossRef] [PubMed]
7. Hoff, H.F.; Bradley, W.A.; Heideman, C.L.; Gaubatz, J.W.; Karagas, M.D.; Gotto, A.M., Jr. Characterization of low density lipoprotein-like particle in the human aorta from grossly normal and atherosclerotic regions. *Biochim. Biophys. Acta* **1979**, *573*, 361–374. [CrossRef]
8. Chu, C.-S.; Law, S.H.; Lenzen, D.; Tan, Y.-H.; Weng, S.-F.; Ito, E.; Wu, J.-C.; Chen, C.-H.; Chan, H.-C.; Ke, L.-Y. Clinical Significance of Electronegative Low-Density Lipoprotein Cholesterol in Atherothrombosis. *Biomedicines* **2020**, *8*, 254. [CrossRef] [PubMed]
9. Ke, L.-Y.; Law, S.H.; Mishra, V.K.; Parveen, F.; Chan, H.-C.; Lu, Y.-H.; Chu, C.-S. Molecular and Cellular Mechanisms of Electronegative Lipoproteins in Cardiovascular Diseases. *Biomedicines* **2020**, *8*, 550. [CrossRef] [PubMed]
10. Odegaard, J.I.; Chawla, A. Pleiotropic actions of insulin resistance and inflammation in metabolic homeostasis. *Science* **2013**, *339*, 172–177. [CrossRef] [PubMed]
11. Takeda, Y.; Matoba, K.; Sekiguchi, K.; Nagai, Y.; Yokota, T.; Utsunomiya, K.; Nishimura, R. Endothelial Dysfunction in Diabetes. *Biomedicines* **2020**, *8*, 182. [CrossRef] [PubMed]
12. Nannelli, G.; Ziche, M.; Donnini, S.; Morbidelli, L. Endothelial Aldehyde Dehydrogenase 2 as a Target to Maintain Vascular Wellness and Function in Ageing. *Biomedicines* **2020**, *8*, 4. [CrossRef] [PubMed]
13. Toma, L.; Stancu, C.S.; Sima, A.V. Endothelial Dysfunction in Diabetes Is Aggravated by Glycated Lipoproteins; Novel Molecular Therapies. *Biomedicines* **2021**, *9*, 18. [CrossRef] [PubMed]
14. Colombo, E.; Signore, A.; Aicardi, S.; Zekiy, A.; Utyuzh, A.; Benedicenti, S.; Amaroli, A. Experimental and Clinical Applications of Red and Near-Infrared Photobiomodulation on Endothelial Dysfunction: A Review. *Biomedicines* **2021**, *9*, 274. [CrossRef] [PubMed]
15. Indolfi, C.; Torella, D.; Coppola, C.; Curcio, A.; Rodriguez, F.; Bilancio, A.; Leccia, A.; Arcucci, O.; Falco, M.; Leosco, D.; et al. Physical training increases eNOS vascular expression and activity and reduces restenosis after balloon angioplasty or arterial stenting in rats. *Circ. Res.* **2002**, *91*, 1190–1197. [CrossRef] [PubMed]
16. Leucker, T.M.; Gerstenblith, G.; Schar, M.; Brown, T.T.; Jones, S.R.; Afework, Y.; Weiss, R.G.; Hays, A.G. Evolocumab, a PCSK9-Monoclonal Antibody, Rapidly Reverses Coronary Artery Endothelial Dysfunction in People Living With HIV and People With Dyslipidemia. *J. Am. Heart Assoc.* **2020**, *9*, e016263. [CrossRef] [PubMed]
17. Li, W.; Park, H.; Guo, E.; Jo, W.; Sim, K.M.; Lee, S.K. Aerobic Exercise Training Inhibits Neointimal Formation via Reduction of PCSK9 and LOX-1 in Atherosclerosis. *Biomedicines* **2020**, *8*, 92. [CrossRef]

Review

Vascular Calcification in Chronic Kidney Disease: Diversity in the Vessel Wall

Prabhatchandra Dube *,†, Armelle DeRiso †, Mitra Patel, Dhanushya Battepati, Bella Khatib-Shahidi, Himani Sharma, Rajesh Gupta, Deepak Malhotra, Lance Dworkin, Steven Haller and David Kennedy *

Department of Medicine, College of Medicine and Life Sciences, University of Toledo, Health Education Building RM 205, 3000 Arlington Ave, Toledo, OH 43614, USA; armelle.deriso@rockets.utoledo.edu (A.D.); mitra.patel@utoledo.edu (M.P.); dhanushya.battepati@rockets.utoledo.edu (D.B.); bella.khatibshahidi@rockets.utoledo.edu (B.K.-S.); himani.sharma@rockets.utoledo.edu (H.S.); rajesh.gupta@utoledo.edu (R.G.); deepak.malhotra@utoledo.edu (D.M.); lance.dworkin@utoledo.edu (L.D.); steven.haller@utoledo.edu (S.H.)
* Correspondence: prabhatchandra.dube@utoledo.edu (P.D.); david.kennedy@utoledo.edu (D.K.)
† These authors contributed equally.

Abstract: Vascular calcification (VC) is one of the major causes of cardiovascular morbidity and mortality in patients with chronic kidney disease (CKD). VC is a complex process expressing similarity to bone metabolism in onset and progression. VC in CKD is promoted by various factors not limited to hyperphosphatemia, Ca/Pi imbalance, uremic toxins, chronic inflammation, oxidative stress, and activation of multiple signaling pathways in different cell types, including vascular smooth muscle cells (VSMCs), macrophages, and endothelial cells. In the current review, we provide an in-depth analysis of the various kinds of VC, the clinical significance and available therapies, significant contributions from multiple cell types, and the associated cellular and molecular mechanisms for the VC process in the setting of CKD. Thus, we seek to highlight the key factors and cell types driving the pathology of VC in CKD in order to assist in the identification of preventative, diagnostic, and therapeutic strategies for patients burdened with this disease.

Keywords: vascular calcification; chronic kidney disease; CKD; uremic toxins; hyperphosphatemia; vascular smooth muscle cells; VSMCs; macrophages; endothelium

1. Introduction

Vascular calcification (VC) is a convoluted process that leads to pathological accumulation of calcium phosphate crystals in the intima and media layers of the vessel wall that worsens the course of atherosclerosis, diabetes, and chronic kidney disease (CKD) [1]. These mineral-enriched plaques induce arterial stiffening, putting patients at risk for fibrosis, inflammation, and oxidative stress on a cellular level. From a clinical perspective, VC is a major problem and is associated with worse outcomes in treatment of coronary artery disease (CAD) and peripheral artery disease (PAD). Coronary calcification is associated with worse outcomes after percutaneous coronary intervention (PCI), and VC is associated with higher risk of amputation after revascularization for lower extremity PAD [2,3]. These associations of CKD can directly increase risks of many clinical complications of VC such as worsening atherosclerosis and increased risk of vascular events such as myocardial infarction, stroke, and vascular occlusive events, which makes it a significant predictor of cardiovascular disease (CVD) and CKD [4–8]. We must give our attention to these patients in the clinical setting to detect early signs of VC in CKD and prepare for prevention and treatment.

VC is now considered as an active and finely regulated process similar to osteogenesis that involves cell-mediated processes and complex interaction between the inhibitor and promoter factors of the calcification process [9,10]. Table 1 lists some major promoter and inhibitor factors involved in VC.

Table 1. Major promoter and inhibitor factors involved in vascular calcification.

Promoters	Inhibitors
Bone morphogenetic protein (BMP) 2, 4, and 6	Matrix Gla protein (MGP)
Osteocalcin	Osteopontin
Alkaline phosphatase	Osteoprotegerin
(Sex determining region Y)-box 9 (SOX9)	Vitamin K
Osterix	Magnesium
Matrix metalloproteinase (MMP) 2, 3, and 7	Bone morphogenetic protein 7 (BMP7)
Runt-related transcription factor 2 (Runx2)	Fetuin
Calcium	Klotho
Phosphate	Parathyroid hormone (PTH)
Glucose	Pyrophosphate
Advanced glycation end products	Carbonic anhydrase
Oxidized low-density lipoproteins	Collagen IV
Collagen I	Inorganic pyrophosphate (PPi)
Receptor activator of nuclear factor-kB ligand (RANKL)	

Moderate to severe calcification manifests in the aorta, cardiac valves, and peripheral vessels, including the tunica intima and tunica medial layers. Intimal calcification is linked to atherosclerosis, while both intimal and medial calcification has been observed in CKD patients [8,11,12].

The pathogenesis of this complication involves an excess build-up of calcium deposits by active and passive means, a surplus in the excretion of osteoid matrix when cells are triggered by toxic stimuli, or an integrated pathway of both processes [13]. These pathways may be attributed to underlying mechanisms at the cellular level, such as iron, calcium, and phosphate metabolism dysfunction. Multiple key processes that attack metabolism regulation include hyperphosphatemia, calcium phosphate imbalances, and a surge in reactive oxidation species (ROS) due to iron misdistribution [13]. Environmental stimuli can trigger these imbalances, transforming vascular smooth muscle cells (VSMCs) into osteoblast-like cells. This results in the build-up of hydroxyapatite in the various layers of the major and minor vessels, inducing calcification in vascular cells [13,14].

Other mechanisms contributing to VC are related to high glucose levels, lipids, and low-density lipoproteins circulating in the endothelial lining [15]. Research efforts demonstrate how a multitude of these combined actions link VC to CKD, a pathophysiological process that develops in response to irregular environmental stimuli [13].

Compared to the general population, patients with CKD are at an alarmingly higher risk of developing cardiovascular morbidity and mortality. About 50% of deaths from CKD are linked to CVD. Patients with chronic renal disease should be paid attention to for markers signaling cardiovascular complications [12,16]. Several studies show that cardiovascular calcifications in patients with renal disease are found to be more progressive and severe compared to non-CKD patients [6,13,17]. The pathogenesis of VC is enhanced in CKD through complex pathways. Modifications in iron, calcium, and phosphate levels due to kidney injury disturb the biochemical equilibrium, affecting bone remodeling in vascular cells [16]. Figure 1 depicts the pathogenesis of VC in CKD and its interconnection with altered bone and mineral homeostasis.

In the presence of risk factors such as VC, CKD patients are more likely to develop pulmonary hypertension (PH), an overlapping complication in patients with renal disease [18]. Additionally, the damaging effects of oxidative stress exacerbate VC in patients with CKD. Antioxidant defenses and free radical generation are balanced under homeostasis. When balance is disturbed, oxidative stress becomes a trigger of the cessation of cellular division. This becomes a marker of chronic and progressive diseases, including CKD and VC [16,19]. The prominence of VC in CKD can be attributed mainly to the combination of CKD traditional risk factors and the underlying uremic-specific mechanisms that induce the cardiovascular condition.

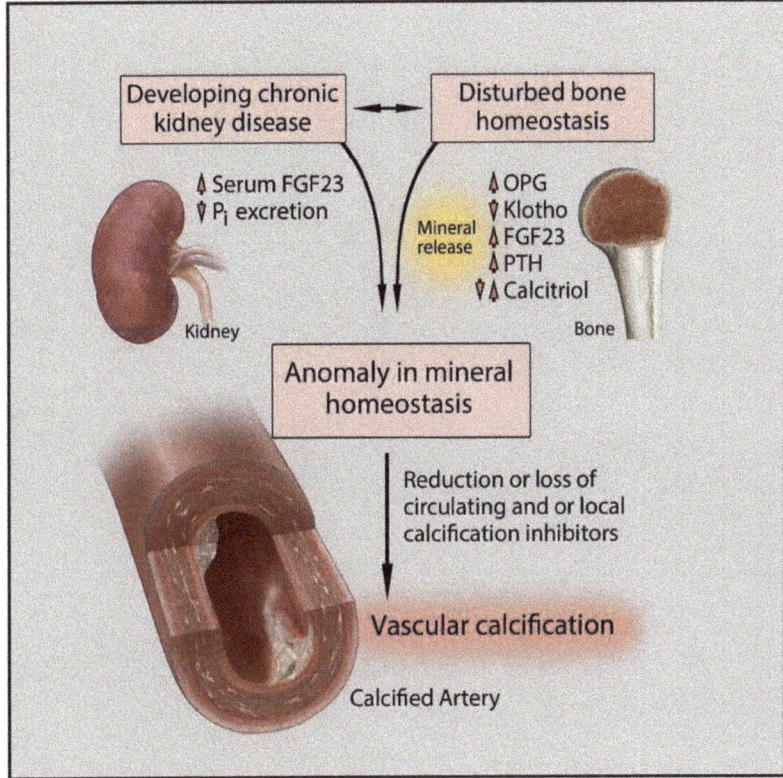

Figure 1. Pathogenesis of vascular calcification in chronic kidney disease (CKD). Altered bone metabolism and mineral homeostasis are commonly found and closely interconnected in CKD patients. The decline in kidney function also leads to elevated serum FGF23 levels and reduced inorganic phosphate excretion. The resulting pathological state is reflected by various altered biomarkers such as OPG, Klotho, PTH, and calcitriol. The disturbed mineral homeostasis also leads to altered serum and tissue levels of Ca, Pi, and Mg causing inflammation and other metabolic disorders. This leads to reduced or complete loss of circulating and/or local calcification inhibitors such as fetuin A, PPi, and MGP, causing vascular calcification. Ca, calcium; FGF23, fibroblast growth factor 23; Mg, magnesium; MGP, matrix Gla protein; OPG, osteoprotegerin; Pi, inorganic phosphate; PPi, inorganic pyrophosphate; PTH, parathyroid hormone. Upper arrow, increase; lower arrow, decrease.

In this review, we discuss the major cell types involved in the VC process, the major cellular and molecular mechanisms involved in the disease, and the important diagnostic and therapeutic interventions that are used clinically to treat patients with VC.

2. Types and Anatomical Presence of VC

VC pathogenesis shares similar mechanisms to that of bone formation. Both processes involve calcium deposition driven by bone matrix proteins, transcription factors, and stenosis. The two major classifications of VC are intimal and medial calcification. Figure 2 is a schematic representation of intimal and medial calcification and their associated pathologies.

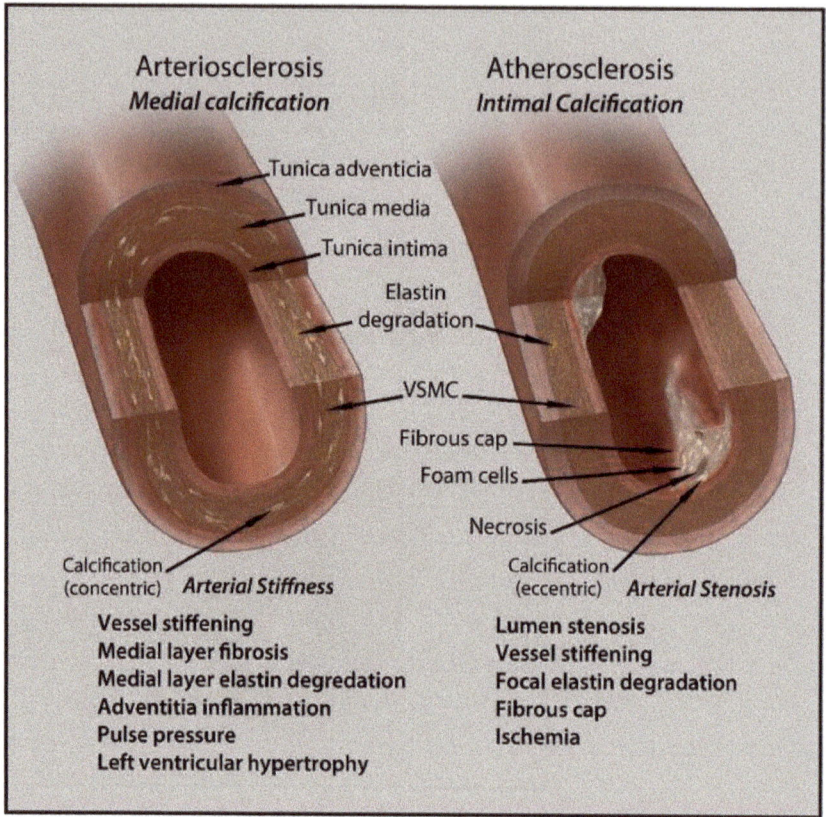

Figure 2. Schematic representation of intimal and medial calcification and their related pathologies.

A less common type of calcification involving calcium accumulation in arterioles is termed calciphylaxis. Intimal calcification shares a significant association with atherosclerosis, chronic inflammation, and the transformation of vascular smooth muscle cells (VSMCs) into osteoblast-like cells [8,20]. Medial calcification is more closely linked to elastin degradation, extracellular matrix remodeling events, and hyperphosphatemia. Diseases associated with this type include CKD, hypertension, and type 2 diabetes mellitus [20–22]. There is little known about the true cause of calciphylaxis, though the rare multifactorial syndrome has been found to have a close relationship with end-stage renal disease [23].

2.1. Intimal Calcification

Intimal calcification develops on the inner layer of vessel walls. This type shares a close link with disturbed lipid metabolism, making it a major indicator of atherosclerosis. Atherosclerosis results from excess lipoprotein depositions in the vessels sub-endothelial walls, resulting in the eccentric obstruction of lumen formation and matrix remodeling activities [20]. Calcification in atherosclerosis is driven by the osteochondrogenic process, which is initially induced by overexpression or inhibition of specific bone-related regulatory factors [7]. Dysfunction in the molecular signaling pathway during regulatory processes contributes to intimal calcification. A core element of the pathogenesis of this disease is inflammation. The osteogenic differentiation of VSMCs is promoted by the stimulus of atherogenic factors, including infiltration of inflammatory cells such as macrophages and recruitment of oxidized low-density lipoproteins and inflammatory oxylipids. These inflammatory factors are abundantly present in calcified vessels and therefore enhance the osteogenic process of VSMCs to osteoblast-like cells [20,21]. The severity and progression of VC can be attributed to the overactivity of inflammatory cytokines at the disease site. Chronic inflammation is linked to lipid accumulation at calcification sites, providing a clear association between the pathologies [7].

Also expressed in atherosclerotic plaques are bone matrix regulatory proteins and transcription factors such as bone morphogenetic protein (BMP) 2, osteopontin (OPN), osteocalcin (OC), and runt-related transcription factor 2 (RunX2). These regulatory factors function in the nucleation of hydroxyapatite minerals, found in plaques as calcified crystals [11,20]. BMP2 activity is strongly linked to VC development, as the protein is involved in the trans-differentiation of VSMCs into osteoblast-like cells, production of ROS, and inflammation. Elevated levels of dephosphorylated OPN have excitatory effects on VC as it can no longer inhibit calcium mineralization. When upregulated, master transcription factor RunX2 induces the transition of VSMCs into osteoblast-like cells during the osteogenic process. The mineralization of matrix vesicles further enhances osteogenic differentiation due to the metaplastic activity of calcifying vascular cells (CVCs) and VSMCs into chondral sites [11,20,21].

Intimal calcification in the aortic valve serves the highest mortality when in conjunction with end-stage renal failure. Atherosclerosis in the setting of renal insufficiency exacerbates the condition in a vicious cycle of calcific plaque development from excess calcium and phosphate precipitates [5]. Other conditions associated with the development of atherosclerotic intimal calcification include hyperlipidemia, hypertension, and osteoporosis [7,21].

2.2. Medial Calcification

The more extensive form of calcification in patients is medial artery calcification. It is significantly associated with cardiovascular complications such as diabetic arteriosclerosis and accounts for significant cardiovascular mortality [7]. Medial calcific sclerosis is a concentric process that develops in the medial layer of vascular walls and is more severe in type 2 diabetes mellitus. Several underlying mechanisms contribute to medial calcification, many of which are similar to the development of intimal VC. An isolated feature of medial calcification involves the degradation of the elastin-rich matrix [20]. Elastin, an extracellular matrix component, is the most abundant protein found in aortic walls. This protein serves the mechanical and elastic integrity of vessel walls, as they are subject to stress from high arterial blood pressure [22]. The formation of calcified vascular crystals obstructs the elastin fibers of arteries and vessels. Loss of this major extracellular constituent incites atherosclerosis, plaque rupture, and osteogenic differentiation [20,22]. The obstruction and destruction of elastin fibers predispose stenosis, ventricular hypertrophy, Marfan syndrome, and other progressive and fatal cardiovascular conditions [20,21].

Like intimal calcification, medial calcification is related to bone-related regulatory proteins and transcription factors, including BMP2, bone alkaline phosphatase (ALP), RunX2, and Msh Homeobox 2 (MSX2). BMP2 activity induces the transformation of smooth muscle cells in the medial layer to osteoblast-like cells. In arterial walls, upregulations in ALP matrix vesicles contribute to elastin degradation [20,21]. Activation of osteogenic transcription factors RunX2, MSX2, and ALP suppresses the activity of calcification inhibitors and promotes extracellular matrix remodeling and release of extracellular vesicles. With reduced levels of calcification inhibition, calcium and phosphate minerals accumulate, accelerating calcific plaque build-up and hyperphosphatemia [11,20]. Exceedingly high levels of circulating inorganic phosphate (Pi) are found to induce apoptosis in VSMCs. Calcium phosphate mineralization is then induced at the sites of smooth muscle cell apoptotic bodies, driving medial artery calcification [20].

Research findings suggest that the molecular mechanism behind hyperphosphatemia contributes to renal insufficiency. In one study using rat models with chronic renal failure, osteogenic factors and cartilage matrix were present in the aorta media at the site of calcified plaques [24]. The calcification process mirrors endochondral bone development, linking medial calcification to renal insufficiency disorders such as CKD [7,25].

2.3. Calciphylaxis

The rarest and most fatal form of VC is calciphylaxis or calcific uremic arteriolopathy. This progressive syndrome leads to extensive tissue and vessel necrosis, thrombosis, and sepsis. Calciphylaxis is found mostly in dialysis patients with end-stage renal disease by means of infection, prompting the high mortality rates [24,26]. Though this rare syndrome's precise etiology is unknown, the condition has been linked to medial calcification, intimal hyperplasia, and other defects involving the calcium phosphate metabolic pathways [20]. Traditional risk factors and comorbidities also play a key role in the pathogenesis of this multifactorial disease. These include smoking, female sex, use of vitamin K antagonists, and overdose on warfarin agents. Comorbidities associated with cardiovascular and renal complications from calcification have been linked to calciphylaxis, such as diabetes mellitus, obesity, hyper- and hypoparathyroidism, hyperphosphatemia, and hypertension, chronic inflammation, and CKD [20,24,26].

Overall, the two major and prevailing calcification types share diverse similarities. Both intimal and medial calcification are involved in the osteochondrogenic process, beginning with calcium deposition at the site of vascular smooth muscle cells [20]. Vessel stiffening is also a common feature of both categories of calcification, whether induced by atherosclerotic intimal plaque development or medial crystal mineralization. The molecular signaling pathways shared by both involve regulatory proteins and transcription factors such as BMP2 and RunX2, each of which drives the osteogenic process [20,21].

On the other hand, calciphylaxis is less commonly found and not as well understood. Diseases such as diabetes, obesity and end-stage renal disease exacerbate it, bringing forth commonalities in its pathogenesis with intimal and medial calcification [26].

3. Major Cell Types of Vascular Calcification

3.1. Vascular Smooth Muscle Cells (VSMCs)

As we already know, VC is an actively driven process that is regulated by VSMCs. VSMCs have a contractile functionality responding to multiple signals such as acetylcholine and norepinephrine to express their phenotype. However, unlike other muscle cells, VSMCs do not terminally differentiate and exhibit phenotypic plasticity [27]. This allows them to vary their phenotypic expression based on environmental contexts, such as the various insults found in CKD. Initially, it was thought that once the initial insult was resolved, the VSMCs returned to their original contractile phenotype; however, recent findings suggest that they can continue to express a spectrum of phenotypes [28]. These phenotypes include osteochondrogenic, adipocytic, and foam cell deposition.

In CKD specifically, there are multiple active triggers that activate the calcification of primarily the medial layer of the vasculature, including inflammatory cytokines, uremic toxins, hypercalcemia, and hyperphosphatemia [29,30]. CKD primarily stimulates an osteochondrogenic phenotypic change in VSMCs. CKD is a chronic inflammatory state, and the accumulation of reactive oxygen species and inflammatory cytokines in the vasculature can precipitate RUNX2 and BMP-2 expression, which are known to stimulate osteocytic VSMC expression in the intima and media leading to VC [31–33]. They are normally balanced out by anti-calcification markers such as MGP, a BMP-2 inhibitor expressed in VSMCs, which are specifically known to be inhibited in CKD. Loss of these anti-calcification inhibitors in CKD has been shown to increase calcification in the vascular media [33,34]. Dysregulation of mineral homeostasis and hyperphosphatemia are considered the leading determinants of VC in CKD [35,36]. Calcium and phosphate both stimulate an osteochondrogenic phenotypic change in VSMCs individually and synergistically. Calcium is a key signaling mediator for VSMCs at physiologic levels [37]. Though the exact mechanism of the effect of hypercalcemia on VSMCs is unknown, calcium induces oxidative stress and can directly deposit in the vasculature. Extracellular calcium deposition can also decrease anti-calcification markers such as MGP, leading to increased calcium deposition. VSMCs express a calcium receptor (CaR), which regulates vascular tone. In vitro, calcified tissue had downregulated expression of this receptor, and complete removal in vitro from VSMCs significantly increased VC [35,37,38]. Mineral cellular phosphate transport is mediated by sodium-dependent phosphate cotransporters PiT-1 and PiT-2 in VSMCs [39]. Knockdown of PiT-1 has also been shown to decrease osteogenic marker expression such as RUNX2 in VSMCs and decrease calcification.

VSMCs not only have an osteochondrogenic phenotypic expression but can differentiate into foam cell-like macrophages, which deposit primarily in the vascular intima. VSMCs were found to be a contributing factor in many atheroma formations in the coronary arteries accounting for high levels of CD48 cells [40]. The differentiation of this phenotypic expression is understudied in CKD patients. However, the suspected trigger is increased oxidative stress as oxidation by reactive oxygen species is an early-stage contributor to atheroma formation [40].

VC in CKD is associated with increased deposition of VSCM-associated extracellular vesicles, which include matrix vesicles and apoptotic bodies [41–43]. Normally, contractile VSMCs release matrix vesicles to maintain homeostasis; however, in pathologic environments such as CKD (hyperphosphatemia, oxidative stress, inflammation), they can transform into a synthetic phenotype and increase secretion of matrix vesicles [44]. These transform target cells into calcified states, which then aggregate to form microcalcifications. They also decrease the expression of calcification inhibitors such as MGP and fetuin-A, like the osteogenic phenotypes [45].

Along with phenotypic changes, VSMC apoptosis and autophagy play a significant role in VC in CKD. Due to increased oxidative and uremic stress in CKD, there is a significant increase in VSMC apoptosis. Apoptotic bodies have significantly high calcium concentrations. Upon apoptosis or necrosis of VSMCs, these vesicles release calcium and DNA, which ultimately deposit on the ECM of the vascular media leading to extensive calcification. Extracellular DNA has been shown to precipitate calcium and phosphate, which may contribute to arterial calcification. This is especially prominent in CKD or ESRD patients undergoing hemodialysis, as it can lead to increased VSMC apoptosis/necrosis through direct membrane contact and complement activation [46]. In normal physiologic conditions, low basal rates of autophagy are required for the removal of unwanted metabolites from the circulation [47]. The increased mineral levels and stress in CKD upregulate these rates and degrade the proteins necessary to maintain the contractile phenotype. This leads to the differentiation of phenotype along with direct calcific deposition in the vasculature [48]. A summary of the major mechanisms involved in VC in CKD is presented in Figure 3, with VSMC-associated osteochondrogenesis playing a central role in the process.

3.2. Macrophages

Monocytes and monocyte-derived macrophages are the crux of our innate immune system. They perform numerous tasks, including host defense, immune regulation, and tissue repair/regeneration [49]. Macrophages secrete many inflammatory substances such as TNF-a, IL-6, and IL-1b. These factors yield several processes that contribute to VC. Firstly, the factors lead to the differentiation of vascular wall cells into chondrocytes and osteoblasts. They can also stimulate the expression of bone morphogenetic protein and increase oxygen free radical production [50]. Calcium burden within the coronary artery is also positively correlated with IL-1b.

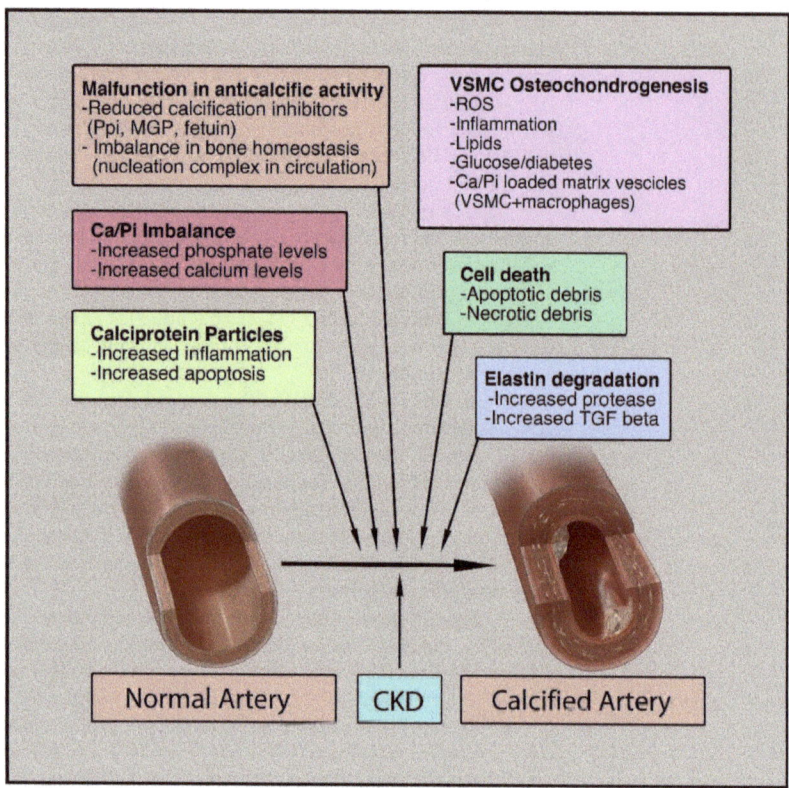

Figure 3. Major mechanisms involved in vascular calcification: Vascular calcification is an active process with multifactorial origins. Significant contributors to this include failure in the anti-calcification process, imbalance in bone homeostasis, osteochondrogenesis of VSMCs, calciprotein particles, cell death, elastin degradation, and extracellular vesicle formation and release. The combination of these factors in the setting of CKD leads to the development of both intimal and medial calcification. VSMC, vascular smooth muscle cell; Ca, calcium; Pi, inorganic phosphate; Ppi, inorganic pyrophosphate; MGP, matrix Gla protein; TGF, transforming growth factor; ROS, reactive oxygen species.

There are different subsets of macrophages that have unique roles in the process of calcification. The two most common subsets that are of interest to investigators are the M1, or "classical activated macrophage", and M2, or "alternative activated macrophage". The M1 type is activated by bacterial lipopolysaccharides and interferon-γ (IFN-γ). These macrophages typically produce reactive oxygen species (ROS), pro-inflammatory cytokines, and reactive nitrogen groups [51]. In certain conditions, and stimuli in various microenvironments, M1 and M2 macrophages display considerable plasticity and can trasnform between one phenotype and another [52]. M1 macrophages can directly release oncostatin M (OSM), which is a cytokine that belongs to the interleukin 6 cytokine group and stimulates vascular smooth muscle cells to take on an osteoblastic phenotype via the JAK-3-STAT3 pathway [53]. This subset of macrophages is also responsible for maintaining a persistent state of chronic inflammation that can interfere with the normal mechanism of VSMCs to differentiating into osteoblasts, leading to interspersed areas of fragmented calcification [54]. This chronic state of inflammation maintained by the M1 macrophages is associated with high levels of TNF-α and IL-6, suggesting that a causal link between macrophage-mediated inflammation and cardiovascular calcification exists in the setting of CKD [55]. Cartilage oligomeric matrix protein (COMP) is an interesting regulator as it facilitates macrophage polarity via integrin β3, which is a COMP-binding protein present on the surface of macrophages. It is known that COMP deficiency results in macrophages transforming into the M1 phenotype (osteogenic phenotype) and at the same time inhibiting macrophages from exhibiting the M2 phenotype, i.e., osteoclast-like cells [56].

The M2 phenotype is activated by Th2 cytokines, and these macrophages have an anti-inflammatory effect. They are involved in parasitic infection resistance, lipid metabolism, allergic reactions, and tumor progression [57]. Ricardo et al. studied how M2 macrophages are able to attenuate VSMCs from differentiating into osteogenic cells via the use of co-cultures of macrophages and smooth muscle cells where they are not in direct contact with each other [58]. Ricardo et al. also demonstrated that the inhibitory effect of M2 is related to increased secretion of adenosine triphosphate (ATP) secretion and the synthesis of pyrophosphoric acid (PPi) via fatty acid β-oxidation [58]. Another study illustrates that extracellular PPi functions as an endogenous inhibitor of vascular calcification both in vivo and in vitro [59]. PPi is generated from extracellular ATP by ectonucleotide pyrophosphatase/phosphodiesterase 1 (ENPP1), and a study by Johnson et al. shows that deleting ENPPI yields aortic calcification in mice [60].

Other functional substances secreted by macrophages include osteogenic genes such as tissue nonspecific alkaline phosphatase (TNAP), osteoprotegerin (OPN), and runt-related transcription factor 2 (Runx2), which further promote the osteogenic process [61]. Additionally, a recent study by Dube et al. showcased that the M1 subtype of macrophages differentiated from mice bone marrow-derived macrophage (BMDM) exhibits osteogenic properties via the constitutive activation of BMP-2 signaling [62]. Dube and colleagues have also discovered that deleting transient potential classical receptor 3 (TRPC3), a nonselective calcium channel in macrophages, leads to a reduction in apoptosis of macrophages triggered by endoplasmic reticulum stress and downregulates the expression of Runx2 and BMP2, leading to a reduction in calcification in advanced atherosclerotic plaques [63–65].

The formation of Ca/P nanocrystals in the vessels as a result of CKD stimulates macrophages to secrete pro-inflammatory cytokines, which in turn exacerbates VC [66]. In addition to the secretion of pro-inflammatory markers, high concentrations of Ca/P also trigger macrophages to release matrix vesicles (MVs). The MVs that are released have a proteomic profile similar to the MVs released by bone osteoblasts [67]. In the early stages of the development of VC, macrophages release an abundance of calcifying matrix vesicles (MVs), which contain the phosphatidylserine-annexin V-S100A9 complex [68]. Sophie et al. further studied this and confirmed that phosphatidylserine forms complexes with annexin V and S100A9 on the surface of macrophage-derived MV membranes. This complex then allows the entrance of calcium ions into the vesicles. The calcium and phosphate ions that enter the MVs via phosphate channels form calcium phosphorus complexes, which accumulate, forming the initial hydroxyapatite crystals [61]. These hydroxyapatite crystals continue to grow, rupturing the membrane, and keep growing to form calcified nodules. Studies have shown that the pro-inflammatory mediators mentioned earlier, such as IL-6, TNF-a, IL-1b, and oncostatin M, are not only involved in the osteochondrogenic transition of vascular/valvular cells but also the release of calcifying MVs [69].

The accumulation of uremic toxins in the circulation of CKD patients is the main driver of VC through the recruitment of monocytes and modulation of cells' inflammatory capabilities [70]. The high levels of uremic toxins lead to the development of atherosclerotic calcification and Monckeberg's medial calcification. There are two notable uremic toxins: indoxyl sulfate (IS) and paracresyl sulfate (pCS). IS and pCS are correlated with the presence of inflammatory markers such as TNF-a, IL-6, and IL-1B in CKD patients [71,72].

Experiments by Six et al. determined a high expression of adhesion molecules such as VCAM-1 and ICAM-1 when CKD mice were fed a high-phosphate diet [73]. This study essentially suggests when the expression of adhesion molecules by endothelial cells and monocytes is promoted, exposure to uremic toxins such as phosphate and IS leads to monocyte extravasation into cardiovascular tissues and thus inflammation-induced VC [70]. Evidence from several experimental studies suggest phosphate (Pi)/IS-induced monocyte recruitment, inflammation, lipid accumulation, and fibrosis are the major drivers of atherosclerotic plaque development and calcific aortic valve disease in CKD patients [70]. On the other hand, IS can also stimulate monocytes and macrophages to differentiate into the M2 phenotype [74]. The M2 phenotype leads to the expression of markers such as IL-10 and TGF-B, which yield an increase in profibrotic inflammatory macrophages [70]. The M2 macrophages demonstrate anti-calcific properties, which suggests that there is a compensatory mechanism. This mechanism protects tissues from developing pathologic calcifications associated with elevated serum phosphate levels [74].

pCS is another uremic toxin that is produced by intestinal bacteria. Various studies suggest that there is an association between increased pCS levels and renal function deterioration, atherosclerosis, and inflammation [70]. Experiments by Jing and colleagues demonstrate that when endothelial cells and macrophages are cultured with pCS in vitro, there is an increase in expression of inflammatory factors such as TNF-a and MCP-1, adhesion molecules [75]. Their studies show that the high levels of pCS seen in CKD patients could potentially be responsible for medial and intimal calcification [70]. The production of uric acid (UA) comes from the metabolism of nucleotides and ATP. It is the end product of purine metabolism [70,76]. It is seen that in patients undergoing hemodialysis, intimal and medial calcification are independently associated with blood uric acid levels. Geraci et al. suggest that in non-CKD patients with asymptomatic hyperuricemia, elevated serum UA levels are associated with carotid-intima media thickness [77]. The high UA levels are also associated with coronary artery calcification [78–80]. Nevertheless, the exact mechanism of how hyperuricemia leads to VC is unclear.

In supporting the findings of UA's effects on macrophages, one study outlined the effects of allopurinol in reducing vascular inflammation. Allopurinol is a xanthine oxidase inhibitor used in treating gout and uric acid kidney stones. Xanthine oxidase is the enzyme responsible for the production of uric acid during purine metabolism. A study in mice showed that administering allopurinol led to reduced arterial stiffness, macrophage accumulation, decrease in vascular oxidative stress, and macrophage polarization to M1 [81]. This study outlines the integral role that UA plays in vascular inflammation, even though the exact causal link is yet to be determined. One key finding from Andrews and colleagues shows that UA may have an even more significant role in VC progression as the administration of allopurinol among stage 3 CKD patients did not impact the carotid-intima media thickness [82]. Among the many roles monocytes and macrophages are involved in, the complex array of phenotypes these cells exhibit is remarkable, especially in the development of VC in the context of CKD. These macrophages secrete many pro-inflammatory markers as outlined, differentiate into osteoclast-like cells, and also demonstrate protective mechanisms against inflammation and calcification. From the supporting studies and experiments, it is evident that the intermediate monocyte subtype plays a key role in the phenomenon of chronic inflammation leading to VC in patients with CKD [83]. Regardless of the many data and experiments that suggest the role of uremic toxins in promoting monocyte/macrophages to exhibit procalcific properties, the exact mechanism by which this event occurs is still unknown. Understanding the role of macrophages in the development of VC in CKD patients is critical. Given the diverse properties these cells exhibit in promoting this chronic disease, it is critical to understand the mechanisms whereby macrophages contribute to the calcification process for future studies regarding drug targets and treatments.

3.3. Endothelial Cells

The development of vascular pathogenesis is largely facilitated by the stability of the endothelium. Lack of integrity in this structure is a driving force of many vascular diseases, including VC [7,84]. Studies show that endothelial cells secrete soluble factors that play a key role in the calcification of the VSMCs [85–87]. The endothelial progenitor cells like macrophages are also known to be involved in the VC in CKD by expressing osteogenic factor osteocalcin [88]. Figure 4 highlights a central role of cytokines and osteogenic factors produced by both endothelial cells and macrophages that leads to calcification in VSMCs.

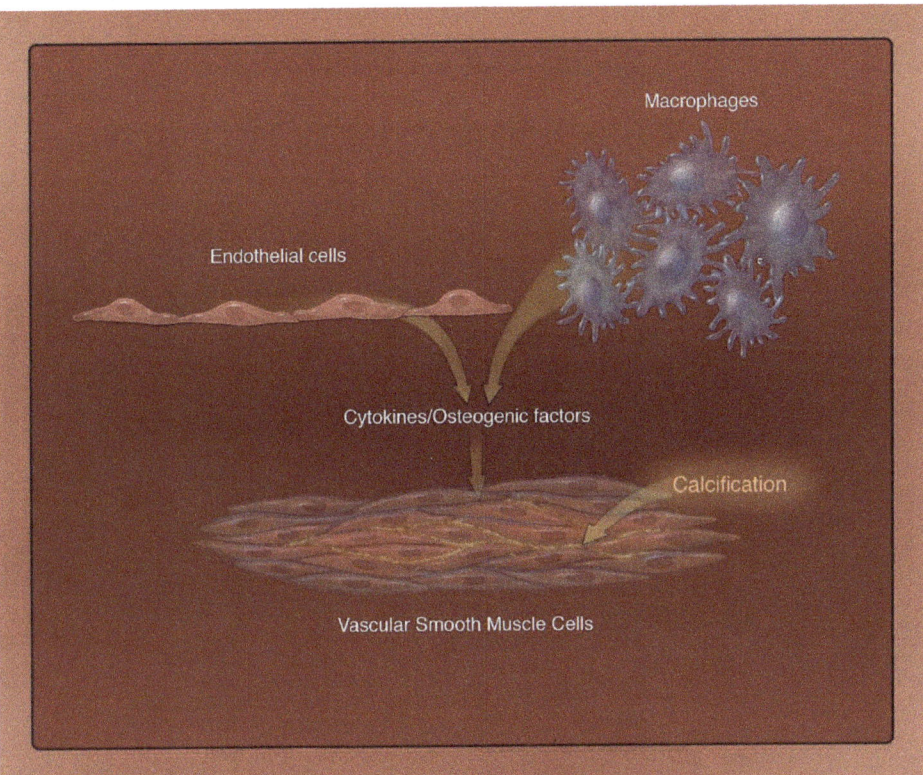

Figure 4. The cellular trifecta of vascular calcification: Endothelial cells and macrophages can secrete various cytokines and osteogenic factors that lead to the activation and progression of calcification in vascular smooth muscle cells.

Endothelial cells also have the capacity to transform into mesenchymal stem cells to attain multipotency before they can differentiate into various cell types. This transformation is known as endothelial–mesenchymal transition (EndMT) [89]. EndMT is a vital mechanism for endothelial cells to undergo osteo/chondrogenesis and secrete factors involved in VC [87,90,91]. Studies by Min Wu et al. show that the use of calcimimetic suppresses the upregulation of mesenchymal markers and the downregulation of the endothelial marker, leading to reduced VC in the aortic samples of uremic rats [92].

Shear and mechanical stress are other factors that disrupt the endothelium and lead to worsening of VC. Vascular wall mechanical stress may contribute to endothelial and interstitial cell proliferation, altering the expression of calcification-related genes in cultured endothelial cells and fibroblasts [93]. Mechanical stress triggers the differentiation of preosteoblasts into mature osteoblasts and can drive the progenitors down the osteogenic lineage [94]. Mechanical stretch can upregulate the pro-osteogenic factors BMP-2 and Sprouty-1 [93,95], and both of these factors modulate the basal expression of osteogenic factors in untreated vascular fibroblasts. Additionally, the application of steady tension to fibroblasts results in downregulation of anti-calcification factors periostin and osteopontin [96,97]. Furthermore, mechanical stress leads to elevation of osteogenic genes thrombospondin and BMP-2 as well as other calcification-related genes [92,97].

Vascular endothelial cells are exposed to shear stress induced from patterns of blood flow [92]. Pulsatile endothelial shear stress can be very pronounced depending on the location and shape of the vessel [98]. This contributes to very turbulent blood flow and shear stress, where bifurcations in arterial tress are the most critical areas for calcification [99,100]. However, disrupted blood flow, with irregular flow and recirculation, can contribute to the development of atherosclerosis [94,99]. The hemodynamic forces, like fluid shear stretch and hydrostatic pressure, caused by pulsatile blood flow and pressure, cause significant stress on endothelial cells [94,100]. Shear stress is a critical modulator of endothelial phenotypes, activating mechanosensors on the endothelial cytoskeleton, which triggers the phosphorylation and activation of genes leading to increased endothelial distress [94]. Furthermore, the stress-induced expression of factors that leads to proinflammatory, procoagulant, proliferative, and proapoptotic functions can contribute to worsening atherosclerosis and VC [94].

Specifically, in CKD uremic vascular calcification occurs in part due to endothelial dysfunction. In a study by Yi-Chou Hou et al., vitamin D is discussed as having a preventative role in VC, especially for those with CKD. Vitamin D deficiency is common in CKD, in part due to proteinuria, the lower therapeutic active dose of vitamin D, and the decline of the glomerular filtration rate [101]. Supplementation of vitamin D has been shown to improve the dysfunction of the endothelium and protect against VC. The effect of vitamin D supplementation has on the endothelium in the vasculature is related to its suppression of the renin–angiotensin–aldosterone system (RAAS), decreased insulin resistance, decreased cholesterol via fewer foam cells, and vascular regeneration [101]. Interestingly, this study suggested that coronary artery calcification progression in those with diabetes mellitus and preserved renal function can be estimated with microalbuminuria.

Hyperphosphatemia, another condition prevalent in CKD due to the high concentration of inorganic phosphate, suppresses the mammalian target of rapamycin (mTOR) signaling. This leads to an increase in the protective autophagic process for endothelial cells by counteracting the reactive-oxygen-species-induced VC [102]. LC3, a protein associated with the autophagosome process, was increased in the endothelial cells of the CKD rat model compared to the controls. During autophagy, LC3 positive vesicles join with ubiquitin and p62, autophagy adaptor. Next, the proteins combine with the lysosomal marker LAMP1 to complete the autophagy process. The end result is calcium and phosphate build-up, contributing to VC [103].

Furthermore, a study by Bouabdallah, et al. shows that Pi and IS together fostered the secretion of interleukin-8 (IL-8) from the endothelial cells of human aortic smooth muscle cells, which induced calcification. It was proposed that IL-8 blocked the production of the potent calcification inhibitor, osteopontin. When antibodies to IL-8 were used, this prevented aortic calcification [104].

Hypertension and VC usually occur together and are ultimately related to one another. Both diseases share endothelial dysfunction. Meng, et al. looked at the aortic calcification in hypertensive rats. They observed a correlation with the expression of the calcium-promoting proteins, MMP-2, and MMP-9 levels in endothelial cells and vascular smooth muscle cell calcification [105]. This is yet another example of endothelial cell dysfunction leading to VC.

In conclusion, endothelial damage leading to VC stems mainly from a variety of reasons, which cause oxidative stress and an increase in an inflammatory response to the endothelium. Crucial factors that lead to endothelial dysfunction include CKD, dyslipidemia, protein-bound uremic toxins, and hypertension.

4. Clinical Implications and Diagnosis of Vascular Calcification in CKD

Vascular disease is the most common cause of death in patients with CKD. In CKD, there are multiple contributing factors to increased VC, such as hyperphosphatemia, hypercalcemia, and increased levels of parathyroid hormone (PTH) [37,106]. High levels of these electrolytes and hormones cause increased cellular activity that increases matrix mineralization, deposition of hydroxyapatite, and inflammation in the vascular intimal and medial walls [107–109].

Amongst CKD patients, VC has been proven to increase cardiovascular mortality and morbidity; however, its predictive value is not clear. Additionally, the increased amount of VC increases one's risk for heart failure, limb ischemia, stroke, and myocardial infarction. Because no data suggest that early detection benefits patients, we generally do not screen CKD patients for VC. Moreover, it is controversial whether interventions that delay the progression of calcification can clinically improve patient outcomes [110].

VC can be diagnosed, detected, and monitored through multiple modalities of imaging, including ultrasound, plain radiographs, and computed tomography (CT) scan. Ultrasound is usually limited to large superficial vessels such as the carotid and femoral arteries, but the information received is qualitative at best and not good for monitoring [111]. Plain radiographs have been used more often for the detection and quantitation of VC. This is normally seen when assessing larger arteries such as the aorta, femoral, and iliac arteries on both anteroposterior and lateral films. The limitation of this imaging modality is that it is difficult to measure VC progression over short periods of time and can be subjective based on the reader [111]. It is also semiquantitative and not as sensitive as a CT scan, which is a superior modality of assessment. VC is detected in >80% of patients undergoing dialysis, which are mostly late CKD and early end-stage renal disease (ESRD) patients. The prevalence amongst non-dialysis CKD patients is around 47% to 83% [110,112–118]. Multi-slice CT (MSCT) scans are often used for diagnosis and follow up for VC in CKD patients. The high image quality allows for precise, quantitative measurements of VC [110,111]. VC in CKD patients is known to be prominent in the media; however, none of these modalities can definitively differentiate intimal from medial calcification [111]. Currently, there are many controversies on the benefit of screening CKD patients for VC. Even if routine screening is performed, there is limited evidence that modifying risk factors and therapies have a significant impact on clinical outcomes [119]. The Kidney Disease: Improving Global Outcomes (KDIGO) guidelines do weakly recommend screening for VC in CKD patients with either lateral radiographs or CT scan [114,120].

5. Treatment Options for Vascular Calcification in CKD

Treatment options for VC in patients with chronic kidney disease include lifestyle modifications, pharmacologic agents, and surgical interventions. One of the most important lifestyle modifications that decreases mortality rates in these patients is smoking cessation [121].

Medical management in these patients is four-fold. Antiplatelet therapy to reduce platelet aggregation, blood pressure control, anti-lipid therapy for plaque-stabilization, and calcium/phosphate balance to prevent further calcification. Antiplatelet therapy, such as aspirin or clopidogrel, has been shown to reduce major cardiovascular events and overall mortality [122]. Blood pressure control is essential; however, there are no specific studies that show that aggressive blood pressure control has altered the course of peripheral artery disease in patients with CKD. Antilipid therapy such as statins significantly reduces the risk of major atherosclerotic events by stabilizing plaques from rupturing.

Though these three targets are important controls of peripheral vascular disease in CKD, calcium and phosphate balance is essential in preventing the progression of VC. Maintaining a calcium phosphate product of less than 60 is imperative in preventing the rapid progression of VC and complications such as calciphylaxis. Phosphate binders are recommended in all patients in CKD to lower phosphate levels [123]. There are two types of phosphate binders, calcium-based and non-calcium-based. Non-calcium-based phosphate binders such as sevelamer are more frequently used due to providing an all-cause mortality benefit in patients in CKD vs. calcium-based phosphate binders such as calcium acetate/magnesium carbonate [124]. Calcimimetics such as Cinacalcet act on calcium-sensing receptors providing a negative feedback loop to decrease PTH levels in an attempt to decrease serum calcium and phosphate levels. Though effective in reducing serum PTH levels, these medications have a questionable impact on all-cause mortality in patients with CKD [125]. Additional studies still need to be performed for assessment. Vitamin D supplements decrease PTH levels by a negative feedback mechanism. Vitamin D deficiency can cause endothelial dysfunction, which can worsen VC. Preventing this by supplementation can attenuate VC in CKD patients [126]. Currently, the KDIGO guidelines recommend vitamin D supplementation in all patients with CKD stage 4 and 5 [120].

Other options include osteoclast activity inhibitors such as bisphosphonates and denosumab. Bisphosphonates such as alendronate are antiresorptive drugs that inhibit osteoclast activity, preventing the release of calcium and phosphate into the bloodstream. They are usually well tolerated but bisphosphonates can cause worsening nephrotoxicity, focal segmental glomerulosclerosis, and osteonecrosis of the jaw [127,128]. Denosumab is a monoclonal antibody that inhibits NFkB-ligand (RANKL), which blocks its osteoclastic and resorptive properties. Few studies have been performed on the effect on VC in vivo and mortality benefits of both bisphosphonates and denosumab in CKD patients; therefore, their clinical benefit is still unclear, and they are not routinely recommended in these patients [123]. Other options also include magnesium supplementation, which has been shown to decrease phosphate-induced calcification in vitro; however, there are very few studies that have shown a statistically significant impact on reducing VC in CKD patients in vivo [123,129].

Interventions include revascularization or amputation. Revascularization via open surgical bypass or percutaneous angioplasty/stenting (also called endovascular revascularization) is usually used in patients experiencing acute limb ischemia or chronic claudication or critical limb ischemia with non-healing wounds. There are no clear data on which type of revascularization is more beneficial, but many physicians would prefer an endovascular first approach if the anatomy is amenable. Amputation is usually reserved for patients that fail revascularization or have non-viable limbs as suggested by symptoms such as paralysis, severe ulceration, or gangrene. Dialysis patients have an extremely high rate of nontraumatic lower extremity amputation compared with the general population [130]. As noted earlier, VC is associated with worse prognosis after coronary or peripheral vascular revascularization procedures [2,3].

There are few experimental therapies that are being studied to combat VC such as ethylenediaminetetraacetic acid (EDTA) chelation therapy, inducing a certain amount of metabolic acidosis, and the use of autologous osteoclasts [11]. EDTA is an amino acid that can bind calcium potentially decreasing free calcium in the blood. While a small human trial showed regression of VC in patients using EDTA, the study was confounded by the lack of an appropriate control group [131]. A state of metabolic acidosis can activate osteoclasts via RANKL and has been shown to decrease VC in vivo and in vitro; however, there are negative side effects of acidosis, and the risks vs. benefits have not been studied [132,133]. Lastly, localized osteoclast therapy from rat-derived bone marrow has been shown to decrease VC in vitro. This has promising potential as it can be used to treat or even prevent VC in certain areas; however, the clinical application of this approach is speculative [130,134]. Along these lines, there are many experimental treatments under active investigation, but more pre-clinical and clinical data will be necessary to determine the efficacy and safety of these treatments.

6. Animal Models of Vascular Calcification

Various animal models have been evaluated in the investigation of VC mechanisms and complications. These studies contribute to understanding the complications that lead to calcification in humans [135]. One review examined CKD rodent models from multiple trials, where the animals developed varying degrees of vascular calcification in the CKD setting. It was found that a particular rodent model shared strong similarities in the symptom development with CKD patients in the clinical setting [136]. This particular rat model developed CKD through the supplementation of adenine, a renal toxin, which was later found to cause nephrotoxicity as a result of 2,8-hydroxyadenine accumulation in the urinary tract [136]. This symptom overlaps with CKD development in patients, providing a useful animal model for CKD-related VC examinations [135]. Several rodent models present conditions including hyperparathyroidism, hyperphosphatemia, and increased levels of both blood urea nitrogen and plasma creatinine [136]. Differences in calcification severity were attributed to varying diet plans, genotypes, and level of kidney damage. As these CKD-related VC animal models share several similarities with the development of VC in CKD patients, they may help reveal important pre-clinical insights [135,136].

Induction of VC through the modification of calcium phosphate levels, alterations in lipid metabolism, downregulation of calcification inhibitors, and promotion of uremic conditions is another approach to modeling VC in CKD [137]. Following these modifications, the mechanisms of VSMCs in the mineralization of the vessel walls of the models is quite similar to the role of VSMCs in humans. These VSMCs are of particular importance as they are associated with VC across multiple species, thus underscoring the hallmark contribution of this cell type [135]. This point is highlighted by several studies where VSMCs have been demonstrated to play an essential role in VC induction. Inhibition of the matrix-gla protein (MGP) yielded substantial VC where VSMCs formed profound cartilage in vessels through osteogenic differentiation [137]. Deficiency in osteopontin (OPN) is another mediator of calcification that induces mineralization in mouse models [135,137].

Altogether, the results in VC animal models provide valuable insight into the mechanisms of VC development in humans [137]. The many factors of VC regulation are comparable across various species from mice to humans making these animal models appropriate for the study of this complex disease and allowing for the needed pre-clinical development of therapeutic interventions [138].

7. Conclusions

Altogether, an extensive multitude of associated factors relates VC to CKD. A wide range of comorbidities escalates VC pathogenesis, including hypertension, hyperglycemia, hyperphosphatemia, hyperlipidemia, inflammation, oxidative stress, and uremia. Structural factors such as the anatomical location of the affected vessel and histological sites of calcification also lead to the advancement of VC. There likewise exists a variety of molecular mechanisms that play essential roles in the pathogenesis of the vessel's calcification. Conventional and alternative prevention mechanisms such as regular exercise, maintaining a healthy diet, and therapeutic approaches are opportunities to decrease the VC burden in CKD. However, there is limited research demonstrating that these mechanisms effectively improve the severity of VC. By defining the exact pathology of the various forms of VC in CKD, targeted and rationale strategies can be developed to understand the clinical importance and address the potential pathological sequalae in this at-risk population.

Author Contributions: P.D., A.D., M.P., D.B., B.K.-S. and H.S. wrote the manuscript; P.D., R.G., D.M., L.D., S.H. and D.K. substantially revised and finalized the manuscript. All authors have read and agreed to the published version of the manuscript.

Funding: This work was supported in part by National Institutes of Health grants (R01HL137004), the David and Helen Boone Foundation Research Fund, the University of Toledo Women and Philanthropy Genetic Analysis Instrumentation Center, and the University of Toledo Medical Research Society.

Acknowledgments: The authors gratefully acknowledge Roy Schneider in the University of Toledo's Center for Creative Instruction for rendering the medical illustrations in this review.

Conflicts of Interest: The authors declare no conflict of interest.

References

1. Gao, J.; Zhang, K.; Chen, J.; Wang, M.-H.; Wang, J.; Liu, P.; Huang, H. Roles of aldosterone in vascular calcification: An update. *Eur. J. Pharmacol.* **2016**, *786*, 186–193. [CrossRef]
2. Généreux, P.; Redfors, B.; Witzenbichler, B.; Arsenault, M.-P.; Weisz, G.; Stuckey, T.D.; Rinaldi, M.J.; Neumann, F.-J.; Metzger, D.C.; Henry, T.D.; et al. Two-year outcomes after percutaneous coronary intervention of calcified lesions with drug-eluting stents. *Int. J. Cardiol.* **2017**, *231*, 61–67. [CrossRef]
3. Lee, H.Y.; Park, U.J.; Kim, H.T.; Roh, Y.-N. The Effect of Severe Femoropopliteal Arterial Calcification on the Treatment Outcome of Femoropopliteal Intervention in Patients with Ischemic Tissue Loss. *Vasc. Spec. Int.* **2020**, *36*, 96–104. [CrossRef]
4. Lacolley, P.; Regnault, V.; Laurent, S. Mechanisms of Arterial Stiffening. *Arter. Thromb. Vasc. Biol.* **2020**, *40*, 1055–1062. [CrossRef]
5. Lioufas, N.M.; Pedagogos, E.; Hawley, C.M.; Pascoe, E.M.; Elder, G.J.; Badve, S.V.; Valks, A.; Toussaint, N.D.; on behalf of the IMPROVE-CKD Investigators. Aortic Calcification and Arterial Stiffness Burden in a Chronic Kidney Disease Cohort with High Cardiovascular Risk: Baseline Characteristics of the Impact of Phosphate Reduction on Vascular End-Points in Chronic Kidney Disease Trial. *Am. J. Nephrol.* **2020**, *51*, 201–215. [CrossRef]
6. Jagieła, J.; Bartnicki, P.; Rysz, J. Selected cardiovascular risk factors in early stages of chronic kidney disease. *Int. Urol. Nephrol.* **2020**, *52*, 303–314. [CrossRef]
7. Demer, L.L.; Tintut, Y. Vascular Calcification. *Circulation* **2008**, *117*, 2938–2948. [CrossRef]
8. Chen, N.X.; Moe, S.M. Vascular calcification: Pathophysiology and risk factors. *Curr. Hypertens. Rep.* **2012**, *14*, 228–237. [CrossRef]
9. Hénaut, L.; Chillon, J.-M.; Kamel, S.; Massy, Z.A. Updates on the Mechanisms and the Care of Cardiovascular Calcification in Chronic Kidney Disease. *Semin. Nephrol.* **2018**, *38*, 233–250. [CrossRef]
10. Voelkl, J.; Lang, F.; Eckardt, K.-U.; Amann, K.; Kuro-O, M.; Pasch, A.; Pieske, B.; Alesutan, I. Signaling pathways involved in vascular smooth muscle cell calcification during hyperphosphatemia. *Cell. Mol. Life Sci.* **2019**, *76*, 2077–2091. [CrossRef]
11. Wu, M.; Rementer, C.; Giachelli, C.M. Vascular Calcification: An Update on Mechanisms and Challenges in Treatment. *Calcif. Tissue Int.* **2013**, *93*, 365–373. [CrossRef]
12. El Din, U.A.A.S.; Salem, M.M.; Azim, S.E.D.U.A. Vascular calcification: When should we interfere in chronic kidney disease patients and how? *World J. Nephrol.* **2016**, *5*, 398–417. [CrossRef] [PubMed]
13. Nakanishi, T.; Nanami, M.; Kuragano, T. The pathogenesis of CKD complications; Attack of dysregulated iron and phosphate metabolism. *Free Radic. Biol. Med.* **2020**, *157*, 55–62. [CrossRef] [PubMed]
14. Olapoju, S.O.; Adejobi, O.I.; Le Thi, X. Fibroblast growth factor 21; review on its participation in vascular calcification pathology. *Vasc. Pharmacol.* **2020**, *125–126*, 106636. [CrossRef] [PubMed]
15. Opdebeeck, B.; D'Haese, P.C.; Verhulst, A. Molecular and Cellular Mechanisms that Induce Arterial Calcification by Indoxyl Sulfate and P-Cresyl Sulfate. *Toxins* **2020**, *12*, 58. [CrossRef] [PubMed]
16. Huang, M.; Zheng, L.; Xu, H.; Tang, D.; Lin, L.; Zhang, J.; Li, C.; Wang, W.; Yuan, Q.; Tao, L.; et al. Oxidative stress contributes to vascular calcification in patients with chronic kidney disease. *J. Mol. Cell. Cardiol.* **2020**, *138*, 256–268. [CrossRef]
17. Sarnak, M.J. Cardiovascular complications in chronic kidney disease. *Am. J. Kidney Dis.* **2003**, *41*, 11–17. [CrossRef]
18. Sheng, B.; Zhu, T.; Li, J. End-stage renal disease and pulmonary hypertension. *Zhong Nan Da Xue Xue Bao Yi Xue Ban = J. Cent. South Univ. Med. Sci.* **2019**, *44*, 1419–1422.
19. Arefin, S.; Buchanan, S.; Hobson, S.; Steinmetz, J.; Alsalhi, S.; Shiels, P.G.; Kublickiene, K.; Stenvinkel, P. Nrf2 in early vascular ageing: Calcification, senescence and therapy. *Clin. Chim. Acta* **2020**, *505*, 108–118. [CrossRef]
20. Lee, S.J.; Lee, I.-K.; Jeon, J.-H. Vascular Calcification—New Insights into Its Mechanism. *Int. J. Mol. Sci.* **2020**, *21*, 2685. [CrossRef]
21. Thompson, B.; Towler, D.A. Arterial calcification and bone physiology: Role of the bone–vascular axis. *Nat. Rev. Endocrinol.* **2012**, *8*, 529–543. [CrossRef]
22. Xu, J.; Shi, G.-P. Vascular wall extracellular matrix proteins and vascular diseases. *Biochim. Biophys. Acta (BBA) Mol. Basis Dis.* **2014**, *1842*, 2106–2119. [CrossRef]
23. Cucchiari, D.; Torregrosa, J.-V. Calcifilaxis en pacientes con enfermedad renal crónica: Una enfermedad todavía desconcertante y potencialmente mortal. *Nefrología* **2018**, *38*, 579–586. [CrossRef] [PubMed]
24. Izquierdo-Gómez, M.M.; Hernández-Betancor, I.; García-Niebla, J.; Marí-López, B.; Laynez-Cerdeña, I.; Lacalzada-Almeida, J. Valve Calcification in Aortic Stenosis: Etiology and Diagnostic Imaging Techniques. *BioMed Res. Int.* **2017**, *2017*, 5178631. [CrossRef]
25. Neven, E.; Dauwe, S.; De Broe, M.E.; Haese, P.C.D.; Persy, V. Endochondral bone formation is involved in media calcification in rats and in men. *Kidney Int.* **2007**, *72*, 574–581. [CrossRef]
26. Qadri, S.I.; Koratala, A. Calciphylaxis with extensive arterial calcification. *Clin. Case Rep.* **2017**, *5*, 1418–1419. [CrossRef]
27. Sinha, S.; Iyer, D.; Granata, A. Embryonic origins of human vascular smooth muscle cells: Implications for In Vitro modeling and clinical application. *Cell. Mol. Life Sci.* **2014**, *71*, 2271–2288. [CrossRef]
28. Durham, A.L.; Speer, M.Y.; Scatena, M.; Giachelli, C.M.; Shanahan, C.M. Role of smooth muscle cells in vascular calcification: Implications in atherosclerosis and arterial stiffness. *Cardiovasc. Res.* **2018**, *114*, 590–600. [CrossRef]
29. Yamada, S.; Giachelli, C.M. Vascular calcification in CKD-MBD: Roles for phosphate, FGF23, and Klotho. *Bone* **2017**, *100*, 87–93. [CrossRef]
30. Paloian, N.J.; Giachelli, C.M. A current understanding of vascular calcification in CKD. *Am. J. Physiol. Physiol.* **2014**, *307*, F891–F900. [CrossRef]

31. Scholze, A.; Jankowski, J.; Pedraza-Chaverri, J.; Evenepoel, P. Oxidative Stress in Chronic Kidney Disease. *Oxidative Med. Cell. Longev.* **2016**, *2016*, 8375186. [CrossRef]
32. Byon, C.H.; Javed, A.; Dai, Q.; Kappes, J.C.; Clemens, T.L.; Darley-Usmar, V.M.; McDonald, J.M.; Chen, Y. Oxidative Stress Induces Vascular Calcification through Modulation of the Osteogenic Transcription Factor Runx2 by AKT Signaling. *J. Biol. Chem.* **2008**, *283*, 15319–15327. [CrossRef]
33. Montezano, A.C.; Zimmerman, D.; Yusuf, H.; Burger, D.; Chignalia, A.Z.; Wadhera, V.; Van Leeuwen, F.N.; Touyz, R.M. Vascular Smooth Muscle Cell Differentiation to an Osteogenic Phenotype Involves TRPM7 Modulation by Magnesium. *Hypertension* **2010**, *56*, 453–462. [CrossRef]
34. Mehrotra, R. Emerging role for fetuin-A as contributor to morbidity and mortality in chronic kidney disease. *Kidney Int.* **2007**, *72*, 137–140. [CrossRef] [PubMed]
35. Shanahan, C.M.; Crouthamel, M.H.; Kapustin, A.; Giachelli, C.M. Arterial Calcification in Chronic Kidney Disease: Key Roles for Calcium and Phosphate. *Circ. Res.* **2011**, *109*, 697–711. [CrossRef] [PubMed]
36. Schlieper, G.; Schurgers, L.; Brandenburg, V.; Reutelingsperger, C.; Floege, J. Vascular calcification in chronic kidney disease: An update. *Nephrol. Dial. Transplant.* **2016**, *31*, 31–39. [CrossRef]
37. House, S.J.; Potier, M.; Bisaillon, J.; Singer, H.A.; Trebak, M. The non-excitable smooth muscle: Calcium signaling and phenotypic switching during vascular disease. *Pflug. Arch. Eur. J. Physiol.* **2008**, *456*, 769–785. [CrossRef]
38. Molostvov, G.; James, S.; Fletcher, S.; Bennett, J.; Lehnert, H.; Bland, R.; Zehnder, D. Extracellular calcium-sensing receptor is functionally expressed in human artery. *Am. J. Physiol. Physiol.* **2007**, *293*, F946–F955. [CrossRef]
39. Yamada, S.; Leaf, E.M.; Chia, J.J.; Cox, T.C.; Speer, M.Y.; Giachelli, C.M. PiT-2, a type III sodium-dependent phosphate transporter, protects against vascular calcification in mice with chronic kidney disease fed a high-phosphate diet. *Kidney Int.* **2018**, *94*, 716–727. [CrossRef]
40. Allahverdian, S.; Chehroudi, A.C.; McManus, B.M.; Abraham, T.; Francis, G.A. Contribution of Intimal Smooth Muscle Cells to Cholesterol Accumulation and Macrophage-Like Cells in Human Atherosclerosis. *Circulation* **2014**, *129*, 1551–1559. [CrossRef]
41. Akers, J.C.; Gonda, D.; Kim, R.; Carter, B.S.; Chen, C.C. Biogenesis of extracellular vesicles (EV): Exosomes, microvesicles, retrovirus-like vesicles, and apoptotic bodies. *J. Neuro-Oncol.* **2013**, *113*, 1–11. [CrossRef] [PubMed]
42. Aikawa, E. Extracellular vesicles in cardiovascular disease: Focus on vascular calcification. *J. Physiol.* **2016**, *594*, 2877–2880. [CrossRef]
43. Shroff, R.C.; McNair, R.; Skepper, J.N.; Figg, N.; Schurgers, L.J.; Deanfield, J.; Rees, L.; Shanahan, C.M. Chronic Mineral Dysregulation Promotes Vascular Smooth Muscle Cell Adaptation and Extracellular Matrix Calcification. *J. Am. Soc. Nephrol.* **2009**, *21*, 103–112. [CrossRef] [PubMed]
44. Hutcheson, J.D.; Goettsch, C.; Bertazzo, S.S.; Maldonado, N.; Ruiz, J.L.; Goh, W.; Yabusaki, K.; Faits, T.; Bouten, C.C.; Franck, G.; et al. Genesis and growth of extracellular-vesicle-derived microcalcification in atherosclerotic plaques. *Nat. Mater.* **2016**, *15*, 335–343. [CrossRef]
45. Yang, W.; Zou, B.; Hou, Y.; Yan, W.; Chen, T.; Qu, S. Extracellular vesicles in vascular calcification. *Clin. Chim. Acta* **2019**, *499*, 118–122. [CrossRef]
46. Coscas, R.; Bensussan, M.; Jacob, M.-P.; Louedec, L.; Massy, Z.; Sadoine, J.; Daudon, M.; Chaussain, C.; Bazin, D.; Michel, J.-B. Free DNA precipitates calcium phosphate apatite crystals in the arterial wall In Vivo. *Atherosclerosis* **2017**, *259*, 60–67. [CrossRef]
47. Singh, R.; Cuervo, A.M. Autophagy in the Cellular Energetic Balance. *Cell Metab.* **2011**, *13*, 495–504. [CrossRef] [PubMed]
48. Shanahan, C.M. Autophagy and matrix vesicles: New partners in vascular calcification. *Kidney Int.* **2013**, *83*, 984–986. [CrossRef]
49. Cochain, C.; Zernecke, A. Macrophages in vascular inflammation and atherosclerosis. *Pflüg. Arch. Eur. J. Physiol.* **2017**, *469*, 485–499. [CrossRef]
50. Byon, C.H.; Sun, Y.; Chen, J.; Yuan, K.; Mao, X.; Heath, J.M.; Anderson, P.G.; Tintut, Y.; Demer, L.L.; Wang, D.; et al. Runx2-Upregulated Receptor Activator of Nuclear Factor κB Ligand in Calcifying Smooth Muscle Cells Promotes Migration and Osteoclastic Differentiation of Macrophages. *Arter. Thromb. Vasc. Biol.* **2011**, *31*, 1387–1396. [CrossRef]
51. Sica, A.; Invernizzi, P.; Mantovani, A. Macrophage plasticity and polarization in liver homeostasis and pathology. *Hepatology* **2014**, *59*, 2034–2042. [CrossRef]
52. Adamson, S.; Leitinger, N. Phenotypic modulation of macrophages in response to plaque lipids. *Curr. Opin. Lipidol.* **2011**, *22*, 335–342. [CrossRef]
53. Zhang, X.; Li, J.; Qin, J.-J.; Cheng, W.-L.; Zhu, X.; Gong, F.-H.; She, Z.; Huang, Z.; Xia, H.; Li, H. Oncostatin M receptor β deficiency attenuates atherogenesis by inhibiting JAK2/STAT3 signaling in macrophages. *J. Lipid Res.* **2017**, *58*, 895–906. [CrossRef] [PubMed]
54. Kraft, C.T.; Agarwal, S.; Ranganathan, K.; Wong, V.W.; Loder, S.; Li, J.; Delano, M.J.; Levi, B. Trauma-induced heterotopic bone formation and the role of the immune system. *J. Trauma Acute Care Surg.* **2016**, *80*, 156–165. [CrossRef]
55. Zhou, S.; Cai, B.; Zhang, Z.; Zhang, Y.; Wang, L.; Liu, K.; Zhang, H.; Sun, L.; Cai, H.; Lu, G.; et al. CDKN2B Methylation and Aortic Arch Calcification in Patients with Ischemic Stroke. *J. Atheroscler. Thromb.* **2017**, *24*, 609–620. [CrossRef] [PubMed]
56. Fu, Y.; Gao, C.; Liang, Y.; Wang, M.; Huang, Y.; Ma, W.; Li, T.; Jia, Y.; Yu, F.; Zhu, W.; et al. Shift of Macrophage Phenotype Due to Cartilage Oligomeric Matrix Protein Deficiency Drives Atherosclerotic Calcification. *Circ. Res.* **2016**, *119*, 261–276. [CrossRef] [PubMed]

57. Porta, C.; Riboldi, E.; Ippolito, A.; Sica, A. Molecular and epigenetic basis of macrophage polarized activation. *Semin. Immunol.* **2015**, *27*, 237–248. [CrossRef]
58. Villa-Bellosta, R.; Hamczyk, M.R.; Andrés, V. Alternatively activated macrophages exhibit an anticalcifying activity dependent on extracellular ATP/pyrophosphate metabolism. *Am. J. Physiol. Physiol.* **2016**, *310*, C788–C799. [CrossRef]
59. Villa-Bellosta, R.; Sorribas, V. Prevention of Vascular Calcification by Polyphosphates and Nucleotides. *Circ. J.* **2013**, *77*, 2145–2151. [CrossRef]
60. Johnson, K.; Polewski, M.D.; Van Etten, D.; Terkeltaub, R.A. Chondrogenesis Mediated by PP i Depletion Promotes Spontaneous Aortic Calcification in NPP1−/− Mice. *Arter. Thromb. Vasc. Biol.* **2005**, *25*, 686–691. [CrossRef]
61. Li, Y.; Sun, Z.; Zhang, L.; Yan, J.; Shao, C.; Jing, L.; Li, L.; Wang, Z. Role of Macrophages in the Progression and Regression of Vascular Calcification. *Front. Pharmacol.* **2020**, *11*, 661. [CrossRef]
62. Dube, P.R.; Birnbaumer, L.; Vazquez, G. Evidence for constitutive bone morphogenetic protein-2 secretion by M1 macrophages: Constitutive auto/paracrine osteogenic signaling by BMP-2 in M1 macrophages. *Biochem. Biophys. Res. Commun.* **2017**, *491*, 154–158. [CrossRef] [PubMed]
63. Dube, P.R.; Chikkamenahalli, L.L.; Birnbaumer, L.; Vazquez, G. Reduced calcification and osteogenic features in advanced atherosclerotic plaques of mice with macrophage-specific loss of TRPC3. *Atherosclerosis* **2018**, *270*, 199–204. [CrossRef]
64. Solanki, S.; Dube, P.R.; Tano, J.-Y.; Birnbaumer, L.; Vazquez, G. Reduced endoplasmic reticulum stress-induced apoptosis and impaired unfolded protein response in TRPC3-deficient M1 macrophages. *Am. J. Physiol. Physiol.* **2014**, *307*, C521–C531. [CrossRef]
65. Solanki, S.; Dube, P.R.; Birnbaumer, L.; Vazquez, G. Reduced Necrosis and Content of Apoptotic M1 Macrophages in Advanced Atherosclerotic Plaques of Mice with Macrophage-Specific Loss of Trpc3. *Sci. Rep.* **2017**, *7*, 42526. [CrossRef]
66. Kapustin, A.N.; Chatrou, M.L.L.; Drozdov, I.; Zheng, Y.; Davidson, S.M.; Soong, D.; Furmanik, M.; Sanchis, P.; De Rosales, R.T.M.; Alvarez-Hernandez, D.; et al. Vascular Smooth Muscle Cell Calcification Is Mediated by Regulated Exosome Secretion. *Circ. Res.* **2015**, *116*, 1312–1323. [CrossRef]
67. Cozzolino, M.; Ciceri, P.; Galassi, A.; Mangano, M.; Carugo, S.; Capelli, I.; Cianciolo, G. The Key Role of Phosphate on Vascular Calcification. *Toxins* **2019**, *11*, 213. [CrossRef]
68. Wang, B.; Wei, G.; Liu, B.; Zhou, X.; Xiao, H.; Dong, N.; Li, F. The Role of High Mobility Group Box 1 Protein in Interleukin-18-Induced Myofibroblastic Transition of Valvular Interstitial Cells. *Cardiology* **2016**, *135*, 168–178. [CrossRef]
69. Kwon, D.-H.; Kim, Y.-K.; Kook, H. New Aspects of Vascular Calcification: Histone Deacetylases and Beyond. *J. Korean Med. Sci.* **2017**, *32*, 1738–1748. [CrossRef] [PubMed]
70. Hénaut, L.; Candellier, A.; Boudot, C.; Grissi, M.; Mentaverri, R.; Choukroun, G.; Brazier, M.; Kamel, S.; Massy, Z.A. New Insights into the Roles of Monocytes/Macrophages in Cardiovascular Calcification Associated with Chronic Kidney Disease. *Toxins* **2019**, *11*, 529. [CrossRef] [PubMed]
71. Rao, P.V.L.N.S.; Gouroju, S.; Bitla, A.R.; Vinapamula, K.S.; Manohar, S.M.; Vishnubhotla, S. Role of gut-derived uremic toxins on oxidative stress and inflammation in patients with chronic kidney disease. *Indian J. Nephrol.* **2017**, *27*, 359–364. [CrossRef]
72. Castillo-Rodríguez, E.; Pizarro-Sánchez, S.; Sanz, A.B.; Ramos, A.M.; Sanchez-Niño, M.D.; Martin-Cleary, C.; Fernandez-Fernandez, B.; Ortiz, A. Inflammatory Cytokines as Uremic Toxins: "Ni Son Todos Los Que Estan, Ni Estan Todos Los Que Son". *Toxins* **2017**, *9*, 114. [CrossRef] [PubMed]
73. Six, I.; Maizel, J.; Barreto, F.C.; Rangrez, A.Y.; Dupont, S.; Slama, M.; Tribouilloy, C.; Choukroun, G.; Mazière, J.C.; Bode-Boeger, S.; et al. Effects of phosphate on vascular function under normal conditions and influence of the uraemic state. *Cardiovasc. Res.* **2012**, *96*, 130–139. [CrossRef] [PubMed]
74. Villa-Bellosta, R.; Hamczyk, M.R.; Andrés, V. Novel phosphate-activated macrophages prevent ectopic calcification by increasing extracellular ATP and pyrophosphate. *PLoS ONE* **2017**, *12*, e0174998. [CrossRef]
75. Jing, Y.J.; Ni, J.W.; Ding, F.H.; Fang, Y.H.; Wang, X.Q.; Wang, H.B.; Chen, X.N.; Chen, N.; Zhan, W.W.; Lu, L.; et al. p-Cresyl sulfate is associated with carotid arteriosclerosis in hemodialysis patients and promotes atherogenesis in apoE−/− mice. *Kidney Int.* **2016**, *89*, 439–449. [CrossRef]
76. Hahn, K.; Kanbay, M.; Lanaspa, M.A.; Johnson, R.J.; Ejaz, A.A. Serum uric acid and acute kidney injury: A mini review. *J. Adv. Res.* **2017**, *8*, 529–536. [CrossRef] [PubMed]
77. Geraci, G.; Mule', G.; Morreale, M.; Cusumano, C.; Castiglia, A.; Gervasi, F.; D'Ignoto, F.; Mogavero, M.; Geraci, C.; Cottone, S. Association between uric acid and renal function in hypertensive patients: Which role for systemic vascular involvement? *J. Am. Soc. Hypertens.* **2016**, *10*, 559–569. [CrossRef] [PubMed]
78. Andrés, M.; Quintanilla, M.-A.; Sivera, F.; Sánchez-Payá, J.; Pascual, E.; Vela, P.; Ruiz-Nodar, J.-M. Silent Monosodium Urate Crystal Deposits Are Associated with Severe Coronary Calcification in Asymptomatic Hyperuricemia: An Exploratory Study. *Arthritis Rheumatol.* **2016**, *68*, 1531–1539. [CrossRef]
79. Kim, H.; Kim, S.-H.; Choi, A.R.; Kim, S.; Choi, H.Y.; Kim, H.J.; Park, H.-C. Asymptomatic hyperuricemia is independently associated with coronary artery calcification in the absence of overt coronary artery disease. *Medicine* **2017**, *96*, e6565. [CrossRef]
80. Kiss, L.Z.; Bagyura, Z.; Csobay-Novák, C.; Lux, Á.; Polgár, L.; Jermendy, Á.; Soós, P.; Szelid, Z.; Maurovich-Horvat, P.; Becker, D.; et al. Serum Uric Acid Is Independently Associated with Coronary Calcification in an Asymptomatic Population. *J. Cardiovasc. Transl. Res.* **2019**, *12*, 204–210. [CrossRef]

81. Aroor, A.R.; Jia, G.; Habibi, J.; Sun, Z.; Ramirez-Perez, F.I.; Brady, B.; Chen, D.; Martinez-Lemus, L.A.; Manrique, C.; Nistala, R.; et al. Uric acid promotes vascular stiffness, maladaptive inflammatory responses and proteinuria in western diet fed mice. *Metab. Clin. Exp.* **2017**, *74*, 32–40. [CrossRef]
82. Andrews, E.S.; Perrenoud, L.; Nowak, K.L.; You, Z.; Pasch, A.; Chonchol, M.; Kendrick, J.; Jalal, D. Examining the effects of uric acid-lowering on markers vascular of calcification and CKD-MBD; A post-hoc analysis of a randomized clinical trial. *PLoS ONE* **2018**, *13*, e0205831. [CrossRef] [PubMed]
83. Rogacev, K.S.; Seiler, S.; Zawada, A.M.; Reichart, B.; Herath, E.; Roth, D.; Ulrich, C.; Fliser, D.; Heine, G.H. CD14++CD16+ monocytes and cardiovascular outcome in patients with chronic kidney disease. *Eur. Heart J.* **2010**, *32*, 84–92. [CrossRef] [PubMed]
84. Mizobuchi, M.; Towler, D.; Slatopolsky, E. Vascular Calcification: The Killer of Patients with Chronic Kidney Disease. *J. Am. Soc. Nephrol.* **2009**, *20*, 1453–1464. [CrossRef] [PubMed]
85. Hergenreider, E.; Heydt, S.; Tréguer, K.; Boettger, T.; Horrevoets, A.J.G.; Zeiher, A.M.; Scheffer, M.P.; Frangakis, A.S.; Yin, X.; Mayr, M.; et al. Atheroprotective communication between endothelial cells and smooth muscle cells through miRNAs. *Nat. Cell Biol.* **2012**, *14*, 249–256. [CrossRef]
86. Shin, V.; Zebboudj, A.F.; Boström, K. Endothelial Cells Modulate Osteogenesis in Calcifying Vascular Cells. *J. Vasc. Res.* **2004**, *41*, 193–201. [CrossRef]
87. Yao, Y.; Jumabay, M.; Ly, A.; Radparvar, M.; Cubberly, M.R.; Boström, K.I. A Role for the Endothelium in Vascular Calcification. *Circ. Res.* **2013**, *113*, 495–504. [CrossRef] [PubMed]
88. Soriano, S.; Carmona, A.; Triviño, F.; Rodriguez, M.; Alvarez-Benito, M.; Martín-Malo, A.; Alvarez-Lara, M.-A.; Ramírez, R.; Aljama, P.; Carracedo, J. Endothelial damage and vascular calcification in patients with chronic kidney disease. *Am. J. Physiol. Physiol.* **2014**, *307*, F1302–F1311. [CrossRef] [PubMed]
89. Yao, J.; Guihard, P.J.; Blazquez-Medela, A.M.; Guo, Y.; Moon, J.H.; Jumabay, M.; Boström, K.I.; Yao, Y. Serine Protease Activation Essential for Endothelial–Mesenchymal Transition in Vascular Calcification. *Circ. Res.* **2015**, *117*, 758–769. [CrossRef]
90. Boström, K.I.; Yao, J.; Guihard, P.J.; Blazquez-Medela, A.M.; Yao, Y. Endothelial-mesenchymal transition in atherosclerotic lesion calcification. *Atherosclerosis* **2016**, *253*, 124–127. [CrossRef]
91. Hjortnaes, J.; Shapero, K.; Goettsch, C.; Hutcheson, J.D.; Keegan, J.; Kluin, J.; Mayer, J.E.; Bischoff, J.; Aikawa, E. Valvular interstitial cells suppress calcification of valvular endothelial cells. *Atherosclerosis* **2015**, *242*, 251–260. [CrossRef]
92. Wu, M.; Tang, R.-N.; Liu, H.; Pan, M.-M.; Liu, B.-C. Cinacalcet ameliorates aortic calcification in uremic rats via suppression of endothelial-to-mesenchymal transition. *Acta Pharmacol. Sin.* **2016**, *37*, 1423–1431. [CrossRef]
93. Rutkovskiy, A.; Lund, M.; Siamansour, T.S.; Reine, T.M.; Kolset, S.O.; Sand, K.L.; Ignatieva, E.; Gordeev, M.L.; Stensløkken, K.-O.; Valen, G.; et al. Mechanical stress alters the expression of calcification-related genes in vascular interstitial and endothelial cells. *Interact. Cardiovasc. Thorac. Surg.* **2019**, *28*, 803–811. [CrossRef]
94. Chiu, J.-J.; Usami, S.; Chien, S. Vascular endothelial responses to altered shear stress: Pathologic implications for atherosclerosis. *Ann. Med.* **2009**, *41*, 19–28. [CrossRef] [PubMed]
95. Urs, S.; Venkatesh, D.; Tang, Y.; Henderson, T.; Yang, X.; Friesel, R.E.; Rosen, C.J.; Liaw, L. Sprouty1 is a critical regulatory switch of mesenchymal stem cell lineage allocation. *FASEB J.* **2010**, *24*, 3264–3273. [CrossRef] [PubMed]
96. Merle, B.; Garnero, P. The multiple facets of periostin in bone metabolism. *Osteoporos. Int.* **2012**, *23*, 1199–1212. [CrossRef] [PubMed]
97. Canfield, A.E.; Farrington, C.; Dziobon, M.D.; Boot-Handford, R.P.; Heagerty, A.M.; Kumar, S.N.; Roberts, I.S.D. The involvement of matrix glycoproteins in vascular calcification and fibrosis: An immunohistochemical study. *J. Pathol.* **2001**, *196*, 228–234. [CrossRef] [PubMed]
98. Van Thienen, J.V.; Fledderus, J.O.; Dekker, R.J.; Rohlena, J.; Van Ijzendoorn, G.A.; Kootstra, N.A.; Pannekoek, H.; Horrevoets, A.J.G. Shear stress sustains atheroprotective endothelial KLF2 expression more potently than statins through mRNA stabilization. *Cardiovasc. Res.* **2006**, *72*, 231–240. [CrossRef] [PubMed]
99. Chatzizisis, Y.S.; Coskun, A.U.; Jonas, M.; Edelman, E.R.; Feldman, C.L.; Stone, P.H. Role of Endothelial Shear Stress in the Natural History of Coronary Atherosclerosis and Vascular Remodeling. *J. Am. Coll. Cardiol.* **2007**, *49*, 2379–2393. [CrossRef] [PubMed]
100. Chien, S. Mechanotransduction and endothelial cell homeostasis: The wisdom of the cell. *Am. J. Physiol. Circ. Physiol.* **2007**, *292*, H1209–H1224. [CrossRef]
101. Hou, Y.-C.; Liu, W.-C.; Zheng, C.-M.; Zheng, J.-Q.; Yen, T.-H.; Lu, K.-C. Role of Vitamin D in Uremic Vascular Calcification. *BioMed Res. Int.* **2017**, *2017*, 2803579. [CrossRef] [PubMed]
102. Hsu, Y.-J.; Hsu, S.-C.; Huang, S.-M.; Lee, H.-S.; Lin, S.-H.; Tsai, C.-S.; Shih, C.-C.; Lin, C.-Y. Hyperphosphatemia induces protective autophagy in endothelial cells through the inhibition of Akt/mTOR signaling. *J. Vasc. Surg.* **2015**, *62*, 210–221.e2. [CrossRef] [PubMed]
103. Dai, X.-Y.; Zhao, M.-M.; Cai, Y.; Guan, Q.-C.; Zhao, Y.; Guan, Y.; Kong, W.; Zhu, W.-G.; Xu, M.-J.; Wang, X. Phosphate-induced autophagy counteracts vascular calcification by reducing matrix vesicle release. *Kidney Int.* **2013**, *83*, 1042–1051. [CrossRef] [PubMed]
104. Bouabdallah, J.; Zibara, K.; Issa, H.; Lenglet, G.; Kchour, G.; Caus, T.; Six, I.; Choukroun, G.; Kamel, S.; Bennis, Y. Endothelial cells exposed to phosphate and indoxyl sulphate promote vascular calcification through interleukin-8 secretion. *Nephrol. Dial. Transplant.* **2019**, *34*, 1125–1134. [CrossRef] [PubMed]

105. Meng, F.; Zhao, Y.; Wang, B.; Li, B.; Sheng, Y.; Liu, M.; Li, H.; Xiu, R. Endothelial Cells Promote Calcification in Aortic Smooth Muscle Cells from Spontaneously Hypertensive Rats. *Cell. Physiol. Biochem.* **2018**, *49*, 2371–2381. [CrossRef]
106. Cianciolo, G.; Galassi, A.; Capelli, I.; Schillaci, R.; La Manna, G.; Cozzolino, M. Klotho-FGF23, Cardiovascular Disease, and Vascular Calcification: Black or White? *Curr. Vasc. Pharmacol.* **2018**, *16*, 143–156. [CrossRef]
107. Kapustin, A.N.; Davies, J.D.; Reynolds, J.L.; McNair, R.; Jones, G.T.; Sidibe, A.; Schurgers, L.J.; Skepper, J.N.; Proudfoot, D.; Mayr, M.; et al. Calcium Regulates Key Components of Vascular Smooth Muscle Cell–Derived Matrix Vesicles to Enhance Mineralization. *Circ. Res.* **2011**, *109*, e1–e12. [CrossRef]
108. Houben, E.; Neradova, A.; Schurgers, L.J.; Vervloet, M. The influence of phosphate, calcium and magnesium on matrix Gla-protein and vascular calcification: A systematic review. *G. Ital. Nefrol.* **2017**, *33*, 1724–5590.
109. Kidney Disease: Improving Global Outcomes (KDIGO) CKD-MBD Work Group. KDIGO clinical practice guideline for the diagnosis, evaluation, prevention, and treatment of Chronic Kidney Disease-Mineral and Bone Disorder (CKD-MBD). *Kidney Int. Suppl.* **2009**, *113*, S1–S130.
110. Reynolds, J.L.; Joannides, A.J.; Skepper, J.N.; McNair, R.; Schurgers, L.J.; Proudfoot, D.; Jahnen-Dechent, W.; Weissberg, P.L.; Shanahan, C.M. Human Vascular Smooth Muscle Cells Undergo Vesicle-Mediated Calcification in Response to Changes in Extracellular Calcium and Phosphate Concentrations: A Potential Mechanism for Accelerated Vascular Calcification in ESRD. *J. Am. Soc. Nephrol.* **2004**, *15*, 2857–2867. [CrossRef]
111. Disthabanchong, S. Vascular calcification in chronic kidney disease: Pathogenesis and clinical implication. *World J. Nephrol.* **2012**, *1*, 43–53. [CrossRef]
112. Merjanian, R.; Budoff, M.; Adler, S.; Berman, N.; Mehrotra, R. Coronary artery, aortic wall, and valvular calcification in nondialyzed individuals with type 2 diabetes and renal disease. *Kidney Int.* **2003**, *64*, 263–271. [CrossRef] [PubMed]
113. Lamprea-Montealegre, J.A.; McClelland, R.L.; Astor, B.C.; Matsushita, K.; Shlipak, M.; de Boer, I.H.; Szklo, M. Chronic kidney disease, plasma lipoproteins, and coronary artery calcium incidence: The Multi-Ethnic Study of Atherosclerosis. *Arterioscler. Thromb. Vasc. Biol.* **2013**, *33*, 652. [CrossRef] [PubMed]
114. Kramer, H.; Toto, R.; Peshock, R.; Cooper, R.; Victor, R. Association between Chronic Kidney Disease and Coronary Artery Calcification: The Dallas Heart Study. *J. Am. Soc. Nephrol.* **2004**, *16*, 507–513. [CrossRef] [PubMed]
115. Qunibi, W.Y.; AbouZahr, F.; Mizani, M.R.; Nolan, C.R.; Arya, R.; Hunt, K.J. Cardiovascular calcification in Hispanic Americans (HA) with chronic kidney disease (CKD) due to type 2 diabetes. *Kidney Int.* **2005**, *68*, 271–277. [CrossRef]
116. Russo, D.; Palmiero, G.; De Blasio, A.P.; Balletta, M.M.; Andreucci, V.E. Coronary artery calcification in patients with CRF not undergoing dialysis. *Am. J. Kidney Dis.* **2004**, *44*, 1024–1030. [CrossRef] [PubMed]
117. Chiu, Y.-W.; Adler, S.G.; Budoff, M.J.; Takasu, J.; Ashai, J.; Mehrotra, R. Coronary artery calcification and mortality in diabetic patients with proteinuria. *Kidney Int.* **2010**, *77*, 1107–1114. [CrossRef] [PubMed]
118. Bellasi, A.; Raggi, P. Vascular calcification in patients with kidney disease: Techniques and Technologies to Assess Vascular Calcification. *Semin. Dial.* **2007**, *20*, 129–133. [CrossRef]
119. Uhlig, K. There Is No Practical Utility in Routinely Screening Dialysis Patients for Vascular Calcification. *Semin. Dial.* **2010**, *23*, 277–279. [CrossRef]
120. Beto, J.; Bhatt, N.; Gerbeling, T.; Patel, C.; Drayer, D. Overview of the 2017 KDIGO CKD-MBD Update: Practice Implications for Adult Hemodialysis Patients. *J. Ren. Nutr.* **2019**, *29*, 2–15. [CrossRef]
121. Ix, J.H.; Katz, R.; Kestenbaum, B.; Fried, L.F.; Kramer, H.; Stehman-Breen, C.; Shlipak, M.G. Association of Mild to Moderate Kidney Dysfunction and Coronary Calcification. *J. Am. Soc. Nephrol.* **2008**, *19*, 579–585. [CrossRef] [PubMed]
122. Jun, M.; Lv, J.; Perkovic, V.; Jardine, M.J. Managing cardiovascular risk in people with chronic kidney disease: A review of the evidence from randomized controlled trials. *Ther. Adv. Chronic Dis.* **2011**, *2*, 265–278. [CrossRef]
123. Himmelsbach, A.; Ciliox, C.; Goettsch, C. Cardiovascular Calcification in Chronic Kidney Disease—Therapeutic Opportunities. *Toxins* **2020**, *12*, 181. [CrossRef] [PubMed]
124. Jamal, S.A.; Vandermeer, B.; Raggi, P.; Mendelssohn, D.C.; Chatterley, T.; Dorgan, M.; Lok, C.E.; Fitchett, D.; Tsuyuki, R.T. Effect of calcium-based versus non-calcium-based phosphate binders on mortality in patients with chronic kidney disease: An updated systematic review and meta-analysis. *Lancet* **2013**, *382*, 1268–1277. [CrossRef]
125. Block, G.A.; Bushinsky, D.A.; Cheng, S.; Cunningham, J.; Dehmel, B.; Drueke, T.B.; Ketteler, M.; KewalRamani, R.; Martin, K.J.; Moe, S.M.; et al. Effect of Etelcalcetide vs. Cinacalcet on Serum Parathyroid Hormone in Patients Receiving Hemodialysis with Secondary Hyperparathyroidism. *JAMA* **2017**, *317*, 156–164. [CrossRef]
126. Chitalia, N.; Recio-Mayoral, A.; Kaski, J.C.; Banerjee, D. Vitamin D deficiency and endothelial dysfunction in non-dialysis chronic kidney disease patients. *Atherosclerosis* **2012**, *220*, 265–268. [CrossRef]
127. Perazella, M.A.; Markowitz, G.S. Bisphosphonate nephrotoxicity. *Kidney Int.* **2008**, *74*, 1385–1393. [CrossRef]
128. Bergner, R.; Diel, I.J.; Henrich, D.; Hoffmann, M.; Uppenkamp, M. Differences in Nephrotoxicity of Intravenous Bisphosphonates for the Treatment of Malignancy Related Bone Disease. *Oncol. Res. Treat.* **2006**, *29*, 534–540. [CrossRef]
129. Louvet, L.; Büchel, J.; Steppan, S.; Passlick-Deetjen, J.; Massy, Z.A. Magnesium prevents phosphate-induced calcification in human aortic vascular smooth muscle cells. *Nephrol. Dial. Transplant.* **2012**, *28*, 869–878. [CrossRef]
130. Sharples, E.J.; Pereira, D.; Summers, S.; Cunningham, J.; Rubens, M.; Goldsmith, D.; Yaqoob, M.M. Coronary artery calcification measured with electron-beam computerized tomography correlates poorly with coronary artery angiography in dialysis patients. *Am. J. Kidney Dis.* **2004**, *43*, 313–319. [CrossRef]

131. Maniscalco, B.S.; Taylor, K.A. Calcification in coronary artery disease can be reversed by EDTA–tetracycline long-term chemotherapy. *Pathophysiology* **2004**, *11*, 95–101. [CrossRef] [PubMed]
132. Krieger, N.S.; Frick, K.K.; Bushinsky, D.A. Mechanism of acid-induced bone resorption. *Curr. Opin. Nephrol. Hypertens.* **2004**, *13*, 423–436. [CrossRef]
133. Mendoza, F.J.; Lopez, I.; Montes de Oca, A.; Perez, J.; Rodriguez, M.; Aguilera-Tejero, E. Metabolicacidosis inhibits soft tissue calcification in uremic rats. *Kidney Int.* **2008**, *73*, 407–414. [CrossRef]
134. Simpson, C.L.; Lindley, S.; Eisenberg, C.; Basalyga, D.M.; Starcher, B.C.; Simionescu, D.T.; Vyavahare, N.R. Toward cell therapy for vascular calcification: Osteoclast-mediated demineralization of calcifiedelastin. *Cardiovasc. Pathol.* **2007**, *16*, 29–37. [CrossRef] [PubMed]
135. Neven, E.; D'Haese, P.C. Vascular Calcification in Chronic Renal Failure. *Circ. Res.* **2011**, *108*, 249–264. [CrossRef] [PubMed]
136. Shobeiri, N.; Adams, M.; Holden, R. Vascular Calcification in Animal Models of CKD: A Review. *Am. J. Nephrol.* **2010**, *31*, 471–481. [CrossRef]
137. Herrmann, J.; Babic, M.; Tölle, M.; Van Der Giet, M.; Schuchardt, M. Research Models for Studying Vascular Calcification. *Int. J. Mol. Sci.* **2020**, *21*, 2204. [CrossRef]
138. Sinha, S.; Eddington, H.; Kalra, P.A. Vascular calcification: Lessons from scientific models. *J. Ren. Care* **2009**, *35*, 51–56. [CrossRef]

Review

Experimental and Clinical Applications of Red and Near-Infrared Photobiomodulation on Endothelial Dysfunction: A Review

Esteban Colombo [1,†], Antonio Signore [1,2,†], Stefano Aicardi [3], Angelina Zekiy [4], Anatoliy Utyuzh [4], Stefano Benedicenti [1] and Andrea Amaroli [1,4,*]

[1] Laser Therapy Centre, Department of Surgical and Diagnostic Sciences, University of Genoa, 16132 Genoa, Italy; esteban.colombo92@gmail.com (E.C.); dr.signore@icloud.com (A.S.); stefano.benedicenti@unige.it (S.B.)

[2] Department of Therapeutic Dentistry, Faculty of Dentistry, First Moscow State Medical University (Sechenov University), 119991 Moscow, Russia

[3] Department for the Earth, Environment and Life Sciences, University of Genoa, 16132 Genoa, Italy; stefano.aicardi94@libero.it

[4] Department of Orthopaedic Dentistry, Faculty of Dentistry, First Moscow State Medical University (Sechenov University), 119991 Moscow, Russia; zekiy82@bk.ru (A.Z.); anatoliy.utyuzh@gmail.com (A.U.)

* Correspondence: andrea.amaroli.71@gmail.com; Tel.: +39-010-3537309

† These authors contributed equally to this work.

Abstract: Background: Under physiological conditions, endothelial cells are the main regulator of arterial tone homeostasis and vascular growth, sensing and transducing signals between tissue and blood. Disease risk factors can lead to their unbalanced homeostasis, known as endothelial dysfunction. Red and near-infrared light can interact with animal cells and modulate their metabolism upon interaction with mitochondria's cytochromes, which leads to increased oxygen consumption, ATP production and ROS, as well as to regulate NO release and intracellular Ca^{2+} concentration. This medical subject is known as photobiomodulation (PBM). We present a review of the literature on the in vitro and in vivo effects of PBM on endothelial dysfunction. Methods: A search strategy was developed consistent with the PRISMA statement. The PubMed, Scopus, Cochrane, and Scholar electronic databases were consulted to search for in vitro and in vivo studies. Results: Fifty out of >12,000 articles were selected. Conclusions: The PBM can modulate endothelial dysfunction, improving inflammation, angiogenesis, and vasodilatation. Among the studies, 808 nm and 18 J (0.2 W, 2.05 cm^2) intracoronary irradiation can prevent restenosis as well as 645 nm and 20 J (0.25 W, 2 cm^2) can stimulate angiogenesis. PBM can also support hypertension cure. However, more extensive randomised controlled trials are necessary.

Keywords: low-level laser therapy; phototherapy; endothelium; vascular disease; healing; angiogenesis; ischemia; hypertension; inflammation; nitric oxide

1. Introduction

An increase in the prosperity and the prevention of childhood mortality through socioeconomic and scientific progress has led to a shift, over the last two decades years, from risks for infancy diseases towards those for adulthood diseases [1,2].

Indeed, the main disease risk factors implicated for worldwide deaths are hypertension (approximately 7–18%), ischemic heart disease (approximately 5–14%), and cerebrovascular disease (approximately 4–14%), which are all threats of cardiovascular illness [3].

The endothelium is a monolayer of endothelial cells cladding the lumen of the vascular beds of the entire cardiovascular system, from the heart to the smallest capillaries.

It is a metabolically and mechanically dynamic organ, separating the vessel wall from the blood- and its components [4]. The endothelium, which was for a long time considered to be a relatively inert cellular monolayer, has recently been recognised as an important modulator of key physiological functions [1].

The endothelium's role is carried out through membrane-bound receptors for hormones, proteins, metabolites, lipid transporting particles, as well as through specific junctional proteins and receptors that regulate cell–cell and cell-matrix interactions [5,6].

Recently, the important regulatory role of mitochondria on endothelial cells' cytoprotective phenomena have been evidenced. In addition, stimuli affecting mitochondrial dynamics of endothelial cells can affect the production of reactive oxygen species (ROS) and adenosine triphosphate (ATP), leading to physiological or pathological response [7]. Additionally, endothelial functions may also depend on changes in intracellular calcium (Ca^{2+}) concentration, depending on depletion of Ca^{2+} stores and in-out-in Ca^{2+} signalling events, as well as multiple transient receptor potential (TRP) channel isoforms activation [8,9].

Under physiological conditions, endothelial cells are the main regulators of arterial tone homeostasis and vascular growth, sensing and transducing signals between tissue and blood [6]. Endothelial cells, through different antiplatelet and anticoagulant mechanisms, counteract thrombosis and, by regulating the expression of binding sites for procoagulant and anticoagulant agents, control the clotting system. Additionally, platelet adhesion to leukocytes represents the initial stage leading to exudation of leukocytes to inflammation or infection areas, followed by a platelet–leukocyte interaction and aggregation, and then vascular occlusion [6]. Lastly, endothelial cells, thanks to several vasoactive substances such as endothelin, angiotensin, nitric oxide (NO), vascular endothelial growth factor (VEGF), for instance, exert a regulatory role in vessel tone and growth, as well as cell proliferation and angiogenesis [1,5,6].

The dynamic features of the endothelium permit rapid response to diverse stimuli such as coagulation proteins, microbial components, shear stress, growth factors, and cytokines. These "activation" responses have evolved for host defence against microorganisms and the repair of tissue injury and are generally localised and beneficial [10].

However, this well-balanced endothelial regulation of vascular function can be affected by several disease risk factors, leading to both unbalanced homeostasis and an unphysiological state of healthy endothelial cells, which is known as endothelial dysfunction. As a consequence, endothelium shows reduced vasodilation induction, experiences a proinflammatory state, and prothrombic properties, which are associated with most forms of cardiovascular diseases, such as coronary artery disease, hypertension, peripheral vascular disease, chronic heart failure, chronic kidney, diabetes, failure, and severe viral infections [10].

However, as carefully reviewed by Daiber et al. [1], the evidence that the progression of endothelial dysfunction can be counteracted and reversed increases the possibility of retarding and, in some cases, preventing the progression of the related diseases and encourages efforts to explore new therapies.

Routinely, treatments applied for patient care have focused on indirect pleiotropic antioxidant properties and modulating NADPH oxidase enzymes and mitochondrial activities to prevent endothelial nitric oxide synthase (eNOS) uncoupling and stimulating anti-inflammatory effects, leading to improved endothelial function. Pharmacological agents can also be employed to equilibrate the oxidatively impaired activity of soluble guanylyl cyclase of smooth muscle [1].

In addition, mitochondria- and calcium-targeted therapeutics to counteract oxidative stress and improve endothelial dysfunction in cardiovascular disease have also been developed [11].

As fully detailed in the next paragraph, light in the red and near-infrared range of wavelength can interact with animal cells and can modulate their metabolism. Indeed, studies conducted by our research team have shown that modulation in oxygen consumption, ATP production, and ROS, after interaction of those wavelengths with mitochondrial cytochromes, as well as light, can regulate the NO release and the intracellular Ca^{2+} concentration [12–15]. This medical topic has been defined as low-level laser therapy, but recently, the more appropriate definition of photobiomodulation (PBM) was introduced [14].

Photobiomodulation can, thereby, counteract inflammation, stimulate growth factor expression, and modulate many cellular pathways [12,14] by different strategies, as described in the paragraph below. Therefore, PBM can support clinicians and their patients to manage and experience faster healing. In this review, we explore the ability of red and near-infrared (NIR) light to modulate cell homeostasis with particular attention to the endothelial cell line, in vitro. Additionally, the in vivo therapeutic effects of PBM on animal and human subjects are discussed. In particular, we present a review from the cell to patient literature on the effect of PBM on endothelial dysfunction.

1.1. Molecular Targets of Red and Near-Infrared Light: Primary and Secondary Effects

Interactions of light at the visible-red and near-infrared wavelengths with non-plant cells have been described [12]. Animal cells interacting with light evolved the ability to use specific light stimuli for vitamin-D production and vision, as examples. Although animal cells did not choose sunlight as a source of energy for their metabolism, a process close to photosynthesis, i.e., light–cell interaction can also take place in cells. This process is referred to as photobiomodulation and describes the ability of photons to interact with atoms and molecules within the cell and induces biological modulation on tissue homeostasis [12]. Specifically, following the standard model of particle physics, i.e., a paradigm of quantum field theory, the four known forces are settled by a fundamental carrier particle, the boson. Photons, belonging to the family of bosons, are the core particle of light and they can carry electromagnetic force [16]. Therefore, they can behave as a power source, where the brightness of light describes the photons' number and their colour is the result of energy contained in each photon. In this way, photons are theoretically able to interact with cells, and then energetically modulate the homeostasis of the tissue [12,16]. However, to make photobiomodulation happen, an energy conversion needs to take place. Indeed, molecules in living systems can absorb photons' energy, reaching an electronically excited state that temporarily modifies their conformation and function.

The uptake of photons' energy by cells needs biomolecules that may undergo an excited state [12]. In the life forms, two types of molecules exist, those specialised to absorb light, such as the photoreceptors, and non-specialised molecules, such as the photoacceptors. The latter are more common than photoreceptors and are part of ubiquitous metabolic pathways not directly related to light processing [17]; this explains how photobiomodulation can affect key cellular pathways of all life forms, from protozoa to humans [12].

Not all molecules feature light-acceptor ability. Organic cofactors or metal ions are fundamental elements that allow excitation of molecules from the ground state to an excited state and their change in conformation and function [18].

Water deserves a separate discussion, as a polar molecule with more hydrogens than atoms, as well as nonlinear and time-dependent chaotic behaviours, because it display extremely complex vibrations. As effectively discussed by Santana-Blank and collaborators, water was considered to be an innocuous medium for a long time [16]. However, recently its key role in physiologic mechanisms has been reconsidered, and therefore the possibility that external radiant energies may route its organisation and selectively lead biological function. For this reason, water in the 600–1100 nm range, can behave as a biological photoacceptor, despite its relatively low coefficient of absorption, and therefore can lead and improve biological reactions [16,19].

Among the different molecules within cells, haemoglobin is a well-known photoacceptor, which reacts with visible and NIR light based on wavelengths and its redox state. Additionally, behaviour such as photoacceptors was also described in other cellular molecules such as catalase, cytochromes, cryptochromes, nitric oxide synthase, nitrosothiols and dinitrosyl iron complexes, and superoxide dismutase [17,18]. Therefore, since many of these photoacceptors were described in the mitochondria, it is not strange that this organelle is considered to be the elective cellular target to explain the light and animal–cell interaction for photobiomodulation. The parallel and convergent evolution of both chloroplasts and mitochondria, from ancestral bacteria, has to be also taken into account to explain mitochondria's ability for responding to light stimuli [12]. Experiments have demonstrated that isolated mitochondria are sensitive to red and NIR spectrum and that the interaction increases ATP synthesis and oxygen consumption, as well as ROS generation, Ca^{2+} modulation, and photodissociation of NO from cytochrome c oxidase [12]. In particular, the cytochrome c oxidase (complex IV) has been shown to be activated in vitro by a red laser (633 nm) [20]. According to metal-ligand systems and absorption spectra, such as 450, 620–680, and 760–895 nm, characteristically different peaks may be related to it [21]. In previous papers, we have shown that 808 nm stimulated complex IV electively and that complex III was excited poorly, while complexes I and II were not affected [12,22]. At 980-nm the interaction behaved as a window effect and it interested complexes III and IV [23]. In addition, by increasing the wavelength to 1064 nm, complexes I, III, and IV were influenced, while the extrinsic mitochondrial membrane complex II seems again to be not receptive to photons at this wavelength [24]. Lastly, red and NIR light exposure induces NO release from S-nitrosothiols and di-nitrosyl iron complexes in a wavelength-dependent modality [25], as well as NO concentration, which can be changed by modulation of eNOS activity and expression after light exposure [13,26]. Lastly, Wang et al. [27] showed, for the parameters tested, that 980 nm affected temperature-gated Ca^{2+} ion channels, also probably via water resonance influence and transient "heating". Altogether, this direct and immediate photochemical change in the photoacceptor's primary targets by light can be considered to be primary effects [12,17] (Figure 1A). They lead to the secondary effects of photobiomodulation, which involve the expression or activation of second messengers, modulation of gene expression and enzyme activities, and then signalling pathways that result in macroscopical physiological consequences [12,17] (Figure 1B). In this way, the PBM, irradiated by intravenous, transmucosal, transcranial, or transcutaneous mode, can affect, after minutes as well as days, the homeostasis of tissue and represents a novel paradigm to treat altered physiological conditions [12,14,17].

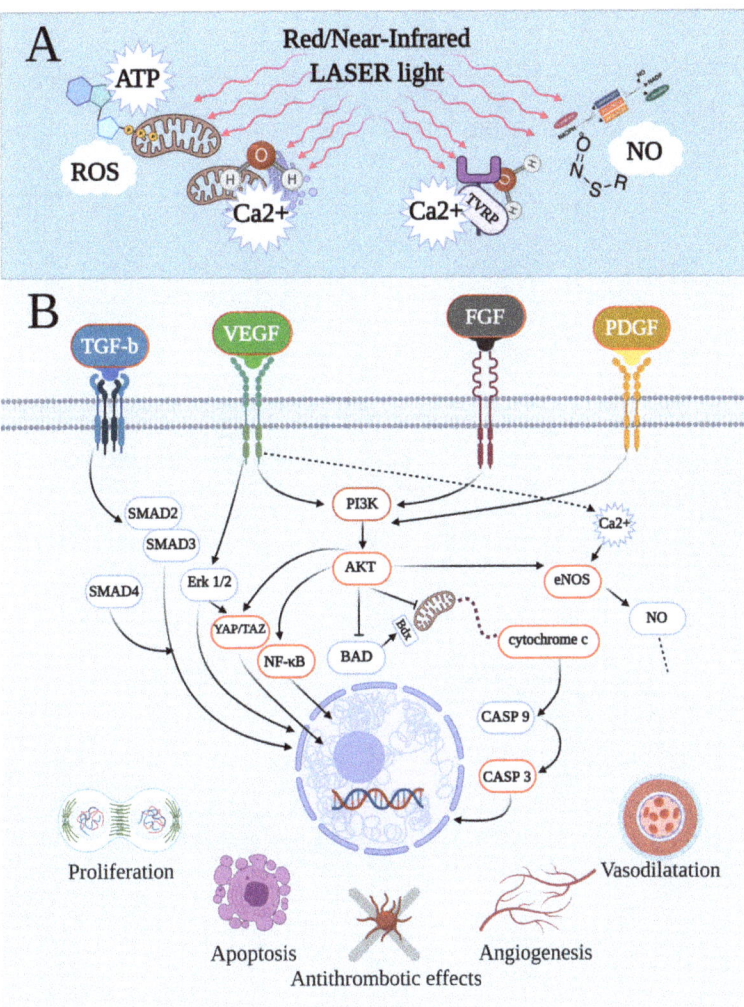

Figure 1. Endothelial-cell pathways modulated by red and near-Infrared laser lights, primary targets and secondary effects. As the first step, both red and near-infrared light interacted with the primary target, inducing immediate photochemical changes in the photoacceptor. This controlled the stimulation of the primary effects (**A**), such as the modulation's levels of adenosine triphosphate (ATP), reactive oxygen species (ROS), intracellular calcium concentration (Ca^{2+}), and nitric oxide (NO). The primary effects lead to the secondary effects (**B**) of photobiomodulation, which involve expression or activation of second messengers, modulation of gene expression and enzyme activities, and then signalling pathways, which resulted in physiological events such as proliferation/healing, apoptosis, antithrombotic effects, angiogenesis, and vasodilatation (**B**), as examples. AKT, alpha serine/threonine kinases; BAD, BCL2 associated agonist of cell death; CASP, caspase; eNOS, endothelial-nitric oxide synthase; Erk, extracellular signal-regulated kinase; FGF, fibroblast growth factor; YAP/TAZ, yes-associated protein/transcriptional coactivator with PDZ-binding motif; Nf-κB, nuclear factor-κB; PI3K, phosphoinositide 3-kinases; PDGF, platelet-derived growth factor; SMAD, small mother against decapentaplegic; TGF-b, transforming growth factor-b; VEGF, vascular endothelial growth factor.

2. Materials and Methods

Our review was carried out in compliance with the PRISMA guidelines (Supplemental Material Figure S1). Papers were independently searched by four authors (A.A., E.C., A.S., and S.A.) on the PubMed, Scopus, Cochrane, and Scholar databases. The following keywords were applied to meet the strategy investigation: "low-level laser therapy" OR "photobiomodulation" OR "laser phototherapy" AND "endothelial OR "endothelial dysfunction" OR "endothelium" OR "vascular" OR "vascular dysfunction". Additional studies were also identified from the references. Articles were listed and duplicates were deleted by all the authors as a consequence of the large number of papers identified. We also initially screened the works by title and abstracts according to inclusion and exclusion criteria. The inclusion criteria included the following: (1) studies published in English in journals with a peer-review process before publication; (2) works published before 1 September 2020; (3) studies that complied with the topic of the review; (4) a clear description of the type of light emitting diode (LED) or laser device and treatment parameters employed; (5) therapies were immediately traceable to PBM; (6) type of articles such as original research, case reports, and short communications; and (7) articles drafted according to "parameter reproducibility in photobiomodulation" by Tunér and Jenkins [28]. The exclusion criteria included the following: (1) in vitro studies on stem cell or cell line not referring to endothelium; (2) LED or laser therapies not adhering to the principles of PBM; (3) studies not focused on the topic of the review; and (4) types of articles such as reviews, abstracts to congress, and patents. The selection process is available in Supplemental Material Figure S1.

3. Results

Fifty out of >12,000 articles selected by PubMed, Scopus, Cochrane, and Scholar, as well as from article references were judged to be eligible for the review (Supplemental Material Figure S1). Actually, research by Scholar was the most dispersive, and no paper after the first 150 papers screened was included in the review. Conversely, PubMed resulted in being the most inclusive and Scopus and Cochrane showed similar useful support.

Following a previous review of PBM on bone socket preservation [14], the most frequent reason for exclusion was the unsuitability with the parameter reproducibility in photobiomodulation described by Tunér and Jenkins (2016) [28]. Four papers out of 54 were rejected from the eligibility because they did not completely fit the inclusion criteria after the full texts were read.

Unfortunately, only 17% of the studies had used a power meter to measure the power at the target, and in many studies, the probe used for irradiation was not clearly described, as well as only a few of the studies clearly described the PBM's parameters in the abstract.

3.1. In Vitro Studies

Concerning the in vitro studies (Table 1), ~50% of the experiments were performed on human umbilical vein endothelial cell lines (HUVEC, 62.5% were umbilical vein primary cell culture and 37.5% were immortalised cells), while 29% of the studies used other primary cell cultures such as human endothelial cells isolated from coronary vessels (HuC EC), neonatal rat ventricular myocytes (RVM-neo), rabbit aorta endothelial cell line (RAEC), human adjacent annulus fibrosus cells (h-AFC), and human pulp fibroblasts HPF. Lastly, ~23% of the experiments were performed on other transformed cell lines such as mouse NCTC clone 929 Clone of strain L (L929), permanent human umbilical vein cell line (EA.hy926), human dermal microvascular endothelial cells (HMVEC-d), and human endothelial cells from the umbilical cord (HECV).

Table 1. In vitro studies on photobiomodulation and endothelial dysfunction, selected after inclusion and exclusion criteria screening. The table shows the schematic design of the experimental setup and the results.

Cell Line	Wavelength	Parameters Irradiated	Methods	Effect of PBM
HuC EC (PCC) [29]	632 nm Laser	Power = 0.0035 W; power density = 0.0017 W/cm^2; time = 0, 60, 180, 300, 600, 900, 1200, 1800, 2400, and 3600 s; spot area = 2 cm^2; energy = from 0 to 12.6 J; fluence = from 0 to 6.3 J/cm^2; mode = continuous wave (CW)	No. of irradiations = 1. Investigation setup: proliferation and human VEGF immunoassay	Stimulation of endothelial cell growth; increment of VEGF secretion by 0.63, 1.05, 2.1, and 4.2 J; no effect by 0.21 and 6.3 J; decrement of VEGF secretion by 8.4 and 12.6 J
RVM-neo (PCC) [26]	670 nm LED	Power = 0.005, 0.025, and 0.05 W; power density = 0.005, 0.025, and 0.05 W/cm^2; time = 300 s; spot area = 1 cm^2; energy = 1.5, 7.5, and 15 J; fluence = 1.5, 7.5, and 15 J/cm^2; mode = CW	No. of irradiations = 1 with incremented energies (1.5, 7.5, and 15 J) in LDH and Casp experiments. 1 with fluence 7.5 J/cm^2 in other experiments. Investigation setup: cell viability; Casp 3 assay; flow cytometer for annexin V, cytochrome c release assay; measurement of intracellular NO, oxygen consumption, ATP synthesis	Decrement of apoptosis indices (casp 3, annexin, cytochrome c); increment of ATP and NO production
RAEC (PCC) [30]	685 nm Laser	Power = 0.02 W, Power density = 0.011 W/cm^2, Time = 720 s, Spot area = 1.8 cm^2, Energy = 14.4 J, Fluence = 8 J/cm^2, Mode = continuous wave (CW)	No. of irradiations = 4 (12, 24, 36, and 48 h after plating). Investigation set-up: proliferation assay; immunohistochemistry for actin filaments	Stimulation of cellular proliferation; changes in the cytoskeleton through the reorganization of actin filaments and neo-formation of stress fibres
h-AFC (PCC) [31]	645 nm Laser	Power = 0.025, 0.010, 0.012 W; power density = 0.009, 0.003, 0.004 W/cm^2; time = 640, 1591, 1257 s to generate 6 J; time = 1278, 3183, 2515 s to generate 32 J; time = 2559, 6366, 5029 s to generate 64 J; spot area = 2.8 cm^2; energy = 16, 32, and 64 J; fluence = 5.76, 11.51, 23.02 J/cm^2; mode = CW	No. of irradiations = 1. Investigation set-up: cell cytotoxicity and LDH release assay. Immunofluorescence and qRT-PCR: TNF-α, IL-1β, MMP1, MMP3, IL-6, IL-8, VEGF-A, VEGF-B, VEGF-C, NGF, and BDNF	Decrement of inflammatory mediators, and catabolic enzymes; 32 J inhibited MMP1, MMP3 BDNF; 64 J inhibited MMP1, MMP3, BDNF, IL-8
HPF (PCC) [32]	660 nm Laser	Power = 0.01 W or 0.015 W; power density = 0.25 or 0.37 W/cm^2; time = 10 s; spot area = 0.04 cm^2; energy = 0.1 or 0.15 J; fluence = 2.5 or 3.7 J/cm^2; mode = CW	No. of irradiations = 1. Investigation setup: viability and proliferation assay; ELISA, VEGF-C, VEGF-A, VEGFR2, FGF-2, PDGF, VEGFR1, PECAM-1, VEGF-D, PLGF, BMP-9	Both energy increased secretion of VEGF-A, VEGF-C, and VEGFR1; upregulation of BMP-9; downregulation of PDGF by both energies; 0.1 J was better than 0.15 J for capillary-like structure formation
HUVEC(PCC) [33]	623.5 nm Laser	Power = 0.0013 W; power density = 0.0001 W/cm^2; time = s; spot area = 9.06 cm^2; energy = 0.78, 2.34, 4.68, 7.02; fluence = 0.086, 0.26, 0.52, 0.77 J/cm^2; mode = CW	No. of irradiations = 1. Investigation setup: nitrate and nitrite Griess Assay; qRT-PCR, eNOS; Western Blot Analysis, eNOS and vinculin; migration assay; tube formation analysis	Upregulation of eNOS expression through PI3K pathway and increment of vinculin protein; cell migration promoted with 2.34 J

Table 1. Cont.

Cell Line	Wavelength	Parameters Irradiated	Methods	Effect of PBM
HUVEC [34]	660, 670, 820, 808 nm Laser	Power = 0.01, 0.02, 0.04, 0.1, 1.5 W; power density = 0.028, 0.057, 0.11, 0.28, 4.28 W/cm^2; time = from 5 to 100 s; spot area = 0.35 cm^2; energy = from 0.05 to 150 J; fluence from 0.14 to 428 J/cm^2; mode = CW	No. of irradiations = 1. Investigation setup: cell proliferation assay	Stimulation of cell proliferation
HUVEC$^{(PCC)}$ [35]	635, 830 nm Laser	635-nm: Power = 0.15 W Power density = 0.00187 W/cm^2 Time = 1066, 2133, 4266 s Spot area = 80 cm^2, Energy = 160, 320, 640 J, Fluence 2, 4, 8 J/cm^2 Mode = CW 830-nm: Power = 0.3 W Power density = 0.00375 W/cm^2 Time = 533, 1066, 2133 s Spot area = 80 cm^2, Energy = 160, 320, 640 J, Fluence 2, 4, 8 J/cm^2 Mode = CW	No. of irradiations = 2 radiations on the day 2 and 4 with one day-break Investigation set-up: proliferation assay; ELISA, VEGF-A, TGF-1	635 nm increased cell proliferation and decreases VEGF-A concentration; 830 nm decreased TGF-1 concentration
HUVEC$^{(PCC)}$ [36]	635 nm Laser	Power = 0.15 W; power density = 0.00187 W/cm^2; time = 1066, 2133, 4266 s; spot area = 80 cm^2; energy = 160, 320, 640 J; fluence 2, 4, 8 J/cm^2; mode = CW	No. of irradiations = 2 radiations on day 2 and 4 with one day break. Investigation setup: proliferation assay; ELISA test, VEGF-A and presence of soluble VEGF receptors (sVEGFR-1 and sVEGFR-2)	Decrement of VEGF-A, sVEGFR-1, and sVEGFR-2; activation of cell proliferation
HUVEC$^{(PCC)}$ hyperglycemia [37]	635, 830 nm Laser	635 nm: Power = 0.15 W; power density = 0.00187 W/cm^2; time = 1066 s; spot area = 80 cm^2; energy = 160 J; fluence 2 J/cm^2; mode = CW 830 nm: Power = 0.3 W; power density = 0.00375 W/cm^2; Time = 533 s; spot area = 80 cm^2; energy = 160 J; fluence 2 J/cm^2; mode = CW	No. of irradiations = 2 on day 5 and 6. Investigation setup: induction of hyperglycemia; proliferation assay; ELISA, TNF-α and IL-6	Increment of proliferation; reduction of inflammation by decrease of TNF-α and IL-6; 830 nm affect more than 635 nm
HUVEC$^{(PCC)}$ hyperglycemia [38]	635, 830 nm Laser	635 nm: Power = 0.15 W; power density = 0.00187 W/cm^2; time = 1066 s; spot area = 80 cm^2; energy = 160 J; fluence 2 J/cm^2; mode = CW 830 nm: Power = 0.3 W; power density = 0.00375 W/cm^2; time = 533 s; spot area = 80 cm^2; energy = 160 J; fluence 2 J/cm^2; mode = CW	No. of irradiations = 2 on day 5 and 6. Investigation setup: induction of hyperglycemia; proliferation assay. ELISA, sE-selectin and sVCAM	Decrement in sE-selectin and sVCAM concentration; increment of proliferation; 830 nm affect more than 635 nm
HUVEC [39]	650 nm Laser	Power = ~2 W; power density = 0.031, 0.011, 0.002 W/cm^2; time = from 16 to 1920 s; spot area = 63.6, 191, 961.6 cm^2; energy = from 32 to 3,840 J; fluence = from 0.5 to 20 J/cm^2; mode = CW	No. of irradiations = 1. Investigation setup: proliferation assay; scratch test; tube formation assay; Western blot analysis, PI3K, p-PI3K, Akt, p-Akt, VEGF-A, eNOS, HIF-1α. ELISA, VEGF-A	Increment of proliferation, migration, and tube formation; activation of by PI3K/Akt signalling pathway

Table 1. Cont.

Cell Line	Wavelength	Parameters Irradiated	Methods	Effect of PBM
HUVEC [40]	660, 780 nm Laser	Power = 0.04 W; power density = 1 W/cm^2; time = 1, 5, 10, 20 s; spot area = 0.04 cm^2; energy = 0.04, 0.2, 0.4, 0.8 J; fluence 1, 5, 10, 20 J/cm^2; mode = CW	No. of irradiations = 1. Investigation setup: cell viability assay; total protein measure	660 nm induced increment of viability and concentration of total proteins; 0.2 and 0.4 J had better effect; 780 nm had inhibitory effect.
L929 [41]	904 nm Laser	Repetition rate = 10 KHz; output power = 50 mW; pulse width = 100 ns; peak power = 50 W; spot area = 0.01 cm^2; active cycle of 0.1%; energy = 200 J or 300 J; fluence = 2 J/cm^2 or 3 J/cm^2	No. of irradiations = 2 for two consecutive days. Investigation setup: qRT-PCR, COL1α1, VEGF	200 J incremented expression of the genes COL1α1 and VEGF; 300 J incremented expression of the genes VEGF
EA.hy926 [42]	660 nm Laser	Power = 1 W; power density = 0.003 W/cm^2; time = 3600 s; spot area = 314 cm^2; energy = 3600 J; fluence = 11.5 J/cm^2; mode = CW	No. of irradiations = 1 for proliferation. 1 + 1 irradiation after 24 h for other experiments. Investigation setup: Western blotting, phospho-ERK, ERK, phosphop38, p38, phospho-JNK, JNK, NF-κB, iNOS, and cleaved caspase 3/8/7/9, PARP	Protection against TNF/CHX-induced apoptosis by inhibition of p38 MAPK and NF-κB signals
HMVEC-d [43]	670 nm Laser	Power = 0.025 W, 0.05 W or 0.1 W; power density = 0.025, 0.05 or 0.1 W/cm^2; time = 30, 60 or 120 s; spot area = 1 cm^2; energy = 0.75, 1.5, 3, 6, or 12 J; fluence = 0.75, 1.5, 3, 6, 12 J/cm^2; mode = CW	No. of irradiations = 1. Investigation setup, NO production	Increment of NO release from bound substances in healthy and diabetic model
HECV-d [44]	808 nm Laser	Power = 0.95 W; power density = 0.95 W/cm^2; time = 60 s; spot area = 1 cm^2; energy = 57 J; fluence = 57 J/cm^2; mode = CW	No. of irradiations = 1 irradiation. Investigation setup: cell viability, lipid peroxidation, scratch-wound healing assay; nitrite/nitrate quantification; Western blotting, NF-κB p65; oxygen consumption measurements; ATP synthase activity assay	Increment of proliferation and migration; moderate increase in ROS production; no increment in oxidative or nitrosative stress and NF-κB p65; shift from anaerobic to aerobic metabolism and increment of ATP

Legend: Akt, alpha serine/threonine kinases; ATP, adenosine triphosphate; BDNF, brain-derived neurotrophic factor; BMP, bone morphogenetic protein; Casp, caspase; COL, collagen type; EA.hy926, permanent human umbilical vein cell line; ELISA, enzyme-linked immunosorbent assay; eNOS, endothelial nitric oxide synthase; ERK, extracellular signal-regulated kinase; FGF, fibroblast growth factor; h-AFC, human adjacent annulus fibrosus cells; HECV, human endothelial cells; HMVEC-d, human dermal microvascular endothelial cells; HIF, hypoxia-inducible factors; HPF, human pulp fibroblasts; HuC EC, human endothelial cells isolated from coronary vessels; HUVEC, human umbilical vein endothelial cells (immortalised); HUVEC$^{(PCC)}$, human umbilical vein endothelial cells from primary cell culture; IL, interleukin; iNOS, inducible nitric oxide synthase; JNK, c-Jun N-terminal kinase; L929, mouse NCTC clone 929 Clone of strain L; LDH, lactate dehydrogenase; MAPK, mitogen-activated protein kinase; MMP, matrix metalloproteinase; NF-κB, nuclear factor-κB; NGF, nerve growth factor; NO, nitric oxide; PARP, poly (ADP-ribose) polymerase; PCC, primary culture cells; PDGF, platelet derived growth factor; PECAM, platelet endothelial cell adhesion molecule; PI3K, phosphoinositide 3-kinases; PLGF, placental growth factor; qRT-PCR, real-time quantitative reverse transcription; RAEC, rabbit aorta endothelial cell line; ROS, reactive oxygen species; RVM-neo, neonatal rat ventricular myocytes; TGF, transforming growth factor; TNF-α, tumour necrosis factor alpha; TNF/CHX, tumour necrosis factor/cycloheximide; VEGF, vascular endothelial growth factor; VEGFR, vascular endothelial growth factor receptor; sVCAM, soluble vascular cell adhesion molecule; YAP/TAZ, yes-associated protein/transcriptional coactivator with PDZ-binding motif.

Except for two studies, the experiments were performed in continuous wave mode of irradiation, and an LED device was used in only one study instead of a diode laser. Lastly, the irradiation was prevalently performed by the visible red light; no study showed an adverse effect.

As described in the previous Section 1.1, PBM is the result of an interaction between light and photoacceptors (primary target, Figure 1A), which is followed by a succession of the primary (Figure 1A) and secondary effects (Figure 1B), and physiological events. Therefore, the results at the cellular level, summarized in Table 1, are considered to be secondary effects (Figure 1B) in response to molecules such as ATP, ROS, Ca^{2+}, and NO (Figure 1A).

The PBM affects the VEGF family and their receptors [29,31,32,35,36,41], as well as anti-inflammatory effects are evidenced by the impact on matrix metalloproteinases (MMP), tumour necrosis factor (TNF-α), interleukin (IL)-6, and IL-8 [30,37]. Additionally, a modulation on NO and ATP production was shown [26,33,43,44] and inhibitory and stimulatory effects on apoptosis [26,42] and proliferation/viability of the cells [30,34,38–40,44] were suggested, respectively.

3.2. In Vivo Preclinical (Animal) Studies

Except for one study on rabbits and one on hamsters, the preclinical experiments were carried out on murine models (Table 2).

Table 2. Preclinical in vivo studies on photobiomodulation and endothelial dysfunction, selected after inclusion and exclusion criteria screening. The table shows the schematic design of the experimental setup on animals and the results.

Animal Model	Wavelength	Parameters Irradiated	Methods	Effect of PBM
Rabbit (ischemia, cardiac disease) [45]	660 nm LED	Power = 0.003 or 0.06 W; power density = 0.003 or 0.06 W/cm^2; spot area = 1 cm^2, energy = 10.8 or 0.54 J; fluence = 10.8 or 0.54 J/cm^2; mode = continuous wave (CW)	No. of irradiations and mode = 1 irradiation and 3 or 5 cycles on 1 point with probe at 25 cm from target. Investigation setup: release of NO from nitrosyl heme proteins	Increment of NO release and cardioprotective effects
Rat (ischemia, long flaps) [46]	810 nm Laser	Power = 0.1 W; power density = 0.0314 W/cm^2; time = 360 s; spot area = 5.28 cm^2; energy = 59.66 J; fluence = 11.30 J/cm^2; mode = CW	No. of irradiations and mode = 1 irradiation for 4, 7, 10, and 14 days, on 1 point with the probe in contact mode. Investigation setup: immuno- and histochemical staining, VEGF, smooth muscle actin, factor VIII	Reduction of inflammation; increment of ischemic flap revascularization and flap viability; decrement of VEGF; increment of smooth muscle actin and factor VIII
Rat (ischemia, infarction) [47]	804 nm Laser	Power = 0.0157, 0.025, 0.037, 0.053 W; power density = 0.005, 0.008, 0.012, 0.017 W/cm^2; time = 120 s; spot area = 3.14 cm^2, energy = 1.88, 3, 4.52, 6.40 J; fluence = 0.6, 0.96, 1.44, 2.04 J/cm^2; mode = CW	No. of irradiations and mode = 1irradiation on 1 point on infarcted heart area. Investigation setup: infarct size and angiogenesis determination; immunoblot analysis, VEGF, iNOS	Increment of angiogenesis and cardioprotection; increment of VEGF and iNOS
Rat (ischemia, coronary) [48]	635 nm Laser	Power = 0.005 W; power density = 0.006 W/cm^2; time = 150 s; spot area = 0.8 cm^2; energy = 0.8 J, fluence = 1 J/cm^2; mode = CW	No. of irradiations and mode = 1 irradiation at 26 mm above the myocardium. Investigation setup: antibody array analysis for cytokines; ELISA, cytokine antibody; echocardiographic assessments	Improvement in ischemic heart disease; modulation of granulocyte-macrophage colony stimulating factor and fractalkine
Rat (ischemia, coronary) [49]	660 nm Laser	Power = 0.015 W, power density = 0.019 W/cm^2, time = 60 s, spot area = 0.785 cm^2, energy = 17.66 J, fluence = 22.5 J/cm^2, mode = CW	No. of irradiations and mode = 1 irradiation at 3 cm from target on 1 point. Investigation setup: biometric data and myocardial size; qRT-PCR and Western blot analysis, interleukins, Mas, kinin B2, and plasma kallikrein; plasma nitric oxide metabolites measurement.	Decrement of myocardium inflammation and infarct size; attenuation of left ventricle dysfunction; decrement of myocardial interleukin-1 beta, interleukin-6 and Mas receptor; increment of kinin B2 and plasma kallikrein; increment of NO derivatives.
Hamster (angiogenesis, mucositis) [50]	660 nm Laser	Power = 0.0328 or 0.0962 W; power density = 10.9 or 32 W/cm^2; time = 16 or 6 s; spot area = 0.003 cm^2; energy = 0.52 or 0.56 J); fluence = 173 or 187 J/cm^2; mode = continuous wave (CW)	No. of irradiations and mode = 1 irradiation on 5 points at days 3, 4, 5, and 6 of the experiment. Investigation setup: immunohistochemistry, COX-2, VEGF, factor VIII	Reduction of mucositis severity. 0.52 J decreased COX-2 and 0.56 J decreased VEGF
Mice HRS/J (angiogenesis, muscle) [51]	780 nm Laser	Power = 0.04 W; power density = 1 W/cm^2; time = 20 s; spot area = 0.04 cm^2; energy = 0.8 J; fluence = 20 J/cm^2; mode = continuous wave (CW)	No. of irradiations and mode = 1 irradiation on 1 point, for 3, 6, or 10 times, on alternate days. Contact mode. Investigation setup: immunoblot, MMP; immunoistochemistry, VEGF, VEGFR-2.	No effect on MMP; decrement of VEGF and VEGFR-2; the effects are visible only after the 10th irradiation

Table 2. Cont.

Animal Model	Wavelength	Parameters Irradiated	Methods	Effect of PBM
Rat (angiogenesis, muscle) [52]	660 nm Laser	Power = 0.02 or 0.04 W; power density = 0.05 or 0.1 W/cm^2; time = 20 s or 50; spot area = 0.4 cm^2; energy = 0.4 or 2 J; fluence = 1 or 5 J/cm^2; mode = CW	No. of irradiations and mode = 1 irradiation on 1 point. Treatments started 48 h post-surgery and were performed five times/week (each 24 h). Contact mode. Investigation setup: histopathological analysis; qRT-PCR, VEGF, COX-2, MyoD, myogenin	Improvement of muscle regeneration; decrement of COX-2; increment of VEGF, MyoD. Myogenin increased with 2 J
Rat (angiogenesis, muscle) [53]	904 nm Laser	Repetition rate = 9.500 Hz; output power r = 0.04 W; pulse width = 60 ns; peak power = 50 W; spot area = 0.1 cm^2; energy = 0.3 J or 0.5 J; fluence = 3 J/cm^2 or 5 J/cm^2	No. of irradiations and mode = 1 irradiation on 5 points, 2, 12, and 24 h after trauma. Probe at 0.5 cm from target. Investigation setup: Griess nitrite, lipid peroxidation, protein carbonylation, glutathione peroxidase, and catalase activity assay; dityrosine autofluorescence determination; qRT-PCR, VEGF, BDNF, IL-6, IL-10	Accelerated recovery; decrement of inflammation; decrement of IL-6; increment of IL-10; 0.3 J prevents thiobarbituric acid-reactive substance, carbonyl, superoxide dismutase, glutathione peroxidase, and catalase increment; BDNF and VEGF are prevented by irradiation
Rat (angiogenesis, skin) [54]	660 nm Laser	Power = 0.03 W; power density = 1.07 W/cm^2; time = 67 s; spot area = 0.028 cm^2; energy = 2 J; fluence = 72 J/cm^2; mode = CW	No. of irradiations and mode = 1 irradiation on 1 point. Alternate days for 14 days. Contact mode. Investigation setup: immunohistochemistry, VEGF, TIMP-2, MMP-3 and -9, collagen I, and III; qRT-PCR: IL-6; ELISA, CINC-1	Accelerated recovery in early stages of tissue repair; modulation of IL-6, CINC-1, VEGF, MMP-3, MMP-9 and TIMP-2; increment of collagen
Rat (angiogenesis, skin) [55]	670 nm Laser	Power = 0.03 W; power density = 0.476 W/cm^2; time = 30 s; spot area = 0.063 cm^2; energy = 0.9 J; fluence = 14.28 J/cm^2; mode = CW	No. of irradiations and mode = 1 irradiatio on 1 point. 15 consecutive days of treatment. Contact mode. Investigation setup: histological analysis; immunohistochemistry, collagen I, TNF-α, VEGF	Accelerated recovery of the cutaneous wound healing; decrement of inflammatory infiltrate and TNF-α; increment of VEGF and collagen type 1.
Rat (angiogenesis - skin) [56]	904 nm Laser	Repetition rate = 100 Hz; output power = 0.00078 W; pulse width = 200 ns; spot area = 1.77 cm^2; time = 600 s; energy = 0.4 J; fluence = 0.2 J/cm^2	No. of irradiations and mode = 1 irradiation on 1 point. Daily for seven days post-burn injury. Contact mode. Investigation setup: assays for oxidative stress and antioxidants markers, ROS, NO, lipid peroxidation, GSH, SOD, catalase, GPx, advanced oxidation protein products; immunoblot, Nrf2, HO-1, Txnrd2	Accelerated recovery of the burn wound healing; decrement of ROS, NO, lipid peroxidation, protein carbonylation, advanced oxidation protein product levels, GSH, and thiol (T-SH, Np-SH, P-SH); increment of endogenous antioxidants levels of SOD, catalase, GPx
Rat (angiogenesis, skin) [57]	660, 780 nm Laser	780 nm: Power = 0.07 W; power density = 1.75 W/cm^2; time = 20 s; spot area = 0.04 cm^2; energy = 1.4 J; fluence = 35 J/cm^2; mode = CW 606 nm: Power = 0.04 W; power density = 1 W/cm^2; time = 20 s; spot area a = 0.04 cm^2; energy = 0.8 J; fluence = 20 J/cm^2; mode = CW	No. of irradiations and mode = 1 irradiation on 2 points for two days. Probe at 1 mm from target. Investigation setup: qRT-PCR, VEGF	Accelerated recovery of wound healing; modulation of expression of VEGF

Table 2. Cont.

Animal Model	Wavelength	Parameters Irradiated	Methods	Effect of PBM
Rat (angiogenesis, skin) [58]	660 nm Laser 635 nm LED	Power = 0.04 W; power density = 0.32 W/cm^2; time = 62 s; spot area = 0.125 cm^2; energy = 2.5 J; fluence = 19.74 J/cm^2; mode = CW	No. of irradiations and mode = 1 irradiation on 4 points for 2, 4, or 6 days. Contact mode. Investigation setup: histology, collagen; immunohistology, TGF-β	Stimulation of angiogenesis; increment of collagen expression; increment of blood vessels formation; TGF-β no stimulated
Rats (angiogenesis, skin) [59]	660 nm Laser	Laser: Power = 0.04 W; power density = 1 W/cm^2; time = 31 or 126 s; spot area = 0.04 cm^2; energy = 0.2 or 0.8 J; fluence = 5 or 20 J/cm^2; mode = CW LED: Power = 0.09 W; power density = 1.06 W/cm^2; time = 17 or 56 s; spot area = 0.085 cm^2; energy = 0.42 or 1.7 J; fluence = 5 or 20 J/cm^2; mode = CW	No. of irradiations and mode = 1 irradiation on 7 points for 2, 6, 13, or 20 days. Contact mode. Investigation setup: histology	Improvement of angiogenesis; light coherence was shown not to be essential to angiogenesis
Rats (angiogenesis, skin) [60]	660 nm Laser	Power = 0.04 W; power density = 1 W/cm^2; time = 4 or 20 s; spot area = 0.04 cm^2; energy = 0.16 or 0.8 J; fluence = 4 or 20 J/cm^2; mode = CW	No. of irradiations and mode = 1 irradiation on 2 points for 14 days. Contact mode. Investigation setup: ELISA, IL-1β and TNF-α; image analysis for micro-vessel density	Improvement of oral wound repair and angiogenesis; increment of IL-1β and TNF-α
Rats (angiogenesis, skin) [61]	660, 780 nm Lasers	Power = 0.04 W; power density = 0.32 W/cm^2; time = 30 or 40 s; spot area = 0.125 cm^2; energy = 1.2 or 1.6 J; fluence = 9.6 or 12.8 J/cm^2; mode = CW	No. of irradiations and mode = 1 irradiation on 24 points for 4 days. Contact mode. Investigation setup: image analysis for micro-vessel density; immunoblotting, HIF-1α; qRT-PCR, VEGF; gelatin zymography, MMP-2 activity	Increment of new vessels formation; Increment of HIF-1α and VEGF; decrement of MMP-2
Rats (angiogenesis, skin) [62]	670 nm Laser	Power = 0.009 W; power density = 0.031 W/cm^2; time = 31 s; spot area = 0.28 cm^2; energy = 0.28 J; fluence = 1 J/cm^2; mode = CW	No. of irradiations and mode = 1 irradiation on 4 points for 4 days. Contact mode. Investigation setup: histomorphometry; immunohistochemistry, VEGF and CD31.	Improvement of late course of healing; increment of collagens and blood vessel; no effect on VEGF
Rats (angiogenesis, skin) [63]	670 nm Laser	Power = 0.009 W; power density = 0.031 W/cm^2; time = 31 s; spot area = 0.28 cm^2; energy = 0.28 J; fluence = 1 J/cm^2; mode = CW	No. of irradiations and mode = 1 irradiation on 4 points for 4 days. Contact mode. Investigation setup: histomorphometry; immunohistochemistry: CD31, NG2, smooth muscle alpha actin, CD8, CD68, Ptch, Gli-2, and Ihh.	Stimulation of later stages of wound healing and angiogenesis; decrement of CD68; increment of CD8
Rats (angiogenesis, tendon rupture) [64]	660 nm Laser	Power = 0.01 or 0.04 W; power density = 0.25 or 1 W/cm^2; time = 10 s; spot area = 0.04 cm^2; energy = 0.1 or 0.4 J; fluence = 2.5 or 10 J/cm^2; mode = CW	No. of irradiations and mode = 1 irradiation on 1 point for 3, 5 and 7 days. Contact mode. Investigation setup: India ink injection	Promotion of neovascularization

Table 2. Cont.

Animal Model	Wavelength	Parameters Irradiated	Methods	Effect of PBM
Aged rats (angiogenesis) [65]	830 nm Laser	Power = 0.05 W; power density = 1.8 W/cm^2; time = 60 s; spot area = 0.028 cm^2; energy = 3 J; fluence = 107 J/cm^2; mode = CW	No. of irradiations and mode = 1 irradiation on 1 point, daily for seven days post injury. Contact mode. Investigation setup: Immunohistochemistry: VEGF, MMP-3, and MMP-9; histochemistry, collagen type I and III	Increment of collagen type I and III production; downregulation of MMP-3 and MMP-9 expression; upregulation of VEGF
SHR rats (blood pressure) [66]	660 nm Laser	Power = 0.1 W; power density = 1.71 W/cm^2; time = 56 s; spot area = 0.058 cm^2; energy = 5.6 J; fluence = 96 J/cm^2; mode = CW	No. of irradiations and mode = 1 irradiation on 6 different points. Transcutaneously, with skin contact at 90° angle. Investigation setup: systolic arterial pressure; NO levels evaluation (abdominal region)	Reduction of systolic arterial pressure; increment of nitric oxide levels; no changement in heart rate
2K rats (blood pressure) [67]	660-nm Laser	Power = 0.1 W, power density = 3.57 W/cm^2, time = 56 s, spot area = 0.028 cm^2, energy = 5.6 J, fluence = 200 J/cm^2, mode = CW	No. of irradiations and mode = 1 irradiation on 6 points. Transcutaneously, with skin contact at 90° angle. Investigation setup: systolic arterial, diastolic arterial, mean arterial pressure, and heart rate were measured; NO levels evaluation	Induction of long lasting hypotensive effect; vasodilation by a NO dependent mechanism
2K-1C rats (blood pressure) [68]	660 nm Laser	Power = 0.1 W; power density = 0.14 W/cm^2; time = from 1 to 186 s; spot area = 0.722 cm^2; energy = 0.1, 0.3, 0.6, 1.2, 2.3, 4.7, 9.3, and 18.6 J; fluence = from 0.14 to 25.76 J/cm^2; mode = CW	No. of irradiations and mode = 1 irradiation on 6 simultaneous points. Transcutaneously, with skin contact at 90° angle. Investigation setup: systolic arterial, diastolic, arterial pressure and heart rate were measured	7.2–55.8 J is the effective therapeutic window to reduce pressure and heart rate and induce a long-lasting hypotensive effect

Legend: BDNF, brain-derived neurotrophic factor; CD, cluster of differentiation; COX, cyclooxygenase; CINC, cytokine-induced neutrophil chemoattractant; ELISA, enzyme-linked immunosorbent assay; GSH, glutathione; GSx, glutathione peroxidase; Gli-2, zinc finger protein GLI2; HIF: hypoxia-inducible factors; HO-1, heme oxygenase; IL, interleukin; Ihh, Indian hedgehog homolog; iNOS, inducible nitric oxide synthase; Mas = MAS-G-protein coupled receptor; MMP: matrix metalloproteinase; MyoD = myoblast determination protein; Nrf2 = nuclear factor erythroid 2-related factor; NO = nitric oxide Ptch = protein patched homolog; qRT-PCR = Real-Time quantitative reverse transcription; ROS = reactive oxygen species; SOD = superoxide dismutase; TGF = transforming growth factor; TIMP = tissue inhibitors of metalloproteinases; TNF-α = tumour necrosis factor alpha; Txnrd2 = thioredoxin reductase; VEGF = vascular endothelial growth factor; VEGFR = vascular endothelial growth factor receptor.

Sixty-six per cent of the studies investigated wound healing and angiogenesis, 21% of the studies were on ischemia, and only 13% of the studies were on blood pressure and micro- and macrocirculation. More than 70% of the studies used visible red light and in 87% of the studies, the light was irradiated with a diode laser. The screening on the literature of preclinical studies confirmed that the evidence shown was from in vitro experiments. No side effect was experienced by animals because of PBM, while angiogenesis was experienced. The data can be catalogued in three macro studies, such as wound healing and angiogenesis, ischemia, and blood features (both circulation and pressure). Basically, modulation of VEGF family proteins [46,47,50–53,55,57,61,65], anti-inflammation [46,49,51,53–55], and nitric oxide production [45,47,49,56,66,67] were described.

3.3. Clinical Studies

Taken together, the clinical studies showed interesting effectiveness of PBM on endothelial dysfunction, not accompanied by adverse effects.

One hundred per cent of humans exposed to PBM showed positive effects (Table 3). Among the nine studies selected, one study was a randomised controlled trial, one study was a randomised trial, and the remaining were pilot studies. In three studies [69–71], patients experienced prevention of prodromal complications in saphenectomy post myocardial revascularization and improvement of recovery after coronary intervention by modulation of endothelin, NO derivates, and transforming-growth-factor-b. In three studies [72–74], inflammation was reduced, as a consequence of antioxidants increment and tumour necrosis factor decrement, as well as in two studies the ischemia damages were counteracted by new microvascular restoration and increased cognitive performances. In one study blood-pressure decreased, and thus hypertension was ameliorated. Lastly, the studies pointed out the angiogenesis effect of PBM through the modulation of the VEGF-family, in accordance with the in vitro and preclinical data. In contrast to the previous in vitro and preclinical studies, five out of nine studies on humans were performed with NIR light, the remaining four studies were conducted with red-light and no study used an LED device.

Table 3. Clinical in vivo studies on photobiomodulation and endothelial dysfunction, selected after inclusion and exclusion criteria screening. The table shows the schematic design of the experimental set-up on patients and the results.

Study/Disease	Wavelength	Parameters Irradiated	Methods	Effect of PBM
91 Patients (RCT) (angioplasty) [69]	808 nm Laser	Power = 0.2 W, power density = 0.1 W/cm^2, time = 90 s, average spot area = 2.05 cm^2, energy = 18 J, fluence = ~9 J/cm^2, mode = continuous wave (CW)	No. of irradiations and mode = 1 intracoronary irradiation during percutaneous coronary interventions. Investigation setup: serum levels of IGF-1, VEGF, TGF and FGF-2 were measured before angioplasty, then, 6 and 12 h and 1 month after the procedure	Smaller neointima formation; IGF-1 and VEGF = no-effect; decrement to FGF-2; increment of TGF-b1
101 Patients (RP) (angioplasty) [70]	808 nm Laser	Power = 0.2 W, power density = 0.1 W/cm^2, time = 90 s, average spot area = 2.05 cm^2, energy = 18 J, fluence = ~9 J/cm^2, mode = CW	No. of irradiations and mode = 1 intracoronary irradiation during percutaneous coronary interventions. Investigation setup: serum levels of NO derivates and endothelin-1 were measured before angioplasty, then, 6 and 12 h and 1 month after the procedure	Improvement of restenosis process; increment of NO derivates; endothelin-1 increased after 6 h but decreased later
14 Patients (PS) (saphenectomy) [71]	780 nm laser	Power = 0.025 W, power density = 0.625 W/cm^2, time = 30 s, average spot area = 0.04 cm^2, energy = 0.75 J, fluence = 19 J/cm^2, mode = CW	No. of irradiations and mode = 1 irradiation surrounding the entire surgical perimeter wound edge. Investigation setup: evaluation of erythema, edema, blister, hematoma, transudation, dehiscence, and pain	Prevention of prodromal complications in saphenectomy post myocardial revascularization
27 Patients (PS) (cerebral ischemia) [75]	633 nm laser	Power = 0.025 or 0.045 W, power density = 0.14 or 0.045 W/cm^2, time = 1200 or 2400 s, average spot area = 0.18 cm^2, energy = 29, 54, 60 or 106 J, fluence = from 161 to 589 J/cm^2, mode = CW	No. of irradiations and mode = 1 intracerebral transcatheter laser irradiation. Investigation setup: restoration of mental and motor functions was detected; rheoencephalography, scintigraphy, computed tomography and magnetic resonance imaging was performed	Restoration of cerebral collateral and capillary blood supply; improvement of microcirculation; restoration of cellular and tissue metabolism; stimulation of neurogenesis and regenerative processes
21 aged Patients (PS) (cerebral ischemia) [73]	1064 nm Laser	Power = 3.4 W, power density = 0.25 W/cm^2, time = 240 s, spot area = 13.6 cm^2, energy = 816 J, fluence = 60 J/cm^2, mode = CW	No. of irradiations and mode = 1 irradiation at the right forehead on 2 points. Investigation setup: prefrontal cortex measures of attention PVT and memory, carotid artery intima-media thickness, electroencephalography, and functional magnetic resonance imaging	Improvement of cognitive performance and both carotid artery and intima-media thickness; increment and improvement of resting-state EEG alpha, beta, and gamma power as well as prefrontal bloodoxygen-evel
7 diabetic Patients (PS) (angiogenesis/healing) [72]	660 nm Laser	Power = 0.1 W, power density = 0.16 W/cm^2, time = 12 s, spot area = 0.6 cm^2, energy = 1.2 J, fluence = 2 J/cm^2, mode = CW	No. of irradiations and mode = 1 irradiation around lesion area; 0.5 cm distant from tissue; points were 2 cm far from each other. Investigation setup: qRT-PCR, IL6, TNF, VEGF and TGF	IL6 not changed; decrement of TNF; increment of VEFG and TGF-β
40 Patients (PS) (angiogenesis/healing) [73]	645 nm Laser	Power = 0.25 W, power density = 0.125 W/cm^2, time = 80 s, spot area = 2 cm^2, energy = 20 J, fluence = 10 J/cm^2, mode = CW	No. of irradiations and mode = 1 irradiation on 2 points at baseline and after 1, 3, and 7 days. Investigation setup: ELISA on crevicular fluid, bradykinin, VEGF and EGF	Improvement of the early phases of the healing and angiogenesis; reduction of bradykinin and VEGF; increment of EGF.
10 Patients (PS) (angiogenesis/healing) [74]	808 nm Laser	Power = 0.05 W, power density = 1.6 W/cm^2, time = 400 s, spot area = 0.031 cm^2, energy = 20 J, fluence = 645 J/cm^2, mode = CW	No. of irradiations and mode = 1 irradiation once a day for three consecutive days. Investigation setup: blood analysis for VEGF, FGF, angiostatin, GSH, symmetric dimethyl-arginine, asymmetric dimethylarginine and L-arginine	No change in VEGF, FGF, SDMA, NO, and ADMA levels; increment of antioxidant and angiogenic potential.
30 Patients (PS) (hypertension) [76]	635 nm Laser	Power = 0.0015 W, Power density = 0.2 W/cm^2, Time = 900 s, spot area = 0.0075 cm^2, Energy = 1.35 J, Fluence = 180 J/cm^2, Mode = CW	No. of irradiations and mode = 1 intravein irradiation for 10 procedures. Investigation setup: endothelium function was evaluated by test with reactive hyperemia	Decrement of cardiac risk; decrement of DayDBP in hyperuricemia group and both DaySBP and DayDBP in the group of patients with AH combined with hyperuricemia

Legend: ADMA, asymmetric dimethylarginine; AH, arterial hypertension; DBP, diastolic blood pressure EEG, electroencephalogram; EGF, epidermal growth factor; ELISA, enzyme-linked immunosorbent assay; FGF, fibroblast growth factors; GSH, glutathione; IGF, insulin-like growth factor; IL, interleukin; NO, nitric oxide; PVT, psychomotor vigilance task; PS, pilot study; qRT-PCR, real-time quantitative reverse transcription; RCT, randomised trial; RT, randomised control; SBP, systolic blood pressure; SDMA, symmetric dimethylarginine; TGF, transforming growth factor; TNF, tumour necrosis factor; VEGF, vascular endothelial growth factor.

4. Discussion
4.1. In Vitro Studies

Concerning endothelial cells, and more generally, the eukaryotic cell, ATP, ROS, Ca^{2+}, and NO molecules are responsible for their metabolic behaviour [12]. Substantially, the modulation of Ca^{2+} homeostasis tends to regulate growth factor release, such as fibroblast growth factors (FGF), transforming growth factor (TGF)-b, VEGF, and platelet-derived growth factor (PDGF) [77,78], as well as ATP, ROS, and NO can act as proangiogenic factors or inducers of cell alteration, according to their concentrations and target [79–81]. The VEGF family and their receptors are influenced by PBM, as evidenced by ELISA, qRT-PCR, and immunoblot investigations on different cell lines [29,31,32,36,41]. In particular, 632 nm and 0.63, 1.05, 2.1, or 4.2 J (0.0035 W, 2 cm^2) increased VEGF secretion, 8.4 and 12.6 J decreased VEGF secretion, while 0.21 and 6.3 J did not have an effect [29]. Additionally, 660 nm and 0.10 J (0.01 W, 0.04 cm^2) and 0.15 J (0.015 W, 0.04 cm^2) increased secretion of VEGF-A, VEGF-C, VEGF-D, and VEGF-receptor (R)1 [32], as well as reduction of VEGF-A, sVEGFR-1, and sVEGFR-2 by irradiation with 635 nm and 160, 320 or 640 J (0.15 W, 80 cm^2) was observed by Góralczyk et al. [36]; 904 nm and 200 or 300 J (50 W, 0.01 cm^2), in a pulsed mode of irradiation, incremented VEGF expression [41]. The VEGF family constitutes five different types of protein signalling molecules, i.e., VEGF-A or VEGF, VEGF-B, VEGF-C, VEGF-D, and placental growth factor (PlGF), which plays a key role in endothelial tissue homeostasis. The link of these molecules with their specific receptors (VEGFR-1, -2, -3) leads to the development of both the cardiovascular system and the angiogenesis, with a beneficial effect on peripheral arterial disease, ischemic heart disease, healing, coagulation, and also the female reproductive cycle [82]. Unfortunately, VEGF is expressed in cancers and VEGF inhibition is considered to be a promising treatment option, stopping tumour metastasis and growth [83]. Therefore, the chance of modulating the VEGF expression by adequate laser therapy parameters could support the clinician for different stimulatory or inhibitory curative approaches on endothelial dysfunction, according to need.

The analysis of the selected literature suggests that the support by PBM to endothelial dysfunction can additionally be traced to an anti-inflammation effect, which leads to an increase in cell viability, similar to the recent description of pre-osteoblast cell line by Hanna et al. [84]. Hwang et al. [31] showed that 645 nm and 32 or 64 J (0.010 or 0.012 W, 2.8 cm^2) inhibited inflammatory mediators and catabolic enzymes, such as matrix metalloproteinase (MMP)-1, MMP-3, and brain-derived neurotrophic factor (BDNF), and MMP-1, MMP-3, BDNF, and interleukin (IL)-8, respectively. Góralczyk et al., [37] described a reduction of inflammation by a decrease in TNF-α and IL-6 after irradiation with 630 nm/160 J (0.15 W, 80 cm^2) or 830 nm/160 J (0.3 W, 80 cm^2).

Nitric oxide and ATP are ambiguous molecules acting as proinflammatory or anti-inflammatory mediators according to their concentration, target, and time of exposure [85,86]. Therefore, they are able to downregulate or upregulate some MMPs synthesis and affect TNF and proinflammatory markers [85–87]. In this context, the ability of laser light to induce NO production or release and to change the energetic cell metabolism, such as the ATP production [25,26,44], can control the proinflammatory molecules. This behaviour may lead to a reduction of apoptosis' indices such as caspase 3, annexin, and cytochrome c [26], as well as protects against TNF/cycloheximide (CHX)-induced apoptosis pathway by inhibiting p38 mitogen-activated protein kinase (MAPK) and nuclear factor kappa-light-chain-enhancer of activated B (NF-κB) signals [42]; also, according to De Nadai et al. [88]. Undoubtedly, the strengths of the in vitro studies selected for our review are the prevalent use of human umbilical vein endothelial cells (HUVEC) cell line and a wide panel of markers also investigated on primary cells. Therefore, despite the fact that different light therapies were administered in the experimental setup, we can try to reveal a pathway of the effectiveness of red and NIR light. The primary interaction between red and NIR light and the photoacceptors into the cell (see [12]) leads to a modulation of the VEGF molecules family, including PDGF, and the FGF and their receptors. The following step appears to involve the phospho-inositide 3-kinase (PI3K) protein, which can interact with the protein kinase B (AKT), a multiple-process cellular protein, able to have a key role in eNOS synthesis [33], NF-kB signal, and inhibition of mitochondrial proapoptotic triggering, via the cytochrome c and caspase 3 pathway. In this way, the cell's fate can be routed through cell proliferation and angiogenesis, or apoptosis, but also an antithrombotic effect can be stimulated (see Table 1). Likewise, the TGF-b signal involving small mother against decapentaplegics (SMADs), a family of structurally similar proteins, may be involved, as well as AKT can take the transcriptional regulator integrating mechanical signal yes-associated protein (YAP)/tafazzin (TAZ) [35], which modulates aspects of cell behaviour, including cell plasticity, proliferation, and stem cell differentiation, essential for tissue regeneration [89]. Additionally, the NO release mediated by red/NIR light through an effect on eNOS expression, the calcium–calmodulin interaction, or the gas dissociation from S-nitrosylated proteins can be involved in an intracellular NO dose-dependent mechanism, as also shown in ex vivo experiments [90]. Lastly, wavelength comparison studies indicate 635 nm as the best effector of endothelial cell proliferation with respect to 808 nm in standard growth condition, while in a like-diabetic hyperglycaemic environment, 808 nm seems to be more effective, probably because of the involvement of different pathways via either NO or mitochondria's metabolism, as suggested by authors [25,42,44]. It is knownthat a cellular redox state has a role in the sensitivity of cells to PBM and can, therefore, explain the different responses to phototherapy by diabetic cells; diabetic pro-oxidant cells are generally differently more sensitive to PBM thanks to oxidation of cytochrome c oxidase and the increased mitochondrial responsiveness [91].

4.2. In Vivo Preclinical (Animal) Studies

In rat model PBM, different wavelengths and laser parameters can improve the healing process in injured muscle and skin; the PBM affects the inflammation process by modulation of its markers.

Basically, 660 nm visible light and 2 J (0.04 W, 0.04 cm^2) or (0.03 W, 0.028 cm^2) decreased cyclooxygenase (Cox)-2 [52] and IL-6, and MMP-3 and -9 [54], respectively.

Infrared light, 904 nm, irradiated in pulsed mode with 3 J on 0.1 cm^2 (output power = 0.04 W) or 0.4 J on 1.77 cm^2 (output power = 0.00078 W), prevented IL-6 [53] formation and has decremented ROS, NO, lipid peroxidation, protein carbonylation, and other proinflammatory markers [56]. Conversely, anti-inflammation indicators such as IL-10 [53] and superoxide dismutase, catalase, and glutathione peroxidase (GPx) were improved [56].

As previously seen through in vitro experiences, different signal pathways influence VEGF family, which is modulated. The same behaviour is observed in the angiogenesis experimental purposes, in which, as described in Table 2, PBM modulated both cytokines and metalloproteinases [51,53–55,61,65], and the collagens and blood vessels took form [54,55,62,65]. The only comparative work pointed out that light coherence was not shown to be essential to angiogenesis [59]. Nevertheless, in only one study, the data contextualised with the conclusion of Hode [92] by which coherence seemed to not be a potentially important factor in the overall efficacy of photobiomodulation, particularly in a clinical setting.

Additionally, the increment of NO production or release observed in cell culture also took place in rats' model and led to a reduction of systolic arterial pressure, induced a long-lasting hypotensive effect, and improved heart rate [66,67]; 660 nm and 5.6 J (0.1 W, 0.058 or 0.028 cm^2) irradiated transcutaneously in six points, seems to represent an effective therapy.

Lastly, modulation of NO and improvement of angiogenesis could also diminish myocardium inflammation, as well as enhanced ischemic revascularisation and angiogenesis, which increased tissue viability and reduced infarct size [46,47,49].

The information experienced by in vitro and preclinical experiments showed coherent results and suggested a promising clinical opportunity for therapeutic approaches on humans.

4.3. Clinical Studies

Thanks to both a randomised controlled trial (RCT) [69] and a randomised trial (RT) study [70] with a six-month follow-up, Derkacz's team strongly demonstrated that 808 nm and 18 J (0.2 W, 2.05 cm^2) intracoronary irradiation, during percutaneous coronary interventions, can counteract restenosis cascade. It has been assumed that growth factors play a role in the restenosis process stimulation. Mitogenic smooth muscle cell expression by VEGF and FGF-2 support, in particular, migration and constitution of neointima [93]. The Darkacz et al. therapy [69,70] stimulated NO release and decreased FGF-2 concentration as well as the expression, in the more advanced stage, of endothelin-1, a protein associated with the process of coronary restenosis after percutaneous transluminal coronary angioplasty [94]; the effect led to smaller neointima formation and prevented the process of restenosis. Despite a different approach and a lower number of patients but with the same six-month follow-up, an improved effect of PBM was also shown by Pinto et al. [71] on saphenectomy post-myocardial revascularisation irradiated with 780 nm/0.75 J (0.025 W, 0.04 cm^2).

Concerning healing and angiogenesis, clinical data was in agreement with the in vitro and on preclinical positive evidence, but the different purposes and therapy parameters impeded a successful comparison among the works.

However, Angiero and co-workers [73] showed interesting results on a high number of patients, and a 21-day follow-up, thanks to irradiation on two points at baseline and after 1, 3, and 7 days with 645 nm and 20 J (0.25 W, 2 cm^2). Patients experienced a reduction of bradykinin (a vasoactive peptide involved in the classical signs of inflammation), local heat, redness, pain, and swelling [95] as well as VEGF, while EGF (a growth factor implicated in wound healing [96]), increased; they concluded that patients with periodontitis had a beneficial effect due to PBM, in the early phases of the revascularisation and healing.

Maksimovich et al. [75] and Vargas et al. [97], through two different wavelengths in the visible (633 nm) and infrared (1064 nm) spectrum, showed in humans the improvement effect of PBM on ischemia, as previously described on rat models. Indeed, PBM on patients affected by cerebral ischemia restored cerebral collateral and capillary blood supply, improved microcirculation, recovered cellular and tissue metabolism, stimulated neurogenesis, and caused regenerative processes, if irradiated by different energy from 29 to 106 J and 633 nm [75]; no side effects were observed in the irradiated patients with follow-up for two-years. In addition, patients with altered neurocognitive function, because of ischemia, incremented cognitive performance, improved carotid artery intima-media thickness, and increased resting-state EEG alpha, beta, and gamma power, by an increment of prefrontal blood oxygen level after irradiation with 1064 nm and 816 J (3.4 W, 13.6 cm^2) [97].

Lastly, although only a paper was selected after inclusion and exclusion criteria as well as research by keywords, PBM could provide support for patients with hypertension. Endothelial dysfunction, which is characterised by an impairment of nitric oxide (NO) bioavailability, is indeed an important risk factor for both hypertension and cardiovascular disease and may represent a major link between such conditions. Mitchell and Mack [98] demonstrated near-infrared light in the form of low-level laser therapy increased NO levels in venous blood draining from the treatment site in healthy subjects. As previously discussed, the preclinical studies on animal model showed effective therapeutic PBM's windows to reduce pressure and heart rate and induce a long-lasting hypotensive effect. Additionally, intravascular laser irradiation of blood discovered by Russian scientists in the 1970s, as well as plasmapheresis of blood irradiated with PBM, showed improvement in patients with hypertension disease by normalisation of plasma levels of stable nitric oxide [99,100]. This was contextualised by the work of Kovalenko et al. [76], where patients affected by hyperuricemia and high blood pressure showed beneficial effects using PBM with 1.35 J, (0.0015 W, 0.0075 cm^2).

5. Conclusions

Light at different wavelengths and intensities is able to improve endothelial function. The effect is correlated to the primary targets of red and NIR light within the cell, which changes leads to secondary cellular pathways activation responsible for modulation of inflammation, angiogenesis, and vasodilatation. However, PBM is a therapy derived by a complex mixture of wavelengths with different frequencies, amplitudes, and energies, which are absorbed, scattered, and reflected by biological material. Therefore, selecting the most inclusive therapy is not easy. However, 808 nm and 18 J (0.2 W, 2.05 cm^2) intracoronary irradiation during percutaneous coronary interventions can support clinicians to prevent restenosis; as well, irradiation on two points at baseline and after 1, 3, and 7 days with 645 nm and 20 J (0.25 W, 2 cm^2) can stimulate angiogenesis. Lastly, the use of PBM to support hypertension cure showed interesting insights that stimulate investigation by more extensive randomised controlled trials.

Supplementary Materials: The following are available online at https://www.mdpi.com/2227-9059/9/3/274/s1, Figure S1. Flow chart demonstrating the selection process.

Author Contributions: Conceptualization, A.A.; methodology, A.A.; software, E.C. and S.A.; validation, A.A, E.C., A.S., S.A., A.Z., A.U., and S.B.; formal analysis, A.A., E.C., and A.S.; data curation, A.A., A.Z., and A.U.; writing—review and editing, A.A., E.C., and A.S.; supervision, A.A. and S.B.; project administration, A.A. All authors have read and agreed to the published version of the manuscript.

Funding: This research received no external funding.

Institutional Review Board Statement: Not applicable.

Informed Consent Statement: Not applicable.

Data Availability Statement: Data available on request from the authors.

Acknowledgments: Authors would like to express special appreciation and thanks to Alberico Benedicenti, for his guidance of our work.

Conflicts of Interest: The authors declare no conflict of interest.

References

1. Daiber, A.; Steven, S.; Weber, A.; Shuvaev, V.V.; Muzykantov, V.R.; Laher, I.; Li, H.; Lamas, S.; Münzel, T. Targeting vascular (endothelial) dysfunction. *Br. J. Pharmacol.* **2017**, *174*, 1591–1619. [CrossRef]
2. Lim, S.S.; Vos, T.; Flaxman, A.D.; Danaei, G.; Shibuya, K.; Adair-Rohani, H.; A AlMazroa, M.; Amann, M.; Anderson, H.R.; Andrews, K.G.; et al. A comparative risk assessment of burden of disease and injury attributable to 67 risk factors and risk factor clusters in 21 regions, 1990–2010: A systematic analysis for the Global Burden of Disease Study 2010. *Lancet* **2012**, *380*, 2224–2260. [CrossRef]
3. Wright, J.T., Jr.; Williamson, J.D.; Whelton, P.K.; Snyder, J.K.; Sink, K.M.; Rocco, M.V.; Reboussin, D.M.; Rahman, M.; Oparil, S.; Lewis, C.E.; et al. A Randomized Trial of Intensive versus Standard Blood-Pressure Control. *N. Engl. J. Med.* **2015**, *373*, 2103–2116.
4. Lerman, A.; Zeiher, A.M. Endothelial function: Cardiac events. *Circulation* **2005**, *111*, 363–368. [CrossRef]
5. Dauphinee, S.M.; Karsan, A. Progress in Inflammation Research. In *Endothelial Dysfunction and Inflammation*; Springer Basel AG: Basel, Switzerland, 2010; p. 243.
6. Rajendran, P.; Rengarajan, T.; Thangavel, J.; Nishigaki, Y.; Sakthisekaran, D.; Sethi, G.; Nishigaki, I. The Vascular Endothelium and Human Diseases. *Int. J. Biol. Sci.* **2013**, *9*, 1057–1069. [CrossRef] [PubMed]
7. Szewczyk, A.; Jarmuszkiewicz, W.; Kozieł, A.; Sobieraj, I.; Nobik, W.; Łukasiak, A.; Skup, A.; Bednarczyk, P.; Drabarek, B.; Dymkowska, D.; et al. Mitochondrial mechanisms of endothelial dysfunction. *Pharmacol. Rep.* **2015**, *67*, 704–710. [CrossRef] [PubMed]
8. Kwan, H.-Y.; Huang, Y.; Yao, X. TRP channels in endothelial function and dysfunction. *Biochim. Biophys. Acta (BBA) Mol. Basis Dis.* **2007**, *1772*, 907–914. [CrossRef] [PubMed]
9. Tran, Q.-K.; Ohashi, K.; Watanabe, H. Calcium signalling in endothelial cells. *Cardiovasc. Res.* **2000**, *48*, 13–22. [CrossRef]
10. Sitia, S.; Tomasoni, L.; Atzeni, F.; Ambrosio, G.; Cordiano, C.; Catapano, A.; Tramontana, S.; Perticone, F.; Naccarato, P.; Camici, P.; et al. From endothelial dysfunction to atherosclerosis. *Autoimmun. Rev.* **2010**, *9*, 830–834. [CrossRef] [PubMed]
11. Kiseleva, R.Y.; Glassman, P.M.; Greineder, C.F.; Hood, E.D.; Shuvaev, V.V.; Muzykantov, V.R. Targeting therapeutics to endothelium: Are we there yet? *Drug Deliv. Transl. Res.* **2018**, *8*, 883–902. [CrossRef] [PubMed]

12. Amaroli, A.; Ferrando, S.; Benedicenti, S. Photobiomodulation Affects Key Cellular Pathways of all Life-Forms: Considerations on Old and New Laser Light Targets and the Calcium Issue. *Photochem. Photobiol.* **2019**, *95*, 455–459. [CrossRef]
13. Amaroli, A.; Benedicenti, A.; Ferrando, S.; Parker, S.; Selting, W.; Gallus, L.; Benedicenti, S. Photobiomodulation by Infrared Diode Laser: Effects on Intracellular Calcium Concentration and Nitric Oxide Production of Paramecium. *Photochem. Photobiol.* **2016**, *92*, 854–862. [CrossRef] [PubMed]
14. Amaroli, A.; Colombo, E.; Zekiy, A.; Aicardi, S.; Benedicenti, S.; De Angelis, N. Interaction between Laser Light and Osteoblasts: Photobiomodulation as a Trend in the Management of Socket Bone Preservation-A Review. *Biology* **2020**, *23*, 409. [CrossRef] [PubMed]
15. Ferrando, S.; Agas, D.; Mirata, S.; Signore, A.; De Angelis, N.; Ravera, S.; Utyuzh, A.S.; Parker, S.; Sabbieti, M.G.; Benedicenti, S.; et al. The 808 nm and 980 nm infrared laser irradiation affects spore germination and stored calcium homeostasis: A comparative study using delivery hand-pieces with standard (Gaussian) or flat-top profile. *J. Photochem. Photobiol. B Biol.* **2019**, *199*, 111627. [CrossRef]
16. Santana-Blank, L.; Rodríguez-Santana, E.; Santana-Rodríguez, J.A.; Santana-Rodríguez, K.E.; Reyes-Barrios, H.; Hamblin, M.R.; De Sousa, M.V.P.; Agrawal, T. Water as a Photoacceptor, Energy Transducer, and Rechargeable Electrolytic Bio-battery in Photobiomodulation. In *Handbook of Low-Level Laser Therapy*; Chapter 8; CRC Press: Boca Raton, FL, USA, 2016; pp. 119–140.
17. Gonzalez-Lima, F.; Rojas, J.C. Low-level light therapy of the eye and brain. *Eye Brain* **2011**, *3*, 49–67. [CrossRef]
18. Stephensa, B.J.; Jonesb, L.R. Tissue Optics. In *Handbook of Low-Level Laser Therapy*; Hamblin, M.R., Pires de Sousa, M.V., Agrawal, T., Eds.; Pan Stanford Publishing: Singapore, 2017; pp. 98–117.
19. Santana-Blank, L.; Rodríguez-Santana, E.; Santana-Rodríguez, K. Theoretic, Experimental, Clinical Bases of the Water Oscillator Hypothesis in Near-Infrared Photobiomodulation. *Photomed. Laser Surg.* **2010**, *28*, S41. [CrossRef]
20. Karu, T.I. Mitochondrial Signaling in Mammalian Cells Activated by Red and Near-IR Radiation. *Photochem. Photobiol.* **2008**, *84*, 1091–1099. [CrossRef]
21. Karu, T.I. Multiple roles of cytochrome c oxidase in mammalian cells under action of red and IR-A radiation. *IUBMB Life* **2010**, *62*, 607–610. [CrossRef] [PubMed]
22. Amaroli, A.; Ravera, S.; Parker, S.; Panfoli, I.; Benedicenti, A.; Benedicenti, S. An 808-nm Diode Laser with a Flat-Top Handpiece Positively Photobiomodulates Mitochondria Activities. *Photomed. Laser Surg.* **2016**, *34*, 564–571. [CrossRef]
23. Amaroli, A.; Pasquale, C.; Zekiy, A.; Utyuzh, A.; Benedicenti, S.; Signore, A.; Ravera, S. Photobiomodulation and oxidative stress: 980nm diode-laser light regulates mitochondria activity and reactive oxygen species production. *Oxidative Med. Cell. Longev.* **2021**, *2021*, 6626286. [CrossRef]
24. Ravera, S.; Ferrando, S.; Agas, D.; De Angelis, N.; Raffetto, M.; Sabbieti, M.G.; Signore, A.; Benedicenti, S.; Amaroli, A. 1064 nm Nd:YAG laser light affects transmembrane mitochondria respiratory chain complexes. *J. Biophotonics* **2019**, *12*, e201900101. [CrossRef]
25. Keszler, A.; Lindemer, B.; Hogg, N.; Weihrauch, D.; Lohr, N.L. Wavelength-dependence of vasodilation and NO release from S-nitrosothiols and dinitrosyl iron complexes by far red/near infrared light. *Arch. Biochem. Biophys.* **2018**, *649*, 47–52. [CrossRef] [PubMed]
26. Zhang, R.; Mio, Y.; Pratt, P.F.; Lohr, N.; Warltier, D.C.; Whelan, H.T.; Zhu, D.; Jacobs, E.R.; Medhora, M.; Bienengraeber, M. Near infrared light protects cardiomyocytes from hypoxia and reoxygenation injury by a nitric oxide dependent mechanism. *J. Mol. Cell. Cardiol.* **2009**, *46*, 4–14. [CrossRef] [PubMed]
27. Wang, Y.; Huang, Y.-Y.; Wang, Y.; Lyu, P.; Hamblin, M.R. Photobiomodulation of human adipose-derived stem cells using 810 nm and 980 nm lasers operates via different mechanisms of action. *Biochim. Biophys. Acta (BBA) Gen. Subj.* **2017**, *1861*, 441–449. [CrossRef] [PubMed]
28. Tunér, J.; Jenkins, P.A. Parameter Reproducibility in Photobiomodulation. *Photomed. Laser Surg.* **2016**, *34*, 91–92. [CrossRef] [PubMed]
29. Kipshidze, N.; Nikolaychik, V.; Keelan, M.H.; Shankar, L.R.; Khanna, A.; Kornowski, R.; Leon, M.; Moses, J. Low-power helium: Neon laser irradiation enhances production of vascular endothelial growth factor and promotes growth of endothelial cells in vitro. *Lasers Surg. Med.* **2001**, *28*, 355–364. [CrossRef] [PubMed]
30. Ricci, R.; Pazos, M.C.; Borges, R.E.; Pacheco-Soares, C. Biomodulation with low-level laser radiation induces changes in en-dothelial cell actin filaments and cytoskeletal organization. *J. Photochem. Photobiol. B Biol.* **2009**, *95*, 6–8. [CrossRef]
31. Hwang, M.H.; Lee, J.W.; Son, H.-G.; Kim, J.; Choi, H. Effects of photobiomodulation on annulus fibrosus cells derived from degenerative disc disease patients exposed to microvascular endothelial cells conditioned medium. *Sci. Rep.* **2020**, *10*, 9655. [CrossRef]
32. Vitor, L.L.R.; Prado, M.T.O.; Neto, N.L.; Oliveira, R.C.; Sakai, V.T.; Santos, C.F.; Dionísio, T.J.; Rios, D.; Cruvinel, T.; Machado, M.A.A.M.; et al. Does photobiomodulation change the synthesis and secretion of angiogenic proteins by different pulp cell lineages? *J. Photochem. Photobiol. B Biol.* **2020**, *203*, 111738. [CrossRef]
33. Chen, C.-H.; Hung, H.-S.; Hsu, S.-H. Low-energy laser irradiation increases endothelial cell proliferation, migration, and eNOS gene expression possibly via PI3K signal pathway. *Lasers Surg. Med.* **2008**, *40*, 46–54. [CrossRef]
34. Lukowicz, M.; Szymańska, J.; Goralczyk, K.; Zając, A.; Rość, D. Effect of low level laser therapy and high intensity laser therapy on endothelial cell proliferation in vitro: Preliminary communication. In Proceedings of the Tenth Symposium on Laser Technology, Szczecin, Poland, 24–28 September 2012. [CrossRef]

35. Szymanska, J.; Goralczyk, K.; Klawe, J.J.; Lukowicz, M.; Michalska, M.; Goralczyk, B.; Zalewski, P.; Newton, J.L.; Gryko, L.; Zajac, A.; et al. Phototherapy with low-level laser influences the proliferation of endothelial cells and vascular endothelial growth factor and transforming growth factor-beta secretion. *J. Physiol. Pharmacol. Off. J. Pol. Physiol. Soc.* 2013, 64, 387–391.
36. Góralczyk, K.; Szymańska, J.; Łukowicz, M.; Drela, E.; Kotzbach, R.; Dubiel, M.; Michalska, M.; Góralczyk, B.; Zając, A.; Rość, D. Effect of LLLT on endothelial cells culture. *Lasers Med. Sci.* 2014, 30, 273–278. [CrossRef]
37. Góralczyk, K.; Szymańska, J.; Szot, K.; Fisz, J.; Rość, D. Low-level laser irradiation effect on endothelial cells under conditions of hyperglycemia. *Lasers Med. Sci.* 2016, 31, 825–831. [CrossRef]
38. Góralczyk, K.; Szymańska, J.; Gryko, Ł.; Fisz, J.; Rość, D. Low-level laser irradiation modifies the effect of hyperglycemia on adhesion molecule levels. *Lasers Med. Sci.* 2018, 33, 1521–1526. [CrossRef] [PubMed]
39. Li, Y.; Xu, Q.; Shi, M.; Gan, P.; Huang, Q.; Wang, A.; Tan, G.; Fang, Y.; Liao, H. Low-level laser therapy induces human umbilical vascular endothelial cell proliferation, migration and tube formation through activating the PI3K/Akt signaling pathway. *Microvasc. Res.* 2020, 129, 103959. [CrossRef]
40. Terena, S.M.L.; Mesquita-Ferrari, R.A.; Araújo, A.M.D.S.; Fernandes, K.P.S.; Fernandes, M.H. Photobiomodulation alters the viability of HUVECs cells. *Lasers Med. Sci.* 2021, 36, 83–90. [CrossRef]
41. Martignago, C.C.; Oliveira, R.F.; Pires-Oliveira, D.A.; Oliveira, P.D.; Pacheco Soares, C.; Monzani, P.S.; Poli-Frederico, R.C. Effect of low-level laser therapy on the gene expression of collagen and vascular endothelial growth factor in a culture of fi-broblast cells in mice. *Lasers Med. Sci.* 2015, 30, 203–208. [CrossRef]
42. Chu, Y.-H.; Chen, S.-Y.; Hsieh, Y.-L.; Teng, Y.-H.; Cheng, Y.-J. Low-level laser therapy prevents endothelial cells from TNF-α/cycloheximide-induced apoptosis. *Lasers Med. Sci.* 2017, 33, 279–286. [CrossRef] [PubMed]
43. Keszler, A.; Lindemer, B.; Weihrauch, D.; Jones, D.; Hogg, N.; Lohr, N.L. Red/near infrared light stimulates release of an endothelium dependent vasodilator and rescues vascular dysfunction in a diabetes model. *Free. Radic. Biol. Med.* 2017, 113, 157–164. [CrossRef] [PubMed]
44. Amaroli, A.; Ravera, S.; Baldini, F.; Benedicenti, S.; Panfoli, I.; Vergani, L. Photobiomodulation with 808-nm diode laser light promotes wound healing of human endothelial cells through increased reactive oxygen species production stimulating mi-tochondrial oxidative phosphorylation. *Lasers Med. Sci.* 2019, 34, 495–504. [CrossRef]
45. Lohr, N.L.; Keszler, A.; Pratt, P.; Bienengraber, M.; Warltier, D.C.; Hogg, N. Enhancement of nitric oxide release from nitrosyl hemoglobin and nitrosyl myoglobin by red/near infrared radiation: Potential role in cardioprotection. *J. Mol. Cell. Cardiol.* 2009, 47, 256–263. [CrossRef]
46. Ma, J.-X.; Yang, Q.-M.; Xia, Y.-C.; Zhang, W.-G.; Nie, F.-F. Effect of 810 nm Near-Infrared Laser on Revascularization of Ischemic Flaps in Rats. *Photomed. Laser Surg.* 2018, 36, 290–297. [CrossRef] [PubMed]
47. Tuby, H.; Maltz, L.; Oron, U. Modulations of VEGF and iNOS in the rat heart by low level laser therapy are associated with cardioprotection and enhanced angiogenesis. *Lasers Surg. Med.* 2006, 38, 682–688. [CrossRef]
48. Yang, Z.; Wu, Y.; Zhang, H.; Jin, P.; Wang, W.; Hou, J.; Wei, Y.; Hu, S. Low-Level Laser Irradiation Alters Cardiac Cytokine Expression Following Acute Myocardial Infarction: A Potential Mechanism for Laser Therapy. *Photomed. Laser Surg.* 2011, 29, 391–398. [CrossRef] [PubMed]
49. Manchini, M.T.; Serra, A.J.; Feliciano, R.D.S.; Santana, E.T.; Antônio, E.L.; Carvalho, P.D.T.C.D.; Montemor, J.; Crajoinas, R.O.; Girardi, A.C.C.; Tucci, P.J.F.; et al. Amelioration of Cardiac Function and Activation of Anti-Inflammatory Vasoactive Peptides Expression in the Rat Myocardium by Low Level Laser Therapy. *PLoS ONE* 2014, 9, e101270. [CrossRef]
50. Lopes, N.N.F.; Plapler, H.; Chavantes, M.C.; Lalla, R.V.; Yoshimura, E.M.; Alves, M.T.S. Cyclooxygenase-2 and vascular endothelial growth factor expression in 5-fluorouracil-induced oral mucositis in hamsters: Evaluation of two low-intensity laser protocols. *Support. Care Cancer* 2009, 17, 1409–1415. [CrossRef] [PubMed]
51. Iyomasa, M.M.; Rizzi, E.C.; Leão, J.C.; Issa, J.P.; Dias, F.J.; Pereira, Y.C.; Fonseca, M.J.; Vicentini, F.T.; Watanabe, I.S. Zymo-graphic and ultrastructural evaluations after low-level laser irradiation on masseter muscle of HRS/J strain mice. *Lasers Med. Sci.* 2013, 28, 777–783. [CrossRef]
52. Rodrigues, N.C.; Brunelli, R.; De Araújo, H.S.S.; Parizotto, N.A.; Renno, A.C.M. Low-level laser therapy (LLLT) (660nm) alters gene expression during muscle healing in rats. *J. Photochem. Photobiol. B Biol.* 2013, 120, 29–35. [CrossRef]
53. Silveira, P.C.L.; Scheffer, D.D.L.; Glaser, V.; Remor, A.P.; Pinho, R.A.; Junior, A.S.A.; Latini, A. Low-level laser therapy attenuates the acute inflammatory response induced by muscle traumatic injury. *Free. Radic. Res.* 2016, 50, 503–513. [CrossRef]
54. Fiorio, F.B.; Dos Santos, S.A.; de Melo Rambo, C.S.; Dalbosco, C.G.; Serra, A.J.; de Melo, B.L.; Leal-Junior, E.C.P.; de Carvalho, P.T.C. Photobiomodulation therapy action in wound repair skin induced in aged rats old: Time course of biomarkers in-flammatory and repair. *Lasers Med. Sci.* 2017, 32, 1769–1782. [CrossRef]
55. Otterço, A.N.; Andrade, A.L.; Brassolatti, P.; Pinto, K.N.Z.; Araújo, H.S.S.; Parizotto, N.A. Photobiomodulation mechanisms in the kinetics of the wound healing process in rats. *J. Photochem. Photobiol. B* 2018, 183, 22–29.
56. Yadav, A.; Verma, S.; Keshri, G.K.; Gupta, A. Role of 904 nm superpulsed laser-mediated photobiomodulation on nitroxidative stress and redox homeostasis in burn wound healing. *Photodermatol. Photoimmunol. Photomed.* 2020, 36, 208–218. [CrossRef]
57. Silva, T.C.; Oliveira, T.M.; Sakai, V.T.; Dionísio, T.J.; Santos, C.F.; Bagnato, V.S.; Machado, M.A. In vivo effects on the ex-pression of vascular endothelial growth factor-A165 messenger ribonucleic acid of an infrared diode laser associated or not with a visible red diode laser. *Photomed. Laser Surg.* 2010, 28, 63–68. [CrossRef]

58. Colombo, F.; Neto, A.A.; Sousa, A.P.; Marchionni, A.M.; Pinheiro, A.L.; Reis, S.R. Effect of low-level laser therapy (λ660 nm) on angiogenesis in wound healing: A immunohistochemical study in a rodent model. *Braz. Dent. J.* **2013**, *24*, 308–312. [CrossRef]
59. Corazza, A.V.; Jorge, J.; Kurachi, C.; Bagnato, V.S. Photobiomodulation on the angiogenesis of skin wounds in rats using different light sources. *Photomed. Laser Surg.* **2007**, *25*, 102–106. [CrossRef]
60. Wagner, V.P.; Curra, M.; Webber, L.P.; Nör, C.; Matte, U.; Meurer, L.; Martins, M.D. Photobiomodulation regulates cytokine release and new blood vessel formation during oral wound healing in rats. *Lasers Med. Sci.* **2016**, *31*, 665–671. [CrossRef]
61. Cury, V.; Moretti, A.I.S.; Assis, L.; Bossini, P.; Crusca, J.D.S.; Neto, C.B.; Fangel, R.; De Souza, H.P.; Hamblin, M.R.; Parizotto, N.A. Low level laser therapy increases angiogenesis in a model of ischemic skin flap in rats mediated by VEGF, HIF-1α and MMP-2. *J. Photochem. Photobiol. B Biol.* **2013**, *125*, 164–170. [CrossRef]
62. Fortuna, T.; Gonzalez, A.C.; Sá, M.F.; Andrade, Z.A.; Reis, S.R.A.; Medrado, A.R.A.P. Effect of 670 nm laser photobiomod-ulation on vascular density and fibroplasia in late stages of tissue repair. *Int. Wound J.* **2018**, *15*, 274–282. [CrossRef] [PubMed]
63. Gonzalez, A.C.; Santos, E.T.; Freire, T.F.C.; Sá, M.F.; Andrade, Z.D.A.; Medrado, A.R.A.P. Participation of the Immune System and Hedgehog Signaling in Neoangiogenesis Under Laser Photobiomodulation. *J. Lasers Med. Sci.* **2019**, *10*, 310–316. [CrossRef] [PubMed]
64. Salate, A.C.; Barbosa, G.; Gaspar, P.; Koeke, P.U.; Parizotto, N.A.; Benze, B.G.; Foschiani, D. Effect of In-Ga-Al-P Diode Laser Irradiation on Angiogenesis in Partial Ruptures of Achilles Tendon in Rats. *Photomed. Laser Surg.* **2005**, *23*, 470–475. [CrossRef] [PubMed]
65. Marques, A.C.; Albertini, R.; Serra, A.J.; da Silva, E.A.; de Oliveira, V.L.; Silva, L.M.; Leal-Junior, E.C.; de Carvalho, P.T. Photobiomodulation therapy on collagen type I and III, vascular endothelial growth factor, and metalloproteinase in experimentally induced tendinopathy in aged rats. *Lasers Med. Sci.* **2016**, *31*, 1915–1923. [CrossRef] [PubMed]
66. Buzinari, T.C.; De Moraes, T.F.; Cárnio, E.C.; Lopes, L.A.; Salgado, H.C.; Rodrigues, G.J. Photobiomodulation induces hypotensive effect in spontaneously hypertensive rats. *Lasers Med. Sci.* **2020**, *35*, 567–572. [CrossRef] [PubMed]
67. Oishi, J.C.; De Moraes, T.F.; Buzinari, T.C.; Cárnio, E.C.; Parizotto, N.A.; Rodrigues, G.J. Hypotensive acute effect of photobiomodulation therapy on hypertensive rats. *Life Sci.* **2017**, *178*, 56–60. [CrossRef] [PubMed]
68. De Moraes, T.F.; Filho, J.C.C.; Oishi, J.C.; Almeida-Lopes, L.; Parizotto, N.A.; Rodrigues, G.J. Energy-dependent effect trial of photobiomodulation on blood pressure in hypertensive rats. *Lasers Med. Sci.* **2020**, *35*, 1041–1046. [CrossRef] [PubMed]
69. Derkacz, A.; Protasiewicz, M.; Rola, P.; Podgórska, K.; Szymczyszyn, A.; Gutherc, R.; Poreba, R.; Doroszko, A. Effects of Intravascular Low-Level Laser Therapy During Coronary Intervention on Selected Growth Factors Levels. *Photomed. Laser Surg.* **2014**, *32*, 582–587. [CrossRef]
70. Derkacz, A.; Szymczyszyn, A.; Szahidewicz-Krupska, E.; Protasiewicz, M.; Poręba, R.; Doroszko, A. Effect of endovascular coronary low-level laser therapy during angioplasty on the release of endothelin-1 and nitric oxide. *Adv. Clin. Exp. Med.* **2017**, *26*, 595–599. [CrossRef] [PubMed]
71. Pinto, N.C.; Pereira, M.H.C.; Tomimura, S.; De Magalhães, A.C.; Pomerantzeff, P.M.; Chavantes, M.C. Low-Level Laser Therapy Prevents Prodromal Signal Complications on Saphenectomy Post Myocardial Revascularization. *Photomed. Laser Surg.* **2014**, *32*, 330–335. [CrossRef]
72. Ruh, A.C.; Frigo, L.; Cavalcanti, M.F.X.B.; Svidnicki, P.; Vicari, V.N.; Lopes-Martins, R.A.B.; Junior, E.C.P.L.; De Isla, N.; Diomede, F.; Trubiani, O.; et al. Laser photobiomodulation in pressure ulcer healing of human diabetic patients: Gene expression analysis of inflammatory biochemical markers. *Lasers Med. Sci.* **2017**, *33*, 165–171. [CrossRef]
73. Angiero, F.; Ugolini, A.; Cattoni, F.; Bova, F.; Blasi, S.; Gallo, F.; Cossellu, G.; Gherlone, E. Evaluation of bradykinin, VEGF, and EGF biomarkers in gingival crevicular fluid and comparison of PhotoBioModulation with conventional techniques in perio-dontitis: A split-mouth randomized clinical trial. *Lasers Med. Sci.* **2020**, *35*, 965–970. [CrossRef]
74. Szymczyszyn, A.; Doroszko, A.; Szahidewicz-Krupska, E.; Rola, P.; Gutherc, R.; Jasiczek, J.; Mazur, G.; Derkacz, A. Effect of the transdermal low-level laser therapy on endothelial function. *Lasers Med. Sci.* **2016**, *31*, 1301–1307. [CrossRef]
75. Maksimovich, I.V. Intracerebral Transcatheter Laser Photobiomodulation Therapy in the Treatment of Binswanger's Disease and Vascular Parkinsonism: Research and Clinical Experience. *Photobiomodul. Photomed. Laser Surg.* **2019**, *37*, 606–614.76. [CrossRef] [PubMed]
76. Kovalenko, Y.L.; A Rudenko, L.; Melekhovets, O.K.; Chepeliuk, A.D.; Melekhovets, I.V. Efficiency of hyperuricemia correction by low level laser therapy in the treatment of arterial hypertension. *Wiad. Lek.* **2018**, *71*, 1310–1315.
77. Martineau, I.; Lacoste, E.; Gagnon, G. Effects of calcium and thrombin on growth factor release from platelet concentrates: Kinetics and regulation of endothelial cell proliferation. *Biomaterials* **2004**, *25*, 4489–4502. [CrossRef] [PubMed]
78. Lacoste, E.; Martineau, I.; Gagnon, G. Platelet Concentrates: Effects of Calcium and Thrombin on Endothelial Cell Proliferation and Growth Factor Release. *J. Periodontol.* **2003**, *74*, 1498–1507. [CrossRef] [PubMed]
79. Stone, W.L.; Leavitt, L.; Varacallo, M. Physiology, Growth Factor. In *StatPearls*; StatPearls Publishing: Treasure Island, FL, USA, 2020.
80. Hu, T.; Ramachandrarao, S.P.; Siva, S.; Valancius, C.; Zhu, Y.; Mahadev, K.; Toh, I.; Goldstein, B.J.; Woolkalis, M.; Sharma, K. Reactive oxygen species production via NADPH oxidase mediates TGF-beta-induced cytoskeletal alterations in endothelial cells. *Am. J. Physiol. Renal. Physiol.* **2005**, *289*, F816–F825. [CrossRef]
81. Pocock, T.M.; Williams, B.; Curry, F.E.; Bates, D.O. VEGF and ATP act by different mechanisms to increase microvascular permeability and endothelial [Ca^{2+}]i. *Am. J. Physiol. Circ. Physiol.* **2000**, *279*, H1625–H1634. [CrossRef]

82. Hoeben, A.; Landuyt, B.; Highley, M.S.; Wildiers, H.; Van Oosterom, A.T.; De Bruijn, E.A. Vascular Endothelial Growth Factor and Angiogenesis. *Pharmacol. Rev.* **2004**, *56*, 549–580. [CrossRef]
83. Zhao, Y.; Adjei, A.A. Targeting Angiogenesis in Cancer Therapy: Moving Beyond Vascular Endothelial Growth Factor. *Oncol.* **2015**, *20*, 660–673. [CrossRef]
84. Hanna, R.; Agas, D.; Benedicenti, S.; Ferrando, S.; Laus, F.; Cuteri, V.; Lacava, G.; Sabbieti, M.G.; Amaroli, A. A Comparative Study Between the Effectiveness of 980 nm Photobiomodulation Delivered by Hand-Piece with Gaussian vs. Flat-Top Profiles on Osteoblasts Maturation. *Front. Endocrinol.* **2019**, *10*, 92. [CrossRef]
85. Cooke, J.P.; Losordo, D.W. Nitric Oxide and Angiogenesis. *Circulation* **2002**, *105*, 2133–2135. [CrossRef]
86. Faas, M.; Sáez, T.; de Vos, P. Extracellular ATP and adenosine: The Yin and Yang in immune responses? *Mol. Asp. Med.* **2017**, *55*, 9–19. [CrossRef]
87. Xia, Z.; Vanhoutte, P.M. Nitric Oxide and Protection against Cardiac Ischemia. *Curr. Pharm. Des.* **2011**, *17*, 1774–1782. [CrossRef]
88. De Nadai, C.; Sestili, P.; Cantoni, O.; Lièvremont, J.-P.; Sciorati, C.; Barsacchi, R.; Moncada, S.; Meldolesi, J.; Clementi, E. Nitric oxide inhibits tumor necrosis factor-alpha -induced apoptosis by reducing the generation of ceramide. *Proc. Natl. Acad. Sci. USA* **2000**, *97*, 5480–5485. [CrossRef] [PubMed]
89. Totaro, A.; Panciera, T.; Piccolo, S. YAP/TAZ upstream signals and downstream responses. *Nat. Cell Biol.* **2018**, *20*, 888–899. [CrossRef] [PubMed]
90. Plass, C.A.; Loew, H.G.; Podesser, B.K.; Prusa, A.M. Light-Induced Vasodilation of Coronary Arteries and Its Possible Clinical Implication. *Ann. Thorac. Surg.* **2012**, *93*, 1181–1186. [CrossRef]
91. Tafur, J.; Mills, P.J. Low-Intensity Light Therapy: Exploring the Role of Redox Mechanisms. *Photomed. Laser Surg.* **2008**, *26*, 323–328. [CrossRef] [PubMed]
92. Hode, T.; Hamblin, M.R.; De Sousa, M.V.P.; Agrawal, T. Is Coherence Important in Photobiomodulation? In *Handbook of Low-Level Laser Therapy*; Chapter 4; CRC Press: Boca Raton, FL, USA, 2016; pp. 51–66.
93. Ashikaga, H.; Ben-Yehuda, O.; Chien, K.R. Chapter 25-Coronary Restenosis. In *Molecular Basis of Cardiovascular Disease*, 2nd ed.; Chien, K.R., Ed.; Elsevier: Amsterdam, The Netherlands, 2004; pp. 455–469.
94. Takase, H.; Sugiyama, M.; Nakazawa, A.; Toriyama, T.; Hayashi, K.; Goto, T.; Sato, K.; Ikeda, K.; Ueda, R.; Dohi, Y. Increased endogenous endothelin-1 in coronary circulation is associated with restenosis after coronary angioplasty. *Can. J. Cardiol.* **2003**, *19*, 902–906.
95. Eddleston, J.; Christiansen, S.C.; Zuraw, B.L. KININS AND NEUROPEPTIDES, Bradykinin. In *Encyclopedia of Respiratory Medicine*; Laurent, G.J., Shapiro, S.D., Eds.; Academic Press: Cambridge, MA, USA, 2006; pp. 502–506.
96. Hardwicke, J.; Schmaljohann, D.; Boyce, D.; Thomas, D. Epidermal growth factor therapy and wound healing—past, present and future perspectives. *Surgeon* **2008**, *6*, 172–177. [CrossRef]
97. Vargas, E.; Barrett, D.W.; Saucedo, C.L.; Huang, L.-D.; Abraham, J.A.; Tanaka, H.; Haley, A.P.; Gonzalez-Lima, F. Beneficial neurocognitive effects of transcranial laser in older adults. *Lasers Med. Sci.* **2017**, *32*, 1153–1162. [CrossRef]
98. Mitchell, U.H.; Mack, G.L. Low-Level Laser Treatment with Near-Infrared Light Increases Venous Nitric Oxide Levels Acutely: A Single-Blind, Randomized Clinical Trial of Efficacy. *Am. J. Phys. Med. Rehabil.* **2013**, *92*, 151–156. [CrossRef] [PubMed]
99. Meneguzzo, D.T.; Ferreira, L.S.; De Carvalho, E.M.; Nakashima, C.F.; Hamblin, M.R.; De Sousa, M.V.P.; Agrawal, T. Intravascular Laser Irradiation of Blood. In *Handbook of Low-Level Laser Therapy*; Chapter 46; CRC Press: Boca Raton, FL, USA, 2016; pp. 933–952.
100. Alizade, I.G.; Karaeva, N.T. Experience in the use of autotransfusions of laser irradiated blood in treating hypertension patients. *Lik. Sprava* **1994**, *5–6*, 29–32.

Article

An Atherogenic Diet Disturbs *Aquaporin 5* Expression in Liver and Adipocyte Tissues of Apolipoprotein E-Deficient Mice: New Insights into an Old Model of Experimental Atherosclerosis

Inês V. da Silva [1,2], Courtney A. Whalen [3], Floyd J. Mattie [3], Cristina Florindo [2], Neil K. Huang [3,4], Sandra G. Heil [5], Thomas Neuberger [6,7], A. Catharine Ross [3], Graça Soveral [1,2,*] and Rita Castro [1,2,3,*]

[1] Research Institute for Medicines (iMed.ULisboa), Faculty of Pharmacy, Universidade de Lisboa, 1649-003 Lisboa, Portugal; imsilva1@campus.ul.pt
[2] Department of Pharmaceutical Sciences and Medicines, Faculty of Pharmacy, Universidade de Lisboa, 1649-003 Lisboa, Portugal; cristinaflorindo@ff.ulisboa.pt
[3] Department of Nutritional Sciences, The Pennsylvania State University, University Park, PA 16802, USA; caw400@psu.edu (C.A.W.); fjm1311@gmail.com (F.J.M.); neil.huang@tufts.edu (N.K.H.); acr6@psu.edu (A.C.R.)
[4] Cardiovascular Nutrition Laboratory, Jean Mayer USDA Human Nutrition Research Center on Aging, Tufts University, Boston, MA 02111, USA
[5] Department of Clinical Chemistry, Medical Center Rotterdam, Erasmus MC University, 3015 GD Rotterdam, The Netherlands; s.heil@erasmusmc.nl
[6] Huck Institutes of the Life Sciences, The Pennsylvania State University, University Park, PA 16802, USA; tun3@psu.edu
[7] Department of Biomedical Engineering, The Pennsylvania State University, University Park, PA 16802, USA
* Correspondence: gsoveral@ff.ulisboa.pt (G.S.); mum689@psu.edu (R.C.)

Abstract: The dysfunction of vascular endothelial cells is profoundly implicated in the pathogenesis of atherosclerosis and cardiovascular disease, the global leading cause of death. Aquaporins (AQPs) are membrane channels that facilitate water and glycerol transport across cellular membranes recently implicated in the homeostasis of the cardiovascular system. Apolipoprotein-E deficient ($apoE^{-/-}$) mice are a common model to study the progression of atherosclerosis. Nevertheless, the pattern of expression of *AQPs* in this atheroprone model is poorly characterized. In this study, $apoE^{-/-}$ mice were fed an atherogenic high-fat (HF) or a control diet. Plasma was collected at multiple time points to assess metabolic disturbances. At the endpoint, the aortic atherosclerotic burden was quantified using high field magnetic resonance imaging. Moreover, the transcriptional levels of several *AQP* isoforms were evaluated in the liver, white adipocyte tissue (WAT), and brown adipocyte tissue (BAT). The results revealed that HF-fed mice, when compared to controls, presented an exacerbated systemic inflammation and atherosclerotic phenotype, with no major differences in systemic methylation status, circulating amino acids, or plasma total glutathione. Moreover, an overexpression of the isoform *AQP5* was detected in all studied tissues from HF-fed mice when compared to controls. These results suggest a novel role for AQP5 on diet-induced atherosclerosis that warrants further investigation.

Keywords: MRI (magnetic resonance imaging); endothelial dysfunction; high-fat diets; plaque burden

1. Introduction

Despite significant advances in the treatment of cardiovascular disease (CVD), it remains the leading cause of mortality and morbidity among adults [1]. The major cause of CVD is atherosclerosis that is elicited by the early impairment of endothelial function and results in a chronic inflammatory condition in which arteries harden through the build-up of lipid-rich plaque in the vessel wall [2].

Aquaporins (AQPs) have been recently proposed to contribute to the homeostasis of the cardiovascular system [3,4]. Aquaporins are channel-forming proteins that facilitate water and small solutes transport across the plasma membrane driven by osmotic or solute gradients [5,6]. These channels play a variety of important physiological roles in mammals [7]. In humans, the 13 isoforms (AQP0-12) are categorized according to structural and functional properties, where the orthodox aquaporins (AQP0, 1, 2, 4, 5, 6, and 8) are mainly water channels; aquaglyceroporins (AQP3, 7, 9, and 10) also transport small uncharged solutes, such as glycerol and urea; and the less well-known S-aquaporins (AQP11 and 12) are under investigation with respect to their subcellular localization and selectivity [8–10]. Recent evidence has shown that AQP3, 5, 8, and 9 also facilitate the permeation of the major reactive oxygen species (ROS), hydrogen peroxide, being thus termed peroxiporins [11–13]. As facilitators of water and glycerol membrane permeation, aquaglyceroporins are crucial for energy production in different organs [14,15]. In circumstances of negative energy balance, glycerol produced by triacylglycerol (TAG) lipolysis in the white adipose tissue (WAT) is released into the bloodstream via AQP7 and used in peripheral tissues as an energy source [16,17]. In the liver, the main organ responsible for whole-body glycerol metabolism, glycerol is taken up via AQP9 to be used for gluconeogenesis [18,19]. The insulin-dependent glucose uptake is also induced by blood glycerol concentration that, by using the AQP7 route, induces pancreatic β-cells to secrete insulin [20]. Thus, the close coordination of aquaglyceroporins in metabolic-related organs is crucial for control of whole-body energy homeostasis and lipid accumulation [15,21]. The balance between the two types of adipose depots, WAT, and brown adipose tissue (BAT) impacts energy homeostasis through their specific metabolic and endocrine functions [22,23]. Adipose tissue releases a large number of adipokines and bioactive mediators that influence not only bodyweight homeostasis but also inflammation, which is a major driver of atherosclerosis [24]. While WAT is an anabolic tissue involved in energy storage, with deleterious consequences for metabolic health, BAT is catabolic and involved in energy production in the form of heat, conferring beneficial effects on adiposity. In consequence, the browning of WAT has been described as a potential strategy to target and control obesity that is causally related to CVD [25]. Our previous work revealed that AQP7 and AQP9 are downregulated along the browning process, which may be related to the physiological role of BAT in heat production, contrasting with the anabolic/catabolic lipid metabolism in white adipose cells [26].

Consistent with the vital role of AQPs in maintaining body water and energy homeostasis, alterations in their physiological functions have been related with the development of cardiometabolic risk factors [22,23]. For instance, the functional importance of AQP1 to maintain endothelial homeostasis and cardiovascular health in humans and mice has been recently reported [4]. Moreover, our prior in vitro study confirmed endothelial AQP1 as a candidate player in the setting of endothelial dysfunction, an early hallmark of atherosclerosis and CVD [14]. Following the emergent importance of AQPs in health and disease [27], these proteins are being viewed as promising novel therapeutic targets for several disorders including the metabolic syndrome, a cluster of atherosclerotic cardiovascular disease risk factors including visceral adiposity, insulin resistance, and dyslipidemia [15].

Apolipoprotein E deficient ($apoE^{-/-}$) mice have been widely used as an animal model to study the pathophysiology of atherosclerosis due to the striking similarities with humans on the molecular mechanisms that lead to endothelial dysfunction and vascular plaque formation [2–4,25,28,29]. Nevertheless, neither the pattern of *AQPs* expression in this atheroprone model is characterized nor the corresponding influence of disease progression. As aforementioned, the functional importance of AQPs on cardiovascular homeostasis has been recently suggested. Thus, we postulated that an atherogenic diet would disturb *AQPs* expression profile, allowing the identification of the *AQPs* isoforms related with vascular toxicity. To investigate this possibility male $apoE^{-/-}$ mice were fed high-fat (HF) or control diets. We assessed systemic inflammation, other metabolic disturbances, and the extent of vascular atherosclerotic lesions formation using biochemical analyses and high-field magnetic resonance imaging (MRI). After confirming a strong atherosclerotic phenotype in the HF-fed mice, we determined the mRNA expressions of the orthodox isoforms, *AQP1* and *AQP5*, and of the aquaglyceroporins *AQP3, AQP7,* and *AQP9*, in different metabolic tissues (BAT, WAT, and liver).

2. Experimental Section

2.1. Animals and Diets

Seven-week-old male $apoE^{-/-}$ mice (Jackson Laboratory, Bar Harbor, ME, USA) were fed one of the following diets prepared based on AIN 93M recommendations but with different levels of fat/cholesterol (Research Diets, New Brunswick, NJ, USA). A control diet (C, 5% w/w fat) or a High Fat diet (HF, 20% w/w fat and 0.15% cholesterol). Macronutrient dietary composition is shown in Table A1. Animals were housed in a room at 22 ± 2 °C with a 12-h light-dark cycle with free access to food and water during 14 (± 2) weeks. Diets were refreshed weekly, at which time animals and their remaining food were weighed. Food consumption was estimated as an average per mouse per day within each cage. All procedures were performed in compliance with the Institutional Animal Care and Use Committee of the Pennsylvania State University, which specifically approved this study.

2.2. Blood Collection

At different time points, blood was collected from the retroorbital sinus of anesthetized animals into heparinized tubes and immediately put on ice. Within 30 min, plasma was isolated by centrifugation at 2000 rpm, 15 min, at 4 °C and stored at -80 °C prior to further biochemical analyses. Due to the limited volume of blood obtained, samples obtained at different time-points were used in the subsequent biochemical analysis, as further detailed.

2.3. Biochemical Analyses

2.3.1. Systemic Methylation Index

The effect of 8 weeks of experimental diets on systemic methylation index was evaluated by the ratio of the plasmatic concentrations of S-adenosylmethionine (AdoMet) to S-adenosylhomocysteine (AdoHcy) (AdoMet/AdoHcy ratio) quantified by liquid chromatography tandem-mass spectrometry (LC-MS/MS), as previously described [30].

2.3.2. Triacylglycerols

The effect of 10 weeks of experimental diets on the fasting plasma concentrations of TAG were measured using a colorimetric assay (Cayman, Ann Arbor, MI, USA) following the manufacturer's protocol.

2.3.3. Cytokines

The effect of 12 weeks of experimental diets on the plasma concentrations of several proinflammatory cytokines/adipokines (interleukin 6, IL-6; macrophage inflammatory protein-1 alpha, MIP-1α; monocyte chemoattractant protein 1, MCP-1; and tumor necrosis factor α, TNF-α) were determined using MSD U-PLEX multiplex assay platforms (Meso Scale Diagnostics, Rockville, MD, USA) following the manufacturer's instructions.

2.3.4. Plasma Amino Acids and Glutathione

Plasma concentrations of amino acids and total glutathione after 12 weeks on each diet were quantified by adequate methodology. Specifically, amino acids were determined by gas chromatography-flame ionization detector (GC-FID) using the Phenomenex EZ:faastTM kit for physiological amino acid analysis (Phenomenex, Torrance, CA, USA) according to the manufacturer's instructions and as previously described in detail [31]. Total glutathione, defined as the total concentration of glutathione after reductive cleavage of all disulfide bonds, was quantified by high-performance liquid chromatography (HPLC) analysis with fluorometric detection, as previously described [32].

2.3.5. Glucose

At the end of each experiment, fasting blood glucose levels were determined using a glucometer (Contour, Bayer, Tarrytown, NY, USA) following the manufacturer's instructions.

2.4. Tissue Collection

At the end-point mice were euthanized by carbon dioxide inhalation and aortas were collected as previously described in detail [33]. Inguinal subcutaneous WAT and interscapular BAT were removed and immediately snap-frozen in liquid nitrogen. Livers were removed, weighted, sampled, and embedded in Optimal Cutting Temperature (OCT) compound (Sakura Finetek, Torrance, CA, USA) or immediately snap-frozen in liquid nitrogen. All samples were stored at $-80\ °C$ before further analyses.

2.5. Oil Red O Staining

Oil Red O staining was used to determine the distribution of lipid droplets in the liver. The OCT-frozen samples were cut into 8-μm-thick sections, fixed in 4% formaldehyde (Thermo Fisher Scientific, Waltham, MA, USA) for 10 min, and rinsed in deionized water and dried at room temperature before Oil Red O staining (Sigma-Aldrich, St. Louis, MO, USA). The sections were stained for 20 min with Oil Red O and rinsed in tap water for 5 min. Hematoxylin (Sigma-Aldrich, St. Louis, MO, USA) was counterstained to assist tissue visualization, and the slides were stained for 1 min and rinsed in tap water for 5 min. All of the representative images are at 400X magnification. The detailed information was described previously [34].

2.6. RNA Extraction

Total RNAs from WAT, BAT, and liver were extracted using the Qiagen RNeasy lipid tissue mini kit and Qiagen RNeasy mini kit, respectively (Qiagen, Germantown, MD, USA) and were stored at $-80\ °C$ prior to analysis. To exclude possible DNA contamination, the optional on-column DNA digestion with the RNase-free DNase (Qiagen) was performed during RNA extraction. All procedures were conducted by following the manufacturer's protocol. The RNA concentrations were determined spectrophotometrically at 260 nm using the NanoDrop 2000c (ThermoFisher Scientific, Waltham, MA, USA). The 260/280 nm ratio was utilized to assess the purity of RNA samples. Only samples with 260/280 nm ratios between 1.8 and 2.2 were used for cDNA synthesis. Additionally, agarose bleach gels were used to qualitatively assess RNA integrity by visualization of the 28S and 18S rRNA bands [35]. To generate cDNA for quantitative PCR, 500 ng of total RNA was reverse-transcribed for 1 h at 42 °C using M-MLV reverse transcriptase (Promega, Madison, WI, USA) and oligo(dT)15 Primer (Promega, Madison, WI, USA).

2.7. Quantitative PCR Analysis

RNA samples were reverse transcribed using M-MLV reverse transcriptase (Promega, Madison, WI, USA) with oligo dT primers (Promega, Madison, WI, USA). Real-time PCR reactions were carried out using a CFX96 Real-Time System C1000 (BioRad, Hercules, CA, USA), the TaqMan Universal PCR Master Mix (Applied Biosystems, Foster City, CA, USA) and the following specific TaqMan pre-designed gene expression primers and probes (Applied Biosystems, Foster City, CA, USA: AQP1 (#Mm00431834_m1), AQP3 (#Mm01208559_m1), AQP5 (#Mm00437578_m1), AQP7 (#Mm00431839_m10), AQP9 (#Mm00508094_m1), and Eef2 (#Mm00833287_g1) as previously described in [26]. The Ct method (2-ΔΔCt) was used for relative quantification of target genes expression after normalization with the Eef2 reference gene [36,37].

2.8. Quantification of Aortic Plaque Volume

Aortas were processed as previously described in detail [33]. Briefly, dissected aortas were equilibrated to 0.1% Magnevist (Bayer HealthCare Pharmaceuticals Inc., Whippany, NJ, USA), 0.25% sodium azide, PBS solution overnight at 4 °C, and plaque volume was determined by MRI using an Agilent 14T micro imaging system (Agilent Technologies, Inc., Santa Clara, CA, USA). After acquisition MR data was reconstructed using Matlab (The Mathworks Inc., Natick, MA, USA) and data segmentation was performed using Avizo 9.5 (Themo Fisher Scientific, Waltham, MA, USA). The lumen of the aorta, the different plaques, and the aorta wall were manually segmented. Quantification of plaque volume was obtained using the material statistics function on the segmented aorta and the results were expressed as the % of plaque volume in relation to the total segmented volume.

2.9. Statistical Analysis

Statistical analyses were performed in GraphPad Prism 7 (GraphPad Software, La Jolla, CA, USA), with statistical significance set to $p < 0.05$. For two-group comparison, an unpaired Student's *t*-test was used. For more than two groups, a one-analysis of variance (ANOVA) was performed, followed by Tukey's.

3. Results and Discussion

To investigate whether a HF diet would disturb the expression of AQPs in an atherosclerosis prone model, we fed *apoE$^{-/-}$* mice HF or control diets. Only male mice were included to control for the known effect of gender on atherosclerosis in this strain [38].

3.1. Diet Consumption, Body and Liver Weights, and Liver Histology

The results showed that HF mice consumed significantly less food than the mice fed a control diet (Figure 1A), nevertheless due to the higher energy density of the HF diet, more calories were consumed by HF-fed mice versus controls (Figure 1B). As expected, HF mice gained more weight compared to controls [39,40] (Figure 1C). No differences were observed in absolute liver weights (g) between the two groups of mice. The values (mean ± SEM, n = 10/group) were 1.17 ± 0.09 and 1.22 ± 0.13, for controls and HF, respectively. The relative liver weights were also similar in the two groups (Figure 1D).

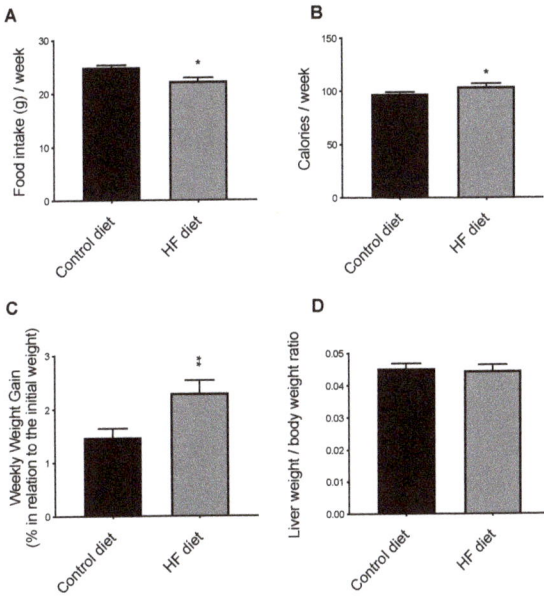

Figure 1. The effect of control and High-Fat (HF) diets on food (**A**) and calories (**B**) intake, and on animal growth (**C**) and relative liver weights (**D**). Data shown are the mean ± SEM (n = 10 per group); * $p < 0.05$; ** $p < 0.01$ control versus HF.

Oil Red O staining showed that livers from HF-fed mice presented more macrovesicular lipid droplets of Oil Red O positive staining than controls (Figure 2). Ballooned hepatocytes and sinusoids capillarization, on the other hand, were also exacerbated in the HF-fed mice, showing more inflated hepatocytes, as compared to the control mice, indicating liver damage by the HF diet. This finding is consistent with a previous study, in which hepatic fat accumulation was reported in $apoE^{-/-}$ mice after seven weeks of a diet containing similar amounts of fat and cholesterol as the HF-diet used in this study [41].

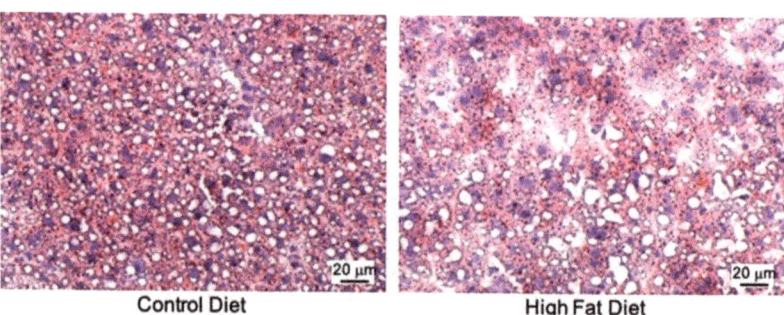

Figure 2. The effect of the experimental diets on liver morphology and hepatic lipid deposition (n = 3–4/group). The images selected as representatives are at 400× magnification. Red: neutral lipid; purple: nuclei.

3.2. Blood Biochemistry

We next evaluated the effect of the HF diet on several biochemical parameters. The results showed that the mice fed a HF diet had significantly increased fasting plasma TAG (Figure 3A) and glucose levels (Figure 3B), thus confirming the major metabolic disturbances elicited by the HF diet previously described by others [31,41,42]. In fact, male $apoE^{-/-}$ mice fed HF-diet are a widely used model of insulin resistance [43], a condition where aquaglyceroporins expression and function are affected [15,44,45].

3.2.1. Systemic Methylation Index

The ratio of the metabolites, S-adenosylmethionine (AdoMet) to S-adenosylhomocysteine (AdoHcy), or AdoMet/AdoHcy measures the cell methylation potential [46–48]. AdoMet is the universal methyl donor to cellular methyltransferases, originating the methylated substrate and the metabolite AdoHcy. Importantly, AdoHcy may negatively regulate the activity of those same methyltransferases. Thus, a decreased AdoMet/AdoHcy ratio reflects a hypomethylating environment. In previous cell studies we have observed that a decrease in AdoMet/AdoHcy ratio downregulated AQP1 expression and promoted an atherogenic endothelial phenotype [14]. Yun et al. have reported that a diminished methylation index was present in wild type mice fed HF diets [49]. These observations led us to measure the systemic concentrations of AdoMet and AdoHcy in this study. Nevertheless, the results revealed similar plasmatic AdoMet/AdoHcy ratios in both groups of animals, thus showing that the systemic methylation index was not affected by the HF diet (Figure 3C) in $apoE^{-/-}$ mice.

3.2.2. Plasma Glutathione

The role of AQPs as facilitators of hydrogen peroxide membrane permeation, a major reactive oxygen species (ROS), has been recently reported [50]. Importantly, ROS build-up is a main pathophysiological mechanism favoring the establishment and progression of atherosclerosis [47]. Glutathione is a major cellular antioxidant that neutralizes hydrogen peroxide to water. Thus, we measured the systemic levels of total glutathione in our animals. The results, shown in Figure 3D, revealed similar plasma levels of glutathione in both groups of animals thereby suggesting that the bioavailability of this tripeptide is intact in the HF-mice. In support of this possibility, the plasma levels of three amino acids that form glutathione, i.e., glutamate (Figure 3E), glycine (Figure 3E), and cysteine (data not shown) did not differ between these two diets. We do acknowledge however that the measurement of the concentrations of both the free and reduced glutathione forms would be necessary to give a functional measure of this major antioxidant system.

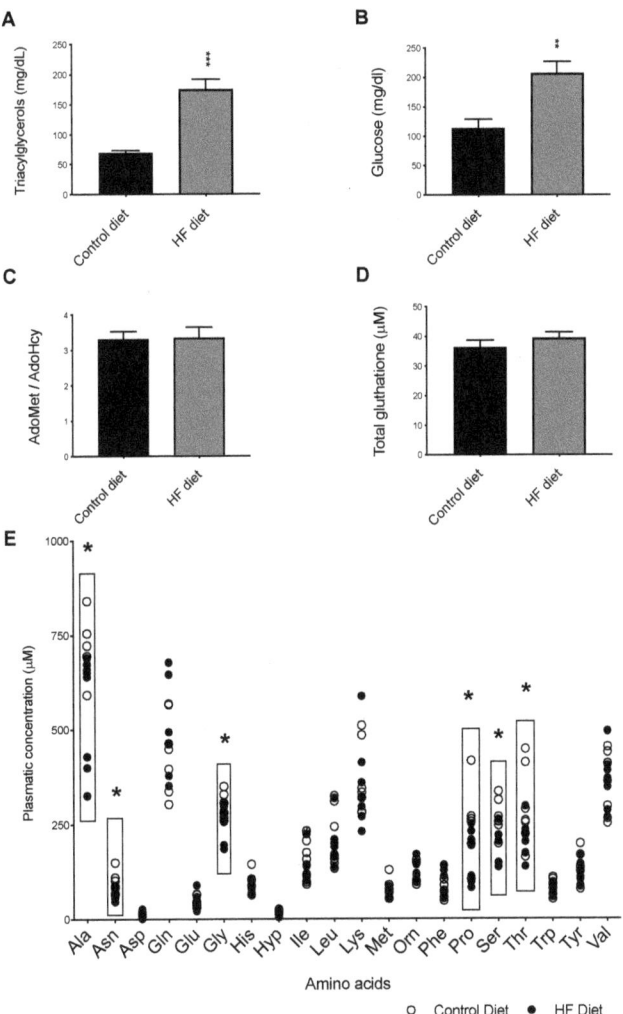

Figure 3. The effect of control and High-Fat (HF) diets on circulating concentrations of triacylglycerols (**A**), glucose (**B**), S-adenosylmethionine to S-adenosylhomocysteine ratio (AdoMet/AdoHcy) (**C**), total glutathione (**D**), and amino acids (**E**). (Ala, alanine; Gly, glycine; Val, valine; Leu, leucine; Ile, isoleucine; Thr, threonine; Ser, serine; Pro, proline; Asn, asparagine; Asp, aspartate; Met, methionine; Hyp, hydroxyproline; Glu, glutamate; Phe, phenylalanine; Gln, glutamine; Orn, ornithine, Lys, lysine; His, histidine; Tyr, tyrosine; Trp, tryptophan). Data shown are the mean ± SEM (n = 6–9 per group); * $p < 0.05$; ** $p < 0.01$; *** $p < 0.001$; HF versus control.

3.2.3. Plasma Amino Acids

Next, we evaluated the plasmatic levels of amino acids to visualize any changes driven by the HF diet that could suggest disturbed metabolic pathways to be further related with diet-induced effects on *AQPs* expression. In fact, recent developments have begun to shed light on associations between compromised cardiometabolic function and altered intermediary metabolism of amino acids [51]. In the present study, the plasma amino acid levels of the essential amino acids were not different between two groups (Figure 3E). Thus, the concentrations of the branched chain amino acids (BCAA), Tyr, and Phe, which were previously implicated in heart failure, a form of CVD [51–53], were similar in both the HF-fed and control groups. As shown in Table A1, both diets contained the same amount of casein, which was the only protein source. Because essential amino acids cannot be synthetized endogenously, this observation suggests that the metabolic pathways responsible for utilization and catabolism of essential amino acids are intact in the HF-fed animals. Among the remaining amino acids, significantly lower levels of alanine (Ala), asparagine (Asn), lysine (Lys), serine (Ser), and proline (Pro) were found in the plasma of the HF-fed mice compared to the control group (Figure 3E), suggesting that the HF diet resulted in a decreased protein turn-over state that is consistent with preferential utilization of the dietary fat as energy fuel [54].

3.2.4. Systemic Inflammation

Finally, we determined the effect of the diets on plasma proinflammatory cytokines, including interferon gamma (IFN-γ), interleukin 6 (IL-6), macrophage inflammatory protein-1 alpha (MIP-1α/CCL3), and tumor necrosis factor alpha (TNF-α) [42,55,56]. IFN-γ is a proinflammatory mediator that is expressed in atherosclerotic lesions [57]. IL-6 is a pleiotropic cytokine with both pro- or antiatherogenic effects but which exacerbates atherosclerosis in murine species [58,59] MCP-1 is an inflammatory chemokine with a critical role in the initiation of atherosclerosis [60]. TNF-α is another major proatherogenic molecule that sustains the progression of the vascular lesions and atherosclerosis [42]. As expected, feeding mice with HF diet significantly elevated the systemic concentration of most cytokines, when compared to the control diet-fed mice (Figure 4). This observation is in agreement with the positive effect of dietary fat on systemic inflammation and with the well-established atherogenic effect of a HF diet in $apoE^{-/-}$ mice [61,62]. In fact, inflammation is central to all stages of atherosclerosis establishment and progression.

3.3. Volume of Atherosclerosis

A method based on ex-vivo MR imaging was used to visualize and quantify the aortic volume of fatty plaque in HF-fed mice versus controls. The results revealed a profound difference of the atherosclerosis burden between the two groups of animals (Figure 5). Plaque volume throughout the whole aorta was significantly increased in the HF group compared to the control group; the atherosclerotic lesions predominantly distributed in the aortic arch and in the areas surrounding the branching points of the major arteries, which includes the brachiocephalic artery (BCA). As a result, the difference between HF-aortas and control aortas in these two highly susceptible regions, aortic arch and BCA, was even more pronounced than in the whole aortas. Interestingly, despite this massive difference of vascular lesions between HF-fed mice and controls both groups of animals presented similar plasma concentrations of the BCAA Val, Leu and Ile, and of Tyr and Phe (Figure 3), which were previously suggested to have a significant role in the pathogenesis of atherosclerosis and CVD [51–53]. The increased systemic inflammation observed in HF-fed mice, however, (Figure 4) is consistent with this augmented arteriosclerotic plaque burden observed in these mice. Taken together, these observations are in agreement with the well-established atherogenic effect of dietary fat [63]. Having induced a strong atherosclerotic phenotype with HF diet in $apoE^{-/-}$ mice, when compared to controls, we determined the expression levels of the orthodox AQP isoforms, *AQP1* and *AQP5*, and of the aquaglyceroporins *AQP3*, *AQP7*, and *AQP9*, in both groups of mice.

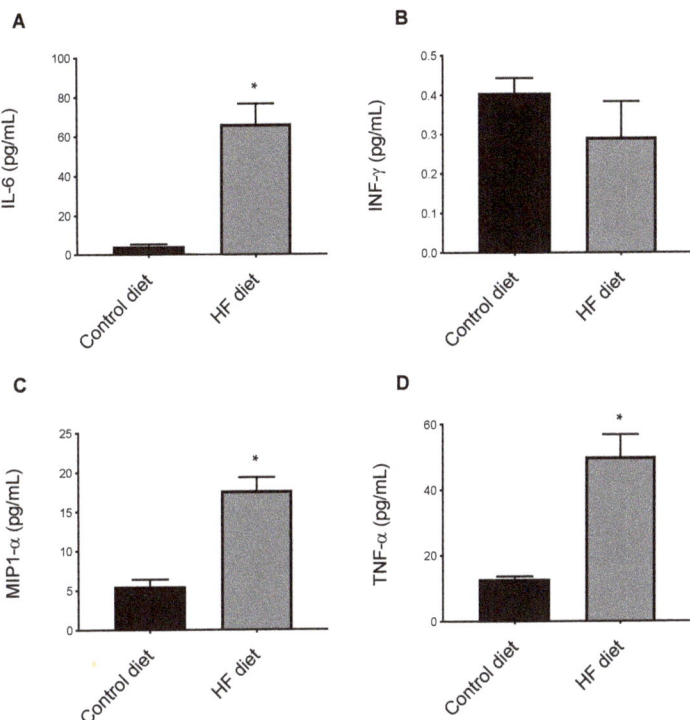

Figure 4. The effect of control and High-Fat (HF) diets on systemic inflammation. Interleukin 6, IL-6 (**A**); interferon gamma, INF-γ (**B**); macrophage inflammatory protein-1 alpha, MIP-1α/CCL3 (**C**); tumor necrosis factor alpha, TNF-α (**D**). Data shown are the mean ± SEM (n = 4 per group). * $p < 0.05$ HF versus control.

3.4. Expression of Aquaporins in Liver and Adipose Tissues

The expression of *AQPs* with impact on metabolism and endothelial function (*AQP1, AQP3, AQP5, AQP7,* and *AQP9*) [14] was evaluated in three distinct metabolic tissues: liver, WAT and BAT. Liver is the organ responsible for the plasma glucose levels maintenance, thus, in a situation of negative energy balance or exercise, glucose is produced in the liver and released to the bloodstream [64]. When the body is in a state of positive energy balance and plasma glucose levels are high (as in the HF-fed mice), energy is stored in WAT in the triacylglycerol form, to be hydrolyzed in case of energy demands. BAT is a tissue that burns excess fat by thermogenesis [65].

All of the investigated *AQP* isoforms were detected in liver, WAT, and BAT, however, each in a tissue-specific profile (Figure 6A,C,E, black bars). In liver, and as previously described, *AQP9* was the most highly expressed isoform [18]. Moreover, *AQP1, AQP3, AQP7,* and *AQP5* were also detected in liver, although at lower levels than *AQP9* (Figure 6A). While AQP9 has been described as responsible for glycerol influx in hepatocytes, no specific function has been attributed to any other hepatic AQP isoform [66]. Concerning the impact of the HF diet on the hepatic *AQPs* expression levels, we observed that *AQP3* was the most affected isoform, with the HF-fed mice presenting levels that were around 80-fold more than controls. Nevertheless, since the overall expression level of *AQP3* was lower than the most abundant isoform *AQP9*, an increase in *AQP3* expression may not have significantly impacted the total glycerol efflux in hepatocytes. Interestingly, however, AQP3 is also a hydrogen peroxide channel [67], and thus its upregulation might represent an increase in oxidative stress in the tissue. This possibility is favored by the fact that the hepatic expression levels of another peroxiporin, AQP5, were also significantly upregulated by the atherogenic diet. In fact, overall oxidative stress is a major driver of the strong atherogenic phenotype observed in these HF-fed animals [46,48]. Additionally, the results showed that hepatic AQP1 was significantly downregulated by the HF-diet whereas the levels of AQP9 remained unaltered (Figure 6B). One limitation to our study is the fact that AQP9 immunolocalization was not performed, which would show whether it is correlated with gene expression or liver pathology. By histopathological analysis we could confirm the presence of hepatic fat accumulation in the HF-fed mice. Nevertheless, the lack of markers of hepatic fibrosis in the present study prevents us to further elucidate the relationship between hepatic AQPs expression and the development of atherosclerosis. However, a study in patients with morbid obesity did not find any relationship between AQP9 expression in the liver and the degree of hepatic steatosis or fibrosis [68]. Therefore, it seems that regulation of fatty liver deposits is not influenced by AQP9 expression. Thus, it is unlikely that animals fed the atherogenic diet and with liver fat deposition show altered hepatic aquaglyceroporin expression profile.

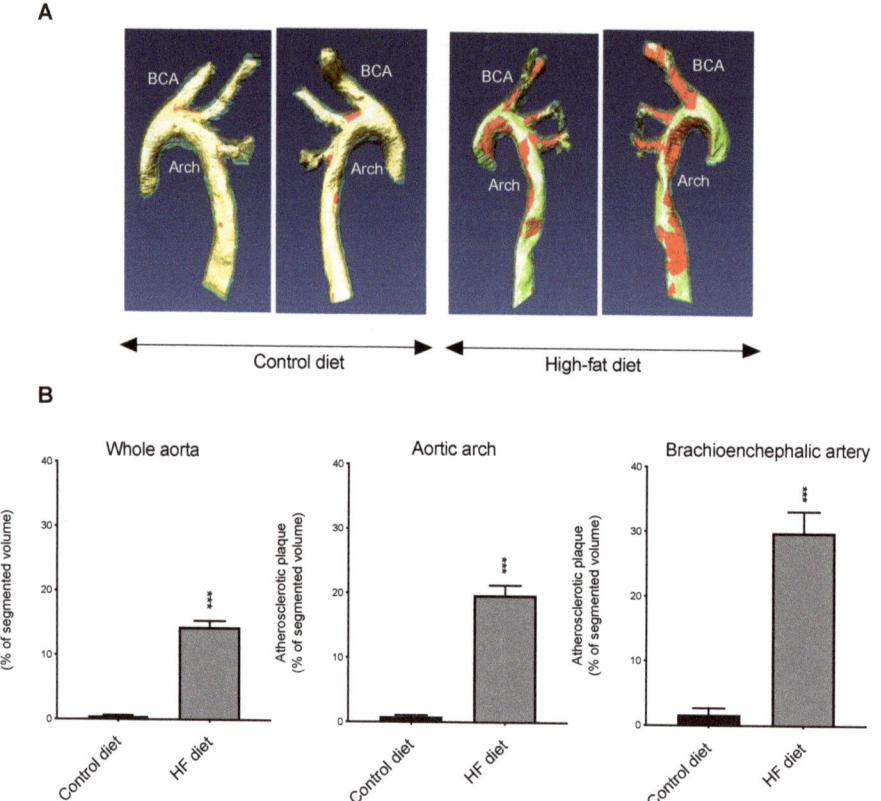

Figure 5. Differences in aortic atherosclerosis burden between mice fed the different diets. (**A**) Representative images (both sides of the same aorta are shown) of the visualization of the arteriosclerotic plaque (colored in red) using 14T-magnetic resonance imaging (MRI) in aortas from mice fed control or high-fat diets (BCA, brachiocephalic artery). (**B**) 14T-MRI volumetric assessment of the atherosclerotic plaque in different aorta segments. Data shown are the mean ± SEM (n = 8–10 per group). *** $p < 0.001$, HF versus control.

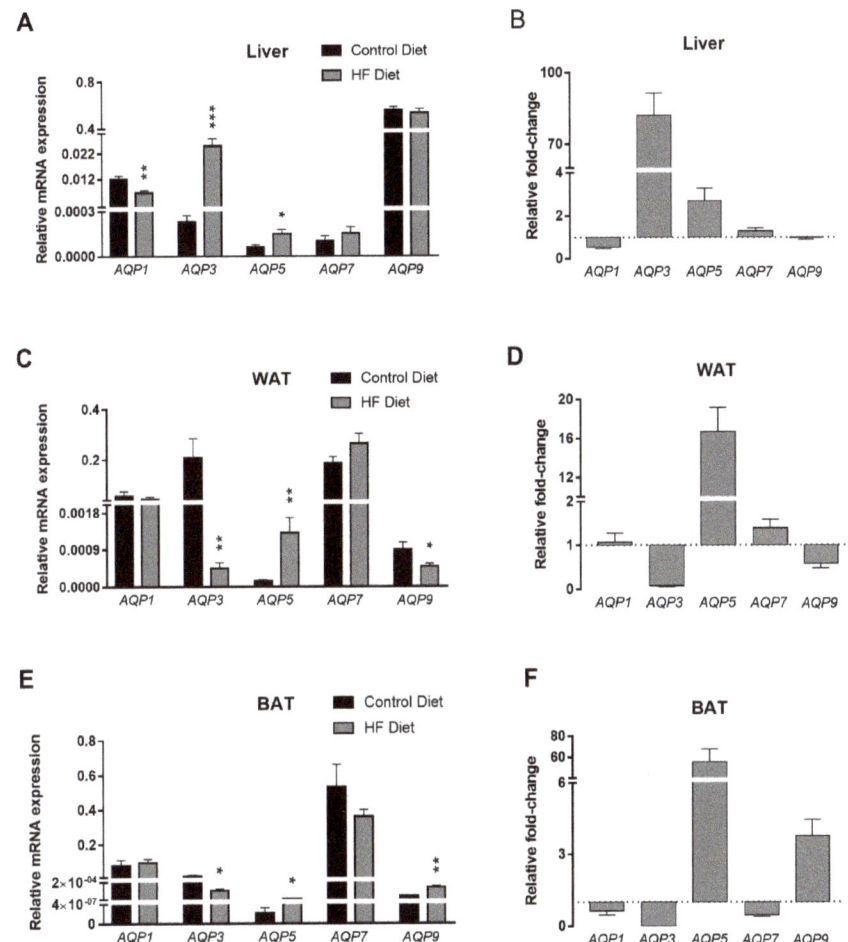

Figure 6. Aquaporins gene expression in $apoE^{-/-}$ mice fed control or atherogenic High-Fat (HF) diets. *AQP1, AQP3, AQP5, AQP7* and *AQP9* expression in liver (**A**), WAT (**C**) and BAT (**E**) in mice fed control or HF diets and respective fold-change in HF-fed animals relative to controls [liver (**B**); WAT (**D**); and BAT (**F**)]. Results are mean ± SEM (n = 8–10 per group). * $p < 0.05$; ** $p < 0.01$; *** $p < 0.001$; all HF versus control.

Interestingly, the association of an inadequate AQP1 function with an atherogenic phenotype has been previously reported albeit in other contexts. For example, we observed that endothelial dysfunction under hypomethylating stress lessened *AQP1* expression in vitro [14], and others reported that the targeted deletion of the *AQP1* gene favored atherosclerosis progression in $apoE^{-/-}$ mice [4]. In the present study, and as discussed above, the AdoMet/AdoHcy ratio was not affected by the HF-diet, so we can exclude a disturbed systemic methylation index as causing the observed HF-induced hepatic *AQP1* downregulation.

In WAT, we have detected abundant levels of *AQP1*, *AQP3* and *AQP7* mRNAs, while *AQP5* and *AQP9* were present in low amounts (Figure 6C) [16,66]. A strong induction of *AQP5* expression of around 15-fold was observed in mice fed HF diet. As aforementioned, *AQP5* is a peroxiporin whose expression has been related with oxidative stress in rodents [69] and humans [13]. We have previously reported dysregulation of endothelial AQP5 expression associated with endothelial dysfunction [14], suggesting AQP5 is closely related to a dysbalanced redox state that favors the atherosclerosis progression. In addition, our present results show that *AQP3* and *AQP9* expression is impaired in WAT from mice fed HF diet (Figure 6D). Knowing that AQP3 and AQP9 are the most abundant aquaglyceroporins in mice adipose membranes [26], their downregulation might represent a cellular strategy to avoid excess glycerol efflux to be used in liver.

BAT presented high levels of *AQP7* gene expression, the most representative isoform in this tissue, and low levels of *AQP1*, *AQP3*, *AQP5*, and *AQP9* mRNAs (Figure 6E). In the BAT of HF-fed mice, *AQP3* was downregulated, *AQP9* and *AQP5* were increased around 4- and 60-fold, respectively, while *AQP1* and *AQP7* expression was unaltered (Figure 6F). The downregulation of *AQP3* in parallel with the dramatic upregulation of *AQP5*, suggests an impairment of glycerol movements coupled to increased sensitivity to oxidative stress, similar to the observed in WAT.

The *AQPs* expression pattern in the three analyzed tissues revealed tissue-specific differences, inherent to their predicted metabolic role. However, *AQP5* expression was consistently induced by HF diet across WAT, BAT, and liver. Since AQP5 facilitates hydrogen peroxide permeation in rodents and humans, its upregulation suggests the presence of an unbalanced redox state [13,69]. Here, we have detected similar plasma levels of total glutathione, the major intracellular antioxidant that neutralizes hydrogen peroxide, in both groups of animals, thereby suggesting that the bioavailability of glutathione was intact in the HF-fed mice. Nevertheless, this similar glutathione bioavailability does not translate into a similar antioxidant capacity and could only be evaluated by determining the concentration of the reduced and oxidized glutathione forms. AQP5 dysfunction has been associated with a vast array of phenotypes, and evidence suggests that AQP5 upregulation promotes tumor cell proliferation [70,71]. Studies correlating AQP5 to obesity are scarce; however, a link between hypothalamus *AQP5* and adipocyte gene expression was reported, indicating a possible regulatory coordination [72]. Interestingly, AQP5 is crucial for mice adipocyte differentiation [73] and *AQP5*-KO mice have lower body weight than controls [74]. Altogether, these data suggest that the increased *AQP5* expression observed in WAT may indicate an increase in adipocyte differentiation to accommodate the excess fat in HF diet.

4. Conclusions

In conclusion, the present study contributes to a better characterization of the well-established $apoE^{-/-}$ mouse model by reporting that the pattern of *AQPs* expression in these mice is disturbed, in a tissue-specific manner, by an atherogenic HF diet. The present report suggests a novel relation between diet, AQP5, and atherosclerosis that warrants further investigation and may ultimately open the door to the development of effective new treatments for CVD.

Author Contributions: Methodology, I.V.d.S., C.A.W., F.J.M., C.F.; investigation, I.V.d.S., C.A.W., F.J.M., C.F.; formal analysis, I.V.d.S., C.A.W., F.J.M., N.K.H., S.G.H., T.N.; original draft preparation, I.V.d.S., G.S., R.C.; reviewing and editing, I.V.d.S., C.A.W., F.J.M., C.F., N.K.H., S.G.H., T.N., G.S., R.C., A.C.R.; conceptualization, N.K.H., T.N., A.C.R., G.S., R.C.; resources, T.N., G.S., R.C.; supervision, A.C.R., G.S., R.C.; laboratory resources, A.C.R. All authors have read and agreed to the published version of the manuscript.

Funding: This work was funded by the Graduate Program in Nutritional Sciences, Penn State High-Field Magnetic Resonance Imaging Facility, and the Huck Institutes of the Life Sciences, all from the Pennsylvania State University, USA, as well as by the Fundação para a Ciência e Tecnologia (FCT), Portugal (grant PTDC/BTM-SAL/28977/2017 and strategic projects UIDB/04138/2020 and UIDP/04138/2020), and publication funds were from The Hershey Company Endowment in Nutrition.

Institutional Review Board Statement: The study was conducted according to the guidelines of the Declaration of Helsinki, and approved by the Institutional Animal Care and Use Committee of the Pennsylvania State University, which specifically approved this study (IACUC #47677) on 31 August 2017.

Informed Consent Statement: Not applicable.

Data Availability Statement: Not applicable.

Acknowledgments: The authors wish to thank Isabel Tavares de Almeida (University of Lisbon, Portugal) for her valuable support.

Conflicts of Interest: The authors declare no conflict of interest.

Appendix A

Table A1. Macronutrient composition of the experimental diets.

	Control Diet		High Fat Diet	
	g	Kcal	g	Kcal
Casein	180	720	184	736
L-Cysteine	3	12	3	12
Corn starch	431	1725	217	868
Maltodextrin 10	155	620	93	372
Sucrose	100	400	102	408
Cellulose	35	0	35	0
Cocoa butter	0	0	155	1395
Primex	25	225	0	0
Corn oil	25	225	25	225
Cholesterol	0	0	11	0

References

1. Virani, S.S.; Alonso, A.; Benjamin, E.J.; Bittencourt, M.S.; Callaway, C.W.; Carson, A.P.; Chamberlain, A.M.; Chang, A.R.; Cheng, S.; Delling, F.N.; et al. Heart Disease and Stroke Statistics-2020 Update: A Report from the American Heart Association. *Circulation* **2020**, *141*, e139–e596. [CrossRef]
2. Sanz, J.; Fayad, Z.A. Imaging of atherosclerotic cardiovascular disease. *Nature* **2008**, *451*, 953–957. [CrossRef]
3. Fontijn, R.D.; Volger, O.L.; van der Pouw-Kraan, T.C.; Doddaballapur, A.; Leyen, T.; Baggen, J.M.; Boon, R.A.; Horrevoets, A.J. Expression of Nitric Oxide-Transporting Aquaporin-1 Is Controlled by KLF2 and Marks Non-Activated Endothelium In Vivo. *PLoS ONE* **2015**, *10*, e0145777. [CrossRef]
4. Wintmo, P.; Johansen, S.H.; Hansen, P.B.L.; Lindholt, J.S.; Urbonavicius, S.; Rasmussen, L.M.; Bie, P.; Jensen, B.L.; Stubbe, J. The water channel AQP1 is expressed in human atherosclerotic vascular lesions and AQP1 deficiency augments angiotensin II-induced atherosclerosis in mice. *Acta Physiol.* **2017**, *220*, 446–460. [CrossRef]
5. Agre, P. Aquaporin water channels (Nobel Lecture). *Angew. Chem. Int. Ed. Engl.* **2004**, *43*, 4278–4290. [CrossRef]
6. Madeira, A.; Moura, T.F.; Soveral, G. Detecting Aquaporin Function and Regulation. *Front. Chem.* **2016**, *4*, 3. [CrossRef]
7. Verkman, A.S.; Anderson, M.O.; Papadopoulos, M.C. Aquaporins: Important but elusive drug targets. *Nat. Rev. Drug Discov.* **2014**, *13*, 259–277. [CrossRef]
8. Ishibashi, K.; Tanaka, Y.; Morishita, Y. The role of mammalian superaquaporins inside the cell. *Biochim. Biophys. Acta* **2014**, *1840*, 1507–1512. [CrossRef]
9. Itoh, T.; Rai, T.; Kuwahara, M.; Ko, S.B.; Uchida, S.; Sasaki, S.; Ishibashi, K. Identification of a novel aquaporin, AQP12, expressed in pancreatic acinar cells. *Biochem. Biophys. Res. Commun.* **2005**, *330*, 832–838. [CrossRef] [PubMed]
10. Madeira, A.; Fernandez-Veledo, S.; Camps, M.; Zorzano, A.; Moura, T.F.; Ceperuelo-Mallafre, V.; Vendrell, J.; Soveral, G. Human aquaporin-11 is a water and glycerol channel and localizes in the vicinity of lipid droplets in human adipocytes. *Obesity* **2014**, *22*, 2010–2017. [CrossRef]
11. Bertolotti, M.; Bestetti, S.; Garcia-Manteiga, J.M.; Medrano-Fernandez, I.; Dal Mas, A.; Malosio, M.L.; Sitia, R. Tyrosine kinase signal modulation: A matter of H_2O_2 membrane permeability? *Antioxid. Redox Signal.* **2013**, *19*, 1447–1451. [CrossRef] [PubMed]
12. Prata, C.; Hrelia, S.; Fiorentini, D. Peroxiporins in Cancer. *Int. J. Mol. Sci.* **2019**, *20*, 1371. [CrossRef]
13. Rodrigues, C.; Pimpao, C.; Mosca, A.F.; Coixao, A.S.; Lopes, D.; da Silva, I.V.; Pedersen, P.A.; Antunes, F.; Soveral, G. Human Aquaporin-5 Facilitates Hydrogen Peroxide Permeation Affecting Adaption to Oxidative Stress and Cancer Cell Migration. *Cancers* **2019**, *11*, 932. [CrossRef]

14. da Silva, I.V.; Barroso, M.; Moura, T.; Castro, R.; Soveral, G. Endothelial Aquaporins and Hypomethylation: Potential Implications for Atherosclerosis and Cardiovascular Disease. *Int. J. Mol. Sci.* **2018**, *19*, 130. [CrossRef]
15. da Silva, I.V.; Rodrigues, J.S.; Rebelo, I.; Miranda, J.P.G.; Soveral, G. Revisiting the metabolic syndrome: The emerging role of aquaglyceroporins. *Cell. Mol. Life Sci. CMLS* **2018**, *75*, 1973–1988. [CrossRef]
16. da Silva, I.V.; Soveral, G. Aquaporins in Obesity. *Adv. Exp. Med. Biol.* **2017**, *969*, 227–238. [CrossRef]
17. Rodriguez, A.; Marinelli, R.A.; Tesse, A.; Fruhbeck, G.; Calamita, G. Sexual Dimorphism of Adipose and Hepatic Aquaglyceroporins in Health and Metabolic Disorders. *Front. Endocrinol.* **2015**, *6*, 171. [CrossRef]
18. Calamita, G.; Gena, P.; Ferri, D.; Rosito, A.; Rojek, A.; Nielsen, S.; Marinelli, R.A.; Fruhbeck, G.; Svelto, M. Biophysical assessment of aquaporin-9 as principal facilitative pathway in mouse liver import of glucogenetic glycerol. *Biol. Cell* **2012**, *104*, 342–351. [CrossRef]
19. Jelen, S.; Wacker, S.; Aponte-Santamaria, C.; Skott, M.; Rojek, A.; Johanson, U.; Kjellbom, P.; Nielsen, S.; de Groot, B.L.; Rutzler, M. Aquaporin-9 protein is the primary route of hepatocyte glycerol uptake for glycerol gluconeogenesis in mice. *J. Biol. Chem.* **2011**, *286*, 44319–44325. [CrossRef] [PubMed]
20. Delporte, C.; Virreira, M.; Crutzen, R.; Louchami, K.; Sener, A.; Malaisse, W.J.; Beauwens, R. Functional role of aquaglyceroporin 7 expression in the pancreatic beta-cell line BRIN-BD11. *J. Cell. Physiol.* **2009**, *221*, 424–429. [CrossRef] [PubMed]
21. Rodriguez, A.; Catalan, V.; Gomez-Ambrosi, J.; Garcia-Navarro, S.; Rotellar, F.; Valenti, V.; Silva, C.; Gil, M.J.; Salvador, J.; Burrell, M.A.; et al. Insulin- and leptin-mediated control of aquaglyceroporins in human adipocytes and hepatocytes is mediated via the PI3K/Akt/mTOR signaling cascade. *J. Clin. Endocrinol. Metab.* **2011**, *96*, E586–E597. [CrossRef]
22. Ravussin, E.; Kozak, L.P. Have we entered the brown adipose tissue renaissance? *Obes. Rev. Off. J. Int. Assoc. Study Obes.* **2009**, *10*, 265–268. [CrossRef]
23. Simopoulos, A.P. The omega-6/omega-3 fatty acid ratio, genetic variation, and cardiovascular disease. *Asia Pac. J. Clin. Nutr.* **2008**, *17* (Suppl. S1), 131–134.
24. Trayhurn, P.; Wood, I.S. Signalling role of adipose tissue: Adipokines and inflammation in obesity. *Biochem. Soc. Trans.* **2005**, *33*, 1078–1081. [CrossRef]
25. Lee, Y.H.; Mottillo, E.P.; Granneman, J.G. Adipose tissue plasticity from WAT to BAT and in between. *Biochim. Biophys. Acta* **2014**, *1842*, 358–369. [CrossRef] [PubMed]
26. da Silva, I.V.; Diaz-Saez, F.; Zorzano, A.; Guma, A.; Camps, M.; Soveral, G. Aquaglyceroporins Are Differentially Expressed in Beige and White Adipocytes. *Int. J. Mol. Sci.* **2020**, *21*, 610. [CrossRef]
27. Soveral, G.; Nielsen, S.; Casini, A. *Aquaporins in Health and Disease: New Molecular Targets for Drug Discovery*; Soveral, G., Nielsen, S., Casini, A., Eds.; CRC Press, Taylor & Francis Group: Boca Raton, FL, USA, 2016.
28. Getz, G.S.; Reardon, C.A. Do the $Apoe^{-/-}$ and $Ldlr^{-/-}$ Mice Yield the Same Insight on Atherogenesis? *Arterioscler. Thromb. Vasc. Biol.* **2016**, *36*, 1734–1741. [CrossRef] [PubMed]
29. Getz, G.S.; Reardon, C.A. ApoE knockout and knockin mice: The history of their contribution to the understanding of atherogenesis. *J. Lipid Res.* **2016**, *57*, 758–766. [CrossRef] [PubMed]
30. Heil, S.G.; Herzog, E.M.; Griffioen, P.H.; van Zelst, B.; Willemsen, S.P.; de Rijke, Y.B.; Steegers-Theunissen, R.P.M.; Steegers, E.A.P. Lower S-adenosylmethionine levels and DNA hypomethylation of placental growth factor (PlGF) in placental tissue of early-onset preeclampsia-complicated pregnancies. *PLoS ONE* **2019**, *14*, e0226969. [CrossRef] [PubMed]
31. Badawy, A.A.; Morgan, C.J.; Turner, J.A. Application of the Phenomenex EZ:faasttrade mark amino acid analysis kit for rapid gas-chromatographic determination of concentrations of plasma tryptophan and its brain uptake competitors. *Amino Acids* **2008**, *34*, 587–596. [CrossRef]
32. Barroso, M.; Rocha, M.S.; Esse, R.; Goncalves, I., Jr.; Gomes, A.Q.; Teerlink, T.; Jakobs, C.; Blom, H.J.; Loscalzo, J.; Rivera, I.; et al. Cellular hypomethylation is associated with impaired nitric oxide production by cultured human endothelial cells. *Amino Acids* **2012**, *42*, 1903–1911. [CrossRef]
33. Whalen, K.J.; Mattie, F.J.; Florindo, C.; van Zelst, B.; Huang, N.K.; Tavares de Almeida, I.; Heil, S.G.; Neuberger, T.; Ross, A.C.; Castro, R. No Effect of Diet-Induced Mild Hyperhomocysteinemia on Vascular Methylating Capacity, Atherosclerosis Progression, and Specific Histone Methylation. *Nutrients* **2020**, *12*, 2182. [CrossRef] [PubMed]
34. Andres-Manzano, M.J.; Andres, V.; Dorado, B. Oil Red O and Hematoxylin and Eosin Staining for Quantification of Atherosclerosis Burden in Mouse Aorta and Aortic Root. *Methods Mol. Biol.* **2015**, *1339*, 85–99. [CrossRef] [PubMed]
35. Aranda, P.S.; LaJoie, D.M.; Jorcyk, C.L. Bleach gel: A simple agarose gel for analyzing RNA quality. *Electrophoresis* **2012**, *33*, 366–369. [CrossRef] [PubMed]
36. Fleige, S.; Pfaffl, M.W. RNA integrity and the effect on the real-time qRT-PCR performance. *Mol. Asp. Med.* **2006**, *27*, 126–139. [CrossRef]
37. Livak, K.J.; Schmittgen, T.D. Analysis of relative gene expression data using real-time quantitative PCR and the 2(-Delta Delta C(T)) Method. *Methods* **2001**, *25*, 402–408. [CrossRef]
38. Caligiuri, G.; Nicoletti, A.; Zhou, X.; Tornberg, I.; Hansson, G.K. Effects of sex and age on atherosclerosis and autoimmunity in apoE-deficient mice. *Atherosclerosis* **1999**, *145*, 301–308. [CrossRef]
39. Jones, N.S.; Watson, K.Q.; Rebeck, G.W. Metabolic Disturbances of a High-Fat Diet Are Dependent on APOE Genotype and Sex. *eNeuro* **2019**, *6*. [CrossRef]

40. Li, Y.; Zhang, C.G.; Wang, X.H.; Liu, D.H. Progression of atherosclerosis in ApoE-knockout mice fed on a high-fat diet. *Eur. Rev. Med. Pharmacol. Sci.* **2016**, *20*, 3863–3867.
41. Schierwagen, R.; Maybuchen, L.; Zimmer, S.; Hittatiya, K.; Back, C.; Klein, S.; Uschner, F.E.; Reul, W.; Boor, P.; Nickenig, G.; et al. Seven weeks of Western diet in apolipoprotein-E-deficient mice induce metabolic syndrome and non-alcoholic steatohepatitis with liver fibrosis. *Sci. Rep.* **2015**, *5*, 12931. [CrossRef]
42. Branen, L.; Hovgaard, L.; Nitulescu, M.; Bengtsson, E.; Nilsson, J.; Jovinge, S. Inhibition of tumor necrosis factor-alpha reduces atherosclerosis in apolipoprotein E knockout mice. *Arterioscler. Thromb. Vasc. Biol.* **2004**, *24*, 2137–2142. [CrossRef] [PubMed]
43. Hansmann, G.; Wagner, R.A.; Schellong, S.; Perez, V.A.; Urashima, T.; Wang, L.; Sheikh, A.Y.; Suen, R.S.; Stewart, D.J.; Rabinovitch, M. Pulmonary arterial hypertension is linked to insulin resistance and reversed by peroxisome proliferator-activated receptor-gamma activation. *Circulation* **2007**, *115*, 1275–1284. [CrossRef] [PubMed]
44. Mendez-Gimenez, L.; Ezquerro, S.; da Silva, I.V.; Soveral, G.; Fruhbeck, G.; Rodriguez, A. Pancreatic Aquaporin-7: A Novel Target for Anti-diabetic Drugs? *Front. Chem.* **2018**, *6*, 99. [CrossRef] [PubMed]
45. da Silva, I.V.; Cardoso, C.; Mendez-Gimenez, L.; Camoes, S.P.; Fruhbeck, G.; Rodriguez, A.; Miranda, J.P.; Soveral, G. Aquaporin-7 and aquaporin-12 modulate the inflammatory phenotype of endocrine pancreatic beta-cells. *Arch. Biochem. Biophys.* **2020**, *691*, 108481. [CrossRef] [PubMed]
46. Barroso, M.; Handy, D.E.; Castro, R. The link between hyperhomocysteinemia and hypomethylation: Implications for cardiovascular disease. *J. Inborn Errors Metab. Screen.* **2017**, *5*, 1–15. [CrossRef]
47. Castro, R.; Rivera, I.; Blom, H.J.; Jakobs, C.; Tavares de Almeida, I. Homocysteine metabolism, hyperhomocysteinaemia and vascular disease: An overview. *J. Inherit. Metab. Dis.* **2006**, *29*, 3–20. [CrossRef]
48. Esse, R.; Barroso, M.; Tavares de Almeida, I.; Castro, R. The Contribution of Homocysteine Metabolism Disruption to Endothelial Dysfunction: State-of-the-Art. *Int. J. Mol. Sci.* **2019**, *20*, 867. [CrossRef]
49. Yun, K.U.; Ryu, C.S.; Oh, J.M.; Kim, C.H.; Lee, K.S.; Lee, C.H.; Lee, H.S.; Kim, B.H.; Kim, S.K. Plasma homocysteine level and hepatic sulfur amino acid metabolism in mice fed a high-fat diet. *Eur. J. Nutr.* **2013**, *52*, 127–134. [CrossRef]
50. Tamma, G.; Valenti, G.; Grossini, E.; Donnini, S.; Marino, A.; Marinelli, R.A.; Calamita, G. Aquaporin Membrane Channels in Oxidative Stress, Cell Signaling, and Aging: Recent Advances and Research Trends. *Oxid. Med. Cell. Longev.* **2018**, *2018*, 1501847. [CrossRef]
51. Adams, S.H. Emerging perspectives on essential amino acid metabolism in obesity and the insulin-resistant state. *Adv. Nutr.* **2011**, *2*, 445–456. [CrossRef]
52. Grajeda-Iglesias, C.; Aviram, M. Specific Amino Acids Affect Cardiovascular Diseases and Atherogenesis via Protection against Macrophage Foam Cell Formation: Review Article. *Rambam Maimonides Med. J.* **2018**, *9*. [CrossRef] [PubMed]
53. Neinast, M.; Murashige, D.; Arany, Z. Branched Chain Amino Acids. *Annu. Rev. Physiol.* **2019**, *81*, 139–164. [CrossRef]
54. Hou, Y.; Yin, Y.; Wu, G. Dietary essentiality of "nutritionally non-essential amino acids" for animals and humans. *Exp. Biol. Med.* **2015**, *240*, 997–1007. [CrossRef] [PubMed]
55. Moss, J.W.; Ramji, D.P. Cytokines: Roles in atherosclerosis disease progression and potential therapeutic targets. *Future Med. Chem.* **2016**, *8*, 1317–1330. [CrossRef] [PubMed]
56. Voloshyna, I.; Littlefield, M.J.; Reiss, A.B. Atherosclerosis and interferon-gamma: New insights and therapeutic targets. *Trends Cardiovasc. Med.* **2014**, *24*, 45–51. [CrossRef]
57. Gupta, S.; Pablo, A.M.; Jiang, X.; Wang, N.; Tall, A.R.; Schindler, C. IFN-gamma potentiates atherosclerosis in ApoE knock-out mice. *J. Clin. Investig.* **1997**, *99*, 2752–2761. [CrossRef] [PubMed]
58. Kleemann, R.; Zadelaar, S.; Kooistra, T. Cytokines and atherosclerosis: A comprehensive review of studies in mice. *Cardiovasc. Res.* **2008**, *79*, 360–376. [CrossRef]
59. Sukovich, D.A.; Kauser, K.; Shirley, F.D.; DelVecchio, V.; Halks-Miller, M.; Rubanyi, G.M. Expression of interleukin-6 in atherosclerotic lesions of male ApoE-knockout mice: Inhibition by 17beta-estradiol. *Arterioscler. Thromb. Vasc. Biol.* **1998**, *18*, 1498–1505. [CrossRef]
60. Gosling, J.; Slaymaker, S.; Gu, L.; Tseng, S.; Zlot, C.H.; Young, S.G.; Rollins, B.J.; Charo, I.F. MCP-1 deficiency reduces susceptibility to atherosclerosis in mice that overexpress human apolipoprotein B. *J. Clin. Investig.* **1999**, *103*, 773–778. [CrossRef]
61. Ishikawa, T. Branched-chain amino acids to tyrosine ratio value as a potential prognostic factor for hepatocellular carcinoma. *World J. Gastroenterol.* **2012**, *18*, 2005–2008. [CrossRef]
62. Mancuso, P. The role of adipokines in chronic inflammation. *Immunotargets Ther.* **2016**, *5*, 47–56. [CrossRef]
63. Duan, Y.; Zeng, L.; Zheng, C.; Song, B.; Li, F.; Kong, X.; Xu, K. Inflammatory Links Between High Fat Diets and Diseases. *Front. Immunol.* **2018**, *9*, 2649. [CrossRef]
64. Rui, L. Energy metabolism in the liver. *Compr. Physiol.* **2014**, *4*, 177–197. [CrossRef]
65. Lowell, B.B.; Flier, J.S. Brown adipose tissue, beta 3-adrenergic receptors, and obesity. *Annu. Rev. Med.* **1997**, *48*, 307–316. [CrossRef]
66. Rodriguez, A.; Catalan, V.; Gomez-Ambrosi, J.; Fruhbeck, G. Aquaglyceroporins serve as metabolic gateways in adiposity and insulin resistance control. *Cell Cycle* **2011**, *10*, 1548–1556. [CrossRef]
67. Miller, E.W.; Dickinson, B.C.; Chang, C.J. Aquaporin-3 mediates hydrogen peroxide uptake to regulate downstream intracellular signaling. *Proc. Natl. Acad. Sci. USA* **2010**, *107*, 15681–15686. [CrossRef] [PubMed]

68. Miranda, M.; Ceperuelo-Mallafre, V.; Lecube, A.; Hernandez, C.; Chacon, M.R.; Fort, J.M.; Gallart, L.; Baena-Fustegueras, J.A.; Simo, R.; Vendrell, J. Gene expression of paired abdominal adipose AQP7 and liver AQP9 in patients with morbid obesity: Relationship with glucose abnormalities. *Metabolism* **2009**, *58*, 1762–1768. [CrossRef]
69. Rodrigues, C.; Mosca, A.F.; Martins, A.P.; Nobre, T.; Prista, C.; Antunes, F.; Cipak Gasparovic, A.; Soveral, G. Rat Aquaporin-5 Is pH-Gated Induced by Phosphorylation and Is Implicated in Oxidative Stress. *Int. J. Mol. Sci.* **2016**, *17*, 2090. [CrossRef]
70. Direito, I.; Madeira, A.; Brito, M.A.; Soveral, G. Aquaporin-5: From structure to function and dysfunction in cancer. *Cell. Mol. Life Sci. CMLS* **2016**, *73*, 1623–1640. [CrossRef] [PubMed]
71. Papadopoulos, M.C.; Saadoun, S. Key roles of aquaporins in tumor biology. *Biochim. Biophys. Acta* **2015**, *1848*, 2576–2583. [CrossRef]
72. Dobrin, R.; Zhu, J.; Molony, C.; Argman, C.; Parrish, M.L.; Carlson, S.; Allan, M.F.; Pomp, D.; Schadt, E.E. Multi-tissue coexpression networks reveal unexpected subnetworks associated with disease. *Genome Biol.* **2009**, *10*, R55. [CrossRef]
73. Madeira, A.; Mosca, A.F.; Moura, T.F.; Soveral, G. Aquaporin-5 is expressed in adipocytes with implications in adipose differentiation. *IUBMB Life* **2015**, *67*, 54–60. [CrossRef]
74. Verkman, A.S.; Yang, B.; Song, Y.; Manley, G.T.; Ma, T. Role of water channels in fluid transport studied by phenotype analysis of aquaporin knockout mice. *Exp. Physiol.* **2000**, *85*, 233s–241s. [CrossRef] [PubMed]

Review

Endothelial Cell Glucose Metabolism and Angiogenesis

Wa Du [1], Lu Ren [1], Milton H. Hamblin [2] and Yanbo Fan [1,3,*]

1. Department of Cancer Biology, University of Cincinnati College of Medicine, Cincinnati, OH 45267, USA; duwa@ucmail.uc.edu (W.D.); renln@ucmail.uc.edu (L.R.)
2. Department of Pharmacology, Tulane University School of Medicine, New Orleans, LA 70112, USA; mhamblin15@hotmail.com
3. Department of Internal Medicine, Division of Cardiovascular Health and Diseases, University of Cincinnati College of Medicine, Cincinnati, OH 45267, USA
* Correspondence: fanyb@ucmail.uc.edu

Abstract: Angiogenesis, a process of new blood vessel formation from the pre-existing vascular bed, is a critical event in various physiological and pathological settings. Over the last few years, the role of endothelial cell (EC) metabolism in angiogenesis has received considerable attention. Accumulating studies suggest that ECs rely on aerobic glycolysis, rather than the oxidative phosphorylation pathway, to produce ATP during angiogenesis. To date, numerous critical regulators of glucose metabolism, fatty acid oxidation, and glutamine metabolism have been identified to modulate the EC angiogenic switch and pathological angiogenesis. The unique glycolytic feature of ECs is critical for cell proliferation, migration, and responses to environmental changes. In this review, we provide an overview of recent EC glucose metabolism studies, particularly glycolysis, in quiescent and angiogenic ECs. We also summarize and discuss potential therapeutic strategies that take advantage of EC metabolism. The elucidation of metabolic regulation and the precise underlying mechanisms could facilitate drug development targeting EC metabolism to treat angiogenesis-related diseases.

Keywords: endothelial cell; glycolysis; metabolism; angiogenesis; pathological angiogenesis; tumor microenvironment

1. Introduction

Vascular endothelial cells (ECs) form a single layer that coats the interior walls of arteries, veins, and capillaries. ECs are necessary for nutrient and oxygen exchanges between the bloodstream and surrounding tissues [1]. In response to proangiogenic stimuli, ECs rapidly change their cellular state from a quiescent to a proliferative and migratory state. Tip cells at the leading edge of sprouting vessel are characterized by a high migratory and matrix-degrading capacity. Stalk cells follow the tip cells, and they are highly proliferative. Compared with the two differentiated EC types, quiescent ECs are less proliferative and migrative [2]. Recently, metabolic pathways have been identified to be critical for many EC functions, including embryonic angiogenesis [3], pathological angiogenesis [4–6], inflammation [7], and barrier function [8]. By means of modern techniques, many new metabolic features of angiogenic ECs have been discovered. Increasing knowledge about the flexibility and adaptability of metabolism during the angiogenic switch will facilitate new therapeutic strategies for patients with angiogenesis-related diseases. Here, we shed light on the remarkable glycolytic features of angiogenic ECs and propose feasible therapeutic approaches targeting EC glucose metabolism.

1.1. Endothelial Cell Glucose Metabolism

For most body cells, carbohydrates, lipids, and proteins ultimately break down into glucose, fatty acids, and amino acids, respectively. The nutrients are used to produce energy through glycolysis or tricarboxylic acid cycle (TCA cycle) pathways. Glucose serves as the primary metabolic fuel to enter tissue cells and be converted to ATP for cellular maintenance. Thus, glucose metabolism is essential not only for energy production but also for metabolic waste removal. The endothelium is a metabolically active organ that maintains both vascular homeostasis and systemic metabolism [9]. In ECs, numerous genes, including insulin receptor substrate 2 (IRS2), peroxisome proliferator-activated receptor-gamma (PPARγ), and fatty acid translocase (FAT)/cluster of differentiation 36 (CD36), have been demonstrated to regulate systemic glucose levels [10–14]. Transcription factor EB (TFEB) is a master regulator of autophagy and lysosomal biogenesis [15]. Recently, we found that EC-specific TFEB overexpression improves systemic glucose tolerance in mice on a high-fat diet. In human primary ECs, TFEB increases glucose uptake and insulin transport across ECs through activation of Akt signaling [16]. It is still a challenge to elucidate detailed mechanisms by which altered EC function affects glucose metabolism in peripheral tissues in vivo. The role and mechanisms of these regulators in crosstalk between ECs and metabolically active tissues remain to be fully explored.

In general, as the first step of glucose metabolism, glucose is transported across the plasma membrane by glucose transporters, especially glucose transporter 1 (GLUT1). As soon as glucose enters the cells, it is phosphorylated to glucose-6-phosphate (G6P) as catalyzed by hexokinase (HK). G6P can be utilized immediately for energy production through the glycolytic pathway. Under aerobic conditions, pyruvate can be fed into mitochondria and fully oxidized to produce ATP. When oxygen is limited, pyruvate is converted to lactate, and glycolysis becomes the primary source of ATP production [17]. In the early 1920s, Otto Warburg first found that tumor cells use the glycolysis pathway as the energy source under aerobic conditions. Until recently, ECs were discovered to have unique features of glucose metabolism [3]. Although glycolysis and oxidative phosphorylation (OXPHOS) are the two major energy-producing pathways in ECs, normally, up to 85% of ATPs are generated through the glycolysis pathway in human umbilical vein endothelial cells (HUVECs) during vessel sprouting [3]. In cultured quiescent aortic ECs, 99% of glucose is catalyzed into lactate [3,18]. In addition, single-cell RNA sequencing data obtained from mouse tumor tissues and choroidal neovasculature revealed that angiogenic ECs (tip, stalk cell) are enriched in gene sets of both glycolysis and OXPHOS when compared with normal ECs [4].

ECs produce and consume energy to fuel cell proliferation and migration, maintain their structures, and adapt to environmental changes during the EC switch from a quiescent to an angiogenic phenotype. The different glycolytic features of tip cells, stalk cells, and quiescent cells are summarized in Figure 1. The unique aerobic glycolytic features of ECs could serve as beneficiary protection for ECs as follows: (1) glycolysis produces ATP with faster kinetics; (2) enabling rapid response to anaerobic conditions to generate energy, especially under nutrient deprivation; (3) glycolysis occurs in the cytosol and it does not require oxygen. ECs can save oxygen for the transendothelial transfer of oxygen to perivascular cells; (4) low mitochondria content (2–6%) in ECs is consistent with the role of mitochondria as an energetic sensor rather than producer [19]; (5) avoiding the production of reactive oxygen species (ROS) and preventing apoptotic cell death in oxidative stress conditions [20]; and (6) glycolysis provides metabolic intermediates to generate amino acids, lipids, and nucleotides [21]. In light of the importance of glycolysis in ECs, we summarize the role and underlying mechanisms of glycolysis in angiogenesis.

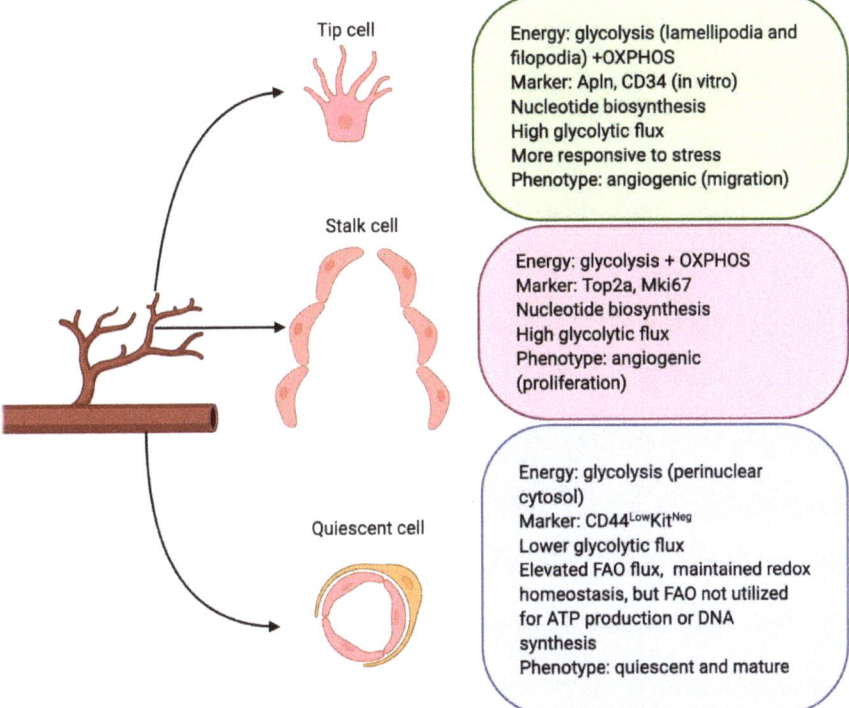

Figure 1. Differential metabolic features in three major endothelial cell (EC) populations. According to the phenotypes of ECs, they can be classified into tip cells, stalk cells, and quiescent cells during angiogenesis. Tip cells grow from the pre-existing vascular bed and are highly responsive to microenvironmental signals for migration. Stalk cells are highly proliferative and follow the tip cells to form a vessel lumen. Quiescent cells maintain vascular homeostasis. Angiogenic ECs show upregulated glycolysis gene signatures during the angiogenic switch to meet their metabolic demands. Quiescent ECs lower their glycolytic flux (35–40%) and use fatty acid oxidation (FAO) flux to maintain energy homeostasis. OXPHOS: oxidative phosphorylation.

1.2. Endothelial Cell Fatty Acid Oxidation

Fatty acids are another source of energy. They are either passively diffused from blood or transported by FAT/CD36 into the cell to fuel the TCA cycle [22]. Through the fatty acid oxidization (FAO) pathway, fatty acids are oxidized into two-carbon acetyl CoA molecules, which can provide twice as much ATP as carbohydrates. Strikingly, a recent study found that in cultured ECs, fatty acids act as a carbon source for deoxynucleoside triphosphate (dNTP) synthesis rather than providing ATP (<5% of total ATP) [23]. Genetic or pharmacological inhibition of carnitine palmitoyltransferase I (CPT1), the rate-limiting enzyme in FAO, causes functional defects in differentiation, proliferation, and barrier function in ECs [8,24,25]. Fatty acid transport protein (FATP) and fatty acid-binding protein (FABP) were also found to regulate EC function when stimulated with vascular endothelial growth factor (VEGF) [26,27]. Loss of fatty acid binding protein 4 (FABP4) in ECs results in decreased proliferation, migration, and sprouting [26]. Taken together, FAO is required for EC dNTP synthesis and proliferation during sprouting angiogenesis.

1.3. Endothelial Cell Glutamine Metabolism

Besides glycolysis and FAO, ECs utilize glutamine as a "conditionally essential" nutrient [28]. In physiological conditions, glutamine is the most abundant free amino acid in plasma and the most important donor of nitrogen atoms for metabolism [29], contributing 30% of the TCA carbons [23]. Nearly 90% of glutamine is transported into ECs through a sodium-dependent transporter system [30] and is oxidized in mitochondria to produce ATP. Highly proliferative ECs utilize glutamine for protein synthesis, the TCA cycle, and redox homeostasis [31,32]. Since ECs have low mitochondrial content, glutamine metabolism showed marginal effects on sprouting angiogenesis. Nevertheless, glutamine could be effectively regulating vascular tone and inflammation [32].

2. Glucose Metabolism in Quiescent ECs

In healthy adult vasculature, the majority of ECs are quiescent [33,34]. A quiescent endothelium is essential to maintaining vascular integrity, suppressing thrombosis, and inhibiting inflammation [34,35]. Laminar shear stress maintains the quiescent state of ECs, inhibits EC glycolysis, and downregulates PFKFB3 in a Krüppel-like factor 2 (KLF2)-dependent manner [36]. Forkhead box O1 (FOXO1), a transcription factor that plays an important role in regulating gluconeogenesis and glycogenolysis, maintains the EC quiescent state and restricts vascular overgrowth by reducing glycolysis. MYC proto-oncogene (MYC), a potent driver of anabolic metabolism, mediates the inhibitory effect of FOXO1 on glycolysis in ECs [37]. Unlike tip cells with compartmentalization of glycolysis in lamellipodia and filopodia, quiescent ECs have glycolysis taking place in the perinuclear cytosol [3]. ECs shift between a proliferative and nonproliferative state based on their metabolic needs. When angiogenesis occurs, EC migration and proliferation rely on glycolysis as a fuel source [38]. Quiescent ECs tend to have lower metabolic rates and reduced metabolic gene transcripts related to glycolysis, the TCA cycle, respiration, and nucleotide synthesis compared with highly activated ECs. Instead, through increasing FAO flux, quiescent ECs utilize FAO to maintain redox homeostasis but do not utilize FAO for ATP production or DNA synthesis [39]. This feature of metabolic adaptation and flexibility reprogram ECs to switch between angiogenic and quiescent states, which significantly impacts vascular disease-related angiogenesis.

3. EC Glucose Metabolism in Pathological Angiogenesis

3.1. Ocular Angiogenesis (Diabetic Retinopathy and Retinal Angiomatous Proliferation)

In diabetic patients, retinal ECs constitutively express GLUT1, which results in elevated ROS in both the cytosol and mitochondria due to high glucose level and insufficient ROS scavenging [40]. As a consequence, ECs lower down their glycolytic flux. Accumulated glycolytic intermediates are directed into other glycolysis branch pathways (e.g., ~3% enters the polyol pathway) and further increase ROS [41]. In the past few years, several studies were carried out to explore glucose metabolism in retinal ECs in the context of diabetic retinopathy. In vitro, loss of EC-*GLUT1* leads to reduced glycolysis, AMP-activated protein kinase (AMPK) activation, and decreased cell proliferation. Conditional deletion of *Glut1* in mouse ECs results in impaired retinal and brain angiogenesis due to tip cell reduction in vivo [42]. Deletion of peroxisome proliferator-activated receptor-alpha (PPARα) in endothelial colony-forming cells (ECFC) decreased mitochondrial oxidation and glycolysis and further exacerbated 4-hydroxynonenal (HNE)-induced mitochondria damage [43].

Recent studies of EC metabolic profiling shed light on the connection of EC metabolism to pathological ocular angiogenesis. Joyal et al. demonstrated that the retina utilized glucose and FAO for ATP production and identified that free fatty acid receptor 1 (FFAR1), a lipid sensor, inhibits glucose uptake when free fatty acids are available [44]. FFAR1 decreases GLUT1 and suppresses glucose uptake in the retinas of very low-density lipoprotein receptor (*Vldlr*) knockout mice. The impaired glucose uptake into photoreceptors results in a dual lipid/glucose fuel shortage and reduction in α-ketoglutarate, an intermediate of the TCA cycle. Low α-ketoglutarate further stabilized hypoxia-inducible factor 1α (HIF-1α) and increased VEGF secretion. As a result, abnormal vessels invaded avascular photoreceptors, which mimicked retinal angiomatous proliferation [44]. In addition, blockade of endothelial carnitine palmitoyltransferase 1A (CPT1A) or glutaminase 1 (GLS1) could reduce ocular neovascularization in mice [23,32]. Fatty acid synthesis is also involved in pathological ocular angiogenesis [45]. EC-specific fatty acid synthase (*Fasn*) knockout or application of the FASN blocker orlistat in vivo impairs angiogenesis and inhibits abnormal ocular neovascularization through malonylation-dependent repression of mammalian target of rapamycin complex 1 (mTORC1) activity. Taken together, the energy sources for ECs rely on both glycolysis and FAO in the pathological process of ocular angiogenesis, as shown in Figure 2.

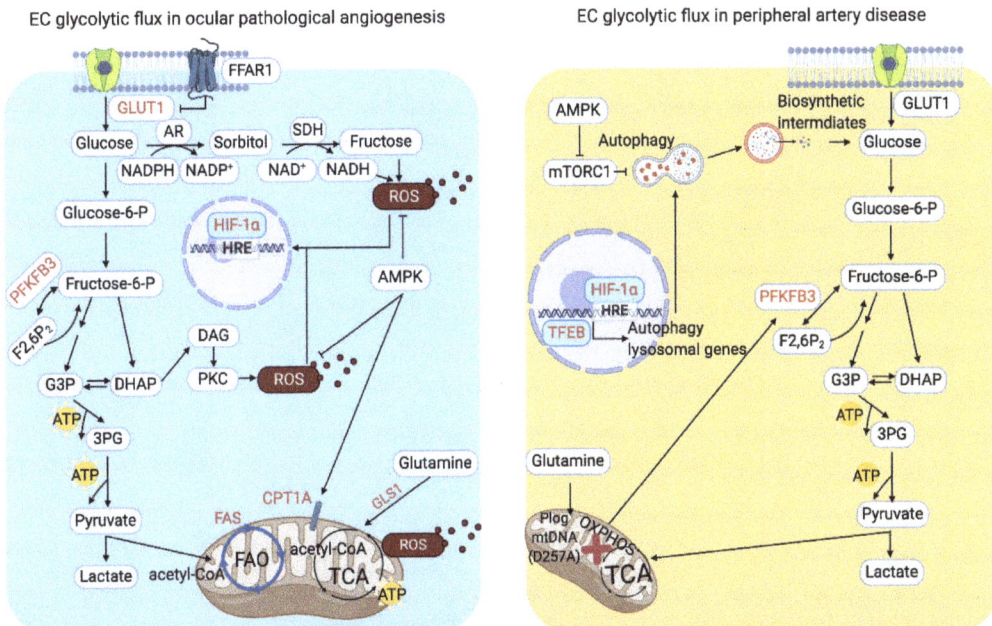

Figure 2. Endothelial cell (EC) glycolytic flux in ocular angiogenesis and peripheral artery disease (PAD). Retinal ECs utilize both glycolysis and fatty acid oxidation (FAO) for adenosine triphosphate (ATP) production. High glucose level leads to overproduction of mitochondrial reactive oxygen species (ROS) in ECs. The accumulated glycolytic intermediates result in lower glycolytic flux, which further increases ROS in both mitochondria and cytosol in diabetic ECs. ECs in PAD show impaired oxidative phosphorylation. Under energy deficiency conditions, phosphofructo-2-kinase/fructose-2,6-biphosphatase 3 (PFKFB3), glycolytic flux, and autophagy have protective effects on ECs in PAD. Red color represents the metabolic genes that positively regulate angiogenesis. GLUT1, glucose transporter 1; AR, aldose reductase; SDH, Sorbitol dehydrogenase; NADPH, nicotinamide adenine dinucleotide phosphate; NADH, nicotinamide adenine dinucleotide; AMPK, AMP-activated protein kinase; G3P, glyceraldehyde 3-phosphate; DHAP, dihydroxyacetone phosphate; DAG, diacylglycerol; PKC, protein kinase C; 3PG, 3-phosphoglyceric acid; FAS, fatty acid synthase; CPT1A, carnitine palmitoyltransferase 1A; GLS1, glutaminase 1; mTORC1, mammalian target of rapamycin complex 1; OXPHOS, oxidative phosphorylation; acetyl-CoA, acetyl coenzyme A; TCA, tricarboxylic acid cycle; F2,6P$_2$, fructose 2,6-bisphosphate; TFEB, transcription factor EB; HIF-1α, hypoxia-inducible factor 1-alpha; HRE, hypoxia response elements; Plog, mitochondrial DNA polymerase gamma; mtDNA, mitochondrial DNA.

3.2. Diabetic Angiogenesis

Hyperglycemia negatively regulates HIF-1α stability and its nuclear translocation by upregulation of prolyl hydroxylase domain protein 2 (PHD2) and PHD3. PHD2 and PHD3 act as oxygen sensors in oxygen-dependent regulation of HIF-1α stability. The hypoxia/VEGF axis is impaired in ECs under high-glucose conditions [46]. Under hypoxia, HIF-1α is also stabilized, and it upregulates glycolytic and glucose uptake-related genes, including glucose transporter 1/3 (GLUT1 and GLUT3), hexokinase 1/2 (HK1/2), phosphoglycerate kinase 1 (PGK1), lactate dehydrogenase A (LDHA), pyruvate kinase M2 (PKM2), phosphofructo-2-kinase/fructose-2,6-biphosphatase 3 (PFKFB3), aldolase A/C (ALDOA/C), glyceraldehyde-3-phosphate dehydrogenase (GAPDH), phosphofructokinase type 1 (PFK1), and pyruvate dehydrogenase kinase 1 (PDK1) [47]. Furthermore, in response to hypoxia, VEGF increases PFKFB3 expression to enhance glycolysis in ECs. However, high glucose reduces PFKFB3 expression in the mouse ECs [48]. These studies suggest that hyperglycemia downregulates two critical promoters of angiogenesis: HIF-1α and PFKFB3 in ECs. Therefore, different from diabetic retinopathy, diabetes leads to insufficient angiogenesis in wound healing [49–51], characterized by decreased angiogenesis and vascular density.

In diabetic rodents and humans, peroxisome proliferator-activated receptor-gamma coactivator 1-α (PGC-1α) expression was elevated in ECs [52]. PGC-1α could be a critical regulator of endothelial activation caused by hyperglycemia [52]. In mice, endothelial PGC-1α inhibits blood flow recovery, exacerbates foot necrosis in the hindlimb ischemia model, and attenuates wound healing. Mechanistically, PGC-1α activates Notch and blocks Rac/Akt/eNOS signaling in ECs [52]. Collectively, as critical metabolic regulators in ECs, PFKFB3, HIF-1α, and PGC-1α could be potential targets to modulate angiogenesis in diabetic conditions.

3.3. Peripheral Arterial Disease (PAD) and EC Glycolytic Flux

PAD is a manifestation of reduced blood supply to the lower extremities that is mostly induced by atherosclerotic obstruction. Ischemia imposes a major energetic challenge on the tissues due to impaired oxidative phosphorylation. Using Polg mtDNA mutator ($Polg^{D257A}$) mice, Ryan et al. observed the remarkable protective effects of glycolytic metabolism and PFKFB3 on hindlimb ischemia. They also collected muscles from patients with critical limb ischemia and found lower PFKFB3 expression and reduced glycolytic flux in patient muscles [53]. Indeed, therapeutic angiogenesis is a promising strategy for the treatment of PAD. Utilizing EC-specific *Tfeb* transgenic and knockout mice, we demonstrated that TFEB promotes angiogenesis and improves blood flow recovery in the mouse hindlimb ischemia model. In ECs, TFEB increases angiogenesis through the activation of AMPKα and upregulation of autophagy [54–56]. As summarized in Figure 2, this finding established a positive relation between TFEB and postischemia angiogenesis. The role of TFEB in glucose metabolism remains to be fully explored.

3.4. Tumor Angiogenesis

3.4.1. Tumor Endothelial Cells (TECs) Adapt Their Metabolism to the Tumor Hypoxic Environment

ECs are more resistant to hypoxia than other cell types, such as cardiomyocytes and neurons [57]. Unlike normal ECs, disorganized TECs are essential for tumor growth characterized by a leaking vascular system, high interstitial fluid pressure, reduced blood flow, poor oxygenation, and acidosis [58]. Hypoxia and ischemia can activate the EC switch from a quiescent to angiogenic phenotype (higher proliferative and migratory abilities). Recently, single-cell RNA sequencing data from tumor tissues revealed that TECs are hyperglycolytic. Their total read counts were 2–4 fold higher than that in normal ECs, which means TECs have high RNA content due to increased metabolic demands of nucleotide biosynthesis and glycolysis [4]. TECs have a distinct metabolic transcriptome signature linked to their angiogenic potential, as shown in Figure 3. Similar to TECs, stroma cells also have metabolic adaptability or flexibility in the tumor microenvironment [59]. TECs showed heterogeneity in cell function and structure, which could be varied in different host organs and tumor types [4,60]. To date, although many antiangiogenic compounds have been identified, antiangiogenesis therapy may not be enough for tumor treatment (metastatic breast cancer and glioma) due to vessel co-option or vasculogenic mimicry [61, 62]. Since tumor cells and tumor stroma cells, including TECs, can adapt their metabolism to survive and proliferate in tumor growth, modulation of cell metabolism could be effective to control not only different phenotypes of TECs at multiple steps of angiogenesis (proliferation, migration, sprouting, and maturation) but also tumor cell growth.

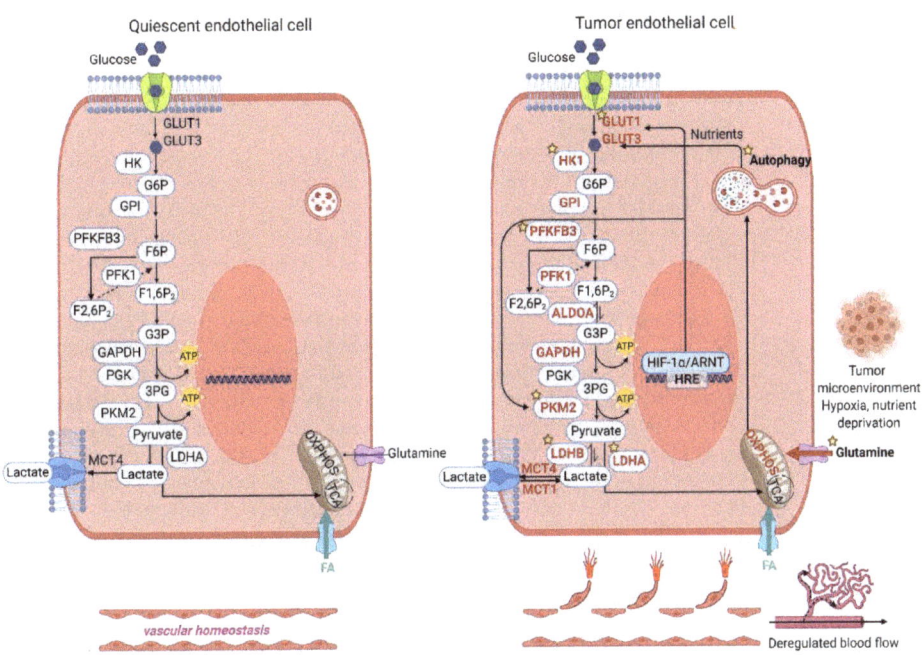

Figure 3. Tumor endothelial cells (TECs) exhibit distinct metabolic transcriptome signatures, which are linked to their angiogenic potential. Compared with quiescent ECs, TECs utilize both glycolysis and OXPHOS for energy production. TECs (tip, stalk, and immature ECs) show upregulated glycolytic genes, including glucose transporter 1 (GLUT1), glucose transporter 3 (GLUT3), hexokinase 1 (HK1), hexokinase 2 (HK2), 6-phosphofructo-2-kinase/fructose-2,6-biphosphatase 3 (PFKFB3), phosphofructokinase 1 (PFK1), aldolase A (ALDOA), glyceraldehyde 3-phosphate dehydrogenase (GAPDH), pyruvate kinase M2 (PKM2), enolase 1 (ENO1), lactate dehydrogenase A (LDHA). TECs can proliferate in a lactate-rich environment. Under hypoxia, hypoxia-inducible factor-1 alpha (HIF-1α) increases the expression of GLUT1 and GLUT3 in TECs. Autophagy is increased to promote TECs to survive and adapt to metabolic needs. The star symbol indicates the steps where chemical compounds are available and the antiangiogenic effects have been tested in preclinical or clinical settings. Red color represents the upregulated metabolic genes. FA, fatty acid; MCT1/4, monocarboxylate transporter 1/4; PGK, phosphoglycerate kinase; GPI, glucose-6-phosphate isomerase.

3.4.2. TECs Release More Lactate and Utilize Lactate for Proliferation

Lactate, the metabolic end-product of glycolysis, is released by ECs in aerobic conditions, and in turn, lactate attenuates HIF-1α degradation by binding and inhibiting HIF prolyl hydroxylase [63]. Hyperglycolytic TECs could be the potential source of lactate in the tumor microenvironment. Both normal ECs and TECs can utilize lactate to support their growth in a dose-dependent manner [63]. In the mouse Lewis lung carcinoma model, suppression of monocarboxylate transporter 1 (MCT1), the main transporter for lactate uptake in ECs, reduces tumor angiogenesis [64]. Lactate dehydrogenase B (LDHB) expression is increased in TECs, which helps re-entered lactate to integrate into the metabolism [65]. Compared with normal ECs, TECs do not only produce more lactate; their growth is also promoted by lactate preferentially. Unlike normal ECs, TECs can proliferate in a lactate-rich environment due to highly expressed carbonic anhydrases II (CAII) that facilitate the transport activity of MCT1/4 [58]. These findings suggest that TECs are more glycolytic than normal ECs, even though normal ECs are already addicted to glycolysis [3]. Collectively, altered TEC glucose metabolism can sustain their proliferation in the tumor microenvironment and survive in an acidic environment.

3.4.3. TECs Display High Glycolytic Flux

Accumulated single-cell RNA sequencing data has revealed numerous transcriptome signatures related to glycolysis in TECs [4]. Compared with normal ECs, TECs showed higher glycolytic flux. The mechanisms by which glycolytic flux is regulated in TECs remain to be fully explored. The first key and irreversible step is the transformation of F6P to fructose-1,6-bisphosphate (F1,6BP) catalyzed by phosphofructokinase 1 (PFK1). PFK1 activity is inhibited by intracellular ATP or citric acid and reactivated by F2,6BP. PFKFB3 has high kinase activity to promote the synthesis of F2,6BP and maintain the increased glycolytic flux [66]. EC-specific deletion of a single *Pfkfb3* allele or administration of the PFKFB3 inhibitor (3-(3-pyridinyl)-1-(4-pyridynyl)-2-propen-1-one, 3PO) reduces tumor cell invasion and metastasis, normalizes tumor vessels, and improves the vascular barrier by reducing VE-cadherin (vascular endothelial cadherin) endocytosis [5]. Augmented glycolysis in TECs fuels multiple metabolic pathways, including the pentose phosphate pathway (PPP), hexosamine biosynthesis pathway (HBP), TCA cycle, and serine biosynthesis pathway [4]. Glycolytic flux is nearly three-fold higher in TECs than in normal ECs, and TECs utilize glucose carbons for biomass production. Additionally, in TECs, hypoxia upregulates glucose transporters (GLUT1 and GLUT3), which are necessary for rapid glucose uptake and increased glycolytic flux [5,67]. Both inter- and intratumor metabolic heterogeneity has been observed within and between the tumors [68]. This would make the strategy of targeting glucose metabolism in TECs more valuable, as tumor cells show high metabolic flexibility. At the same time, TECs are more stable and consistent among various tumor types.

3.4.4. TECs Exhibit Increased Autophagy

Autophagy is a conserved cellular degradation pathway that is critical to maintain cellular homeostasis and to adapt to the metabolic needs to sustain proliferation and survival [69,70]. However, the effects of autophagy in cancer are still controversial because autophagy plays opposite roles in precancerous and malignant tumors [71,72]. The role of autophagy in the vasculature has gained more attention as tumor vessels are involved in both nutrient replenishment and metastasis for starved and stressed tumor cells. In the tumor microenvironment, TECs are subjected to low glucose, starvation, low blood flow, and hypoxia. Autophagy is upregulated in TECs in response to extracellular stresses [69]. Mechanistically, autophagy is controlled by upstream regulators, including mammalian target of rapamycin (mTOR) and AMP-activated protein kinase-α (AMPKα). Accumulated studies suggest that autophagy regulators, including Beclin 1 (BECN1), TFEB, and high-mobility group box protein 1 (HMGB1) modulate angiogenesis [54,65,73]. Compared with normal ECs, TECs may adjust their autophagy/lysosomal activity to mitigate the detrimental effects of hypoxia. ECs maintain glycogen stores during hypoxia but not under low-glucose conditions. Autophagy sustains cell survival in nutrient-deprivation conditions, in which cells use glycogen as a critical backup energy source [74]. In Figure 4, we summarize the glycogen storage and breakdown pathways modulated by autophagy. In a high-glucose environment, ECs store glycogen to prepare energy for extracellular stresses. Upon hypoxia or nutrient deprivation, 90% of glycogen is mobilized and converted into glucose-1-phosphate (GP) catalyzed by glycogen debranching enzymes and glycogen phosphorylase [75]. Then, GP is converted into glucose-6-phosphate (G6P) catalyzed by phosphoglucose mutase. G6P can be directly used for glycolysis to maintain EC proliferation and migration. Alternative autophagy-dependent glycogen breakdown (10%) produces nonphosphorylated glucose catalyzed by lysosomal 1,4-α-glucosidase to meet metabolic requirements [76]. Nonphosphorylated glucose can either be used in glycolysis or stored as glycogen in cells, depending on the metabolic status. In various cell types,

TFEB increases autolysosome numbers and stimulates the fusion between lysosomes and autophagosomes under hypoxia and nutrient-deprivation stress [77,78]. In this scenario, TFEB, together with other autophagy regulators, would be critical in glycogen storage and mobilization through enhancing autophagic flux. The role of autophagy in TEC metabolic reprogramming remains to be fully explored.

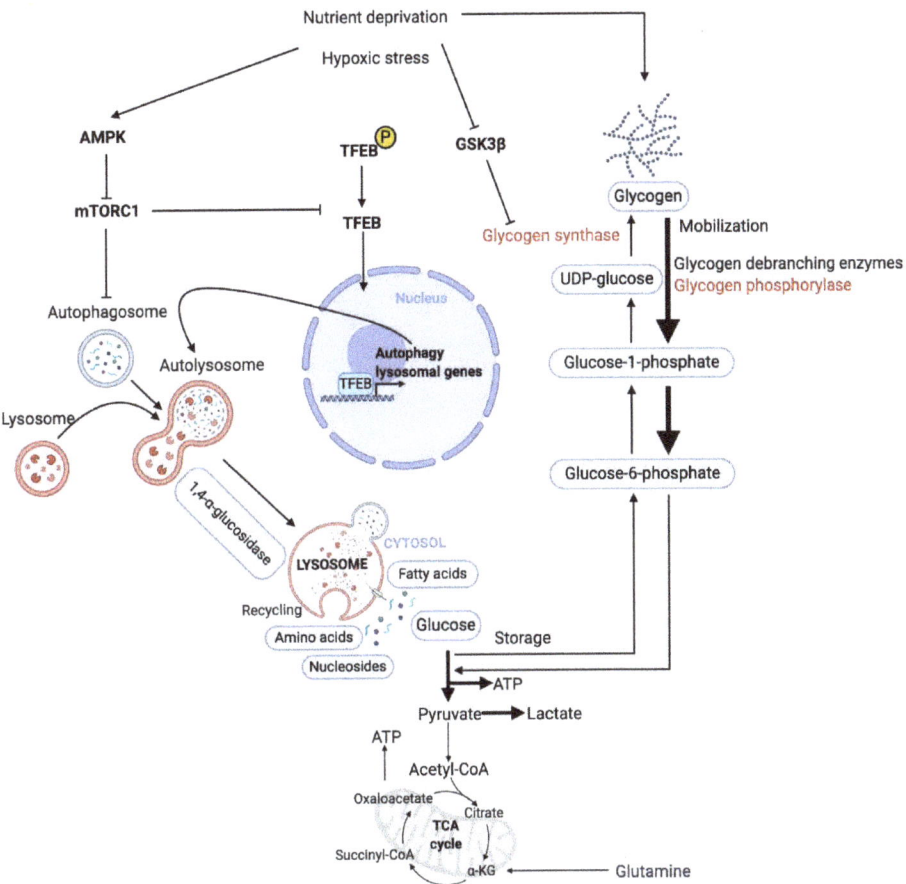

Figure 4. EC glucose and glycogen metabolism in hypoxia and nutrient-deprivation conditions. Under hypoxia and nutrient-deprivation conditions, cellular energy levels are decreased and autophagy is upregulated by the AMPK–mTORC1 pathway. The autophagy–lysosomal pathway promotes the recycling of nutrients, including glucose, for cell survival. In response to environmental changes, ECs use glycogen as a backup energy source. Upon nutrient deprivation, TFEB translocates to the cell nucleus, where it activates target genes involved in lysosomal function and autophagy. Upregulated autophagy/lysosomal activity supports ECs to resist the detrimental effects of hypoxia and nutrient deprivation. Red color represents the key metabolic genes.

4. EC Metabolic Regulators of Antiangiogenesis and Vessel Normalization

In both preclinical and clinical settings, anti-VEGF therapy showed transitory, limited efficacy and acquired resistance [79]. Recent single-cell RNA sequencing data suggested that glycolytic genes were upregulated in tip cells in xenograft tumors after pharmacological inhibition of VEGF and Notch signaling [60]. Abnormal tumor vessels promote tumor growth, metastasis, and resistance to chemotherapy. Tumor vessel normalization has been recognized as a promising strategy for anticancer treatment. Blockade of PFKFB3 improves vessel maturation and perfusion, thereby reducing tumor cell invasion, intravasation, and metastasis and enhancing the efficiency of chemotherapy on tumors [5]. Glycolysis drives EC rearrangement by increasing filopodia formation and reducing intercellular adhesion. PFKFB3 blockade promotes the disturbed EC rearrangement in high-VEGF conditions [80]. The glycolytic enzyme pyruvate kinase M2 (PKM2) regulates cell–cell junctions and migration in ECs. PKM2 knockdown promotes proper VE-cadherin internalization/traffic at endothelial junctions, which may help vessel normalization in tumors [81]. Thus, manipulation of TEC glycolysis for vessel normalization constitutes a potential therapeutic intervention in tumors.

Taken together, targeting TEC glucose metabolism and thereby inhibiting angiogenesis is a promising strategy for cancer treatment. We summarize the compounds that target critical metabolic enzymes in glycolysis and other metabolic pathways in Table 1. Of note, these compounds regulate metabolism in both TECs and cancer cells. For instance, cancer cells also readily use glycolysis for energy metabolism [82]. Therefore, understanding TECs in metabolism and antiangiogenic resistance can help develop novel strategies to treat cancer.

Table 1. Compounds that directly target metabolic enzymes involved in tumor angiogenesis.

Target	Compound	Tumor Type	Status
Glycolysis			
Glucose transporters Glut1	Phloretin	Cervical cancer cell	Preclinical [83]
	Silybin (Silibinin)	Prostate cancer	Clinical phase II [84]
	Canagliflozin	Liver cancer cell	Preclinical [85]
	Curcumin	Multiple cancers	Clinical phase I/II [86]
	Fasentin	Breast cancer	Preclinical [87]
	Genistein	Multiple cancers	Clinical phase I/II [88,89]
Hexokinases	2-DG [1]	Multiple cancers Glioblastoma	Clinical phase I/II [90–92]
	Ketoconazole	Melanoma, breast cancer, glioblastoma, lung cancer, prostate cancer	Preclinical [93]
	Lonidamine		Clinical phase III [94]
	Methyl jasmonate	Gastric cancer	Preclinical [95]
PFKFB3 [2]	3PO [3]	Melanoma, lung carcinoma, pancreatic cancer	Preclinical [5,96,97]
	PFK158	Advanced solid malignancies	Clinical phase I [98]
Pyruvate kinase-M2 (PK-M2)	TLN-232	Metastatic renal cell	Clinical I/II [99]
	Shikonin	Lung carcinoma	Preclinical [100]
Lactate dehydrogenase	PTK787/ZK 222584 (Vatalanib)	Colon cancer, advanced colorectal cancer	Preclinical [101]
	Gossypol	Multiple cancers	Clinical phase I/II [102,103]
	Oxamate	Breast cancer	Preclinical [104]
Lactate	Lonidamine	Prostate cancer	Clinical phase III [94]
	AZD3965	Gastric cancer, prostate cancer lymphoma	Clinical phase I [105]
TCA cycle			
PDK1 [4]	Dichloroacetate (DCA)	Non-small-cell lung cancer, breast cancer	Preclinical [106]
OXPHOS			
Mitochondrial complex I/III	Metformin	Breast cancer	Preclinical [107]
	Phenformin	Cholangiocarcinoma	Clinical phase I [108,109]
	Arsenic trioxide	Gastric cancer cells	Preclinical [110]

Table 1. Cont.

Target	Compound	Tumor Type	Status
FAO			
CPT1 [5]	Etomoxir	Lung carcinoma, prostate cancer cell line	Preclinical [23,111,112]
	Perhexiline	Prostate cancer, glioma	Preclinical [112,113]
Glutamine metabolism			
GLS1 [6]	BPTES	Osteosarcoma, pancreatic cancer	Preclinical [114,115]
	CB-839	Multiple cancers	Clinical phase I/II [32]

[1] 2-DG, 2-deoxyglucose; [2] PFKFB3, fructose-2,6-biphosphatase 3; [3] 3PO, 3-(3-pyridinyl)-1-(4-pyridynyl)-2-propen-1-one; [4] PDK1, pyruvate dehydrogenase kinase 1; [5] CPT1, carnitine palmitoyltransferase 1; [6] GLS1, glutaminase 1.

5. Conclusions and Open Questions

The metabolic regulation of ECs is gaining much attention, especially in pathological angiogenesis within the tissue-specific microenvironment. EC metabolism has been summarized into glucose metabolism, fatty acid oxidation, and glutamine metabolism. In this review, we summarize glucose metabolism within quiescent and angiogenic ECs. Metabolic adaption of ECs during the angiogenic switch and in healthy tissues is well-documented [116]. TEC metabolic reprogramming should be studied as a common and expected feature of metabolism. Many antiangiogenic therapies are designed to inhibit VEGF receptors or VEGF signaling in ECs, which results in the insufficient treatment of tumors and increased tumor metastasis due to elevated hypoxia in the tumor microenvironment [117]. It could be a promising strategy to modulate metabolic regulators in TECs. Targeting TEC metabolism would allow us to design new strategies combined with the classical antiangiogenic strategies to fight cancer.

Glycolysis, but not oxidative phosphorylation (OXPHOS), is chosen for ATP generation in cultured human primary ECs, mouse angiosarcoma, and mouse hemangioma [3]. However, single-cell RNA sequencing data from in vivo tumor tissues suggested that angiogenic ECs (tip cells, proliferating cells, and immature cells) still rely much on both glycolysis and OXPHOS as energy sources [4]. During tumor blood vessel sprouting, whether large amounts of glucose are available is still speculative. Since aerobic glycolysis is upregulated in TECs, they should be classified as oxidative ECs and glycolytic ECs, even within the same tumor. This would be beneficial for further understanding of the metabolic demands in tumor bioenergetics. TECs support tumor progression and affect chemotherapeutic resistance and metastasis. The metabolic switch is not specific to ECs but exists as an example of global adaptation and flexibility to environmental changes.

Unlike tumor cells that carry various mutations, TECs are more genetically stable. Targeting the metabolism of TECs instead of tumor cell metabolism could be a promising strategy against tumor progression. It is expected that metabolic modulators are able to affect different steps of the angiogenic process in ECs. Here, we summarize some questions that remain to be answered: What is the exact role of the autophagy/lysosome pathway in EC metabolism? Is there any glucose competition between ECs and other cell types, including tumor cells, stromal cells, and macrophages, in the tumor microenvironment? How do TECs escape from cell death in hypoxic and nutrient-deprived conditions? Understanding the role and underlying mechanisms of EC metabolism will facilitate new therapeutic approaches to modulate angiogenesis-related diseases.

Author Contributions: W.D. and Y.F. conceived the idea of the review article. W.D. and Y.F. performed the literature search. W.D. and Y.F. wrote the manuscript. W.D. drew the figures and summarized the table. Y.F., L.R., and M.H.H. revised the manuscript. All authors have read and agreed to the published version of the manuscript.

Funding: This work was supported by the National Institutes of Health under R01HL138094 and R01HL145176 (to Y.F.).

Institutional Review Board Statement: Not applicable.

Informed Consent Statement: Not applicable.

Data Availability Statement: Data sharing not applicable.

Acknowledgments: The authors would like to thank all the other studies in this field that were not cited due to space limitation. The figures were created with BioRender.com.

Conflicts of Interest: The authors declare no conflict of interest.

Abbreviations

The following Abbreviations are used in the manuscript:

3PG	3-phosphoglyceric acid
3PO	3-(3-pyridinyl)-1-(4-pyridynyl)-2-propen-1-one
AD	aldose reductase
ALDOA/C	aldolase A/aldolase C
AMPK	AMP-activated protein kinase
ATP	adenosine triphosphate
CPT1A	carnitine palmitoyltransferase 1A
DAG	diacylglycerol
DHAP	dihydroxyacetone phosphate
ECFC	endothelial colony-forming cell
ECs	endothelial cells
ENO1	enolase 1
F6P	fructose 6-phosphate
FABP	fatty acid-binding protein
FAO	fatty acid oxidation
FASN	fatty acid synthase
FAT	fatty acid translocase
FATP	fatty acid transport protein
G3P	glyceraldehyde 3-phosphate
G6P	glucose-6-phosphate
GAPDH	glyceraldehyde 3-phosphate dehydrogenase
GLS1	glutaminase 1
GLUT1/3	glucose transporter 1/3
GPI	glucose-6-phosphate isomerase
HIF-1α	hypoxia-inducible factor 1 alpha
HK1/2	hexokinase 1/2
HMGB1	high-mobility group box 1
HUVECs	human umbilical vein endothelial cells
IRS1/2	insulin receptor substrate 1/2
LDHA/LDHB	lactate dehydrogenase A/lactate dehydrogenase B
MCT1/4	monocarboxylate transporter 1/4
mTOR	mammalian target of rapamycin
NAD	nicotinamide adenine dinucleotide
NADPH	nicotinamide adenine dinucleotide phosphate
OXPHOS	oxidative phosphorylation
PAD	peripheral artery disease
PDK1	pyruvate dehydrogenase kinase 1
PFKFB3	6-phosphofructo-2-kinase/fructose-2,6-biphosphatase 3
PFK1	phosphofructokinase 1
PGK	phosphoglycerate kinase
PHD	prolyl hydroxylase domain protein
PKC	protein kinase C
PKM2	pyruvate kinase M2
PPARγ	peroxisome proliferator-activated receptor gamma

PPARα	peroxisome proliferator-activated receptor alpha
ROS	reactive oxygen species
SDH	sorbitol dehydrogenase
SIRT1	sirtuin 1
TECs	tumor endothelial cells
TFEB	transcription factor EB

References

1. Cines, D.B.; Pollak, E.S.; Buck, C.A.; Loscalzo, J.; Zimmerman, G.A.; McEver, R.P.; Pober, J.S.; Wick, T.M.; Konkle, B.A.; Schwartz, B.S.; et al. Endothelial cells in physiology and in the pathophysiology of vascular disorders. *Blood* **1998**, *91*, 3527–3561.
2. Blancas, A.A.; Wong, L.E.; Glaser, D.E.; McCloskey, K.E. Specialized tip/stalk-like and phalanx-like endothelial cells from embryonic stem cells. *Stem Cells Dev.* **2013**, *22*, 1398–1407. [CrossRef] [PubMed]
3. De Bock, K.; Georgiadou, M.; Schoors, S.; Kuchnio, A.; Wong, B.W.; Cantelmo, A.R.; Quaegebeur, A.; Ghesquiere, B.; Cauwenberghs, S.; Eelen, G.; et al. Role of PFKFB3-driven glycolysis in vessel sprouting. *Cell* **2013**, *154*, 651–663. [CrossRef]
4. Rohlenova, K.; Goveia, J.; Garcia-Caballero, M.; Subramanian, A.; Kalucka, J.; Treps, L.; Falkenberg, K.D.; de Rooij, L.; Zheng, Y.; Lin, L.; et al. Single-Cell RNA Sequencing Maps Endothelial Metabolic Plasticity in Pathological Angiogenesis. *Cell Metab.* **2020**, *31*, 862–877.e14. [CrossRef] [PubMed]
5. Cantelmo, A.R.; Conradi, L.C.; Brajic, A.; Goveia, J.; Kalucka, J.; Pircher, A.; Chaturvedi, P.; Hol, J.; Thienpont, B.; Teuwen, L.A.; et al. Inhibition of the Glycolytic Activator PFKFB3 in Endothelium Induces Tumor Vessel Normalization, Impairs Metastasis, and Improves Chemotherapy. *Cancer Cell* **2016**, *30*, 968–985. [CrossRef]
6. Cohen, E.B.; Geck, R.C.; Toker, A. Metabolic pathway alterations in microvascular endothelial cells in response to hypoxia. *PLoS ONE* **2020**, *15*, e0232072. [CrossRef] [PubMed]
7. Wang, K.C.; Yeh, Y.T.; Nguyen, P.; Limqueco, E.; Lopez, J.; Thorossian, S.; Guan, K.L.; Li, Y.J.; Chien, S. Flow-dependent YAP/TAZ activities regulate endothelial phenotypes and atherosclerosis. *Proc. Natl. Acad. Sci. USA* **2016**, *113*, 11525–11530. [CrossRef]
8. Patella, F.; Schug, Z.T.; Persi, E.; Neilson, L.J.; Erami, Z.; Avanzato, D.; Maione, F.; Hernandez-Fernaud, J.R.; Mackay, G.; Zheng, L.; et al. Proteomics-based metabolic modeling reveals that fatty acid oxidation (FAO) controls endothelial cell (EC) permeability. *Mol. Cell. Proteom.* **2015**, *14*, 621–634. [CrossRef]
9. Theodorou, K.; Boon, R.A. Endothelial Cell Metabolism in Atherosclerosis. *Front. Cell Dev. Biol.* **2018**, *6*, 82. [CrossRef] [PubMed]
10. Graupera, M.; Claret, M. Endothelial Cells: New Players in Obesity and Related Metabolic Disorders. *Trends Endocrinol. Metab.* **2018**, *29*, 781–794. [CrossRef]
11. Kubota, T.; Kubota, N.; Kadowaki, T. The role of endothelial insulin signaling in the regulation of glucose metabolism. *Rev. Endocr. Metab. Disord.* **2013**, *14*, 207–216. [CrossRef] [PubMed]
12. Kubota, T.; Kubota, N.; Kumagai, H.; Yamaguchi, S.; Kozono, H.; Takahashi, T.; Inoue, M.; Itoh, S.; Takamoto, I.; Sasako, T.; et al. Impaired insulin signaling in endothelial cells reduces insulin-induced glucose uptake by skeletal muscle. *Cell Metab.* **2011**, *13*, 294–307. [CrossRef] [PubMed]
13. Kanda, T.; Brown, J.D.; Orasanu, G.; Vogel, S.; Gonzalez, F.J.; Sartoretto, J.; Michel, T.; Plutzky, J. PPARgamma in the endothelium regulates metabolic responses to high-fat diet in mice. *J. Clin. Investig.* **2009**, *119*, 110–124. [CrossRef] [PubMed]
14. Son, N.H.; Basu, D.; Samovski, D.; Pietka, T.A.; Peche, V.S.; Willecke, F.; Fang, X.; Yu, S.Q.; Scerbo, D.; Chang, H.R.; et al. Endothelial cell CD36 optimizes tissue fatty acid uptake. *J. Clin. Investig.* **2018**, *128*, 4329–4342. [CrossRef] [PubMed]
15. Bala, S.; Szabo, G. TFEB, a master regulator of lysosome biogenesis and autophagy, is a new player in alcoholic liver disease. *Dig. Med. Res.* **2018**, *1*. [CrossRef]
16. Sun, J.; Lu, H.; Liang, W.; Zhao, G.; Ren, L.; Hu, D.; Chang, Z.; Liu, Y.; Garcia-Barrio, M.T.; Zhang, J.; et al. Endothelial TFEB (Transcription Factor EB) Improves Glucose Tolerance via Upregulation of IRS (Insulin Receptor Substrate) 1 and IRS2. *Arter. Thromb. Vasc. Biol.* **2020**. [CrossRef]
17. Gatenby, R.A.; Gillies, R.J. Why do cancers have high aerobic glycolysis? *Nat. Rev. Cancer* **2004**, *4*, 891–899. [CrossRef]
18. Krutzfeldt, A.; Spahr, R.; Mertens, S.; Siegmund, B.; Piper, H.M. Metabolism of exogenous substrates by coronary endothelial cells in culture. *J. Mol. Cell. Cardiol.* **1990**, *22*, 1393–1404. [CrossRef]
19. Oldendorf, W.H.; Cornford, M.E.; Brown, W.J. The large apparent work capability of the blood-brain barrier: A study of the mitochondrial content of capillary endothelial cells in brain and other tissues of the rat. *Ann. Neurol.* **1977**, *1*, 409–417. [CrossRef]
20. Yetkin-Arik, B.; Vogels, I.M.C.; Nowak-Sliwinska, P.; Weiss, A.; Houtkooper, R.H.; Van Noorden, C.J.F.; Klaassen, I.; Schlingemann, R.O. The role of glycolysis and mitochondrial respiration in the formation and functioning of endothelial tip cells during angiogenesis. *Sci. Rep.* **2019**, *9*, 12608. [CrossRef]
21. Zhu, J.; Thompson, C.B. Metabolic regulation of cell growth and proliferation. *Nat. Rev. Mol. Cell Biol.* **2019**, *20*, 436–450. [CrossRef] [PubMed]
22. Li, X.; Kumar, A.; Carmeliet, P. Metabolic Pathways Fueling the Endothelial Cell Drive. *Annu. Rev. Physiol.* **2019**, *81*, 483–503. [CrossRef] [PubMed]
23. Schoors, S.; Bruning, U.; Missiaen, R.; Queiroz, K.C.; Borgers, G.; Elia, I.; Zecchin, A.; Cantelmo, A.R.; Christen, S.; Goveia, J.; et al. Fatty acid carbon is essential for dNTP synthesis in endothelial cells. *Nature* **2015**, *520*, 192–197. [CrossRef] [PubMed]

24. Xiong, J.; Kawagishi, H.; Yan, Y.; Liu, J.; Wells, Q.S.; Edmunds, L.R.; Fergusson, M.M.; Yu, Z.X.; Rovira, I.I.; Brittain, E.L.; et al. A Metabolic Basis for Endothelial-to-Mesenchymal Transition. *Mol. Cell* **2018**, *69*, 689–698.e7. [CrossRef] [PubMed]
25. Wong, B.W.; Wang, X.; Zecchin, A.; Thienpont, B.; Cornelissen, I.; Kalucka, J.; Garcia-Caballero, M.; Missiaen, R.; Huang, H.; Bruning, U.; et al. The role of fatty acid beta-oxidation in lymphangiogenesis. *Nature* **2017**, *542*, 49–54. [CrossRef] [PubMed]
26. Elmasri, H.; Ghelfi, E.; Yu, C.W.; Traphagen, S.; Cernadas, M.; Cao, H.; Shi, G.P.; Plutzky, J.; Sahin, M.; Hotamisligil, G.; et al. Endothelial cell-fatty acid binding protein 4 promotes angiogenesis: Role of stem cell factor/c-kit pathway. *Angiogenesis* **2012**, *15*, 457–468. [CrossRef]
27. Harjes, U.; Bridges, E.; Gharpure, K.M.; Roxanis, I.; Sheldon, H.; Miranda, F.; Mangala, L.S.; Pradeep, S.; Lopez-Berestein, G.; Ahmed, A.; et al. Antiangiogenic and tumour inhibitory effects of downregulating tumour endothelial FABP4. *Oncogene* **2017**, *36*, 912–921. [CrossRef]
28. Altman, B.J.; Stine, Z.E.; Dang, C.V. From Krebs to clinic: Glutamine metabolism to cancer therapy. *Nat. Rev. Cancer* **2016**, *16*, 619–634. [CrossRef]
29. Coloff, J.L.; Murphy, J.P.; Braun, C.R.; Harris, I.S.; Shelton, L.M.; Kami, K.; Gygi, S.P.; Selfors, L.M.; Brugge, J.S. Differential Glutamate Metabolism in Proliferating and Quiescent Mammary Epithelial Cells. *Cell Metab.* **2016**, *23*, 867–880. [CrossRef]
30. Pan, M.; Wasa, M.; Ryan, U.; Souba, W. Inhibition of pulmonary microvascular endothelial glutamine transport by glucocorticoids and endotoxin. *JPEN J. Parenter. Enter. Nutr.* **1995**, *19*, 477–481. [CrossRef]
31. Kim, B.; Li, J.; Jang, C.; Arany, Z. Glutamine fuels proliferation but not migration of endothelial cells. *EMBO J.* **2017**, *36*, 2321–2333. [CrossRef] [PubMed]
32. Huang, H.; Vandekeere, S.; Kalucka, J.; Bierhansl, L.; Zecchin, A.; Bruning, U.; Visnagri, A.; Yuldasheva, N.; Goveia, J.; Cruys, B.; et al. Role of glutamine and interlinked asparagine metabolism in vessel formation. *EMBO J.* **2017**, *36*, 2334–2352. [CrossRef] [PubMed]
33. Eelen, G.; de Zeeuw, P.; Treps, L.; Harjes, U.; Wong, B.W.; Carmeliet, P. Endothelial Cell Metabolism. *Physiol. Rev.* **2018**, *98*, 3–58. [CrossRef] [PubMed]
34. Chistiakov, D.A.; Orekhov, A.N.; Bobryshev, Y.V. Effects of shear stress on endothelial cells: Go with the flow. *Acta Physiol.* **2017**, *219*, 382–408. [CrossRef]
35. Incalza, M.A.; D'Oria, R.; Natalicchio, A.; Perrini, S.; Laviola, L.; Giorgino, F. Oxidative stress and reactive oxygen species in endothelial dysfunction associated with cardiovascular and metabolic diseases. *Vasc. Pharmacol.* **2018**, *100*, 1–19. [CrossRef] [PubMed]
36. Doddaballapur, A.; Michalik, K.M.; Manavski, Y.; Lucas, T.; Houtkooper, R.H.; You, X.; Chen, W.; Zeiher, A.M.; Potente, M.; Dimmeler, S.; et al. Laminar shear stress inhibits endothelial cell metabolism via KLF2-mediated repression of PFKFB3. *Arterioscler. Thromb. Vasc. Biol.* **2015**, *35*, 137–145. [CrossRef]
37. Wilhelm, K.; Happel, K.; Eelen, G.; Schoors, S.; Oellerich, M.F.; Lim, R.; Zimmermann, B.; Aspalter, I.M.; Franco, C.A.; Boettger, T.; et al. FOXO1 couples metabolic activity and growth state in the vascular endothelium. *Nature* **2016**, *529*, 216–220. [CrossRef]
38. De Bock, K.; Georgiadou, M.; Carmeliet, P. Role of endothelial cell metabolism in vessel sprouting. *Cell Metab.* **2013**, *18*, 634–647. [CrossRef]
39. Kalucka, J.; Bierhansl, L.; Conchinha, N.V.; Missiaen, R.; Elia, I.; Bruning, U.; Scheinok, S.; Treps, L.; Cantelmo, A.R.; Dubois, C.; et al. Quiescent Endothelial Cells Upregulate Fatty Acid beta-Oxidation for Vasculoprotection via Redox Homeostasis. *Cell Metab.* **2018**, *28*, 881–894.e13. [CrossRef]
40. Brandes, R.P.; Weissmann, N.; Schroder, K. Redox-mediated signal transduction by cardiovascular Nox NADPH oxidases. *J. Mol. Cell. Cardiol.* **2014**, *73*, 70–79. [CrossRef]
41. Du, X.; Matsumura, T.; Edelstein, D.; Rossetti, L.; Zsengeller, Z.; Szabo, C.; Brownlee, M. Inhibition of GAPDH activity by poly(ADP-ribose) polymerase activates three major pathways of hyperglycemic damage in endothelial cells. *J. Clin. Investig.* **2003**, *112*, 1049–1057. [CrossRef]
42. Veys, K.; Fan, Z.; Ghobrial, M.; Bouche, A.; Garcia-Caballero, M.; Vriens, K.; Conchinha, N.V.; Seuwen, A.; Schlegel, F.; Gorski, T.; et al. Role of the GLUT1 Glucose Transporter in Postnatal CNS Angiogenesis and Blood-Brain Barrier Integrity. *Circ. Res.* **2020**, *127*, 466–482. [CrossRef] [PubMed]
43. Shao, Y.; Chen, J.; Dong, L.J.; He, X.; Cheng, R.; Zhou, K.; Liu, J.; Qiu, F.; Li, X.R.; Ma, J.X. A Protective Effect of PPARalpha in Endothelial Progenitor Cells Through Regulating Metabolism. *Diabetes* **2019**, *68*, 2131–2142. [CrossRef] [PubMed]
44. Joyal, J.S.; Sun, Y.; Gantner, M.L.; Shao, Z.; Evans, L.P.; Saba, N.; Fredrick, T.; Burnim, S.; Kim, J.S.; Patel, G.; et al. Retinal lipid and glucose metabolism dictates angiogenesis through the lipid sensor Ffar1. *Nat. Med.* **2016**, *22*, 439–445. [CrossRef] [PubMed]
45. Bruning, U.; Morales-Rodriguez, F.; Kalucka, J.; Goveia, J.; Taverna, F.; Queiroz, K.C.S.; Dubois, C.; Cantelmo, A.R.; Chen, R.; Loroch, S.; et al. Impairment of Angiogenesis by Fatty Acid Synthase Inhibition Involves mTOR Malonylation. *Cell Metab.* **2018**, *28*, 866–880.e15. [CrossRef] [PubMed]
46. Tan, J.T.; Prosser, H.C.; Dunn, L.L.; Vanags, L.Z.; Ridiandries, A.; Tsatralis, T.; Lecce, L.; Clayton, Z.E.; Yuen, S.C.; Robertson, S.; et al. High-Density Lipoproteins Rescue Diabetes-Impaired Angiogenesis via Scavenger Receptor Class B Type I. *Diabetes* **2016**, *65*, 3091–3103. [CrossRef] [PubMed]
47. Wong, B.W.; Marsch, E.; Treps, L.; Baes, M.; Carmeliet, P. Endothelial cell metabolism in health and disease: Impact of hypoxia. *EMBO J.* **2017**, *36*, 2187–2203. [CrossRef]

48. Rudnicki, M.; Abdifarkosh, G.; Nwadozi, E.; Ramos, S.V.; Makki, A.; Sepa-Kishi, D.M.; Ceddia, R.B.; Perry, C.G.; Roudier, E.; Haas, T.L. Endothelial-specific FoxO1 depletion prevents obesity-related disorders by increasing vascular metabolism and growth. *Elife* **2018**, *7*. [CrossRef]
49. Dinh, T.; Veves, A. Microcirculation of the diabetic foot. *Curr. Pharm. Des.* **2005**, *11*, 2301–2309. [CrossRef]
50. Lin, C.J.; Lan, Y.M.; Ou, M.Q.; Ji, L.Q.; Lin, S.D. Expression of miR-217 and HIF-1alpha/VEGF pathway in patients with diabetic foot ulcer and its effect on angiogenesis of diabetic foot ulcer rats. *J. Endocrinol. Investig.* **2019**, *42*, 1307–1317. [CrossRef]
51. Wetterau, M.; George, F.; Weinstein, A.; Nguyen, P.D.; Tutela, J.P.; Knobel, D.; Cohen Ba, O.; Warren, S.M.; Saadeh, P.B. Topical prolyl hydroxylase domain-2 silencing improves diabetic murine wound closure. *Wound Repair Regen.* **2011**, *19*, 481–486. [CrossRef]
52. Sawada, N.; Jiang, A.; Takizawa, F.; Safdar, A.; Manika, A.; Tesmenitsky, Y.; Kang, K.T.; Bischoff, J.; Kalwa, H.; Sartoretto, J.L.; et al. Endothelial PGC-1alpha mediates vascular dysfunction in diabetes. *Cell Metab.* **2014**, *19*, 246–258. [CrossRef]
53. Ryan, T.E.; Schmidt, C.A.; Tarpey, M.D.; Amorese, A.J.; Yamaguchi, D.J.; Goldberg, E.J.; Inigo, M.M.; Karnekar, R.; O'Rourke, A.; Ervasti, J.M.; et al. PFKFB3-mediated glycolysis rescues myopathic outcomes in the ischemic limb. *JCI Insight* **2020**, *5*. [CrossRef]
54. Fan, Y.; Lu, H.; Liang, W.; Garcia-Barrio, M.T.; Guo, Y.; Zhang, J.; Zhu, T.; Hao, Y.; Zhang, J.; Chen, Y.E. Endothelial TFEB (Transcription Factor EB) Positively Regulates Postischemic Angiogenesis. *Circ. Res.* **2018**, *122*, 945–957. [CrossRef]
55. Lu, H.; Sun, J.; Hamblin, M.H.; Chen, Y.E.; Fan, Y. Transcription factor EB regulates cardiovascular homeostasis. *EBioMedicine* **2021**, *63*, 103207. [CrossRef]
56. Lu, H.; Sun, J.; Liang, W.; Chang, Z.; Rom, O.; Zhao, Y.; Zhao, G.; Xiong, W.; Wang, H.; Zhu, T.; et al. Cyclodextrin Prevents Abdominal Aortic Aneurysm via Activation of Vascular Smooth Muscle Cell Transcription Factor EB. *Circulation* **2020**, *142*, 483–498. [CrossRef]
57. Long, X.; Boluyt, M.O.; Hipolito, M.L.; Lundberg, M.S.; Zheng, J.S.; O'Neill, L.; Cirielli, C.; Lakatta, E.G.; Crow, M.T. p53 and the hypoxia-induced apoptosis of cultured neonatal rat cardiac myocytes. *J. Clin. Investig.* **1997**, *99*, 2635–2643. [CrossRef]
58. Annan, D.A.; Maishi, N.; Soga, T.; Dawood, R.; Li, C.; Kikuchi, H.; Hojo, T.; Morimoto, M.; Kitamura, T.; Alam, M.T.; et al. Carbonic anhydrase 2 (CAII) supports tumor blood endothelial cell survival under lactic acidosis in the tumor microenvironment. *Cell Commun. Signal.* **2019**, *17*, 169. [CrossRef]
59. Schworer, S.; Vardhana, S.A.; Thompson, C.B. Cancer Metabolism Drives a Stromal Regenerative Response. *Cell Metab.* **2019**, *29*, 576–591. [CrossRef]
60. Zhao, Q.; Eichten, A.; Parveen, A.; Adler, C.; Huang, Y.; Wang, W.; Ding, Y.; Adler, A.; Nevins, T.; Ni, M.; et al. Single-Cell Transcriptome Analyses Reveal Endothelial Cell Heterogeneity in Tumors and Changes following Antiangiogenic Treatment. *Cancer Res.* **2018**, *78*, 2370–2382. [CrossRef]
61. Qian, C.N.; Tan, M.H.; Yang, J.P.; Cao, Y. Revisiting tumor angiogenesis: Vessel co-option, vessel remodeling, and cancer cell-derived vasculature formation. *Chin. J. Cancer* **2016**, *35*, 10. [CrossRef]
62. Andonegui-Elguera, M.A.; Alfaro-Mora, Y.; Caceres-Gutierrez, R.; Caro-Sanchez, C.H.S.; Herrera, L.A.; Diaz-Chavez, J. An Overview of Vasculogenic Mimicry in Breast Cancer. *Front. Oncol.* **2020**, *10*, 220. [CrossRef]
63. Hunt, T.K.; Aslam, R.S.; Beckert, S.; Wagner, S.; Ghani, Q.P.; Hussain, M.Z.; Roy, S.; Sen, C.K. Aerobically derived lactate stimulates revascularization and tissue repair via redox mechanisms. *Antioxid. Redox Signal.* **2007**, *9*, 1115–1124. [CrossRef]
64. Sonveaux, P.; Copetti, T.; De Saedeleer, C.J.; Vegran, F.; Verrax, J.; Kennedy, K.M.; Moon, E.J.; Dhup, S.; Danhier, P.; Frerart, F.; et al. Targeting the lactate transporter MCT1 in endothelial cells inhibits lactate-induced HIF-1 activation and tumor angiogenesis. *PLoS ONE* **2012**, *7*, e33418. [CrossRef]
65. Van Beijnum, J.R.; Dings, R.P.; van der Linden, E.; Zwaans, B.M.; Ramaekers, F.C.; Mayo, K.H.; Griffioen, A.W. Gene expression of tumor angiogenesis dissected: Specific targeting of colon cancer angiogenic vasculature. *Blood* **2006**, *108*, 2339–2348. [CrossRef]
66. Sakakibara, R.; Kato, M.; Okamura, N.; Nakagawa, T.; Komada, Y.; Tominaga, N.; Shimojo, M.; Fukasawa, M. Characterization of a human placental fructose-6-phosphate, 2-kinase/fructose-2,6-bisphosphatase. *J. Biochem.* **1997**, *122*, 122–128. [CrossRef]
67. Yeh, W.L.; Lin, C.J.; Fu, W.M. Enhancement of glucose transporter expression of brain endothelial cells by vascular endothelial growth factor derived from glioma exposed to hypoxia. *Mol. Pharmacol.* **2008**, *73*, 170–177. [CrossRef]
68. Loponte, S.; Lovisa, S.; Deem, A.K.; Carugo, A.; Viale, A. The Many Facets of Tumor Heterogeneity: Is Metabolism Lagging Behind? *Cancers* **2019**, *11*, 1574. [CrossRef]
69. Filippi, I.; Saltarella, I.; Aldinucci, C.; Carraro, F.; Ria, R.; Vacca, A.; Naldini, A. Different Adaptive Responses to Hypoxia in Normal and Multiple Myeloma Endothelial Cells. *Cell. Physiol. Biochem.* **2018**, *46*, 203–212. [CrossRef]
70. Schaaf, M.B.; Houbaert, D.; Mece, O.; Agostinis, P. Autophagy in endothelial cells and tumor angiogenesis. *Cell Death Differ.* **2019**, *26*, 665–679. [CrossRef]
71. Galluzzi, L.; Pietrocola, F.; Bravo-San Pedro, J.M.; Amaravadi, R.K.; Baehrecke, E.H.; Cecconi, F.; Codogno, P.; Debnath, J.; Gewirtz, D.A.; Karantza, V.; et al. Autophagy in malignant transformation and cancer progression. *EMBO J.* **2015**, *34*, 856–880. [CrossRef]
72. Rouschop, K.M.; van den Beucken, T.; Dubois, L.; Niessen, H.; Bussink, J.; Savelkouls, K.; Keulers, T.; Mujcic, H.; Landuyt, W.; Voncken, J.W.; et al. The unfolded protein response protects human tumor cells during hypoxia through regulation of the autophagy genes MAP1LC3B and ATG5. *J. Clin. Investig.* **2010**, *120*, 127–141. [CrossRef]
73. Lee, S.J.; Kim, H.P.; Jin, Y.; Choi, A.M.; Ryter, S.W. Beclin 1 deficiency is associated with increased hypoxia-induced angiogenesis. *Autophagy* **2011**, *7*, 829–839. [CrossRef]

74. Vizan, P.; Sanchez-Tena, S.; Alcarraz-Vizan, G.; Soler, M.; Messeguer, R.; Pujol, M.D.; Lee, W.N.; Cascante, M. Characterization of the metabolic changes underlying growth factor angiogenic activation: Identification of new potential therapeutic targets. *Carcinogenesis* **2009**, *30*, 946–952. [CrossRef]
75. Pilar Lopez, M.; Gomez-Lechon, M.J.; Castell, J.V. Role of glucose, insulin, and glucagon in glycogen mobilization in human hepatocytes. *Diabetes* **1991**, *40*, 263–268. [CrossRef]
76. Zois, C.E.; Harris, A.L. Glycogen metabolism has a key role in the cancer microenvironment and provides new targets for cancer therapy. *J. Mol. Med.* **2016**, *94*, 137–154. [CrossRef]
77. Hong, J.; Wuest, T.R.; Min, Y.; Lin, P.C. Oxygen Tension Regulates Lysosomal Activation and Receptor Tyrosine Kinase Degradation. *Cancers* **2019**, *11*, 1653. [CrossRef]
78. Mansueto, G.; Armani, A.; Viscomi, C.; D'Orsi, L.; De Cegli, R.; Polishchuk, E.V.; Lamperti, C.; Di Meo, I.; Romanello, V.; Marchet, S.; et al. Transcription Factor EB Controls Metabolic Flexibility during Exercise. *Cell Metab.* **2017**, *25*, 182–196. [CrossRef]
79. Bergers, G.; Hanahan, D. Modes of resistance to anti-angiogenic therapy. *Nat. Rev. Cancer* **2008**, *8*, 592–603. [CrossRef]
80. Cruys, B.; Wong, B.W.; Kuchnio, A.; Verdegem, D.; Cantelmo, A.R.; Conradi, L.C.; Vandekeere, S.; Bouche, A.; Cornelissen, I.; Vinckier, S.; et al. Glycolytic regulation of cell rearrangement in angiogenesis. *Nat. Commun.* **2016**, *7*, 12240. [CrossRef]
81. Gomez-Escudero, J.; Clemente, C.; Garcia-Weber, D.; Acin-Perez, R.; Millan, J.; Enriquez, J.A.; Bentley, K.; Carmeliet, P.; Arroyo, A.G. PKM2 regulates endothelial cell junction dynamics and angiogenesis via ATP production. *Sci. Rep.* **2019**, *9*, 15022. [CrossRef]
82. Warburg, O. On the origin of cancer cells. *Science* **1956**, *123*, 309–314. [CrossRef]
83. Hsiao, Y.H.; Hsieh, M.J.; Yang, S.F.; Chen, S.P.; Tsai, W.C.; Chen, P.N. Phloretin suppresses metastasis by targeting protease and inhibits cancer stemness and angiogenesis in human cervical cancer cells. *Phytomedicine* **2019**, *62*, 152964. [CrossRef]
84. Yang, S.H.; Lin, J.K.; Chen, W.S.; Chiu, J.H. Anti-angiogenic effect of silymarin on colon cancer LoVo cell line. *J. Surg. Res.* **2003**, *113*, 133–138. [CrossRef]
85. Kaji, K.; Nishimura, N.; Seki, K.; Sato, S.; Saikawa, S.; Nakanishi, K.; Furukawa, M.; Kawaratani, H.; Kitade, M.; Moriya, K.; et al. Sodium glucose cotransporter 2 inhibitor canagliflozin attenuates liver cancer cell growth and angiogenic activity by inhibiting glucose uptake. *Int. J. Cancer* **2018**, *142*, 1712–1722. [CrossRef]
86. Fu, Z.; Chen, X.; Guan, S.; Yan, Y.; Lin, H.; Hua, Z.C. Curcumin inhibits angiogenesis and improves defective hematopoiesis induced by tumor-derived VEGF in tumor model through modulating VEGF-VEGFR2 signaling pathway. *Oncotarget* **2015**, *6*, 19469–19482. [CrossRef]
87. Ocana, M.C.; Martinez-Poveda, B.; Mari-Beffa, M.; Quesada, A.R.; Medina, M.A. Fasentin diminishes endothelial cell proliferation, differentiation and invasion in a glucose metabolism-independent manner. *Sci. Rep.* **2020**, *10*, 6132. [CrossRef]
88. Jia, Z.; Zhen, W.; Velayutham Anandh Babu, P.; Liu, D. Phytoestrogen genistein protects against endothelial barrier dysfunction in vascular endothelial cells through PKA-mediated suppression of RhoA signaling. *Endocrinology* **2013**, *154*, 727–737. [CrossRef]
89. Su, S.J.; Yeh, T.M.; Chuang, W.J.; Ho, C.L.; Chang, K.L.; Cheng, H.L.; Liu, H.S.; Cheng, H.L.; Hsu, P.Y.; Chow, N.H. The novel targets for anti-angiogenesis of genistein on human cancer cells. *Biochem. Pharmacol.* **2005**, *69*, 307–318. [CrossRef]
90. Singh, S.; Pandey, S.; Chawla, A.S.; Bhatt, A.N.; Roy, B.G.; Saluja, D.; Dwarakanath, B.S. Dietary 2-deoxy-D-glucose impairs tumour growth and metastasis by inhibiting angiogenesis. *Eur. J. Cancer* **2019**, *123*, 11–24. [CrossRef]
91. Huang, C.C.; Wang, S.Y.; Lin, L.L.; Wang, P.W.; Chen, T.Y.; Hsu, W.M.; Lin, T.K.; Liou, C.W.; Chuang, J.H. Glycolytic inhibitor 2-deoxyglucose simultaneously targets cancer and endothelial cells to suppress neuroblastoma growth in mice. *Dis. Models Mech.* **2015**, *8*, 1247–1254. [CrossRef]
92. Merchan, J.R.; Kovacs, K.; Railsback, J.W.; Kurtoglu, M.; Jing, Y.; Pina, Y.; Gao, N.; Murray, T.G.; Lehrman, M.A.; Lampidis, T.J. Antiangiogenic activity of 2-deoxy-D-glucose. *PLoS ONE* **2010**, *5*, e13699. [CrossRef]
93. Agnihotri, S.; Mansouri, S.; Burrell, K.; Li, M.; Mamatjan, Y.; Liu, J.; Nejad, R.; Kumar, S.; Jalali, S.; Singh, S.K.; et al. Ketoconazole and Posaconazole Selectively Target HK2-expressing Glioblastoma Cells. *Clin. Cancer Res.* **2019**, *25*, 844–855. [CrossRef]
94. Del Bufalo, D.; Trisciuoglio, D.; Scarsella, M.; D'Amati, G.; Candiloro, A.; Iervolino, A.; Leonetti, C.; Zupi, G. Lonidamine causes inhibition of angiogenesis-related endothelial cell functions. *Neoplasia* **2004**, *6*, 513–522. [CrossRef]
95. Zheng, L.; Li, D.; Xiang, X.; Tong, L.; Qi, M.; Pu, J.; Huang, K.; Tong, Q. Methyl jasmonate abolishes the migration, invasion and angiogenesis of gastric cancer cells through down-regulation of matrix metalloproteinase 14. *BMC Cancer* **2013**, *13*, 74. [CrossRef]
96. Schoors, S.; De Bock, K.; Cantelmo, A.R.; Georgiadou, M.; Ghesquiere, B.; Cauwenberghs, S.; Kuchnio, A.; Wong, B.W.; Quaegebeur, A.; Goveia, J.; et al. Partial and transient reduction of glycolysis by PFKFB3 blockade reduces pathological angiogenesis. *Cell Metab.* **2014**, *19*, 37–48. [CrossRef]
97. Clem, B.; Telang, S.; Clem, A.; Yalcin, A.; Meier, J.; Simmons, A.; Rasku, M.A.; Arumugam, S.; Dean, W.L.; Eaton, J.; et al. Small-molecule inhibition of 6-phosphofructo-2-kinase activity suppresses glycolytic flux and tumor growth. *Mol. Cancer Ther.* **2008**, *7*, 110–120. [CrossRef]
98. Clem, B.F.; O'Neal, J.; Tapolsky, G.; Clem, A.L.; Imbert-Fernandez, Y.; Kerr, D.A., 2nd; Klarer, A.C.; Redman, R.; Miller, D.M.; Trent, J.O.; et al. Targeting 6-phosphofructo-2-kinase (PFKFB3) as a therapeutic strategy against cancer. *Mol. Cancer Ther.* **2013**, *12*, 1461–1470. [CrossRef]
99. Porporato, P.E.; Dhup, S.; Dadhich, R.K.; Copetti, T.; Sonveaux, P. Anticancer targets in the glycolytic metabolism of tumors: A comprehensive review. *Front. Pharmacol.* **2011**, *2*, 49. [CrossRef]

100. Komi, Y.; Suzuki, Y.; Shimamura, M.; Kajimoto, S.; Nakajo, S.; Masuda, M.; Shibuya, M.; Itabe, H.; Shimokado, K.; Oettgen, P.; et al. Mechanism of inhibition of tumor angiogenesis by beta-hydroxyisovalerylshikonin. *Cancer Sci.* **2009**, *100*, 269–277. [CrossRef]
101. Koukourakis, M.I.; Giatromanolaki, A.; Sivridis, E.; Gatter, K.C.; Trarbach, T.; Folprecht, G.; Shi, M.M.; Lebwohl, D.; Jalava, T.; Laurent, D.; et al. Prognostic and predictive role of lactate dehydrogenase 5 expression in colorectal cancer patients treated with PTK787/ZK 222584 (vatalanib) antiangiogenic therapy. *Clin. Cancer Res.* **2011**, *17*, 4892–4900. [CrossRef]
102. Ulus, G.; Koparal, A.T.; Baysal, K.; Yetik Anacak, G.; Karabay Yavasoglu, N.U. The anti-angiogenic potential of (+/-) gossypol in comparison to suramin. *Cytotechnology* **2018**, *70*, 1537–1550. [CrossRef]
103. Pang, X.; Wu, Y.; Wu, Y.; Lu, B.; Chen, J.; Wang, J.; Yi, Z.; Qu, W.; Liu, M. (-)-Gossypol suppresses the growth of human prostate cancer xenografts via modulating VEGF signaling-mediated angiogenesis. *Mol. Cancer Ther.* **2011**, *10*, 795–805. [CrossRef]
104. El-Sisi, A.E.; Sokar, S.S.; Abu-Risha, S.E.; El-Mahrouk, S.R. Oxamate potentiates taxol chemotherapeutic efficacy in experimentally-induced solid ehrlich carcinoma (SEC) in mice. *Biomed. Pharmacother.* **2017**, *95*, 1565–1573. [CrossRef]
105. Zhang, J.; Muri, J.; Fitzgerald, G.; Gorski, T.; Gianni-Barrera, R.; Masschelein, E.; D'Hulst, G.; Gilardoni, P.; Turiel, G.; Fan, Z.; et al. Endothelial Lactate Controls Muscle Regeneration from Ischemia by Inducing M2-like Macrophage Polarization. *Cell Metab.* **2020**, *31*, 1136–1153.e7. [CrossRef]
106. Sutendra, G.; Dromparis, P.; Kinnaird, A.; Stenson, T.H.; Haromy, A.; Parker, J.M.; McMurtry, M.S.; Michelakis, E.D. Mitochondrial activation by inhibition of PDKII suppresses HIF1a signaling and angiogenesis in cancer. *Oncogene* **2013**, *32*, 1638–1650. [CrossRef]
107. Dallaglio, K.; Bruno, A.; Cantelmo, A.R.; Esposito, A.I.; Ruggiero, L.; Orecchioni, S.; Calleri, A.; Bertolini, F.; Pfeffer, U.; Noonan, D.M.; et al. Paradoxic effects of metformin on endothelial cells and angiogenesis. *Carcinogenesis* **2014**, *35*, 1055–1066. [CrossRef]
108. Jaidee, R.; Kongpetch, S.; Senggunprai, L.; Prawan, A.; Kukongviriyapan, U.; Kukongviriyapan, V. Phenformin inhibits proliferation, invasion, and angiogenesis of cholangiocarcinoma cells via AMPK-mTOR and HIF-1A pathways. *Naunyn-Schmiedeberg Arch. Pharmacol.* **2020**, *393*, 1681–1690. [CrossRef]
109. Navarro, P.; Bueno, M.J.; Zagorac, I.; Mondejar, T.; Sanchez, J.; Mouron, S.; Munoz, J.; Gomez-Lopez, G.; Jimenez-Renard, V.; Mulero, F.; et al. Targeting Tumor Mitochondrial Metabolism Overcomes Resistance to Antiangiogenics. *Cell Rep.* **2016**, *15*, 2705–2718. [CrossRef]
110. Zhang, L.; Liu, L.; Zhan, S.; Chen, L.; Wang, Y.; Zhang, Y.; Du, J.; Wu, Y.; Gu, L. Arsenic Trioxide Suppressed Migration and Angiogenesis by Targeting FOXO3a in Gastric Cancer Cells. *Int. J. Mol. Sci.* **2018**, *19*, 3739. [CrossRef]
111. Agius, L.; Meredith, E.J.; Sherratt, H.S. Stereospecificity of the inhibition by etomoxir of fatty acid and cholesterol synthesis in isolated rat hepatocytes. *Biochem. Pharmacol.* **1991**, *42*, 1717–1720. [CrossRef]
112. Ashrafian, H.; Horowitz, J.D.; Frenneaux, M.P. Perhexiline. *Cardiovasc. Drug Rev.* **2007**, *25*, 76–97. [CrossRef]
113. Iwamoto, H.; Abe, M.; Yang, Y.; Cui, D.; Seki, T.; Nakamura, M.; Hosaka, K.; Lim, S.; Wu, J.; He, X.; et al. Cancer Lipid Metabolism Confers Antiangiogenic Drug Resistance. *Cell Metab.* **2018**, *28*, 104–117.e5. [CrossRef]
114. Elgogary, A.; Xu, Q.; Poore, B.; Alt, J.; Zimmermann, S.C.; Zhao, L.; Fu, J.; Chen, B.; Xia, S.; Liu, Y.; et al. Combination therapy with BPTES nanoparticles and metformin targets the metabolic heterogeneity of pancreatic cancer. *Proc. Natl. Acad. Sci. USA* **2016**, *113*, E5328–5336. [CrossRef]
115. Ren, L.; Ruiz-Rodado, V.; Dowdy, T.; Huang, S.; Issaq, S.H.; Beck, J.; Wang, H.; Tran Hoang, C.; Lita, A.; Larion, M.; et al. Glutaminase-1 (GLS1) inhibition limits metastatic progression in osteosarcoma. *Cancer Metab.* **2020**, *8*, 4. [CrossRef]
116. Rohlenova, K.; Veys, K.; Miranda-Santos, I.; De Bock, K.; Carmeliet, P. Endothelial Cell Metabolism in Health and Disease. *Trends Cell Biol.* **2018**, *28*, 224–236. [CrossRef]
117. Ribatti, D.; Annese, T.; Ruggieri, S.; Tamma, R.; Crivellato, E. Limitations of Anti-Angiogenic Treatment of Tumors. *Transl. Oncol.* **2019**, *12*, 981–986. [CrossRef]

Review

Endothelial Dysfunction in Pulmonary Hypertension: Cause or Consequence?

Kondababu Kurakula [1,†], Valérie F. E. D. Smolders [2,†], Olga Tura-Ceide [3,4,5], J. Wouter Jukema [6], Paul H. A. Quax [2] and Marie-José Goumans [1,*]

1. Department of Cell and Chemical Biology, Laboratory for CardioVascular Cell Biology, Leiden University Medical Center, 2300 RC Leiden, The Netherlands; K.B.Kurakula@lumc.nl
2. Department of Surgery, Einthoven Laboratory for Experimental Vascular Medicine, Leiden University Medical Center, 2300 RC Leiden, The Netherlands; v.f.e.d.smolders@lumc.nl (V.F.E.D.S.); P.H.A.Quax@lumc.nl (P.H.A.Q.)
3. Department of Pulmonary Medicine, Hospital Clínic-Institut d'Investigacions Biomèdiques August Pi i Sunyer (IDIBAPS), University of Barcelona, 08036 Barcelona, Spain; olgaturac@gmail.com
4. Department of Pulmonary Medicine, Dr. Josep Trueta University Hospital de Girona, Santa Caterina Hospital de Salt and the Girona Biomedical Research Institut (IDIBGI), 17190 Girona, Catalonia, Spain
5. Biomedical Research Networking Centre on Respiratory Diseases (CIBERES), 28029 Madrid, Spain
6. Department of Cardiology, Leiden University Medical Center, 2300 RC Leiden, The Netherlands; j.w.jukema@lumc.nl
* Correspondence: M.J.T.H.Goumans@lumc.nl
† The authors contributed equally.

Abstract: Pulmonary arterial hypertension (PAH) is a rare, complex, and progressive disease that is characterized by the abnormal remodeling of the pulmonary arteries that leads to right ventricular failure and death. Although our understanding of the causes for abnormal vascular remodeling in PAH is limited, accumulating evidence indicates that endothelial cell (EC) dysfunction is one of the first triggers initiating this process. EC dysfunction leads to the activation of several cellular signalling pathways in the endothelium, resulting in the uncontrolled proliferation of ECs, pulmonary artery smooth muscle cells, and fibroblasts, and eventually leads to vascular remodelling and the occlusion of the pulmonary blood vessels. Other factors that are related to EC dysfunction in PAH are an increase in endothelial to mesenchymal transition, inflammation, apoptosis, and thrombus formation. In this review, we outline the latest advances on the role of EC dysfunction in PAH and other forms of pulmonary hypertension. We also elaborate on the molecular signals that orchestrate EC dysfunction in PAH. Understanding the role and mechanisms of EC dysfunction will unravel the therapeutic potential of targeting this process in PAH.

Keywords: pulmonary hypertension; endothelial dysfunction; vasoactive factors; EndoMT; inflammation; TGF-β; epigenetics

Citation: Kurakula, K.; Smolders, V.F.E.D.; Tura-Ceide, O.; Jukema, J.W.; Quax, P.H.A.; Goumans, M.-J. Endothelial Dysfunction in Pulmonary Hypertension: Cause or Consequence? *Biomedicines* 2021, 9, 57. https://doi.org/10.3390/biomedicines9010057

Received: 24 November 2020
Accepted: 3 January 2021
Published: 9 January 2021

Publisher's Note: MDPI stays neutral with regard to jurisdictional claims in published maps and institutional affiliations.

Copyright: © 2021 by the authors. Licensee MDPI, Basel, Switzerland. This article is an open access article distributed under the terms and conditions of the Creative Commons Attribution (CC BY) license (https://creativecommons.org/licenses/by/4.0/).

1. Introduction

Pulmonary hypertension (PH) is a condition that is defined by a mean pulmonary arterial pressure of more than 20 mmHg at rest and 30 mmHg during exercise. The range of genetic, molecular, and humoral causes that can lead to this increase in pressure is extensive. Therefore, PH is grouped into different classes that are based on clinical and pathological findings as well as therapeutic interventions [1,2]. The World Health Organization (WHO) classifies PH into five groups, namely: 1. Pulmonary arterial hypertension (PAH), 2. Pulmonary hypertension due to left heart disease (PH-LHD), 3. Pulmonary hypertension due to lung disease (PH-LD), 4. Chronic thromboembolic pulmonary hypertension (CTEPH), and 5. Pulmonary hypertension due to unclear and/or multifactorial mechanisms [1,3,4]. PH is increasingly becoming a global health issue due to the ageing population. Although PH-LHD and PH-LD are the most prevalent PH groups, research and drug development mainly focus on PAH and CTEPH, which are rarer diseases that mainly affect younger people [5]. This review will focus mostly on PAH because of the amount of research conducted in PAH as compared to the other four groups.

PAH is characterized by remodeling of distal pulmonary arteries, causing a progressive increase in vascular resistance. Vascular remodeling is associated with alterations in vasoconstriction, pulmonary artery- endothelial cells (PAECs) and -smooth muscle cells (PASMCs) cell proliferation, inflammation, apoptosis, angiogenesis, and thrombosis, which leads to the muscularization and occlusion of the lumen of pulmonary arteries by the formation of vascular lesions. Some of the lesions found in PAH are plexiform lesions, which are characterized by enhanced endothelial cell (EC) proliferation, thrombotic lesions and neointima formation, the formation of a layer of myofibroblasts, and extracellular matrix between the endothelium and the external elastic lamina [6,7]. One of the first triggers for development of PAH is thought to be EC injury triggering the activation of cellular signaling pathways that are not yet completely understood.

In normal conditions, the endothelium is in a quiescent and genetically stable state. When activated, the endothelium secretes different growth factors and cytokines that affect EC and SMC proliferation, apoptosis, coagulation, attract inflammatory cells, and/or affect vasoactivity in order to restore homeostasis. Prolonged or chronic activation of the endothelium leads to EC dysfunction, the loss of homeostatic functions, leading to pathological changes, and it is crucial in the development of cardiovascular diseases and so too in PAH [8,9]. Many different factors have been suggested to be triggers of EC dysfunction in PAH, like shear stress, hypoxia, inflammation, cilia length, and genetic factors (Figure 1) [6,10–12]. As a consequence, the endothelium switches from a quiescent to an overactive state, where it starts to secrete vasoconstrictive factors, like endothelin-1 (ET-1) [13] and thromboxane [14], and proliferative factors, like vascular endothelial growth factor (VEGF), fibroblast growth factor 2 (FGF2) [15], CXCL12 [16], and reduce the secretion of vasodilators, like nitric oxide (NO) and prostacyclin, which indicates that EC dysfunction might play a central role in the pathogenesis of PAH. Whether EC dysfunction is the primary cause or rather the consequence of changes in environmental factors remains to be resolved [8,17].

The purpose of this review is to provide a state-of-the-art overview on the features and driving forces of EC dysfunction in PAH and highlight the current progress made in understanding this phenomenon. Finally, this review discusses several models for studying EC dysfunction in PH and explores possible molecular targets and drugs for restoring EC function in PH.

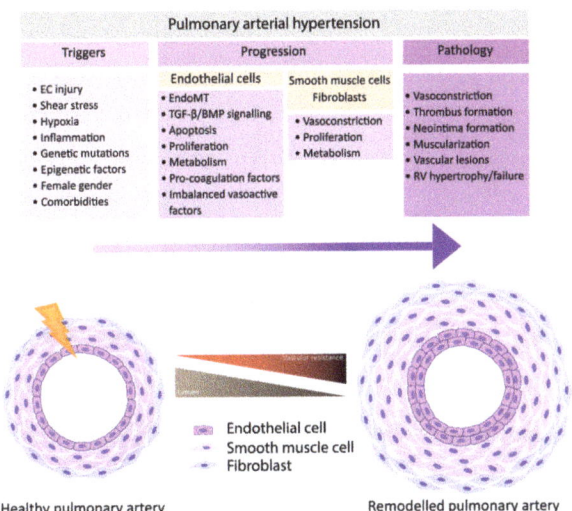

Figure 1. Pulmonary artery remodeling, vascular resistance and pulmonary arterial hypertension (PAH) development. PAH results from a progressive increase in vascular resistance caused by pulmonary vascular remodeling. Molecular mechanisms behind the process of vascular remodeling are still not fully elucidated but endothelial cell (EC) injury is thought to be one of the early triggers. EC injury can be caused by shear stress, hypoxia and inflammation. Host factors such as genetic mutations and gender but also epigenetic factors and comorbidities are thought to play an important role in EC dysfunction. EC dysfunction leads to altered cell signaling that induces cellular processes such as EndoMT, apoptosis, and proliferation. In addition, changes are found in cell metabolism and in the secretion of vasoactive, coagulation and thrombotic factors. Additionally, vascular smooth muscle cells and fibroblasts are found to display a diseased cellular phenotype. EC dysfunction eventually promotes vasoconstriction, thrombus formation, neointima formation, muscularization, and the development of vascular lesions. As lumen size decreases, pulmonary vascular resistance increases and induces right ventricle (RV) hypertrophy, with eventual RV failure.

2. Factors contributing to EC Dysfunction in PH

Approximately 80% of familial PAH (hPAH) and 20% of idiopathic cases of PAH (iPAH) are associated with mutations in the bone morphogenic type 2 receptor (BMPR2), but a penetrance of 20–30% suggests secondary stimuli, such as inflammation and thrombosis, as important contributors to EC dysfunction and PAH development [18–21]. More recently, alterations in endothelial metabolic functions in the pulmonary vasculature are emerging as important regulators of endothelial dysfunction.

2.1. Bone Morphogenic Type 2 Receptor

BMPR2 encodes for a transmembrane serine/threonine kinase receptor belonging to the transforming growth factor-β (TGFβ) family of signaling proteins (Figure 2) [22]. BMPR2 modulates cellular growth, apoptosis, inflammation, and differentiation via the binding of bone morphogenetic proteins (BMPs) to a heteromeric complex of a BMP type-I receptor and BMPR2, in a time, concentration, and cell type dependent manner [23]. BMPs are secreted cytokines that play important roles in vascular development and homeostasis. Alterations in the functions of BMPs are associated with severe developmental disorders and diverse human disease [23–25]. BMPR2 promotes the survival of PAECs depending on the localization in the vascular bed, and it has an anti-proliferative effect on PASMCs [26–28].

To date, over 380 PAH related mutations in *BMPR2* are known, mostly loss of function mutations [29,30]. The low penetrance of disease development associated with *BMPR2* mutations observed in humans has also been confirmed in experimental models of PH, where *BMPR2* deletion alone does not induce PAH in the majority of the cases [31–33]. Interestingly, reduced levels of BMPR2 have also been found in PH patients without *BMPR2* mutations, which suggests the additional involvement of genetic modifiers or environmental factors reducing BMPR2 dependent signaling [34–37].

BMPR2 is predominantly present in ECs lining the vascular lumen in the lung and expression is reduced in ECs from PH lung. Therefore, mutated *BMPR2* is postulated to play a significant role in EC dysfunction in PAH [30,34]. Association between endothelial BMPR2 expression levels and PAH development was further supported by the observation that mice with endothelial specific deletion of *BMPR2* were prone to developing PAH [38,39]. PAECs overexpressing a kinase-inactive BMPR2 mutant show increased susceptibility to apoptosis and conditioned medium from these PAECs stimulated proliferation of PASMCs via increased release of TGFβ1 and fibroblast growth factor (FGF)-2 [40]. More recently, mutations in *GDF2*, the gene encoding the BMP9 ligand, have been identified in PAH patients and associated with reduced circulating levels of both BMP9 and BMP10 [41]. The presence of these PAH-linked mutations in the endothelial BMPR2/ligand axis provide additional genetic evidence to support a critical role for endothelial dysfunction in the pathobiology of PAH. Moreover, BMP9 administration selectively enhanced endothelial BMPR2 signaling in PAECs and reversed PH in both MCT and SuHx rats [42]. Based on this knowledge, one might speculate a causal role for these mutations in EC dysfunction and subsequent PAH development.

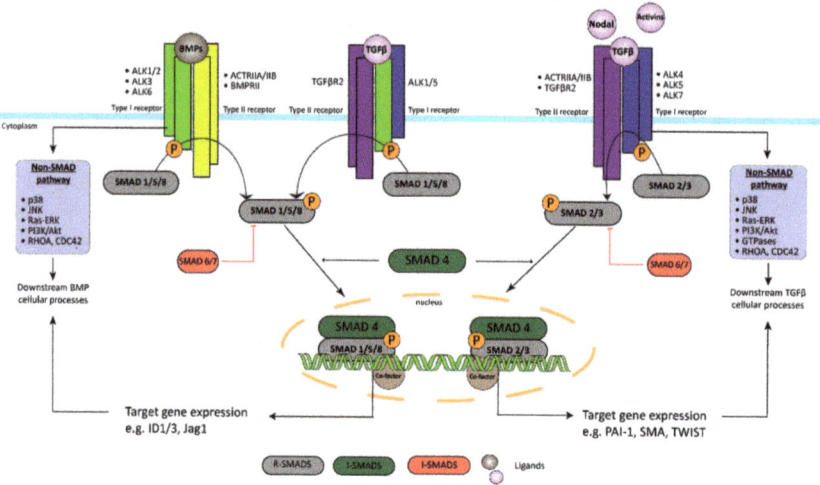

Figure 2. Transforming growth factor-β (TGF-β) superfamily signaling in PAH. The bone morphogenetic protein (BMP)/TGF-β signaling pathway is an important factor in the existence of EC dysfunction in PAH. Decreased expression of BMPR2 but more importantly various mutations in the BMPR2/BMP-ligand axis are associated with specific changes in EC behavior such as increased proliferation and migration, but also structural changes that cause the loss of the protective EC barrier. In addition, the TGF-β superfamily signaling also plays an important role the initiation of EndoMT by triggering the overexpression of genes, like TWIST1, αSMA, and phospho-vimentin. Receptor-regulated Smads (R-Smads); Common mediator Smad (Co-Smad); Inhibitory Smads (I-Smads).

Further evidence in the association between BMPR2 and EC dysfunction comes from studies showing that BMPR2 deficiency in iPAH PAECs is associated with the loss of DNA damage control via reduced DNA repair related genes, such as BRCA1 [43]. In addition, transcriptome analysis of PAECs from iPAH patients revealed a correlation between reduced BMPR2 levels and the downregulation of β-catenin, resulting in reduced Collagen-4 (COL4) and ephrinA1 (EFNA1) expression [44]. COL4 and EFNA1 both perform intertwining roles in endothelium structure. Moreover, siRNA mediated silencing of *BMPR2* in PAECs resulted in increased PAEC proliferation, migration, and the disruption of cytoskeletal architecture. One of the changes observed was an increase in Ras/Raf/ERK signaling, and Ras inhibitors, like nintedanib [45], reversed the enhanced proliferation and hypermotility of BMPR2 silencing in PAECs [46].

2.2. Inflammation

Mutations in *BMPR2* are known to predispose patients to developing PAH, but low penetrance and the time of disease onset suggest that a second hit required developing PAH. Pulmonary inflammation is such a plausible second hit that puts patients with BMPR2 mutations at risk of developing PAH. Exposure of *Bmpr2* mutant rats to 5-lipoxygenase, inducer of lung inflammation, induced severe PAH pathology with an endothelial transformation that required TGF-β signaling [47]. However, the administration of only IL-6 to rats and overexpression of IL-6 in transgenic mice also led to the occlusion of pulmonary arteries and RV hypertrophy without a silent mutation of BMPR2 [48,49]. Accordingly, it has also been found that pro-inflammatory cytokine TNFα in vitro downregulates the expression of BMPR2 via NOTCH signaling in ECs [50]. Altogether, this could suggest that sustained inflammation is an important trigger in PH development, potentially through the induction of EC dysfunction. Pulmonary arteries of PAH patients showed the infiltration of macrophages, dendritic cells, and lymphocytes into the plexiform lesions and an increased migration of monocytes [10,51]. Increased levels of pro-inflammatory cytokines and chemokines, such as IL-1β, TNFα, and IL-6, which are known activators of vascular endothelium, were found (Figure 3) [52–54]. Hence, has been found that IL-1β stimulates endothelial ET-1 production [55].

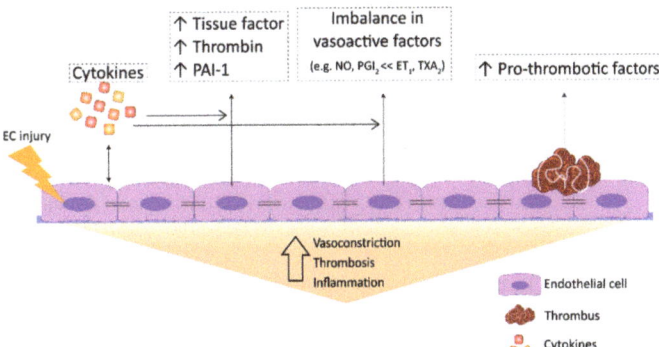

Figure 3. Endothelial dysfunction in PAH. PAH is characterized by endothelial dysfunction that causes an imbalance in the production of several endothelial-specific factors. The endothelium presents a pro-inflammatory phenotype with an increased expression of cytokines, a pro-thrombotic surface due to changes in the expression of clotting factors (e.g., TF) and increased expression of pro-thrombotic factors, and an imbalanced production of vasoactive factors that promote vasoconstriction. Upon endothelial cell injury, pulmonary artery- endothelial cells (PAECs) become dysfunctional and alter their secretion of cytokines and other factors that regulate coagulation, thrombosis, and vascular tone. A failure of PAECs in maintaining vessel homeostasis promotes vasoconstriction, thrombosis and inflammation that initiate PAH disease progression.

2.3. Thrombosis in PAH

The presence of thrombotic lesions in the pulmonary vasculature is a common pathological finding in PAH [56]. However, the role of thrombosis in PAH remains controversial. Few studies demonstrated that coagulation factors, such as proteases, tissue factor (TF), factor Xa, and thrombin, activate the coagulation cascade, which leads to the formation of fibrin clots that obstruct/narrow the lumen, could promote EC dysfunction, and can eventually contribute to vascular remodeling in PAH (Figure 3) [57]. In contrast, some studies support the hypothesis that thrombosis is an epiphenomenon of vascular remodeling in PAH [58]. Thus, it is still unknown whether thrombosis contributes to the pathogenesis of PAH or acts as a bystander.

Although altered platelet activation has been reported in PAH patients, their exact role in PAH remains controversial. Only platelets and ECs express and release von Willebrand Factor (vWF) upon activation, which facilitates the interaction between each other. Circulating vWF levels are significantly increased in PAH patients, which suggests the potential involvement of platelets in EC dysfunction in PAH [59]. CD40L, a proinflammatory mediator, is expressed on the surface of activated platelets. Upon activation, CD40L is cleaved into its soluble form (sCD40L), which is known to be greatly increased in PAH patients [60]. sCD40L interacts with its receptor CD40, expressed on ECs, and may lead to EC dysfunction and eventually contributes to vascular remodeling in PAH. Altogether implicating the role of platelets in EC dysfunction and thrombosis in PAH. Although there is considerable evidence to suggest that platelets contribute to the EC dysfunction and the pathogenesis of PAH, the molecular mechanisms have yet to be delineated.

2.4. Coagulation in PAH

Under physiological conditions, transmembrane glycoprotein TF is expressed at low levels in the pulmonary vessel wall, but its expression is significantly increased in pulmonary vascular lesions of PAH patients [61–63]. Increased TF/thrombin signaling contributes to vascular remodeling and the formation of plexiform lesions in PAH by inducing the proliferation and migration of SMCs and mediating the migration and angiogenesis of ECs. Furthermore, ECs from PAH patients release enhanced TF-expressing microparticles, further implicating TF as a crucial mediator in the vascular remodeling in PAH [64]. PAH patients exhibit a hypercoagulable state, consistent with the increased TF expression [65]. PAH patients have higher levels of fibrinopeptide-A (FPA), plasminogen activator inhibitor-1 (PAI), and thrombin, and lower levels of thrombomodulin [66]. Although all of the factors involved in coagulation cascade are increased in PAH, the relative contribution of EC dysfunction to their increase remains to be elucidated.

2.5. EC Metabolism

ECs in PAH have a metabolic phenotype that is similar to that seen in cancer. ECs in PAH have a metabolic phenotype similar to that seen in cancer, namely a metabolic reprogramming towards increased glycolytic metabolism which renders ECs with a pro-survival advantage and higher proliferation [67,68]. This metabolic shift is thought to be driven through the upregulation of glycolytic enzymes PFKFB3, hexokinase, and lactate dehydrogenase, and mitochondrial enzyme pyruvate dehydrogenase kinase (PDK) [67,69]. Therefore, the concept of targeting EC metabolism to treat PAH is emerging and raised great scientific interest. Based on a recent study in rodents, one such potential target could be PFKFB3. The blockage of endothelial PFKFB3 has shown to attenuate PH development in rats that were treated with SuHx [70]. Moreover, dichloroacetate (DCA), which is an inhibitor of the mitochondrial enzyme PDK, has been found to improve patient hemodynamics and functional capacity in genetically susceptible PAH patients [71]. Despite promising results and being based on metabolomic heterogeneity of PAH [72], comprehensive metabolic characterization of ECs still needs further investigation to further expand our understanding of the complex pathobiology of PAH.

2.6. Shear Stress

Abundant evidence demonstrates that shear stress is altered in the pulmonary vasculature in PAH. PAH is strongly associated with increased main pulmonary artery diameter and reduced main pulmonary artery flow rate, which suggests that the shear stress is lower globally and, thus, leads to a reduction in NO release from the endothelium [73]. Several studies found 2–3-fold lower shear stress in PAH patients when compared to control subjects, and such a reduction has a correlation with a reduction in NO bioavailability in PAH patients. This implies that the pruning of the distal pulmonary vasculature in PAH may be a way for the lung to preserve microvascular perfusion by increasing microvascular resistance and elevating shear stress [74]. However, like congenital heart disease, the microvasculature in PAH may also experience high shear stress or high oscillations in flow, sue to increased stiffness in the pulmonary arteries [75]. Despite the lower shear stress in the main pulmonary arteries, the pulsatility may elevate in the microvasculature and the stiffness of the arteries increases, which explains the coupling of microvascular dysfunction with macrovascular dysfunction.

Interestingly, decreasing the pulmonary flow via banding prevented the development of plexiform lesions in a rat model of PAH, which suggests a causative role for increased force transmission in the initiation and development of PAH [76,77]. However, pulmonary artery banding in rats induced right ventricle dysfunction [78]. Furthermore, PAH patients treated with vasodilators have shown increased survival, suggesting that dampening microvascular shear stress or pulsatile flow may improve PAH.

Using microvascular ECs derived from PAH patients, Szulcek et al. demonstrated that PAH ECs show a delayed shear adaptation and, thus, promoted shear induced endothelial dysfunction and abnormal vascular remodeling [79]. In another study, pulmonary artery ECs were subjected to high pulsatile flow, but the same mean shear stress displayed exacerbated inflammation and increased cell elongation, which could all be normalized by stabilization of microtubules [80]. Future research should focus on decoupling the microvascular shear stress, pulsatile flow, oscillation index, and right ventricular function using in vitro and in vivo models to better understand the contribution of shear stress to the EC dysfunction and development of PAH.

3. Features of EC Dysfunction

PAH is characterized by a dysfunctional endothelium, of which the balance between vasodilation and vasoconstriction, but also the growth factor production and cell survival are altered (Figure 3). In addition, ECs undergo endothelial to mesenchymal transition (EndoMT), which, all together, causes perturbations in pulmonary vascular homeostasis that promote vascular remodeling (Figure 4).

3.1. Perturbations in Vasoactivity

Reduced vasorelaxation in PAH mainly contributes to the altered expression of the vasodilators NO and prostacyclin. NO is a fast-reacting endogenous free radical that is produced by endothelial NO Synthase (eNOS). NO is essential for vasorelaxation via PASMCs, but it also has antithrombotic effects and controls EC differentiation and growth [81–83]. NO has long been implicated in the pathogenesis of PAH, and the lungs of PAH patients have reduced NO expression [84] (Figure 3). Whole exome sequencing has identified that mutations in Caveolin-1 are associated with PAH. Caveolin-1 is highly expressed in ECs and, interestingly, the C-terminus of caveolin-1 directly interacts with eNOS, which may result in the disruption in NO levels, ultimately triggering PAH [85]. However, other studies reported contradictory results and some PH patients even show an increase in eNOS expression [84]. Furthermore, eNOS$^{-/-}$ mice show reduced vascular remodeling after chronic hypoxia that is caused by reduced vascular proliferation [86], pointing out the complexity of its role in PAH. Prostacyclin, also produced by EC with additional antithrombotic and antiproliferative properties [8,87–89], is synthesized from arachidonic acid, by prostacyclin synthase, and cyclo-oxygenase (COX) [90]. Decreased prostacyclin levels are measured in various patients with different forms of PAH, like iPAH and HIV-associated PAH [8,91], explaining, in part, the increase in pulmonary vasoconstriction, SMC proliferation, and coagulation occurring in these patients. Interestingly, in experimental PH models, mice overexpressing prostacyclin synthase are protected from developing chronic hypoxia-induced PAH [92].

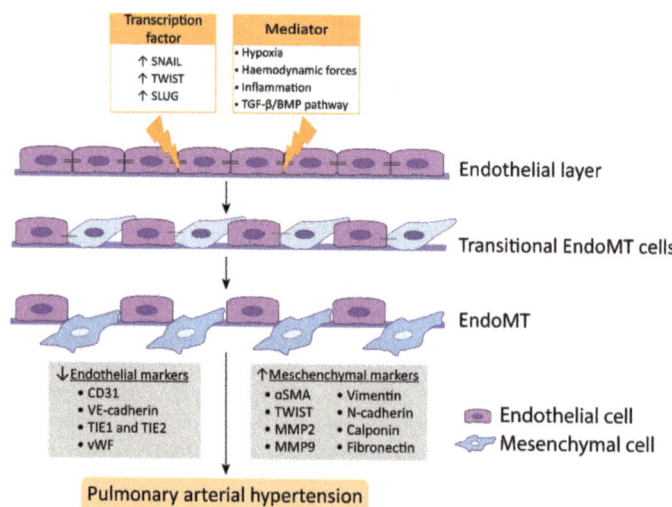

Figure 4. Endothelial to mesenchymal transition (EndoMT) in PAH. EndoMT in PAH is thought to be an important process contributing to vascular remodeling. Activation by transcriptional factors, hypoxia, haemodynamic forces, inflammation, and TGF-β/BMP pathway signaling pulmonary endothelial cells (PAECs) undergo cellular transition to a mesenchymal phenotype, in which PAECs lose endothelial markers and gain mesenchymal markers, such as αSMA and TWIST. These mesenchymal-like cells present an invasive character and hence contribute to vascular remodeling in PAH.

ET-1, on the other hand, is a potent vasoconstrictor, which is mainly synthesized in EC and the lungs show the highest level of ET-1 in the entire body [93]. ET-1 exhibits its effects by binding to the ET_A and ET_B receptors, which activate signalling pathways in vSMCs regulating proliferation, vasorelaxation and vasoconstriction [89,93]. ET_A is predominantly expressed on vSMCs and is involved in vasoconstriction and proliferation of these cells, while ET_B is expressed on vSMCs and PAECs, and is involved in stimulating the release of vasodilators, like NO and prostacyclin, and the inhibition of apoptosis [55,89,93–95]. The expression of ET-1 and its receptors is increased in lungs of PAH patients and experimental PH models (Figure 3) [96–99]. Furthermore, a correlation exists between the expression of ET-1 and an increase in pulmonary resistance in PAH [98]. The increased synthesis of endothelial ET-1, accompanied with an increase in expression of ET_A on PASMCs, likely contributes to the increased vasoconstriction and vascular remodelling observed in PAH [88,99,100]. Another vasoconstrictor, thromboxane A_2, which is produced by ECs and platelets, but is also an inducer of platelet aggregation and a vSMCs mitogen, is increased in PAH [8,14], creating an imbalance that might contribute to excessive platelet aggregation and vascular remodeling observed in PAH [14] (Figure 3).

At last, the expression of the growth factor vascular endothelial growth factor (VEGF) and its receptor VEGF receptor 2 (VEGFR2) are found to be increased in ECs from plexiform lesions from iPAH patients. Additionally, the plasma levels of VEGF are found to be elevated in PH patients [101,102]. The relation between PAH and increased VEGF expression is still poorly understood. It is suggested that VEGF levels in PAECs are elevated in early stages of PAH as a protective response, while, during disease progression, VEGF keeps promoting the growth of PAECs, causing the formation of plexiform lesions [8].

3.2. Endothelial to Mesenchymal Transition

EndoMT is a phenomenon where ECs acquire a mesenchymal-like phenotype that is accompanied with a loss of endothelial markers and increase of mesenchymal markers. In addition, ECs lose cell-cell contact, change their morphology, and adopt a highly migratory and invasive phenotype, thereby losing features of a healthy endothelium (Figure 4) [103,104]. In the lungs of human PAH patients and monocrotaline (MCT) and Sugen/hypoxia (SuHx) experimental PH rat models, EndoMT was observed, whereby cells express high levels of α-SMA and activated phospho-vimentin and VE-cadherin, indicating their endothelial origin [105–107]. Moreover, TWIST1, which is a key transcription factor in inducing EndoMT, is highly expressed in human PAH lungs as compared to healthy lungs [106] (Figure 4).

TGFβ treatment of PAECs induces the expression of the EndoMT transcription factors TWIST1 and SNAIL1 [103,108] and the mesenchymal markers α-SMA and phospho-vimentin [109] (Figure 4). TWIST1 increases the expression of TGFβ, leading to enhanced TGFβ signaling [110]. In addition, reduced BMPR2 signaling promotes EndoMT via the upregulation of the High Mobility Group AT-hook 1 and its target gene SLUG, independent of TGFβ signaling [111]. More interestingly, BMP-7, a protein previously described as having anti-inflammatory and anti-tumor effects in several diseases, was attenuated by hypoxia-induced EndoMT in PAECs both in vivo and in vitro by inhibiting the m-TORC1 signaling pathway [112]. BMPR2 loss favors EndoMT, allowing for cells of myo-fibroblastic character to create a vicious feed-forward process, leading to hyperactivated TGFβ signaling [113]. In summary, alterations in TGFβ/BMP signaling are linked to the process of EndoMT that was observed in PAH [114].

Hypoxia is also an inducer of EndoMT through hypoxia-inducible transcription factor-1α (HIF-1α) and HIF-2α, and both transcription factors are increased in PAH [115,116] (Figure 4). PAH ECs display an increased expression of HIF-2α, leading to SNAIL upregulation [107]. In addition, HIF-1α knockdown alone effectively blocks hypoxia-induced EndoMT, but also the knockdown of its downstream target gene TWIST1 showed the effective blockage of hypoxia-induced EndoMT in microvascular ECs (MVECs); however, it was less pronounced [117]. Nonetheless, it is important to realize that microvascular endothelium may differ from arterial endothelial function. Finally, in addition to transcription factors, microRNAs, such as miR-181b, have been shown to be implicated in EndoMT in PAH. The overexpression of miR-181b in rat pulmonary arterial ECs (rPAECs) attenuated inflammation-induced EndoMT by inhibiting the expression of TGF-βR1 and circulating proteoglycan endocan [118].

3.3. Apoptosis

EC apoptosis may also play a role in PH development via vascular dropout and selection pressure on ECs, contributing to the apoptosis-resistant phenotype of ECs in vascular lesions [119]. Several attempts were made in order to elucidate the molecular pathways that are involved in the regulation of PAEC apoptosis. The hypothesis is that disturbed responses to VEGF signaling, in combination with hypoxia, cause an initial increase in apoptosis in PAECs, leading to the emergence of aggressive apoptosis resistant and hyperproliferative ECs that cause the formation of intimal lesions [120–122]. A possible explanation for the initial increase in apoptosis of PAECs is that the loss of BMPR2 signaling promotes mitochondrial dysfunction and subsequent PAEC apoptosis [123]. White et al., interestingly, proposes a model in which the pro-apoptotic factor programmed cell death-4 (PDCD4) activates the cleavage of caspase-3, inducing PAEC apoptosis. Interestingly, they show that reducing PDCD4 levels in vivo by overexpressing miRNA-21 prevents PH development in SuHx rats [124]. Besides an initial increase in apoptosis, PAH is also characterized by PAECs that are hyperproliferative and apoptosis resistant [122]. PAECs from iPAH patients showed an increased expression of pro-survival factors IL-15,

BCL-2, and Mcl-1, together with persistent activation of the pro-survival STAT3 signaling pathway [122]. Furthermore, Notch1 was elevated in lungs from iPAH patients and from SuHx rats. Notch1 contributes to PAH pathogenesis by increasing EC proliferation and inhibiting apoptosis via p21 downregulation and regulating BCL-2 and survivin expression. Furthermore, HIF1α expression promotes Notch signaling human PAECs [125]. In contrast, Miyagawa et al., demonstrated that contact-mediated communication between SMC and EC activates EC derived Notch1 and alters the cells epigenome in order to regulate Notch1-dependent genes that maintain endothelial integrity and prevent pulmonary vascular remodeling in a murine model of hypoxia-induced pulmonary hypertension [126]. Therefore, the role of Notch1 is complex and controversial in PAH and warrants more research to delineate the molecular mechanisms.

4. Epigenetics

In recent years, epigenetics has become a growing field of interest in PAH research. Currently, the main focus of study for targeting PAH is the following three mechanisms of epigenetic regulation: DNA methylation, histone modifications, and RNA interference (Figure 5) [17].

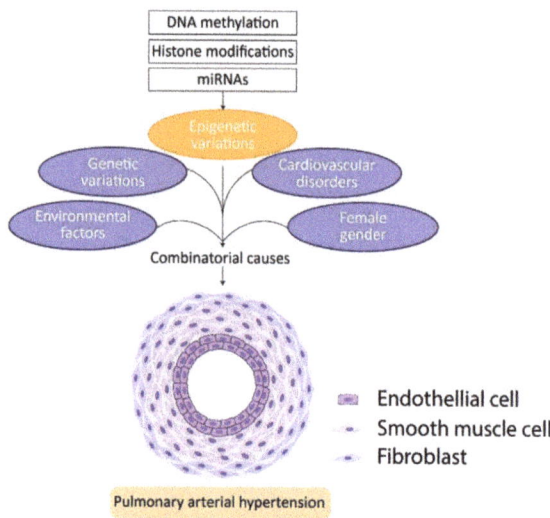

Figure 5. Epigenetics in PAH. In addition to genetic variations and other risk factors, such as gender, comorbidities, and environmental factors, epigenetic variations in PAH gain interest. Differences in DNA methylation profiles, increased histone acetylation and dysregulated miRNA expression in PAH patients point out a growing field in PAH research that provides better understanding of disease pathology.

DNA methylation profiling of PAECs from iPAH and hPAH patients revealed differences in the expression of several genes that are involved in inflammatory processes, remodeling, and lipid metabolism when compared to the controls [127]. Among those genes, ABCA1 was found to be most differently methylated/downregulated in the discrimination between PAH and controls. ABCA1 belongs to the family of ATP binding cassette (ABC) transporters that are important for pulmonary homeostasis [127]. Furthermore, ABCA1 is linked to PAH pathophysiology in a MCT animal model of PAH, where the activation of ABCA1 improved RV hypertrophy and pulmonary haemodynamics [17,127].

Increased histone acetylation through histone-deacetylases (HDAC) is associated with vascular remodeling found in PAH [128,129]. In humans, HDAC enzymes are divided into four classes: class-1 HDACs (HDAC-1, -2, -3, and -8), class-2a HDACs (HDAC-4, -5, -7, and -9), class-2b HDACs (HDAC-6 and -10), class-3 HDACs (Sir2-like proteins), and class-4 HDACs (HDAC-11) [130]. HDAC-1 and -5 show increased expression in both lungs of iPAH patients and chronic hypoxic rats whereas HDAC-4 was only increased in human iPAH lungs [129]. More recently, HDAC-6 has been linked to PAH pathogenesis, possibly through the upregulation of HSP90 [131]. HDAC-6 was overexpressed in PAECs and PASMCs of PAH patients and PH experimental models [132]. In the SuHx and MCT rat model pharmacological HDAC-6 inhibition improved PH [132]. Several other studies showed that class-1 HDAC inhibitors attenuate PAH by suppressing arterial remodeling in a chronic hypoxia model and by reducing inflammation in PH-fibroblasts [129,133,134]. In PAECs, class-2a HDAC inhibitors restore the levels of myocyte-enhancer-factor-2 and attenuate PAH in both the MCT and SuHx PAH rat models [135].

The epigenetic regulator bromodomain-containing-protein-4 (BRD4) is linked to the pathogenesis of PAH [136]. BRD4 is a member of the Bromodomain and Extra-Terminal (BET) motif family, which binds histones to influence gene expression [137]. BRD4 is overexpressed in the lungs of PAH patients in a miR-204 dependent manner. It inhibits apoptosis by sending cell survival signals [136,138], and stimulates the proliferation of PAEC and PASMC proliferation at these sites [17,138]. The selective inhibition of BRD4 with RVX-208 restored EC function, reversed PAH in the MCT and SuHx rat models, and supported the RV function in pulmonary artery banding model of PAH [136].

5. EC Dysfunction in Other PH Groups

Patients with PAH, which are classified as group 1, are just a proportion of the five broad groups of patients suffering from PH. The remaining groups, group 2 (PH due to left-sided heart disease), group 3 (PH due to lung disease), group 4 (PH due to chronic thromboembolic disease), and group 5 (PH due to unclear and/or multifactorial mechanisms), also present signs of EC dysfunction.

5.1. Group 2 PH

Group 2 PH is due to a complication of left heart disease and it is most common in patients with heart failure (HF). Therefore, research in group 2 PH mostly focuses on left ventricular dysfunction and not so much the lung vasculature. However, features of EC dysfunction are also observed in PH-LHD [139]. An experimental model of chronic HF showed reduced NO activity and responsiveness to NO in pulmonary arteries [140]. Moreover, ET-1 is elevated in certain PH-LHD phenotypes and ET-1 activity is increased in plasma of patients with chronic HF. Blocking the ET_A receptor caused pulmonary vasodilation in these patients [141,142]. Furthermore, polymorphisms that are found in eNOS also contribute to PH development in patients with LHD [143]. Despite the presence of similar perturbations in vasoactivity between PAH and PH-LHD, treating PH-LHD patients with drugs used to treat PAH patients was not beneficial and even harmful [139,144].

5.2. Group 3 PH

Chronic obstructive lung disease (COPD) associated PH is the best described form of PH in group 3. The main trigger of COPD is considered to be cigarette smoke, which causes chronic inflammation in the lung that subsequently triggers EC dysfunction and leads to PH [145]. Cigarette smoke decreases eNOS and prostacyclin expression in PAECs [146,147]. COPD patients can show the overexpression of VEGF and ET-1 in pulmonary arteries [148,149]. A role for HIF1α and EndoMT has also been suggested in COPD [150,151]. Although there are similarities in EC dysfunction, the drugs used to treat PAH are currently not recommended for group 3 PH, due to a lack of evidence how these drugs may influence PH progression in combination with the underlying lung diseases [152].

5.3. Group 4 PH

CTEPH develops as a result of a pulmonary embolism (PE) that does not resolve [153]. These organized pulmonary thrombi in the lungs are associated with distal vascular remodeling of non-occluded vessels similar to the remodeling observed in PAH lungs [153]. Whether patients develop CTEPH due to primary EC dysfunction or as a consequence of PE remains to be resolved. Nevertheless, evidence supports that features of EC dysfunction, which are similar to those observed in PAH, are present in these patients and could play a causal role in CTEPH development. Activated platelets with a hyper-responsiveness to thrombin are likely to contribute to the CTEPH pathogenesis and progression via enhancing inflammatory responses of pulmonary ECs [154]. EC dysfunction-associated vascular remodeling has been suggested as a common mechanism between CTEPH and PAH [153,155]. Primary cell cultures that were isolated from endarterectomized tissue co-expressed both EC and SMC markers, suggesting a role for EndoMT in intimal remodeling/lesion development in CTEPH [156]. The existence of endothelial dysfunction in CTEPH pathogenesis is further supported by the fact that conditioned medium from CTEPH-derived PAECs, containing high levels of growth factors and inflammatory cytokines, increased PASMC proliferation and monocyte migration [157]. In addition, PAECs from CTEPH patients show an increased proliferation, altered angiogenic potential and metabolism, and apoptosis resistance [158–162]. Increased levels of soluble intracellular adhesion molecule-1 (ICAM1) in PAECs from CTEPH patients and in endarterectomy may contribute to EC proliferation and apoptosis resistance through its effect on cell survival pathways [161]. Additionally, FoxO1, in a PI3K/Akt dependent manner, is a possible contributor to the loss of balance between cell survival and death and it was downregulated after PE in a rat model of CTEPH [163]. A recent study reported that decreased levels of ADAMTS13 and increased levels of vWF levels were observed in plasma of CTEPH patients, suggesting the role of the ADAMTS13–vWF axis in CTEPH pathobiology. However, it remains unknown as to whether this axis plays a role in EC dysfunction in CTEPH [164]. Finally, PAECs isolated from CTEPH patients showed a significant rise in basal calcium levels, which is an important regulatory molecule for EC function [165]. This imbalance in calcium homeostasis is caused by angiostatic factors, such as PF4, IP-10, and collagen type 1, which are formed in the microenvironment that is created by the unresolved clot and eventually leads to EC dysfunction [165]. So far, a soluble guanylate cyclase stimulator (Riociguat) is the only PAH based therapy that has been approved in patients with CTEPH that are not eligible for surgery [166].

6. Current and Future Perspectives

Although much progress has been made to understand EC dysfunction in PAH, to date there is still no definitive cure and patients only have a median survival rate of 2.8 years [167]. Current therapies for PAH, which consist of calcium channel blockers, ET-1 receptor antagonists, phosphodiesterase type 5 inhibitors, prostacyclin-derivatives, and, more recently, also Riociguat, focus on restoring the imbalanced endothelial vasoactive factor production to promote SMC relaxation, but with limited or no effect on other features of EC dysfunction and subsequent progressive pulmonary vascular remodeling [168–170]. Therefore, research on EC dysfunction and its stimuli to target structural changes that narrow lumen size in PAH is vital to find a cure.

A first step towards reversing vascular remodeling in PAH is the use of apoptosis-inducing drugs, such as anthracyclines and proteasome inhibitors. They are already used in combination with cardio-protectants, such as p53 inhibitors, to reduce pulmonary pressure and restore blood flow in the experimental models of PAH [171,172]. The combinatorial use is essential in circumventing the lack of cell-type/organ specificity of cell-killing drugs. Cancer patients, but also experimental PAH animals treated with only cell-killing drugs, show signs of cardiotoxicity that should be prevented in PAH patients that already suffer from reduced right heart function [171,173–175].

Another way to target progressive pulmonary vascular remodeling focuses on restoring signaling pathways and EC function, e.g., using selective TGF-β ligand traps [176] or TGF-β synthesis inhibitors, like kallistatin, which are known to improve hemodynamics, remodeling, and survival in experimental PH models, and to inhibit EndoMT in HUVECs, stimulate eNOS expression, and prevent TGF-β induced miRNA-21 synthesis, respectively [176,177]. However, blocking inflammation to restore normal EC function in PAH was not successful. One explanation might be the complexity of the immune system and, by inhibiting the bad side, one also suppresses beneficial inflammatory pathways [178,179].

Modulating BMPR2 has also been proposed as a therapeutic approach to reverse endothelial dysfunction in PAH. A recent study comparing human induced pluripotent stem cell-derived ECs (iPSC-ECs) from unaffected BMPR2-mutation carriers with iPSC-ECs from BMPR2-mutation carriers that present PAH identified several BMPR2 modifiers and differentially expressed genes in unaffected iPSC-ECs. These BMPR2 modifiers exert a protective response against PAH by improving downstream signaling, which compensates against BMPR2 mutation-induced EC dysfunction and offers insights towards new strategies for rescuing BMPR2 signaling [180]. A potential therapy for stimulating BMPR2 signaling is through pharmaceuticals [181]. Direct enhancement of endothelial BMPR2 signaling using recombinant BMP9 protein prevents and reverses the established experimental PAH [42]. However, in contrast to Long et al., Tu et al. (2019) showed that the deletion or inhibition of BMP9 protects against experimental PH via its effect on endothelial production of ET-1, apelin, and adrenomedullin [182]. In line with this, we have recently shown that BMP9-induced aberrant EndoMT in PAH pulmonary ECs is dependent on exacerbated pro-inflammatory signaling mediated through IL6 [54]. These studies show the BMP receptor family complexity as therapeutics in PAH. More recently, ACTRIIA-Fc, an activin and growth and differentiation factor (GDF) ligand trap, prevented and reversed existing PH in experimental PAH models. ACTRIIA-Fc inhibited SMAD2/3 activation and restored a favorable balance of BMP signaling versus TGF-β/activin/GDF signaling. ACTRIIA-Fc is currently tested in a phase-2 clinical trial for efficacy and safety in PAH patients (NCT03496207) [183]. However, a recent study shows that TGF-β/SMAD signaling is regulated differently in PH animal models compared to PAH patients [184]. Therefore, more research should be performed on this complex TGF-β/activin/GDF signaling. Spiekerkoetter et al. uncovered a molecular mechanism, where FK506 (tacrolimus) restores defective BMPR2 signaling in PAECs from iPAH patients and reverses severe PAH in several rat models [181]. Based on improvements in clinical parameters and the stabilization of cardiac function of end-stage PAH patients in a phase-2a clinical trial, a low dose of FK506 was proposed as potentially beneficial in the treatment of end-stage PAH [185]. These findings open-up an area, in which correcting BMPR2 mutations in combination with other therapies might be more successful in curing PAH. A proposed hypothesis to cure PAH describes collecting iPSCs from PAH patients, restoring the BMPR2 mutation with CRISPR/Cas9, and reinjecting those iPSCs in the patient to normalize EC function and signaling along with administration of drugs that could restore the protective gene expression profile of unaffected BMPR2 mutation carriers [186]. 6-Mercaptopurine (MP), which is a well-established immunosuppressive drug, inhibits EC dysfunction and reverses development of PH in the SuHx rat model by restoring BMP signaling through the upregulation of nuclear receptor Nur77 [187]. A recent proof-of-concept study with MP in a small group of PAH patients showed a significant reduction pulmonary vascular resistance, accompanied by increased BMPR2 mRNA expression in the patients' peripheral blood mononuclear cells. However, unexpected severe side-effects require further dose optimization and/or the use of other thiopurine analogues [188]. Next to a role for BMPR2, the loss of KCNK3 function/expression is a hallmark of PAH. A recent study shows that the loss of KCNK3 is inducing EC dysfunction by promoting the metabolic shift and apoptosis resistance in PAECs. Therefore, targeting KCNK3 might restore EC function; however, the mechanisms remain unknown [189]. The transplantation of mesenchymal cells in rats from the SuHx model improved haemodynamic parameters, but, more interestingly, reduced EndoMT (partially) through the modulation of HIF2α expression [190]. Furthermore, mesenchymal stem cells are also suggested to reduce inflammation through the secretion of paracrine factors and attenuate vascular remodeling by lowering collagen deposition [190–192]. However, the underlying mechanisms for this observation remain unclear [190]. Several recent studies demonstrate the role of endothelial HIF-2a in the pathogenesis of PAH, and therapeutic targeting of HIF-2a with small molecule inhibitors, such as PT2567, have showed a beneficial effect in PAH in vivo [193,194]. A recent study demonstrates that human pulmonary ECs of patients with PAH are more vulnerable to cellular senescence, a process that is associated with EC dysfunction. Interestingly, targeting senescence while using the senolytic drug ABT 263 reversed established PH in a MCT+shunt induced PAH rat model by specifically inhibiting senescent vascular cells [77]. However, more research should be performed on the safety and efficacy of senolytics in patients.

Finally, epigenetic modulation has received growing interest as potential therapeutic intervention. Especially, specific HDAC inhibition shows great promise in reversing pulmonary remodeling and pressure [129]. A problem with broad-spectrum HDAC drugs is that they show severe side effects on the right ventricle, which can have fatal consequences in PAH patients with RV failure [133,195,196]. Therefore, searches for more selective HDAC inhibitors that do not show cardiotoxicity are still being done. One example is MGCD0103, which is a HDAC inhibitor that selectively inhibits class-1 HDACs that has been tested in a chronic hypoxia rat model. This inhibitor showed improved hemodynamics, reduced wall thickening, while RV function was maintained [133]. Additionally, BET inhibitors, such as RVX208, seem to be promising in the treatment of PAH through its beneficial effect on reducing the apoptosis-resistant and pro-inflammatory phenotype in PASMCs and MVECs isolated from PAH patients, but also on vascular remodeling and the RV in several experimental models of PH [136]. Finally, miRNA-21 has been associated with multiple pathogenic features, such as TGF-β signalling, EndoMT, and apoptosis, which are central to PAH. Therefore, therapeutic modulation of miRNA-21 may be an important issue for future research to restore pathogenic signaling.

7. Conclusions

To date, we still do not fully understand what triggers the onset and progression of PAH. We do know that BMPR2 mutations, epigenetics, physiological conditions, and inflammation are important triggers. EC dysfunction plays a central role in all of this, through EC proliferation, EndoMT, and a misbalanced production of vasoactive factors, resulting in the disorganized growth of PASMCs. Although several preclinical studies demonstrate that EC dysfunction is a cause rather than a consequence of PAH, more research should be performed in PAH patients in order to better understand this. For example, non-carriers of BMPR2 mutation along with carriers of BMPR2 mutation from the same family should be followed up for several years to understand whether EC dysfunction or other triggers are a cause or consequence. Despite advancements that have been made in treating this disease, very few therapies have little or no direct impact on EC dysfunction. Therefore, successful treatments should focus on multiple aspects of EC dysfunction and not solely on its effect on SMCs and fibroblasts in PAH. A better understanding of the molecular mechanisms that are involved in EC dysfunction in PAH is of utmost importance for developing successful therapies to save the lung as well as the heart, and perhaps cure PAH in the future.

Author Contributions: Drafted or substantively revised the manuscript: K.K., V.F.E.D.S., O.T.-C., J.W.J., P.H.A.Q., M.-J.G.; Has approved the final version of the manuscript: K.K., V.F.E.D.S., O.T.-C., J.W.J., P.H.A.Q., M.-J.G. All authors have read and agreed to the published version of the manuscript.

Funding: This research was funded by the European Commission Horizon 2020 research and innovation program under the MOGLYNET H2020-MSCA-ITN-EJD grant (agreement No 675527), a Miguel Servet grant from the Instituto de Salud Carlos III (CP17/00114), research grants PI15/00582 and PI18/00960 from the Institute of Health Carlos III Spain, Catalan Society of Pneumology (SOCAP 2019), a DCVA consortium grant PHAEDRA-IMPACT. We also acknowledge support for KK by the Dutch Lung Foundation (Longfonds) grant number-5.2.17.198J0 and by the Leiden University Foundation grant (W18378-2-32).

Institutional Review Board Statement: Not applicable.

Informed Consent Statement: Not applicable.

Data Availability Statement: Not applicable.

Conflicts of Interest: The authors declare no conflict of interest.

References

1. Dumitrescu, D.; Hager, A.; Held, M.; Sinning, C.; Greiner, S.; Kruck, I.; Meyer, J.; Pabst, S.; Köhler, T.; Kovacs, G.; et al. Definition, clinical classification and initial diagnosis of pulmonary hypertension: Updated recommendations from the Cologne Consensus Conference 2018. *Int. J. Cardiol.* **2018**, *272*, 11–19.
2. Simonneau, G.; Hoeper, M.M. The revised definition of pulmonary hypertension: Exploring the impact on patient management. *Eur. Heart J. Suppl. J. Eur. Soc. Cardiol.* **2019**, *21*, K4–K8. [CrossRef]
3. Simonneau, G.; Montani, D.; Celermajer, D.S.; Denton, C.P.; Gatzoulis, M.A.; Krowka, M.; Williams, P.G.; Souza, R. Haemodynamic definitions and updated clinical classification of pulmonary hypertension. *Eur. Respir. J.* **2019**, *53*, 1801913. [CrossRef] [PubMed]
4. Vonk Noordegraaf, A.; Groeneveldt, J.A.; Bogaard, H.J. Pulmonary hypertension. *Eur. Respir. Rev.* **2016**, *25*, 4. [CrossRef] [PubMed]
5. Hoeper, M.M.; Humbert, M.; Souza, R.; Idrees, M.; Kawut, S.M.; Sliwa-Hahnle, K.; Jing, Z.C.; Gibbs, J.S.R. A global view of pulmonary hypertension. *Lancet Respir. Med.* **2016**, *4*, 306–322. [CrossRef]
6. Humbert, M.; Morrell, N.W.; Archer, S.L.; Stenmark, K.R.; MacLean, M.R.; Lang, I.M.; Christman, B.W.; Weir, E.K.; Eickelberg, O.; Voelkel, N.F.; et al. Cellular and molecular pathobiology of pulmonary arterial hypertension. *J. Am. Coll. Cardiol.* **2004**, *43*, 13S–24S. [CrossRef] [PubMed]
7. Tuder, R.M.; Groves, B.; Badesch, D.B.; Voelkel, N.F. Exuberant endothelial cell growth and elements of inflammation are present in plexiform lesions of pulmonary hypertension. *Am. J. Pathol.* **1994**, *144*, 275–285.
8. Budhiraja, R.; Tuder, R.M.; Hassoun, P.M. Endothelial Dysfunction in Pulmonary Hypertension. *Circulation* **2004**, *109*, 159–165. [CrossRef]
9. Hadi, H.A.; Carr, C.S.; Al Suwaidi, J. Endothelial dysfunction: Cardiovascular risk factors, therapy, and outcome. *Vasc. Health Risk Manag.* **2005**, *1*, 183–198.
10. Humbert, M.; Montani, D.; Perros, F.; Dorfmüller, P.; Adnot, S.; Eddahibi, S. Endothelial cell dysfunction and cross talk between endothelium and smooth muscle cells in pulmonary arterial hypertension. *Vasc. Pharmacol.* **2008**, *49*, 113–118. [CrossRef]
11. Nicod, L.P. The endothelium and genetics in pulmonary arterial hypertension. *Swiss Med. Wkly.* **2007**, *137*, 437–442. [PubMed]
12. Dummer, A.; Rol, N.; Szulcek, R.; Kurakula, K.; Pan, X.; Visser, B.I.; Bogaard, H.J.; DeRuiter, M.C.; Goumans, M.J.; Hierck, B.P. Endothelial dysfunction in pulmonary arterial hypertension: Loss of cilia length regulation upon cytokine stimulation. *Pulm. Circ.* **2018**, *8*. [CrossRef] [PubMed]
13. Stewart, D.J.; Levy, R.D.; Cernacek, P.; Langleben, D. Increased plasma endothelin-1 in pulmonary hypertension: Marker or mediator of disease? *Ann. Intern. Med.* **1991**, *114*, 464–469. [CrossRef] [PubMed]
14. Christman, B.W.; McPherson, C.D.; Newman, J.H.; King, G.A.; Bernard, G.R.; Groves, B.M.; Loyd, J.E. An Imbalance between the Excretion of Thromboxane and Prostacyclin Metabolites in Pulmonary Hypertension. *N. Engl. J. Med.* **1992**, *327*, 70–75. [CrossRef]
15. Tu, L.; Dewachter, L.; Gore, B.; Fadel, E.; Dartevelle, P.; Simonneau, G.; Humbert, M.; Eddahibi, S.; Guignabert, C. Autocrine fibroblast growth factor-2 signaling contributes to altered endothelial phenotype in pulmonary hypertension. *Am. J. Respir. Cell Mol. Biol.* **2011**, *45*, 311–322. [CrossRef]
16. Dai, Z.; Zhu, M.M.; Peng, Y.; Jin, H.; Machireddy, N.; Qian, Z.; Zhang, X.; Zhao, Y.Y. Endothelial and Smooth Muscle Cell Interaction via FoxM1 Signaling Mediates Vascular Remodeling and Pulmonary Hypertension. *Am. J. Respir. Crit. Care Med.* **2018**, *198*, 788–802. [CrossRef]
17. Ranchoux, B.; Harvey, L.D.; Ayon, R.J.; Babicheva, A.; Bonnet, S.; Chan, S.Y.; Yuan, J.X.J.; Perez, V.J. Endothelial dysfunction in pulmonary arterial hypertension: An evolving landscape (2017 Grover Conference Series). *Pulm. Circ.* **2018**, *8*. [CrossRef]
18. Orriols, M.; Gomez-Puerto, M.C.; Ten Dijke, P. BMP type II receptor as a therapeutic target in pulmonary arterial hypertension. *Cell. Mol. Life Sci.* **2017**, *74*, 2979–2995. [CrossRef]
19. Newman, J.H.; Wheeler, L.; Lane, K.B.; Loyd, E.; Gaddipati, R.; Phillips, J.A., 3rd; Loyd, J.E. Mutation in the gene for bone morphogenetic protein receptor II as a cause of primary pulmonary hypertension in a large kindred. *N. Engl. J. Med.* **2001**, *345*, 319–324. [CrossRef]
20. Larkin, E.K.; Newman, J.H.; Austin, E.D.; Hemnes, A.R.; Wheeler, L.; Robbins, I.M.; West, J.D.; Phillips, J.A., 3rd; Hamid, R.; Loyd, J.E. Longitudinal analysis casts doubt on the presence of genetic anticipation in heritable pulmonary arterial hypertension. *Am. J. Respir. Crit. Care Med.* **2012**, *186*, 892–896. [CrossRef]

21. Morrell, N.W.; Aldred, M.A.; Chung, W.K.; Elliott, C.G.; Nichols, W.C.; Soubrier, F.; Trembath, R.C.; Loyd, J.E. Genetics and genomics of pulmonary arterial hypertension. *Eur. Respir. J.* **2019**, *53*, D13–D21. [CrossRef] [PubMed]
22. Liu, F.; Ventura, F.; Doody, J.; Massagué, J. Human type II receptor for bone morphogenic proteins (BMPs): Extension of the two-kinase receptor model to the BMPs. *Mol. Cell. Biol.* **1995**, *15*, 3479–3486. [CrossRef] [PubMed]
23. Goumans, M.J.; Zwijsen, A.; Ten Dijke, P.; Bailly, S. Bone Morphogenetic Proteins in Vascular Homeostasis and Disease. *Cold Spring Harb. Perspect. Biol.* **2018**, *10*, a031989. [CrossRef] [PubMed]
24. Sanchez-Duffhues, G.; Williams, E.; Goumans, M.J.; Heldin, C.H.; Ten Dijke, P. Bone morphogenetic protein receptors: Structure, function and targeting by selective small molecule kinase inhibitors. *Bone* **2020**, *138*, 115472. [CrossRef] [PubMed]
25. Kurakula, K.; Goumans, M.J.; Ten Dijke, P. Regulatory RNAs controlling vascular (dys)function by affecting TGF-ß family signalling. *EXCLI J.* **2015**, *14*, 832–850. [PubMed]
26. Yang, X.; Long, L.; Southwood, M.; Rudarakanchana, N.; Upton, P.D.; Jeffery, T.K.; Atkinson, C.; Chen, H.; Trembath, R.C.; Morrell, N.W. Dysfunctional Smad signaling contributes to abnormal smooth muscle cell proliferation in familial pulmonary arterial hypertension. *Circ. Res.* **2005**, *96*, 1053–1063. [CrossRef]
27. Teichert-Kuliszewska, K.; Kutryk, M.J.B.; Kuliszewski, M.A.; Karoubi, G.; Courtman, D.W.; Zucco, L.; Granton, J.; Stewart, D.J. Bone morphogenetic protein receptor-2 signaling promotes pulmonary arterial endothelial cell survival: Implications for loss-of-function mutations in the pathogenesis of pulmonary hypertension. *Circ. Res.* **2006**, *98*, 209–217. [CrossRef]
28. Zhang, S.; Fantozzi, I.; Tigno, D.D.; Yi, E.S.; Platoshyn, O.; Thistlethwaite, P.A.; Kriett, J.M.; Yung, G.; Rubin, L.J.; Yuan, J.X. Bone morphogenetic proteins induce apoptosis in human pulmonary vascular smooth muscle cells. *Am. J. Physiol. Lung Cell. Mol. Physiol.* **2003**, *285*, L740–L754. [CrossRef]
29. Gräf, S.; Haimel, M.; Bleda, M.; Hadinnapola, C.; Southgate, L.; Li, W.; Hodgson, J.; Liu, B.; Salmon, R.M.; Southwood, M.; et al. Identification of rare sequence variation underlying heritable pulmonary arterial hypertension. *Nat. Commun.* **2018**, *9*, 1416. [CrossRef]
30. Frump, A.; Prewitt, A.; de Caestecker, M.P. BMPR2 mutations and endothelial dysfunction in pulmonary arterial hypertension (2017 Grover Conference Series). *Pulm. Circ.* **2018**, *8*. [CrossRef]
31. Soon, E.; Crosby, A.; Southwood, M.; Yang, P.; Tajsic, T.; Toshner, M.; Appleby, S.; Shanahan, C.M.; Bloch, K.D.; Pepke-Zaba, J.; et al. Bone morphogenetic protein receptor type II deficiency and increased inflammatory cytokine production: A gateway to pulmonary arterial hypertension. *Am. J. Respir. Crit. Care Med.* **2015**. [CrossRef] [PubMed]
32. Liu, D.; Wang, J.; Kinzel, B.; Müeller, M.; Mao, X.; Valdez, R.; Liu, Y.; Li, E. Dosage-dependent requirement of BMP type II receptor for maintenance of vascular integrity. *Blood* **2007**, *110*, 1502–1510. [CrossRef] [PubMed]
33. Long, L.; MacLean, M.R.; Jeffery, T.K.; Morecroft, I.; Yang, X.; Rudarakanchana, N.; Southwood, M.; James, V.; Trembath, R.C.; Morrell, N.W. Serotonin increases susceptibility to pulmonary hypertension in BMPR2-deficient mice. *Circ. Res.* **2006**, *98*, 818–827. [CrossRef] [PubMed]
34. Atkinson, C.; Stewart, S.; Upton, P.D.; Machado, R.; Thomson, J.R.; Trembath, R.C.; Morrell, N.W. Primary pulmonary hypertension is associated with reduced pulmonary vascular expression of type II bone morphogenetic protein receptor. *Circulation* **2002**, *105*, 1672–1678. [CrossRef] [PubMed]
35. Brock, M.; Trenkmann, M.; Gay, R.E.; Michel, B.A.; Gay, S.; Fischler, M.; Ulrich, S.; Speich, R.; Huber, L.C. Interleukin-6 modulates the expression of the bone morphogenic protein receptor type II through a novel STAT3-microRNA cluster 17/92 pathway. *Circ. Res.* **2009**, *104*, 1184–1191. [CrossRef] [PubMed]
36. Andruska, A.; Spiekerkoetter, E. Consequences of BMPR2 Deficiency in the Pulmonary Vasculature and Beyond: Contributions to Pulmonary Arterial Hypertension. *Int. J. Mol. Sci.* **2018**, *19*, 2499. [CrossRef]
37. Happé, C.; Kurakula, K.; Sun, X.Q.; da Silva Goncalves Bos, D.; Rol, N.; Guignabert, C.; Tu, L.; Schalij, I.; Wiesmeijer, K.C.; Tura-Ceide, O.; et al. The BMP Receptor 2 in Pulmonary Arterial Hypertension: When and Where the Animal Model Matches the Patient. *Cells* **2020**, *9*, 1422. [CrossRef]
38. Hong, K.H.; Lee, Y.J.; Lee, E.; Park, S.O.; Han, C.; Beppu, H.; Li, E.; Raizada, M.K.; Bloch, K.D.; Oh, S.P. Genetic ablation of the BMPR2 gene in pulmonary endothelium is sufficient to predispose to pulmonary arterial hypertension. *Circulation* **2008**, *118*, 722–730. [CrossRef]
39. Majka, S.; Hagen, M.; Blackwell, T.; Harral, J.; Johnson, J.A.; Gendron, R.; Paradis, H.; Crona, D.; Loyd, J.E.; Nozik-Grayck, E.; et al. Physiologic and molecular consequences of endothelial Bmpr2 mutation. *Respir. Res.* **2011**, *12*, 84. [CrossRef]
40. Yang, X.; Long, L.; Reynolds, P.N.; Morrell, N.W. Expression of Mutant BMPR-II in Pulmonary Endothelial Cells Promotes Apoptosis and a Release of Factors that Stimulate Proliferation of Pulmonary Arterial Smooth Muscle Cells. *Pulm. Circ.* **2011**, *1*, 103–110. [CrossRef]
41. Hodgson, J.; Swietlik, E.M.; Salmon, R.M.; Hadinnapola, C.; Nikolic, I.; Wharton, J.; Guo, J.; Liley, J.; Haimel, M.; Bleda, M.; et al. Characterization of GDF2 Mutations and Levels of BMP9 and BMP10 in Pulmonary Arterial Hypertension. *Am. J. Respir. Crit. Care Med.* **2020**, *201*, 575–585. [CrossRef] [PubMed]
42. Long, L.; Ormiston, M.L.; Yang, X.; Southwood, M.; Gräf, S.; Machado, R.D.; Mueller, M.; Kinzel, B.; Yung, L.M.; Wilkinson, J.M.; et al. Selective enhancement of endothelial BMPR-II with BMP9 reverses pulmonary arterial hypertension. *Nat. Med.* **2015**, *21*, 777–785. [CrossRef] [PubMed]
43. Li, M.; Vattulainen, S.; Aho, J.; Orcholski, M.; Rojas, V.; Yuan, K.; Helenius, M.; Taimen, P.; Myllykangas, S.; De Jesus Perez, V.; et al. Loss of bone morphogenetic protein receptor 2 is associated with abnormal DNA Repair in pulmonary arterial hypertension. *Am. J. Respir. Cell Mol. Biol.* **2014**, *50*, 1118–1128. [CrossRef] [PubMed]

44. Rhodes, C.J.; Im, H.; Cao, A.; Hennigs, J.K.; Wang, L.; Sa, S.; Chen, P.I.; Nickel, N.P.; Miyagawa, K.; Hopper, R.K.; et al. RNA Sequencing Analysis Detection of a Novel Pathway of Endothelial Dysfunction in Pulmonary Arterial Hypertension. *Am. J. Respir. Crit. Care Med.* **2015**, *192*, 356–366. [CrossRef]
45. Rol, N.; de Raaf, M.A.; Sun, X.Q.; Kuiper, V.P.; da Silva Gonçalves Bos, D.; Happé, C.; Kurakula, K.; Dickhoff, C.; Thuillet, R.; Tu, L.; et al. Nintedanib improves cardiac fibrosis but leaves pulmonary vascular remodelling unaltered in experimental pulmonary hypertension. *Cardiovasc. Res.* **2019**, *115*, 432–439. [CrossRef]
46. Awad, K.S.; Elinoff, J.M.; Wang, S.; Gairhe, S.; Ferreyra, G.A.; Cai, R.; Sun, J.; Solomon, M.A.; Danner, R.L. Raf/ERK drives the proliferative and invasive phenotype of BMPR2-silenced pulmonary artery endothelial cells. *Am. J. Physiol. Lung Cell. Mol. Physiol.* **2016**, *310*, L187–L201. [CrossRef]
47. Tian, W.; Jiang, X.; Sung, Y.K.; Shuffle, E.; Wu, T.H.; Kao, P.N.; Tu, A.B.; Dorfmüller, P.; Cao, A.; Wang, L.; et al. Phenotypically Silent Bone Morphogenetic Protein Receptor 2 Mutations Predispose Rats to Inflammation-Induced Pulmonary Arterial Hypertension by Enhancing the Risk for Neointimal Transformation. *Circulation* **2019**, *140*, 1409–1425. [CrossRef]
48. Miyata, M.; Sakuma, F.; Yoshimura, A.; Ishikawa, H.; Nishimaki, T.; Kasukawa, R. Pulmonary hypertension in rats. 2. Role of interleukin-6. *Int. Arch. Allergy Immunol.* **1995**, *108*, 287–291. [CrossRef]
49. Steiner, M.K.; Syrkina, O.L.; Kolliputi, N.; Mark, E.J.; Hales, C.A.; Waxman, A.B. Interleukin-6 overexpression induces pulmonary hypertension. *Circ. Res.* **2009**, *104*, 236–244. [CrossRef]
50. Hurst, L.A.; Dunmore, B.J.; Long, L.; Crosby, A.; Al-Lamki, R.; Deighton, J.; Southwood, M.; Yang, X.; Nikolic, M.Z.; Herrera, B.; et al. TNFα drives pulmonary arterial hypertension by suppressing the BMP type-II receptor and altering NOTCH signalling. *Nat. Commun.* **2017**, *8*, 14079. [CrossRef]
51. Dorfmüller, P.; Perros, F.; Balabanian, K.; Humbert, M. Inflammation in pulmonary arterial hypertension. *Eur. Respir. J.* **2003**, *22*, 358–363. [CrossRef] [PubMed]
52. Jasiewicz, M.; Knapp, M.; Waszkiewicz, E.; Ptaszynska-Kopczynska, K.; Szpakowicz, A.; Sobkowicz, B.; Musial, W.J.; Kaminski, K.A. Enhanced IL-6 trans-signaling in pulmonary arterial hypertension and its potential role in disease-related systemic damage. *Cytokine* **2015**, *76*, 187–192. [CrossRef] [PubMed]
53. Groth, A.; Vrugt, B.; Brock, M.; Speich, R.; Ulrich, S.; Huber, L.C. Inflammatory cytokines in pulmonary hypertension. *Respir. Res.* **2014**, *15*, 47. [CrossRef] [PubMed]
54. Szulcek, R.; Sanchez-Duffhues, G.; Rol, N.; Pan, X.; Tsonaka, R.; Dickhoff, C.; Yung, L.M.; Manz, X.D.; Kurakula, K.; Kiełbasa, S.M.; et al. Exacerbated inflammatory signaling underlies aberrant response to BMP9 in pulmonary arterial hypertension lung endothelial cells. *Angiogenesis* **2020**, *23*, 699–714. [CrossRef] [PubMed]
55. Veyssier-Belot, C.; Cacoub, P. Role of endothelial and smooth muscle cells in the physiopathology and treatment management of pulmonary hypertension. *Cardiovasc. Res.* **1999**, *44*, 274–282. [CrossRef]
56. Fuster, V.; Steele, P.M.; Edwards, W.D.; Gersh, B.J.; McGoon, M.D.; Frye, R.L. Primary pulmonary hypertension: Natural history and the importance of thrombosis. *Circulation* **1984**, *70*, 580–587. [CrossRef] [PubMed]
57. Johnson, S.R.; Granton, J.T.; Mehta, S. Thrombotic arteriopathy and anticoagulation in pulmonary hypertension. *Chest* **2006**, *130*, 545–552. [CrossRef]
58. Berger, G.; Azzam, Z.S.; Hoffman, R.; Yigla, M. Coagulation and anticoagulation in pulmonary arterial hypertension. *Isr. Med. Assoc. J.* **2009**, *11*, 376–379.
59. Kawut, S.M.; Horn, E.M.; Berekashvili, K.K.; Widlitz, A.C.; Rosenzweig, E.B.; Barst, R.J. Von Willebrand factor independently predicts long-term survival in patients with pulmonary arterial hypertension. *Chest* **2005**, *128*, 2355–2362. [CrossRef]
60. Damås, J.K.; Otterdal, K.; Yndestad, A.; Aass, H.; Solum, N.O.; Frøland, S.S.; Simonsen, S.; Aukrust, P.; Andreassen, A.K. Soluble CD40 ligand in pulmonary arterial hypertension: Possible pathogenic role of the interaction between platelets and endothelial cells. *Circulation* **2004**, *110*, 999–1005. [CrossRef]
61. Kroone, C.; Vos, M.; Rademakers, T.; Kuijpers, M.; Hoogenboezem, M.; van Buul, J.; Heemskerk, J.W.M.; Ruf, W.; van Hylckama Vlieg, A.; Versteeg, H.H.; et al. LIM-only protein FHL2 attenuates vascular tissue factor activity, inhibits thrombus formation in mice and FHL2 genetic variation associates with human venous thrombosis. *Haematologica* **2020**, *105*, 1677–1685. [CrossRef] [PubMed]
62. Kurakula, K.; Koenis, D.S.; Herzik, M.A., Jr.; Liu, Y.; Craft, J.W., Jr.; van Loenen, P.B.; Vos, M.; Tran, M.K.; Versteeg, H.H.; Goumans, M.T.H.; et al. Structural and cellular mechanisms of peptidyl-prolyl isomerase Pin1-mediated enhancement of Tissue Factor gene expression, protein half-life, and pro-coagulant activity. *Haematologica* **2018**, *103*, 1073–1082. [CrossRef] [PubMed]
63. White, R.J.; Meoli, D.F.; Swarthout, R.F.; Kallop, D.Y.; Galaria, I.I.; Harvey, J.L.; Miller, C.M.; Blaxall, B.C.; Hall, C.M.; Pierce, R.A.; et al. Plexiform-like lesions and increased tissue factor expression in a rat model of severe pulmonary arterial hypertension. *Am. J. Physiol. Lung Cell. Mol. Physiol.* **2007**, *293*, L583–L590. [CrossRef] [PubMed]
64. Bakouboula, B.; Morel, O.; Faure, A.; Zobairi, F.; Jesel, L.; Trinh, A.; Zupan, M.; Canuet, M.; Grunebaum, L.; Brunette, A.; et al. Procoagulant membrane microparticles correlate with the severity of pulmonary arterial hypertension. *Am. J. Respir. Crit. Care Med.* **2008**, *177*, 536–543. [CrossRef] [PubMed]
65. Tournier, A.; Wahl, D.; Chaouat, A.; Max, J.P.; Regnault, V.; Lecompte, T.; Chabot, F. Calibrated automated thrombography demonstrates hypercoagulability in patients with idiopathic pulmonary arterial hypertension. *Thromb. Res.* **2010**, *126*, e418–e422. [CrossRef] [PubMed]

66. Huber, K.; Beckmann, R.; Frank, H.; Kneussl, M.; Mlczoch, J.; Binder, B.R. Fibrinogen, t-PA, and PAI-1 plasma levels in patients with pulmonary hypertension. *Am. J. Respir. Crit. Care Med.* **1994**, *150*, 929–933. [CrossRef]
67. Boucherat, O.; Vitry, G.; Trinh, I.; Paulin, R.; Provencher, S.; Bonnet, S. The cancer theory of pulmonary arterial hypertension. *Pulm. Circ.* **2017**, *7*, 285–299. [CrossRef]
68. Smolders, V.F.; Zodda, E.; Quax, P.H.A.; Carini, M.; Barberà, J.A.; Thomson, T.M.; Tura-Ceide, O.; Cascante, M. Metabolic Alterations in Cardiopulmonary Vascular Dysfunction. *Front. Mol. Biosci.* **2018**, *5*, 120. [CrossRef]
69. Maron, B.A.; Leopold, J.A. Emerging Concepts in the Molecular Basis of Pulmonary Arterial Hypertension: Part II: Neurohormonal Signaling Contributes to the Pulmonary Vascular and Right Ventricular Pathophenotype of Pulmonary Arterial Hypertension. *Circulation* **2015**, *131*, 2079–2091. [CrossRef]
70. Cao, Y.; Zhang, X.; Wang, L.; Yang, Q.; Ma, Q.; Xu, J.; Wang, J.; Kovacs, L.; Ayon, R.J.; Liu, Z.; et al. PFKFB3-mediated endothelial glycolysis promotes pulmonary hypertension. *Proc. Natl. Acad. Sci. USA* **2019**, *116*, 13394–13403. [CrossRef]
71. Michelakis, E.D.; Gurtu, V.; Webster, L.; Barnes, G.; Watson, G.; Howard, L.; Cupitt, J.; Paterson, I.; Thompson, R.B.; Chow, K.; et al. Inhibition of pyruvate dehydrogenase kinase improves pulmonary arterial hypertension in genetically susceptible patients. *Sci. Transl. Med.* **2017**, *9*. [CrossRef] [PubMed]
72. Zhao, Y.; Peng, J.; Lu, C.; Hsin, M.; Mura, M.; Wu, L.; Chu, L.; Zamel, R.; Machuca, T.; Waddell, T.; et al. Metabolomic heterogeneity of pulmonary arterial hypertension. *PLoS ONE* **2014**, *9*, e88727. [CrossRef] [PubMed]
73. Schäfer, M.; Kheyfets, V.O.; Schroeder, J.D.; Dunning, J.; Shandas, R.; Buckner, J.K.; Browning, J.; Hertzberg, J.; Hunter, K.S.; Fenster, B.E. Main pulmonary arterial wall shear stress correlates with invasive hemodynamics and stiffness in pulmonary hypertension. *Pulm. Circ.* **2016**, *6*, 37–45. [CrossRef] [PubMed]
74. Allen, R.P.; Schelegle, E.S.; Bennett, S.H. Diverse forms of pulmonary hypertension remodel the arterial tree to a high shear phenotype. *Am. J. Physiol. Heart Circ. Physiol.* **2014**, *307*, H405–H417. [CrossRef]
75. Gatzoulis, M.A.; Alonso-Gonzalez, R.; Beghetti, M. Pulmonary arterial hypertension in paediatric and adult patients with congenital heart disease. *Eur. Respir. Rev.* **2009**, *18*, 154–161. [CrossRef]
76. Abe, K.; Shinoda, M.; Tanaka, M.; Kuwabara, Y.; Yoshida, K.; Hirooka, Y.; McMurtry, I.F.; Oka, M.; Sunagawa, K. Haemodynamic unloading reverses occlusive vascular lesions in severe pulmonary hypertension. *Cardiovasc. Res.* **2016**, *111*, 16–25. [CrossRef]
77. van der Feen, D.E.; Bossers, G.P.L.; Hagdorn, Q.A.J.; Moonen, J.R.; Kurakula, K.; Szulcek, R.; Chappell, J.; Vallania, F.; Donato, M.; Kok, K.; et al. Cellular senescence impairs the reversibility of pulmonary arterial hypertension. *Sci. Transl. Med.* **2020**, *12*. [CrossRef]
78. Hirata, M.; Ousaka, D.; Arai, S.; Okuyama, M.; Tarui, S.; Kobayashi, J.; Kasahara, S.; Sano, S. Novel Model of Pulmonary Artery Banding Leading to Right Heart Failure in Rats. *BioMed Res. Int.* **2015**, *2015*, 753210. [CrossRef]
79. Szulcek, R.; Happé, C.M.; Rol, N.; Fontijn, R.D.; Dickhoff, C.; Hartemink, K.J.; Grünberg, K.; Tu, L.; Timens, W.; Nossent, G.D.; et al. Delayed Microvascular Shear Adaptation in Pulmonary Arterial Hypertension. Role of Platelet Endothelial Cell Adhesion Molecule-1 Cleavage. *Am. J. Respir. Crit. Care Med.* **2016**, *193*, 1410–1420. [CrossRef]
80. Li, M.; Tan, Y.; Stenmark, K.R.; Tan, W. High Pulsatility Flow Induces Acute Endothelial Inflammation through Overpolarizing Cells to Activate NF-κB. *Cardiovasc. Eng. Technol.* **2013**, *4*, 26–38. [CrossRef]
81. Förstermann, U.; Sessa, W.C. Nitric oxide synthases: Regulation and function. *Eur. Heart J.* **2012**, *33*, 829–837. [CrossRef] [PubMed]
82. Ziche, M.; Morbidelli, L.; Masini, E.; Amerini, S.; Granger, H.J.; Maggi, C.A.; Geppetti, P.; Ledda, F. Nitric oxide mediates angiogenesis In Vivo and endothelial cell growth and migration In Vitro promoted by substance P. *J. Clin. Investig.* **1994**, *94*, 2036–2044. [CrossRef] [PubMed]
83. Babaei, S.; Teichert-Kuliszewska, K.; Monge, J.C.; Mohamed, F.; Bendeck, M.P.; Stewart, D.J. Role of nitric oxide in the angiogenic response In Vitro to basic fibroblast growth factor. *Circ. Res.* **1998**, *82*, 1007–1015. [CrossRef] [PubMed]
84. Giaid, A.; Saleh, D. Reduced Expression of Endothelial Nitric Oxide Synthase in the Lungs of Patients with Pulmonary Hypertension. *N. Engl. J. Med.* **1995**, *333*, 214–221. [CrossRef]
85. Austin, E.D.; Ma, L.; LeDuc, C.; Berman Rosenzweig, E.; Borczuk, A.; Phillips, J.A., 3rd; Palomero, T.; Sumazin, P.; Kim, H.R.; Talati, M.H.; et al. Whole exome sequencing to identify a novel gene (caveolin-1) associated with human pulmonary arterial hypertension. *Circ. Cardiovasc. Genet.* **2012**, *5*, 336–343. [CrossRef]
86. Quinlan, T.R.; Li, D.; Laubach, V.E.; Shesely, E.G.; Zhou, N.; Johns, R.A. eNOS-deficient mice show reduced pulmonary vascular proliferation and remodeling to chronic hypoxia. *Am. J. Physiol. Lung Cell. Mol. Physiol.* **2000**. [CrossRef]
87. Epstein, F.H.; Vane, J.R.; Änggård, E.E.; Botting, R.M. Regulatory Functions of the Vascular Endothelium. *N. Engl. J. Med.* **2010**, *323*, 27–36. [CrossRef]
88. Chen, Y.F.; Oparil, S. Endothelial Dysfunction in the Pulmonary Vascular Bed. *Am. J. Med. Sci.* **2000**, *320*, 223–232. [CrossRef]
89. Humbert, M.; Sitbon, O. Treatment of Pulmonary Arterial Hypertension. *N. Engl. J. Med.* **2004**, *351*, 1425–1436. [CrossRef]
90. Mitchell, J.A.; Ahmetaj-Shala, B.; Kirkby, N.S.; Wright, W.R.; Mackenzie, L.S.; Reed, D.M.; Mohamed, N. Role of prostacyclin in pulmonary hypertension. *Glob. Cardiol. Sci. Pract.* **2014**, *2014*, 382–393. [CrossRef]
91. Tuder, R.M.; Cool, C.D.; Geraci, M.W.; Wang, J.; Abman, S.H.; Wright, L.; Badesch, D.; Voelkel, N.F. Prostacyclin synthase expression is decreased in lungs from patients with severe pulmonary hypertension. *Am. J. Respir. Crit. Care Med.* **1999**. [CrossRef] [PubMed]

92. Geraci, M.W.; Gao, B.; Shepherd, D.C.; Moore, M.D.; Westcott, J.Y.; Fagan, K.A.; Alger, L.A.; Tuder, R.M.; Voelkel, N.F. Pulmonary prostacyclin synthase overexpression in transgenic mice protects against development of hypoxic pulmonary hypertension. *J. Clin. Investig.* **1999**, *103*, 1509–1515. [CrossRef] [PubMed]
93. Chester, A.H.; Yacoub, M.H. The role of endothelin-1 in pulmonary arterial hypertension. *Glob. Cardiol. Sci. Pract.* **2014**, *2014*, 62–78. [CrossRef] [PubMed]
94. Shao, D.; Park, J.E.S.; Wort, S.J. The role of endothelin-1 in the pathogenesis of pulmonary arterial hypertension. *Pharm. Res.* **2011**, *63*, 504–511. [CrossRef] [PubMed]
95. Shichiri, M.; Kato, H.; Marumo, F.; Hirata, Y. Endothelin-1 as an autocrine/paracrine apoptosis survival factor for endothelial cells. *Hypertension* **1997**, *30*, 1198–1203. [CrossRef]
96. Giaid, A.; Yanagisawa, M.; Langleben, D.; Michel, R.P.; Levy, R.; Shennib, H.; Kimura, S.; Masaki, T.; Duguid, W.P.; Stewart, D.J. Expression of Endothelin-1 in the Lungs of Patients with Pulmonary Hypertension. *N. Engl. J. Med.* **1993**, *328*, 1732–1739. [CrossRef]
97. Li, H.B.; Chen, S.J.; Chen, Y.F.; Meng, Q.C.; Durand, J.; Oparil, S.; Elton, T.S. Enhanced Endothelin-1 and Endothelin Receptor Gene-Expression in Chronic Hypoxia. *J. Appl. Physiol.* **1994**, *77*, 1451–1459. [CrossRef]
98. Frasch, H.F.; Marshall, C.; Marshall, B.E. Endothelin-1 is elevated in monocrotaline pulmonary hypertension. *Am. J. Physiol. Lung Cell. Mol. Physiol.* **1999**, *276*, L304–L310. [CrossRef]
99. Davie, N.; Haleen, S.J.; Upton, P.D.; Polak, J.M.; Yacoub, M.H.; Morrell, N.W.; Wharton, J. ET(A) and ET(B) receptors modulate the proliferation of human pulmonary artery smooth muscle cells. *Am. J. Respir. Crit. Care Med.* **2002**, *165*, 398–405. [CrossRef]
100. Galié, N.; Manes, A.; Branzi, A. The endothelin system in pulmonary arterial hypertension. *Cardiovasc. Res.* **2004**, *61*, 227–237. [CrossRef]
101. Tuder, R.M.; Chacon, M.; Alger, L.; Wang, J.; Taraseviciene-Stewart, L.; Kasahara, Y.; Cool, C.D.; Bishop, A.E.; Geraci, M.; Semenza, G.L.; et al. Expression of angiogenesis-related molecules in plexiform lesions in severe pulmonary hypertension: Evidence for a process of disordered angiogenesis. *J. Pathol.* **2001**, *195*, 367–374. [CrossRef] [PubMed]
102. Säleby, J.; Bouzina, H.; Ahmed, S.; Lundgren, J.; Rådegran, G. Plasma receptor tyrosine kinase RET in pulmonary arterial hypertension diagnosis and differentiation. *ERJ Open Res.* **2019**, *5*, 00037–02019. [CrossRef] [PubMed]
103. Sánchez-Duffhues, G.; García de Vinuesa, A.; Ten Dijke, P. Endothelial-to-mesenchymal transition in cardiovascular diseases: Developmental signaling pathways gone awry. *Dev. Dyn.* **2018**, *247*, 492–508. [CrossRef] [PubMed]
104. Medici, D.; Kalluri, R. Endothelial-mesenchymal transition and its contribution to the emergence of stem cell phenotype. *Semin. Cancer Biol.* **2012**, *22*, 379–384. [CrossRef] [PubMed]
105. Ranchoux, B.; Antigny, F.; Rucker-Martin, C.; Hautefort, A.; Péchoux, C.; Bogaard, H.J.; Dorfmüller, P.; Remy, S.; Lecerf, F.; Planté, S.; et al. Endothelial-to-mesenchymal transition in pulmonary hypertension. *Circulation* **2015**, *131*, 1006–1018. [CrossRef]
106. Good, R.B.; Gilbane, A.J.; Trinder, S.L.; Denton, C.P.; Coghlan, G.; Abraham, D.J.; Holmes, A.M. Endothelial to Mesenchymal Transition Contributes to Endothelial Dysfunction in Pulmonary Arterial Hypertension. *Am. J. Pathol.* **2015**, *185*, 1850–1858. [CrossRef]
107. Tang, H.; Babicheva, A.; McDermott, K.M.; Gu, Y.; Ayon, R.J.; Song, S.; Wang, Z.; Gupta, A.; Zhou, T.; Sun, X.; et al. Endothelial HIF-2α Contributes to Severe Pulmonary Hypertension by Inducing Endothelial-to-Mesenchymal Transition. *Am. J. Physiol. Lung Cell. Mol. Physiol.* **2017**, *314*, L256–L275. [CrossRef]
108. Goumans, M.J.; van Zonneveld, A.J.; ten Dijke, P. Transforming growth factor beta-induced endothelial-to-mesenchymal transition: A switch to cardiac fibrosis? *Trends Cardiovasc. Med.* **2008**, *18*, 293–298. [CrossRef]
109. Ursoli Ferreira, F.; Eduardo Botelho Souza, L.; Hassibe Thomé, F.; Tomazini Pinto, M.; Origassa, C.; Salustiano, S.; Marcel Faça, V.; Olsen Câmara, N.; Kashima, S.; Tadeu Covas, D. Endothelial Cells Tissue-Specific Origins Affects Their Responsiveness to TGF-β2 during Endothelial-to-Mesenchymal Transition. *Int. J. Mol. Sci.* **2019**, *20*, 458. [CrossRef]
110. Mammoto, T.; Muyleart, M.; Konduri, G.G.; Mammoto, A. Twist1 in Hypoxia-induced Pulmonary Hypertension through Transforming Growth Factor-β–Smad Signaling. *Am. J. Respir. Cell Mol. Biol.* **2018**, *58*, 194–207. [CrossRef]
111. Hopper, R.K.; Moonen, J.R.A.J.; Diebold, I.; Cao, A.; Rhodes, C.J.; Tojais, N.F.; Hennigs, J.K.; Gu, M.; Wang, L.; Rabinovitch, M. In pulmonary arterial hypertension, reduced bmpr2 promotes endothelial-to-Mesenchymal transition via hmga1 and its target slug. *Circulation* **2016**, *133*, 1783–1794. [CrossRef] [PubMed]
112. Zhang, H.; Liu, Y.; Yan, L.; Du, W.; Zhang, X.; Zhang, M.; Chen, H.; Zhang, Y.; Zhou, J.; Sun, H.; et al. Bone morphogenetic protein-7 inhibits endothelial-mesenchymal transition in pulmonary artery endothelial cell under hypoxia. *J. Cell. Physiol.* **2018**. [CrossRef] [PubMed]
113. Hiepen, C.; Jatzlau, J.; Hildebrandt, S.; Kampfrath, B.; Goktas, M.; Murgai, A.; Cuellar Camacho, J.L.; Haag, R.; Ruppert, C.; Sengle, G.; et al. BMPR2 acts as a gatekeeper to protect endothelial cells from increased TGFβ responses and altered cell mechanics. *PLoS Biol.* **2019**, *17*, e3000557. [CrossRef] [PubMed]
114. Rol, N.; Kurakula, K.B.; Happé, C.; Bogaard, H.J.; Goumans, M.J. TGF-β and BMPR2 Signaling in PAH: Two Black Sheep in One Family. *Int. J. Mol. Sci.* **2018**, *19*, 2585. [CrossRef] [PubMed]
115. Lei, W.; He, Y.; Shui, X.; Li, G.; Yan, G.; Zhang, Y.; Huang, S.; Chen, C.; Ding, Y. Expression and analyses of the HIF-1 pathway in the lungs of humans with pulmonary arterial hypertension. *Mol. Med. Rep.* **2016**, *14*, 4383–4390. [CrossRef]
116. Dai, Z.; Zhu, M.M.; Peng, Y.; Machireddy, N.; Evans, C.E.; Machado, R.; Zhang, X.; Zhao, Y.Y. Therapeutic Targeting of Vascular Remodeling and Right Heart Failure in Pulmonary Arterial Hypertension with a HIF-2α Inhibitor. *Am. J. Respir. Crit. Care Med.* **2018**, *198*, 1423–1434. [CrossRef]
117. Zhang, B.; Niu, W.; Dong, H.Y.; Liu, M.L.; Luo, Y.; Li, Z.C. Hypoxia induces endothelial-mesenchymal transition in pulmonary vascular remodeling. *Int. J. Mol. Med.* **2018**, *42*, 270–278. [CrossRef]

118. Zhao, H.; Wang, Y.; Zhang, X.; Guo, Y.; Wang, X. miR-181b-5p inhibits endothelial-mesenchymal transition in monocrotaline-induced pulmonary arterial hypertension by targeting endocan and TGFBR1. *Toxicol. Appl. Pharmacol.* **2020**, *386*, 114827. [CrossRef]
119. Tuder, R.M.; Archer, S.L.; Dorfmüller, P.; Erzurum, S.C.; Guignabert, C.; Michelakis, E.; Rabinovitch, M.; Schermuly, R.; Stenmark, K.R.; Morrell, N.W. Relevant issues in the pathology and pathobiology of pulmonary hypertension. *J. Am. Coll. Cardiol.* **2013**, *62*, D4–D12. [CrossRef]
120. Taraseviciene-Stewart, L.; Kasahara, Y.; Alger, L.; Hirth, P.; Mc Mahon, G.; Waltenberger, J.; Voelkel, N.F.; Tuder, R.M. Inhibition of the VEGF receptor 2 combined with chronic hypoxia causes cell death-dependent pulmonary endothelial cell proliferation and severe pulmonary hypertension. *FASEB J.* **2001**, *15*, 427–438. [CrossRef]
121. Sakao, S.; Taraseviciene-Stewart, L.; Lee, J.D.; Wood, K.; Cool, C.D.; Voelkel, N.F. Initial apoptosis is followed by increased proliferation of apoptosis-resistant endothelial cells. *FASEB J. Off. Publ. Fed. Am. Soc. Exp. Biol.* **2005**, *19*, 1178–1180. [CrossRef] [PubMed]
122. Masri, F.A.; Xu, W.; Comhair, S.A.; Asosingh, K.; Koo, M.; Vasanji, A.; Drazba, J.; Anand-Apte, B.; Erzurum, S.C. Hyperproliferative apoptosis-resistant endothelial cells in idiopathic pulmonary arterial hypertension. *Am. J. Physiol. Lung Cell. Mol. Physiol.* **2007**, *293*, 548–554. [CrossRef] [PubMed]
123. Diebold, I.; Hennigs, J.K.; Miyagawa, K.; Li, C.G.; Nickel, N.P.; Kaschwich, M.; Cao, A.; Wang, L.; Reddy, S.; Chen, P.I.; et al. BMPR2 preserves mitochondrial function and DNA during reoxygenation to promote endothelial cell survival and reverse pulmonary hypertension. *Cell Metab.* **2015**, *21*, 596–608. [CrossRef] [PubMed]
124. White, K.; Dempsie, Y.; Caruso, P.; Wallace, E.; McDonald, R.A.; Stevens, H.; Hatley, M.E.; Van Rooij, E.; Morrell, N.W.; Maclean, M.R.; et al. Endothelial apoptosis in pulmonary hypertension is controlled by a microRNA/programmed cell death 4/caspase-3 axis. *Hypertension* **2014**, *64*, 185–194. [CrossRef]
125. Dabral, S.; Tian, X.; Kojonazarov, B.; Savai, R.; Ghofrani, H.A.; Weissmann, N.; Florio, M.; Sun, J.; Jonigk, D.; Maegel, L.; et al. Notch1 signalling regulates endothelial proliferation and apoptosis in pulmonary arterial hypertension. *Eur. Respir. J.* **2016**, *48*, 1137–1149. [CrossRef]
126. Miyagawa, K.; Shi, M.; Chen, P.I.; Hennigs, J.K.; Zhao, Z.; Wang, M.; Li, C.G.; Saito, T.; Taylor, S.; Sa, S.; et al. Smooth Muscle Contact Drives Endothelial Regeneration by BMPR2-Notch1-Mediated Metabolic and Epigenetic Changes. *Circ. Res.* **2019**, *124*, 211–224. [CrossRef]
127. Hautefort, A.; Chesné, J.; Preussner, J.; Pullamsetti, S.S.; Tost, J.; Looso, M.; Antigny, F.; Girerd, B.; Riou, M.; Eddahibi, S.; et al. Pulmonary endothelial cell DNA methylation signature in pulmonary arterial hypertension. *Oncotarget* **2017**, *8*, 52995–53016. [CrossRef]
128. Cavasin, M.A.; Stenmark, K.R.; McKinsey, T.A. Emerging Roles for Histone Deacetylases in Pulmonary Hypertension and Right Ventricular Remodeling (2013 Grover Conference series). *Pulm. Circ.* **2015**, *5*, 63–72. [CrossRef]
129. Zhao, L.; Chen, C.N.; Hajji, N.; Oliver, E.; Cotroneo, E.; Wharton, J.; Wang, D.; Li, M.; McKinsey, T.A.; Stenmark, K.R.; et al. Histone Deacetylation Inhibition in Pulmonary Hypertension: Therapeutic of Valproic Acid and SuPotentialberoylanilide Hydroxamic Acid. *Circulation* **2012**, *126*, 455–467. [CrossRef]
130. Seto, E.; Yoshida, M. Erasers of histone acetylation: The histone deacetylase enzymes. *Cold Spring Harb. Perspect. Biol.* **2014**, *6*, a018713. [CrossRef]
131. Chabot, S.; Boucherat, O.; Ruffenach, G.; Breuils-Bonnet, S.; Tremblay, E.; Provencher, S.; Bonnet, S. HDAC6-HSP90 interplay in pulmonary arterial hypertension. *FASEB J.* **2016**, *30*, 774.4.
132. Boucherat, O.; Chabot, S.; Paulin, R.; Trinh, I.; Bourgeois, A.; Potus, F.; Lampron, M.C.; Lambert, C.; Breuils-Bonnet, S.; Nadeau, V.; et al. HDAC6: A Novel Histone Deacetylase Implicated in Pulmonary Arterial Hypertension. *Sci. Rep.* **2017**, *7*, 4546. [CrossRef] [PubMed]
133. Cavasin, M.A.; Demos-Davies, K.; Horn, T.R.; Walker, L.A.; Lemon, D.D.; Birdsey, N.; Weiser-Evans, M.C.M.; Harral, J.; Irwin, D.C.; Anwar, A.; et al. Selective class i histone deacetylase inhibition suppresses hypoxia-induced cardiopulmonary remodeling through an antiproliferative mechanism. *Circ. Res.* **2012**, *110*, 739–748. [CrossRef] [PubMed]
134. Li, M.; Riddle, S.R.; Frid, M.G.; El Kasmi, K.C.; McKinsey, T.A.; Sokol, R.J.; Strassheim, D.; Meyrick, B.; Yeager, M.E.; Flockton, A.R.; et al. Emergence of Fibroblasts with a Proinflammatory Epigenetically Altered Phenotype in Severe Hypoxic Pulmonary Hypertension. *J. Immunol.* **2011**, *187*, 2711–2722. [CrossRef] [PubMed]
135. Kim, J.; Hwangbo, C.; Hu, X.; Kang, Y.; Papangeli, I.; Mehrotra, D.; Park, H.; Ju, H.; McLean, D.L.; Comhair, S.A.; et al. Restoration of Impaired Endothelial MEF2 Function Rescues Pulmonary Arterial Hypertension. *Circulation* **2015**, *131*, 190–199. [CrossRef] [PubMed]
136. Feen, D.E.V.D.; Kurakula, K.; Tremblay, E.; Boucherat, O.; Bossers, G.P.L. Multicenter preclinical validation of BET inhibition for the treatment of pulmonary arterial hypertension. *Am. J. Respir. Crit. Care Med.* **2019**, *200*, 910–920. [CrossRef] [PubMed]
137. Devaiah, B.N.; Gegonne, A.; Singer, D.S. Bromodomain 4: A cellular Swiss army knife. *J. Leukoc. Biol.* **2016**, *100*, 679–686. [CrossRef] [PubMed]
138. Meloche, J.; Potus, F.; Vaillancourt, M.; Bourgeois, A.; Johnson, I.; Deschamps, L.; Chabot, S.; Ruffenach, G.; Henry, S.; Breuils-Bonnet, S.; et al. Bromodomain-Containing Protein 4. *Circ. Res.* **2015**, *117*, 525–535. [CrossRef]
139. Fernández, A.I.; Yotti, R.; González-Mansilla, A.; Mombiela, T.; Gutiérrez-Ibanes, E.; Pérez Del Villar, C.; Navas-Tejedor, P.; Chazo, C.; Martínez-Legazpi, P.; Fernández-Avilés, F.; et al. The Biological Bases of Group 2 Pulmonary Hypertension. *Int. J. Mol. Sci.* **2019**, *20*, 5884. [CrossRef]
140. Ontkean, M.; Gay, R.; Greenberg, B. Diminished endothelium-derived relaxing factor activity in an experimental model of chronic heart failure. *Circ. Res.* **1991**, *69*, 1088–1096. [CrossRef]

141. Givertz, M.M.; Colucci, W.S.; LeJemtel, T.H.; Gottlieb, S.S.; Hare, J.M.; Slawsky, M.T.; Leier, C.V.; Loh, E.; Nicklas, J.M.; Lewis, B.E. Acute endothelin A receptor blockade causes selective pulmonary vasodilation in patients with chronic heart failure. *Circulation* **2000**, *101*, 2922–2927. [CrossRef] [PubMed]
142. Meoli, D.F.; Su, Y.R.; Brittain, E.L.; Robbins, I.M.; Hemnes, A.R.; Monahan, K. The transpulmonary ratio of endothelin 1 is elevated in patients with preserved left ventricular ejection fraction and combined pre- and post-capillary pulmonary hypertension. *Pulm. Circ.* **2018**, *8*. [CrossRef] [PubMed]
143. Duarte, J.D.; Kansal, M.; Desai, A.A.; Riden, K.; Arwood, M.J.; Yacob, A.A.; Stamos, T.D.; Cavallari, L.H.; Zamanian, R.T.; Shah, S.J.; et al. Endothelial nitric oxide synthase genotype is associated with pulmonary hypertension severity in left heart failure patients. *Pulm. Circ.* **2018**, *8*. [CrossRef] [PubMed]
144. Vachiéry, J.L.; Tedford, R.J.; Rosenkranz, S.; Palazzini, M.; Lang, I.; Guazzi, M.; Coghlan, G.; Chazova, I.; De Marco, T. Pulmonary hypertension due to left heart disease. *Eur. Respir. J.* **2019**, *53*, 1801897. [CrossRef]
145. Szucs, B.; Szucs, C.; Petrekanits, M.; Varga, J.T. Molecular Characteristics and Treatment of Endothelial Dysfunction in Patients with COPD: A Review Article. *Int. J. Mol. Sci.* **2019**, *20*, 4329. [CrossRef]
146. Barberà, J.A.; Peinado, V.I.; Santos, S.; Ramirez, J.; Roca, J.; Rodriguez-Roisin, R. Reduced Expression of Endothelial Nitric Oxide Synthase in Pulmonary Arteries of Smokers. *Am. J. Respir. Crit. Care Med.* **2001**, *164*, 709–713. [CrossRef]
147. Nana-Sinkam, S.P.; Jong, D.L.; Sotto-Santiago, S.; Stearman, R.S.; Keith, R.L.; Choudhury, Q.; Cool, C.; Parr, J.; Moore, M.D.; Bull, T.M.; et al. Prostacyclin prevents pulmonary endothelial cell apoptosis induced by cigarette smoke. *Am. J. Respir. Crit. Care Med.* **2007**. [CrossRef]
148. Santos, S.; Peinado, V.I.; Ramírez, J.; Morales-Blanhir, J.; Bastos, R.; Roca, J.; Rodriguez-Roisin, R.; Barberà, J.A. Enhanced expression of vascular endothelial growth factor in pulmonary arteries of smokers and patients with moderate chronic obstructive pulmonary disease. *Am. J. Respir. Crit. Care Med.* **2003**. [CrossRef]
149. Carratu, P.; Scoditti, C.; Maniscalco, M.; Seccia, T.; Di Gioia, G.; Gadaleta, F.; Cardone, R.; Dragonieri, S.; Pierucci, P.; Spanevello, A.; et al. Exhaled and arterial levels of endothelin-1 are increased and correlate with pulmonary systolic pressure in COPD with pulmonary hypertension. *BMC Pulm. Med.* **2008**, *8*, 20. [CrossRef]
150. Xiong, P.Y.; Potus, F.; Chan, W.; Archer, S.L. Models and Molecular Mechanisms of World Health Organization Group 2 to 4 Pulmonary Hypertension. *Hypertension* **2018**, *71*, 34–55. [CrossRef]
151. Reimann, S.; Fink, L.; Wilhelm, J.; Hoffmann, J.; Bednorz, M.; Seimetz, M.; Dessureault, I.; Troesser, R.; Ghanim, B.; Klepetko, W.; et al. Increased S100A4 expression in the vasculature of human COPD lungs and murine model of smoke-induced emphysema. *Respir. Res.* **2015**, *16*, 127. [CrossRef] [PubMed]
152. Olschewski, H.; Behr, J.; Bremer, H.; Claussen, M.; Douschan, P.; Halank, M.; Held, M.; Hoeper, M.M.; Holt, S.; Klose, H.; et al. Pulmonary hypertension due to lung diseases: Updated recommendations from the Cologne Consensus Conference 2018. *Int. J. Cardiol.* **2018**, *272*, 63–68. [CrossRef]
153. Simonneau, G.; Torbicki, A.; Dorfmüller, P.; Kim, N. The pathophysiology of chronic thromboembolic pulmonary hypertension. *Eur. Respir. Rev.* **2017**, *26*, 160112. [CrossRef] [PubMed]
154. Yaoita, N.; Shirakawa, R.; Fukumoto, Y.; Sugimura, K.; Miyata, S.; Miura, Y.; Nochioka, K.; Miura, M.; Tatebe, S.; Aoki, T.; et al. Platelets are highly activated in patients of chronic thromboembolic pulmonary hypertension. *Arterioscler. Thromb. Vasc. Biol.* **2014**, *34*, 2486–2494. [CrossRef] [PubMed]
155. Humbert, M. Pulmonary arterial hypertension and chronic thromboembolic pulmonary hypertension: Pathophysiology. *Eur. Respir. Rev.* **2010**, *19*, 59–63. [CrossRef] [PubMed]
156. Sakao, S.; Hao, H.; Tanabe, N.; Kasahara, Y.; Kurosu, K.; Tatsumi, K. Endothelial-like cells in chronic thromboembolic pulmonary hypertension: Crosstalk with myofibroblast-like cells. *Respir. Res.* **2011**, *12*, 109. [CrossRef] [PubMed]
157. Mercier, O.; Arthur Ataam, J.; Langer, N.B.; Dorfmüller, P.; Lamrani, L.; Lecerf, F.; Decante, B.; Dartevelle, P.; Eddahibi, S.; Fadel, E. Abnormal pulmonary endothelial cells may underlie the enigmatic pathogenesis of chronic thromboembolic pulmonary hypertension. *J. Heart Lung Transpl.* **2017**, *36*, 305–314. [CrossRef]
158. Tura-Ceide, O.; Aventín, N.; Piccari, L.; Morén, C.; Guitart-Mampel, M.; Garrabou, G.; García-Lucio, J.; Chamorro, N.; Blanco, I.; Peinado, V.; et al. Endothelial dysfunction in patients with chronic thromboembolic pulmonary hypertension (CTEPH). *Eur. Respir. Soc.* **2016**, *48*, PA3606.
159. Naito, A.; Sakao, S.; Lang, I.M.; Voelkel, N.F.; Jujo, T.; Ishida, K.; Sugiura, T.; Matsumiya, G.; Yoshino, I.; Tanabe, N.; et al. Endothelial cells from pulmonary endarterectomy specimens possess a high angiogenic potential and express high levels of hepatocyte growth factor. *BMC Pulm. Med.* **2018**, *18*, 197. [CrossRef]
160. Quarck, R.; Wynants, M.; Verbeken, E.; Meyns, B.; Delcroix, M. Contribution of inflammation and impaired angiogenesis to the pathobiology of chronic thromboembolic pulmonary hypertension. *Eur. Respir. J.* **2015**, *46*, 431–443. [CrossRef]
161. Arthur Ataam, J.; Mercier, O.; Lamrani, L.; Amsallem, M.; Ataam, J.A.; Ataam, S.A.; Guihaire, J.; Lecerf, F.; Capuano, V.; Ghigna, M.R.; et al. ICAM-1 promotes the abnormal endothelial cell phenotype in chronic thromboembolic pulmonary hypertension. *J. Heart Lung Transplant.* **2019**, *38*, 982–996. [CrossRef] [PubMed]
162. Smolders, V.; Rodríguez, C.; Morén, C.; Blanco, I.; Osorio, J.; Piccari, L.; Bonjoch, C.; Quax, P.H.A.; Peinado, V.I.; Castellà, M.; et al. Decreased Glycolysis as Metabolic Fingerprint of Endothelial Cells in Chronic Thromboembolic Pulmonary Hypertension. *Am. J. Respir. Cell Mol. Biol.* **2020**, *63*, 710–713. [CrossRef] [PubMed]

163. Deng, C.; Zhong, Z.; Wu, D.; Chen, Y.; Lian, N.; Ding, H.; Zhang, Q.; Lin, Q.; Wu, S. Role of FoxO1 and apoptosis in pulmonary vascular remolding in a rat model of chronic thromboembolic pulmonary hypertension. *Sci. Rep.* **2017**, *7*, 2210. [CrossRef] [PubMed]
164. Newnham, M.; South, K.; Bleda, M.; Auger, W.R.; Barberà, J.A.; Bogaard, H.; Bunclark, K.; Cannon, J.E.; Delcroix, M.; Hadinnapola, C.; et al. The ADAMTS13-VWF axis is dysregulated in chronic thromboembolic pulmonary hypertension. *Eur. Respir. J.* **2019**, *53*, 1801805. [CrossRef] [PubMed]
165. Zabini, D.; Nagaraj, C.; Stacher, E.; Lang, I.M.; Nierlich, P.; Klepetko, W.; Heinemann, A.; Olschewski, H.; Bálint, Z.; Olschewski, A. Angiostatic factors in the pulmonary endarterectomy material from chronic thromboembolic pulmonary hypertension patients cause endothelial dysfunction. *PLoS ONE* **2012**, *7*, e43793. [CrossRef] [PubMed]
166. Conole, D.; Scott, L.J. Riociguat: First global approval. *Drugs* **2013**, *73*, 1967–1975. [CrossRef]
167. Prins, K.W.; Thenappan, T. WHO Group I Pulmonary Hypertension: Epidemiology and Pathophysiology. *Cardiol. Clin.* **2016**, *34*, 363–374. [CrossRef]
168. Hoeper, M.M.; Apitz, C.; Grünig, E.; Halank, M.; Ewert, R.; Kaemmerer, H.; Kabitz, H.J.; Kähler, C.; Klose, H.; Leuchte, H.; et al. Targeted therapy of pulmonary arterial hypertension: Updated recommendations from the Cologne Consensus Conference 2018. *Int. J. Cardiol.* **2018**, *272S*, 37–45. [CrossRef]
169. Lan, N.S.H.; Massam, B.D.; Kulkarni, S.S.; Lang, C.C. Pulmonary Arterial Hypertension: Pathophysiology and Treatment. *Diseases* **2018**, *6*, 38. [CrossRef]
170. Humbert, M.; Ghofrani, H.A. The molecular targets of approved treatments for pulmonary arterial hypertension. *Thorax* **2016**, *71*, 73–83. [CrossRef]
171. Suzuki, Y.J.; Ibrahim, Y.F.; Shults, N.V. Apoptosis-based therapy to treat pulmonary arterial hypertension. *J. Rare Dis. Res. Treat.* **2016**, *1*, 17–24. [PubMed]
172. Ibrahim, Y.F.; Wong, C.M.; Pavlickova, L.; Liu, L.; Trasar, L.; Bansal, G.; Suzuki, Y.J. Mechanism of the susceptibility of remodeled pulmonary vessels to drug-induced cell killing. *J. Am. Heart Assoc.* **2014**, *3*, e000520. [CrossRef] [PubMed]
173. Kim, S.Y.; Lee, J.H.; Huh, J.W.; Kim, H.J.; Park, M.K.; Ro, J.Y.; Oh, Y.M.; Lee, S.D.; Lee, Y.S. Bortezomib alleviates experimental pulmonary arterial hypertension. *Am. J. Respir. Cell Mol. Biol.* **2012**, *47*, 698–708. [CrossRef] [PubMed]
174. Jain, D.; Russell, R.R.; Schwartz, R.G.; Panjrath, G.S.; Aronow, W. Cardiac Complications of Cancer Therapy: Pathophysiology, Identification, Prevention, Treatment, and Future Directions. *Curr. Cardiol. Rep.* **2017**, *19*, 36. [CrossRef] [PubMed]
175. Voelkel, N.F.; Quaife, R.A.; Leinwand, L.A.; Barst, R.J.; McGoon, M.D.; Meldrum, D.R.; Dupuis, J.; Long, C.S.; Rubin, L.J.; Smart, F.W.; et al. Right ventricular function and failure: Report of a National Heart, Lung, and Blood Institute working group on cellular and molecular mechanisms of right heart failure. *Circulation* **2006**, *114*, 1883–1891. [CrossRef] [PubMed]
176. Yung, L.M.; Nikolic, I.; Paskin-Flerlage, S.D.; Pearsall, R.S.; Kumar, R.; Yu, P.B. A Selective Transforming Growth Factor-β Ligand Trap Attenuates Pulmonary Hypertension. *Am. J. Respir. Crit. Care Med.* **2016**, *194*, 1140–1151. [CrossRef]
177. Guo, Y.; Li, P.; Bledsoe, G.; Yang, Z.R.; Chao, L.; Chao, J. Kallistatin inhibits TGF-β-induced endothelial-mesenchymal transition by differential regulation of microRNA-21 and eNOS expression. *Exp. Cell Res.* **2015**, *337*, 103–110. [CrossRef]
178. Marsh, L.M.; Jandl, K.; Grünig, G.; Foris, V.; Bashir, M.; Ghanim, B.; Klepetko, W.; Olschewski, H.; Olschewski, A.; Kwapiszewska, G. The inflammatory cell landscape in the lungs of patients with idiopathic pulmonary arterial hypertension. *Eur. Respir. J.* **2018**, *51*, 1701214. [CrossRef]
179. Kumar, R.; Graham, B. How does inflammation contribute to pulmonary hypertension? *Eur. Respir. J.* **2018**, *51*, 1702403. [CrossRef]
180. Gu, M.; Shao, N.Y.; Sa, S.; Li, D.; Termglinchan, V.; Ameen, M.; Karakikes, I.; Sosa, G.; Grubert, F.; Lee, J.; et al. Patient-Specific iPSC-Derived Endothelial Cells Uncover Pathways that Protect against Pulmonary Hypertension in BMPR2 Mutation Carriers. *Cell Stem Cell* **2017**, *20*, 490–504. [CrossRef]
181. Spiekerkoetter, E.; Tian, X.; Cai, J.; Hopper, R.K.; Sudheendra, D.; Li, C.G.; El-Bizri, N.; Sawada, H.; Haghighat, R.; Chan, R.; et al. FK506 activates BMPR2, rescues endothelial dysfunction, and reverses pulmonary hypertension. *J. Clin. Investig.* **2013**, *123*, 3600–3613. [CrossRef] [PubMed]
182. Tu, L.; Desroches-Castan, A.; Mallet, C.; Guyon, L.; Cumont, A.; Phan, C.; Robert, F.; Thuillet, R.; Bordenave, J.; Sekine, A.; et al. Selective BMP-9 Inhibition Partially Protects Against Experimental Pulmonary Hypertension. *Circ. Res.* **2019**, *124*, 846–855. [CrossRef] [PubMed]
183. Yung, L.M.; Yang, P.; Joshi, S.; Augur, Z.M.; Kim, S.S.J.; Bocobo, G.A.; Dinter, T.; Troncone, L.; Chen, P.S.; McNeil, M.E.; et al. ACTRIIA-Fc rebalances activin/GDF versus BMP signaling in pulmonary hypertension. *Sci. Transl. Med.* **2020**, *12*. [CrossRef] [PubMed]
184. Sanada, T.J.; Sun, X.-Q.; Happé, C.; Guignabert, C.; Tu, L.; Schalij, I.; Bogaard, H.-J.; Goumans, M.-J.; Kurakula, K. Altered TGFβ/SMAD Signaling in Human and Rat Models of Pulmonary Hypertension: An Old Target Needs Attention. *Cells* **2021**, *10*, 84. [CrossRef]
185. Spiekerkoetter, E.; Sung, Y.K.; Sudheendra, D.; Bill, M.; Aldred, M.A.; van de Veerdonk, M.C.; Vonk Noordegraaf, A.; Long-Boyle, J.; Dash, R.; Yang, P.C.; et al. Low-Dose FK506 (Tacrolimus) in End-Stage Pulmonary Arterial Hypertension. *Am. J. Respir. Crit. Care Med.* **2015**, *192*, 254–257. [CrossRef] [PubMed]
186. Quarck, R.; Perros, F. Rescuing BMPR2-driven endothelial dysfunction in PAH: A novel treatment strategy for the future? *Stem Cell Investig.* **2017**, *4*, 56. [CrossRef]
187. Kurakula, K.; Sun, X.Q.; Happé, C.; da Silva Goncalves Bos, D.; Szulcek, R.; Schalij, I.; Wiesmeijer, K.C.; Lodder, K.; Tu, L.; Guignabert, C.; et al. 6-mercaptopurine, an agonist of Nur77, reduces progression of pulmonary hypertension by enhancing BMP signalling. *Eur. Respir. J.* **2019**, *54*, 1802400. [CrossRef]

188. Botros, L.; Szulcek, R.; Jansen, S.M.; Kurakula, K.; Goumans, M.T.; van Kuilenburg, A.B.P.; Vonk Noordegraaf, A.; de Man, F.S.; Aman, J.; Bogaard, H.J. The Effects of Mercaptopurine on Pulmonary Vascular Resistance and BMPR2 Expression in Pulmonary Arterial Hypertension. *Am. J. Respir. Crit. Care Med.* **2020**. [CrossRef]
189. Le Ribeuz, H.; Dumont, F.; Ruellou, G.; Lambert, M.; Balliau, T.; Quatredeniers, M.; Girerd, B.; Cohen-Kaminsky, S.; Mercier, O.; Yen-Nicolaÿ, S.; et al. Proteomic Analysis of KCNK3 Loss of Expression Identified Dysregulated Pathways in Pulmonary Vascular Cells. *Int. J. Mol. Sci.* **2020**, *21*, 7400. [CrossRef]
190. Huang, J.; Lu, W.; Ouyang, H.; Chen, Y.; Zhang, C.; Luo, X.; Li, M.; Shu, J.; Zheng, Q.; Chen, H.; et al. Transplantation of Mesenchymal Stem Cells Attenuates Pulmonary Hypertension by Normalizing the EndMT. *Am. J. Respir. Cell Mol. Biol.* **2019**, *62*, 49–60. [CrossRef]
191. de Mendonça, L.; Felix, N.S.; Blanco, N.G.; Da Silva, J.S.; Ferreira, T.P.; Abreu, S.C.; Cruz, F.F.; Rocha, N.; Silva, P.M.; Martins, V.; et al. Mesenchymal stromal cell therapy reduces lung inflammation and vascular remodeling and improves hemodynamics in experimental pulmonary arterial hypertension. *Stem Cell Res. Ther.* **2017**, *8*, 220. [CrossRef] [PubMed]
192. Martire, A.; Bedada, F.B.; Uchida, S.; Pöling, J.; Krüger, M.; Warnecke, H.; Richter, M.; Kubin, T.; Herold, S.; Braun, T. Mesenchymal stem cells attenuate inflammatory processes in the heart and lung via inhibition of TNF signaling. *Basic Res. Cardiol.* **2016**, *111*, 54. [CrossRef] [PubMed]
193. Macias, D.; Moore, S.; Crosby, A.; Southwood, M.; Du, X.; Tan, H.; Xie, S.; Vassallo, A.; Wood, A.J.; Wallace, E.M.; et al. Targeting HIF2α-ARNT hetero-dimerisation as a novel therapeutic strategy for Pulmonary Arterial Hypertension. *Eur. Respir. J.* **2020**. [CrossRef] [PubMed]
194. Hu, C.J.; Poth, J.M.; Zhang, H.; Flockton, A.; Laux, A.; Kumar, S.; McKeon, B.; Mouradian, G.; Li, M.; Riddle, S.; et al. Suppression of HIF2 signalling attenuates the initiation of hypoxia-induced pulmonary hypertension. *Eur. Respir. J.* **2019**, *54*, 541900378. [CrossRef]
195. Bogaard, H.J.; Mizuno, S.; Al Hussaini, A.A.; Toldo, S.; Abbate, A.; Kraskauskas, D.; Kasper, M.; Natarajan, R.; Voelkel, N.F. Suppression of histone deacetylases worsens right ventricular dysfunction after pulmonary artery banding in rats. *Am. J. Respir. Crit. Care Med.* **2011**, *183*, 1402–1410. [CrossRef]
196. Wang, Y.; Yan, L.; Zhang, Z.; Prado, E.; Fu, L.; Xu, X.; Du, L. Epigenetic Regulation and Its Therapeutic Potential in Pulmonary Hypertension. *Front. Pharmacol.* **2018**, *9*, 241. [CrossRef]

Review

Endothelial Dysfunction in Diabetes Is Aggravated by Glycated Lipoproteins; Novel Molecular Therapies

Laura Toma, Camelia Sorina Stancu and Anca Volumnia Sima *

Lipidomics Department, Institute of Cellular Biology and Pathology "Nicolae Simionescu" of the Romanian Academy, 8, B.P. Hasdeu Street, 050568 Bucharest, Romania; laura.toma@icbp.ro (L.T.); camelia.stancu@icbp.ro (C.S.S.)
* Correspondence: anca.sima@icbp.ro

Abstract: Diabetes and its vascular complications affect an increasing number of people. This disease of epidemic proportion nowadays involves abnormalities of large and small blood vessels, all commencing with alterations of the endothelial cell (EC) functions. Cardiovascular diseases are a major cause of death and disability among diabetic patients. In diabetes, EC dysfunction (ECD) is induced by the pathological increase of glucose and by the appearance of advanced glycation end products (AGE) attached to the plasma proteins, including lipoproteins. AGE proteins interact with their specific receptors on EC plasma membrane promoting activation of signaling pathways, resulting in decreased nitric oxide bioavailability, increased intracellular oxidative and inflammatory stress, causing dysfunction and finally apoptosis of EC. Irreversibly glycated lipoproteins (AGE-Lp) were proven to have an important role in accelerating atherosclerosis in diabetes. The aim of the present review is to present up-to-date information connecting hyperglycemia, ECD and two classes of glycated Lp, glycated low-density lipoproteins and glycated high-density lipoproteins, which contribute to the aggravation of diabetes complications. We will highlight the role of dyslipidemia, oxidative and inflammatory stress and epigenetic risk factors, along with the specific mechanisms connecting them, as well as the new promising therapies to alleviate ECD in diabetes.

Keywords: diabetes; hyperglycemia; glycated lipoproteins; glycated LDL; glycated HDL; endothelial cell dysfunction; molecular mechanisms; epigenetic factors; therapeutic approaches

1. Introduction

The prevalence of diabetes mellitus (DM) is rapidly increasing worldwide [1]. The decreased quality of life of diabetic patients and the social and economic burden of this disease emphasize the need to establish the causative mechanisms of DM that will finally allow the identification of new therapies to cure diabetes and its associated vascular complications. Cardiovascular diseases (CVD) are the clinical manifestations of atherosclerosis, which represents one of the main vascular threats of diabetes. Published data show that the risk of acute cardiovascular events (such as stroke or myocardial infarction) is seven to ten times higher in diabetic patients compared to non-diabetic subjects [2]. In addition, microvascular afflictions, including retinopathy, nephropathy, neuropathy and limb ischemia, occur at a very high rate in diabetic patients compared to non-diabetic individuals [2].

The primary cause of the pathophysiologic alterations of the diabetic patient's vasculature is the exposure to high levels of blood glucose. It is well known that high glucose (HG) can induce vascular complications in diabetic patients by affecting the normal function of the vessel wall's cells. Unfortunately, large-scale clinical studies have shown that despite good glycemic control, the vascular complications persist and even evolve [3,4]. This phenomenon is known as the "metabolic memory" of the cells [5]. The first cells of the vessel wall exposed to plasma HG are the endothelial cells (EC). Constant plasma hyperglycemia or intermittent HG due to poor glycemic control induces EC dysfunction (ECD) [6–8]. ECD is considered a critical step in the initiation and evolution of atherosclerosis [9].

It favors an increased trans-endothelial transport of plasma proteins and lipoproteins (Lp), stimulates the adhesion and sub-endothelial transmigration of blood monocytes, supports the migration and proliferation of vascular smooth muscle cells (SMC) from the media to the intima and impedes the fibrinolytic processes, finally increasing the risk of cardiovascular events in diabetic patients [10]. Prolonged plasma HG induces also the formation of advanced glycation end products (AGEs), which by non-enzymatic attachment to proteins compromise their proper functioning. The interaction between the receptor for AGE (RAGE) and AGE proteins activates numerous signaling pathways and represents a powerful determinant of ECD [11]. Glycated lipoproteins (gLp), which are formed in excess in the plasma of diabetic patients, are ligands for RAGE and contribute substantially to ECD.

The aim of the present review is to select and summarize the molecular mechanisms that determine ECD due to cells' interaction with gLp and to present new therapeutic strategies to alleviate CVD in diabetes. Special attention is given to the interaction of EC with glycated low-density lipoproteins (gLDL) and glycated high-density lipoproteins (gHDL) as important players in the accelerated-atherosclerotic process in diabetes.

2. Structural and Biochemical Alterations of Proteins and Lp Induced by High Glucose

2.1. Generation of Advanced Glycation End Products

Diabetes is a metabolic disorder affecting people worldwide, its major complication being vascular diseases, especially accelerated atherosclerosis. Characteristic of diabetes is the increased levels of blood glucose. Hyperglycemia induces the non-enzymatic glycation of blood proteins, resulting in AGE formation [12]. Extracellular glycation of different molecules can be the result of Maillard reactions or of prolonged oxidative stress. In Maillard reactions, AGEs formation starts with the condensing reaction between the carbonyl group from glucose (or other reducing sugars) and the primary amino groups of proteins, lipoproteins, nucleic acids or other molecules. The reversible Schiff bases formed are transformed into Amadori products and then into various irreversible, crosslinked, fluorescent and chemically reactive adducts [13]. The rearrangement of Amadori products to form AGEs occurs via oxidative or non-oxidative processes. The rearrangement of Schiff bases takes place at an alkaline pH, while the Amadori rearrangements start at a low pH, having a lower reaction rate [14].

Another source of AGEs is the advanced lipoxidation end products (ALEs) that result from successive oxidation cascades. The formation of ALEs starts with the peroxidation of lipids (such as polyunsaturated fatty acids from cellular membranes), leading to reactive carbonyl species (RCS), such as malondialdehyde (MDA) or 4-hydroxy-trans-2-nonenal (4-HNE), or to α-oxoaldehydes, such as glyoxal, methylglyoxal, 3-deoxyglucuson and acrolein [15]. After further arrangements, RCS can generate AGEs such as Nε-carboxyethyllysine (CEL), arginine pyrimidine, pentosidine, pyrralin, glyoxal lysine and Nε-carboxymethyllysine (CML) [14].

Although protein glycation takes place in vivo in tissues and fluids even under physiological conditions, in diabetes, this reaction takes place at a faster rate because of the increased availability of glucose and its catabolic aldehydes, such as glycoaldehyde and methylglyoxal and increased oxidative stress [16,17]. Most of the time, the AGEs and ALEs coexist in vivo, acting together in macromolecular complexes, such as Lp, and contributing to the aggravation of the vascular damage in diabetes.

2.2. Irreversible Glycation of Lipoproteins

The formation of AGEs represents an important mechanism in the propagation of the atherosclerotic process in diabetes. AGEs accumulate on long-lived proteins, affecting the cellular membranes, cytosolic proteins, the components of the basal lamina and of the extracellular matrix [18]. Plasma Lp can also be modified by excessive glucose, resulting in non-enzymatically glycated Lp (AGE-Lp).

Two classes of Lp, the low-density Lp (LDL) with pro-atherogenic potential and the protective, anti-atherogenic high-density Lp (HDL), are mainly involved in the atherosclerotic process. LDLs are the main cholesterol carriers in human plasma and, in pathological conditions, LDLs accumulate in the subendothelium as extracellular modified Lp, contributing to the atherosclerotic process by releasing pro-inflammatory bioactive lipids (such as oxidized phospholipids), inducing immune responses (by promoting the recruitment of immuno-inflammatory cells, such as monocytes, neutrophils, lymphocytes, or dendritic cells) or by being taken up by macrophages and determining the formation of foam cells [19]. Apart from LDL, lipoprotein (a) (Lp(a)) has received increasing interest in recent years and is now recognized as an important independent risk factor for CVD in patients with or without type 2 DM (T2DM) [20]. Lp(a) is composed of an LDL-like particle that binds to apoprotein(a) (apo(a)) via plasminogen-like domains, has structural similarity to plasminogen and tissue plasminogen activator (t-PA) and thereby may induce diminished fibrinolysis and even thrombogenesis due to the stimulation of the secretion of plasminogen activator inhibitor-1 (PAI-1). Further, Lp(a) carries cholesterol and binds atherogenic proinflammatory oxidized phospholipids, thus impairing the endothelial proper function and stimulating the attraction of inflammatory cells [21]. HDL plays important roles in maintaining the homeostasis of the vascular system, mainly by exerting antioxidant and anti-inflammatory effects and by participating in the reverse cholesterol transport process [19].

In LDL, glycosylation affects the apolipoprotein B (apoB), the main protein of this class of Lp. The irreversible glycation of native LDL (nLDL) starts with the non-enzymatic addition of reducing sugars to the positively charged arginine and/or lysine residues of apoB and continues with the formation of sugar-amino acid adducts, collectively known as AGE (Figure 1).

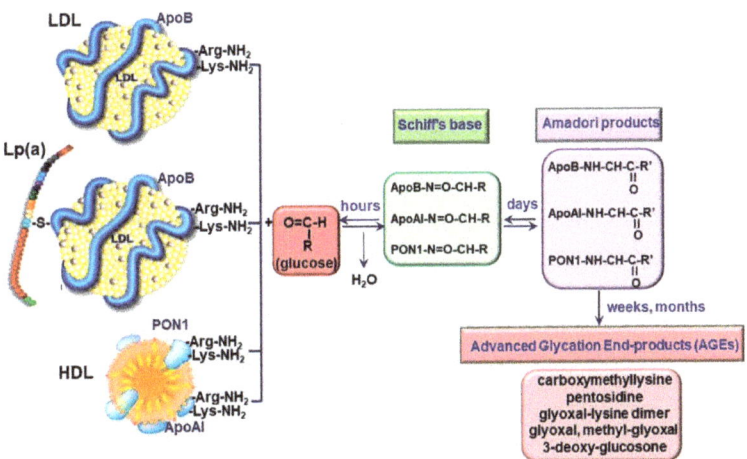

Figure 1. Formation of irreversibly glycated lipoproteins by Maillard reactions.

Since the arginine and lysine residues from apoB are important for the specific recognition by the LDL receptor (LDLR), an impaired LDLR-mediated uptake, a decreased gLDL clearance and an increased gLDL mean lifetime in the plasma of diabetic patients occur [16,22]. Furthermore, gLDL has an increased susceptibility towards oxidative modification, a critical step in atherogenesis [17]. It was reported that irreversible glycated LDL (AGE-LDL) has a 5-fold increased mean level of lipid peroxides, a reduction of the free amino groups of apoB and a higher gel electrophoresis mobility, reflecting the increase in the overall negative charge and the loss of positive charges [23]. Later, Deleanu et al. showed that AGE-LDL has an increased content of conjugated dienes, 4-HNE, MDA and

7-ketocholesterol compared to nLDL. In addition, the composition in free fatty acids is different in AGE-LDL compared to nLDL, with a decrease of linoleic acid, dihomogamma-linolenic acid and arachidonic acid content being observed [24].

The levels of circulating Lp(a) are increased in T2DM and type 1 DM (T1DM) patients and positively correlated with blood concentrations of glycated hemoglobulin A1c. In addition, the levels of glycated Lp(a) are elevated in diabetic patients [25].

In the plasma of diabetic subjects, HDL main proteins become glycated; thus HDL lose their protective function. Godfrey et al. showed that apoAI, the main protein of HDL, is glycated in HDL isolated from diabetic patients. The in vitro glycation of apoAI appears at sites involved in membrane fusion and ligand binding and induces structural alterations of HDL particles. These modifications determine the decrease in HDL stability and functionality in plasma [26]. Kashyap et al. reported that the glycation of apoAI at Lys residues (Lys-12, Lys-96, Lys-133, and Lys-205) favors the appearance of apoAI crosslinking and is associated with a decreased apoAI half-life compared to its non-glycated form [27]. Proteomic analysis revealed changes in 7 of the 45 identified proteins in HDL from T2DM patients, including apolipoprotein (apo) A-II, apoE and PON-1. The data demonstrate early changes in the lipid and protein composition of specific HDL subspecies in adolescents with T2DM that are related to early markers of arterial disease [28].

2.3. Receptors for AGE-Proteins

The interaction of AGE-proteins with the cells is mediated by specific receptors. These include the type I cell surface receptor for AGEs (RAGE), AGE-receptor complexes (AGE-R1/OST-48, AGE-R2/80K-H, AGE-R3/galectin-3) and some members of the scavenger receptors family (SR-A; CD36, SR-BI, LOX-1; FEEL-1; FEEL-2) [14]. Unlike AGE-receptor complexes, which take up AGEs and participate in their clearance from circulation, RAGE interaction with AGEs leads to the activation of the intracellular signaling pathways, leading to increased oxidative and inflammatory stress [11,29]. RAGE belongs to the immunoglobulin (Ig) superfamily and is expressed on various cell types: EC, SMC, monocytes/macrophages, T-lymphocytes, dendritic cells, fibroblasts, neuronal cells, glia cells, chondrocytes, keratinocytes. Besides AGE, RAGE recognizes a large number of ligands, including HMGB1, S100/calgranulin protein, amyloid β peptide and lipopolysaccharides (LPS) [13]. Essential for RAGE signaling is Diaphanous1 (DIAPH1)/mammalian diaphanous-related formin (mDia1), a protein involved in the cytoskeleton organization that interacts directly with the cytoplasmic domain of RAGE. Key down-effectors of DIAPH 1 include the activation of Src kinase, Rho GTP-ases (including cdc42 and Rac), glycogen synthase kinase3b (GSK3b), AKT and Rho-associated, coiled-coil-containing protein kinase (ROCK). Activation of these signaling pathways was linked with different pathological situations including myocardial ischemia, diabetes-associated nephropathy, retinopathy and inflammation [30,31]. Besides mDia1, it was reported that the RAGE cytoplasmic domain can interact also with ERK and the adaptor protein for toll-like receptors (TIRAP) [32,33].

Multiple alternative splice forms of RAGE with partial functionality were also discovered. Of them, three isoforms are most important: (1) the N-truncated RAGE lacking the extracellular V-domain, thus preventing AGE binding; (2) the dominant-negative RAGE that lacks an intracellular domain, remaining anchored to the cell surface and permitting AGE interaction, but without stimulating the intracellular signaling; and (3) the endogenous secreted RAGE (esRAGE) that lacks both the transmembrane region and intracellular domain. Besides these isoforms, another soluble form of RAGE is formed by the actions of MMP-9 and ADAM-10 matrix metalloproteinases that cleave the cytosolic domain of the RAGE protein [34]. Together with esRAGE, this form of truncated RAGE provides the soluble form of RAGE (sRAGE). Since sRAGE is lacking the transmembrane and cytoplasmic domains, the ligand binding to sRAGE is unable to trigger the intracellular signaling cascades. These features make sRAGE an anti-inflammatory, effective decoy and competitive inhibitor for full-length RAGE, being involved in the scavenging and clearance of AGEs from circulation [11]. The pathophysiological role of RAGE will be discussed in the next sections.

3. Endothelial Cell Dysfunction in Diabetes

ECs are instrumental for maintaining the homeostasis of the vascular system due to their multiple functions. It is known that ECs participate in the regulation of the vascular tone by secreting different vasodilators (such as nitric oxide, prostacyclin) and vasoconstrictors (endothelin, thromboxanes). They act as a selective barrier to control the exchange of macromolecules between the blood and tissues, control the extravasation and the traffic of pro-inflammatory leucocytes by regulating the expression of the cell adhesion molecules and cytokines and keep the balance between the pro-thrombotic and pro-fibrinolytic factors [35,36].

In diabetes, the primary metabolic modification is the chronically elevated blood glucose. ECs are the first cells of the vascular wall that interact with the blood-increased glycemia and suffer structural and functional alterations [18,35]. The structural modifications of EC in diabetes start with the switch to a secretory phenotype, as demonstrated by the overdevelopment of the rough endoplasmic reticulum (RER) and Golgi complexes, the enrichment of the intermediary filaments and Weibel–Palade bodies, the enlargement of the inter-endothelial junctions, and the increase in the number of plasmalemmal vesicles favoring the formation of transendothelial channels [37,38]. These modifications determine the formation of a hyperplasic basal lamina and the increase in EC permeability, favoring the subendothelial accumulation of native and modified LDLs, which contribute to atheroma formation [18].

ECD in diabetes is well documented and is regarded as an important player in the pathogenesis of CVD [39]. Dysfunctional ECs suffer a shift to a vasoconstrictor, prothrombotic and pro-inflammatory phenotypes. RAGE plays an important role in the development of the vascular complications associated with diabetes. It was demonstrated that deletion of RAGE in different animal models determines a significant attenuation of the atherosclerotic process [40,41]. The specific deletion of the cytoplasmic domain of RAGE of EC in transgenic mice was associated with a decrease in the inflammatory stress, revealing a prominent role for RAGE in ECD [40]. Studies in cultured EC have demonstrated that the interaction of different AGEs with RAGE determines the development of oxidative stress by activation of NADPH oxidase or induction of mitochondrial dysfunction [11,42,43] and lowers the bioavailability of nitric oxide (NO) [35]. In addition, series of intracellular phosphorylation reactions leading to the activation of MAPK (such as ERK1/2, p38), the enhancing of Jak/Stat signaling pathway and the activation of nuclear factor kappa B (NF-kB) are induced [14]. These effects result in the stimulation of the synthesis of pro-inflammatory cytokines and chemokines, including interleukin-6 (IL-6), monocyte chemoattractant protein 1 (MCP-1), tumor necrosis factor α (TNFα) and transforming growth factor β (TGF-β) and the overexpression of adhesion molecules, such as vascular cell adhesion molecule (VCAM-1) or intracellular cell adhesion molecule (ICAM-1), exacerbating the atherosclerotic process in diabetes [29,44,45]. Of great importance, it was demonstrated that AGEs increase the endothelial hyper-permeability by dissociating the adherens junctions through RAGE-mDia1 binding [46]. Interestingly, active NF-kB is also involved in the transcription of RAGE and of some of its ligands (such as HMGB1). Thus, the primary activation of RAGE unfortunately generates a positive feedback loop of self-sustained activation cycle, through NF-kB, amplifying the deleterious effects of AGE/RAGE interactions [47] (Figure 2).

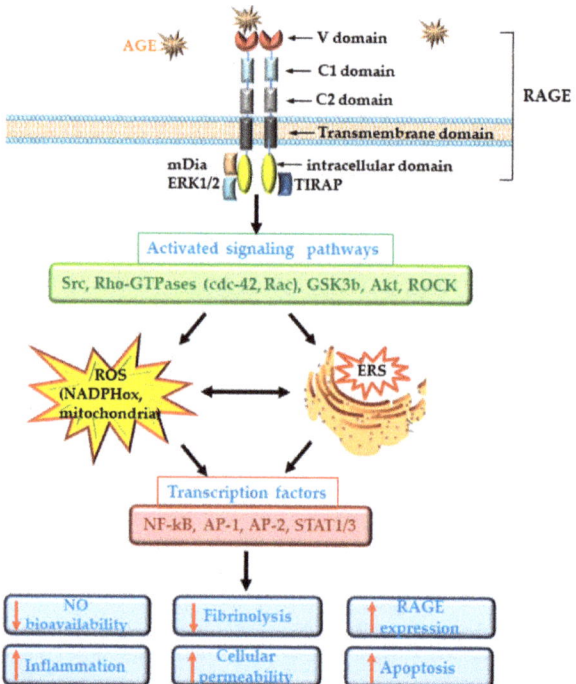

Figure 2. Schematic representation of advanced glycation end products (AGEs)/receptor for AGE (RAGE) interactions. The figure depicts the main signaling pathways, cellular processes and transcription factors involved in the generation of cellular dysfunction determined by AGEs. The black, thick arrows indicate the succesive activation of different signaling pathways and transcription factors stimulated by AGEs/RAGE interaction; the two headed black arrows indicate the interconnection between oxidative stress (ROS) and endoplasmic reticulum stress (ERS); the red arrows indicate the stimulation (up-headed arrows) or inhibition (down-headed arrows) of cellular processes which determine the dysfunction of endothelial cells.

In diabetic patients, the alteration of EC function was measured as: decreased forearm blood flow [48,49], increased levels of soluble adhesion molecules such as E-selectin, soluble VCAM-1 or soluble ICAM-1 [50,51], elevated plasma levels of von Willebrand factor (vWF) and PAI-1 [49,52,53]. More than being just a consequence of diabetes, ECD plays an important role in the development of microvascular (nephropathy, retinopathy, neuropathy) and macrovascular (ischemic heart disease, stroke, peripheral vascular disease) complications of diabetes [39].

4. Glycated Lipoproteins Detrimental Actions in Endothelial Cells

4.1. Reduction of Nitric Oxide Bioavailability

NO is essential in maintaining the functionality of the vascular system, due to its wide pleiotropic beneficial actions, having antiplatelet adhesion, anti-inflammatory, antioxidant and anti-apoptotic properties [35]. In diabetic conditions, the NO bioavailability is decreased due to the reduced endothelial NO synthesis or due to the interactions of NO with the excess pro-oxidant molecules (such as superoxide anion). In EC, NO is synthesized from L-arginine and O_2 by the constitutive endothelial nitric oxide synthase (eNOS or NOS3) in a reaction that produces stoichiometric amounts of l-citrulline and NO. Besides L-arginine, the presence of cofactors such as heme, tetrahydrobiopterin (BH_4), flavin adenine mononucleotide (FMN), flavin adenine dinucleotide (FAD) and NADPH is also required. ENOS functions as a homodimer, and its regulation is complex, involving transcriptional, post-transcriptional and post-translational mechanisms; any alteration of these mechanisms resulting in the decrease of NO production. Interaction with calmodulin, palmitoylation and myristoylation, phosphorylation by protein kinase B (Akt) or AMP-activated protein kinase (AMPK) at Ser1177 (Ser1179), and its localization in the cell's membrane, determine eNOS activation. In contrast, the interactions with caveolin or the phosphorylation at Thr495/Tyr657 by the redox-active kinases (protein kinase C, PKC) promotes the inactivation of eNOS and the subsequent decrease of NO production [54]. The reactive oxygen species (ROS) can decrease NO synthesis through an interesting molecular mechanism: ROS lowers the levels of BH_4 or L-arginine, and eNOS becomes "uncoupled", generating superoxide instead of NO, thus exacerbating the oxidative stress [35]. In addition, the reaction between the superoxide molecule and NO determines the formation of peroxynitrites (ONOO-), highly reactive nitrogen species that exacerbate the oxidative stress and decrease the bioavailability of NO. An important participant in the peroxynitrite formation is the inducible NO synthase (iNOS) that can be stimulated by the interaction of AGE with RAGE [11]. Unlike eNOS, which synthesizes NO in small quantities over long periods of time, iNOS synthesizes large amounts of NO in short periods of time, thus stimulating the formation of peroxynitrites in a pro-oxidant medium and amplifying the oxidative stress [35,54].

The specific effects of gLp on NO synthesis and bioavailability (synthesized as shown in Figure 3a) were studied mainly in cultured ECs from various sources. Rabini et al. reported that the incubation for a short period of time of human aortic EC (HAEC) with LDL isolated from T1DM patients stimulates the production of NO in parallel with that of peroxynitrites [55]. Comparing the effects of gLDL obtained in vitro with those isolated from diabetic patients, Artwolh et al. demonstrated that both gLDLs determine the reduction of eNOS expression [56]. In agreement with this, Toma et al. showed that a 24 h incubation of cultured EC with 100 µg/mL AGE-LDL induces a slight inhibition of eNOS expression while increasing iNOS expression and ROS production [57]. Nair et al. confirmed and added to these results by showing that physiological concentrations of gLDL reduce the abundance of eNOS protein and its activity by a mechanism involving the stimulation of RAGE/H-Ras signaling pathway and of endoplasmic reticulum stress (ERS) [58]. The contribution of AGE-LDL to the alteration of the vascular reactivity was evidenced by using the myograph technique applied to mesenteric arteries from normal hamsters. Compared to nLDL, AGE-LDL decreased by 60% the relaxation of normal mesenteric arteries when stimulated with acetylcholine and by 40% when stimulated with sodium nitroprusside, demonstrating that AGE-LDLs affect both EC- and SMC-dependent relaxation of the arteries [59]. An interesting mechanism that might explain eNOS inhibition in cultured EC exposed to glycated and oxidized LDL (glyc-oxLDL) was described by Dong et al. [60]. They observed that the decrease in eNOS protein levels induced by glyc-oxLDL was accompanied by an increase in intracellular Ca(2+) levels, ROS production, and Ca(2+)-dependent calpain activity. By using specific pharmacologic inhibitors and silencing RNA, Dong et al. demonstrated that eNOS protein levels are decreased due to the activation of Ca(2+)-dependent calpain protease that upregulates the degradation of eNOS in EC exposed to glyc-oxLDL [60]. These studies prove without a doubt that gLDLs determine the decrease of NO bioavailability by affecting various molecular mechanisms, thus providing insights related to the development of cardiovascular complications in diabetes. The mechanisms by which gLDL induce ECD are summarized in Figure 3a.

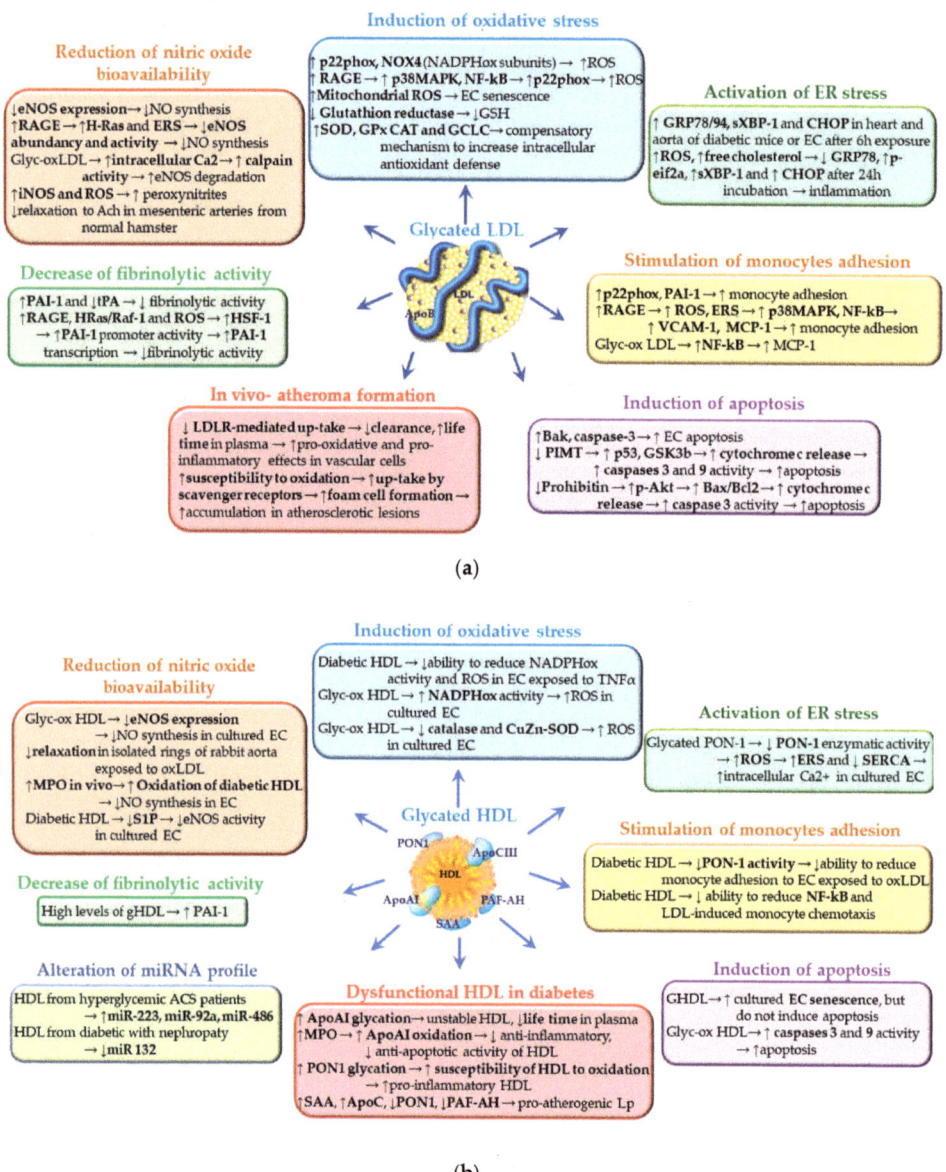

Figure 3. Molecular mechanisms of gLDL (**a**) or gHDL (**b**) contributing to endothelial cell dysfunction. The blue arrows indicate the detrimental actions of gLp in endothelial cells; the small black arrows indicate the specific molecular mechanisms by which gLp determine the dysfunction of endothelial cells as showed by experimental data.

It is known that one of the anti-atherosclerotic properties of native HDL is the stimulation of NO production in EC, determining the endothelium-dependent vasorelaxation [61]. Unfortunately, in diabetic patients, HDL is largely glycated [62]. The effects of gHDL on NO bioavailability (Figure 3b) were analyzed in a few studies. Matsunaga et al. demonstrated that 48h exposure of HAEC to 100 μg/mL glycated and oxidized HDL determines the decrease in eNOS expression and NO production [63]. Persegol et al. reported for the first time that HDL isolated from T2DM patients does not preserve the endothelium-dependent vasorelaxation function in isolated rings of rabbit aorta incubated with oxidized LDL [64]. In good agreement, in a randomized controlled trial, Sorrentino et al. showed that HDLs isolated from T2DM patients lose their ability to stimulate NO production from cultured HAEC and NO-mediated vasodilation of intact arterial segments. These effects were accompanied by an increase in gHDL peroxidation and HDL-associated myeloperoxidase (MPO) protein and activity, suggesting a mechanism in which MPO plays an important role [65]. More recently, it was shown that HDL isolated from T2DM loses its ability to stimulate eNOS activity and its capacity to suppress the NF-κB-mediated inflammatory response in TNFα-exposed EC. Interestingly, the loss of HDL's ability to stimulate eNOS activity was correlated with a decrease of sphingosine-1-phosphate (S1P) levels in the plasma of T2DM; the authors suggested that the loss of S1P is a possible mechanism to explain the inability of HDL to exert its protective functions, contributing to the generation of vascular complications in diabetes [66].

The presented studies converge on the idea that gHDLs reduce NO bioavailability; however, more studies are needed for further clarification of the involved mechanisms. The mechanisms by which gHDLs induce ECD are summarized in Figure 3b.

4.2. Induction of Oxidative Stress

Under physiological conditions, ROS act as essential mediators of the redox signaling pathways involved in various cellular responses including proliferation, migration, differentiation or gene expression. In pathological situations, ROS is generated in excess, becoming toxic for cells, which leads to oxidation of molecules, enhancement of the inflammatory response, cellular aging or apoptosis [67]. Oxidative stress results from the imbalance between ROS production and ROS detoxification. In EC, the main sources of ROS are: the NADPH oxidases complex (specifically Nox2 and Nox4), the mitochondria and the uncoupled eNOS [35]. The cellular antioxidant defense system comprises a set of antioxidant enzymes, superoxide dismutase (SOD), catalase (CAT), glutathione peroxidase (GPx) and the thioredoxin system, which work together to decompose ROS, in particular the superoxide, to water and oxygen [67,68]. In EC, the antioxidant enzymes are stimulated by oxidative stress and participate actively in the adaptive response of the cells to ROS [69]. They have the ability to regenerate the proteins inactivated by oxidative stress and play an important role in EC survival by inhibiting the activation of c-Jun N-terminal kinase (JNK) and p38 mitogen-activated protein kinase (p38 MAPK) [67].

4.2.1. Upregulation of the Main EC Pro-Oxidant Proteins by gLp

The induction of oxidative and inflammatory stress by gLp has been well documented, especially in the case of gLDL. Toma et al. reported that AGE-LDL stimulates NADPH oxidase activity by upregulating the p22phox and NOX-4 subunits and determines the increase of MCP-1 released in EC culture medium [70]. The upregulation of p22phox is p38 kinase and NF-kB-dependent, due to RAGE/AGE-LDL interaction. In addition, the authors showed that all these effects are significantly augmented in the presence of HG in the culture media. Interestingly, exposure of EC to nLDL in HG culture medium compared to normal glucose determines the stimulation of RAGE and NADPH subunits (p22phox, NOX4, and p67phox), supporting the concerted pathogenic potential of hyperglycemia and dyslipidemia in T2DM patients [70]. These results were confirmed by other groups, who reported that gLDLs increase the oxidative stress in cultured EC by stimulating the RAGE/NADPH oxidase axis [44,71]. Recent studies [44,45,72] show that stimulation of the oxidative stress was followed by the upregulation of the pro-thrombotic PAI-1 or of various pro-inflammatory proteins (C reactive protein, MCP-1, VCAM-1) that exacerbate the inflammatory state and stimulate monocyte adhesion to EC (Figure 3a).

Mitochondria play an important role in ROS generation, calcium homeostasis and cell survival. Increasing evidence demonstrates that functional alterations of mitochondria are involved in the promotion of ECD in diabetes [73]. Few studies focusing on the understanding of the effects of gLp on mitochondrial functioning have been done. It was reported that compared to nLDL, gLDL stimulates mitochondrial ROS generation in ECs [74,75]. ROS generation was accompanied by the attenuation of the activity of key enzymes from the mitochondrial electron transport chain, impaired mitochondrial oxygen consumption and the reduction of mitochondrial membrane potential [74]. These results were confirmed by a recent study [76] showing that the L5 subfraction of electronegative LDL augments mitochondrial free-radical production in cultured HAEC, leading to premature vascular endothelial senescence (Figure 3a).

HDLs isolated from diabetic patients, in contrast to HDL isolated from healthy subjects, lose their ability to reduce superoxide production and NADPH oxidase activity in TNFα-stimulated ECs [65]. Matsunaga et al. showed that oxidation of gHDL stimulates the increase of H_2O_2 by activating NADPH oxidase [63] (Figure 3b). Therefore, innovative therapies designed to increase endogenous antioxidants, in particular those associated with HDLs, will be valuable anti-diabetic treatment in addition to the existing ones.

4.2.2. Modulation of the Activity of the Cellular Antioxidant Defense System by gLp

The cellular antioxidant defense system plays an important role in ROS detoxification, being critical in maintaining the cellular proper function. Studying the impact of gLp on the endothelial antioxidant system, Zhao et al. observed that both nLDLs and gLDLs induce stimulation of SOD, GPx and CAT activity after 24 h incubation with cultured ECs. Interestingly, the lowering of glutathione reductase, the enzyme responsible for the restoration of reduced glutathione (GSH) pool, was observed in gLDL-exposed ECs versus nLDL. In agreement with this result, the authors showed that GSH levels in EC are reduced by gLDLs [77]. Another antioxidant protein shown to be affected by gLDL was the catalytic subunit of glutamate cysteine ligase (GCLC), the first rate-limiting enzyme for GSH synthesis. Toma et al. reported an increase of GCLC gene expression in gLDL-exposed EC compared to cells exposed to nLDLs. Interestingly, using specific ERS inhibitors, the authors observed the decrease of GCLC levels to normal values. This result suggests that GCLC is increased as an attempt of the cells to restore their intracellular antioxidant protection by re-establishing the GSH pool consumed for the proper folding of proteins accumulated in the ER when the cells are exposed to gLDL [45].

The exposure of EC to glyc-oxHDL determined the down-regulation of catalase and Cu(2+), Zn(2+)-superoxide dismutase (CuZn-SOD) expression resulting in the formation of increased levels of H_2O_2 [63]. No specific studies regarding the effect of gHDL on ECs' antioxidant defense have been published. The results are presented summarized in Figure 3a,b.

4.3. Activation of Endoplasmic Reticulum Stress

The ER is an organelle with multiple functions, involved in maintaining homeostasis, function and survival of the cells. The main role of the ER is the synthesis of proteins and lipids, the ER lumen being the place of residence for many foldases and chaperones (e.g., GRP78, GRP94) needed for proper protein folding. Due to its importance in cellular homeostasis, the ER functions are highly regulated. However, in pathological situations, in the lumen of the ER, a high amount of misfolded proteins may accumulate, inducing ERS [78]. To resolve ERS, cells initiate the unfolded protein response (UPR), a set of three signaling pathways meant to restore ER homeostasis, as follows: (1) the protein kinase-like ER kinase (PERK)/eukaryotic initiation factor (eIF)-2α branch to attenuate the novo translation of proteins; (2) the activating transcription factor (ATF)-6 upregulates the transcription of molecular chaperones needed to increase the capacity of ER folding; (3) the inositol-requiring enzyme 1 alpha (IRE1α)/spliced X-box binding protein 1 (sXBP-1) simulates the transcription of chaperones and activates the ER-associated protein degradation (ERAD) machinery. However, if ERS is prolonged and the cells do not restore the ER proper functions, the UPR generates the activation of the pro-inflammatory signaling cascades by the activation of NF-κB, p38MAPK and c-Jun N-terminal Kinase (JNK), finally leading to cellular apoptosis through the activation of CCAAT/enhancer-binding protein homologous protein (CHOP) and inhibition of anti-apoptotic Bcl-2, in parallel with the stimulation of the pro-apoptotic Bim [79].

It was reported that stressors such as high glucose, ROS or accumulated free cholesterol can be triggers of ERS in EC [80]. It is known that long-term dyslipidemia induces insulin resistance, a possible mechanism being the enhancement of ERS [81]. Although ERS was frequently associated with diabetes and vascular complications, the involvement of gLp in promoting ERS is not very well documented. Zhao et al. showed that the hearts and ascending aortae from diabetic mice present an increased level of UPR markers [82]. They showed that a 6 h incubation of cultured EC with gLDL stimulates ERS by the increase of GRP78/94, sXBP-1 and CHOP [82]. Toma et al. complemented these studies by demonstrating that a 24 h exposure of EC to gLDL increases ERS, probably due to an increase in accumulated free cholesterol in EC membranes induced by gLDL [45]. Unlike Zhao et al., Toma observed no modification of sXBP-1 levels, but a decrease in GRP78 levels. The dissimilarities might originate in the different exposure times to gLDL, being known that the IRE1α branch of UPR is attenuated after 8 h, despite the persistence of ERS [83]. Furthermore, the study demonstrated that ROS levels induced by gLDL were decreased by inhibitors of ERS and that ERS was alleviated by ROS inhibitors, showing an interconnection between ROS and ERS in gLDL-exposed HEC [45] (summarized in Figure 3a).

In normal conditions, native HDL exerts an anti-oxidant and anti-inflammatory action in EC activated by pro-inflammatory proteins or loaded with lipids. To study the effect of gHDL in EC, Yu et al. used a recombinant, non-enzymatically glycated PON-1. The glycated PON-1 presented a diminished enzymatic activity, and its incubation with ECs induced ERS and reduced the activity of sarco/ER Ca2+-ATPase (SERCA), thus increasing the intracellular levels of calcium by a mechanism involving the oxidative stress [84] (Figure 3b).

4.4. Stimulation of Monocytes Adhesion to Endothelial Cells

Monocytes' adhesion and their subsequent transmigration into the subendothelium are instrumental for atherosclerosis inception and progression. The redox-sensitive NF-kB, the activator protein-1 (AP-1) and the redox-sensitive MAP kinases play important roles in the pathophysiology of diabetes, all being involved in the transcription of different pro-inflammatory cytokines, chemokines or cell adhesion molecules [85]. The overexpression of adhesion molecules (such as ICAM-1, VCAM-1, ninjurin-1) on EC plasma membrane stimulates monocytes adhesion and transmigration into the subendothelial space, contributing to atherosclerosis initiation and progression in diabetes.

Both gLDL and gHDL can activate ECs and stimulate the adhesion of monocytes to ECs through different molecular mechanisms [44,45,72]. Zhao et al. showed that incubation of cultured EC with gLDL determines an increased monocytes adhesion. Very interesting, they showed that the transfection of ECs with siRNA specific for p22phox or PAI-1 prevented the gLDL-induced monocytes adhesion to ECs, suggesting that the oxidative stress and the fibrinolytic modulators participate in the inflammatory processes in diabetes [44]. Toma et al. reported that monocytes adhesion is increased in ECs exposed to gLDLs due to the increase in VCAM-1 expression through signaling pathways involving RAGE, ROS, ERS stimulation and subsequent activation of p38MAP kinase and NF-kB [45]. Later, Toma et al. confirmed and added to the previous reports by evidencing the involvement of gLDLs in the secretion of MCP-1 and C reactive protein through mechanisms involving stimulation of RAGE, oxidative stress and ERS [72] (Figure 3a).

The effect of gHDLs on EC function was reported by Hedrick et al., who showed that gHDLs, compared to native HDLs, do not inhibit monocytes adhesion to HAEC exposed to oxidized LDL. The study concluded that the inability of gHDLs to exert anti-inflammatory protection is due to the significant reduction of PON-1 activity observed after HDL glycation [86]. HDLs isolated from diabetic patients did not inhibit the phosphorylation of NF-kB p65 subunit in TNFα-exposed EC compared to HDLs from normoglycemic subjects [66,87]. Moreover, HDL from T2DM patients had an impaired ability to inhibit LDL oxidation and LDL-induced monocyte chemotaxis [88] (Figure 3b). These studies demonstrate that the irreversible glycation of LDL and HDL participates substantially in the development of a pro-inflammatory state in EC, contributing to the major complications of diabetes.

4.5. Generation of Fibrinolytic Regulators

Intravascular thrombosis resulting from imbalances between the coagulation and the fibrinolytic processes plays a critical role in the diabetic vascular complications [89]. Plasmin, the main active product of the fibrinolytic system, maintains the fluidity of blood in the vasculature by breaking down the fibrin clots. Plasmin is generated from plasminogen through the balanced action of its activators, t-PA, urokinase plasminogen activators (u-PA) and the PAI-1, the physiological inhibitor of t-PA and u-PA. Vascular ECs synthesize both activators and inhibitors of fibrinolysis, playing a critical role in the homeostasis of fibrinolytic activity in the blood [89].

In diabetic patients, an attenuated fibrinolytic activity was determined [90,91], predisposing the patients to vascular accidents. Published data indicate that gLps have an important contribution to the alteration of the fibrinolytic system equilibrium in diabetes. Shen et al. extensively described the effects of gLps on the generation of fibrinolytic regulators from ECs. They were the first to demonstrate that gLDL amplifies the production of PAI-1 in parallel with the reduction of de novo synthesis of t-PA in HUVEC [92]. Similar results were obtained for LDL isolated from T1DM or T2DM patients or VLDL from T2DM patients [90]. Comparing the effects of oxidized LDL and gLDL on the fibrinolytic factors, Ma et al. showed that both modified Lp increase, PAI-promoter activity, PAI-1 mRNA level and its release from EC in a similar manner. Using specific inhibitors, the authors demonstrated that an intact Golgi apparatus is required for PAI-1 generation in ECs [93]. Zhao et al. reported that PAI-1 is upregulated by gLDL through a mechanism depending on the heat shock factor-1 (HSF-1) binding to PAI-1 promoter, which stimulates PAI-1 transcription [94]. Sangle showed later on that the endothelial RAGE, oxidative stress, and HRas/Raf-1 signaling pathways participate in the upregulation of HSF-1 or PAI-1 in EC exposed to gLDL [95] (summarized in Figure 3a).

Zhang et al. observed that glycation of Lp(a) enhances the production of PAI-1 and further decreases the generation of t-PA from HUVEC and human coronary artery EC [25]. Thus, glycation of Lp(a) attenuates the fibrinolytic activity in blood and contributes to the increased incidence of cardiovascular complications in diabetic patients with hyperlipoprotein(a).

Studies regarding the effects of gHDL on the fibrinolytic regulators expression in EC are scarce. Ren S et al. showed that although gHDLs do not significantly modify t-PA in EC, a high concentration of gHDLs (>/=100 μg/mL) moderately increases the release of PAI-1 (Figure 3b). However, it is interesting that HDL glycation does not affect the ability of HDL to reduce PAI-1 and to restore t-PA generation in ECs exposed to gLDL [91,96].

4.6. Induction of Endothelial Cell Apoptosis

Cellular apoptosis or programmed cell death is a physiological process important in the development of embryos and continuing throughout adult life. Apoptosis can be induced by two main separate pathways: one controlled by B-cell lymphoma 2 (BCL2), a family of proteins including anti-apoptotic BCL2 and bcl-XL and pro-apoptotic Bad, Bax, Bid or Bak, and the other one controlled by the so-called "death receptors", including Fas, TNFα receptor 1 (TNFR1), death receptor 3 (DR3) or TNFα-related apoptosis-inducing ligand (TRAIL). Both these pathways converge to the activation of inactive initiator caspases (caspase-8 and -9) and the downstream effector caspases (e.g., caspases-3 and -7), finally determining the DNAse activation and cell's death [97].

In diabetic conditions, an exacerbated apoptotic process was observed in ECs. Published data indicate hyperglycemia as the main causal factor for the development of EC apoptosis, a specific role being played by gLDL. Artwolh et al. demonstrated that 48 h exposure of HUVEC to 100 µg/mL of gLDLs isolated from diabetic patients determines the apoptosis of cultured ECs due to the increase in pro-apoptotic Bak and caspase-3 [56]. Similar results were obtained in ECs using the electronegative LDL subfraction L5 isolated from T2DM patients [98]. Li et al. showed that gLDLs decrease EC survival in a dose-dependent manner by promoting apoptosis [99]. The authors demonstrated that gLDLs increase p53 nuclear transcription factor levels and activate the glycogen synthase kinase 3 (GSK3b), determining the increase of the cytochrome c release from the mitochondria, which in turn successively activates caspase-9 and the effector caspase-3. Interestingly, Li et al. revealed that all these processes are abolished by the overexpression of Protein L-isoaspartyl methyltransferase (PIMT), a protein that plays a role in the repair and/or degradation of damaged proteins, whose levels are decreased by gLDL in an ERK1/2-dependent manner [99]. Another interesting mechanism by which gLDLs may induce EC apoptosis was described by Yin et al., who showed that HUVEC exposure to gLDL determines a significant increase in the Bax/Bcl-2 ratio, the release of cytochrome c from the mitochondria in the cytosol and stimulation of caspase-3 activity, determining cellular apoptosis [100]. In addition, the authors observed that gLDLs decrease the expression of prohibitin, a chaperone involved in the stabilization of mitochondrial proteins, involved in maintaining normal mitochondrial morphology and function. Prohibitin overexpression reversed the apoptotic process by inhibiting Akt phosphorylation, decreasing Bax/Bcl2 ratio, cytochrome c release and caspase 3 activity, showing that prohibitin plays a critical role in gLDL-induced EC apoptosis [100] (summarized in Figure 3a). In conclusion, the presented studies clearly demonstrate that gLDLs can and will stimulate the mitochondrial apoptotic pathways in EC.

The effect of gHDLs on EC apoptosis was also investigated. Matsunaga et al. showed that 100 µg/mL gHDL do not stimulate the apoptosis of HAEC. In contrast, if gHDL is additionally oxidized (glyc-oxHDL), a significant increase of caspase 3 activity and expression and caspase 9 expression, as well as HAEC apoptosis, are induced, suggesting that the oxidation process, rather than glycation, is detrimental for HDL function or pro-apoptotic effects [101]. Although HDL glycation is not sufficient to induce EC apoptosis, the glycation modifications suffered by gHDLs can determine EC senescence [102]. Park et al. reported that exposure of HUVECS to fructated ApoAI determines the increase in cellular senescence along with a lysosomal enlargement (Figure 3b).

5. Glycated Lipoproteins in Diabetes

5.1. Glycated LDL Participate in Atheroma Formation

Glycation determines the crosslinking of proteins that become more resistant to proteolysis and accumulate in different tissues, thus affecting their function [12]. In atheroma from diabetic patients, LDLs, AGE-proteins and oxidation products (such as 4-HNE-proteins) co-localize, being present intracellularly (in foam cells derived from macrophages or in SMC of the fibrous cap), as well as extracellularly, in the necrotic areas formed near the internal elastic lamina [103,104]. Tames et al. showed that glycated apoB levels are about two-fold higher in DM patients compared to healthy subjects, reaching a concentration of 9.3 mg/dl [105]. It was reported that LDL glycation determines structural alterations in Lp, inhibiting their recognition and uptake by the specific LDLR, allowing an enhanced uptake by scavenger receptors present in macrophages, promoting foam cells and atherosclerotic plaque formation [16]. Using an original model of hamsters with T1DM generated by injection with streptozotocin, and made hyperlipidemic by feeding a high-fat diet for 12 weeks, Simionescu et al. showed that levels of gLDL doubled in hamsters' plasma compared to normal ones [106]. Using the model of streptozotocin-injected hamster with/without hyperlipidemia, Sima and Stancu showed that the aortic intima contains focal deposits of AGE-proteins and modified LDL, distributed either diffusely in the extracellular space or associated with SMC or macrophage-derived foam cells [104]. MDA-Lys and HNE-Lys adducts have been previously detected in atherosclerotic plaques of diabetic or diabetic-hyperlipidemic hamsters [107] or in atheroma from diabetic patients [104]. Using different methods to block the effects of reactive carbonyl species, it was reported that these compounds have robust pro-inflammatory properties, thus contributing to atherosclerosis development [108].

5.2. Dysfunctional HDL Are Pro-Atherogenic Particles in Diabetes

Functional HDLs have been reported to mediate different beneficial actions in normal physiological conditions by (i) participating in the glycemic control by interacting with pancreatic beta cells [109], (ii) inhibiting LDL glycation and oxidation [110], (iii) maintaining the function of EC through stimulation of NO production, (iv) reducing monocytes adhesion to EC and (v) participating in the reverse cholesterol transfer process [111]. Unfortunately, the metabolic disturbances characteristic for the diabetic conditions determine alterations in HDL composition and function, favoring the appearance of the dysfunctional HDL [112]. Dysfunctional HDLs, characterized by an increased content of MDA, MPO and ceruloplasmin, in parallel with decreased PON-1 protein and activity, were found in plasma from patients with coronary artery disease and/or diabetes and were demonstrated to play pro-inflammatory effects in EC [87].

In T2DM patients, Godfrey et al. observed an increase of methylglyoxal and dicarbonyl groups in ApoAI from HDL that was associated with the alteration of HDL functions [26]. HDL isolated from T2DM patients was shown to be dysfunctional, being unable to inhibit the inflammatory stress in TNFα-exposed EC [66] or LDL oxidation [88]. ApoA-I was demonstrated to also be a target of MPO, affecting the anti-inflammatory and anti-apoptotic activity of HDL in ECs or impairing the reverse cholesterol transport mediated by ABCA-1 [113,114]. Another protein associated with HDL and known to be affected in diabetes is the anti-oxidant enzyme PON-1. PON1 levels and activity were demonstrated to be low in T1DM or T2DM patients [115,116] determining an impaired antioxidant capacity of HDL [116,117]. Trying to decipher the mechanism of PON-1 regulation, Stancu et al. used the model of the hyperlipidemic hamster that in time develops hyperglycemia, similar to T2DM. These pathological conditions were associated with a decrease in PON-1 protein and activity in the plasma, small intestine and liver through a mechanism regulated by LXR and PPAR gamma [81]. Besides glycation and oxidation, other alterations in HDL composition were observed in diabetes. Proteomic studies showed that diabetic HDL is enriched in acute-phase reactant serum amyloid A (SAA) and apolipoprotein C-III, while the activity of PON-1 and platelet-activating factor-acetylhydrolase (PAF-AH) is decreased, causing the shift of HDL from anti-atherosclerotic to pro-atherogenic Lp [112,118]. HDL subclasses' distribution was also altered in diabetes. Dullaart et al. reported that in the plasma of T2DM patients, large and medium-size HDL particles' number decreased, while small HDL particles were increased [119]. In good agreement, epidemiological studies showed that HDL2, the larger, less dense subfraction of HDL, is negatively correlated with the development of T2DM [120].

In T1DM patients, HDLs are dysfunctional, presenting a low antioxidant protection [121] and a decreased capacity to participate in the reverse cholesterol transport [122]. In parallel, the increase was reported sof the pro-inflammatory SAA in HDL2 and HDL3 subfractions of HDL isolated from T1DM patients with poor glycemic control [123].

5.3. Hyperglycemia Alters miRNAs Profiles in Plasma and Lipoproteins

It is known that Lps are carriers for microRNAs (miRNAs) in plasma [124–126]. MiRNAs are small noncoding RNAs formed by approximately 22 nucleotides, demonstrated to regulate different target genes by post-transcriptional mechanisms. Using different experimental models, it was shown that miRNAs expression is modulated by diabetic conditions, being reported in the few existing studies to correlate with the presence of diabetic complications. Simionescu et al. demonstrated that circulating miR-125a-5p and miR-146a levels are positively correlated with hyperglycemia in the serum of coronary artery of affected patients [127]. In agreement with this, Huang et al. determined a 5-

fold increase in miR-146a and miR-155 in the kidney samples obtained from patients with diabetic nephropathy and showed that these miRNAs induce inflammatory stress in glomerular EC in culture through the stimulation of TNFα secretion and NF-kB activation [128]. Published data indicate that miRNAs are mostly associated with HDL and can be transferred to cultured cells, thus affecting cells' functionality by modulating their genes' expression [124,125]. For instance, it was demonstrated that HDL transfers a functional miR-223 to EC, thereby downregulating the expression of ICAM-1, thus exerting an anti-inflammatory response [129]. In a study that evaluated miRNAs in the plasma and HDL from diabetic subjects with stable (SA) or unstable angina (ACS), Simionescu and Niculescu et al. demonstrated that miR-223, miR-92a and miR-486 were increased in HDL from hyperglycemic ACS patients compared to normoglycemic ones. Interesting, the distribution of these analyzed miRNAs in HDL was statistically different in the two groups of patients, discriminating between SA and ACS patients [130]. Later on, Florijn et al. demonstrated an altered distribution of miRNAs in HDL isolated from diabetic patients with nephropathy (DN), namely a decreased level of miR-132 was found in HDL from DN patients compared with healthy subjects. Trying to find the pathological relevance of these results, Florijn et al. tested the functional impact of HDL-associated miRNAs in ECs and demonstrated that HDL-miR-132 significantly stimulated the angiogenic capacity of ECs, suggesting that HDL enrichment with miR-132 represents a causal mechanism for the advancement of DN [131].

6. Promising Therapies to Reduce ECD in Diabetes

6.1. In Vitro Approaches to Decrease the Effects of gLp in EC

Considering the importance of EC in maintaining the homeostasis of the vascular system, the finding of therapeutic strategies to reverse ECD is of great importance.

The beneficial effects of some compounds (amlodipine, caffeic acid, cyanidin-3-glucoside, procyanidin B2) to restore the proper function of EC affected by gLp was evaluated in a few in vitro studies. It was shown that amlodipine, a calcium channel blocker, increases the bioavailability of NO in ECs exposed to gLDL, as demonstrated by the increase in the NO/peroxynitrite ratio in the culture media of the treated cells. This beneficial action was the result of the increased expression of eNOS, in parallel with the reduction in iNOS and NADPHox-dependent ROS induced by amlodipine. In addition, amlodipine reduced the activation of the pro-inflammatory NF-kB and p38 MAPK determining the decrease of monocyte adhesion to ECs, in part through downregulating MCP-1 and VCAM-1 expression [57]. Caffeic acid, a phenolic acid present in normal diets, was successfully tested for its antioxidant and anti-inflammatory effects in HECs exposed to gLDL. Caffeic acid reduced the secretion of CRP, VCAM-1, and MCP-1 in gLDL-exposed HEC by inhibiting RAGE expression and NADPHox-dependent ROS and attenuating the gLDL-activated ERS. In addition, it exerted anti-apoptotic actions, as demonstrated by CHOP inhibition and the restoration of the mitochondrial transmembrane potential [72]. Cyanidin-3-glucoside (C3G), an anthocyanin present in dark-skinned berries, was shown to reduce the intracellular superoxide production and to improve the viability of EC exposed to gLDL [71]. It was proven that the beneficial effects of C3G are the result of the downregulation of RAGE, thus determining the inhibition of NOX4 expression and normalization of mitochondrial-dependent ROS [71]. An interesting protective mechanism for EC function was described by Li X et al. for grape seed procyanidin B2 (GSPB2). They demonstrated that GSPB2 inhibits the apoptosis of ECs exposed to gLDL by increasing the expression of the protein PIMT, a repair enzyme that controls the release of cytochrome c and activation of caspase 3 and 9 [99]. Another mechanism by which GSPB2 reduced apoptosis, increasing the viability of ECs exposed to gLDL, was described later by Yin et al. [100]. In cultured ECs, the authors demonstrated that GSPB2 increases the expression of prohibitin, a protein implicated in cellular survival and apoptosis that is decreased by gLDL exposure. The increase in prohibitin by GSP2 reduced the cytochrome c release into the cellular cytosol, Bax/Bcl-2 ratio and attenuated the caspase-3 activity, thus protecting against apoptosis [100].

ERS decrease could be another approach for the diminution of gLp effects on ECs in diabetes. ERS inhibition by 4-phenylbutyric acid (PBA) effectively repressed the activation of eIF2α and CHOP signals and reversed AGEs-induced cell apoptosis in mesangial cells in vitro [132]. Stimulation of the ER folding capacity through chemical chaperones, such as tauro-urso-deoxycholate and PBA, promises to be a novel and valuable therapeutic attempt for preventing the effects of diabetes in the vascular cells [133].

6.2. Therapies Used to Alleviate the Vascular Disorders in Diabetes

In diabetic patients, several therapies have been used to reduce the vascular complications of this disease. These include changes in the lifestyle and/or administration of oral hypoglycemic drugs, vitamins, antioxidants, statins, etc., to target the pathological mechanisms that drive the development of diabetes and its complications. Unfortunately, although having beneficial properties for human health, some of them failed to reduce the major vascular complications in diabetes. For example, data from the large-scale Look AHEAD (Action for Health in Diabetes) trial showed that achieving changes of lifestyle and weight loss confers health benefits for DM patients (including decreasing sleep apnea, reducing the need for diabetes medication, maintaining physical mobility), but the cardiovascular risk remaine unchanged [134].

Since AGEs are important players in the progression of diabetic complications, the inhibition of the mechanisms that promote their formation is desirable. As highlighted, hyperglycemia and oxidative stress are the main contributors to AGE formation. The ability of different classes of therapeutic compounds with hypoglycemiant and anti-oxidant properties to ameliorate diabetes and its vascular complications was tested in different clinical studies and will be further presented.

6.2.1. The Use of Hypoglycemiant Compounds

Oral hypoglycemiant compounds are an important class of drugs used in the treatment of diabetes. The compounds used at present act either by increasing the production of insulin (sulphonylureas, glinides and, more recently, incretin mimetics) or stimulating insulin sensitivity in the target tissue (insulin sensitizers such as metformin and thiazolidinediones) [134]. Metformin has been successfully used for long-term treatment as first-line therapy for T2DM to improve glycemic control in patients, being associated with a low risk of hypoglycemic events [135]. In a recent cohort study of USA veterans with T2DM, Wang et al. showed that metformin reduced the rate of CVD events [136]. In a meta-analysis, Lee et al. demonstrated that metformin is associated with fewer major adverse cardiac events in T2DM patients from Taiwan [137]. The beneficial cardio-metabolic effects of metformin were studied also in T1DM patients in a large double-blind randomized, placebo-controlled trial, the conclusion being that this drug reduces atherosclerosis progression [138]. In recent years, new antidiabetic drug classes, such as glucagon-like peptide 1 (GLP-1) receptor agonists (RAs), sodium-glucose co-transporter-2 (SGLT-2) inhibitors and dipeptidyl peptidase-4 (DPP-4) inhibitors, were used as add-on therapies secondary to metformin. In a recent meta-analysis, Fey et al. compared the effects of these compounds on cardiovascular outcomes in diabetic patients. They reported that compared with other antidiabetic drugs, SGLT-2 inhibitors are superior in reducing the cardiovascular mortality, hospitalization for heart failure and the appearance of renal damages, while the effects of DPP-4 inhibitors on cardiovascular and renal outcomes are comparable to placebo [139].

6.2.2. Therapeutic Compounds to Reduce Formation of AGE

Data from literature clearly support the involvement of AGE products (from exogenous or endogenous sources) in the generation of micro- and macrovascular complications of diabetes. Notably, it was stated that AGE are possible mediators of the metabolic memory of cells in diabetes [5], their formation determining the evolution of diabetic complications even when glycemia is under strict control. Thus, the inhibition of AGE formation is one of the possible therapeutic approaches to reduce the diabetic burden.

Few therapeutic compounds were reported to block the formation of AGE, including metformin, derivates of vitamin B and statins. Metformin was reported to prevent AGE formation in vitro [140] and to reduce the levels of CML, while increasing plasma soluble RAGE in patients with metabolic syndrome [141].

Derivates of vitamin B6, such as pyridoxamine, were shown to be beneficial in reducing AGE levels and the vascular complications of diabetes. Recently, Pereira et al. reported that pyridoxamine significantly improved the function of EC in the aorta and mesenteric arteries of an animal model of non-obese T2DM by decreasing the vascular oxidative damage and AGE levels [142]. Using different animal models of diabetes, other studies showed that pyridoxamine can reduce the apoptosis of cardiomyoblasts in cardiac ischemia [143], impede the development of atherosclerotic lesions and improve the cardiac ejection fraction in diabetic mice with myocardial infarction [144], or reduce the arterial stiffening by inhibiting the glycation of aortic collagen [145]. Another derivate of vitamin B6, 2-aminomethyphenol pyridoxamine, was shown to reduce the CML levels, triglycerides and total cholesterol in the plasma of diabetic STZ-induced rats [146].

Lipid-lowering statins were constantly reported to inhibit the AGE/RAGE axis. Clinical trials indicate that atorvastatin, simvastatin, pravastatin and pitavastatin have potent AGE-lowering effects, independent of the glycemic control of the patients [147–149]. Several mechanisms to explain the AGE-lowering effects of statins have been described. First, it was shown that statins stimulate ADAM-10 mediated RAGE shedding, increasing sRAGE and the subsequent AGE clearance [150]. Since it is known that oxidative stress is an important participant in the formation of advanced glycoxidation and lipoxidation end products, it is reasonable to accept that the antioxidant properties of the statins decrease AGE. In addition, a third mechanism can be proposed, related to the lipid-lowering capacity of statins: the decrease in the cholesterol-rich LDL due to statins administration results in a reduced formation of gLDL. Besides reducing AGE, the pleiotropic effects of statin include antioxidant, anti-inflammatory, immunomodulatory, anti-proliferative and endothelial protective effects, supporting their use in preventing the CVD risk in diabetic patients [151]. Their beneficial effects were reported in several cohort studies and clinical trials. The Heart Protection Study (HPS) shows that administration of simvastatin substantially reduces the major vascular events in diabetic subjects [152]. Beneficial effects were observed also for atorvastatin in the Anglo-Scandinavian Cardiac Outcomes Trial-Lipid Lowering Arm (ASCOT-LLA) or the Collaborative Atorvastatin Diabetes Study (CARDS), both studies reporting a decreased risk of nonfatal myocardial infarction and fatal coronary heart disease (CHD) in hypertensive diabetic patients without a previous history of CHD (ASCOT-LLA) and a significant reduction of acute CVD events (including stroke) in T2DM patients with normal LDL cholesterol levels [153]. Recently, Ramos et al. reported that statin therapy significantly reduces the risk of CVD in older patients (>75 years) with T2DM [154].

6.2.3. Antioxidants to Decrease CVD in Diabetes

Knowing the important involvement of ROS in diabetes, compounds with antioxidant properties such as vitamin C, E, alpha-lipoic acid, allopurinol and statins [155–158] were used to improve the health of diabetic patients.

The obtained data are contradictory: the results from HOPE (Heart Outcomes Prevention Evaluation) trial show that antioxidants, like vitamin E, have no effect on cardiovascular outcomes in DM patients [159], while a recent meta-analysis based on published data indicates that antioxidant vitamins (in particular vitamin E) have benefits, such as reduction of glycated hemoglobin levels, enhancing antioxidant capacity and improvement of EC function in non-obese T2DM patients, supporting their use in protecting against the complications of diabetes [155,156,160](Akbar S et al., 2011). In a very recent study, Baziar et al. showed that alpha lipoic acid (ALA) may decrease the CVD risk by reducing the level of oxidized LDL in the plasma of T2DM patients without affecting their lipid profiles or glycemic indices [157]. Altunina et al. showed that ALA reduced the systemic inflammation (measured as decreased CRP, IL-6 and TNF-α) in T2DM patients with a history of non-Q-myocardial infarction [161].

6.3. New Promising Therapies to Alleviate ECD in Diabetes

In diabetic-associated vascular diseases, LDL are modified (glycated and/or oxidized) and play a key role in accelerating the progression of atherosclerosis. Chronic hyperglycemia enhances glucose-induced LDL oxidation and/or glycation, increasing their pro-atherogenic properties [17,107] and promoting vascular injury. A correlation between arterial tissue AGE and circulating AGE-apoB, and the contribution of AGE-specific receptors (RAGE) in atheroma formation was reported [162]. Thus, inhibition of RAGE is a promising therapy to promote vascular healing in diabetes.

Together with reduction of AGEs, the alteration of miRNAs profiles in the plasma of diabetic subjects is indicated to be involved in the promotion of diabetic complications even under strict glycemic control [10], suggesting that miRNA-based therapies could be efficient in the reduction of CVD in diabetes.

In the next section, the progress of therapies based on RAGE inhibition, or on reestablishment of a miRNAs profile to reduce ECD and diabetes complications will be presented. Future therapeutic strategies such as reduction of Lp(a) or the use of CRISP/Cas technology to increase the anti-inflammatory AGE-receptor complexes will also be discussed.

6.3.1. RAGE Inhibitors

A new class of compounds proposed to reduce the diabetic burden is the RAGE inhibitors. The fact that knocking out the RAGE gene in mice does not distress their health and does not seem to affect their growth suggests that RAGE inhibition could be a safe in vivo therapeutic approach [163]. Small-molecular inhibitors targeting either the extracellular ligand-binding site or, more recently, the intracellular signaling domain of RAGE have been developed [163]. TTP488, also known as PF-04494700 or azeliragon, is an orally bioavailable small-molecule inhibitor for the extracellular domain of RAGE. TTP488 inhibits the binding of multiple RAGE ligands, including AGEs, HMGB1, S100B and Aβ. Unfortunately, no data regarding the effects of TTP488 in diabetes exist, as the effects of TTP488 are documented mainly in the context of Alzheimer's disease where preclinical studies show that TTP488 slows the cognitive decline at small concentrations [163]. Another extracellular RAGE inhibitor named FPS-ZMI did not cause toxic side effects in mice, even at doses as high as 500 mg/kg [164] and was demonstrated to have beneficial effects, such as reducing the cardiac hypertrophy and inflammation in cardiac tissues in a mouse model of heart failure [165]. A set of intracellular inhibitors of the cytoplasmatic domain of RAGE was identified by Manigrasso et al. [166]. It consists of 13 small molecules able to inhibit competitively the interaction between the cytoplasmic domain of RAGE and DIAPH1 [166]. These compounds exhibited anti-migratory effects in cultured SMC and anti-inflammatory effects in THP-1 macrophages exposed to CML-AGE [166]. In addition, a subset of these inhibitors was able to reduce the ischemia/reperfusion injury in the hearts of a streptozotocin-induced diabetic mouse [31,166].

AGE proteins induce dysfunction of EC along the vascular tree in diabetes. While AGE-receptor complexes act as scavengers for glycated proteins, AGE-RAGE/TLRs interaction induces intracellular signaling pathways, leading to increased oxidative and inflammatory stress. Therefore, genome-editing technology would be a promising tool to upregulate AGE-receptor complexes and down-regulate RAGE/TLRs in order to counteract the harmful effects of AGE-proteins in EC [167]. Due to its high efficacy, good repeatability, simple design and low cost, the CRISPR/Cas9 technology is successfully used to develop innovative organ-specific therapies for better management of diabetic-associated vascular diseases [168].

6.3.2. MiRNA Based Therapies

The identification of epigenetic mechanisms that affect the function of EC in diabetes opened also the pathway for developing new promising therapeutic interventions. MiRNAs–based therapies are centered on two main approaches: the miRNA inhibitory therapy that aims to downregulate the aberrantly over-expressed miRNAs and the miRNA replacement therapy, which aims to restore the activity of beneficial miRNAs downregulated by pathologic conditions [10]. Different in vitro and in vivo studies indicate that miRNA-targeted therapies could be beneficial in alleviating micro- and macro-complications of diabetes. In vitro, it was reported, for example, that upregulation of miR-149-5p reduced ECD induced by high glucose-exposure by increasing the level of NO and the expression of eNOS, while decreasing the levels of ET-1, vWF and ICAM-1 [169]. In vivo, it was shown that the knockout of miR-21 alleviated the microvascular damage, inflammation, and cell apoptosis in the retina of db/db mice through the decrease in PPARα [170]. Icli et al. showed that delivery of miR-135a-3p inhibitors to wounds of diabetic db/db mice speeded the wound closure and angiogenesis and increased the thickness of the granulation tissue through the activation of p38 MAPK signaling in EC [171]. In addition, the delivery of miR-126 significantly improved neurological and cognitive functions in T2DM-stroke mice and induced the polarization of macrophages towards the anti-inflammatory M2 phenotype in the ischemic zone [172]. Although far from being used in clinical practice, the strategies that target the epigenetic mechanisms used in combination with standard anti-diabetic treatments might represent an additional opportunity to reduce the complications of diabetes.

6.3.3. Inhibition of Glycated-Lp(a)

To date, no therapeutic approaches have been described to reduce the level of glycated Lp(a) in diabetic patients. A reasonable way to do this is to decrease the Lp(a) synthesis in parallel with the hypoglycemic therapies. Analysis of trials investigating evolocumab effects in CVD patients reported significant dose-related decreases in Lp(a) levels, in parallel with the decrease in LDL-C and apolipoprotein B levels [173]. In the last decade, an increased interest in antisense therapy has been noticed. This approach consists of inhibiting the synthesis of specific proteins by using complementary oligonucleotides that bind to their mRNA in the nucleus. Data from a clinical study that investigated an antisense oligonucleotide that selectively reduces the synthesis of apo(a) in the liver, and consequently Lp(a) plasma levels, also show a reduction in oxidized phospholipids on apo(a) and a reduced monocyte inflammatory activation [173]. Conjugation of antisense oligonucleotide with a GalNAc3 complex guides the drug to the hepatocytes via the asialoglycoprotein receptor, making it more potent than the parent antisense oligonucleotide. Thus, the necessary dose of the drug will be reduced and its tolerability improved [174].

7. Discussion

Diabetes is a complex disease, affecting numerous organs and the entire vascular tree. Accelerated atherosclerosis represents a major complication of the macro-vasculature of diabetic patients that can cause major acute cardiovascular events. Endothelial dysfunction has been demonstrated in both the peripheral and coronary arteries circulation. Alterations in lipid metabolism are at the core of T2DM phenotypes and probably greatly contribute to the increased risk of cardiovascular disease associated with diabetes. An important player in the induction and progression of atherosclerosis in diabetes is the interaction between the glycated and/or oxidized Lp and EC. In this context, more documented is the interaction between EC and gLDL or gHDL, while data regarding gLp(a) are scarce and require in-depth investigation. The mechanisms by which gLDL and gHDL induce ECD are complex and interconnected. Although gLDL can interact with different receptors, the harmful effects of gLDL on EC are induced especially by the interaction with RAGE, a process that triggers the activation of various signaling pathways and affects all EC functions, determining decreased NO production, the dysregulation of oxidants–antioxidants balance or alteration of the proper function of ER. The extended presence of gLDL in diabetes induces a pro-inflammatory status of EC and finally apoptosis. Notably, the combined exposure to high glucose concentrations exacerbates the effects of gLDL. Since the gLDL/RAGE interaction is the key step in generating ECD, the receptors for AGE-proteins are promising therapeutic targets to be addressed by novel gene-editing technology. These approaches could be complemented by the administration of molecules that restore the mitochondrial and ER functions in EC.

At the same time, glycation of HDL renders them dysfunctional, unable to exert anti-atherosclerotic actions due to the modified composition and activity of the associated enzymes and a decreased half-life in the plasma of diabetic patients. The gHDL not only lose their anti-atherogenic potential due to the glycation of its main proteins (apoAI, apoAII, apoE, PON-1) but become pro-atherogenic by enrichment with pro-oxidant and pro-inflammatory molecules, such as the pro-oxidant enzyme MPO and the pro-inflammatory protein SAA. However, many other issues regarding the effects of gHDL in EC (such as modulation of the intracellular antioxidant system or of the fibrinolytic system) and the precise mechanism by which gHDL affects the function of EC are still missing, so more studies are needed.

Certain compounds with anti-oxidant properties were effective to alleviate ECD in vitro. Alas, the administration of exogenous anti-oxidants to humans failed to improve the oxidative status and HDL function and its ensuing protective role in EC. The future employment of the gene-editing technology [CRISPR/(d)-CAS9] in vivo to upregulate endogenous anti-oxidant proteins, in particular those associated with HDL, could be a promising approach to improve EC function in diabetes. Taking into account the important impact of gLp on EC function, any new therapy that will be developed in the future to improve EC function in diabetes needs to be accompanied by a hypoglycemic and/or lipid-lowering therapy to ensure a less aggressive environment and to inhibit the formation of AGE-Lp.

Identification of LDL and HDL as important carriers of miRNAs with a dynamic profile under pathological conditions may result in the development of new therapies to improve EC function by delivery of specific protective miRNAs or inhibitors for the pro-atherogenic ones. However, the overall efficacy of these strategies needs to be further examined in pre-clinical and clinical trials.

Reversal of endothelial dysfunction is an open and exciting field of investigation, of fundamental relevance for many diseases, and in particular for diabetes and atherosclerosis. The progress of the biotechnology field will permit the use of targeted nanosystems such as nanoliposomes or nanoemulsions, or the specific regulation of different proteins based on the new CRISP/Cas9 technology to reverse endothelial dysfunction in diabetes, as well as other pathologies. In the future, these and other promising therapies focused on EC-based translational approaches will provide powerful tools to increase the quality of life of diabetic patients.

Author Contributions: Conceptualization, L.T., A.V.S. and C.S.S.; writing—original draft preparation, L.T. and A.V.S.; writing—review and editing, A.V.S., C.S.S. and L.T.; visualization, L.T. and C.S.S.; supervision, A.V.S. All authors have read and agreed to the published version of the manuscript.

Funding: This research was funded by the Romanian Ministry of National Education (grant number PNII #41-067/2007), PN-II-PT-PCCA-2011-3.1-0184 project (grant number PCCA-127/2012) and PN-II-RU-TE-2014-4-0506 project (grant number TE 11/2015) and by the Romanian Academy.

Acknowledgments: The authors thank Gabriela M. Sanda for the help in editing the present paper.

Conflicts of Interest: The authors declare no conflict of interest.

References

1. Oguntibeju, O.O. Type 2 diabetes mellitus, oxidative stress and inflammation: Examining the links. *Int. J. Physiol. Pathophysiol. Pharmacol.* **2019**, *11*, 45–63.
2. Gregg, E.W.; Williams, D.E.; Geiss, L. Changes in diabetes-related complications in the United States. *N. Engl. J. Med.* **2014**, *371*, 286–287. [CrossRef]
3. Nathan, D.M.; Cleary, P.A.; Backlund, J.Y.; Genuth, S.M.; Lachin, J.M.; Orchard, T.J.; Raskin, P.; Zinman, B.; Diabetes, C.; Complications Trial/Epidemiology of Diabetes, I.; et al. Intensive diabetes treatment and cardiovascular disease in patients with type 1 diabetes. *N. Engl. J. Med.* **2005**, *353*, 2643–2653. [CrossRef]
4. Holman, R.R.; Paul, S.K.; Bethel, M.A.; Matthews, D.R.; Neil, H.A. 10-year follow-up of intensive glucose control in type 2 diabetes. *N. Engl. J. Med.* **2008**, *359*, 1577–1589. [CrossRef]
5. Ceriello, A. The emerging challenge in diabetes: The "metabolic memory". *Vasc. Pharmacol.* **2012**, *57*, 133–138. [CrossRef]
6. An, H.; Wei, R.; Ke, J.; Yang, J.; Liu, Y.; Wang, X.; Wang, G.; Hong, T. Metformin attenuates fluctuating glucose-induced endothelial dysfunction through enhancing GTPCH1-mediated eNOS recoupling and inhibiting NADPH oxidase. *J. Diabetes Its Complicat.* **2016**, *30*, 1017–1024. [CrossRef]
7. Liu, T.; Gong, J.; Chen, Y.; Jiang, S. Periodic vs constant high glucose in inducing pro-inflammatory cytokine expression in human coronary artery endothelial cells. *Inflamm. Res.* **2013**, *62*, 697–701. [CrossRef]
8. Liu, T.S.; Pei, Y.H.; Peng, Y.P.; Chen, J.; Jiang, S.S.; Gong, J.B. Oscillating high glucose enhances oxidative stress and apoptosis in human coronary artery endothelial cells. *J. Endocrinol. Investig.* **2014**, *37*, 645–651. [CrossRef]
9. Widlansky, M.E.; Hill, R.B. Mitochondrial regulation of diabetic vascular disease: An emerging opportunity. *Transl. Res. J. Lab. Clin. Med.* **2018**, *202*, 83–98. [CrossRef]
10. Coco, C.; Sgarra, L.; Potenza, M.A.; Nacci, C.; Pasculli, B.; Barbano, R.; Parrella, P.; Montagnani, M. Can Epigenetics of Endothelial Dysfunction Represent the Key to Precision Medicine in Type 2 Diabetes Mellitus? *Int. J. Mol. Sci.* **2019**, *20*, 2949. [CrossRef]
11. Fishman, S.L.; Sonmez, H.; Basman, C.; Singh, V.; Poretsky, L. The role of advanced glycation end-products in the development of coronary artery disease in patients with and without diabetes mellitus: A review. *Mol. Med.* **2018**, *24*, 59. [CrossRef]
12. Kosmopoulos, M.; Drekolias, D.; Zavras, P.D.; Piperi, C.; Papavassiliou, A.G. Impact of advanced glycation end products (AGEs) signaling in coronary artery disease. *Biochim. Biophys. Acta Mol. Basis Dis.* **2019**, *1865*, 611–619. [CrossRef]
13. Cepas, V.; Collino, M.; Mayo, J.C.; Sainz, R.M. Redox Signaling and Advanced Glycation Endproducts (AGEs) in Diet-Related Diseases. *Antioxidants* **2020**, *9*, 142. [CrossRef]
14. Ott, C.; Jacobs, K.; Haucke, E.; Navarrete Santos, A.; Grune, T.; Simm, A. Role of advanced glycation end products in cellular signaling. *Redox Biol.* **2014**, *2*, 411–429. [CrossRef]
15. Mol, M.; Degani, G.; Coppa, C.; Baron, G.; Popolo, L.; Carini, M.; Aldini, G.; Vistoli, G.; Altomare, A. Advanced lipoxidation end products (ALEs) as RAGE binders: Mass spectrometric and computational studies to explain the reasons why. *Redox Biol.* **2019**, *23*, 101083. [CrossRef]
16. Alique, M.; Luna, C.; Carracedo, J.; Ramirez, R. LDL biochemical modifications: A link between atherosclerosis and aging. *Food Nutr. Res.* **2015**, *59*, 29240. [CrossRef]
17. Younis, N.; Sharma, R.; Soran, H.; Charlton-Menys, V.; Elseweidy, M.; Durrington, P.N. Glycation as an atherogenic modification of LDL. *Curr. Opin. Lipidol.* **2008**, *19*, 378–384. [CrossRef]
18. Simionescu, M.; Popov, D.; Sima, A. Endothelial dysfunction in diabetes. In *Vascular Involvement in Diabetes—Clinical, Experimental and Beyond*; Cheta, D.M., Ed.; Romanian Academy Publishing House and Karger: Bucharest, Basel, 2005; pp. 15–34.
19. Boren, J.; Chapman, M.J.; Krauss, R.M.; Packard, C.J.; Bentzon, J.F.; Binder, C.J.; Daemen, M.J.; Demer, L.L.; Hegele, R.A.; Nicholls, S.J.; et al. Low-density lipoproteins cause atherosclerotic cardiovascular disease: Pathophysiological, genetic, and therapeutic insights: A consensus statement from the European Atherosclerosis Society Consensus Panel. *Eur. Heart J.* **2020**, *41*, 2313–2330. [CrossRef]
20. Cybulska, B.; Klosiewicz-Latoszek, L.; Penson, P.E.; Banach, M. What do we know about the role of lipoprotein(a) in atherogenesis 57 years after its discovery? *Prog. Cardiovasc. Dis.* **2020**, *63*, 219–227. [CrossRef]
21. Schmidt, K.; Noureen, A.; Kronenberg, F.; Utermann, G. Structure, function, and genetics of lipoprotein (a). *J. Lipid Res.* **2016**, *57*, 1339–1359. [CrossRef]
22. Witztum, J.L.; Fisher, M.; Pietro, T.; Steinbrecher, U.P.; Elam, R.L. Nonenzymatic glucosylation of high-density lipoprotein accelerates its catabolism in guinea pigs. *Diabetes* **1982**, *31*, 1029–1032. [CrossRef]
23. Sima, A.V.; Botez, G.M.; Stancu, C.S.; Manea, A.; Raicu, M.; Simionescu, M. Effect of irreversibly glycated LDL in human vascular smooth muscle cells: Lipid loading, oxidative and inflammatory stress. *J. Cell. Mol. Med.* **2010**, *14*, 2790–2802. [CrossRef]
24. Deleanu, M.; Sanda, G.M.; Stancu, C.S.; Popa, M.E.; Sima, A.V. Profiles of Fatty Acids and the Main Lipid Peroxidation Products of Human Atherogenic Low Density Lipoproteins. *Rev. Chim.* **2016**, *67*, 8–12.
25. Zhang, J.; Ren, S.; Shen, G.X. Glycation amplifies lipoprotein(a)-induced alterations in the generation of fibrinolytic regulators from human vascular endothelial cells. *Atherosclerosis* **2000**, *150*, 299–308. [CrossRef]
26. Godfrey, L.; Yamada-Fowler, N.; Smith, J.; Thornalley, P.J.; Rabbani, N. Arginine-directed glycation and decreased HDL plasma concentration and functionality. *Nutr. Diabetes* **2014**, *4*, e134. [CrossRef]

27. Kashyap, S.R.; Osme, A.; Ilchenko, S.; Golizeh, M.; Lee, K.; Wang, S.; Bena, J.; Previs, S.F.; Smith, J.D.; Kasumov, T. Glycation Reduces the Stability of ApoAI and Increases HDL Dysfunction in Diet-Controlled Type 2 Diabetes. *J. Clin. Endocrinol. Metab.* **2018**, *103*, 388–396. [CrossRef]
28. Gordon, S.M.; Davidson, W.S.; Urbina, E.M.; Dolan, L.M.; Heink, A.; Zang, H.; Lu, L.J.; Shah, A.S. The effects of type 2 diabetes on lipoprotein composition and arterial stiffness in male youth. *Diabetes* **2013**, *62*, 2958–2967. [CrossRef]
29. Schmidt, A.M.; Yan, S.D.; Wautier, J.L.; Stern, D. Activation of receptor for advanced glycation end products: A mechanism for chronic vascular dysfunction in diabetic vasculopathy and atherosclerosis. *Circ. Res.* **1999**, *84*, 489–497. [CrossRef]
30. Lu, Q.; Lu, L.; Chen, W.; Chen, H.; Xu, X.; Zheng, Z. RhoA/mDia-1/profilin-1 signaling targets microvascular endothelial dysfunction in diabetic retinopathy. *Graefe's Arch. Clin. Exp. Ophthalmol. Albrecht Graefes Arch. Klin. Exp. Ophthalmol.* **2015**, *253*, 669–680. [CrossRef]
31. Egana-Gorrono, L.; Lopez-Diez, R.; Yepuri, G.; Ramirez, L.S.; Reverdatto, S.; Gugger, P.F.; Shekhtman, A.; Ramasamy, R.; Schmidt, A.M. Receptor for Advanced Glycation End Products (RAGE) and Mechanisms and Therapeutic Opportunities in Diabetes and Cardiovascular Disease: Insights From Human Subjects and Animal Models. *Front. Cardiovasc. Med.* **2020**, *7*, 37. [CrossRef]
32. Ishihara, K.; Tsutsumi, K.; Kawane, S.; Nakajima, M.; Kasaoka, T. The receptor for advanced glycation end-products (RAGE) directly binds to ERK by a D-domain-like docking site. *FEBS Lett.* **2003**, *550*, 107–113. [CrossRef]
33. Sakaguchi, M.; Murata, H.; Yamamoto, K.; Ono, T.; Sakaguchi, Y.; Motoyama, A.; Hibino, T.; Kataoka, K.; Huh, N.H. TIRAP, an adaptor protein for TLR2/4, transduces a signal from RAGE phosphorylated upon ligand binding. *PLoS ONE* **2011**, *6*, e23132. [CrossRef]
34. Zhang, L.; Bukulin, M.; Kojro, E.; Roth, A.; Metz, V.V.; Fahrenholz, F.; Nawroth, P.P.; Bierhaus, A.; Postina, R. Receptor for advanced glycation end products is subjected to protein ectodomain shedding by metalloproteinases. *J. Biol. Chem.* **2008**, *283*, 35507–35516. [CrossRef]
35. Meza, C.A.; La Favor, J.D.; Kim, D.H.; Hickner, R.C. Endothelial Dysfunction: Is There a Hyperglycemia-Induced Imbalance of NOX and NOS? *Int. J. Mol. Sci.* **2019**, *20*, 3775. [CrossRef]
36. Yau, J.W.; Teoh, H.; Verma, S. Endothelial cell control of thrombosis. *BMC Cardiovasc. Disord.* **2015**, *15*, 130. [CrossRef]
37. Mompeo, B.; Popov, D.; Sima, A.; Constantinescu, E.; Simionescu, M. Diabetes-induced structural changes of venous and arterial endothelium and smooth muscle cells. *J. Submicrosc. Cytol. Pathol.* **1998**, *30*, 475–484.
38. Suganya, N.; Bhakkiyalakshmi, E.; Sarada, D.V.; Ramkumar, K.M. Reversibility of endothelial dysfunction in diabetes: Role of polyphenols. *Br. J. Nutr.* **2016**, *116*, 223–246. [CrossRef]
39. Shi, Y.; Vanhoutte, P.M. Macro- and microvascular endothelial dysfunction in diabetes. *J. Diabetes* **2017**, *9*, 434–449. [CrossRef]
40. Harja, E.; Bu, D.X.; Hudson, B.I.; Chang, J.S.; Shen, X.; Hallam, K.; Kalea, A.Z.; Lu, Y.; Rosario, R.H.; Oruganti, S.; et al. Vascular and inflammatory stresses mediate atherosclerosis via RAGE and its ligands in apoE-/- mice. *J. Clin. Investig.* **2008**, *118*, 183–194. [CrossRef]
41. Sun, L.; Ishida, T.; Yasuda, T.; Kojima, Y.; Honjo, T.; Yamamoto, Y.; Yamamoto, H.; Ishibashi, S.; Hirata, K.; Hayashi, Y. RAGE mediates oxidized LDL-induced pro-inflammatory effects and atherosclerosis in non-diabetic LDL receptor-deficient mice. *Cardiovasc. Res.* **2009**, *82*, 371–381. [CrossRef]
42. Dobi, A.; Bravo, S.B.; Veeren, B.; Paradela-Dobarro, B.; Alvarez, E.; Meilhac, O.; Viranaicken, W.; Baret, P.; Devin, A.; Rondeau, P. Advanced glycation end-products disrupt human endothelial cells redox homeostasis: New insights into reactive oxygen species production. *Free Radic. Res.* **2019**, *53*, 150–169. [CrossRef]
43. Wang, C.C.; Lee, A.S.; Liu, S.H.; Chang, K.C.; Shen, M.Y.; Chang, C.T. Spironolactone ameliorates endothelial dysfunction through inhibition of the AGE/RAGE axis in a chronic renal failure rat model. *BMC Nephrol.* **2019**, *20*, 351. [CrossRef]
44. Zhao, R.; Ren, S.; Moghadasain, M.H.; Rempel, J.D.; Shen, G.X. Involvement of fibrinolytic regulators in adhesion of monocytes to vascular endothelial cells induced by glycated LDL and to aorta from diabetic mice. *J. Leukoc. Biol.* **2014**, *95*, 941–949. [CrossRef]
45. Toma, L.; Sanda, G.M.; Deleanu, M.; Stancu, C.S.; Sima, A.V. Glycated LDL increase VCAM-1 expression and secretion in endothelial cells and promote monocyte adhesion through mechanisms involving endoplasmic reticulum stress. *Mol. Cell. Biochem.* **2016**, *417*, 169–179. [CrossRef]
46. Zhou, X.; Weng, J.; Xu, J.; Xu, Q.; Wang, W.; Zhang, W.; Huang, Q.; Guo, X. Mdia1 is Crucial for Advanced Glycation End Product-Induced Endothelial Hyperpermeability. *Cell. Physiol. Biochem. Int. J. Exp. Cell. Physiol. Biochem. Pharmacol.* **2018**, *45*, 1717–1730. [CrossRef]
47. Li, J.; Schmidt, A.M. Characterization and functional analysis of the promoter of RAGE, the receptor for advanced glycation end products. *J. Biol. Chem.* **1997**, *272*, 16498–16506. [CrossRef]
48. Caballero, A.E.; Arora, S.; Saouaf, R.; Lim, S.C.; Smakowski, P.; Park, J.Y.; King, G.L.; LoGerfo, F.W.; Horton, E.S.; Veves, A. Microvascular and macrovascular reactivity is reduced in subjects at risk for type 2 diabetes. *Diabetes* **1999**, *48*, 1856–1862. [CrossRef]
49. Carrizzo, A.; Izzo, C.; Oliveti, M.; Alfano, G.; Virtuoso, N.; Capunzo, M.; Di Pietro, P.; Calabrese, M.; De Simone, E.; Sciarretta, S.; et al. The Main Determinants of Diabetes Mellitus Vascular Complications: Endothelial Dysfunction and Platelet Hyperaggregation. *Int. J. Mol. Sci.* **2018**, *19*, 2968. [CrossRef]
50. Thorand, B.; Baumert, J.; Chambless, L.; Meisinger, C.; Kolb, H.; Doring, A.; Lowel, H.; Koenig, W.; Group, M.K.S. Elevated markers of endothelial dysfunction predict type 2 diabetes mellitus in middle-aged men and women from the general population. *Arterioscler. Thromb. Vasc. Biol.* **2006**, *26*, 398–405. [CrossRef]

51. Lim, S.C.; Caballero, A.E.; Smakowski, P.; LoGerfo, F.W.; Horton, E.S.; Veves, A. Soluble intercellular adhesion molecule, vascular cell adhesion molecule, and impaired microvascular reactivity are early markers of vasculopathy in type 2 diabetic individuals without microalbuminuria. *Diabetes Care* **1999**, *22*, 1865–1870. [CrossRef]
52. Festa, A.; D'Agostino, R., Jr.; Tracy, R.P.; Haffner, S.M.; Insulin Resistance Atherosclerosis, S. Elevated levels of acute-phase proteins and plasminogen activator inhibitor-1 predict the development of type 2 diabetes: The insulin resistance atherosclerosis study. *Diabetes* **2002**, *51*, 1131–1137. [CrossRef]
53. Vischer, U.M.; Emeis, J.J.; Bilo, H.J.; Stehouwer, C.D.; Thomsen, C.; Rasmussen, O.; Hermansen, K.; Wollheim, C.B.; Ingerslev, J. von Willebrand factor (vWf) as a plasma marker of endothelial activation in diabetes: Improved reliability with parallel determination of the vWf propeptide (vWf:AgII). *Thromb. Haemost.* **1998**, *80*, 1002–1007.
54. Siragusa, M.; Fleming, I. The eNOS signalosome and its link to endothelial dysfunction. *Pflug. Archiv Eur. J. Physiol.* **2016**, *468*, 1125–1137. [CrossRef]
55. Rabini, R.A.; Vignini, A.; Salvolini, E.; Staffolani, R.; Martarelli, D.; Moretti, N.; Mazzanti, L. Activation of human aortic endothelial cells by LDL from Type 1 diabetic patients: An in vitro study. *Atherosclerosis* **2002**, *165*, 69–77. [CrossRef]
56. Artwohl, M.; Graier, W.F.; Roden, M.; Bischof, M.; Freudenthaler, A.; Waldhausl, W.; Baumgartner-Parzer, S.M. Diabetic LDL triggers apoptosis in vascular endothelial cells. *Diabetes* **2003**, *52*, 1240–1247. [CrossRef]
57. Toma, L.; Stancu, C.S.; Sanda, G.M.; Sima, A.V. Anti-oxidant and anti-inflammatory mechanisms of amlodipine action to improve endothelial cell dysfunction induced by irreversibly glycated LDL. *Biochem. Biophys. Res. Commun.* **2011**, *411*, 202–207. [CrossRef]
58. Mohanan Nair, M.; Zhao, R.; Xie, X.; Shen, G.X. Impact of glycated LDL on endothelial nitric oxide synthase in vascular endothelial cells: Involvement of transmembrane signaling and endoplasmic reticulum stress. *J. Diabetes Its Complicat.* **2016**, *30*, 391–397. [CrossRef]
59. Stancu, C.S.; Georgescu, A.; Toma, L.; Sanda, G.M.; Sima, A.V. Glycated low density lipoproteins alter vascular reactivity in hyperlipidemic hyperglycemic hamsters. *Ann. Rom. Soc. Cell Biol.* **2012**, *17*, 9–15.
60. Dong, Y.; Wu, Y.; Wu, M.; Wang, S.; Zhang, J.; Xie, Z.; Xu, J.; Song, P.; Wilson, K.; Zhao, Z.; et al. Activation of protease calpain by oxidized and glycated LDL increases the degradation of endothelial nitric oxide synthase. *J. Cell. Mol. Med.* **2009**, *13*, 2899–2910. [CrossRef]
61. Nofer, J.R.; van der Giet, M.; Tolle, M.; Wolinska, I.; von Wnuck Lipinski, K.; Baba, H.A.; Tietge, U.J.; Godecke, A.; Ishii, I.; Kleuser, B.; et al. HDL induces NO-dependent vasorelaxation via the lysophospholipid receptor S1P3. *J. Clin. Investig.* **2004**, *113*, 569–581. [CrossRef]
62. Curtiss, L.K.; Witztum, J.L. Plasma apolipoproteins AI, AII, B, CI, and E are glucosylated in hyperglycemic diabetic subjects. *Diabetes* **1985**, *34*, 452–461. [CrossRef] [PubMed]
63. Matsunaga, T.; Nakajima, T.; Miyazaki, T.; Koyama, I.; Hokari, S.; Inoue, I.; Kawai, S.; Shimomura, H.; Katayama, S.; Hara, A.; et al. Glycated high-density lipoprotein regulates reactive oxygen species and reactive nitrogen species in endothelial cells. *Metab. Clin. Exp.* **2003**, *52*, 42–49. [CrossRef] [PubMed]
64. Persegol, L.; Verges, B.; Foissac, M.; Gambert, P.; Duvillard, L. Inability of HDL from type 2 diabetic patients to counteract the inhibitory effect of oxidised LDL on endothelium-dependent vasorelaxation. *Diabetologia* **2006**, *49*, 1380–1386. [CrossRef] [PubMed]
65. Sorrentino, S.A.; Besler, C.; Rohrer, L.; Meyer, M.; Heinrich, K.; Bahlmann, F.H.; Mueller, M.; Horvath, T.; Doerries, C.; Heinemann, M.; et al. Endothelial-vasoprotective effects of high-density lipoprotein are impaired in patients with type 2 diabetes mellitus but are improved after extended-release niacin therapy. *Circulation* **2010**, *121*, 110–122. [CrossRef]
66. Vaisar, T.; Couzens, E.; Hwang, A.; Russell, M.; Barlow, C.E.; DeFina, L.F.; Hoofnagle, A.N.; Kim, F. Type 2 diabetes is associated with loss of HDL endothelium protective functions. *PLoS ONE* **2018**, *13*, e0192616. [CrossRef]
67. Scioli, M.G.; Storti, G.; D'Amico, F.; Rodriguez Guzman, R.; Centofanti, F.; Doldo, E.; Cespedes Miranda, E.M.; Orlandi, A. Oxidative Stress and New Pathogenetic Mechanisms in Endothelial Dysfunction: Potential Diagnostic Biomarkers and Therapeutic Targets. *J. Clin. Med.* **2020**, *9*, 1995. [CrossRef]
68. He, L.; He, T.; Farrar, S.; Ji, L.; Liu, T.; Ma, X. Antioxidants Maintain Cellular Redox Homeostasis by Elimination of Reactive Oxygen Species. *Cell. Physiol. Biochem. Int. J. Exp. Cell. Physiol. Biochem. Pharmacol.* **2017**, *44*, 532–553. [CrossRef]
69. Sena, C.M.; Leandro, A.; Azul, L.; Seica, R.; Perry, G. Vascular Oxidative Stress: Impact and Therapeutic Approaches. *Front. Physiol.* **2018**, *9*, 1668. [CrossRef]
70. Toma, L.; Stancu, C.S.; Botez, G.M.; Sima, A.V.; Simionescu, M. Irreversibly glycated LDL induce oxidative and inflammatory state in human endothelial cells; added effect of high glucose. *Biochem. Biophys. Res. Commun.* **2009**, *390*, 877–882. [CrossRef]
71. Xie, X.; Zhao, R.; Shen, G.X. Impact of cyanidin-3-glucoside on glycated LDL-induced NADPH oxidase activation, mitochondrial dysfunction and cell viability in cultured vascular endothelial cells. *Int. J. Mol. Sci.* **2012**, *13*, 15867–15880. [CrossRef]
72. Toma, L.; Sanda, G.M.; Niculescu, L.S.; Deleanu, M.; Stancu, C.S.; Sima, A.V. Caffeic acid attenuates the inflammatory stress induced by glycated LDL in human endothelial cells by mechanisms involving inhibition of AGE-receptor, oxidative, and endoplasmic reticulum stress. *BioFactors* **2017**, *43*, 685–697. [CrossRef] [PubMed]
73. Popov, D.L. Mitochondrial Dysfunction Signature in Diabetic Vascular Endothelium. *J. Clin. Exp. Pathol.* **2018**, *8*. [CrossRef]
74. Sangle, G.V.; Chowdhury, S.K.; Xie, X.; Stelmack, G.L.; Halayko, A.J.; Shen, G.X. Impairment of mitochondrial respiratory chain activity in aortic endothelial cells induced by glycated low-density lipoprotein. *Free Radic. Biol. Med.* **2010**, *48*, 781–790. [CrossRef] [PubMed]

75. Xie, X.; Chowdhury, S.R.; Sangle, G.; Shen, G.X. Impact of diabetes-associated lipoproteins on oxygen consumption and mitochondrial enzymes in porcine aortic endothelial cells. *Acta Biochim. Pol.* **2010**, *57*, 393–398. [CrossRef]
76. Wang, Y.C.; Lee, A.S.; Lu, L.S.; Ke, L.Y.; Chen, W.Y.; Dong, J.W.; Lu, J.; Chen, Z.; Chu, C.S.; Chan, H.C.; et al. Human electronegative LDL induces mitochondrial dysfunction and premature senescence of vascular cells in vivo. *Aging Cell* **2018**, *17*, e12792. [CrossRef]
77. Zhao, R.; Shen, G.X. Functional modulation of antioxidant enzymes in vascular endothelial cells by glycated LDL. *Atherosclerosis* **2005**, *179*, 277–284. [CrossRef]
78. Pathomthongtaweechai, N.; Chutipongtanate, S. AGE/RAGE signaling-mediated endoplasmic reticulum stress and future prospects in non-coding RNA therapeutics for diabetic nephropathy. *Biomed. Pharmacother. Biomed. Pharmacother.* **2020**, *131*, 110655. [CrossRef]
79. Maamoun, H.; Abdelsalam, S.S.; Zeidan, A.; Korashy, H.M.; Agouni, A. Endoplasmic Reticulum Stress: A Critical Molecular Driver of Endothelial Dysfunction and Cardiovascular Disturbances Associated with Diabetes. *Int. J. Mol. Sci.* **2019**, *20*, 1658. [CrossRef]
80. Basha, B.; Samuel, S.M.; Triggle, C.R.; Ding, H. Endothelial dysfunction in diabetes mellitus: Possible involvement of endoplasmic reticulum stress? *Exp. Diabetes Res.* **2012**, *2012*, 481840. [CrossRef] [PubMed]
81. Stancu, C.S.; Carnuta, M.G.; Sanda, G.M.; Toma, L.; Deleanu, M.; Niculescu, L.S.; Sasson, S.; Simionescu, M.; Sima, A.V. Hyperlipidemia-induced hepatic and small intestine ER stress and decreased paraoxonase 1 expression and activity is associated with HDL dysfunction in Syrian hamsters. *Mol. Nutr. Food Res.* **2015**, *59*, 2293–2302. [CrossRef]
82. Zhao, R.; Xie, X.; Le, K.; Li, W.; Moghadasian, M.H.; Beta, T.; Shen, G.X. Endoplasmic reticulum stress in diabetic mouse or glycated LDL-treated endothelial cells: Protective effect of Saskatoon berry powder and cyanidin glycans. *J. Nutr. Biochem.* **2015**, *26*, 1248–1253. [CrossRef] [PubMed]
83. Lin, J.H.; Li, H.; Yasumura, D.; Cohen, H.R.; Zhang, C.; Panning, B.; Shokat, K.M.; Lavail, M.M.; Walter, P. IRE1 signaling affects cell fate during the unfolded protein response. *Science* **2007**, *318*, 944–949. [CrossRef] [PubMed]
84. Yu, W.; Liu, X.; Feng, L.; Yang, H.; Yu, W.; Feng, T.; Wang, S.; Wang, J.; Liu, N. Glycation of paraoxonase 1 by high glucose instigates endoplasmic reticulum stress to induce endothelial dysfunction in vivo. *Sci. Rep.* **2017**, *7*, 45827. [CrossRef] [PubMed]
85. Suryavanshi, S.V.; Kulkarni, Y.A. NF-kappabeta: A Potential Target in the Management of Vascular Complications of Diabetes. *Front. Pharmacol.* **2017**, *8*, 798. [CrossRef] [PubMed]
86. Hedrick, C.C.; Thorpe, S.R.; Fu, M.X.; Harper, C.M.; Yoo, J.; Kim, S.M.; Wong, H.; Peters, A.L. Glycation impairs high-density lipoprotein function. *Diabetologia* **2000**, *43*, 312–320. [CrossRef]
87. Carnuta, M.G.; Stancu, C.S.; Toma, L.; Sanda, G.M.; Niculescu, L.S.; Deleanu, M.; Popescu, A.C.; Popescu, M.R.; Vlad, A.; Dimulescu, D.R.; et al. Dysfunctional high-density lipoproteins have distinct composition, diminished anti-inflammatory potential and discriminate acute coronary syndrome from stable coronary artery disease patients. *Sci. Rep.* **2017**, *7*, 7295. [CrossRef]
88. Morgantini, C.; Natali, A.; Boldrini, B.; Imaizumi, S.; Navab, M.; Fogelman, A.M.; Ferrannini, E.; Reddy, S.T. Anti-inflammatory and antioxidant properties of HDLs are impaired in type 2 diabetes. *Diabetes* **2011**, *60*, 2617–2623. [CrossRef]
89. Kearney, K.; Tomlinson, D.; Smith, K.; Ajjan, R. Hypofibrinolysis in diabetes: A therapeutic target for the reduction of cardiovascular risk. *Cardiovasc. Diabetol.* **2017**, *16*, 34. [CrossRef]
90. Ren, S.; Lee, H.; Hu, L.; Lu, L.; Shen, G.X. Impact of diabetes-associated lipoproteins on generation of fibrinolytic regulators from vascular endothelial cells. *J. Clin. Endocrinol. Metab.* **2002**, *87*, 286–291. [CrossRef]
91. Shen, G.X. Impact and mechanism for oxidized and glycated lipoproteins on generation of fibrinolytic regulators from vascular endothelial cells. *Mol. Cell. Biochem.* **2003**, *246*, 69–74. [CrossRef]
92. Zhang, J.; Ren, S.; Sun, D.; Shen, G.X. Influence of glycation on LDL-induced generation of fibrinolytic regulators in vascular endothelial cells. *Arterioscler. Thromb. Vasc. Biol.* **1998**, *18*, 1140–1148. [CrossRef] [PubMed]
93. Ma, G.M.; Halayko, A.J.; Stelmack, G.L.; Zhu, F.; Zhao, R.; Hillier, C.T.; Shen, G.X. Effects of oxidized and glycated low-density lipoproteins on transcription and secretion of plasminogen activator inhibitor-1 in vascular endothelial cells. *Cardiovasc. Pathol. Off. J. Soc. Cardiovasc. Pathol.* **2006**, *15*, 3–10. [CrossRef] [PubMed]
94. Zhao, R.; Shen, G.X. Involvement of heat shock factor-1 in glycated LDL-induced upregulation of plasminogen activator inhibitor-1 in vascular endothelial cells. *Diabetes* **2007**, *56*, 1436–1444. [CrossRef]
95. Sangle, G.V.; Zhao, R.; Mizuno, T.M.; Shen, G.X. Involvement of RAGE, NADPH oxidase, and Ras/Raf-1 pathway in glycated LDL-induced expression of heat shock factor-1 and plasminogen activator inhibitor-1 in vascular endothelial cells. *Endocrinology* **2010**, *151*, 4455–4466. [CrossRef] [PubMed]
96. Ren, S.; Shen, G.X. Impact of antioxidants and HDL on glycated LDL-induced generation of fibrinolytic regulators from vascular endothelial cells. *Arterioscler. Thromb. Vasc. Biol.* **2000**, *20*, 1688–1693. [CrossRef] [PubMed]
97. Watson, E.C.; Grant, Z.L.; Coultas, L. Endothelial cell apoptosis in angiogenesis and vessel regression. *Cell. Mol. Life Sci. CMLS* **2017**, *74*, 4387–4403. [CrossRef] [PubMed]
98. Yang, C.Y.; Chen, H.H.; Huang, M.T.; Raya, J.L.; Yang, J.H.; Chen, C.H.; Gaubatz, J.W.; Pownall, H.J.; Taylor, A.A.; Ballantyne, C.M.; et al. Pro-apoptotic low-density lipoprotein subfractions in type II diabetes. *Atherosclerosis* **2007**, *193*, 283–291. [CrossRef]

99. Li, X.L.; Li, B.Y.; Cheng, M.; Yu, F.; Yin, W.B.; Cai, Q.; Zhang, Z.; Zhang, J.H.; Wang, J.F.; Zhou, R.H.; et al. PIMT prevents the apoptosis of endothelial cells in response to glycated low density lipoproteins and protective effects of grape seed procyanidin B2. *PLoS ONE* **2013**, *8*, e69979. [CrossRef]
100. Yin, W.; Li, B.; Li, X.; Yu, F.; Cai, Q.; Zhang, Z.; Wang, J.; Zhang, J.; Zhou, R.; Cheng, M.; et al. Critical role of prohibitin in endothelial cell apoptosis caused by glycated low-density lipoproteins and protective effects of grape seed procyanidin B2. *J. Cardiovasc. Pharmacol.* **2015**, *65*, 13–21. [CrossRef]
101. Matsunaga, T.; Iguchi, K.; Nakajima, T.; Koyama, I.; Miyazaki, T.; Inoue, I.; Kawai, S.; Katayama, S.; Hirano, K.; Hokari, S.; et al. Glycated high-density lipoprotein induces apoptosis of endothelial cells via a mitochondrial dysfunction. *Biochem. Biophys. Res. Commun.* **2001**, *287*, 714–720. [CrossRef] [PubMed]
102. Park, K.H.; Kim, J.Y.; Choi, I.; Kim, J.R.; Won, K.C.; Cho, K.H. Fructated apolipoprotein A-I exacerbates cellular senescence in human umbilical vein endothelial cells accompanied by impaired insulin secretion activity and embryo toxicity. *Biochem. Cell Biol. Biochim. Biol. Cell.* **2016**, *94*, 337–345. [CrossRef] [PubMed]
103. Bucala, R.; Makita, Z.; Koschinsky, T.; Cerami, A.; Vlassara, H. Lipid advanced glycosylation: Pathway for lipid oxidation in vivo. *Proc. Natl. Acad. Sci. USA* **1993**, *90*, 6434–6438. [CrossRef] [PubMed]
104. Sima, A.; Stancu, C. Modified lipoproteins accumulate in human coronary atheroma. *J. Cell. Mol. Med.* **2002**, *6*, 110–111. [CrossRef] [PubMed]
105. Tames, F.J.; Mackness, M.I.; Arrol, S.; Laing, I.; Durrington, P.N. Non-enzymatic glycation of apolipoprotein B in the sera of diabetic and non-diabetic subjects. *Atherosclerosis* **1992**, *93*, 237–244. [CrossRef]
106. Simionescu, M.; Popov, D.; Sima, A.; Hasu, M.; Costache, G.; Faitar, S.; Vulpanovici, A.; Stancu, C.; Stern, D.; Simionescu, N. Pathobiochemistry of combined diabetes and atherosclerosis studied on a novel animal model. The hyperlipemic-hyperglycemic hamster. *Am. J. Pathol.* **1996**, *148*, 997–1014. [PubMed]
107. Sima, A.; Popov, D.; Starodub, O.; Stancu, C.; Cristea, C.; Stern, D.; Simionescu, M. Pathobiology of the heart in experimental diabetes: Immunolocalization of lipoproteins, immunoglobulin G, and advanced glycation endproducts proteins in diabetic and/or hyperlipidemic hamster. *Lab. Investig. J. Tech. Methods Pathol.* **1997**, *77*, 3–18.
108. Hwang, S.W.; Lee, Y.M.; Aldini, G.; Yeum, K.J. Targeting Reactive Carbonyl Species with Natural Sequestering Agents. *Molecules* **2016**, *21*, 280. [CrossRef]
109. von Eckardstein, A.; Widmann, C. High-density lipoprotein, beta cells, and diabetes. *Cardiovasc. Res.* **2014**, *103*, 384–394. [CrossRef]
110. Younis, N.H.; Soran, H.; Charlton-Menys, V.; Sharma, R.; Hama, S.; Pemberton, P.; Elseweidy, M.M.; Durrington, P.N. High-density lipoprotein impedes glycation of low-density lipoprotein. *Diabetes Vasc. Dis. Res.* **2012**, *10*, 152–160. [CrossRef]
111. Meneses, M.J.; Silvestre, R.; Sousa-Lima, I.; Macedo, M.P. Paraoxonase-1 as a Regulator of Glucose and Lipid Homeostasis: Impact on the Onset and Progression of Metabolic Disorders. *Int. J. Mol. Sci.* **2019**, *20*, 4049. [CrossRef]
112. Femlak, M.; Gluba-Brzozka, A.; Cialkowska-Rysz, A.; Rysz, J. The role and function of HDL in patients with diabetes mellitus and the related cardiovascular risk. *Lipids Health Dis.* **2017**, *16*, 207. [CrossRef] [PubMed]
113. Undurti, A.; Huang, Y.; Lupica, J.A.; Smith, J.D.; DiDonato, J.A.; Hazen, S.L. Modification of high density lipoprotein by myeloperoxidase generates a pro-inflammatory particle. *J. Biol. Chem.* **2009**, *284*, 30825–30835. [CrossRef] [PubMed]
114. Lu, N.; Xie, S.; Li, J.; Tian, R.; Peng, Y.Y. Myeloperoxidase-mediated oxidation targets serum apolipoprotein A-I in diabetic patients and represents a potential mechanism leading to impaired anti-apoptotic activity of high density lipoprotein. *Clin. Chim. Acta Int. J. Clin. Chem.* **2015**, *441*, 163–170. [CrossRef] [PubMed]
115. Craciun, E.C.; Leucuta, D.C.; Rusu, R.L.; David, B.A.; Cret, V.; Dronca, E. Paraoxonase-1 activities in children and adolescents with type 1 diabetes mellitus. *Acta Biochim. Pol.* **2016**, *63*, 511–515. [CrossRef]
116. Mastorikou, M.; Mackness, B.; Liu, Y.; Mackness, M. Glycation of paraoxonase-1 inhibits its activity and impairs the ability of high-density lipoprotein to metabolize membrane lipid hydroperoxides. *Diabet. Med. J. Br. Diabet. Assoc.* **2008**, *25*, 1049–1055. [CrossRef]
117. Boemi, M.; Leviev, I.; Sirolla, C.; Pieri, C.; Marra, M.; James, R.W. Serum paraoxonase is reduced in type 1 diabetic patients compared to non-diabetic, first degree relatives; influence on the ability of HDL to protect LDL from oxidation. *Atherosclerosis* **2001**, *155*, 229–235. [CrossRef]
118. Rosenson, R.S.; Brewer, H.B., Jr.; Ansell, B.J.; Barter, P.; Chapman, M.J.; Heinecke, J.W.; Kontush, A.; Tall, A.R.; Webb, N.R. Dysfunctional HDL and atherosclerotic cardiovascular disease. *Nat. Rev. Cardiol.* **2016**, *13*, 48–60. [CrossRef]
119. Dullaart, R.P.; Otvos, J.D.; James, R.W. Serum paraoxonase-1 activity is more closely related to HDL particle concentration and large HDL particles than to HDL cholesterol in Type 2 diabetic and non-diabetic subjects. *Clin. Biochem.* **2014**, *47*, 1022–1027. [CrossRef]
120. Tabara, Y.; Arai, H.; Hirao, Y.; Takahashi, Y.; Setoh, K.; Kawaguchi, T.; Kosugi, S.; Ito, Y.; Nakayama, T.; Matsuda, F.; et al. Different inverse association of large high-density lipoprotein subclasses with exacerbation of insulin resistance and incidence of type 2 diabetes: The Nagahama study. *Diabetes Res. Clin. Pract.* **2017**, *127*, 123–131. [CrossRef]
121. Manjunatha, S.; Distelmaier, K.; Dasari, S.; Carter, R.E.; Kudva, Y.C.; Nair, K.S. Functional and proteomic alterations of plasma high density lipoproteins in type 1 diabetes mellitus. *Metab. Clin. Exp.* **2016**, *65*, 1421–1431. [CrossRef]

122. Machado-Lima, A.; Iborra, R.T.; Pinto, R.S.; Sartori, C.H.; Oliveira, E.R.; Nakandakare, E.R.; Stefano, J.T.; Giannella-Neto, D.; Correa-Giannella, M.L.; Passarelli, M. Advanced glycated albumin isolated from poorly controlled type 1 diabetes mellitus patients alters macrophage gene expression impairing ABCA-1-mediated reverse cholesterol transport. *Diabetes/Metab. Res. Rev.* **2013**, *29*, 66–76. [CrossRef] [PubMed]
123. McEneny, J.; Daniels, J.A.; McGowan, A.; Gunness, A.; Moore, K.; Stevenson, M.; Young, I.S.; Gibney, J. A Cross-Sectional Study Demonstrating Increased Serum Amyloid A Related Inflammation in High-Density Lipoproteins from Subjects with Type 1 Diabetes Mellitus and How this Association Was Augmented by Poor Glycaemic Control. *J. Diabetes Res.* **2015**, *2015*, 351601. [CrossRef] [PubMed]
124. Vickers, K.C.; Palmisano, B.T.; Shoucri, B.M.; Shamburek, R.D.; Remaley, A.T. MicroRNAs are transported in plasma and delivered to recipient cells by high-density lipoproteins. *Nat. Cell Biol.* **2011**, *13*, 423–433. [CrossRef] [PubMed]
125. Wagner, J.; Riwanto, M.; Besler, C.; Knau, A.; Fichtlscherer, S.; Roxe, T.; Zeiher, A.M.; Landmesser, U.; Dimmeler, S. Characterization of levels and cellular transfer of circulating lipoprotein-bound microRNAs. *Arterioscler. Thromb. Vasc. Biol.* **2013**, *33*, 1392–1400. [CrossRef]
126. Niculescu, L.S.; Simionescu, N.; Sanda, G.M.; Carnuta, M.G.; Stancu, C.S.; Popescu, A.C.; Popescu, M.R.; Vlad, A.; Dimulescu, D.R.; Simionescu, M.; et al. MiR-486 and miR-92a Identified in Circulating HDL Discriminate between Stable and Vulnerable Coronary Artery Disease Patients. *PLoS ONE* **2015**, *10*, e0140958. [CrossRef]
127. Simionescu, N.; Niculescu, L.S.; Sanda, G.M.; Margina, D.; Sima, A.V. Analysis of circulating microRNAs that are specifically increased in hyperlipidemic and/or hyperglycemic sera. *Mol. Biol. Rep.* **2014**, *41*, 5765–5773. [CrossRef]
128. Huang, Y.; Liu, Y.; Li, L.; Su, B.; Yang, L.; Fan, W.; Yin, Q.; Chen, L.; Cui, T.; Zhang, J.; et al. Involvement of inflammation-related miR-155 and miR-146a in diabetic nephropathy: Implications for glomerular endothelial injury. *BMC Nephrol.* **2014**, *15*, 142. [CrossRef]
129. Tabet, F.; Vickers, K.C.; Cuesta Torres, L.F.; Wiese, C.B.; Shoucri, B.M.; Lambert, G.; Catherinet, C.; Prado-Lourenco, L.; Levin, M.G.; Thacker, S.; et al. HDL-transferred microRNA-223 regulates ICAM-1 expression in endothelial cells. *Nat. Commun.* **2014**, *5*, 3292. [CrossRef]
130. Simionescu, N.; Niculescu, L.S.; Carnuta, M.G.; Sanda, G.M.; Stancu, C.S.; Popescu, A.C.; Popescu, M.R.; Vlad, A.; Dimulescu, D.R.; Simionescu, M.; et al. Hyperglycemia Determines Increased Specific MicroRNAs Levels in Sera and HDL of Acute Coronary Syndrome Patients and Stimulates MicroRNAs Production in Human Macrophages. *PLoS ONE* **2016**, *11*, e0161201. [CrossRef]
131. Florijn, B.W.; Duijs, J.; Levels, J.H.; Dallinga-Thie, G.M.; Wang, Y.; Boing, A.N.; Yuana, Y.; Stam, W.; Limpens, R.; Au, Y.W.; et al. Diabetic Nephropathy Alters the Distribution of Circulating Angiogenic MicroRNAs Among Extracellular Vesicles, HDL, and Ago-2. *Diabetes* **2019**, *68*, 2287–2300. [CrossRef]
132. Chiang, C.K.; Wang, C.C.; Lu, T.F.; Huang, K.H.; Sheu, M.L.; Liu, S.H.; Hung, K.Y. Involvement of Endoplasmic Reticulum Stress, Autophagy, and Apoptosis in Advanced Glycation End Products-Induced Glomerular Mesangial Cell Injury. *Sci. Rep.* **2016**, *6*, 34167. [CrossRef] [PubMed]
133. Hetz, C.; Axten, J.M.; Patterson, J.B. Pharmacological targeting of the unfolded protein response for disease intervention. *Nat. Chem. Biol.* **2019**, *15*, 764–775. [CrossRef] [PubMed]
134. Roberts, A.C.; Porter, K.E. Cellular and molecular mechanisms of endothelial dysfunction in diabetes. *Diabetes Vasc. Dis. Res.* **2013**, *10*, 472–482. [CrossRef] [PubMed]
135. Inzucchi, S.E.; Bergenstal, R.M.; Buse, J.B.; Diamant, M.; Ferrannini, E.; Nauck, M.; Peters, A.L.; Tsapas, A.; Wender, R.; Matthews, D.R.; et al. Management of hyperglycemia in type 2 diabetes: A patient-centered approach: Position statement of the American Diabetes Association (ADA) and the European Association for the Study of Diabetes (EASD). *Diabetes Care* **2012**, *35*, 1364–1379. [CrossRef]
136. Wang, C.P.; Lorenzo, C.; Habib, S.L.; Jo, B.; Espinoza, S.E. Differential effects of metformin on age related comorbidities in older men with type 2 diabetes. *J. Diabetes Its Complicat.* **2017**, *31*, 679–686. [CrossRef]
137. Lee, K.T.; Yeh, Y.H.; Chang, S.H.; See, L.C.; Lee, C.H.; Wu, L.S.; Liu, J.R.; Kuo, C.T.; Wen, M.S. Metformin is associated with fewer major adverse cardiac events among patients with a new diagnosis of type 2 diabetes mellitus: A propensity score-matched nationwide study. *Medicine* **2017**, *96*, e7507. [CrossRef]
138. Petrie, J.R.; Chaturvedi, N.; Ford, I.; Brouwers, M.; Greenlaw, N.; Tillin, T.; Hramiak, I.; Hughes, A.D.; Jenkins, A.J.; Klein, B.E.K.; et al. Cardiovascular and metabolic effects of metformin in patients with type 1 diabetes (REMOVAL): A double-blind, randomised, placebo-controlled trial. *Lancet Diabetes Endocrinol.* **2017**, *5*, 597–609. [CrossRef]
139. Fei, Y.; Tsoi, M.F.; Cheung, B.M.Y. Cardiovascular outcomes in trials of new antidiabetic drug classes: A network meta-analysis. *Cardiovasc. Diabetol.* **2019**, *18*, 112. [CrossRef]
140. Rahbar, S.; Natarajan, R.; Yerneni, K.; Scott, S.; Gonzales, N.; Nadler, J.L. Evidence that pioglitazone, metformin and pentoxifylline are inhibitors of glycation. *Clin. Chim. Acta Int. J. Clin. Chem.* **2000**, *301*, 65–77. [CrossRef]
141. Haddad, M.; Knani, I.; Bouzidi, H.; Berriche, O.; Hammami, M.; Kerkeni, M. Plasma Levels of Pentosidine, Carboxymethyl-Lysine, Soluble Receptor for Advanced Glycation End Products, and Metabolic Syndrome: The Metformin Effect. *Dis. Markers* **2016**, *2016*, 6248264. [CrossRef]
142. Pereira, A.; Fernandes, R.; Crisostomo, J.; Seica, R.M.; Sena, C.M. The Sulforaphane and pyridoxamine supplementation normalize endothelial dysfunction associated with type 2 diabetes. *Sci. Rep.* **2017**, *7*, 14357. [CrossRef] [PubMed]

143. Almeida, F.; Santos-Silva, D.; Rodrigues, T.; Matafome, P.; Crisostomo, J.; Sena, C.; Goncalves, L.; Seica, R. Pyridoxamine reverts methylglyoxal-induced impairment of survival pathways during heart ischemia. *Cardiovasc. Ther.* **2013**, *31*, e79–e85. [CrossRef] [PubMed]
144. Watson, A.M.; Soro-Paavonen, A.; Sheehy, K.; Li, J.; Calkin, A.C.; Koitka, A.; Rajan, S.N.; Brasacchio, D.; Allen, T.J.; Cooper, M.E.; et al. Delayed intervention with AGE inhibitors attenuates the progression of diabetes-accelerated atherosclerosis in diabetic apolipoprotein E knockout mice. *Diabetologia* **2011**, *54*, 681–689. [CrossRef] [PubMed]
145. Chang, K.C.; Liang, J.T.; Tsai, P.S.; Wu, M.S.; Hsu, K.L. Prevention of arterial stiffening by pyridoxamine in diabetes is associated with inhibition of the pathogenic glycation on aortic collagen. *Br. J. Pharmacol.* **2009**, *157*, 1419–1426. [CrossRef] [PubMed]
146. Degenhardt, T.P.; Alderson, N.L.; Arrington, D.D.; Beattie, R.J.; Basgen, J.M.; Steffes, M.W.; Thorpe, S.R.; Baynes, J.W. Pyridoxamine inhibits early renal disease and dyslipidemia in the streptozotocin-diabetic rat. *Kidney Int.* **2002**, *61*, 939–950. [CrossRef]
147. Tam, H.L.; Shiu, S.W.; Wong, Y.; Chow, W.S.; Betteridge, D.J.; Tan, K.C. Effects of atorvastatin on serum soluble receptors for advanced glycation end-products in type 2 diabetes. *Atherosclerosis* **2010**, *209*, 173–177. [CrossRef]
148. Fukushima, Y.; Daida, H.; Morimoto, T.; Kasai, T.; Miyauchi, K.; Yamagishi, S.; Takeuchi, M.; Hiro, T.; Kimura, T.; Nakagawa, Y.; et al. Relationship between advanced glycation end products and plaque progression in patients with acute coronary syndrome: The JAPAN-ACS sub-study. *Cardiovasc. Diabetol.* **2013**, *12*, 5. [CrossRef]
149. Cuccurullo, C.; Iezzi, A.; Fazia, M.L.; De Cesare, D.; Di Francesco, A.; Muraro, R.; Bei, R.; Ucchino, S.; Spigonardo, F.; Chiarelli, F.; et al. Suppression of RAGE as a basis of simvastatin-dependent plaque stabilization in type 2 diabetes. *Arterioscler. Thromb. Vasc. Biol.* **2006**, *26*, 2716–2723. [CrossRef]
150. Quade-Lyssy, P.; Kanarek, A.M.; Baiersdorfer, M.; Postina, R.; Kojro, E. Statins stimulate the production of a soluble form of the receptor for advanced glycation end products. *J. Lipid Res.* **2013**, *54*, 3052–3061. [CrossRef]
151. Bahrambeigi, S.; Rahimi, M.; Yousefi, B.; Shafiei-Irannejad, V. New potentials for 3-hydroxy-3-methyl-glutaryl-coenzymeA reductase inhibitors: Possible applications in retarding diabetic complications. *J. Cell. Physiol.* **2019**, *234*, 19393–19405. [CrossRef]
152. Collins, R.; Armitage, J.; Parish, S.; Sleigh, P.; Peto, R.; Heart Protection Study Collaborative, G. MRC/BHF Heart Protection Study of cholesterol-lowering with simvastatin in 5963 people with diabetes: A randomised placebo-controlled trial. *Lancet* **2003**, *361*, 2005–2016. [CrossRef] [PubMed]
153. Colhoun, H.M.; Betteridge, D.J.; Durrington, P.N.; Hitman, G.A.; Neil, H.A.; Livingstone, S.J.; Thomason, M.J.; Mackness, M.I.; Charlton-Menys, V.; Fuller, J.H.; et al. Primary prevention of cardiovascular disease with atorvastatin in type 2 diabetes in the Collaborative Atorvastatin Diabetes Study (CARDS): Multicentre randomised placebo-controlled trial. *Lancet* **2004**, *364*, 685–696. [CrossRef]
154. Ramos, R.; Comas-Cufi, M.; Marti-Lluch, R.; Ballo, E.; Ponjoan, A.; Alves-Cabratosa, L.; Blanch, J.; Marrugat, J.; Elosua, R.; Grau, M.; et al. Statins for primary prevention of cardiovascular events and mortality in old and very old adults with and without type 2 diabetes: Retrospective cohort study. *Bmj* **2018**, *362*, k3359. [CrossRef] [PubMed]
155. Balbi, M.E.; Tonin, F.S.; Mendes, A.M.; Borba, H.H.; Wiens, A.; Fernandez-Llimos, F.; Pontarolo, R. Antioxidant effects of vitamins in type 2 diabetes: A meta-analysis of randomized controlled trials. *Diabetol. Metab. Syndr.* **2018**, *10*, 18. [CrossRef] [PubMed]
156. Chen, J.; Wu, J.; Kong, D.; Yang, C.; Yu, H.; Pan, Q.; Liu, W.; Ding, Y.; Liu, H. The Effect of Antioxidant Vitamins on Patients With Diabetes and Albuminuria: A Meta-Analysis of Randomized Controlled Trials. *J. Ren. Nutr. Off. J. Counc. Ren. Nutr. Natl. Kidney Found.* **2020**, *30*, 101–110. [CrossRef] [PubMed]
157. Baziar, N.; Nasli-Esfahani, E.; Djafarian, K.; Qorbani, M.; Hedayati, M.; Mishani, M.A.; Faghfoori, Z.; Ahmaripour, N.; Hosseini, S. The Beneficial Effects of Alpha Lipoic Acid Supplementation on Lp-PLA2 Mass and Its Distribution between HDL and apoB-Containing Lipoproteins in Type 2 Diabetic Patients: A Randomized, Double-Blind, Placebo-Controlled Trial. *Oxidative Med. Cell. Longev.* **2020**, *2020*, 5850865. [CrossRef] [PubMed]
158. Alem, M.M. Allopurinol and endothelial function: A systematic review with meta-analysis of randomized controlled trials. *Cardiovasc. Ther.* **2018**, *36*, e12432. [CrossRef]
159. Lonn, E.; Yusuf, S.; Hoogwerf, B.; Pogue, J.; Yi, Q.; Zinman, B.; Bosch, J.; Dagenais, G.; Mann, J.F.; Gerstein, H.C.; et al. Effects of vitamin E on cardiovascular and microvascular outcomes in high-risk patients with diabetes: Results of the HOPE study and MICRO-HOPE substudy. *Diabetes Care* **2002**, *25*, 1919–1927. [CrossRef]
160. Montero, D.; Walther, G.; Stehouwer, C.D.; Houben, A.J.; Beckman, J.A.; Vinet, A. Effect of antioxidant vitamin supplementation on endothelial function in type 2 diabetes mellitus: A systematic review and meta-analysis of randomized controlled trials. *Obes. Rev. An Off. J. Int. Assoc. Study Obes.* **2014**, *15*, 107–116. [CrossRef]
161. Altunina, N.V.; Lizogub, V.G.; Bondarchuk, O.M. Alpha-Lipoic Acid as a Means of Influence on Systemic Inflammation in Type 2 Diabetes Mellitus Patients with Prior Myocardial Infarction. *J. Med. Life* **2020**, *13*, 32–36. [CrossRef]
162. Stitt, A.W.; He, C.; Friedman, S.; Scher, L.; Rossi, P.; Ong, L.; Founds, H.; Li, Y.M.; Bucala, R.; Vlassara, H. Elevated AGE-modified ApoB in sera of euglycemic, normolipidemic patients with atherosclerosis: Relationship to tissue AGEs. *Mol. Med.* **1997**, *3*, 617–627. [CrossRef] [PubMed]
163. Hudson, B.I.; Lippman, M.E. Targeting RAGE Signaling in Inflammatory Disease. *Annu. Rev. Med.* **2018**, *69*, 349–364. [CrossRef] [PubMed]
164. Deane, R.; Singh, I.; Sagare, A.P.; Bell, R.D.; Ross, N.T.; LaRue, B.; Love, R.; Perry, S.; Paquette, N.; Deane, R.J.; et al. A multimodal RAGE-specific inhibitor reduces amyloid beta-mediated brain disorder in a mouse model of Alzheimer disease. *J. Clin. Investig.* **2012**, *122*, 1377–1392. [CrossRef] [PubMed]

165. Liu, Y.; Yu, M.; Zhang, Z.; Yu, Y.; Chen, Q.; Zhang, W.; Zhao, X. Blockade of receptor for advanced glycation end products protects against systolic overload-induced heart failure after transverse aortic constriction in mice. *Eur. J. Pharmacol.* **2016**, *791*, 535–543. [CrossRef]
166. Manigrasso, M.B.; Pan, J.; Rai, V.; Zhang, J.; Reverdatto, S.; Quadri, N.; DeVita, R.J.; Ramasamy, R.; Shekhtman, A.; Schmidt, A.M. Small Molecule Inhibition of Ligand-Stimulated RAGE-DIAPH1 Signal Transduction. *Sci. Rep.* **2016**, *6*, 22450. [CrossRef]
167. Watanabe, M.; Toyomura, T.; Wake, H.; Liu, K.; Teshigawara, K.; Takahashi, H.; Nishibori, M.; Mori, S. Differential contribution of possible pattern-recognition receptors to advanced glycation end product-induced cellular responses in macrophage-like RAW264.7 cells. *Biotechnol. Appl. Biochem.* **2020**, *67*, 265–272. [CrossRef]
168. Xu, Y.; Li, Z. CRISPR-Cas systems: Overview, innovations and applications in human disease research and gene therapy. *Comput. Struct. Biotechnol. J.* **2020**, *18*, 2401–2415. [CrossRef]
169. Yuan, J.; Chen, M.; Xu, Q.; Liang, J.; Chen, R.; Xiao, Y.; Fang, M.; Chen, L. Effect of the Diabetic Environment On the Expression of MiRNAs in Endothelial Cells: Mir-149-5p Restoration Ameliorates the High Glucose-Induced Expression of TNF-alpha and ER Stress Markers. *Cell. Physiol. Biochem. Int. J. Exp. Cell. Physiol. Biochem. Pharmacol.* **2017**, *43*, 120–135. [CrossRef]
170. Chen, Q.; Qiu, F.; Zhou, K.; Matlock, H.G.; Takahashi, Y.; Rajala, R.V.S.; Yang, Y.; Moran, E.; Ma, J.X. Pathogenic Role of microRNA-21 in Diabetic Retinopathy Through Downregulation of PPARalpha. *Diabetes* **2017**, *66*, 1671–1682. [CrossRef]
171. Icli, B.; Wu, W.; Ozdemir, D.; Li, H.; Haemmig, S.; Liu, X.; Giatsidis, G.; Cheng, H.S.; Avci, S.N.; Kurt, M.; et al. MicroRNA-135a-3p regulates angiogenesis and tissue repair by targeting p38 signaling in endothelial cells. *FASEB J. Off. Publ. Fed. Am. Soc. Exp. Biol.* **2019**, *33*, 5599–5614. [CrossRef]
172. Venkat, P.; Cui, C.; Chopp, M.; Zacharek, A.; Wang, F.; Landschoot-Ward, J.; Shen, Y.; Chen, J. MiR-126 Mediates Brain Endothelial Cell Exosome Treatment-Induced Neurorestorative Effects After Stroke in Type 2 Diabetes Mellitus Mice. *Stroke* **2019**, *50*, 2865–2874. [CrossRef] [PubMed]
173. Gencer, B.; Kronenberg, F.; Stroes, E.S.; Mach, F. Lipoprotein(a): The revenant. *Eur. Heart J.* **2017**, *38*, 1553–1560. [CrossRef] [PubMed]
174. Viney, N.J.; van Capelleveen, J.C.; Geary, R.S.; Xia, S.; Tami, J.A.; Yu, R.Z.; Marcovina, S.M.; Hughes, S.G.; Graham, M.J.; Crooke, R.M.; et al. Antisense oligonucleotides targeting apolipoprotein(a) in people with raised lipoprotein(a): Two randomised, double-blind, placebo-controlled, dose-ranging trials. *Lancet* **2016**, *388*, 2239–2253. [CrossRef]

Review

Can Metformin Exert as an Active Drug on Endothelial Dysfunction in Diabetic Subjects?

Teresa Salvatore [1], Pia Clara Pafundi [2], Raffaele Galiero [2], Luca Rinaldi [2], Alfredo Caturano [2], Erica Vetrano [2], Concetta Aprea [2], Gaetana Albanese [2], Anna Di Martino [2], Carmen Ricozzi [2], Simona Imbriani [2] and Ferdinando Carlo Sasso [2],*

[1] Department of Precision Medicine, University of Campania Luigi Vanvitelli, Via De Crecchio 7, I-80138 Naples, Italy; teresa.salvatore@unicampania.it

[2] Department of Advanced Medical and Surgical Sciences, University of Campania Luigi Vanvitelli, Piazza Luigi Miraglia 2, I-80138 Naples, Italy; piaclara.pafundi@unicampania.it (P.C.P.); raffaele.galiero@unicampania.it (R.G.); luca.rinaldi@unicampania.it (L.R.); alfredo.caturano@unicampania.it (A.C.); erica.vetrano@gmail.com (E.V.); concetta.aprea27@outlook.it (C.A.); gaetanaalbanese@hotmail.it (G.A.); annadimarti@alice.it (A.D.M.); carmenricozzi28@gmail.com (C.R.); simo.imbriani@gmail.com (S.I.)

* Correspondence: ferdinando.sasso@unicampania.it; Tel.: +39-081-566-5010

Abstract: Cardiovascular mortality is a major cause of death among in type 2 diabetes (T2DM). Endothelial dysfunction (ED) is a well-known important risk factor for the development of diabetes cardiovascular complications. Therefore, the prevention of diabetic macroangiopathies by preserving endothelial function represents a major therapeutic concern for all National Health Systems. Several complex mechanisms support ED in diabetic patients, frequently cross-talking each other: uncoupling of eNOS with impaired endothelium-dependent vascular response, increased ROS production, mitochondrial dysfunction, activation of polyol pathway, generation of advanced glycation end-products (AGEs), activation of protein kinase C (PKC), endothelial inflammation, endothelial apoptosis and senescence, and dysregulation of microRNAs (miRNAs). Metformin is a milestone in T2DM treatment. To date, according to most recent EASD/ADA guidelines, it still represents the first-choice drug in these patients. Intriguingly, several extraglycemic effects of metformin have been recently observed, among which large preclinical and clinical evidence support metformin's efficacy against ED in T2DM. Metformin seems effective thanks to its favorable action on all the aforementioned pathophysiological ED mechanisms. AMPK pharmacological activation plays a key role, with metformin inhibiting inflammation and improving ED. Therefore, aim of this review is to assess metformin's beneficial effects on endothelial dysfunction in T2DM, which could preempt development of atherosclerosis.

Keywords: metformin; endothelial dysfunction; diabetes; CV risk

1. Introduction

Type 2 diabetes mellitus (T2DM) has been recognized for long a disease of the cardiovascular system, so that, at the beginning of the 1980s, someone has established it as a "cardiovascular disease diagnosed by glycemia". This anecdotal definition preempted the scientific evidence produced by Haffner in 1998, who demonstrated a similar mortality risk from Coronary Heart Disease (CHD) in a diabetic patient without previous myocardial infarction (MI) as compared to a non-diabetic subject who had suffered from a previous heart ischemic accident [1].

Metformin is the drug of choice for T2DM treatment and most of patients are usually treated first with this drug, and then with other anti-hyperglycemic agents in add-on to their therapeutic regimen as required [2]. Large evidence from the literature, both from preclinical and clinical studies, strengthen the anti-atherosclerotic properties of this drug [3].

In this review, we address the impact of metformin's administration on macrovascular complications of diabetes. In particular, we analyze metformin's beneficial effects on that distinctive pathophysiological condition named endothelial dysfunction (ED), which preempts the early development of atherosclerosis.

1.1. Endothelial Function

The endothelium is a thin monolayer covering the inner surface of blood vessels. It represents a barrier between circulating blood and all tissues and secretes a plethora of bioactive mediators of vascular tone [4].

The most important factor for the maintenance of vascular homeostasis is nitric oxide (NO), derived from the oxidation of L-arginine in the catalysis of endothelial nitric oxide synthase (eNOS), an enzyme constitutively expressed in endothelial cells (ECs) [5]. Once produced, NO rapidly moves to vascular smooth muscle cells (VSMCs), where it activates the soluble guanylate cyclase system which, in turn, increases cyclic guanosine-3′,5 mono phosphate (cGMP) and determines VSMCs relaxation [6]. Other vasodilators released by endothelial cells include endothelium derived hyperpolarization (EDH) factor of VSMCs, prostacyclin I_2 (PGI_2), bradykinin, histamine, serotonin, and substance P. As opposed to vasodilators, endothelium secretes a number of vasoconstrictors, especially cyclooxygenase-derived prostanoids, endothelin-1 (ET-1), angiotensin II (ANG II) and reactive oxygen species (ROS), usually associated with ED [7,8]. The relative contribution of vasodilator signals to the endothelium-dependent relaxation depends on blood vessels size, with NO dominant in conduit arteries and EDH factor as the diameter of the arteries decreases [9].

The fine equilibrium between these opposite factors is crucial to maintain a normal arterial patency. Conversely, an unbalanced production in favor of vasoconstrictor signals compromises the vascular auto-regulation and the functional and structural integrity of the endothelium, thus originating the ED [7].

1.2. Endothelial Dysfunction

ED is a condition of altered metabolism and function of endothelium inducing vascular injury and defective repair. Functionally, ED can be defined as a reduced bioavailability of NO, which affects the impaired response to an endothelium-dependent vasodilator such as acetylcholine.

Endothelium-derived NO not only keeps blood flow, though it also acts as a negative modulator of platelet aggregation, pro-inflammatory gene expression, ICAM-1 (intercellular adhesion molecule 1) and VCAM-1 (vascular cell adhesion molecule 1) production, E-selectin expression, ET-1 synthesis, VSMC proliferation, and lipoprotein oxidation [10]. Thus, ED is characterized by a series of features which goes beyond the hemodynamic dysregulation, including excess production of reactive oxygen species (ROS), enhanced expression of adhesion molecules and inflammatory mediators [11], and increased permeability of vascular endothelium. All of these promote both beginning and progression of atherogenesis [12,13].

The predictive role of ED on the cardiovascular risk has been largely documented in clinical studies by non-invasive, semi-invasive and invasive techniques measuring ED in humans in situ [14,15]. Besides ED functional measures, circulating levels of adhesion molecules and proinflammatory cytokines have also been used as surrogate markers of endothelial activation and cardiovascular risk [16].

2. Endothelial Dysfunction in Diabetes

The literature extensively supports ED as an important risk factor for the development of T2DM cardiovascular complications [17].

Based on a state of insulin resistance (IR), the interaction of three pathological conditions frequently associated with diabetes (hypertension, dyslipidemia and hyperglycemia), plays a pivotal role in the pathogenesis of the atherosclerotic process [18]. In this scenario hyperglycemia, playing a key role in any complication of diabetes [19], is a leading actor.

Acute hyperglycemia, achieved by intra-arterial infusion of dextrose, has been documented to impair endothelium-dependent vasodilation in healthy humans [20]. Likewise, the acute increase in plasma glucose after administration of oral glucose tolerance test (OGTT) determines, within a 1–2 h time period, a reduction of flow-mediated vasodilation in non-diabetic subjects, with a higher response in individuals with impaired glucose tolerance (IGT), and even more in those with diabetes [21]. A similar harmful effect is likely expected from prolonged and repeated post-prandial hyperglycemias, as it may routinely happen in T2DM. These hyperglycemic spikes may exert a dramatic and long-lasting epigenetic "memory" effect on the endothelial function, as reported in ECs cultured in high glucose and then restored to normoglycemia [22], which suggests transient hyperglycemia as a potential HbA1c–independent risk factor for diabetic complications [23]. A recent study in small mesenteric arteries from healthy and diabetic *db/db* mice has demonstrated that both acute and chronic exposure to high glucose interfere with local and conducted vasodilation in the resistance vasculature mediated by EDH [24].

Strong accumulating evidence suggests oxidative stress, defined as increased formation of ROS, reactive nitrogen species (RNS), and/or decreased antioxidant potentials, as the cornerstone of ED in the development of diabetic complications [25]. This condition triggers the production of pro-inflammatory cytokines and adhesion molecules responsible of intimal lesions formation [26,27]. Indirectly, some downstream processes (e.g., insulin resistance, formation of oxidized-low density lipoprotein (ox-LDL), inhibition of AMP-protein kinase (AMPK), and adiponectin) contribute to inflammation during the progression of atherosclerosis [28]. In turn, inflammation enhances ROS production, with a consequent arise of a variety of vicious cycles which intertwine each other, thus featuring the pathogenic complexity of the diabetes-accelerated atherosclerosis [29,30]. Moreover, endothelial damage increases albuminuria, both an independent and strong marker of CV risk [31,32]. The mechanisms by which hyperglycemia induces endothelial dysfunction are summarized in Figure 1 and are described in detail in the following paragraphs.

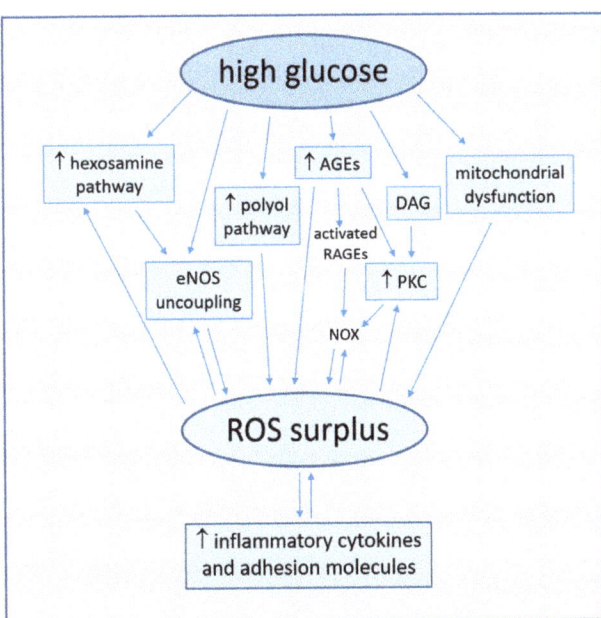

Figure 1. Main mechanisms of high glucose-induced endothelial dysfunction (direct arrows indicate the direction of the pathway, whilst double arrow stands for bidirectional pathway).

2.1. Increased ROS Production

Oxidative stress plays a major role in the pathophysiology of diabetic vascular disease [28,33]. Such a role is consistent with large evidence that increased concentrations of glucose in cultured endothelial cells induce an overproduction of ROS, with the subsequent activation of intracellular signal transduction pathways leading to ED [34,35]. High glucose concentration has been well established to cause endothelial cell damage by both an overproduction of ROS in mitochondria and by multiple biochemical pathways.

2.1.1. Uncoupling of eNOS

The deep reduction in endothelium-dependent vasodilatation associated with T2DM can be linked to changes in eNOS phosphorylation and desensitization induced by signal transduction pathways activated by ROS surplus. As an example, oxidative stress can activate the hexosamine biosynthetic pathway under diabetic and hyperglycemic conditions. This activation is further accompanied by an increase in O-linked N-acetylglucosamine modification of eNOS and a decrease in O-linked serine phosphorylation at residue 1177 [36].

The functional disturbance of the enzyme results in the production of superoxide anion ($O_2^-\cdot$) rather than NO, a phenomenon named eNOS uncoupling [37,38].

The ability of eNOS to generate NO can be disabled by the deficiency of tetrahydrobiopterin (BH4), an essential enzyme co-factor, which transforms eNOS into an oxidant-producing enzyme of $O_2^-\cdot$ [39,40]. ROS may induce oxidative changes of BH4 to dihydrobiopterin (BH2), a BH4 competing compound ineffective as eNOS co-factor. BH2/BH4 competition results in the dissociation of dimeric eNOS to the monomeric form, which acts through its oxygenase domain as an NADPH oxidase, further enhancing ROS generation, in a harmful perpetuation of a vicious circle [10,41–43]. Interestingly, the hyperglycemia-induced ED in normal subjects may be prevented by pre-treatment with the BH4 active isomer, 6R-BH4, whilst not by its inactive stereoisomer, 6S-BH4 [44]. In addition, the oral treatment with sepiapterin, a stable precursor of BH4, reduced oxidative stress and improved acetylcholine-mediated endothelium-dependent vasodilation in small mesenteric resistance arteries from *db/db* obese diabetic mice [45].

GTP cyclohydrolase I (GTPCH I) is the first enzyme in the BH4 biosynthetic pathway, constitutively expressed in endothelial cells and critical for the maintenance of NO synthesis [46]. Studies in HUVECs exposed to high glucose and in streptozotocin-injected diabetic mice have found that hyperglycemia may trigger BH4 deficiency by increasing 26S proteasome-mediated degradation of GTPCH I [47]. This degradation could be either prevented or improved by AMPK overexpression or activation [48].

NO derived from dimeric eNOS and $O_2^-\cdot$ from monomeric eNOS induces the formation of peroxynitrite ($ONOO^-\cdot$). This may facilitate the release of zinc from the zinc-thiolate cluster of eNOS, which is useful to maintain the dimeric structure of the enzyme, thus resulting in a further enhancement of eNOS uncoupling. Since loss of zinc and eNOS uncoupling activity have been both observed in ECs cells exposed to elevated glucose and in tissues of a diabetic mice model, we may hypothesize a significance of this process under in vivo conditions in diabetes [10,49].

The functions of many proteins may be affected by increased oxidant levels. As an example, a characteristic reaction of $ONOO^-\cdot$ is the nitration of protein-bound tyrosine residues to generate 3-nitrotyrosine–positive proteins [50]. Some researchers have suggested that an increased nitration of PGI_2 synthase (PGIS), more likely via dysfunctional eNOS, may characterize the diabetic disease. Such a hypothesis stands on observations that exposure of isolated bovine coronary arteries to high glucose switched angiotensin II–stimulated PGI_2-dependent relaxation into a persistent vasoconstriction [51]. As well, a significant suppression of PGIS activity, along with increased $O_2^-\cdot$ and PGIS-nitration, was also observed in aortas of streptozotocin-treated diabetic mice [51].

2.1.2. Mitochondrial Dysfunction

The mitochondrial electron transport chain (ETC) is the primary source of hyperglycemia-induced ROS production via a greater oxygen use, increased redox potential and shift of O_2 transport towards the respiratory chain complex II [25,29]. Other mechanisms of mitochondrial dysfunction include increased NADH/FADH2 ratio [52] and mitochondrial fission, which triggers an accumulation of fragmented mitochondria with impaired ETC activity [53].

2.1.3. Activation of the Polyol Pathway

Increased intracellular glucose levels overload ETC and are shunted into alternative pathways, in turn generating ROS. In the polyol pathway, accounting for >30% of glucose metabolism during hyperglycemia [54], glucose is converted by NADPH-dependent aldose-reductase to the sugar alcohol sorbitol, and sorbitol to fructose by sorbitol-dehydrogenase. The oxidative stress generated by these reactions depends on the consumption of NADPH, a cofactor required to regenerate the ROS scavenger glutathione (GSH), and on NAD+ reduction to NADH, which is subsequently oxidized by NADH oxidase, with consequent production of superoxide ions [55]. Aldose-reductase has been indeed implied in the increased expression of inflammatory cytokines [56,57].

2.1.4. Generation of Advanced Glycation End-Products (AGEs)

In conditions of hyperglycemia, the nonenzymatic fragmentation of the glycolytic intermediate triose phosphate produces methylglyoxal, precursor of the majority of AGE products formed by a nonenzymatic reaction of either ketones or aldehydes and the amino groups of proteins, during which large amounts of ROS are generated [25].

AGEs can interact with two types of cell surface receptors, scavengers involved in AGE removal and receptors for AGE (RAGEs), which initiate detrimental cellular signals, promoting inflammation and atherogenesis [29,58,59]. As an example, AGEs dose-dependently activate oxidative stress-mediated P38 activation of mitogen-activated protein kinase (MAPK) signaling in endothelial cells, which enhances NO synthesis inhibition by AGEs [60].

Both AGEs and methylglyoxal also promote the expression of RAGEs ligands. In particular, oxidized AGEs activate RAGEs to stimulate NADPH oxidase (NOX) [61], another important source of ROS production. NOX, which in healthy state determines ROS production, in pathological conditions may be hyper-expressed and hyperactive, as observed in cultured mice microvascular endothelial cells (MMECs) and human umbilical artery endothelial cells (HUAECs) exposed to high glucose [62,63]. Cells exposed to glucose fluctuations produce higher levels of NOX-derived ROS as compared to cells steadily exposed to high glucose, thus indicating the detrimental effect on vascular health of acute glycemic variations [64].

2.1.5. Activation of Protein Kinase C (PKC)

PKC is a serine/threonine related protein kinase acting in a wide variety of biological systems and regulating cell growth and proliferation, senescence, and apoptosis. The enzyme, once activated, induces many atherogenic processes, like ROS overproduction, endothelial dysfunction, increased vascular permeability, and inhibited angiogenesis [33,65].

In particular, NOX PKC-dependent activation is considered among the major sources of high glucose-induced ROS production, even more than mitochondrion [66,67].

In either a hyperglycemic or diabetic environment, PKC is activated by oxidative stress and AGEs and by diacylglycerol (DAG), whose levels increase in endothelial cells due to the shunting of glycolytic intermediates to dihydroxyacetone phosphate [65,68]. DAG-PKC is among the several cellular pathways activating when oxidative stress causes DNA fragmentation and stimulation of the DNA repair enzyme, nuclear poly ADP ribose polymerase (PARP). This enzyme inhibits the glyceraldehyde-3-phosphate dehydrogenase

(GAPDH), shunting early glycolytic intermediates into pathogenic signaling pathways, including AGE, polyol, DAG-PKC, and hexosamine pathways [25].

2.2. Endothelial Apoptosis and Senescence

Endothelial cell apoptosis and senescence are pivotal processes for the development of atherosclerosis, due to their activation by a plethora of pathways sharing the common pathophysiological mechanism of oxidative stress [69–71].

Studies on cultured ECs have shown that the promotion of senescence features (e.g., shortening of telomere length, elevated DNA damage, increase genomic instability and growth arrest) can be modulated by two factors intrinsically related to diabetes, high glucose [72], and AGE products [73], thus enhancing the intracellular levels of oxidative stress [74–76]. The implied cellular signals are diverse. As observed in high glucose exposed umbilical vein endothelial cells (HUVECs), Bax protein expression increases in the absence of Bcl-2 modifications, producing an elevated Bax/Bcl-2 ratio which activates the cleavage of procaspase-3 into active caspase-3, a crucial mediator of apoptosis [77]. As well, also the high-glucose induced NF-kB-dependent activation of c-Jun N-terminal kinase (JNK) and ROS-dependent Akt dephosphorylation may be involved [78].

Intriguingly, carbonic anhydrase, overexpressed in endothelial cells of diabetic ischemic heart, determines endothelial cell apoptosis in vitro, thus playing a key role in the remodeling process [79].

2.3. Other Pathogenetic Mechanisms of Vascular Dysfunction

A dysregulation of microRNAs (miRNAs), small non-coding RNAs, may contribute to the progression of atherosclerosis and diabetes-induced vascular dysfunction. As an example, a reduction in miRNA-126 levels has been associated with an increased leucocyte adherence to ECs and impairment of peripheral angiogenesis in T2DM [80]. Moreover, miR-29c and miR-204 were significantly dysregulated in atherosclerotic plaques from patients with DM [81].

T2DM has been proven as characterized by an imbalance of gut microbiota, which can directly promote atherogenesis by oxidative stress, inflammation, and changes in some metabolites, even though the bacteria possibly associated with progression of diabetes-accelerated atherosclerosis have not been identified yet [29].

3. Metformin Promotes Cardiovascular Health

Targeting and reduction of ED, an earlier phenomenon among the vascular abnormalities induced by cardiovascular risk factors, may represent a way to slow down diabetes-associated macrovasculopathy.

Since hyperglycemia-induced ROS may be the factor primarily involved in endothelial damage in diabetes, a protective action for correction of oxidative stress could be predicted. However, intervention studies in humans using orally administered antioxidants such as vitamins E and C have not been proven effective [82].

On the other hand, any anti-hyperglycemic drug achieving a rigorous glycemic control should mitigate the deleterious impact of diabetes on endothelium. However, despite antihyperglycemic effectiveness, not all these agents are able to reduce the CVD risk. Some drugs have been reported as independently associated with an increased risk (e.g., heart failure for rosiglitazone) [83], whilst others, likely provided of additional pleiotropic actions, resulted protective for the cardiovascular system. In this context, metformin, GLP-1 agonists (GLP1RA) and SGLT2 inhibitors (SGLT2i) obtained the strongest evidence for a beneficial effect on the endothelial function [84]. GLP1RA and SGLT2i have been approved for diabetes therapy in the most recent years. Therefore, it does not surprise the larger data on cardiovascular benefits available only for metformin which, after a 60-years history supporting its use, remains the first-choice agent for most T2DM patients.

The ability of metformin to reduce the diabetes-related CV risk arises from direct effects on the endothelium regardless, at least to some extent, of an improvement in metabolic disturbances (i.e., insulin resistance and hyperglycemia), and commonly associated risk factors (i.e., dyslipidemia and hypertension) [85–87].

3.1. Overview on Metformin

3.1.1. Historical Notes

Metformin (1,1-Dimethylbiguanide) is a synthetic derivative of galegine, a compound of French lilac tested as a glucose-lowering agent in humans in the 1920s, but soon discarded due to its toxicity [88,89]. Its anti-hyperglycemic effectiveness has been demonstrated more than half a century ago by the French medical doctor Jean Sterne and the drug has been first used the UK in 1958 under the trade name Glucophage R ('glucose eater'). The Food and Drug Administration (FDA) approved it for T2DM treatment only in 1994, after 20 years of use in Europe [88].

Despite long history and large clinical experience, metformin mechanism of action still remains not fully understood and even controversial, as it often happens with drugs of herbal origins not primarily designed for a specific cellular target.

3.1.2. Pharmacological Effects on Glucose Metabolism

Metformin primarily regulates glucose homeostasis. Specifically, it inhibits liver glucose production by the downregulation of hepatic gluconeogenesis and glycogenolysis. Metformin also alleviates IR, with an enhancement of peripheral glucose uptake via GLUT4 transport and subsequent significant reduction of plasma insulin levels [90,91]. Most recent evidence reports an important contribution about the beneficial metabolic responses to metformin before drug absorption, due to the interaction with gut microbiota and the modulation of incretin axis [92,93], thus supporting the role in the relationship between glycemic index and cardiometabolic diseases [94].

Over the past few decades, metformin has realistically emerged as a drug acting not only on specific targets of metabolism, though also on a series of other mechanisms and signaling pathways [95,96], some of which involved in the atherosclerosis prevention [97].

3.1.3. Activation of AMPK

In the literature there is a general consensus about the key role of AMK activation on metformin's cellular actions, in particular at level of liver and skeletal muscle [98]. AMPK is a heterotrimeric serine/threonine protein kinase containing one catalytic α subunit and two non-catalytic subunits, scaffold β and regulatory γ subunits. Each subunit has two isoforms ($\alpha 1$, $\alpha 2$, $\beta 1$, $\beta 2$, $\gamma 1$, $\gamma 2$, and $\gamma 3$), widely expressed in different tissues and subcellular sites [99].

AMPK is a major regulator of cellular energy homeostasis coordinating the enzymes involved in carbohydrate and fat metabolism to enable ATP conservation and synthesis. Conditions of increased AMP:ATP ratio (exercise, metabolic stress, and hypoxia) activate AMPK, which switches off the ATP-consuming pathways and on the ATP-generating ones [98].

Increasing evidence suggests that the role of AMPK goes beyond energy metabolism control, as the enzyme may regulate a very wide range of cell functions accounting for a variety of metformin pleiotropic actions [100,101]. As an example, AMPK stimulates eNOS production [102], thus supporting a protective role of this kinase on the endothelium, as demonstrated in a study on obese rats [103,104].

The phosphorylation at Thr1172 of the α-subunit activates AMPK, whilst AMP and/or adenosine diphosphate (ADP) binding to the γ-subunit protects the enzyme against dephosphorylation [105]. Upon ATP depletion, AMPK is phosphorylated and activated by upstream kinases such as liver kinase B-1 (LKB1), constitutively expressed in most cell types [106], and calcium/calmodulin-dependent protein kinase-beta, activated by intracellular calcium and expressed only in certain cell types, including ECs [107,108]. Intriguingly, AMPK has been found dysregulated in experimental animal models and in humans with either metabolic syndrome or T2DM [109].

Metformin and AICAR (5-amino-imidazole carboxamide riboside) are the two most commonly used AMPK activators. AICAR is an analog of AMP directly activating the enzyme, but not suitable for human use [110]. On the contrary, metformin is not a specific activator of AMP, but it can be used in humans. How exactly metformin activates AMPK is still unclear [111]. The drug might increase the phosphorylation of AMPK catalytic α subunit at Thr1172, as reported by studies on primary hepatocytes [112], or inhibit AMP deaminase [113].

On the other side, there is large evidence that enhanced AMPK expression is secondary to the increased intracellular ADP/ATP and AMP/ADP ratios resulting from a mild, transient and specific inhibitory action of metformin on ETC's mitochondrial complex I (NADH: ubiquinone oxidoreductase) [90,93]. Even this mechanism is debated and the extent to which it is physiologically relevant is still uncertain, as the required concentration seems about 500–1000 times than the highest attained therapeutically [114].

Beyond ETC's complex I inhibition, other mitochondrial actions have been described, including a direct binding of metformin to mitochondrial copper ions [115] and a non-competitive inhibition of mitochondrial glycerol 3-phosphate dehydrogenase shuttle, producing impaired respiration, reduced cytoplasmic NAD+/NADH ratio and undermined glucose production from both glycerol and lactate [116]. The physiological relevance of these mechanisms is unclear. Incidentally, whether metformin can access mitochondria to a sufficiently high concentration to inhibit complex 1 or exert other actions is still object of debate [117].

Considering the plurality of cellular sites of metformin action, not all the effects of the drug are necessarily mediated via either the direct or indirect activation of AMPK. For instance, Foretz et al. reported that metformin was able to inhibit liver gluconeogenesis in transgenic mice lacking AMPK subunits and LKB1 [118], whilst Buse et al. showed that a significant component of the anti-hyperglycemic effects of metformin resided in microbiome [92].

3.2. Metformin Reduces Cardiovascular Mortality in Diabetes

Publication in 1998 of the United Kingdom Prospective Diabetes Study (UKPDS), a trial designed to assess whether intensive blood-glucose control reduced the risk of macrovascular or microvascular complications in T2DM patients, represented the event which has changed metformin's history. Remarkably, UKPDS findings attributed to metformin the role of first choice anti-hyperglycemic drug after demonstrating, in overweight patients randomized to metformin as compared to conventional dietary measures, a risk reduction of 39% for nonfatal myocardial infarct, 42% for diabetes-related death, and 36% for all-cause mortality [119].

A Cochrane meta-analysis supports the benefits of metformin as compared to other antidiabetic drugs, proving a reduced all-causes mortality [120].

Two further meta-analysis have strengthened this result. The first showed metformin as the only antidiabetic agent able to improve all-cause mortality without causing any harm in diabetic patients with heart failure [121]. The other instead reported significantly lower all-cause mortality rates in diabetic individuals taking metformin as compared either to non-diabetics or diabetics receiving non-metformin therapies [122]. On the contrary, an evaluation of 35 clinical trials including over 18,000 participants found a significant benefit for metformin versus placebo/no therapy, but not versus active-comparators [123]. Another metanalysis of randomized trials has left doubts about whether metformin reduces risk of cardiovascular disease in T2DM or not [124]. Moreover, an observational study using the REACH Registry showed an association between metformin's use in secondary prevention and a decreased mortality [86]. In a retrospective Danish cohort study on T2DM patients admitted with myocardial infarction and not treated with emergent percutaneous coronary intervention, monotherapy with sulfonylureas was associated with increased cardiovascular risk compared with metformin monotherapy [125].

4. Protective Properties of Metformin on Endothelium

Metformin displays multiple beneficial effects against CVD, among the most relevant those exerted on vascular endothelial function [8,114,117,126]. A 4.3-year clinical trial has shown a metformin-associated reduction of several plasma ED biomarkers (e.g., vWF, sVCAM-1, t-PA, PAI-1, and sICAM-1), regardless of changes in HbA1c, insulin dose, and body weight. The authors reported that ED improvement explained about 34% of the reduced cardiovascular risk associated with biguanide treatment [127].

The endothelial protection exerted by metformin may not represent the product of a single pharmacological action, though rather the result of concurrent multiple mechanisms involving endothelium-dependent vascular response, oxidative stress, leukocyte-endothelium interactions, mitochondrial function, and others. Literature data highlight the role of hyperglycemia in ED pathogenesis [128], even though metformin therapeutic concentrations may improve vascular endothelial reactivity in non-diabetic patients, regardless of glucose levels [129]. More likely, metformin exerts both anti-hyperglycemic-mediated and direct actions on endothelial function.

We will now discuss metformin's impact on endothelium and possible underlying cellular and biochemical mechanisms observed in human investigations and in preclinical studies.

4.1. Metformin Improves Endothelium-Dependent Vascular Response

Almost 30 years ago, Marfella et al. demonstrated that metformin improved hemodynamic and rheological responses to infusion of l-arginine, the natural precursor of NO, in newly diagnosed T2DM patients without micro- and macrovascular complications [130].

At the dawn of the third millennium, when NO has begun to emerge as a protective CV factor [131], the analysis of vascular response to metformin in T2DM patients by a direct measurement with forearm strain-gauge plethysmography, proved an improvement of endothelium-dependent vasodilation after a 12-week treatment, indicating the endothelium as the primary site for dysfunctional blood flow. Notably, ED improvement has been associated with a reduction in whole-body IR [132].

4.1.1. Role of Insulin Resistance Correction

Insulin is known to promote NO production by activating the PI3K/Akt/eNOS signaling pathway, which results in vasodilation and vascular protection [133,134].

Once IR develops, pathway-specific impairment in PI3K-dependent signaling may cause imbalance between production of NO and secretion of ET-1, thus leading to endothelial dysfunction [135].

The aforementioned study by Mather et al. supported the conclusion that endothelium-dependent vascular response correction by metformin was more likely secondary to improved insulin signaling [132] (Figure 2), consistently with previous reports on these subjects [136]. Steinberg et al. had already demonstrated that excessive exposure of endothelium to free fatty acids (FFAs) increased $O_2^-\cdot$ production, impaired NO activity, and reduced endothelium-dependent vasodilation [137]. IR is characterized, along with the involvement of numerous other systems [138], by sustained elevations in serum FFAs and failure of appropriate suppression following meals, due to a compromised ability of insulin-resistant adipocytes to store and retain FFAs [139]. The link between FFA excess and ED may lie in a sequential process starting from the increased de novo synthesis of DAG, which activates PKC, in turn responsible for endothelial $O_2^-\cdot$ overproduction via NOX stimulation [67], eNOS inhibition [140], and activation of a vicious worsening of insulin signaling in the endothelial cells [141].

Preclinical studies in mesenteric arteries and aortas from insulin-resistant rats support an improvement of ACh-induced vasodilation by treatment with the insulin sensitizer metformin [142,143]. Indeed, the relationship between IR and ED in humans is not so clear. On the one hand, metformin improves endothelial function in non-diabetic insulin resistant populations [144]. Otherwise, troglitazone, a ligand of nuclear receptor peroxisome proliferator-activated receptor (PPAR)-γ with insulin-sensitizing actions, administered to obese subjects, determined an improvement in insulin sensitivity but no effects on both endothelium-dependent and independent vascular responses [145]. Accordingly, in a study on T2DM patients treated with sulfonylureas, the improvement in ED with the addition of either metformin or pioglitazone did not seem associated neither with a better glycemic control nor with insulin sensitivity [146]. Moreover, a pilot trial in uncomplicated T1DM patients showed a significant improvement of ED, irrespective of glycemic control and body weight, after a 6-months metformin treatment in add-on to basal-bolus insulin regimen [147].

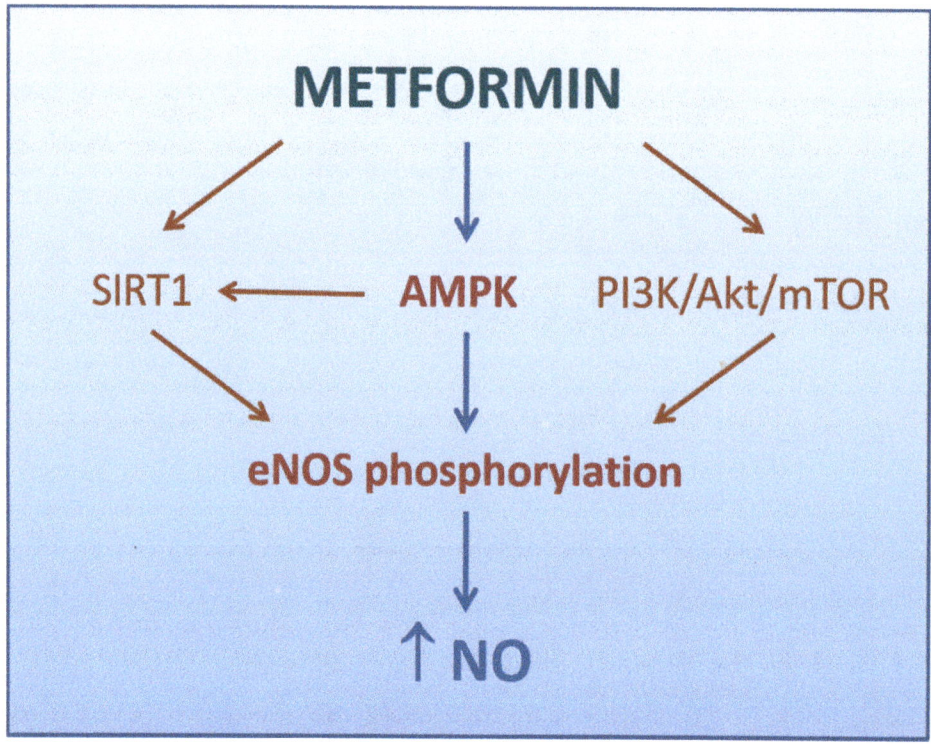

Figure 2. Mechanisms by which metformin promotes NO production (direct arrows indicate the direction of the pathway).

4.1.2. Role of AMPK Activation

An attractive hypothesis of how metformin enhances endothelium-dependent vasodilation may reside in the activation of AMPK [148]. It has been extensively demonstrated that several stimuli, not last metformin, may induce AMPK-dependent eNOS phosphorylation, thus resulting in increased NO production and vasodilation in conduit arteries [102,149,150] (Figure 2). Matsumoto and colleagues reported an improvement of the endothelium-dependent responses by metformin even in the resistance arteries of diabetic rats, thanks to the suppression of prostanoid signaling [151]. Later, a study on mice with endothelium-specific deficiency of α-catalytic subunit of AMPK, demonstrated eAMPK α1 as the main upstream enzyme that mediates EDH responses of microvessels, thus regulating blood pressure and coronary flow responses in vivo [152]. Since these findings have not been confirmed, the contribution of AMPK in the tone regulation at level of microvasculature, where EDH signaling plays a more prominent role, still remains controversial [153].

4.1.3. Other Mechanisms

Based on the evidence that Sirtuin-1 (SIRT1), a NAD-dependent deacetylase with antiaging activities, enhances the activity of eNOS with NO generation and endothelial-dependent vascular relaxation [154], we can speculate that metformin indirectly increases eNOS activity by directly inducing SIRT1 expression and/or activation (Figure 2). This hypothesis is supported by the observation that a 72-h exposure to metformin may reduce hyperglycemia-induced endothelial senescence and apoptosis via a SIRT1-dependent process [155].

Ghosh et al. demonstrated that a brief exposure of aortic tissue and microvascular endothelial cells to metformin can either reverse or reduce the high glucose-induced ED via mechanisms linked to increased phosphorylation of eNOS and Akt, a cytosolic protein involved in the intracellular signaling pathway PI3K/Akt/mTOR regulating the cell cycle. Of note, this response was not accompanied by changes either in AMPK phosphorylation or SIRT1 expression [156].

4.2. Metformin Promotes Antioxidation

ROS are strongly involved in ED occurrence, due to their vasoconstrictor action and the reactivity with NO to produce $ONOO^-\cdot$, with further reduction of NO bioavailability [157].

Large evidence supports metformin inhibitory effect on oxidative stress, in vitro in hyperglycemic environments [158,159] as well as in vivo in high fructose-fed rats [160] and T2DM patients [161].

Experiments in BAECs and HUVECs in the presence of either NOX inhibitor apocynin or ETC inhibitor rotenone, report that metformin inhibits ROS formation from both respiratory mitochondrial chain and NOX [162,163]. PKC-NOX pathway inhibition by metformin was later confirmed in human aortic endothelial cells [164]. In rats exposed to the prooxidant rotenone, metformin's co-treatment is able to correct redox imbalance and toxicity of erythrocytes [165]. Metformin has been also reported to prevent the rise in lipid peroxides and oxidized proteins and the fall of mitochondrial aconitase activity, a sensitive parameter for the mitochondrial generation of ROS inside in aortic tissue, heart and kidney of diabetic Goto-Kakizaki rats [166] and the DNA damage related to oxidative stress in lymphocytes from elderly subjects [167]. The significant reductions in NO release and the pronounced increase in nitroxidative stress observed in obese Zucker rats significantly reverted with metformin treatment, as a result of improved eNOS coupling and bioavailable NO, and other mechanisms regulating endothelial function beyond glucose control [168].

The underlying mechanisms of these antioxidant properties of metformin still remain controversial (Figure 3). The scavenging direct capacity of trapping free radicals is negligible [169].

More likely, metformin enhances the endogenous antioxidant defense by preventing the hyperglycemia-related inhibition of glucose-6-phosphate-dehydrogenase (G6PDH), which would either hamper the regeneration of reduced GSH [170] or increase superoxide dismutase-1 [171]. AMPK pathway has been proven to potentially reduce the intracellular ROS level by activating the fork-head transcription factor 3 (FOXO3), subsequently upregulating thioredoxin expression, a major component of an important endogenous antioxidant system, which promotes the reduction of proteins by cysteine thiol-disulfide exchange [172]. This pathway seems responsible for the attenuation of intracellular ROS levels induced by metformin in primary human aortic endothelial cells exposed to palmitic acid [173].

Otherwise, metformin decreases ROS cellular production. Several experiments have proven metformin's capacity to downregulate NOX, among the major cellular producers of ROS [174–177]. Accordingly, a study in cultured HUVEC and murine aortas isolated from AMPK-α2 deficient mice demonstrated that this enzyme acts as a physiological suppressor of NOX and ROS production in endothelial cells [178]. Since oxidative stress is proportional to the accumulation of AGEs in diabetic animals [179], the antioxidant activity of metformin may be partially due to the inhibition of glycation, a process directly related to free-radical production.

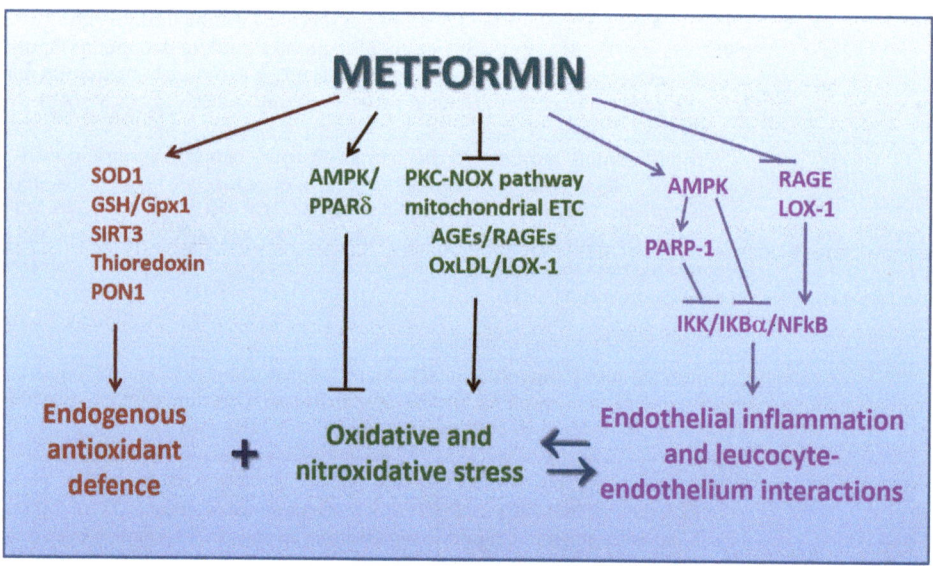

Figure 3. Schematic picture of the mechanisms by which metformin exerts a protective action against oxidative stress and endothelial dysfunction (direct arrows indicate the direction of the pathway, whilst blocked arrows stands for inhibition of that specific pathway. Double arrow stands instead for a bidirectional reaction).

Other mechanisms may be overexpression of SIRT3, a NAD+-dependent deacetylase specifically located in the mitochondria, and glutathione peroxidase 1, which protects leukocytes against oxidative stress by reducing hydroperoxides [180,181]. Metformin may also inhibit endoplasmic reticulum stress and oxidative stress by activating AMPK/PPARδ pathway, as reported in a study on aortae from obese diabetic mice [182] (Figure 3).

4.3. Metformin Counteracts the Pro-Atherogenic Role of oxLDL and LOX-1

Oxidized low-density lipoprotein (OxLDL), as well as class E scavenger lectin-like oxidized receptor 1 (LOX-1) mediating OxLDL uptake by vascular cells, are involved in events critical in atherosclerosis development from ED until plaque instability and rupture [183]. OxLDL is a product of chronic oxidative stress which, in parallel, can act as pro-oxidant by stimulating NOX and ROS generation [184]. On the other hand, LOX-1 may bind with high affinity a broad spectrum of structurally distinct ligands besides OxLDL, among which AGEs which, in turn, upregulate LOX-1 expression in diabetes [185].

Metformin has been proven to inhibit the expression of both RAGEs and LOX-1, more likely through a modulation of redox-sensible nuclear factors, including NF-kB, which are involved in such receptor cell expression [186].

Exposition of cultured endothelial cells to oxidized and glycated LDL (HOG-LDL) causes aberrant ER stress via enhanced sarcoplasmic/endoplasmic reticulum Ca^{2+} AT-Pase oxidation, significantly mitigated by either pharmacological (included metformin) or genetic activation of AMPK, which results in an improved endothelium-dependent relaxation [187].

OxLDL signals mainly activate via LOX-1 diverse cellular second messengers, including NF-κB and AP-1, two oxidative stress-responsive transcription factors involved in the regulation of cytokines, chemokines, and adhesion molecules in endothelial cells. In turn, some of the induced cytokines activate NF-κB and AP-1, thus reinforcing the inflammatory signaling cascade [188].

A study on human primary coronary artery endothelial cells showed for the first time that OxLDL induced ED, cell death, and impaired vasorelaxation, partially via TRAF3IP2, a redox-sensitive cytoplasmic adapter protein and an upstream regulator of IKK/NF-κB and JNK/AP-1. Moreover, while native HDL3 inhibited, oxidatively-modified HDL3 potentiated OxLDL-induced TRAF3IP2 expression. AMPK activators (adiponectin, AICAR and metformin), through AMPK-dependent Akt activation, antagonized the pro-apoptotic effects of OxLDL-induced TRAF3IP2 expression [189].

A study on HUVECs showed that SIRT1 and AMPK silencing decreased the protective function of metformin against OxLDL-increased LOX-1 expression and OxLDL-collapsed AKT/eNOS levels [190].

Both in diabetic rats [191], and newly diagnosed diabetic patients [192], metformin has been shown to restore the activity of paraoxonase 1, an antioxidant associated with circulating HDL that hydrolyzes lipid peroxides in LDL.

4.4. Metformin Inhibits Endothelial Inflammation and Leukocyte-Endothelium Interactions

Beyond its anti-oxidative properties, metformin also exerts eminent anti-inflammatory effects, as expected from an AMPK activator [193].

In a study published in 2003, treatment of human ECs with AGEs for up to 12 h has been shown to significantly increase human monocyte adhesion, an effect prevented by the presence of metformin in incubation medium [194]. Incidentally, the drug also prevented monocyte differentiation into macrophages and foam cell, a process that metformin regulates via AMPK-mediated inhibition of STAT3 activation [195].

These findings were later extended. In fact, metformin has been demonstrated to suppress the cytokine-induced activation of NF-kB in HUVECs. As a consequence, NF-kB-regulated gene expression of various inflammatory and cell adhesion molecules was inhibited. This effect was determined via the AMPK-dependent inhibition of the IKK/IKBα/NF-κB pathway [196].

An excessive and sustained oxidative stress can cause overactivation of poly (ADP-ribose) polymerase-1 (PARP-1), which worsens the oxidative stress and stimulates proinflammatory and necrotic responses [197]. An investigation on HUVECs and in vivo on mice has demonstrated a possible metformin involvement in a pathway linking AMPK, PARP-1, and B-cell lymphoma–6 protein (Bcl-6) in the prevention of monocyte adhesion to endothelial cells and attenuation of endothelial inflammation. PARP-1 binding to Bcl-6 intron 1 has been proven to suppress the transcription of Bcl-6, a corepressor for inflammatory mediators recruiting monocytes to vascular endothelial cells upon inflammation. Conversely, phosphorylation of PARP-1 at Ser-177 by activated-AMPK decreased its binding to Bcl-6 intron 1, with subsequent transcriptional up-regulation of Bcl-6 and co-repression of VCAM-1, MCP-1, and MCP-3 to finally result in an anti-inflammatory phenotype [198]. A later report further confirmed that vascular protection of metformin partially occurs through the activation of the AMPK-PARP-1 cascade [199].

In a study on retinal endothelial cells under hyperglycemic conditions, SIRT1 activation by metformin significantly attenuated ROS mediated activation of PARP through the upregulation of LKB1/AMPK, with the subsequent suppression of NF-kB, as well as of proapoptotic gene Bax [200]. All these mechanisms are summarized in Figure 3.

4.5. Metformin Attenuates the Apoptosis, Senescence, and Death of Endothelial Cells

Mitochondria are the powerhouse of the cell, providing over 90% of ATP consumed by the cell, but they also play an important role in the commitment to cell death [201]. Several intermembrane space proteins have no pro-apoptotic activity when persisting inside mitochondria, though they promote cell death once released into the cytosol by opening an inner membrane channel, the so-called permeability transition pore (PTP) [202]. Metformin has been found to prevent the PTP opening determined by the high glucose-induced oxidative stress in several endothelial cell types [203], and the biguanide given at the time of reperfusion may reduce myocardial infarct size in the heart of both non-diabetic and diabetic rats [204].

SIRT1 plays a central role in the regulation of endothelial cell growth, senescence, and apoptosis, as well as in atherosclerosis development [205]. Metformin may be considered either a direct or LKB-1/AMPK-mediated modulator of SIRT1 expression, able to alleviate hyperglycemia-caused endothelial senescence and cell death (Figure 4). Similarly to the results by Zheng et al. (see Section 4.4) [200], in a study on MMECs hyperglycemia has been proven to accelerate endothelial apoptosis and senescence via changes in SIRT1 expression and downstream signaling targets FoxO-1/p53, whereas metformin prevents these detrimental effects attenuating hyperglycemia-induced oxidative stress and upregulating SIRT1 expression [206].

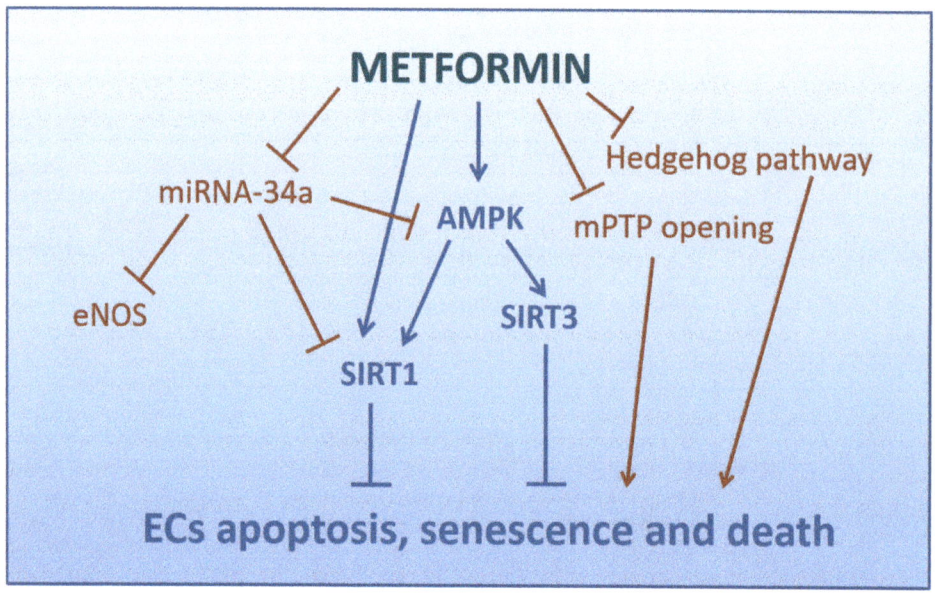

Figure 4. Mechanisms by which Metformin protect Endothelial Cells from apoptosis, senescence, and death (direct arrows indicate the direction of the pathway, whilst blocked arrows stands for inhibition of that specific pathway).

miRNA-34a has been reported as highly expressed in ECs and it may directly bind to SIRT1, the so-called anti-ageing gene, thus inhibiting sirtuin1 expression and regulating apoptosis via the sirtuin1-p53 pathway [207]. In HUVECs, miRNA-34a overexpression down-regulates sirtuin1 expression and induces ECs senescence, whereas miRNA-34a knock-down enhances sirtuin1 expression and attenuates endothelial senescence [208]. A study in MMCs has reported that hyperglycemia-mediated induction of miRNA-34a results in impaired angiogenesis, a defect revertible by therapeutic intervention with metformin, likely through the modulation of miRNA-34a levels which, in turn, regulates sirtuin1, AMPK and eNOS activity [206].

Using a H_2O_2-induced senescence model of human and murine fibroblast and HUVECs, autophagic dysfunction and decline in NAD+ synthesis have been shown as two features of senescent cells induced by oxidative stress, both restored by metformin through AMPK activation [209].

Two recent investigations have identified novel molecular mechanisms for metformin-mediated age-delaying effects on endothelium (Figure 4). The first is AMPK-mediated and lies on the regulation of mitochondrial biogenesis/function and senescence by H3K79me acting through SIRT3 [210]. Indeed, the second is AMPK-independent and consists in the downregulation of autophagy via the Hedgehog pathway, a signaling critically involved in adult tissue maintenance, renewal, and regeneration [211].

4.6. Metformin Inhibits Mitochondrial Fission

Mitochondria form a complex and dynamic network undergoing continuous cycles of fusion and fission events which are crucial to maintain organelle homeostasis [212]. Mitochondrial fusion seems beneficial as it distributes metabolites, proteins, and DNA throughout the mitochondrial population. In contrast, excessive mitochondrial fission may be detrimental due to accumulation of fragmented mitochondria with ETC impairment and mitochondrial ROS increase, as it occurs after cell exposure to high glucose concentrations [213]. In endothelial cells, mitochondrial fission contributes to the reduction in eNOS-derived NO bioavailability [214], impairment of angiogenesis [215], and induction of apoptosis [216]. An increased mitochondrial fission has been reported in different tissues of T2DM patients, more remarked in those with poor glycemic control [217,218].

AMPK activation by metformin may slow atherosclerosis development in diabetes by reducing the mitochondrial fission and its detrimental consequences. Using streptozotocin (STZ)-induced diabetic ApoE2/2 mice, a well-established model for the study of human atherosclerosis metformin has be found to reduce dynamin-related protein 1 (Drp1) expression and Drp1-mediated mitochondrial fission in an AMPK-dependent manner. Concomitantly, mitochondrial-derived superoxide release was mitigated, endothelial-dependent vasodilation improved, vascular inflammation inhibited, and atherosclerotic lesions suppressed [219].

Mitochondrial biogenesis is a response of stress adaptation to improve efficiency of cellular energy and preserve the cellular integrity [220]. The process has been frequently associated with the activation of AMPK by not well-defined mechanisms. A study by Le et al. has demonstrated that AICAR in endothelial cells induces mitochondrial biogenesis and stress adaptation via an AMPK/eNOS/mTORC1 pathway [221].

4.7. Other Protective Vascular Actions by Metformin

Several studies have shown that Ang II binding to Ang II type 1 receptor (AT1R) is involved in the progression of cardiovascular diseases, including atherosclerosis, hypertension, cardiac hypertrophy, and heart failure. Metformin has been shown to potentially decrease AT1R expression in mice aortas and attenuate vascular senescence and atherosclerosis induced by a high-fat diet, thus suggesting that AT1R downregulation may, at least partially, mediate the protective effect of metformin in the vascular system [171].

It has been further reported that AMPK pharmacological activation with metformin (as well as salicylate, resveratrol, and AICAR), inhibited inflammation in perivascular adipose tissue and improved ED against inflammatory insult in an AMPK/SIRT1-interdependent manner [222].

The loss of glycocalyx, a proteoglycan-rich hydrogel which separates blood from endothelium, represents an early event in the development of endothelial dysfunction [223]. A study demonstrated that metformin's treatment, preserving glycocalix may restore the blunted hyperemic response in myocardial microvascular perfusion in rats challenged with a high-fat diet [224].

The proliferation and migration of human aortal smooth muscle cells, a well-known etiological factor of atherosclerosis, restenosis, and pulmonary hypertension, can be significantly inhibited by metformin through AMPK activation, even though this result has been obtained at very high drug concentrations, precluded to achieve in vivo [225].

The endothelial-to-mesenchymal transition (EndoMT), a cellular process involved in ED and vascular disease pathogenesis, is characterized by the loss of endothelial features and gaining of mesenchymal ones by ECs [226]. A study on HUVECs described that high glucose could induce EndoMT and suppress the endothelial protective axis of Kruppel-like factor 4 (KLF4), a master transcription factor maintaining vascular homeostasis, and Ch25h, a promoter of reverse cholesterol efflux. Metformin inhibited these effects by increasing Ch25h expression not only through KLF4, though also epigenetic changes, including DNA methylation and active histone modification [227].

5. Conclusions

Diabetes is a serious and global health problem affecting about 500 million people worldwide, a number expected to grow along with the associated high burden of premature and accelerated atherosclerosis impact on both life quality and expectancy. Since cardiovascular mortality is a major cause of death among individuals with T2DM, the prevention of macroangiopathies by preserving endothelial function represents a major therapeutic concern in this population.

Over its 60-year old history of use, multiple advantages of metformin have been proven, being inexpensive, mildly weight-lowering, relatively free of side effects other than gastrointestinal-related, with a very low risk of hypoglycemia and especially of so feared lactic acidosis [228]. Above all, an extensive pre-clinical and clinical literature details its vascular benefits.

Metformin as first choice treatment for T2DM patients is currently the most widely prescribed oral anti-hyperglycemic agent worldwide, with nearly 150 million annual prescriptions [229,230].

A US study has calculated that approximately 1 of 12 adults has a combination of pre-diabetes and risk factors which may justify the introduction of metformin as indicated by the American Diabetes Association guidelines [231]. Therefore, a higher proportion of relatively healthy individuals might benefit from metformin's treatment to either prevent or delay both diabetes and cardiovascular events, even in secondary prevention, as recently demonstrated by a prospective study with a 24-month follow-up in pre-DM patients with stable angina and nonobstructive coronary stenosis [232].

However, a much wider use of this drug can be implemented as a viable cardiovascular preventive strategy, starting with the many millions of non-diabetic insulin resistant individuals with metabolic syndrome, until considering even the elderly population with its burden of several comorbidities [228,233], and those suffering from some common rheumatologic diseases closely associated with cardiovascular risk such as rheumatoid arthritis and gout [234].

The wide range of possible indications and the well documented benefits associated with its use fully deserve to metformin the attribute of "wonder drug" or "aspirin" of current times recently coined.

Funding: This research received no external funding.

Conflicts of Interest: The authors declare no conflict of interest.

References

1. Haffner, S.M.; Lehto, S.; Rönnemaa, T.; Pyörälä, K.; Laakso, M. Mortality from Coronary Heart Disease in Subjects with Type 2 Diabetes and in Nondiabetic Subjects with and without Prior Myocardial Infarction. *New Engl. J. Med.* **1998**, *339*, 229–234. [CrossRef] [PubMed]
2. American Diabetes Association. 9. Pharmacologic Approaches to Glycemic Treatment: Standards of Medical Care in Diabetes—2019. *Diabetes Care* **2019**, *42* (Suppl. 1), S90–S102. [CrossRef] [PubMed]
3. Luo, F.; Das, A.; Chen, J.; Wu, P.; Li, X.; Fang, Z. Metformin in patients with and without diabetes: A paradigm shift in cardiovascular disease management. *Cardiovasc. Diabetol.* **2019**, *18*, 1–9. [CrossRef] [PubMed]
4. Chistiakov, D.A.; Revin, V.V.; Sobenin, I.A.; Orekhov, A.N.; Bobryshev, Y.V. Vascular Endothelium: Functioning in Norm, Changes in Atherosclerosis and Current Dietary Approaches to Improve Endothelial Function. *Mini-Rev. Med. Chem.* **2015**, *15*, 338–350. [CrossRef]
5. Nishida, K.; Harrison, D.G.; Navas, J.P.; Fisher, A.A.; Dockery, S.P.; Uematsu, M.; Nerem, R.M.; Alexander, R.W.; Murphy, T.J. Molecular cloning and characterization of the constitutive bovine aortic endothelial cell nitric oxide synthase. *J. Clin. Investig.* **1992**, *90*, 2092–2096. [CrossRef]
6. Rapoport, R.M.; Murad, F. Agonist-induced endothelium-dependent relaxation in rat thoracic aorta may be mediated through cGMP. *Circ. Res.* **1983**, *52*, 352–357. [CrossRef]

7. Vanhoutte, P.M.; Shimokawa, H.; Feletou, M.; Tang, E.H.C. Endothelial dysfunction and vascular disease —A 30th anniversary update. *Acta Physiol.* **2017**, *219*, 22–96. [CrossRef]
8. Nafisa, A.; Gray, S.G.; Cao, Y.; Wang, T.; Xu, S.; Wattoo, F.H.; Barras, M.; Cohen, N.D.; Kamato, D.; Little, P.J. Endothelial function and dysfunction: Impact of metformin. *Pharmacol. Ther.* **2018**, *192*, 150–162. [CrossRef]
9. Shimokawa, H.; Yasutake, H.; Fujii, K.; Owada, M.K.; Nakaike, R.; Fukumoto, Y.; Takayanagi, T.; Nagao, T.; Egashira, K.; Fujishima, M.; et al. The Importance of the Hyperpolarizing Mechanism Increases as the Vessel Size Decreases in Endothelium-Dependent Relaxations in Rat Mesenteric Circulation. *J. Cardiovasc. Pharmacol.* **1996**, *28*, 703–711. [CrossRef]
10. Triggle, C.R.; Ding, H.; Marei, I.; Anderson, T.J.; Hollenberg, M.D. Why the endothelium? The endothelium as a target to reduce diabetes-associated vascular disease. *Can. J. Physiol. Pharmacol.* **2020**, *98*, 415–430. [CrossRef]
11. Esposito, K.; Ciotola, M.; Sasso, F.C.; Cozzolino, D.; Saccomanno, F.; Assaloni, R.; Ceriello, A.; Giugliano, D. Effect of a single high-fat meal on endothelial function in patients with the metabolic syndrome: Role of tumor necrosis factor-α. *Nutr. Metab. Cardiovasc. Dis.* **2007**, *17*, 274–279. [CrossRef] [PubMed]
12. Gimbrone, M.A., Jr.; García-Cardeña, G. Endothelial Cell Dysfunction and the Pathobiology of Atherosclerosis. *Circ. Res.* **2016**, *118*, 620–636. [CrossRef] [PubMed]
13. Marfella, R.; Ferraraccio, F.; Rizzo, M.R.; Portoghese, M.; Barbieri, M.; Basilio, C.; Nersita, R.; Siniscalchi, L.I.; Sasso, F.C.; Ambrosino, I.; et al. Innate Immune Activity in Plaque of Patients with Untreated andl-Thyroxine-Treated Subclinical Hypothyroidism. *J. Clin. Endocrinol. Metab.* **2011**, *96*, 1015–1020. [CrossRef] [PubMed]
14. Heitzer, T.; Schlinzig, T.; Krohn, K.; Meinertz, T.; Munzel, T. Endothelial Dysfunction, Oxidative Stress, and Risk of Cardiovascular Events in Patients with Coronary Artery Disease. *Circulation* **2001**, *104*, 2673–2678. [CrossRef] [PubMed]
15. Anderson, T.J.; Phillips, S.A. Assessment and Prognosis of Peripheral Artery Measures of Vascular Function. *Prog. Cardiovasc. Dis.* **2015**, *57*, 497–509. [CrossRef] [PubMed]
16. Zhang, J.; DeFelice, A.F.; Hanig, J.P.; Colatsky, T. Biomarkers of Endothelial Cell Activation Serve as Potential Surrogate Markers for Drug-induced Vascular Injury. *Toxicol. Pathol.* **2010**, *38*, 856–871. [CrossRef] [PubMed]
17. Shi, Y.; Vanhoutte, P.M. Macro- and microvascular endothelial dysfunction in diabetes. *J. Diabetes* **2017**, *9*, 434–449. [CrossRef]
18. Kim, J.-A.; Montagnani, M.; Koh, K.K.; Quon, M.J. Reciprocal Relationships Between Insulin Resistance and Endothelial Dysfunction. *Circulation* **2006**, *113*, 1888–1904. [CrossRef]
19. Sasso, F.C.; Salvatore, T.; Tranchino, G.; Cozzolino, D.; Caruso, A.A.; Persico, M.; Gentile, S.; Torella, D.; Torella, R. Cochlear dysfunction in type 2 diabetes: A complication independent of neuropathy and acute hyperglycemia. *Metabolism* **1999**, *48*, 1346–1350. [CrossRef]
20. Williams, S.B.; Goldfine, A.B.; Timimi, F.K.; Ting, H.H.; Roddy, M.-A.; Simonson, D.C.; Creager, M.A. Acute Hyperglycemia Attenuates Endothelium-Dependent Vasodilation in Humans In Vivo. *Circulation* **1998**, *97*, 1695–1701. [CrossRef]
21. Kawano, H.; Motoyama, T.; Hirashima, O.; Hirai, N.; Miyao, Y.; Sakamoto, T.; Kugiyama, K.; Ogawa, H.; Yasue, H. Hyperglycemia rapidly suppresses flow-mediated endothelium- dependent vasodilation of brachial artery. *J. Am. Coll. Cardiol.* **1999**, *34*, 146–154. [CrossRef]
22. El-Osta, A.; Brasacchio, D.; Yao, D.; Pocai, A.; Jones, P.L.; Roeder, R.G.; Cooper, M.E.; Brownlee, M. Transient high glucose causes persistent epigenetic changes and altered gene expression during subsequent normoglycemia. *J. Exp. Med.* **2008**, *205*, 2409–2417. [CrossRef] [PubMed]
23. Marfella, R.; Sasso, F.C.; Siniscalchi, M.; Paolisso, P.; Rizzo, M.R.; Ferraro, F.; Stabile, E.; Sorropago, G.; Calabrò, P.; Carbonara, O.; et al. Peri-Procedural Tight Glycemic Control during Early Percutaneous Coronary Intervention Is Associated with a Lower Rate of In-Stent Restenosis in Patients with Acute ST-Elevation Myocardial Infarction. *J. Clin. Endocrinol. Metab.* **2012**, *97*, 2862–2871. [CrossRef] [PubMed]
24. Lemmey, H.A.L.; Ye, X.; Ding, H.; Triggle, C.R.; Garland, C.J.; Dora, K.A. Hyperglycaemia disrupts conducted vasodilation in the resistance vasculature of db/db mice. *Vasc. Pharmacol.* **2018**, 29–35. [CrossRef]
25. Brownlee, M. Biochemistry and molecular cell biology of diabetic complications. *Nat. Cell Biol.* **2001**, *414*, 813–820. [CrossRef]
26. Galkina, E.; Ley, K. Vascular Adhesion Molecules in Atherosclerosis. *Arter. Thromb. Vasc. Biol.* **2007**, *27*, 2292–2301. [CrossRef]
27. Forbes, J.M.; Cooper, M.E. Mechanisms of Diabetic Complications. *Physiol. Rev.* **2013**, *93*, 137–188. [CrossRef]
28. Förstermann, U.; Xia, N.; Li, H. Roles of Vascular Oxidative Stress and Nitric Oxide in the Pathogenesis of Atherosclerosis. *Circ. Res.* **2017**, *120*, 713–735. [CrossRef]
29. Yuan, T.; Yang, T.; Chen, H.; Fu, D.; Hu, Y.; Wang, J.; Yuan, Q.; Yu, H.; Xu, W.; Xie, X. New insights into oxidative stress and inflammation during diabetes mellitus-accelerated atherosclerosis. *Redox Biol.* **2019**, *20*, 247–260. [CrossRef]
30. Marfella, R.; D'Amico, M.; Di Filippo, C.; Siniscalchi, M.; Sasso, F.C.; Ferraraccio, F.; Rossi, F.; Paolisso, G. The possible role of the ubiquitin proteasome system in the development of atherosclerosis in diabetes. *Cardiovasc. Diabetol.* **2007**, *6*, 35. [CrossRef]
31. Minutolo, R.; Sasso, F.C.; Chiodini, P.; Cianciaruso, B.; Carbonara, O.; Zamboli, P.; Tirino, G.; Pota, A.; Torella, R.; Conte, G.; et al. Management of cardiovascular risk factors in advanced type 2 diabetic nephropathy: A comparative analysis in nephrology, diabetology and primary care settings. *J. Hypertens.* **2006**, *24*, 1655–1661. [CrossRef] [PubMed]
32. Minutolo, R.; Gabbai, F.B.; Provenzano, M.; Chiodini, P.; Borrelli, S.; Garofalo, C.; Sasso, F.C.; Santoro, D.; Bellizzi, V.; Conte, G.; et al. Cardiorenal prognosis by residual proteinuria level in diabetic chronic kidney disease: Pooled analysis of four cohort studies. *Nephrol. Dial. Transplant.* **2018**, *33*, 1942–1949. [CrossRef] [PubMed]

33. Beckman, J.S.; Koppenol, W.H. Nitric oxide, superoxide, and peroxynitrite: The good, the bad, and ugly. *Am. J. Physiol. Physiol.* **1996**, *271*, C1424–C1437. [CrossRef] [PubMed]
34. Shah, M.S.; Brownlee, M. Molecular and Cellular Mechanisms of Cardiovascular Disorders in Diabetes. *Circ. Res.* **2016**, *118*, 1808–1829. [CrossRef] [PubMed]
35. Ceriello, A.; Russo, P.D.; Amstad, P.; Cerutti, P. High Glucose Induces Antioxidant Enzymes in Human Endothelial Cells in Culture: Evidence Linking Hyperglycemia and Oxidative Stress. *Diabetes* **1996**, *45*, 471–477. [CrossRef] [PubMed]
36. Du, X.L.; Edelstein, D.; Dimmeler, S.; Ju, Q.; Sui, C.; Brownlee, M. Hyperglycemia inhibits endothelial nitric oxide synthase activity by posttranslational modification at the Akt site. *J. Clin. Investig.* **2001**, *108*, 1341–1348. [CrossRef]
37. Förstermann, U.; Munzel, T. Endothelial Nitric Oxide Synthase in Vascular Disease: From marvel to menace. *Circulation* **2006**, *113*, 1708–1714. [CrossRef]
38. Aljofan, M.; Ding, H. High glucose increases expression of cyclooxygenase-2, increases oxidative stress and decreases the generation of nitric oxide in mouse microvessel endothelial cells. *J. Cell. Physiol.* **2009**, *222*, 669–675. [CrossRef]
39. Pannirselvam, M.; Verma, S.; Anderson, T.J.; Triggle, C.R. Cellular basis of endothelial dysfunction in small mesenteric arteries from spontaneously diabetic (db/db−/−) mice: Role of decreased tetrahydrobiopterin bioavailability. *Br. J. Pharmacol.* **2002**, *136*, 255–263. [CrossRef]
40. Moens, A.L.; Kass, D.A. Tetrahydrobiopterin and Cardiovascular Disease. *Arter. Thromb. Vasc. Biol.* **2006**, *26*, 2439–2444. [CrossRef]
41. Wever, R.M.; Van Dam, T.; Van Rijn, H.J.; De Groot, F.; Rabelink, T.J. Tetrahydrobiopterin Regulates Superoxide and Nitric Oxide Generation by Recombinant Endothelial Nitric Oxide Synthase. *Biochem. Biophys. Res. Commun.* **1997**, *237*, 340–344. [CrossRef] [PubMed]
42. Crabtree, M.J.; Tatham, A.L.; Hale, A.B.; Alp, N.J.; Channon, K. Critical Role for Tetrahydrobiopterin Recycling by Dihydrofolate Reductase in Regulation of Endothelial Nitric-oxide Synthase Coupling: Relative importance of the de novo biopterin synthesis versus salvage pathways. *J. Biol. Chem.* **2009**, *284*, 28128–28136. [CrossRef] [PubMed]
43. Channon, K.M. Tetrahydrobiopterin and Nitric Oxide Synthase Recouplers. *Bone Regul. Osteoporos. Ther.* **2020**, 1–14. [CrossRef]
44. Ihlemann, N.; Rask-Madsen, C.; Perner, A.; Dominguez, H.; Hermann, T.; Køber, L.; Torp-Pedersen, C. Tetrahydrobiopterin restores endothelial dysfunction induced by an oral glucose challenge in healthy subjects. *Am. J. Physiol. Circ. Physiol.* **2003**, *285*, H875–H882. [CrossRef]
45. Pannirselvam, M.; Simon, V.; Verma, S.; Anderson, T.; Triggle, C.R. Chronic oral supplementation with sepiapterin prevents endothelial dysfunction and oxidative stress in small mesenteric arteries from diabetic (db/db) mice. *Br. J. Pharmacol.* **2003**, *140*, 701–706. [CrossRef]
46. Franscini, N.; Bächli, E.; Blau, N.; Fischler, M.; Walter, R.B.; Schaffner, A.; Schoedon, G. Functional Tetrahydrobiopterin Synthesis in Human Platelets. *Circulation* **2004**, *110*, 186–192. [CrossRef]
47. Xu, J.; Wu, Y.; Song, P.; Zhang, M.; Wang, S.; Zou, M.-H. Proteasome-Dependent Degradation of Guanosine 5′-Triphosphate Cyclohydrolase I Causes Tetrahydrobiopterin Deficiency in Diabetes Mellitus. *Circulation* **2007**, *116*, 944–953. [CrossRef]
48. Wang, S.; Xu, J.; Song, P.; Viollet, B.; Zou, M.-H. In Vivo Activation of AMP-Activated Protein Kinase Attenuates Diabetes-Enhanced Degradation of GTP Cyclohydrolase I. *Diabetes* **2009**, *58*, 1893–1901. [CrossRef]
49. Zou, M.-H.; Shi, C.; Cohen, R.A. Oxidation of the zinc-thiolate complex and uncoupling of endothelial nitric oxide synthase by peroxynitrite. *J. Clin. Investig.* **2002**, *109*, 817–826. [CrossRef]
50. Srinivasan, S.; Hatley, M.E.; Bolick, D.T.; Palmer, L.A.; Edelstein, D.; Brownlee, M.; Hedrick, C.C. Hyperglycaemia-induced superoxide production decreases eNOS expression via AP-1 activation in aortic endothelial cells. *Diabetologia* **2004**, *47*, 1727–1734. [CrossRef]
51. Wu, N.; Shen, H.; Liu, H.-N.; Wang, Y.; Bai, Y.; Wu, N. Acute blood glucose fluctuation enhances rat aorta endothelial cell apoptosis, oxidative stress and pro-inflammatory cytokine expression in vivo. *Cardiovasc. Diabetol.* **2016**, *15*, 1–13. [CrossRef] [PubMed]
52. He, C.; Hart, P.C.; Germain, D.; Bonini, M.G. SOD2 and the Mitochondrial UPR: Partners Regulating Cellular Phenotypic Transitions. *Trends Biochem. Sci.* **2016**, *41*, 568–577. [CrossRef] [PubMed]
53. Jheng, H.-F.; Tsai, P.-J.; Guo, S.-M.; Kuo, L.-H.; Chang, C.-S.; Su, I.-J.; Tsai, Y.-S. Mitochondrial Fission Contributes to Mitochondrial Dysfunction and Insulin Resistance in Skeletal Muscle. *Mol. Cell. Biol.* **2011**, *32*, 309–319. [CrossRef] [PubMed]
54. Yabe-Nishimura, C. Aldose reductase in glucose toxicity: A potential target for the prevention of diabetic complications. *Pharmacol. Rev.* **1998**, *50*, 21–34.
55. Tang, W.H.; Martin, K.A.; Hwa, J. Aldose Reductase, Oxidative Stress, and Diabetic Mellitus. *Front. Pharmacol.* **2012**, *3*, 87. [CrossRef]
56. Reddy, A.B.M.; Ramana, K.V.; Srivastava, S.; Bhatnagar, A.; Srivastava, S.K. Aldose Reductase Regulates High Glucose-Induced Ectodomain Shedding of Tumor Necrosis Factor (TNF)-α via Protein Kinase C-δ and TNF-α Converting Enzyme in Vascular Smooth Muscle Cells. *Endocrinology* **2008**, *150*, 63–74. [CrossRef]
57. Yadav, U.C.; Ramana, K.V.; Srivastava, S.K. Aldose reductase inhibition suppresses airway inflammation. *Chem. Interact.* **2011**, *191*, 339–345. [CrossRef]

58. Byun, K.; Yoo, Y.; Son, M.; Lee, J.; Jeong, G.-B.; Park, Y.M.; Salekdeh, G.H.; Lee, B. Advanced glycation end-products produced systemically and by macrophages: A common contributor to inflammation and degenerative diseases. *Pharmacol. Ther.* **2017**, *177*, 44–55. [CrossRef]
59. Bongarzone, S.; Savickas, V.; Luzi, F.; Gee, A.D. Targeting the Receptor for Advanced Glycation Endproducts (RAGE): A Medicinal Chemistry Perspective. *J. Med. Chem.* **2017**, *60*, 7213–7232. [CrossRef]
60. Shen, C.; Li, Q.; Zhang, Y.C.; Ma, G.; Feng, Y.; Zhu, Q.; Dai, Q.; Chen, Z.; Yao, Y.; Chen, L.; et al. Advanced glycation endproducts increase EPC apoptosis and decrease nitric oxide release via MAPK pathways. *Biomed. Pharmacother.* **2010**, *64*, 35–43. [CrossRef]
61. Chen, J.; Jing, J.; Yu, S.; Song, M.; Tan, H.; Cui, B.; Huang, L. Advanced glycation endproducts induce apoptosis of endothelial progenitor cells by activating receptor RAGE and NADPH oxidase/JNK signaling axis. *Am. J. Transl. Res.* **2016**, *8*, 2169–2178. [PubMed]
62. Ding, H.; Aljofan, M.; Triggle, C.R. Oxidative stress and increased eNOS and NADPH oxidase expression in mouse microvessel endothelial cells. *J. Cell. Physiol.* **2007**, *212*, 682–689. [CrossRef] [PubMed]
63. Taye, A.; Saad, A.H.; Kumar, A.H.; Morawietz, H. Effect of apocynin on NADPH oxidase-mediated oxidative stress-LOX-1-eNOS pathway in human endothelial cells exposed to high glucose. *Eur. J. Pharmacol.* **2010**, *627*, 42–48. [CrossRef] [PubMed]
64. Quagliaro, L.; Piconi, L.; Assaloni, R.; Martinelli, L.; Motz, E.; Ceriello, A. Intermittent High Glucose Enhances Apoptosis Related to Oxidative Stress in Human Umbilical Vein Endothelial Cells: The Role of Protein Kinase C and NAD(P)H-Oxidase Activation. *Diabetes* **2003**, *52*, 2795–2804. [CrossRef] [PubMed]
65. Geraldes, P.; King, G.L. Activation of Protein Kinase C Isoforms and Its Impact on Diabetic Complications. *Circ. Res.* **2010**, *106*, 1319–1331. [CrossRef] [PubMed]
66. Inoguchi, T.; Sonta, T.; Tsubouchi, H.; Etoh, T.; Kakimoto, M.; Sonoda, N.; Sato, N.; Sekiguchi, N.; Kobayashi, K.; Sumimoto, H.; et al. Protein Kinase C-Dependent Increase in Reactive Oxygen Species (ROS) Production in Vascular Tissues of Diabetes: Role of Vascular NAD(P)H Oxidase. *J. Am. Soc. Nephrol.* **2003**, *14*, 227S–232S. [CrossRef]
67. Inoguchi, T.; Li, P.; Umeda, F.; Yu, H.Y.; Kakimoto, M.; Imamura, M.; Aoki, T.; Etoh, T.; Hashimoto, T.; Naruse, M.; et al. High glucose level and free fatty acid stimulate reactive oxygen species production through protein kinase C–dependent activation of NAD(P)H oxidase in cultured vascular cells. *Diabetes* **2000**, *49*, 1939–1945. [CrossRef]
68. Xia, P.; Inoguchi, T.; Kern, T.S.; Engerman, R.L.; Oates, P.J.; King, G.L. Characterization of the Mechanism for the Chronic Activation of Diacylglycerol-Protein Kinase C Pathway in Diabetes and Hypergalactosemia. *Diabetes* **1994**, *43*, 1122–1129. [CrossRef]
69. Hayashi, T.; Matsui-Hirai, H.; Miyazaki-Akita, A.; Fukatsu, A.; Funami, J.; Ding, Q.-F.; Kamalanathan, S.; Hattori, Y.; Ignarro, L.J.; Iguchi, A. Endothelial cellular senescence is inhibited by nitric oxide: Implications in atherosclerosis associated with menopause and diabetes. *Proc. Natl. Acad. Sci. USA* **2006**, *103*, 17018–17023. [CrossRef]
70. Erusalimsky, J.D. Vascular endothelial senescence: From mechanisms to pathophysiology. *J. Appl. Physiol.* **2009**, *106*, 326–332. [CrossRef]
71. Peng, N.; Meng, N.; Wang, S.; Zhao, F.; Zhao, J.; Su, L.; Zhang, S.; Zhang, Y.; Zhao, B.; Miao, J. An activator of mTOR inhibits oxLDL-induced autophagy and apoptosis in vascular endothelial cells and restricts atherosclerosis in apolipoprotein E-/- mice. *Sci. Rep.* **2015**, *4*, 5519. [CrossRef] [PubMed]
72. Yokoi, T.; Fukuo, K.; Yasuda, O.; Hotta, M.; Miyazaki, J.; Takemura, Y.; Kawamoto, H.; Ichijo, H.; Ogihara, T. Apoptosis Signal-Regulating Kinase 1 Mediates Cellular Senescence Induced by High Glucose in Endothelial Cells. *Diabetes* **2006**, *55*, 1660–1665. [CrossRef]
73. Chen, J.; Brodsky, S.V.; Goligorsky, D.M.; Hampel, D.J.; Li, H.; Gross, S.S.; Goligorsky, M.S. Glycated Collagen I Induces Premature Senescence-Like Phenotypic Changes in Endothelial Cells. *Circ. Res.* **2002**, *90*, 1290–1298. [CrossRef] [PubMed]
74. Yu, T.; Sheu, S.-S.; Robotham, J.L.; Yoon, Y. Mitochondrial fission mediates high glucose-induced cell death through elevated production of reactive oxygen species. *Cardiovasc. Res.* **2008**, *79*, 341–351. [CrossRef] [PubMed]
75. Zhong, W.; Zou, G.; Gu, J.; Zhang, J. L-arginine attenuates high glucose-accelerated senescence in human umbilical vein endothelial cells. *Diabetes Res. Clin. Pr.* **2010**, *89*, 38–45. [CrossRef] [PubMed]
76. Matsui-Hirai, H.; Hayashi, T.; Yamamoto, S.; Ina, K.; Maeda, M.; Kotani, H.; Iguchi, A.; Ignarro, L.J.; Hattori, Y. Dose-Dependent Modulatory Effects of Insulin on Glucose-Induced Endothelial Senescence In Vitro and In Vivo: A Relationship between Telomeres and Nitric Oxide. *J. Pharmacol. Exp. Ther.* **2011**, *337*, 591–599. [CrossRef]
77. Yang, Z.-H.; Mo, X.; Gong, Q.; Pan, Q.; Yang, X.; Cai, W.; Li, C.; Ma, J.-X.; He, Y.; Gao, G. Critical effect of VEGF in the process of endothelial cell apoptosis induced by high glucose. *Apoptosis* **2008**, *13*, 1331–1343. [CrossRef]
78. Ho, F.M.; Lin, W.-W.; Chen, B.C.; Chao, C.M.; Yang, C.-R.; Lin, L.Y.; Lai, C.C.; Liu, S.; Liau, C.S. High glucose-induced apoptosis in human vascular endothelial cells is mediated through NF-κB and c-Jun NH2-terminal kinase pathway and prevented by PI3K/Akt/eNOS pathway. *Cell. Signal.* **2006**, *18*, 391–399. [CrossRef]
79. Torella, D.; Ellison, G.M.; Torella, M.; Vicinanza, C.; Aquila, I.; Iaconetti, C.; Scalise, M.; Marino, F.; Henning, B.J.; Lewis, F.C.; et al. Carbonic Anhydrase Activation Is Associated with Worsened Pathological Remodeling in Human Ischemic Diabetic Cardiomyopathy. *J. Am. Heart Assoc.* **2014**, *3*, e000434. [CrossRef]
80. Tang, N.; Jiang, S.; Yang, Y.; Liu, S.; Ponnusamy, M.; Xin, H.; Yu, T. Noncoding RNAs as therapeutic targets in atherosclerosis with diabetes mellitus. *Cardiovasc. Ther.* **2018**, *36*, e12436. [CrossRef]

81. Torella, D.; Iaconetti, C.; Tarallo, R.; Marino, F.; Giurato, G.; Veneziano, C.; Aquila, I.; Scalise, M.; Mancuso, T.; Cianflone, E.; et al. miRNA Regulation of the Hyperproliferative Phenotype of Vascular Smooth Muscle Cells in Diabetes. *Diabetes* **2018**, *67*, 2554–2568. [CrossRef] [PubMed]
82. Kinlay, S.; Behrendt, D.; Fang, J.C.; Delagrange, D.; Morrow, J.; Witztum, J.L.; Rifai, N.; Selwyn, A.P.; Creager, M.A.; Ganz, P. long-term effect of combined vitamins e and c on coronary and peripheral endothelial function. *J. Am. Coll. Cardiol.* **2004**, *43*, 629–634. [CrossRef] [PubMed]
83. Selvin, E.; Bolen, S.; Yeh, H.-C.; Wiley, C.; Wilson, L.M.; Marinopoulos, S.S.; Feldman, L.; Vassy, J.L.; Wilson, R.; Bass, E.B.; et al. Cardiovascular Outcomes in Trials of Oral Diabetes Medications: A systematic review. *Arch. Intern. Med.* **2008**, *168*, 2070–2080. [CrossRef] [PubMed]
84. Yandrapalli, S.; Jolly, G.; Horblitt, A.; Sanaani, A.; Aronow, W. Cardiovascular benefits and safety of non-insulin medications used in the treatment of type 2 diabetes mellitus. *Postgrad. Med.* **2017**, *129*, 811–821. [CrossRef] [PubMed]
85. Roumie, C.L.; Hung, A.M.; Greevy, R.A.; Grijalva, C.G.; Liu, X.; Murff, H.J.; Elasy, T.A.; Griffin, M.R. Comparative Effectiveness of Sulfonylurea and Metformin Monotherapy on Cardiovascular Events in Type 2 Diabetes Mellitus: A cohort study. *Ann. Intern. Med.* **2012**, *157*, 601–610. [CrossRef] [PubMed]
86. Roussel, R.; Travert, F.; Pasquet, B.; Wilson, P.W.; Smith, S.C.; Goto, S.; Ravaud, P.; Marre, M.; Porath, A.; Bhatt, D.L.; et al. Reduction of Atherothrombosis for Continued Health (REACH) Registry Investigators. Metformin use and mortality among patients with diabetes and atherothrombosis. *Arch. Intern. Med.* **2010**, *170*, 1892–1899. [CrossRef]
87. Scheen, A.J.; Paquot, N. Metformin revisited: A critical review of the benefit-risk balance in at-risk patients with type 2 diabetes. *Diabetes Metab.* **2013**, *39*, 179–190. [CrossRef]
88. Bailey, C.J. Metformin: Historical overview. *Diabetologia* **2017**, *60*, 1566–1576. [CrossRef]
89. Ríos, J.L.; Francini, F.; Schinella, G.R. Natural Products for the Treatment of Type 2 Diabetes Mellitus. *Planta Med.* **2015**, *81*, 975–994. [CrossRef]
90. Rena, G.; Hardie, D.G.; Pearson, E.R. The mechanisms of action of metformin. *Diabetologia* **2017**, *60*, 1577–1585. [CrossRef]
91. Eva, A.-R.; Ranal-Muino, E.; Fernandez-Fernandez, C.; Pazos-Garcia, C.; Vila-Altesor, M. Metabolic Effects of Metformin in Humans. *Curr. Diabetes Rev.* **2019**, *15*, 328–339. [CrossRef]
92. Buse, J.B.; DeFronzo, R.A.; Rosenstock, J.; Kim, T.; Burns, C.; Skare, S.; Baron, A.; Fineman, M. The Primary Glucose-Lowering Effect of Metformin Resides in the Gut, Not the Circulation. Results From Short-term Pharmacokinetic and 12-Week Dose-Ranging Studies. *Diabetes Care* **2016**, *39*, 198–205. [CrossRef] [PubMed]
93. Foretz, M.; Guigas, B.; Viollet, B. Understanding the glucoregulatory mechanisms of metformin in type 2 diabetes mellitus. *Nat. Rev. Endocrinol.* **2019**, *15*, 569–589. [CrossRef] [PubMed]
94. Sasso, F.C.; Rinaldi, L.; Lascar, N.; Marrone, A.; Pafundi, P.C.; Adinolfi, L.E.; Marfella, R. Role of Tight Glycemic Control during Acute Coronary Syndrome on CV Outcome in Type 2 Diabetes. *J. Diabetes Res.* **2018**, *2018*, 1–8. [CrossRef] [PubMed]
95. Wiernsperger, N.F. Metformin as a cellular protector; a synoptic view of modern evidences. *J. Nephropharmacol.* **2015**, *4*, 31–36. [PubMed]
96. Nath, M.; Bhattacharjee, K.; Choudhury, Y. Pleiotropic effects of anti-diabetic drugs: A comprehensive review. *Eur. J. Pharmacol.* **2020**, *884*, 173349. [CrossRef] [PubMed]
97. Jenkins, A.J.; Welsh, P.; Petrie, J.R. Metformin, lipids and atherosclerosis prevention. *Curr. Opin. Lipidol.* **2018**, *29*, 346–353. [CrossRef]
98. Garcia, D.; Shaw, R.J. AMPK: Mechanisms of Cellular Energy Sensing and Restoration of Metabolic Balance. *Mol. Cell* **2017**, *66*, 789–800. [CrossRef]
99. Hardie, D.G.; Ross, F.A.; Hawley, S.A. AMPK: A nutrient and energy sensor that maintains energy homeostasis. *Nat. Rev. Mol. Cell Biol.* **2012**, *13*, 251–262. [CrossRef]
100. Hardie, D.G. AMP-activated protein kinase—An energy sensor that regulates all aspects of cell function. *Genes Dev.* **2011**, *25*, 1895–1908. [CrossRef]
101. Mihaylova, M.M.; Shaw, R.J. The AMPK signalling pathway coordinates cell growth, autophagy and metabolism. *Nat. Cell Biol.* **2011**, *13*, 1016–1023. [CrossRef] [PubMed]
102. Chen, Z.-P.; Mitchelhill, K.I.; Michell, B.J.; Stapleton, D.; Rodriguez-Crespo, I.; Witters, L.A.; Power, D.A.; De Montellano, P.R.O.; Kemp, B.E. AMP-activated protein kinase phosphorylation of endothelial NO synthase. *FEBS Lett.* **1999**, *443*, 285–289. [CrossRef]
103. Lee, W.J.; Lee, I.K.; Kim, H.S.; Kim, Y.M.; Koh, E.H.; Won, J.C.; Han, S.M.; Kim, M.-S.; Jo, I.; Oh, G.T.; et al. α-Lipoic Acid Prevents Endothelial Dysfunction in Obese Rats via Activation of AMP-Activated Protein Kinase. *Arter. Thromb. Vasc. Biol.* **2005**, *25*, 2488–2494. [CrossRef] [PubMed]
104. Zou, M.-H.; Wu, Y. Amp-activated protein kinase activation as a strategy for protecting vascular endothelial function. *Clin. Exp. Pharmacol. Physiol.* **2007**, *35*, 535–545. [CrossRef]
105. Jeon, S.-M. Regulation and function of AMPK in physiology and diseases. *Exp. Mol. Med.* **2016**, *48*, e245. [CrossRef]
106. Shaw, R.J.; Kosmatka, M.; Bardeesy, N.; Hurley, R.L.; Witters, L.A.; Depinho, R.A.; Cantley, L.C. The tumor suppressor LKB1 kinase directly activates AMP-activated kinase and regulates apoptosis in response to energy stress. *Proc. Natl. Acad. Sci. USA* **2004**, *101*, 3329–3335. [CrossRef]
107. Hawley, S.A.; Pan, D.A.; Mustard, K.J.; Ross, L.; Bain, J.; Edelman, A.M.; Frenguelli, B.G.; Hardie, D.G. Calmodulin-dependent protein kinase kinase-β is an alternative upstream kinase for AMP-activated protein kinase. *Cell Metab.* **2005**, *2*, 9–19. [CrossRef]

108. Stahmann, N.; Woods, A.; Carling, D.; Heller, R. Thrombin Activates AMP-Activated Protein Kinase in Endothelial Cells via a Pathway Involving Ca2+/Calmodulin-Dependent Protein Kinase Kinase β. *Mol. Cell. Biol.* **2006**, *26*, 5933–5945. [CrossRef]
109. Herzig, S.; Shaw, R.J. AMPK: Guardian of metabolism and mitochondrial homeostasis. *Nat. Rev. Mol. Cell Biol.* **2018**, *19*, 121–135. [CrossRef]
110. Corton, J.M.; Gillespie, J.G.; Hawley, S.A.; Hardie, D.G. 5-Aminoimidazole-4-Carboxamide Ribonucleoside. A Specific Method for Activating AMP-Activated Protein Kinase in Intact Cells? *JBIC J. Biol. Inorg. Chem.* **1995**, *229*, 558–565. [CrossRef]
111. Viollet, B.; Guigas, B.; Garcia, N.S.; Leclerc, J.; Foretz, M.; Andreelli, F. Cellular and molecular mechanisms of metformin: An overview. *Clin. Sci.* **2011**, *122*, 253–270. [CrossRef] [PubMed]
112. Zhou, G.; Myers, R.; Li, Y.; Chen, Y.; Shen, X.; Fenyk-Melody, J.; Wu, M.; Ventre, J.; Doebber, T.; Fujii, N.; et al. Role of AMP-activated protein kinase in mechanism of metformin action. *J. Clin. Investig.* **2001**, *108*, 1167–1174. [CrossRef] [PubMed]
113. Ouyang, J.; Parakhia, R.A.; Ochs, R.S. Metformin Activates AMP Kinase through Inhibition of AMP Deaminase. *J. Biol. Chem.* **2010**, *286*, 1–11. [CrossRef] [PubMed]
114. Ding, H.; Ye, K.; Triggle, C.R. Impact of currently used anti-diabetic drugs on myoendothelial communication. *Curr. Opin. Pharmacol.* **2019**, *45*, 1–7. [CrossRef]
115. Logie, L.; Harthill, J.; Patel, K.; Bacon, S.; Hamilton, D.L.; Macrae, K.; McDougall, G.; Wang, H.-H.; Xue, L.; Jiang, H.; et al. Cellular Responses to the Metal-Binding Properties of Metformin. *Diabetes* **2012**, *61*, 1423–1433. [CrossRef]
116. Madiraju, A.K.; Erion, D.M.; Rahimi, Y.; Zhang, X.-M.; Braddock, D.T.; Albright, R.A.; Prigaro, B.J.; Wood, J.L.; Bhanot, S.; Macdonald, M.J.; et al. Metformin suppresses gluconeogenesis by inhibiting mitochondrial glycerophosphate dehydrogenase. *Nat. Cell Biol.* **2014**, *510*, 542–546. [CrossRef]
117. Kinaan, M.; Ding, H.; Triggle, C.R. Metformin: An Old Drug for the Treatment of Diabetes but a New Drug for the Protection of the Endothelium. *Med Princ. Pr.* **2015**, *24*, 401–415. [CrossRef]
118. Foretz, M.; Hébrard, S.; Leclerc, J.; Zarrinpashneh, E.; Soty, M.; Mithieux, G.; Sakamoto, K.; Andreelli, F.; Viollet, B. Metformin inhibits hepatic gluconeogenesis in mice independently of the LKB1/AMPK pathway via a decrease in hepatic energy state. *J. Clin. Investig.* **2010**, *120*, 2355–2369. [CrossRef]
119. Effect of intensive blood-glucose control with metformin on complications in overweight patients with type 2 diabetes (UKPDS 34). UK Prospective Diabetes Study (UKPDS) Group. *Lancet* **1998**, *352*, 854–865. [CrossRef]
120. Saenz, A.; Fernandez-Esteban, I.; Mataix, A.; Segura, M.A.; Roque, M.; Moher, D. Metformin monotherapy for type 2 diabetes mellitus. *Cochrane Database Syst. Rev.* **2005**, *20*, CD002966. [CrossRef]
121. Eurich, D.T.; McAlister, F.A.; Blackburn, D.F.; Majumdar, S.R.; Tsuyuki, R.T.; Varney, J.; Johnson, J.A. Benefits and harms of antidiabetic agents in patients with diabetes and heart failure: Systematic review. *BMJ* **2007**, *335*, 497. [CrossRef] [PubMed]
122. Campbell, J.M.; Bellman, S.M.; Stephenson, M.D.; Lisy, K. Metformin reduces all-cause mortality and diseases of ageing independent of its effect on diabetes control: A systematic review and meta-analysis. *Ageing Res. Rev.* **2017**, *40*, 31–44. [CrossRef] [PubMed]
123. Lamanna, C.; Monami, M.; Marchionni, N.; Mannucci, E. Effect of metformin on cardiovascular events and mortality: A meta-analysis of randomized clinical trials. *Diabetes Obes. Metab.* **2011**, *13*, 221–228. [CrossRef] [PubMed]
124. Griffin, S.; Leaver, J.K.; Irving, G.J. Impact of metformin on cardiovascular disease: A meta-analysis of randomised trials among people with type 2 diabetes. *Diabetologia* **2017**, *60*, 1620–1629. [CrossRef] [PubMed]
125. Jørgensen, C.H.; Gislason, G.; Andersson, C.; Ahlehoff, O.; Charlot, M.; Schramm, T.K.; Vaag, A.; Abildstrøm, S.Z.; Torp-Pedersen, C.; Hansen, T.W. Effects of oral glucose-lowering drugs on long term outcomes in patients with diabetes mellitus following myocardial infarction not treated with emergent percutaneous coronary intervention—A retrospective nationwide cohort study. *Cardiovasc. Diabetol.* **2010**, *9*, 54. [CrossRef]
126. Triggle, C.R.; Ding, H. Metformin is not just an antihyperglycaemic drug but also has protective effects on the vascular endothelium. *Acta Physiol.* **2017**, *219*, 138–151. [CrossRef]
127. De Jager, J.; Kooy, A.; Schalkwijk, C.; Van Der Kolk, J.; Lehert, P.; Bets, D.; Wulffelé, M.G.; Donker, A.J.; Stehouwer, C.D.A. Long-term effects of metformin on endothelial function in type 2 diabetes: A randomized controlled trial. *J. Intern. Med.* **2014**, *275*, 59–70. [CrossRef]
128. Chait, A.; Bornfeldt, K.E. Diabetes and atherosclerosis: Is there a role for hyperglycemia? *J. Lipid Res.* **2008**, *50*, S335–S339. [CrossRef]
129. De Aguiar, L.G.K.; Kraemer-Aguiar, L.G.; Bahia, L.R.; Villela, N.; Laflor, C.; Sicuro, F.; Wiernsperger, N.F.; Bottino, D.; Bouskela, E. Metformin Improves Endothelial Vascular Reactivity in First-Degree Relatives of Type 2 Diabetic Patients With Metabolic Syndrome and Normal Glucose Tolerance. *Diabetes Care* **2006**, *29*, 1083–1089. [CrossRef]
130. Marfella, R.; Acampora, R.; Verrazzo, G.; Ziccardi, P.; De Rosa, N.; Giunta, R.; Giugliano, D. Metformin Improves Hemodynamic and Rheological Responses to L-Arginine in NIDDM Patients. *Diabetes Care* **1996**, *19*, 934–939. [CrossRef]
131. Kapur, N. Nitric oxide: An emerging role in cardioprotection? *Heart* **2001**, *86*, 368–372. [CrossRef]
132. Mather, K.J.; Verma, S.; Anderson, T.J. Improved endothelial function with metformin in type 2 diabetes mellitus. *J. Am. Coll. Cardiol.* **2001**, *37*, 1344–1350. [CrossRef]
133. Steinberg, H.O.; Brechtel, G.; Johnson, A.; Fineberg, N.; Baron, A.D. Insulin-mediated skeletal muscle vasodilation is nitric oxide dependent. A novel action of insulin to increase nitric oxide release. *J. Clin. Investig.* **1994**, *94*, 1172–1179. [CrossRef] [PubMed]

134. Zeng, G.; Quon, M.J. Insulin-stimulated production of nitric oxide is inhibited by wortmannin. Direct measurement in vascular endothelial cells. *J. Clin. Investig.* **1996**, *98*, 894–898. [CrossRef]
135. Muniyappa, R.; Sowers, J.R. Role of insulin resistance in endothelial dysfunction. *Rev. Endocr. Metab. Disord.* **2013**, *14*, 5–12. [CrossRef]
136. Steinberg, H.O.; Chaker, H.; Leaming, R.; Johnson, A.; Brechtel, G.; Baron, A.D. Obesity/insulin resistance is associated with endothelial dysfunction. Implications for the syndrome of insulin resistance. *J. Clin. Investig.* **1996**, *97*, 2601–2610. [CrossRef]
137. Steinberg, H.O.; Tarshoby, M.; Monestel, R.; Hook, G.; Cronin, J.; Johnson, A.; Bayazeed, B.; Baron, A.D. Elevated circulating free fatty acid levels impair endothelium-dependent vasodilation. *J. Clin. Investig.* **1997**, *100*, 1230–1239. [CrossRef]
138. Cozzolino, D.; Sessa, G.; Salvatore, T.; Sasso, F.C.; Giugliano, D.; Lefèbvre, P.J.; Torella, R. The involvement of the opioid system in human obesity: A study in normal weight relatives of obese people. *J. Clin. Endocrinol. Metab.* **1996**, *81*, 713–718. [CrossRef]
139. Balletshofer, B.; Rittig, K.; Maerker, E.; Häring, H.; Volk, A.; Jacob, S.; Rett, K. Impaired Non-Esterified Fatty Acid Suppression is Associated with Endothelial Dysfunction in Insulin Resistant Subjects. *Horm. Metab. Res.* **2001**, *33*, 428–431. [CrossRef]
140. Michell, B.J.; Chen, Z.-P.; Tiganis, T.; Stapleton, D.; Katsis, F.; Power, D.A.; Sim, A.T.; Kemp, B.E. Coordinated Control of Endothelial Nitric-oxide Synthase Phosphorylation by Protein Kinase C and the cAMP-dependent Protein Kinase. *J. Biol. Chem.* **2001**, *276*, 17625–17628. [CrossRef]
141. Ruderman, N.B.; Cacicedo, J.M.; Itani, S.; Yagihashi, N.; Saha, A.K.; Ye, J.M.; Chen, K.; Zou, M.; Carling, D.; Boden, G.; et al. Malonyl-CoA and AMP-activated protein kinase (AMPK): Possible links between insulin resistance in muscle and early endothelial cell damage in diabetes. *Biochem. Soc. Trans.* **2003**, *31*, 202–206. [CrossRef] [PubMed]
142. Katakam, P.V.G.; Ujhelyi, M.R.; Hoenig, M.; Miller, A.W. Metformin Improves Vascular Function in Insulin-Resistant Rats. *Hypertension* **2000**, *35*, 108–112. [CrossRef] [PubMed]
143. Sena, C.; Matafome, P.; Louro, T.; Nunes, E.; Fernandes, R.; Seiça, R.M. Metformin restores endothelial function in aorta of diabetic rats. *Br. J. Pharmacol.* **2011**, *163*, 424–437. [CrossRef] [PubMed]
144. Vitale, C.; Mercuro, G.; Cornoldi, A.; Fini, M.; Volterrani, M.; Rosano, G. Metformin improves endothelial function in patients with metabolic syndrome. *J. Intern. Med.* **2005**, *258*, 250–256. [CrossRef]
145. Tack, C.J.J.; Ong, M.K.E.; Lutterman, J.A.; Smits, P. Insulin-induced vasodilatation and endothelial function in obesity/insulin resistance. Effects of troglitazone. *Diabetologia* **1998**, *41*, 569–576. [CrossRef]
146. Naka, K.K.; Papathanassiou, K.; Bechlioulis, A.; Pappas, K.; Kazakos, N.; Kanioglou, C.; Kostoula, A.; Vezyraki, P.; Makriyiannis, D.; Tsatsoulis, A.; et al. Effects of pioglitazone and metformin on vascular function in patients with type 2 diabetes treated with sulfonylureas. *Diabetes Vasc. Dis. Res.* **2011**, *9*, 52–58. [CrossRef]
147. Pitocco, D.; Zaccardi, F.; Tarzia, P.; Milo, M.; Scavone, G.; Rizzo, P.; Pagliaccia, F.; Nerla, R.; Di Franco, A.; Manto, A.; et al. Metformin improves endothelial function in type 1 diabetic subjects: A pilot, placebo-controlled randomized study. *Diabetes Obes. Metab.* **2013**, *15*, 427–431. [CrossRef]
148. Salt, I.P.; Hardie, D.G. AMP-Activated Protein Kinase: An Ubiquitous Signaling Pathway With Key Roles in the Cardiovascular System. *Circ. Res.* **2017**, *120*, 1825–1841. [CrossRef]
149. Davis, B.J.; Xie, Z.; Viollet, B.; Zou, M.-H. Activation of the AMP-Activated Kinase by Antidiabetes Drug Metformin Stimulates Nitric Oxide Synthesis In Vivo by Promoting the Association of Heat Shock Protein 90 and Endothelial Nitric Oxide Synthase. *Diabetes* **2006**, *55*, 496–505. [CrossRef]
150. Bosselaar, M.; Boon, H.; Van Loon, L.J.C.; Broek, P.H.H.V.D.; Smits, P.; Tack, C.J. Intra-arterial AICA-riboside administration induces NO-dependent vasodilation in vivo in human skeletal muscle. *Am. J. Physiol. Metab.* **2009**, *297*, E759–E766. [CrossRef]
151. Matsumoto, T.; Noguchi, E.; Ishida, K.; Kobayashi, T.; Yamada, N.; Kamata, K. Metformin normalizes endothelial function by suppressing vasoconstrictor prostanoids in mesenteric arteries from OLETF rats, a model of type 2 diabetes. *Am. J. Physiol. Circ. Physiol.* **2008**, *295*, H1165–H1176. [CrossRef] [PubMed]
152. Enkhjargal, B.; Godo, S.; Sawada, A.; Suvd, N.; Saito, H.; Noda, K.; Satoh, K.; Shimokawa, H. Endothelial AMP-Activated Protein Kinase Regulates Blood Pressure and Coronary Flow Responses Through Hyperpolarization Mechanism in Mice. *Arter. Thromb. Vasc. Biol.* **2014**, *34*, 1505–1513. [CrossRef] [PubMed]
153. Chen, H.; Vanhoutte, P.M.; Leung, S.W.S. Vascular adenosine monophosphate-activated protein kinase: Enhancer, brake or both? *Basic Clin. Pharmacol. Toxicol.* **2019**, *127*, 81–91. [CrossRef] [PubMed]
154. Mattagajasingh, I.; Kim, C.-S.; Naqvi, A.; Yamamori, T.; Hoffman, T.A.; Jung, S.-B.; DeRicco, J.; Kasuno, K.; Irani, K. SIRT1 promotes endothelium-dependent vascular relaxation by activating endothelial nitric oxide synthase. *Proc. Natl. Acad. Sci. USA* **2007**, *104*, 14855–14860. [CrossRef] [PubMed]
155. Arunachalam, G.; Samuel, S.M.; Marei, I.; Ding, H.; Triggle, C.R. Metformin modulates hyperglycaemia-induced endothelial senescence and apoptosis through SIRT1. *Br. J. Pharmacol.* **2013**, *171*, 523–535. [CrossRef] [PubMed]
156. Ghosh, S.; Lakshmanan, A.P.; Hwang, M.J.; Kubba, H.; Mushannen, A.; Triggle, C.R.; Ding, H. Metformin improves endothelial function in aortic tissue and microvascular endothelial cells subjected to diabetic hyperglycaemic conditions. *Biochem. Pharmacol.* **2015**, *98*, 412–421. [CrossRef] [PubMed]
157. Tang, E.H.C.; Vanhoutte, P.M. Endothelial dysfunction: A strategic target in the treatment of hypertension? *Pflügers Archiv.* **2010**, *459*, 995–1004. [CrossRef] [PubMed]

158. Zou, M.-H.; Kirkpatrick, S.S.; Davis, B.J.; Nelson, J.S.; Wiles, W.G.; Schlattner, U.; Neumann, D.; Brownlee, M.; Freeman, M.B.; Goldman, M.H. Withdrawal: Activation of the AMP-activated protein kinase by the anti-diabetic drug metformin in vivo: Role of mitochondrial reactive nitrogen species. *J. Biol. Chem.* **2019**, *294*, 13525. [CrossRef]
159. Bellin, C.; De Wiza, D.H.; Wiernsperger, N.F.; Rosen, P. Generation of Reactive Oxygen Species by Endothelial and Smooth Muscle Cells: Influence of Hyperglycemia and Metformin. *Horm. Metab. Res.* **2006**, *38*, 732–739. [CrossRef]
160. Faure, P.; Rossini, E.; Wiernsperger, N.; Richard, M.J.; Favier, A.; Halimi, S. An insulin sensitizer improves the free radical defense system potential and insulin sensitivity in high fructose-fed rats. *Diabetes* **1999**, *48*, 353–357. [CrossRef]
161. Gargiulo, P.; Caccese, D.; Pignatelli, P.; Brufani, C.; De Vito, F.; Marino, R.; Lauro, R.; Violi, F.; Di Mario, U.; Sanguigni, V. Metformin decreases platelet superoxide anion production in diabetic patients. *Diabetes/Metab. Res. Rev.* **2002**, *18*, 156–159. [CrossRef] [PubMed]
162. Ouslimani, N.; Peynet, J.; Bonnefont-Rousselot, D.; Thérond, P.; Legrand, A.; Beaudeux, J.-L. Metformin decreases intracellular production of reactive oxygen species in aortic endothelial cells. *Metabolism* **2005**, *54*, 829–834. [CrossRef] [PubMed]
163. Batandier, C.; Guigas, B.; Detaille, D.; El-Mir, M.-Y.; Fontaine, E.; Rigoulet, M.; Leverve, X.M. The ROS Production Induced by a Reverse-Electron Flux at Respiratory-Chain Complex 1 is Hampered by Metformin. *J. Bioenerg. Biomembr.* **2006**, *38*, 33–42. [CrossRef] [PubMed]
164. Batchuluun, B.; Inoguchi, T.; Sonoda, N.; Sasaki, S.; Inoue, T.; Fujimura, Y.; Miura, D.; Takayanagi, R. Metformin and liraglutide ameliorate high glucose-induced oxidative stress via inhibition of PKC-NAD(P)H oxidase pathway in human aortic endothelial cells. *Atherosclerosis* **2014**, *232*, 156–164. [CrossRef] [PubMed]
165. Tripathi, S.S.; Singh, A.K.; Akhtar, F.; Chaudhary, A.; Rizvi, S.I. Metformin protects red blood cells against rotenone induced oxidative stress and cytotoxicity. *Arch. Physiol. Biochem.* **2019**, 1–10. [CrossRef] [PubMed]
166. Rosen, P.; Wiernsperger, N.F. Metformin delays the manifestation of diabetes and vascular dysfunction in Goto–Kakizaki rats by reduction of mitochondrial oxidative stress. *Diabetes/Metab. Res. Rev.* **2006**, *22*, 323–330. [CrossRef]
167. Sultuybek, G.K.; Ozdas, S.B.; Curgunlu, A.; Tezcan, V.; Onaran, I. Does metformin prevent short term oxidant-induced DNA damage? In vitro study on lymphocytes from aged subjects. *J. Basic Clin. Physiol. Pharmacol.* **2007**, *18*, 129–140. [CrossRef]
168. Sambe, T.; Mason, R.P.; Dawoud, H.; Sherratt, S.C.; Malinski, T. Metformin treatment decreases nitroxidative stress, restores nitric oxide bioavailability and endothelial function beyond glucose control. *Biomed. Pharmacother.* **2018**, *98*, 149–156. [CrossRef]
169. Khouri, H.; Collin, F.; Bonnefont-Rousselot, D.; Legrand, A.; Jore, D.; Gardes-Albert, M. Radical-induced oxidation of metformin. *JBIC J. Biol. Inorg. Chem.* **2004**, *271*, 4745–4752. [CrossRef]
170. Mithieux, G.; Guignot, L.; Bordet, J.-C.; Wiernsperger, N. Intrahepatic mechanisms underlying the effect of metformin in decreasing basal glucose production in rats fed a high-fat diet. *Diabetes* **2002**, *51*, 139–143. [CrossRef]
171. Forouzandeh, F.; Salazar, G.; Patrushev, N.; Xiong, S.; Hilenski, L.; Fei, B.; Alexander, R.W. Metformin Beyond Diabetes: Pleiotropic Benefits of Metformin in Attenuation of Atherosclerosis. *J. Am. Heart Assoc.* **2014**, *3*, e001202. [CrossRef] [PubMed]
172. Li, X.-N.; Song, J.; Zhang, L.; Lemaire, S.A.; Hou, X.; Zhang, C.; Coselli, J.S.; Chen, L.; Wang, X.L.; Zhang, Y.; et al. Activation of the AMPK-FOXO3 Pathway Reduces Fatty Acid-Induced Increase in Intracellular Reactive Oxygen Species by Upregulating Thioredoxin. *Diabetes* **2009**, *58*, 2246–2257. [CrossRef] [PubMed]
173. Hou, X.; Song, J.; Li, X.-N.; Zhang, L.; Wang, X.; Chen, L.; Shen, Y.H. Metformin reduces intracellular reactive oxygen species levels by upregulating expression of the antioxidant thioredoxin via the AMPK-FOXO3 pathway. *Biochem. Biophys. Res. Commun.* **2010**, *396*, 199–205. [CrossRef] [PubMed]
174. Cheng, G.; Lanza-Jacoby, S. Metformin decreases growth of pancreatic cancer cells by decreasing reactive oxygen species: Role of NOX4. *Biochem. Biophys. Res. Commun.* **2015**, *465*, 41–46. [CrossRef] [PubMed]
175. Sato, N.; Takasaka, N.; Yoshida, M.; Tsubouchi, K.; Minagawa, S.; Araya, J.; Saito, N.; Fujita, Y.; Kurita, Y.; Kobayashi, K.; et al. Metformin attenuates lung fibrosis development via NOX4 suppression. *Respir. Res.* **2016**, *17*, 1–12. [CrossRef] [PubMed]
176. Guang, W.; Wei, R.; Ke, J.; Yang, J.; Liu, Y.; Wang, X.; Wang, G.; Hong, T. Metformin attenuates fluctuating glucose-induced endothelial dysfunction through enhancing GTPCH1-mediated eNOS recoupling and inhibiting NADPH oxidase. *J. Diabetes Complicat.* **2016**, *30*, 1017–1024. [CrossRef]
177. Bułdak, Ł.; Łabuzek, K.; Bułdak, R.J.; Machnik, G.; Bołdys, A.; Basiak, M.; Bogusław, O. Metformin reduces the expression of NADPH oxidase and increases the expression of antioxidative enzymes in human monocytes/macrophages cultured in vitro. *Exp. Ther. Med.* **2016**, *13*, 794. [CrossRef]
178. Wang, S.; Zhang, M.; Liang, B.; Xu, J.; Xie, Z.; Liu, C.; Viollet, B.; Yan, D.; Zou, M.-H. AMPKα2 Deletion Causes Aberrant Expression and Activation of NAD(P)H Oxidase and Consequent Endothelial Dysfunction In Vivo. *Circ. Res.* **2010**, *106*, 1117–1128. [CrossRef]
179. Forbes, J.M.; Cooper, M.E.; Thallas, V.; Burns, W.; Thomas, M.C.; Brammar, G.C.; Lee, F.; Grant, S.L.; Burrell, L.M.; Jerums, G.; et al. Reduction of the Accumulation of Advanced Glycation End Products by ACE Inhibition in Experimental Diabetic Nephropathy. *Diabetes* **2002**, *51*, 3274–3282. [CrossRef]
180. Diaz-Morales, N.; Rovira-Llopis, S.; Bañuls, C.; Lopez-Domenech, S.; Escribano-Lopez, I.; Veses, S.; Jover, A.; Rocha, M.; Hernandez-Mijares, A.; Victor, V.M. Does Metformin Protect Diabetic Patients from Oxidative Stress and Leukocyte-Endothelium Interactions? *Antioxid. Redox Signal.* **2017**, *27*, 1439–1445. [CrossRef]

181. Javadipour, M.; Rezaei, M.; Keshtzar, E.; Khodayar, M.J. Metformin in contrast to berberine reversed arsenic-induced oxidative stress in mitochondria from rat pancreas probably via Sirt3-dependent pathway. *J. Biochem. Mol. Toxicol.* **2019**, *33*, e22368. [CrossRef] [PubMed]
182. Cheang, W.S.; Tian, X.Y.; Wong, W.T.; Lau, C.-W.; Lee, S.S.-T.; Chen, Z.-Y.; Yao, X.; Wang, N.; Huang, Y. Metformin Protects Endothelial Function in Diet-Induced Obese Mice by Inhibition of Endoplasmic Reticulum Stress Through 5′ Adenosine Monophosphate–Activated Protein Kinase–Peroxisome Proliferator–Activated Receptor δ Pathway. *Arter. Thromb. Vasc. Biol.* **2014**, *34*, 830–836. [CrossRef] [PubMed]
183. Kanuri, S.H.; Mehta, J.L. Role of Ox-LDL and LOX-1 in Atherogenesis. *Curr. Med. Chem.* **2019**, *26*, 1693–1700. [CrossRef]
184. Heinloth, A.; Heermeier, K.; Raff, U.; Wanner, C.; Galle, J. Stimulation of NADPH oxidase by oxidized low-density lipoprotein induces proliferation of human vascular endothelial cells. *J. Am. Soc. Nephrol.* **2000**, *11*, 1819–1825.
185. Shiu, S.W.M.; Wong, Y.; Tan, K.C.B. Effect of advanced glycation end products on lectin-like oxidized low density lipoprotein receptor-1 expression in endothelial cells. *J. Atheroscler. Thromb.* **2012**, *19*, 1083–1092. [CrossRef]
186. Ouslimani, N.; Mahrouf, M.; Peynet, J.; Bonnefont-Rousselot, D.; Cosson, C.; Legrand, A.; Beaudeux, J.-L. Metformin reduces endothelial cell expression of both the receptor for advanced glycation end products and lectin-like oxidized receptor 1. *Metabolism* **2007**, *56*, 308–313. [CrossRef]
187. Dong, Y.; Zhang, M.; Wang, S.; Liang, B.; Zhao, Z.; Liu, C.; Wu, M.; Choi, H.C.; Lyons, T.J.; Zou, M.-H. Activation of AMP-Activated Protein Kinase Inhibits Oxidized LDL-Triggered Endoplasmic Reticulum Stress In Vivo. *Diabetes* **2010**, *59*, 1386–1396. [CrossRef]
188. Xu, S.; Ogura, S.; Chen, J.; Little, P.J.; Moss, J.; Liu, P. LOX-1 in atherosclerosis: Biological functions and pharmacological modifiers. *Cell. Mol. Life Sci.* **2013**, *70*, 2859–2872. [CrossRef]
189. Valente, A.J.; Irimpen, A.M.; Siebenlist, U.; Chandrasekar, B. OxLDL induces endothelial dysfunction and death via TRAF3IP2: Inhibition by HDL3 and AMPK activators. *Free. Radic. Biol. Med.* **2014**, *70*, 117–128. [CrossRef]
190. Hung, C.-H.; Chan, S.-H.; Chu, P.-M.; Lin, H.-C.; Tsai, K.-L. Metformin regulates oxLDL-facilitated endothelial dysfunction by modulation of SIRT1 through repressing LOX-1-modulated oxidative signaling. *Oncotarget* **2016**, *7*, 10773–10787. [CrossRef]
191. Roxo, D.F.; Arcaro, C.A.; Gutierres, V.O.; Costa, M.C.; Oliveira, J.O.; Lima, T.F.O.; Assis, R.P.; Brunetti, I.L.; Baviera, A.M. Curcumin combined with metformin decreases glycemia and dyslipidemia, and increases paraoxonase activity in diabetic rats. *Diabetol. Metab. Syndr.* **2019**, *11*, 1–8. [CrossRef] [PubMed]
192. Mirmiranpoor, H.; Mousavizadeh, M.; Noshad, S.; Ghavami, M.; Ebadi, M.; Ghasemiesfe, M.; Nakhjavani, M.; Esteghamati, A. Comparative effects of pioglitazone and metformin on oxidative stress markers in newly diagnosed type 2 diabetes patients: A randomized clinical trial. *J. Diabetes its Complicat.* **2013**, *27*, 501–507. [CrossRef] [PubMed]
193. Salt, I.P.; Palmer, T.M. Exploiting the anti-inflammatory effects of AMP-activated protein kinase activation. *Expert Opin. Investig. Drugs* **2012**, *21*, 1155–1167. [CrossRef] [PubMed]
194. Mamputu, M.; Wiernsperger, N.; Renier, G. Metformin inhibits monocyte adhesion to endothelial cells and foam cell formation. *Br. J. Diabetes Vasc. Dis.* **2003**, *3*, 302–310. [CrossRef]
195. Vasamsetti, S.B.; Karnewar, S.; Kanugula, A.K.; Thatipalli, A.R.; Kumar, J.M.; Kotamraju, S. Metformin Inhibits Monocyte-to-Macrophage Differentiation via AMPK-Mediated Inhibition of STAT3 Activation: Potential Role in Atherosclerosis. *Diabetes* **2015**, *64*, 2028–2041. [CrossRef]
196. Hattori, Y.; Suzuki, K.; Hattori, S.; Kasai, K. Metformin Inhibits Cytokine-Induced Nuclear Factor κB Activation Via AMP-Activated Protein Kinase Activation in Vascular Endothelial Cells. *Hypertension* **2006**, *47*, 1183–1188. [CrossRef]
197. Jagtap, P.; Szabó, C. Poly(ADP-ribose) polymerase and the therapeutic effects of its inhibitors. *Nat. Rev. Drug Discov.* **2005**, *4*, 421–440. [CrossRef]
198. Gongol, B.; Marin, T.; Peng, I.-C.; Woo, B.; Martin, M.; King, S.; Sun, W.; Johnson, D.A.; Chien, S.; Shyy, J.Y.-J. AMPKα2 exerts its anti-inflammatory effects through PARP-1 and Bcl-6. *Proc. Natl. Acad. Sci. USA* **2013**, *110*, 3161–3166. [CrossRef]
199. Shang, F.; Zhang, J.; Li, Z.; Zhang, J.; Yin, Y.; Wang, Y.; Marin, T.L.; Gongol, B.; Xiao, H.; Zhang, Y.-Y.; et al. Cardiovascular Protective Effect of Metformin and Telmisartan: Reduction of PARP1 Activity via the AMPK-PARP1 Cascade. *PLoS ONE* **2016**, *11*, e0151845. [CrossRef]
200. Zheng, Z.; Chen, H.; Li, J.; Li, T.; Zheng, B.; Zheng, Y.; Jin, H.; He, Y.; Gu, Q.; Xu, X. Sirtuin 1-Mediated Cellular Metabolic Memory of High Glucose Via the LKB1/AMPK/ROS Pathway and Therapeutic Effects of Metformin. *Diabetes* **2011**, *61*, 217–228. [CrossRef]
201. Nieminen, A.-L. Apoptosis and necrosis in health and disease: Role of mitochondria. *Adv. Clin. Chem.* **2003**, *224*, 29–55. [CrossRef]
202. Grimm, S.; Brdiczka, D. The permeability transition pore in cell death. *Apoptosis* **2007**, *12*, 841–855. [CrossRef] [PubMed]
203. Detaille, D.; Guigas, B.; Chauvin, C.; Batandier, C.; Fontaine, E.; Wiernsperger, N.; Leverve, X. Metformin Prevents High-Glucose-Induced Endothelial Cell Death Through a Mitochondrial Permeability Transition-Dependent Process. *Diabetes* **2005**, *54*, 2179–2187. [CrossRef] [PubMed]
204. Bhamra, G.S.; Hausenloy, D.J.; Davidson, S.M.; Carr, R.D.; Paiva, M.; Wynne, A.M.; Mocanu, M.M.; Yellon, D.M. Metformin protects the ischemic heart by the Akt-mediated inhibition of mitochondrial permeability transition pore opening. *Basic Res. Cardiol.* **2008**, *103*, 274–284. [CrossRef]

205. Bai, B.; Liang, Y.; Xu, C.; Lee, M.Y.; Xu, A.; Wu, N.; Vanhoutte, P.M.; Wang, Y. Cyclin-Dependent Kinase 5–Mediated Hyperphosphorylation of Sirtuin-1 Contributes to the Development of Endothelial Senescence and Atherosclerosis. *Circulation* **2012**, *126*, 729–740. [CrossRef]
206. Arunachalam, G.; Lakshmanan, A.P.; Samuel, S.M.; Triggle, C.; Ding, H. Molecular Interplay between microRNA-34a and Sirtuin1 in Hyperglycemia-Mediated Impaired Angiogenesis in Endothelial Cells: Effects of Metformin. *J. Pharmacol. Exp. Ther.* **2016**, *356*, 314–323. [CrossRef]
207. Qin, B.; Yang, H.; Xiao, B. Role of microRNAs in endothelial inflammation and senescence. *Mol. Biol. Rep.* **2011**, *39*, 4509–4518. [CrossRef]
208. Ito, T.; Yagi, S.; Yamakuchi, M. MicroRNA-34a regulation of endothelial senescence. *Biochem. Biophys. Res. Commun.* **2010**, *398*, 735–740. [CrossRef]
209. Han, X.; Tai, H.; Wang, X.; Wang, Z.; Zhou, J.; Wei, X.; Ding, Y.; Gong, H.; Mo, C.; Zhang, J.; et al. AMPK activation protects cells from oxidative stress-induced senescence via autophagic flux restoration and intracellular NAD + elevation. *Aging Cell* **2016**, *15*, 416–427. [CrossRef]
210. Karnewar, S.; Neeli, P.K.; Panuganti, D.; Kotagiri, S.; Mallappa, S.; Jain, N.; Jerald, M.K.; Kotamraju, S. Metformin regulates mitochondrial biogenesis and senescence through AMPK mediated H3K79 methylation: Relevance in age-associated vascular dysfunction. *Mol. Basis Dis.* **2018**, *1864*, 1115–1128. [CrossRef]
211. Niu, C.; Chen, Z.; Kim, K.T.; Sun, J.; Xue, M.; Chen, G.; Li, S.; Shen, Y.; Zhu, Z.; Wang, X.; et al. Metformin alleviates hyperglycemia-induced endothelial impairment by downregulating autophagy via the Hedgehog pathway. *Autophagy* **2019**, *15*, 843–870. [CrossRef] [PubMed]
212. Westermann, B. Mitochondrial fusion and fission in cell life and death. *Nat. Rev. Mol. Cell Biol.* **2010**, *11*, 872–884. [CrossRef] [PubMed]
213. Yu, T.; Robotham, J.L.; Yoon, Y. Increased production of reactive oxygen species in hyperglycemic conditions requires dynamic change of mitochondrial morphology. *Proc. Natl. Acad. Sci. USA* **2006**, *103*, 2653–2658. [CrossRef] [PubMed]
214. Shenouda, S.M.; Widlansky, M.E.; Chen, K.; Xu, G.; Holbrook, M.; Tabit, C.E.; Hamburg, N.M.; Frame, A.A.; Caiano, T.L.; Kluge, M.A.; et al. Altered Mitochondrial Dynamics Contributes to Endothelial Dysfunction in Diabetes Mellitus. *Circulation* **2011**, *124*, 444–453. [CrossRef] [PubMed]
215. Lugus, J.J.; Ngoh, G.A.; Bachschmid, M.M.; Walsh, K. Mitofusins are required for angiogenic function and modulate different signaling pathways in cultured endothelial cells. *J. Mol. Cell. Cardiol.* **2011**, *51*, 885–893. [CrossRef] [PubMed]
216. Wang, W.; Wang, Y.; Long, J.; Wang, J.; Haudek, S.B.; Overbeek, P.; Chang, B.H.; Schumacker, P.T.; Danesh, F.R. Mitochondrial Fission Triggered by Hyperglycemia Is Mediated by ROCK1 Activation in Podocytes and Endothelial Cells. *Cell Metab.* **2012**, *15*, 186–200. [CrossRef] [PubMed]
217. Yoon, Y.; Galloway, C.A.; Jhun, B.S.; Yu, T. Mitochondrial Dynamics in Diabetes. *Antioxid. Redox Signal.* **2011**, *14*, 439–457. [CrossRef] [PubMed]
218. Diaz-Morales, N.; Rovira-Llopis, S.; Bañuls, C.; Escribano-Lopez, I.; De Marañon, A.M.; Lopez-Domenech, S.; Orden, S.; Torres, I.R.; Alvarez, A.; Veses, S.; et al. Are Mitochondrial Fusion and Fission Impaired in Leukocytes of Type 2 Diabetic Patients? *Antioxid. Redox Signal.* **2016**, *25*, 108–115. [CrossRef] [PubMed]
219. Wang, Q.; Zhang, M.; Torres, G.; Wu, S.; Ouyang, C.; Xie, Z.; Zou, M.-H. Metformin Suppresses Diabetes-Accelerated Atherosclerosis via the Inhibition of Drp1-Mediated Mitochondrial Fission. *Diabetes* **2017**, *66*, 193–205. [CrossRef] [PubMed]
220. Kluge, M.A.; Fetterman, J.L.; Vita, J.A. Mitochondria and Endothelial Function. *Circ. Res.* **2013**, *112*, 1171–1188. [CrossRef]
221. Li, C.; Reif, M.M.; Craige, S.M.; Kant, S.; Keaney, J.F. Endothelial AMPK activation induces mitochondrial biogenesis and stress adaptation via eNOS-dependent mTORC1 signaling. *Nitric Oxide* **2016**, *55*, 45–53. [CrossRef]
222. Sun, Y.; Li, J.; Xiao, N.; Wang, M.; Kou, J.; Qi, L.; Huang, F.; Liu, B.; Liu, K. Pharmacological activation of AMPK ameliorates perivascular adipose/endothelial dysfunction in a manner interdependent on AMPK and SIRT1. *Pharmacol. Res.* **2014**, *89*, 19–28. [CrossRef]
223. Nieuwdorp, M.; Meuwese, M.C.; Vink, H.; Hoekstra, J.B.L.; Kastelein, J.J.P.; Stroes, E.S.G. The endothelial glycocalyx: A potential barrier between health and vascular disease. *Curr. Opin. Lipidol.* **2005**, *16*, 507–511. [CrossRef] [PubMed]
224. Van Haare, J.; Kooi, M.; Van Teeffelen, J.W.G.E.; Vink, H.; Slenter, J.; Cobelens, H.; Strijkers, G.J.; Koehn, D.; Post, M.J.; Van Bilsen, M. Metformin and sulodexide restore cardiac microvascular perfusion capacity in diet-induced obese rats. *Cardiovasc. Diabetol.* **2017**, *16*, 1–13. [CrossRef] [PubMed]
225. Hao, B.; Xiao, Y.; Song, F.; Long, X.; Huang, J.; Tian, M.; Deng, S.; Wu, Q. Metformin-induced activation of AMPK inhibits the proliferation and migration of human aortic smooth muscle cells through upregulation of p53 and IFI16. *Int. J. Mol. Med.* **2017**, *41*, 1365–1376. [CrossRef] [PubMed]
226. Chen, P.-Y.; Qin, L.; Baeyens, N.; Li, G.; Afolabi, T.; Budatha, M.; Tellides, G.; Schwartz, M.A.; Simons, M. Endothelial-to-mesenchymal transition drives atherosclerosis progression. *J. Clin. Investig.* **2015**, *125*, 4514–4528. [CrossRef] [PubMed]
227. Yu, B.; Wu, Y.; Li, Z. KLF4/Ch25h axis activated by metformin suppresses EndoMT in human umbilical vein endothelial cells. *Biochem. Biophys. Res. Commun.* **2020**, *522*, 838–844. [CrossRef]
228. Kuan, I.H.S.; Savage, R.L.; Duffull, S.B.; Walker, R.J.; Wright, D.F. The Association between Metformin Therapy and Lactic Acidosis. *Drug Saf.* **2019**, *42*, 1449–1469. [CrossRef]
229. He, L.; Wondisford, F.E. Metformin Action: Concentrations Matter. *Cell Metab.* **2015**, *21*, 159–162. [CrossRef]

230. Caturano, A.; Galiero, R.; Pafundi, P.C. Metformin for Type 2 Diabetes. *JAMA* **2019**, *322*, 1312. [CrossRef]
231. Rhee, M.K.; Herrick, K.A.; Ziemer, D.C.; Vaccarino, V.; Weintraub, W.S.; Narayan, K.V.; Kolm, P.; Twombly, J.G.; Phillips, L.S. Many Americans Have Pre-Diabetes and Should Be Considered for Metformin Therapy. *Diabetes Care* **2009**, *33*, 49–54. [CrossRef] [PubMed]
232. Sardu, C.; Paolisso, P.; Sacra, C.; Mauro, C.; Minicucci, F.; Portoghese, M.; Rizzo, M.R.; Barbieri, M.; Sasso, F.C.; D'Onofrio, N.; et al. Effects of Metformin Therapy on Coronary Endothelial Dysfunction in Patients With Prediabetes With Stable Angina and Nonobstructive Coronary Artery Stenosis: The CODYCE Multicenter Prospective Study. *Diabetes Care* **2019**, *42*, 1946–1955. [CrossRef] [PubMed]
233. Della Corte, C.M.; Ciaramella, V.; Di Mauro, C.; Castellone, M.D.; Papaccio, F.; Fasano, M.; Sasso, F.C.; Martinelli, E.; Troiani, T.; De Vita, F.; et al. Metformin increases antitumor activity of MEK inhibitors through GLI1 downregulation in LKB1 positive human NSCLC cancer cells. *Oncotarget* **2015**, *7*, 4265–4278. [CrossRef] [PubMed]
234. Salvatore, T.; Pafundi, P.C.; Galiero, R.; Gjeloshi, K.; Masini, F.; Acierno, C.; Di Martino, A.; Albanese, G.; Alfano, M.; Rinaldi, L.; et al. Metformin: A Potential Therapeutic Tool for Rheumatologists. *Pharmaceuticals* **2020**, *13*, 234. [CrossRef] [PubMed]

Review

Endothelial to Mesenchymal Transition in Pulmonary Vascular Diseases

Eunsik Yun [1], Yunjin Kook [1], Kyung Hyun Yoo [1,2], Keun Il Kim [1,2], Myeong-Sok Lee [1,2], Jongmin Kim [1,2,*] and Aram Lee [1,*]

1. Division of Biological Sciences, Sookmyung Women's University, Seoul 04310, Korea; yes951212@naver.com (E.Y.); jenny1267@naver.com (Y.K.); khryu@sookmyung.ac.kr (K.H.Y.); kikim@sookmyung.ac.kr (K.I.K.); mslee@sookmyung.ac.kr (M.-S.L.)
2. Research Institute for Women's Health, Sookmyung Women's University, Seoul 04310, Korea
* Correspondence: jkim@sookmyung.ac.kr (J.K.); aram0918@sookmyung.ac.kr (A.L.); Tel.: +82-2-710-9553 (J.K. & A.L.); Fax: +82-2-2077-7322 (J.K. & A.L.)

Received: 24 November 2020; Accepted: 17 December 2020; Published: 21 December 2020

Abstract: Lung diseases, such as pulmonary hypertension and pulmonary fibrosis, are life-threatening diseases and have common features of vascular remodeling. During progression, extracellular matrix protein deposition and dysregulation of proteolytic enzymes occurs, which results in vascular stiffness and dysfunction. Although vasodilators or anti-fibrotic therapy have been mainly used as therapy owing to these characteristics, their effectiveness does not meet expectations. Therefore, a better understanding of the etiology and new therapeutic approaches are needed. Endothelial cells (ECs) line the inner walls of blood vessels and maintain vascular homeostasis by protecting vascular cells from pathological stimuli. Chronic stimulation of ECs by various factors, including pro-inflammatory cytokines and hypoxia, leads to ECs undergoing an imbalance of endothelial homeostasis, which results in endothelial dysfunction and is closely associated with vascular diseases. Emerging studies suggest that endothelial to mesenchymal transition (EndMT) contributes to endothelial dysfunction and plays a key role in the pathogenesis of vascular diseases. EndMT is a process by which ECs lose their markers and show mesenchymal-like morphological changes, and gain mesenchymal cell markers. Despite the efforts to elucidate these molecular mechanisms, the role of EndMT in the pathogenesis of lung disease still requires further investigation. Here, we review the importance of EndMT in the pathogenesis of pulmonary vascular diseases and discuss various signaling pathways and mediators involved in the EndMT process. Furthermore, we will provide insight into the therapeutic potential of targeting EndMT.

Keywords: lung disease; endothelial to mesenchymal transition; pulmonary hypertension; pulmonary fibrosis

1. Introduction

Endothelial cells (ECs), a monolayer composed of the inner cellular lining of the vascular lumen, play an important role in various physiological processes to maintain vascular homeostasis [1–3]. These cells are involved in the regulation of vascular tone, permeability, and inflammatory responses [4]. However, endothelial injury by stimuli, such as hypoxia, pro-inflammatory cytokines and abnormal mechanical forces, can induce endothelial-to-mesenchymal transition (EndMT), resulting in endothelial dysfunction and destruction of homeostasis [2,5]. EndMT is the process by which ECs lose their cellular features and acquire mesenchymal characteristics [6]. EndMT-derived cells gain migration potential by losing endothelial markers, such as cluster of differentiation 31 (CD31) and vascular endothelial cadherin (VE-cadherin), which are involved in cell-to-cell contact [7,8]. Concomitantly, the expressions of mesenchymal markers, such as fibronectin, alpha-smooth muscle actin (SMAα), smooth muscle protein 22 alpha, vimentin, and neural cadherin (N-cadherin), are upregulated [7,8]. The morphology of ECs undergoing EndMT changes from a cobblestone monolayer to an elongated phenotype [9]. This phenomenon mainly occurs during embryonic cardiac development, but is also involved in various lung diseases, such as pulmonary arterial hypertension (PAH) and pulmonary fibrosis (PF) (Figure 1) [7,10–13].

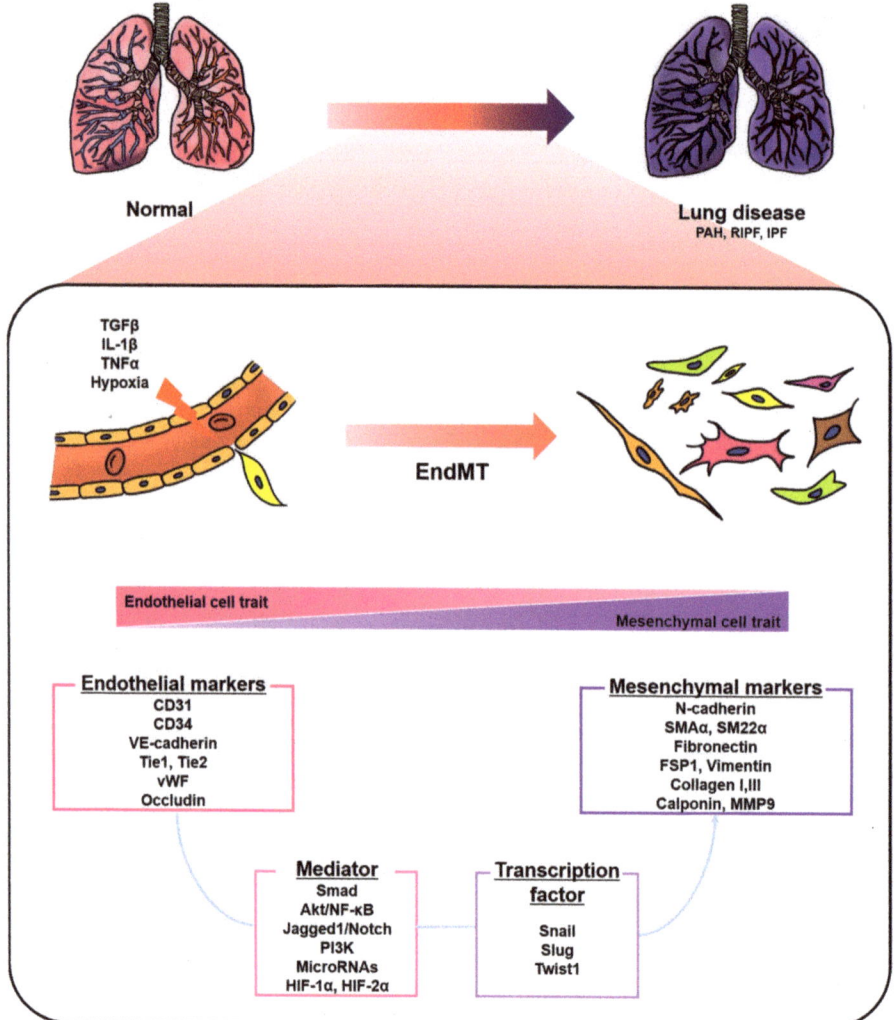

Figure 1. A schematic representation of endothelial-to-mesenchymal transition (EndMT) involved in lung diseases. Endothelial cells stimulated by transforming growth factor-β (TGFβ), interleukin 1 beta (IL-1β), tumor necrosis factor alpha (TNFα), and hypoxia undergo EndMT. EndMT is characterized by phenotypic change from a cobblestone into an elongated shape, loss of endothelial markers, and the acquisition of mesenchymal markers. EndMT contributes to the pathogenesis of lung diseases, including pulmonary arterial hypertension (PAH), radiation-induced pulmonary fibrosis (RIPF), and idiopathic pulmonary fibrosis (IPF). Various mediators and transcription factors are identified in this process.

Pulmonary hypertension (PH) is categorized into five groups: PAH, PH due to left heart disease, PH due to lung diseases and/or hypoxia, PH due to pulmonary arterial obstructions, and PH with unclear and/or multifactorial mechanisms [14–16]. PAH has been defined as pulmonary artery pressure (PAP) ≥ 25 mmHg at rest and occurs as a result of multiple causes, such as heritable factors (mainly bone morphogenic protein receptor-2 (BMPR2) mutations), drugs and toxins, as well as association with other diseases; however, PAH without known causes is known as idiopathic PAH (IPAH) [14,17]. Vascular remodeling in PAH is characterized by the aberrant proliferation of pulmonary arterial ECs (PAECs) and smooth muscle cells (SMCs), which form occlusive neointima and vascular structural changes [18–20]. These progressive changes cause excess vasoconstriction and right ventricle hypertrophy and, ultimately, death [18–20]. Endothelial dysfunction is a key player in the pathogenesis of PAH [21]. Growing evidence suggests that EndMT potentially contributes to endothelial dysfunction and the vascular remodeling of PAH [7,11,22,23]. Indeed, many studies have demonstrated that various signaling pathways and mediators, including transforming growth factor beta (TGFβ), nuclear factor kappa B (NF-κB), Notch, and microRNA, are involved in the EndMT of PAH [24,25]. It has been reported that the endothelial-specific loss of BMPR2, known as the principal mutation factor of heritable PAH, induces EndMT in vitro and in vivo [7,11,23]. In addition, exposure to hypoxia or chronic stimulation with proinflammatory cytokines or TGFβ also induce EndMT in vitro and in vivo [26–30]. However, the contribution of EndMT to disease progression is not fully understood [2]. Current therapies for PAH, such as phosphodiesterase-5 inhibitors, prostacyclin analogues, and endothelin receptor antagonists, can help relieve symptoms and slow progression, but there is no effective treatment [21,31]. Thus, targeting EndMT is emerging as a novel therapeutic approach by alleviating vascular remodeling and the PAH phenotype in vitro and in vivo [7,29,32–36].

Idiopathic PF (IPF) is chronic, progressive, and the most common interstitial lung disease without a definite etiology [37,38]. Various cell types, such as epithelial cells, pneumocytes, ECs, pericytes, fibrocytes, resident fibroblasts, and mesenchymal cells, are associated with the pathogenesis of IPF [25]. The injured epithelial cells, through aging, genetic susceptibility and repetitive microinjury, release fibrogenic factors and cytokines, resulting in the recruitment of contractile myofibroblasts, which are key cellular mediators of fibrosis [38]. Recruited myofibroblasts undergoing activation and proliferation induce extracellular matrix expansion, which consequently results in aberrant vascular remodeling in the lung [38]. The myofibroblasts are derived not only from the proliferation of resident mesenchymal cells, circulating fibrocytes, lung interstitium pericytes, epithelial–mesenchymal transition, but also EndMT. [38–40]. Many studies have demonstrated that EndMT occurs in the lung tissue of IPF patients and animal models, suggesting EndMT may play an important role in pathological processes in PF [25,41,42]. In addition, emerging evidence indicates that inhibiting EndMT can also be a therapeutic strategy in PF in vivo [41,43–45].

This review highlights the role of EndMT associated with pulmonary diseases, such as PAH and PF. Moreover, this review discusses molecular mechanisms, epigenetic modulation, and recent clinical relevance in lung diseases.

2. EndMT in Pulmonary Hypertension

PH is characterized by the muscularization of arterioles, medial thickening, plexiform region formation, intimal fibrosis, and the hyperproliferation of ECs and SMCs [15,16,46,47]. Most studies have identified EndMT by analyzing the co-expression of endothelial markers and mesenchymal markers in the lung tissue of patients and experimental PH animal models. EndMT has been observed in pathological lesions in the lungs of PH patients [7,30,32,48,49]. Endothelial (CD31, CD34, and VE-cadherin) and mesenchymal marker (SMAα) double-positive cells were observed in intimal and plexiform lesions in the lung tissue of PAH patients [7]. Another group also demonstrated that neointimal and plexiform lesions in the lung tissue of human PAH patients contain endothelial markers, CD31 or von Willebrand factor (vWF), and SMAα co-expressing cells [48]. Isobe et al. reported that the CD44 spliced variant form (CD44v) results from EndMT, and its positive cells also expressed vWF and SMAα in neointimal lesions of IPAH patients [32]. The 4 ± 1% of pulmonary arterioles in systemic sclerosis (SSc)-PAH patients showed vWF/SMAα co-localization [30]. CD31 and SMAα co-expressing cells were detected in endarterectomized tissues from patients with chronic thromboembolic pulmonary hypertension (CTEPH) [49].

In addition to performing the double staining of endothelial and mesenchymal markers, ECs isolated from the lung have also been used for studying EndMT [49,50]. Endothelial-like cells isolated from the vascular tissue of patients with CTEPH underwent disruption of the endothelial monolayer and abnormal growth even after sorting with CD31 [49]. In addition, conditioned media from myofibroblast-like cells isolated from CTEPH patients induced phenotypic changes and mesenchymal marker expression in pulmonary microvascular ECs (PMVECs) [49]. Pulmonary vascular ECs (PVECs) isolated from patients with IPAH exhibited molecular characteristics of EndMT and a spindle-shaped morphology, which was similar to that of normal PVECs treated with TGFβ1, a well-known factor of EndMT [50]. Pulmonary arteries isolated from PAH patients showed increased mRNA levels of mesenchymal markers and EndMT-related factors, which also supports EndMT [7].

Animal models have also been used to demonstrate EndMT. Monocrotaline (MCT) injection causes endothelial injury and pulmonary vascular remodeling, and is commonly used to induce severe PH [50,51]. Several groups observed the reduction of endothelial markers and the induction of mesenchymal markers, as well as the co-staining of SMAα and endothelial marker (CD31 or CD34), in the lung tissue of MCT-induced PH rats [7,28,29,50,52]. Zhang et al. found that changes in endothelial and mesenchymal cell marker expressions occurred in a time-dependent manner during MCT-induced PAH development [51]. Chronic hypoxia also contributes to the vascular remodeling of small pulmonary arteries [27,53]. With this, it has been demonstrated that three weeks of hypoxia induces EndMT in the pulmonary arteries of rats and mice [26,53]. EndMT was further identified within the intimal layer of small pulmonary arteries, but not in large arteries, in chronic hypoxia-induced PH rats [27]. The combination of SU5416, a vascular endothelial growth factor receptor antagonist, and a chronic hypoxia model (SuHx) has been used for severe PH owing to the similarity of pathological lesions to plexiform lesions of human PAH [53]. In the lung of the SuHx model that had over 80 mmHg of right ventricular systolic pressure (RVSP), transitions of vWF+ vimentin− ECs to vWF− vimentin high mesenchymal-like cells were observed in pulmonary vascular lesions [7]. Tie2+ vimentin+ and Tie2+ SMAα+ cells were also found in occlusive lesions [7]. In addition, 6 ± 1% of pulmonary vessels had vWF/SMAα double-positive ECs, which contrasts with normal tissues having only 1% transitional EndMT cells in SuHx mice [30].

In general, endothelial and mesenchymal marker double-positive cells are considered EndMT-induced cells. However, this approach has the limitation of not being able to distinguish complete EndMT (cEndMT), where there are lost endothelial markers, and partial EndMT (pEndMT) cells. To overcome this problem, several studies have used endothelial-specific fluorescence transgenic animals [48,54]. Qiao et al. established VE-cadherin Cre or Tie2 Cre-mTomato/mGFP lineage-tracing mice [48]. Histological analysis identified SMAα-expressing neointima in an experimental PH animal model derived from the endothelium in VE-cadherin Cre or Tie2 Cre-mTomato/mGFP lineage-tracing mice [48]. Furthermore, cEndMT cells isolated from SuHx-induced Cdh5-Cre/CAG-GFP double-transgenic mice showed a spindle-like morphology and were characterized by mesenchymal-like functions, such as high proliferation and migration ability [54]. Additionally, conditioned media from cEndMT had a paracrine effect on the proliferation and migration of non-endothelial mesenchymal cells, suggesting that EndMT contributes directly and indirectly to the vascular remodeling of PAH [54].

3. EndMT in Pulmonary Fibrosis

IPF characterizes matrix deposition and fibrotic tissue remodeling, and it has been demonstrated that fibroblasts are involved in pathogenesis; thus, efforts to identify the origin of fibroblasts have been made. [42,55]. In the lung tissue of radiation-induced pulmonary fibrosis (RIPF) patients and radiation-exposed mouse models, the co-localization of CD31 and SMAα was significantly elevated compared to that of the control group, indicating EndMT [41]. The same group also reported endothelial heat shock protein beta 1 (HSPB1)-dependent EndMT in the PF of lung cancer [45]. The bleomycin-induced PF in animal models is the most commonly used model to study human IPF by causing damage to epithelial cells and alveolar inflammation [56,57]. Another group reported significant alterations of EC markers in the lungs of bleomycin-treated endothelial-specific autophagy-related 7 (ATG7) knockout mice compared to bleomycin-treated WT mice [58]. Hashimoto et al. established a Tie2-Cre/CAG-CAT-LacZ double transgenic mice model to track endothelial-derived fibroblasts in bleomycin-induced PF [42]. The 16.2% of lung fibroblasts isolated from bleomycin-treated mice were X-gal-staining-positive and 14.8% of X-gal-positive cells were SMAα- and Collagen I-double positive (myofiboblast), while the other 85.2% were SMAα-negative and Collagen I-positive, suggesting that a significant number of fibroblasts are EC-derived [42]. Suzuki et al. demonstrated that PVECs isolated from lipopolysaccharide (LPS)-induced mouse lungs undergo EndMT using the double staining of CD31 and SMAα or S100A4 [59]. Flow cytometry analysis showed that the number of SMAα + PVECs and S100A4 + PVECs increased, while the total number of PVECs decreased [59].

Taken together, EndMT may play a key role in the pathogenesis of lung diseases. Many studies describe EndMT based on the evidence of co-expression of EC markers and mesenchymal markers in the lung tissue of animal disease models or human patients, which has a primary limitation because EndMT is a switching process; thus, the underlying molecular mechanisms are not yet fully understood. The methods to clarify partial and complete EndMT processes have been improved using endothelial-specific fluorescence transgenic mice; however, further investigation with human samples is needed. Thus, the clinical relevance of EndMT should be thoroughly assessed.

4. Key Signaling Pathways and Mediators during EndMT in Lung Diseases

The understanding of the key molecular mechanisms and mediators during EndMT is an important step toward finding how to develop EndMT inhibitors that can be applied to vascular disease therapy. Inflammatory stress contributes to endothelial dysfunction in the pathogenesis of lung diseases [2,60]. The combination of proinflammatory cytokines, including interleukin 1 beta (IL-1β), tumor necrosis factor alpha (TNFα), and TGFβ, is a powerful EndMT inducer [28–30,32,34]. Good et al. found that the combination of IL-1β, TNFα, and TGFβ1 for six days induces EndMT in PAECs (I-EndMT) [30]. I-EndMT PAECs and lung fibroblasts isolated from patients with SSc-PAH showed elevated levels of cytokines, such as IL-6, IL-8, IL-13, and TNFα [30]. In addition, a cocktail of IL-1β, TNFα, and TGFβ1 induces CD44v and EndMT in PAECs [32]. CD44v-positive EndMT-induced PAECs showed upregulations of proinflammatory cytokines and chemokines, such as TNFα, IL-1β, IL-6, and CXCL12 [32]. The combination of TGFβ1 and IL-1β induces EndMT through Smad2/3 and ERK1/2 phosphorylation, which means that both Smad and non-Smad signaling are involved in this process [28]. Moreover, it has been demonstrated that cytokine levels, such as TGFβ1, IL-1β, IL-6, and TNFα, are increased in the lung tissue of MCT-induced PH rats [28,29]. Therefore, inflammatory cytokines induce EndMT and also induce cells to exhibit proinflammatory characteristics.

Among the various signaling pathways involved in EndMT, TGFβ signaling is known to be a major regulator of EndMT [61]. TGFβ upregulates EndMT-associated transcription factors, such as Snail, Slug, and Twist1, which leads to the upregulation of mesenchymal markers [2,61]. Although TGFβ induces EndMT mainly through the Smad-dependent canonical signaling pathway, Smad-independent non-canonical TGFβ signaling is also involved [61]. Non-canonical TGFβ signaling includes phosphatidylinositol 3-kinase (PI3K), mitogen-activated protein kinase (MAPK), and extracellular signal-regulated kinase (ERK) [62]. One group reported that *ATG7* knockdown promotes EndMT in PAECs [58]. During this process, mRNA and protein levels of TGFβ1 and its receptors, TGFβR1 and TGFβR2, are increased, and this elevates the phosphorylation of Smad2/3, leading to the upregulation of Slug and pro-fibrotic genes, connective tissue growth factor, and Collagen I [58]. Sabbineni et al. reported that the endothelial loss of *Akt1* increased TGFβ2 expression, which in turn elevated the phosphorylation of p38-MAPK and Smad2/3, resulting in EndMT [53]. These results indicate that both canonical and non-canonical signaling pathways are involved in TGFβ2-induced EndMT [53]. In addition, the inhibitor of beta catenin (β-catenin) suppressed the expression of mesenchymal markers and ameliorated vascular thickening in a SuHx PH model, suggesting that Akt1-mediated β-catenin signaling is a novel pathway for inducing EndMT [53]. Caveolin-1 plays an important role in the internalization of the TGFβ receptor [63]. The expression levels of Snail, Slug, SMAα and Collagen I were higher in pulmonary ECs isolated from *Caveolin-1* knockout mice than in WT mice [63]. Moreover, TGFβ1 treatment further increased SMAα and Collagen I expression in *Caveolin-1* knockout cells [63]. The phosphorylated Twist1 and vimentin were elevated in the lungs of PAH patients and MCT-induced PH rats, and TGFβ treatment increases Twist1 expression [7,64]. Mammoto et al. reported that the overexpression of Twist1 induces EndMT through TGFβR2-Smad2 signaling, and the phosphorylation of Twist1 Ser42 is required during hypoxia-induced EndMT [65].

BMPR2 is a member of the TGF receptor superfamily and is highly expressed on the pulmonary vascular endothelium [36,66–68]. BMPR2 mutations and low expression levels are closely associated with PAH [36,66–69]. Roughly 70–80% of familial PAH and 10–20% of sporadic cases of IPAH patients have BMPR2 mutations [36,66]. Several studies have demonstrated that BMPR2 expression is associated with EndMT in the lung vasculature of PAH animal models and patients [7,23,50,54]. Dysfunction of BMPR2 signaling induces EndMT through high mobility group AT-hook 1 (HMGA1) upregulation [23]. The knockdown of *HMGA1* or *Slug* prevented *BMPR2* silencing-induced SMAα expression [23]. In addition, pulmonary ECs isolated from endothelial-specific *BMPR2* knockout mice also showed EndMT with elevated HMGA1 and its target, Slug, expression [23]. *BMPR2*-deficient (BMPR2Δ140Ex1/+) rats exhibit spontaneous pulmonary vascular remodeling [7]. A recent study reported that *BMPR2* knockdown leads to the switch of cell junction protein from VE-cadherin to N-cadherin and increases Slug and Twist [11]. During this process, the heteromerization of BMP and TGFβ receptors was facilitated, leading to increased lateral TGFβ signaling responses [11]. Reynolds et al. reported that adenoviral BMPR2 delivery attenuates vascular remodeling in PAH animal models and treatment with BMPR2 ligands ameliorates TGFβ1-induced EndMT in vitro [36]. Accumulating evidence indicates that altered BMPR2 signaling is closely related to EndMT, and restoration of BMPR2 signaling can be a strategy for inhibiting EndMT.

Hypoxia contributes to EndMT in pulmonary ECs [26,27,70]. The hypoxia-inducible factor (HIF) family consists of HIF-1, HIF-2, and HIF-3, which are key regulators in maintaining oxygen homeostasis [50,71]. It has been reported that PVECs from PAH patients show EndMT with higher HIF-2α levels compared to control [50]. Although HIF-2α is degraded by prolyl hydroxylase domain protein 2 (PHD2) under normoxia, PHD2 was decreased in PVECs from IPAH, leading to the upregulation of Snail and Slug [50,71]. This suggests that PHD2 and HIF-2α are closely associated with EndMT. *PHD2* endothelial-specific knockout mice showed a severe PH phenotype, even under normoxia, while endothelial-specific *HIF-2α* knockout mice prevented developing hypoxia-induced PH [50]. Hypoxia upregulates HIF-1α, which acts as an upstream regulator of Twist1 by binding to its promoter and leads to EndMT [27]. Choi et al. described that HIF-1α is elevated in EndMT-derived cells in the lung tissue of radiation-induced fibrosis mice and human RIPF patients [41]. In addition, this study demonstrated that HIF-1α mediates TGFβ receptor/Smad signaling in radiation-induced EndMT [41]. These studies reflect the critical role of the HIF family in hypoxia-induced EndMT.

Notch is a family of transmembrane receptors and consists of Notch 1, 2, 3, and 4 [72,73]. Notch is activated by ligands, Jagged 1, Jagged 2, and Delta-like 1, 3, and 4, and produces the intracellular domain of Notch by proteolytic processing [72,73]. Notch signaling pathways have been associated with epithelial-to-mesenchymal transition and EndMT [72–74]. Noseda et al. demonstrated that activated Notch (Notch4IC, Notch1IC) and Jagged 1 lead to EndMT in human microvascular ECs [72]. Zhang et al. identified that Galectin-3 (Gal-3) is increased in the lung vasculature of patients with PAH and in the experimental animal model [33]. In vitro, Gal-3 treatment activated Jagged 1/Notch1 pathway, leading to EndMT [33]. The activated Jagged 1/Notch1 pathway was also identified in the PMVECs of bleomycin-induced PF rats [73]. *Jagged1* knockdown resulted in the downregulation of SMAα and NF-κB expression in bleomycin-treated rat PMVECs [73]. In addition, expressions of SMAα and Jagged 1/Notch1 were positively correlated [73].

NF-κB signaling is known to play a critical role in EndMT [51]. NF-κB transcriptionally regulates Snail, which is a transcription factor for promoting EndMT [51]. Several groups have shown the activation of NF-κB-Snail signaling in TGFβ1-induced ECs and MCT-rat models [51,75]. In addition, the NOD1 agonist, g-dglutamyl-meso-diaminopimelic acid (iE-DAP), induces EndMT via Akt/NF-κB signaling [76]. Taken together, Jagged 1/Notch signaling and NF-κB signaling are vital during EndMT.

MicroRNAs (miRNAs) are 22-nucleotide, small, non-coding RNAs and important regulators of EndMT in many diseases [77,78]. It has been demonstrated that miRNAs, such as miR-21, miR-27a, miR-126a-5p, miR-130a, miR-139-5p, and miR-199a-5p, are involved in EndMT in vitro and in vivo. Parikh et al. used a network-based bioinformatic method to identify PH-modifying miRNAs and found that miR-21 is upregulated in the pulmonary vessels of PH animal models and human PAH patients [79]. In PAECs, hypoxia, inflammation, and BMPR2-dependent signaling induced miR-21 and suppressed its target, RhoB [79]. Another group provided evidence that miR-21 levels, Akt phosphorylation/activation, Snail expression, and NF-κB signaling were elevated in TGFβ1-induced EndMT [75]. Our group determined that iE-DAP downregulates miR-139-5p and activates Akt/NF-κB signaling, which leads to EndMT [76]. In addition, the overexpression of miR-139-5p reversed the nuclear translocation of NF-κB, resulting in the inhibition of iE-DAP-induced EndMT [76]. Li et al. reported that the mouse lung tissue of MCT-induced PAH showed increased miR-130a expression, and its regulation was NF-κB-dependent [80]. Moreover, the overexpression of miR-130a induced EndMT in PMVECs [80]. Further, a lung tissue microarray identified that miR-126a-5p is upregulated in a neonatal PH rat model [81]. Hypoxia induced the expression of miR-126a-5p and led to EndMT through PI3K/Akt signaling in primary cultured rat PMVECs [81]. Furthermore, circulating miR-126a-5p levels were increased in the sera of PAH patients [81]. Several studies have reported that miR-27a is increased in pulmonary arteries of PAH. [26,82]. Moreover, there is upregulation of miR-27a in pulmonary arteries of PAH rats and hypoxia-induced PAECs [26]. MiR-27a acts as an EndMT inducer through the suppression of Smad5 and the upregulation of Snail and Twist [26]. In addition, the contribution of Snail-induced miR-199a-5p to radiation-induced EndMT has been evaluated previously [83].

In conclusion, many studies have demonstrated the interplay of various signaling pathways in the process of EndMT (Figure 2). However, better knowledge of how they engage in crosstalk with one another and what other mediators are involved is required for developing therapeutic strategies.

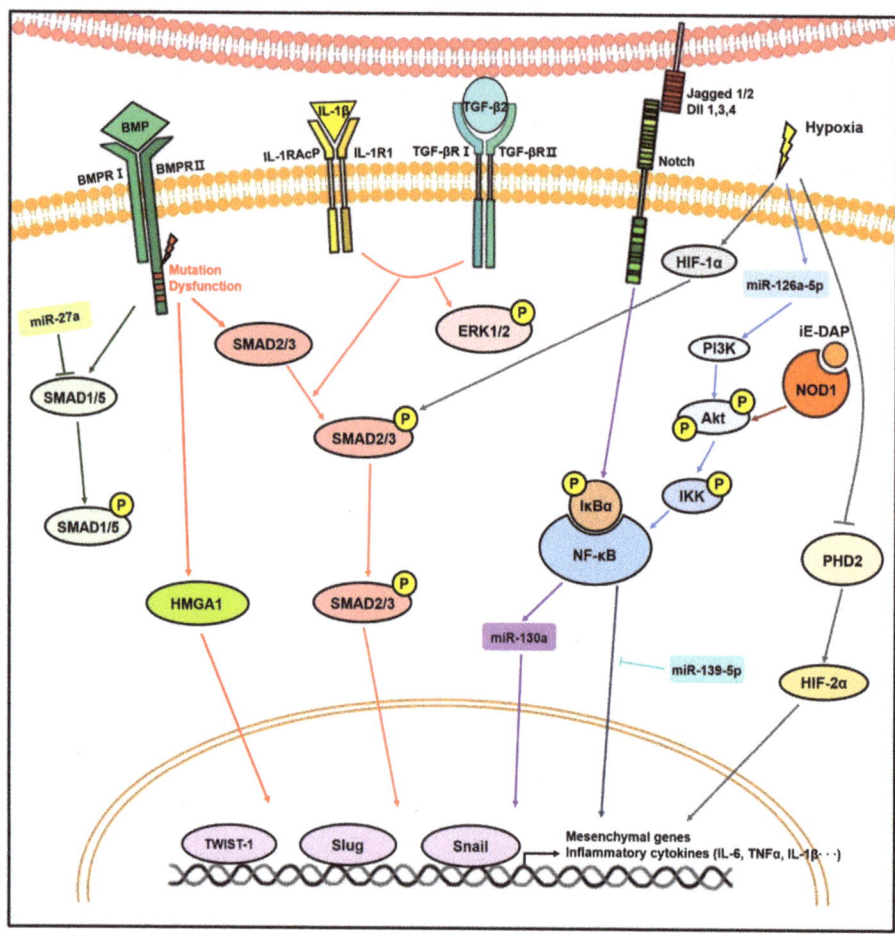

Figure 2. Molecular signaling pathways involved in endothelial-to-mesenchymal transition (EndMT). Stimulation with transforming growth factor-β (TGFβ), bone morphogenic protein (BMP), Notch ligands, inflammatory stress, and hypoxia induce expression of transcription factors, such as Twist1, Slug, and Snail, resulting in EndMT. During this process, mediators including microRNAs (miRNAs), Smad, Akt/nuclear factor kappa B (NF-κB), and hypoxia-inducible factor (HIF) play important roles in EndMT.

5. Targeting EndMT for Potential Therapeutic Applications

Many studies have reported potential therapeutic strategies to alleviate PF and PAH by targeting EndMT [7,33,34,41,52]. Adenoviral BMPR2 administration inhibited vascular remodeling and improved cardiac function in chronic hypoxia or MCT-induced PH rats [36]. In addition, recombinant BMP2 or BMP7 treatment reversed TGFβ1-induced EndMT in PMVECs [36]. One study suggested that modulating miRNAs may be a therapeutic strategy for PAH. When the MCT-exposed rats were infected with lentivirus-overexpressing miR-181b, RVSP, mean PAP, the decrease in pulmonary arterial wall thickness and the overexpression of miR-181b inhibited EndMT in rat PAECs by negatively regulating endocan and TGFβR1 [29]. Our group demonstrated that ginsenoside Rg3 attenuates iE-DAP-induced EndMT by upregulating miR-139-5p [76]. Nintedanib, a tyrosine kinase inhibitor, significantly decreased the phosphorylation of platelet-derived growth factor and fibroblast growth factor receptors, which were increased in the pulmonary artery of PAH, resulting in the improvement of PAH in vivo [34]. In addition, nintedanib downregulated the expression of mesenchymal markers in PMVECs [34]. One study suggested that hydrogen sulfide inhibits the NF-κB-Snail pathway and EndMT, resulting in a therapeutic effect on PAH [51]. In addition, treatment with the prostaglandin E2 receptor 4 (EP4) agonist decreases EndMT in vitro and reduces right ventricular fibrosis in a rat model of PAH [64]. The expression of Gal-3 was increased in PAH patients and a hypoxia-induced PAH rat model, and treatment with Gal-3 inhibitor, N-Lac, recovered RVSP, pulmonary artery acceleration, and pulmonary arterial velocity time integral [33]. Salvianolic acid A restores pulmonary vascular remodeling and improves vascular relaxation by upregulating Nrf2/HO-1 signaling while reducing TGFβ1 and EndMT in MCT-induced PAH rat models [35]. CD44v binds to and stabilizes the cystine transporter subunit (x-CT) when PAECs undergo EndMT. Sulfasalazine inhibits x-CT and restores EndMT in vitro and in vivo [32]. Rapamycin increased the expression of p120-catenin, a cytoplasmic scaffold protein that regulates cell-to-cell adhesion by binding VE-cadherin, and led to the downregulation of Twist1 expression in the lungs of MCT-induced PH rats [7]. Moreover, rapamycin inhibited the migration and proliferation of PAECs isolated from normal and PAH patients [7]. As such, rapamycin has curative effects on EndMT in MCT-induced PH rat models [7].

CD26/dipeptidyl peptidase 4 (DPP-4) is widely distributed in various cell types of the lung and promotes TGFβ signaling and EndMT [44,84]. Vildagliptin, a DPP-4 inhibitor, inhibits LPS-induced EndMT in PMVECs and attenuates the LPS-induced PF mouse model [44]. Another study confirmed that treatment with a DPP-4 inhibitor, sitagliptin, reduces pulmonary arterial remodeling and alleviates EndMT in PH-induced rats [52]. In addition, treatment with atazanavir sulphate, an antiretroviral protease inhibitor, ameliorates EndMT in cobalt chloride ($CoCl_2$)-induced hypoxic PMVECs and decreases fibrotic lesions in a bleomycin-induced rat PF model, suggesting a potential therapeutic effect of atazanavir sulphate on EndMT [43]. Further, the overexpression of HSPB1, which protects against cellular stress, decreased radiation-induced EndMT in vitro and in vivo [45]. The HIF-1α inhibitor reduced the expression of HIF-1α, the phosphorylation of Smad2/3, and EndMT in radiation-induced PAECs; furthermore, it decreased the SMAα+ CD31+ double-positive cells and fibrosis area in the lungs of the RIPF mouse model [41].

Taken together, EndMT plays a critical part in vascular diseases, especially PF and PAH, and many studies on various therapeutic strategies targeting EndMT have been conducted by modulating various signaling pathways and epigenetic factors (Figure 3). However, further studies are still necessary to clarify the mechanism between EndMT and pulmonary vascular diseases to find more effective therapeutic agents that can treat vascular diseases, such as PF or PAH, by targeting EndMT. Tables 1 and 2 each list a summary of studies on EndMT in lung diseases and targeting EndMT as a therapeutic strategy.

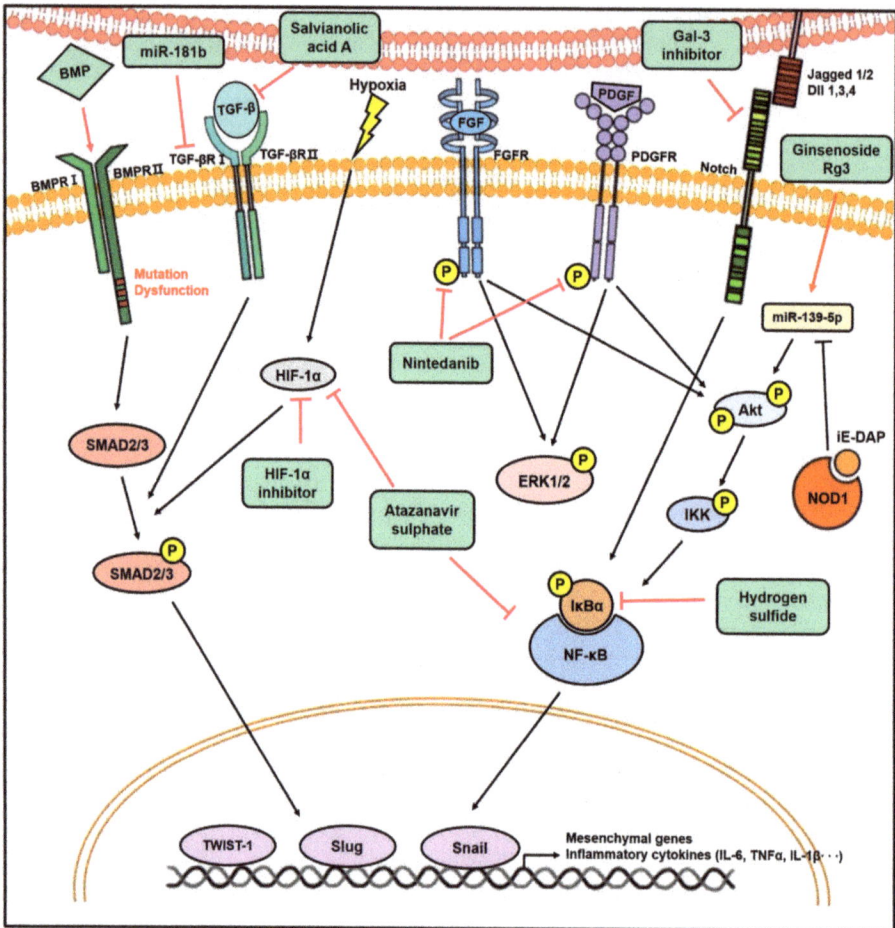

Figure 3. The novel therapeutic approach for lung diseases by targeting EndMT. The modulation of signaling pathways involved in EndMT by miRNAs, inhibitors, and agonists has therapeutic effects in vitro and in vivo. Targeting EndMT reduces mesenchymal marker expression and pulmonary vascular remodeling, which ultimately ameliorates the hemodynamic phenotypes in the animal model of PH. In addition, the inhibition of EndMT decreases fibrotic lesion in various PF animal models. This provides insight into the therapeutic potential of targeting EndMT.

Table 1. Summary of the main studies on EndMT in lung diseases.

Lung Diseases	In Vivo Model	In Vitro Model	Cell Line	Endothelial Markers	Mesenchymal Markers	EndMT Mediators	Reference
PH	SuHx, MCT rats	BMPR2 deficiency	PAECs	CD31, VE-cad, CD34, Tie2	SMAα, Vimentin, p-Vimentin	Twist1	[7]
PH	BMPR2 KO mice	BMPR2 deficiency	EAhy926	VE-cad	N-cad	Slug, Twist	[11]
PH	EC-specific BMPR2 KO mice	BMPR2 deficiency	PAECs	vWF, CD31, VE-cad	SM22α, SMAα, p-Vimentin	HMGA1, Snail, Slug	[23]
PH	Hypoxia-exposed rats	Hypoxia	Rat PMVECs	CD31	SMAα, Collagen I, III	HIF-1α, Twist1	[27]
PH	SuHx mice	Combination of TGFβ1, TNFα, and IL-1β	PAECs	vWF, CD31, VE-cad, Occudin	SMAα, Calponin, Collagen I	Inflammatory cytokines	[30]
PH	VE-cad Cre or Tie2 Cre-mTomato/mGFP lineage tracing mice			vWF, CD31, VE-cad, Tie2	SMAα	Not determined	[48]
PH	MCT rats, EC-specific phd2, egln1, KO mice	Hypoxia, TGFβ1	PVECs	CD31, VE-cad	SM22α, Vimentin, FN, SMAα, FSP1	HIF-2α, Snail, Slug, PHD2	[50]
PF	Radiation-exposed mice	Radiation	PAECs, PMVECs	CD31, VE-cad	SMAα, Vimentin, FSP1, Collagen, MMP9	TGFβ-RI, Smad2/3, HIF-1α, Snail	[41]
PF	Bleomycin-treated Tie2-Cre/CAG-CAT-Lac mice	Combination of Ras and TGFβ	MS-1	CD31, VE-cad, CD34, Tie2,	SMAα, FN, Collagen I	Not determined	[42]
PF	Radiation-exposed mice	HSPB1 deficiency	PAECs, PMVECs	CD31, VE-cad	SMAα	Inflammatory cytokines	[45]

PH, pulmonary hypertension; PF, pulmonary fibrosis; SuHx, SU-5416/hypoxia model; MCT, monocrotaline; BMPR2, bone morphogenic protein receptor-2; KO, knockout; EC, endothelial cell; TGFβ, transforming growth factor beta; IL-1β, interleukin-1 beta; PHD2, prolyl hydroxylase domain protein 2; HSPB1, endothelial heat shock protein beta 1; PAECs, pulmonary arterial endothelial cells; PVECs, pulmonary vascular endothelial cells; PMVECs, pulmonary microvascular endothelial cells; MS-1, mouse microvascular endothelial cell line; CD31, cluster of differentiation 31; CD31, cluster of differentiation 34; VE-cad, vascular endothelial cadherin; vWF, von willebrand factor; SMAα, α-smooth muscle actin; N-cad, neural cadherin; SM22α, smooth muscle protein 22-α; FN, fibronectin; FSP1, fibroblast-specific protein 1; MMP, matrix metallopeptidase 9; HIF, hypoxia-inducible factor; HMGA1, high-mobility group AT-hook 1; p-Vimentin, phosphorylated vimentin.

Table 2. Key studies targeting EndMT as a therapeutic strategy in PH and PF.

Clinical Relevance	In Vitro Model	In Vivo Model	Negative Regulator of EndMT	Reference
PH		MCT rat	Rapamycin	[7]
PH	Combination of TGFβ1, TNFα, and IL-1β-treated rat PAECs	MCT rat	miR-181b	[29]
PH	Combination of TGFβ1, TNFα, and IL-1β-treated PAECs	SuHx mice	Sulfasalazine	[32]
PH		Hypoxia-exposed rat	Galectin-3 inhibitor	[33]
PH	Combination of TGFβ2, TNFα and IL-1β-treated PMVECs	SuHx rat	Nintedanib	[34]
PH	TGFβ1-treated PAECs	MCT rat	Salvianolic acid A	[35]
PH	TGFβ1-treated PAECs	MCT rat	Hydrogen sulfide	[51]
PH		MCT rat	Sitagliptin	[52]
PH	TGFβ-treated HUVECs	MCT rat	EP4 agonist	[64]
PH	TGFβ1-treated PMVECs	MCT rat, Hypoxia-exposed rat	BMPR2, rhBMP2, rhBMP7	[36]
PF	Radiation-exposed PAECs	Radiation-exposed mice	HIF-1α inhibitor	[41]
PF	CoCl$_2$-treated PMVECs	Bleomycin-treated rat	Atazanavir sulphate	[43]
PF	LPS-treated PMVECs	LPS-treated mice	Vildagliptin	[44]
PF	Radiation-exposed PMVECs	Radiation-exposed EC conditionally overexpressed HSPB1 mice	HSPB1	[45]

PH, pulmonary hypertension; PF, pulmonary fibrosis; SuHx, SU-5416/hypoxia model; MCT, monocrotaline; EC, endothelial cell; LPS, lipopolysaccharides; CoCl$_2$, cobalt chloride; TGFβ, transforming growth factor beta; TNFα, tumor necrosis factor alpha; IL-1β, interleukin 1beta; HSPB1, heat shock protein beta 1; HIF1α, hypoxia-inducible factor 1 alpha; BMPR2, bone morphogenic protein receptor-2; rhBMP, recombinant human bone morphogenic protein; EP4, prostaglandin E2 receptor 4; PAECs, pulmonary arterial endothelial cells; PMVECs, pulmonary microvascular endothelial cells; HUVECs, human umbilical vein endothelial cells.

6. Conclusions and Perspective

Here, we review the role of EndMT and its downstream pathways, and the therapeutic implications of targeting EndMT for lung diseases. As described earlier, many studies have demonstrated that EndMT is associated with the pathogenesis of PAH and PF in both in vitro and in vivo models, as well as in the lung tissue of humans. Thus, targeting EndMT can be a new therapeutic approach to treating lung diseases given the fact that there are no drugs to cure PAH or PF. Indeed, emerging evidence suggests that pharmacological approaches to inhibiting EndMT have the potential to treat lung diseases in vitro and in vivo. EndMT can also contribute to other human diseases, such as cancer, atherosclerosis, neointima formation, vascular calcification, and cerebral cavernous malformations. Therefore, inhibiting EndMT might represent a broadly applicable therapeutic strategy for the treatment of not only pulmonary vascular diseases, but also many diseases associated with endothelial dysfunction. Although it is now clear that EndMT is closely associated with the pathogenesis of multiple diseases, the EndMT-targeting therapeutic approach needs to be carefully evaluated before clinical application. Given that ECs display organ-specific heterogeneity in function and phenotype, in health and disease, and in their response to environmental stimuli, it is possible that certain vessel types and vascular beds are more sensitive to EndMT-inducing signals. Therefore, it is important to understand the exact molecular mechanisms related to the EndMT process in the context of the heterogeneity of ECs. Importantly, EndMT may be a reversible biological process, what is called mesenchymal-to-endothelial transition, suggesting that exploration of the regulatory mechanism of the reversible process of EndMT will provide new insights into the prevention and treatment of various human diseases, and may be applied to tissue engineering. In conclusion, we believe that studies of EndMT in the context of endothelial heterogeneity will provide us with better insights into the molecular mechanisms of a broad variety of human diseases, and will help to develop novel vascular bed-specific therapies.

Author Contributions: Conceptualization, J.K., A.L., E.Y.; investigation, J.K., A.L., E.Y., Y.K.; writing-review and editing, J.K., A.L., E.Y.; visualization, Y.K., E.Y.; supervision, K.H.Y., K.I.K., M.-S.L. All authors have read and agreed to the published version of the manuscript.

Funding: This research was supported by the Basic Science Research Program through the National Research Foundation of Korea (NRF) funded by the minister of Education, Science and Technology (NRF-2016R1A5A1011974 and NRF-2019R1A2C4069815 to J.K.).

Conflicts of Interest: The authors declare no conflict of interest.

References

1. Potenta, S.; Zeisberg, E.; Kalluri, R. The role of endothelial-to-mesenchymal transition in cancer progression. *Br. J. Cancer* **2008**, *99*, 1375–1379. [CrossRef] [PubMed]
2. Cho, J.G.; Lee, A.; Chang, W.; Lee, M.S.; Kim, J. Endothelial to mesenchymal transition represents a key link in the interaction between inflammation and endothelial dysfunction. *Front. Immunol.* **2018**, *9*, 294. [CrossRef]
3. Kim, S.M.; Huh, J.W.; Kim, E.Y.; Shin, M.K.; Park, J.E.; Kim, S.W.; Lee, W.; Choi, B.; Chang, E.J. Endothelial dysfunction induces atherosclerosis: Increased aggrecan expression promotes apoptosis in vascular smooth muscle cells. *BMB Rep.* **2019**, *52*, 145–150.
4. Sena, C.M.; Pereira, A.M.; Seica, R. Endothelial dysfunction—A major mediator of diabetic vascular disease. *Biochim. Biophys. Acta* **2013**, *1832*, 2216–2231. [CrossRef]
5. Krenning, G.; Barauna, V.G.; Krieger, J.E.; Harmsen, M.C.; Moonen, J.R. Endothelial plasticity: Shifting phenotypes through force feedback. *Stem Cells Int.* **2016**, *2016*, 9762959. [CrossRef]
6. Sabbineni, H.; Verma, A.; Somanath, P.R. Isoform-specific effects of transforming growth factor beta on endothelial-to-mesenchymal transition. *J. Cell. Physiol.* **2018**, *233*, 8418–8428. [CrossRef]
7. Ranchoux, B.; Antigny, F.; Rucker-Martin, C.; Hautefort, A.; Pechoux, C.; Bogaard, H.J.; Dorfmuller, P.; Remy, S.; Lecerf, F.; Plante, S.; et al. Endothelial-to-mesenchymal transition in pulmonary hypertension. *Circulation* **2015**, *131*, 1006–1018. [CrossRef]
8. Ursoli Ferreira, F.; Eduardo Botelho Souza, L.; Hassibe Thome, C.; Tomazini Pinto, M.; Origassa, C.; Salustiano, S.; Marcel Faca, V.; Olsen Camara, N.; Kashima, S.; Tadeu Covas, D. Endothelial cells tissue-specific origins affects their responsiveness to TGF-beta2 during endothelial-to-mesenchymal transition. *Int. J. Mol. Sci.* **2019**, *20*, 458. [CrossRef]
9. Sanchez-Duffhues, G.; Garcia de Vinuesa, A.; Ten Dijke, P. Endothelial-to-mesenchymal transition in cardiovascular diseases: Developmental signaling pathways gone awry. *Dev. Dyn. Off. Publ. Am. Assoc. Anat.* **2018**, *247*, 492–508. [CrossRef]
10. Guan, S.; Zhou, J. CXCR7 attenuates the TGF-beta-induced endothelial-to-mesenchymal transition and pulmonary fibrosis. *Mol. Biosyst.* **2017**, *13*, 2116–2124. [CrossRef] [PubMed]
11. Hiepen, C.; Jatzlau, J.; Hildebrandt, S.; Kampfrath, B.; Goktas, M.; Murgai, A.; Cuellar Camacho, J.L.; Haag, R.; Ruppert, C.; Sengle, G.; et al. BMPR2 acts as a gatekeeper to protect endothelial cells from increased TGFbeta responses and altered cell mechanics. *PLoS Biol.* **2019**, *17*, e3000557. [CrossRef]
12. Zeisberg, E.M.; Tarnavski, O.; Zeisberg, M.; Dorfman, A.L.; McMullen, J.R.; Gustafsson, E.; Chandraker, A.; Yuan, X.; Pu, W.T.; Roberts, A.B.; et al. Endothelial-to-mesenchymal transition contributes to cardiac fibrosis. *Nature Med.* **2007**, *13*, 952–961. [CrossRef] [PubMed]
13. Thuan, D.T.B.; Zayed, H.; Eid, A.H.; Abou-Saleh, H.; Nasrallah, G.K.; Mangoni, A.A.; Pintus, G. A potential link between oxidative stress and endothelial-to-mesenchymal transition in systemic sclerosis. *Front. Immunol.* **2018**, *9*, 1985. [CrossRef] [PubMed]
14. Simonneau, G.; Montani, D.; Celermajer, D.S.; Denton, C.P.; Gatzoulis, M.A.; Krowka, M.; Williams, P.G.; Souza, R. Haemodynamic definitions and updated clinical classification of pulmonary hypertension. *Eur. Respir. J.* **2019**, *53*. [CrossRef]
15. Kim, J.D.; Lee, A.; Choi, J.; Park, Y.; Kang, H.; Chang, W.; Lee, M.S.; Kim, J. Epigenetic modulation as a therapeutic approach for pulmonary arterial hypertension. *Exp. Mol. Med.* **2015**, *47*, e175. [CrossRef]
16. Lee, A.; McLean, D.; Choi, J.; Kang, H.; Chang, W.; Kim, J. Therapeutic implications of microRNAs in pulmonary arterial hypertension. *BMB Rep.* **2014**, *47*, 311–317. [CrossRef]
17. Roos-Hesselink, J.W.; Zijlstra, F. Pulmonary hypertension, how to diagnose and who to treat? *Neth. Heart J.* **2011**, *19*, 493–494. [CrossRef]
18. Lyle, M.A.; Davis, J.P.; Brozovich, F.V. Regulation of pulmonary vascular smooth muscle contractility in pulmonary arterial hypertension: Implications for therapy. *Front. Physiol.* **2017**, *8*. [CrossRef]

19. Pietra, G.G.; Edwards, W.D.; Kay, J.M.; Rich, S.; Kernis, J.; Schloo, B.; Ayres, S.M.; Bergofsky, E.H.; Brundage, B.H.; Detre, K.M.; et al. Histopathology of primary pulmonary hypertension. A qualitative and quantitative study of pulmonary blood vessels from 58 patients in the National Heart, Lung, and Blood Institute, primary pulmonary hypertension registry. *Circulation* **1989**, *80*, 1198–1206. [CrossRef]
20. Rabinovitch, M. Molecular pathogenesis of pulmonary arterial hypertension. *J. Clin. Investig.* **2012**, *122*, 4306–4313. [CrossRef]
21. Ranchoux, B.; Harvey, L.D.; Ayon, R.J.; Babicheva, A.; Bonnet, S.; Chan, S.Y.; Yuan, J.X.; Perez, V.J. Endothelial dysfunction in pulmonary arterial hypertension: An evolving landscape (2017 Grover Conference Series). *Pulm. Circ.* **2018**, *8*. [CrossRef] [PubMed]
22. McLaughlin, V.V.; Archer, S.L.; Badesch, D.B.; Barst, R.J.; Farber, H.W.; Lindner, J.R.; Mathier, M.A.; McGoon, M.D.; Park, M.H.; Rosenson, R.S.; et al. ACCF/AHA 2009 expert consensus document on pulmonary hypertension. A report of the American College of Cardiology foundation task force on expert consensus documents and the American Heart Association developed in collaboration with the American College of Chest physicians; American Thoracic Society, Inc.; and the Pulmonary Hypertension Association. *J. Am. Coll. Cardiol.* **2009**, *53*, 1573–1619. [CrossRef] [PubMed]
23. Hopper, R.K.; Moonen, J.R.A.J.; Diebold, I.; Cao, A.Q.; Rhodes, C.J.; Tojais, N.F.; Hennigs, J.K.; Gu, M.X.; Wang, L.L.; Rabinovitch, M. In pulmonary arterial hypertension, reduced BMPR2 promotes endothelial-to-mesenchymal transition via HMGA1 and its target slug. *Circulation* **2016**, *133*, 1783–1794. [CrossRef]
24. Gorelova, A.; Berman, M.; Al Ghouleh, I. Endothelial-to-mesenchymal transition in pulmonary arterial hypertension. *Antioxid. Redox Signal.* **2020**. [CrossRef]
25. Gaikwad, A.V.; Eapen, M.S.; McAlinden, K.D.; Chia, C.; Larby, J.; Myers, S.; Dey, S.; Haug, G.; Markos, J.; Glanville, A.R.; et al. Endothelial to mesenchymal transition (EndMT) and vascular remodeling in pulmonary hypertension and idiopathic pulmonary fibrosis. *Expert Rev. Respir. Med.* **2020**, *14*, 1027–1043. [CrossRef] [PubMed]
26. Liu, T.; Zou, X.Z.; Huang, N.; Ge, X.Y.; Yao, M.Z.; Liu, H.; Zhang, Z.; Hu, C.P. MiR-27a promotes endothelial-mesenchymal transition in hypoxia-induced pulmonary arterial hypertension by suppressing BMP signaling. *Life Sci.* **2019**, *227*, 64–73. [CrossRef] [PubMed]
27. Zhang, B.; Niu, W.; Dong, H.Y.; Liu, M.L.; Luo, Y.; Li, Z.C. Hypoxia induces endothelialmesenchymal transition in pulmonary vascular remodeling. *Int. J. Mol. Med.* **2018**, *42*, 270–278. [CrossRef] [PubMed]
28. Wang, J.; Yu, M.; Xu, J.; Cheng, Y.; Li, X.; Wei, G.; Wang, H.; Kong, H.; Xie, W. Glucagon-like peptide-1 (GLP-1) mediates the protective effects of dipeptidyl peptidase IV inhibition on pulmonary hypertension. *J. Biomed. Sci.* **2019**, *26*, 6. [CrossRef] [PubMed]
29. Zhao, H.; Wang, Y.; Zhang, X.; Guo, Y.; Wang, X. miR-181b-5p inhibits endothelial-mesenchymal transition in monocrotaline-induced pulmonary arterial hypertension by targeting endocan and TGFBR1. *Toxicol. Appl. Pharmacol.* **2020**, *386*, 114827. [CrossRef] [PubMed]
30. Good, R.B.; Gilbane, A.J.; Trinder, S.L.; Denton, C.P.; Coghlan, G.; Abraham, D.J.; Holmes, A.M. Endothelial to mesenchymal transition contributes to endothelial dysfunction in pulmonary arterial hypertension. *Am. J. Pathol.* **2015**, *185*, 1850–1858. [CrossRef]
31. Shah, S.J. Pulmonary hypertension. *JAMA* **2012**, *308*, 1366–1374. [CrossRef]
32. Isobe, S.; Kataoka, M.; Endo, J.; Moriyama, H.; Okazaki, S.; Tsuchihashi, K.; Katsumata, Y.; Yamamoto, T.; Shirakawa, K.; Yoshida, N.; et al. Endothelial-mesenchymal transition drives expression of CD44 variant and xCT in pulmonary hypertension. *Am. J. Respir. Cell. Mol. Biol.* **2019**, *61*, 367–379. [CrossRef] [PubMed]
33. Zhang, L.; Li, Y.M.; Zeng, X.X.; Wang, X.Y.; Chen, S.K.; Gui, L.X.; Lin, M.J. Galectin-3-mediated transdifferentiation of pulmonary artery endothelial cells contributes to hypoxic pulmonary vascular remodeling. *Cell. Physiol. Biochem.* **2018**, *51*, 763–777. [CrossRef] [PubMed]
34. Tsutsumi, T.; Nagaoka, T.; Yoshida, T.; Wang, L.; Kuriyama, S.; Suzuki, Y.; Nagata, Y.; Harada, N.; Kodama, Y.; Takahashi, F.; et al. Nintedanib ameliorates experimental pulmonary arterial hypertension via inhibition of endothelial mesenchymal transition and smooth muscle cell proliferation. *PLoS ONE* **2019**, *14*, e0214697. [CrossRef] [PubMed]
35. Chen, Y.C.; Yuan, T.Y.; Zhang, H.F.; Yan, Y.; Wang, D.S.; Fang, L.H.; Lu, Y.; Du, G.H. Activation of Nrf2 attenuates pulmonary vascular remodeling via inhibiting endothelial-to-mesenchymal transition: An insight from a plant polyphenol. *Int. J. Biol. Sci.* **2017**, *13*, 1067–1081. [CrossRef]

36. Reynolds, A.M.; Holmes, M.D.; Danilov, S.M.; Reynolds, P.N. Targeted gene delivery of BMPR2 attenuates pulmonary hypertension. *Eur. Respir. J.* **2012**, *39*, 329–343. [CrossRef]
37. King, T.E., Jr.; Pardo, A.; Selman, M. Idiopathic pulmonary fibrosis. *Lancet* **2011**, *378*, 1949–1961. [CrossRef]
38. Richeldi, L.; Collard, H.R.; Jones, M.G. Idiopathic pulmonary fibrosis. *Lancet* **2017**, *389*, 1941–1952. [CrossRef]
39. Selman, M.; Pardo, A. Alveolar epithelial cell disintegrity and subsequent activation a key process in pulmonary fibrosis. *Am. J. Respir. Crit. Care Med.* **2012**, *186*, 119–121. [CrossRef]
40. Wynn, T.A. Cellular and molecular mechanisms of fibrosis. *J. Pathol.* **2008**, *214*, 199–210. [CrossRef]
41. Choi, S.H.; Hong, Z.Y.; Nam, J.K.; Lee, H.J.; Jang, J.; Yoo, R.J.; Lee, Y.J.; Lee, C.Y.; Kim, K.H.; Park, S.; et al. A hypoxia-induced vascular endothelial-to-mesenchymal transition in development of radiation-induced pulmonary fibrosis. *Clin. Cancer Res. Off. J. Am. Assoc. Cancer Res.* **2015**, *21*, 3716–3726. [CrossRef] [PubMed]
42. Hashimoto, N.; Phan, S.H.; Imaizumi, K.; Matsuo, M.; Nakashima, H.; Kawabe, T.; Shimokata, K.; Hasegawa, Y. Endothelial-mesenchymal transition in bleomycin-induced pulmonary fibrosis. *Am. J. Resp. Cell. Mol.* **2010**, *43*, 161–172. [CrossRef] [PubMed]
43. Song, S.; Ji, Y.; Zhang, G.; Zhang, X.; Li, B.; Li, D.; Jiang, W. Protective effect of atazanavir sulphate against pulmonary fibrosis in vivo and in vitro. *Basic Clin. Pharmacol. Toxicol.* **2018**, *122*, 199–207. [CrossRef] [PubMed]
44. Suzuki, T.; Tada, Y.; Gladson, S.; Nishimura, R.; Shimomura, I.; Karasawa, S.; Tatsumi, K.; West, J. Vildagliptin ameliorates pulmonary fibrosis in lipopolysaccharide-induced lung injury by inhibiting endothelial-to-mesenchymal transition. *Respir. Res.* **2017**, *18*, 177. [CrossRef]
45. Choi, S.H.; Nam, J.K.; Kim, B.Y.; Jang, J.; Jin, Y.B.; Lee, H.J.; Park, S.; Ji, Y.H.; Cho, J.; Lee, Y.J. HSPB1 inhibits the endothelial-to-mesenchymal transition to suppress pulmonary fibrosis and lung tumorigenesis. *Cancer Res.* **2016**, *76*, 1019–1030. [CrossRef]
46. Lu, X.; Gong, J.; Dennery, P.A.; Yao, H. Endothelial-to-mesenchymal transition: Pathogenesis and therapeutic targets for chronic pulmonary and vascular diseases. *Biochem. Pharm.* **2019**, *168*, 100–107. [CrossRef]
47. Kim, J.; Kang, Y.; Kojima, Y.; Lighthouse, J.K.; Hu, X.; Aldred, M.A.; McLean, D.L.; Park, H.; Comhair, S.A.; Greif, D.M.; et al. An endothelial apelin-FGF link mediated by miR-424 and miR-503 is disrupted in pulmonary arterial hypertension. *Nat. Med.* **2013**, *19*, 74–82. [CrossRef]
48. Qiao, L.; Nishimura, T.; Shi, L.; Sessions, D.; Thrasher, A.; Trudell, J.R.; Berry, G.J.; Pearl, R.G.; Kao, P.N. Endothelial fate mapping in mice with pulmonary hypertension. *Circulation* **2014**, *129*, 692–703. [CrossRef]
49. Sakao, S.; Hao, H.; Tanabe, N.; Kasahara, Y.; Kurosu, K.; Tatsumi, K. Endothelial-like cells in chronic thromboembolic pulmonary hypertension: crosstalk with myofibroblast-like cells. *Respir. Res.* **2011**, *12*, 109. [CrossRef]
50. Tang, H.; Babicheva, A.; McDermott, K.M.; Gu, Y.; Ayon, R.J.; Song, S.; Wang, Z.; Gupta, A.; Zhou, T.; Sun, X.; et al. Endothelial HIF-2alpha contributes to severe pulmonary hypertension due to endothelial-to-mesenchymal transition. *Am. J. Physiol. Lung Cell Mol. Physiol.* **2018**, *314*, L256–L275. [CrossRef]
51. Zhang, H.; Lin, Y.; Ma, Y.; Zhang, J.; Wang, C.; Zhang, H. Protective effect of hydrogen sulfide on monocrotalineinduced pulmonary arterial hypertension via inhibition of the endothelial mesenchymal transition. *Int. J. Mol. Med.* **2019**. [CrossRef] [PubMed]
52. Xu, J.; Wang, J.J.; He, M.Y.; Han, H.H.; Xie, W.P.; Wang, H.; Kong, H. Dipeptidyl peptidase IV (DPP-4) inhibition alleviates pulmonary arterial remodeling in experimental pulmonary hypertension. *Lab Invest.* **2018**, *98*, 1333–1346. [CrossRef] [PubMed]
53. Sabbineni, H.; Verma, A.; Artham, S.; Anderson, D.; Amaka, O.; Liu, F.; Narayanan, S.P.; Somanath, P.R. Pharmacological inhibition of beta-catenin prevents EndMT in vitro and vascular remodeling in vivo resulting from endothelial Akt1 suppression. *Biochem. Pharmacol.* **2019**, *164*, 205–215. [CrossRef] [PubMed]
54. Suzuki, T.; Carrier, E.J.; Talati, M.H.; Rathinasabapathy, A.; Chen, X.; Nishimura, R.; Tada, Y.; Tatsumi, K.; West, J. Isolation and characterization of endothelial-to-mesenchymal transition cells in pulmonary arterial hypertension. *Am. J. Physiol. Lung Cell Mol. Physiol.* **2018**, *314*, L118–L126. [CrossRef] [PubMed]
55. Nataraj, D.; Ernst, A.; Kalluri, R. Idiopathic Pulmonary Fibrosis is Associated with Endothelial to Mesenchymal Transition. *Am. J. Resp. Cell Mol.* **2010**, *43*, 129–130. [CrossRef] [PubMed]
56. Zhou, Y.; Li, P.; Duan, J.X.; Liu, T.; Guan, X.X.; Mei, W.X.; Liu, Y.P.; Sun, G.Y.; Wan, L.; Zhong, W.J.; et al. Aucubin Alleviates Bleomycin-Induced Pulmonary Fibrosis in a Mouse Model. *Inflammation* **2017**, *40*, 2062–2073. [CrossRef]

57. Mouratis, M.A.; Aidinis, V. Modeling pulmonary fibrosis with bleomycin. *Curr. Opin. Pulm. Med.* **2011**, *17*, 355–361. [CrossRef]
58. Singh, K.K.; Lovren, F.; Pan, Y.; Quan, A.; Ramadan, A.; Matkar, P.N.; Ehsan, M.; Sandhu, P.; Mantella, L.E.; Gupta, N.; et al. The essential autophagy gene ATG7 modulates organ fibrosis via regulation of endothelial-to-mesenchymal transition. *J. Biol. Chem.* **2015**, *290*, 2547–2559. [CrossRef]
59. Suzuki, T.; Tada, Y.; Nishimura, R.; Kawasaki, T.; Sekine, A.; Urushibara, T.; Kato, F.; Kinoshita, T.; Ikari, J.; West, J.; et al. Endothelial-to-mesenchymal transition in lipopolysaccharide-induced acute lung injury drives a progenitor cell-like phenotype. *Am. J. Physiol. Lung Cell Mol. Physiol.* **2016**, *310*, L1185–L1198. [CrossRef]
60. Groth, A.; Vrugt, B.; Brock, M.; Speich, R.; Ulrich, S.; Huber, L.C. Inflammatory cytokines in pulmonary hypertension. *Respir. Res.* **2014**, *15*, 47. [CrossRef]
61. Kovacic, J.C.; Dimmeler, S.; Harvey, R.P.; Finkel, T.; Aikawa, E.; Krenning, G.; Baker, A.H. Endothelial to Mesenchymal Transition in Cardiovascular Disease: JACC State-of-the-Art Review. *J. Am. Coll. Cardiol.* **2019**, *73*, 190–209. [CrossRef]
62. Piera-Velazquez, S.; Jimenez, S.A. Endothelial to Mesenchymal Transition: Role in Physiology and in the Pathogenesis of Human Diseases. *Physiol. Rev.* **2019**, *99*, 1281–1324. [CrossRef]
63. Li, Z.; Wermuth, P.J.; Benn, B.S.; Lisanti, M.P.; Jimenez, S.A. Caveolin-1 deficiency induces spontaneous endothelial-to-mesenchymal transition in murine pulmonary endothelial cells in vitro. *Am. J. Pathol.* **2013**, *182*, 325–331. [CrossRef] [PubMed]
64. Lai, Y.J.; Chen, I.C.; Li, H.H.; Huang, C.C. EP4 Agonist L-902,688 Suppresses EndMT and Attenuates Right Ventricular Cardiac Fibrosis in Experimental Pulmonary Arterial Hypertension. *Int. J. Mol. Sci.* **2018**, *19*, 727. [CrossRef] [PubMed]
65. Mammoto, T.; Muyleart, M.; Konduri, G.G.; Mammoto, A. Twist1 in Hypoxia-induced Pulmonary Hypertension through Transforming Growth Factor-beta-Smad Signaling. *Am. J. Respir. Cell Mol. Biol.* **2018**, *58*, 194–207. [CrossRef] [PubMed]
66. Morrell, N.W.; Aldred, M.A.; Chung, W.K.; Elliott, C.G.; Nichols, W.C.; Soubrier, F.; Trembath, R.C.; Loyd, J.E. Genetics and genomics of pulmonary arterial hypertension. *Eur. Respir. J.* **2019**, *53*. [CrossRef]
67. Kim, M.J.; Park, S.Y.; Chang, H.R.; Jung, E.Y.; Munkhjargal, A.; Lim, J.S.; Lee, M.S.; Kim, Y. Clinical significance linked to functional defects in bone morphogenetic protein type 2 receptor, BMPR2. *BMB Rep.* **2017**, *50*, 308–317. [CrossRef]
68. Mathew, R.; Huang, J.; Iacobas, S.; Iacobas, D.A. Pulmonary Hypertension Remodels the Genomic Fabrics of Major Functional Pathways. *Genes* **2020**, *11*, 126. [CrossRef]
69. Atkinson, C.; Stewart, S.; Upton, P.D.; Machado, R.; Thomson, J.R.; Trembath, R.C.; Morrell, N.W. Primary pulmonary hypertension is associated with reduced pulmonary vascular expression of type II bone morphogenetic protein receptor. *Circulation* **2002**, *105*, 1672–1678. [CrossRef]
70. Pinto, M.T.; Covas, D.T.; Kashima, S.; Rodrigues, C.O. Endothelial Mesenchymal Transition: Comparative Analysis of Different Induction Methods. *Biol. Proced. Online* **2016**, *18*, 10. [CrossRef]
71. Appelhoff, R.J.; Tian, Y.M.; Raval, R.R.; Turley, H.; Harris, A.L.; Pugh, C.W.; Ratcliffe, P.J.; Gleadle, J.M. Differential function of the prolyl hydroxylases PHD1, PHD2, and PHD3 in the regulation of hypoxia-inducible factor. *J. Biol. Chem.* **2004**, *279*, 38458–38465. [CrossRef]
72. Noseda, M.; McLean, G.; Niessen, K.; Chang, L.; Pollet, I.; Montpetit, R.; Shahidi, R.; Dorovini-Zis, K.; Li, L.; Beckstead, B.; et al. Notch activation results in phenotypic and functional changes consistent with endothelial-to-mesenchymal transformation. *Circ. Res.* **2004**, *94*, 910–917. [CrossRef] [PubMed]
73. Yin, Q.; Wang, W.; Cui, G.; Yan, L.; Zhang, S. Potential role of the Jagged1/Notch1 signaling pathway in the endothelial-myofibroblast transition during BLM-induced pulmonary fibrosis. *J. Cell. Physiol.* **2018**, *233*, 2451–2463. [CrossRef] [PubMed]
74. Chang, A.C.; Fu, Y.; Garside, V.C.; Niessen, K.; Chang, L.; Fuller, M.; Setiadi, A.; Smrz, J.; Kyle, A.; Minchinton, A.; et al. Notch initiates the endothelial-to-mesenchymal transition in the atrioventricular canal through autocrine activation of soluble guanylyl cyclase. *Dev. Cell* **2011**, *21*, 288–300. [CrossRef] [PubMed]
75. Guo, Y.; Li, P.; Bledsoe, G.; Yang, Z.R.; Chao, L.; Chao, J. Kallistatin inhibits TGF-beta-induced endothelial-mesenchymal transition by differential regulation of microRNA-21 and eNOS expression. *Exp. Cell Res.* **2015**, *337*, 103–110. [CrossRef]

76. Lee, A.; Yun, E.; Chang, W.; Kim, J. Ginsenoside Rg3 protects against iE-DAP-induced endothelial-to-mesenchymal transition by regulating the miR-139-5p-NF-kappaB axis. *J. Ginseng Res.* **2020**, *44*, 300–307. [CrossRef]
77. Kim, J. MicroRNAs as critical regulators of the endothelial to mesenchymal transition in vascular biology. *BMB Rep.* **2018**, *51*, 65–72. [CrossRef]
78. Salvi, V.; Gianello, V.; Tiberio, L.; Sozzani, S.; Bosisio, D. Cytokine Targeting by miRNAs in Autoimmune Diseases. *Front. Immunol.* **2019**, *10*, 15. [CrossRef]
79. Parikh, V.N.; Jin, R.C.; Rabello, S.; Gulbahce, N.; White, K.; Hale, A.; Cottrill, K.A.; Shaik, R.S.; Waxman, A.B.; Zhang, Y.Y.; et al. MicroRNA-21 integrates pathogenic signaling to control pulmonary hypertension: results of a network bioinformatics approach. *Circulation* **2012**, *125*, 1520–1532. [CrossRef]
80. Li, L.; Kim, I.K.; Chiasson, V.; Chatterjee, P.; Gupta, S. NF-kappaB mediated miR-130a modulation in lung microvascular cell remodeling: Implication in pulmonary hypertension. *Exp. Cell Res.* **2017**, *359*, 235–242. [CrossRef]
81. Xu, Y.P.; He, Q.; Shen, Z.; Shu, X.L.; Wang, C.H.; Zhu, J.J.; Shi, L.P.; Du, L.Z. MiR-126a-5p is involved in the hypoxia-induced endothelial-to-mesenchymal transition of neonatal pulmonary hypertension. *Hypertens Res* **2017**, *40*, 552–561. [CrossRef] [PubMed]
82. Kang, B.Y.; Park, K.K.; Green, D.E.; Bijli, K.M.; Searles, C.D.; Sutliff, R.L.; Hart, C.M. Hypoxia mediates mutual repression between microRNA-27a and PPARgamma in the pulmonary vasculature. *PLoS ONE* **2013**, *8*, e79503. [CrossRef] [PubMed]
83. Yi, M.; Liu, B.; Tang, Y.; Li, F.; Qin, W.; Yuan, X. Irradiated Human Umbilical Vein Endothelial Cells Undergo Endothelial-Mesenchymal Transition via the Snail/miR-199a-5p Axis to Promote the Differentiation of Fibroblasts into Myofibroblasts. *BioMed Res. Int.* **2018**, *2018*, 4135806. [CrossRef] [PubMed]
84. Srivastava, S.P.; Goodwin, J.E.; Kanasaki, K.; Koya, D. Inhibition of Angiotensin-Converting Enzyme Ameliorates Renal Fibrosis by Mitigating DPP-4 Level and Restoring Antifibrotic MicroRNAs. *Genes* **2020**, *11*, 211. [CrossRef]

Publisher's Note: MDPI stays neutral with regard to jurisdictional claims in published maps and institutional affiliations.

© 2020 by the authors. Licensee MDPI, Basel, Switzerland. This article is an open access article distributed under the terms and conditions of the Creative Commons Attribution (CC BY) license (http://creativecommons.org/licenses/by/4.0/).

Article

The Inhibition of Prolyl Oligopeptidase as New Target to Counteract Chronic Venous Insufficiency: Findings in a Mouse Model

Giovanna Casili, Marika Lanza, Sarah Adriana Scuderi, Salvatore Messina, Irene Paterniti, Michela Campolo and Emanuela Esposito *

Department of Chemical, Biological, Pharmaceutical and Environmental Sciences, University of Messina, 98166 Messina, Italy; gcasili@unime.it (G.C.); mlanza@unime.it (M.L.); sarascud@outlook.it (S.A.S.); smessina23@gmail.com (S.M.); ipaterniti@unime.it (I.P.); campolom@unime.it (M.C.)
* Correspondence: eesposito@unime.it

Received: 20 November 2020; Accepted: 10 December 2020; Published: 13 December 2020

Abstract: (1) Background: Chronic venous insufficiency (CVI) is a common disorder related to functional and morphological abnormalities of the venous system. Inflammatory processes and angiogenesis alterations greatly concur to the onset of varicose vein. KYP-2047 is a selective inhibitor of prolyl oligopeptidase (POP), a serine protease involved in the release of pro-angiogenic molecules. The aim of the present study is to evaluate the capacity of KYP-2047 to influence the angiogenic and inflammatory mechanisms involved in the pathophysiology of CVI. (2) Methods: An in vivo model of CVI-induced by saphene vein ligation (SVL) and a tissue block culture study were performed. Mice were subjected to SVL followed by KYP-2047 treatment (intraperitoneal, 10 mg/kg) for 7 days. Histological analysis, Masson's trichrome, Van Gieson staining, and mast cells evaluation were performed. Release of cytokines, nitric oxide synthase production, TGF-beta, VEGF, α-smooth muscle actin, PREP, Endoglin, and IL-8 quantification were investigated. (3) Results: KYP-2047 treatment ameliorated the histological abnormalities of the venous wall, reduced the collagen increase and modulated elastin content, lowered cytokines levels and prevented mast degranulation. Moreover, a decreased expression of TGF-beta, eNOS, VEGF, α-smooth muscle actin, IL-8, and PREP was observed in in vivo study; also a reduction in VEGF and Endoglin expression was confirmed in tissue block culture study. (4) Conclusions: For the first time, this research, highlighting the importance of POP as new target for vascular disorders, revealed the therapeutic potential of KYP-2047 as a helpful treatment for the management of CVI.

Keywords: chronic venous insufficiency; prolyl oligopeptidase (POP); inflammation; angiogenesis; endothelial disfunction

1. Introduction

Chronic venous insufficiency (CVI) is a debilitating condition whose manifestations are extremely prevalent and reported to affect up to 80% of the population [1], with a varying percentage of incidences from 2 to 56% in men and 1 to 60% in women [2,3]; advanced age, obesity, and a positive family history represent the most important risk factors for developing this chronic venous disease [4]. The pathophysiology is complex, with a wide clinical spectrum, ranging from asymptomatic but cosmetic problems to severe symptoms; this includes telangiectases, reticular veins, varicose veins, edema, pigmentation and/or eczema, lipodermatosclerosis, and venous ulceration [5]. Genetic susceptibility and environmental factors affect CVI, generating a concerning symptomatology: pain, dermal irritation, swelling, skin changes, associated with a real risk of developing debilitating venous ulceration. The increase in venous pressure referred to as venous hypertension symbolizes the main signature of this venous pathology [2]. The symptomatology correlated to CVI results in various clinical signs that may severely compromise life quality, carrying out the importance of a speedy recognition to provide symptomatic relief and prevent pathology progression [6]. CVI is a relatively common health problem but is often ignored by healthcare providers because of an underappreciation of the magnitude and impact of the problem, as well as incomplete recognition of the various presenting manifestations of primary and secondary venous disorders [7]. There is a wide range of clinical options—both conservative and invasive—for the treatment of CVI, united by a single goal, which is to improve the symptomatology and to prevent sequelae and complications, promoting the ulcer healing [8–10]. Despite great therapeutic advances, there is to date no intervention that can definitively prevent or resolve the recurrence of this chronic venous disease. Hence there is a need to better understand the focal points of the pathogenetic process in order to find better therapeutic strategies.

Angiogenesis is crucial in the formation of collateral vessels as part of an adaptive response to vascular occlusion and ischemia, playing an important role in conditions including vascular diseases; conversely, excessive pathological angiogenesis driven by inflammation is a key contributor to the development and progression of cardiovascular-related disorders, representing an unfavorable process in venous system disorders [11,12]. It is known that angiogenesis inhibition could significantly impact the tissue breakdown and could hereby enable the formation of venous ulcerations [13]. In turn, activated leukocytes can release a large amount of elastase and other proteinases associated with tissue injury and lipodermatosclerotic skin remodeling. Specifically, the phenomenon of the "leucocyte trap" is of great interest, suggesting the fundamental importance of inflammation in the pathogenesis of venous insufficiency [14,15].

Prolyl oligopeptidase (POP), also known as propyl endopeptidase, is a serine protease involved in the release of pro-angiogenic and anti-fibrogenic molecules. Particularly, POP is present in all organs, localized in specific cells and cell layers across the brain and peripheral tissues (as skeletal muscle, testis, liver, kidney, lung, renal cortex, gut) and it is implicated in the hydrolysis of proline-containing bioactive peptides, such as angiotensins, arginine-vasopressin, substance P, and neurotensin [16,17]; it has also been shown to be involved in several other physiological and pathological functions such as inflammation [18].

Several potent substrate-like POP inhibitors have been developed and 4-phenyl-butanoyl-L-prolyl-2(S)-cyanopyrrolidine (KYP-2047) appears to be the most potent and widely studied in vitro and in vivo model. The aim of the present study was to evaluate the capacity of KYP-2047 to influence the angiogenic and inflammatory mechanisms involved in the pathophysiology of CVI, to restore normal vascular blood flow. Particularly, anti-inflammatory and anti-angiogenic activities of treatment with KYP-2047 were evaluated through an in vivo mouse model of CVI performed by saphene-vein ligation and confirmed in a tissue block culture study.

2. Materials and Methods

2.1. Animals

Male NMRI mice (12 weeks old) were obtained from Charles River and housed under specific pathogen-free conditions with the approval of the local animal ethics and welfare committee. The animals were maintained in a 12-h light–dark cycle and were provided with food and water at libitum. Animal experiments are in compliance with Italian regulations on protection of animals used for experimental and other scientific purposes (DM 116192) as well as EU regulations (OJ of EC L 358/1 12/18/1986) and ARRIVE guidelines. This study was approved by the University of Messina Review Board for the care of animals in compliance with Italian regulations on protection of animals (n 499/2018-PR released on 23 February 2018).

2.2. Saphene Vein Ligation (SVL) Model

All animals were treated under general anesthesia and oxygenated with a mechanical respirator. The animal was placed in the lateral supine position on the operating table. The hind legs were shaved, the skin was disinfected with iodine, and sterile draping was applied. A small transverse skin incision at the level of the ankle was performed, to exhibit underlying vasculature the lateral saphenous vein; it was chosen because of similarity to the human saphenous vein regarding diameter and length. The ligation was performed as previously described [19]. Proximally and distally the lateral saphenous vein was ligated with a Vicryl 3-0 suture (Johnson & Johnson, New Brunswick, NJ, USA). Long ends of the suture were used for identification of the proximal end. Large side branches were ligated with a similar suture or with a titanium clip. The veins were surgically exposed over the total length of treatment and monitored macroscopically for occlusion or any complication (perforation, rupture, or vein wall hematoma).

2.3. Experimental Groups

Mice were divided into six experimental groups:

Group 1: Sham + vehicle, control group to which the saphenous vein ligation was not performed, orally administered with saline for 7 days (n = 8);

Group 2: Sham + Simvastatin, group to which the saphenous vein ligation was not performed, orally administered with Simvastatin (20 mg/kg) for 7 days (n = 8);

Group 3: Sham + KYP-2047 group to which the saphenous vein ligation was not performed; KYP-2047 (10 mg/kg) was intraperitoneal (i.p.) administered for 7 days (n = 8);

Group 4: Saphene vein ligation (SVL), group subjected to ligation of the saphenous vein, orally administered with saline for 7 days (n = 8);

Group 5: SVL + Simvastatin, group subjected to ligation of the saphenous vein, orally administered with Simvastatin (20 mg/kg), 30 min after saphene vein ligation, for 7 days (n = 8);

Group 6: SVL + KYP-2047, subjected to ligation of the saphenous vein, i.p. administered with KYP-2047 (10 mg/kg), 30 min after saphene vein ligation, for 7 days (n = 8)

KYP-2047 was dissolved in saline containing 5% of Tween 80 and administered at the recommended dose of 10 mg/kg, according to the bibliography [20]. Moreover, mice were treated with Simvastatin because statins represent a potential pharmacological treatment option suitable to prevent growth of varicose veins and to limit the formation of recurrence after varicose vein surgery [21]. Based on animal weight and ingestion, the applied dose of Simvastatin was 20 mg/kg per day.

2.4. Histological Analysis

Seven days after the surgery, mice were sacrificed and lateral saphene veins were collected for histopathological examination and standard hematoxylin and eosin (H&E) staining was performed. Briefly, lateral saphene veins were before fixed in 10% (w/v) PBS-buffered formalin and then 7-μm sections were prepared from paraffin-fixed tissues. Following dehydration with ethanol and embedding with paraffin, 7 μm sections were made, followed by H&E staining. Based on the knowledge of the histopathology of varicose vein disorder [22], the following morphological criteria were considered to perform the histological score: score = 0, no structural or morphological damage to the three layers of the vessel wall; score = 1, slight morphological alteration; score = 2, dilation of the vessel lumen and structural alteration of the vessel wall; score = 3, hypertrophy of the tunica media, followed by slight neutrophilic accumulation; score = 4, structural alteration of the tunica adventitia and formation of focal epithelial edema with narrowing of the vessel lumen; score = 5, high presence of neutrophilic infiltrate and collapse of the whole vessel wall. The sections were evaluated by computer-assisted color image investigation (Leica QWin V3, Cambridge, UK). The histological results were showed 20× (50 μm of the Bar scale).

2.5. Masson Trichrome Staining

To evaluate the degree of fibrosis, tissue sections from saphene veins were stained with Masson trichrome according to the manufacturer's protocol (Bio-Optica, Milan, Italy), as previously described [23]. Particularly, the entire area represented by the three tunicae, surrounding the veins, was considered for the quantification of the collagen in each section.

2.6. Van Gieson Staining

To detect the elastic fibers, tissue sections from saphene veins were stained with Elastica van Gieson staining kit, according to the manufacturer's protocol (#115974, Sigma-Aldrich, St. Louis, MO, USA), as previous described [24].

2.7. Toluidine Blue Staining

For evaluation of the number and degranulation of mast cells, tissue sections from saphene veins were stained with toluidine blue (#05-M23001, Bio-Optica, Milan, Italy) as previously described [25].

2.8. Western Blot Analysis for Interleukin 1β (IL-1β), Tumor Necrosis Factor α (TNF-α), Transforming Growth Factor β (TGFβ1), Vascular Endothelial Growth Factor (VEGF), α-Smooth Muscle Actin (αSMA) and Prolyl Endopeptidase

Total cytosolic and nuclear extracts were prepared, as previously described [26], on saphene veins. The following primary antibodies were used: anti- IL-1β (Santa Cruz Biotechnology, Dallas, TX, USA, 1:500 #sc12742, D.B.A, Milan, Italy), anti-TNF-α (Santa Cruz Biotechnology; 1:500 #sc52746), anti-TGFβ1 (Santa Cruz Biotechnology, 1:500 #sc130348, D.B.A, Milan, Italy), anti-VEGF (Santa Cruz Biotechnology; 1:1000 #sc7269), anti- αSMA (Santa Cruz Biotechnology; 1:500 #sc53015) and anti-prolyl endopeptidase (Abcam; 1:1000,#ab58988) in 1× phosphate-buffer saline (Biogenerica srl, Catania, Italy), 5% w/v non-fat dried milk, 0.1% Tween-20 at 4 °C overnight. Membranes were incubated with peroxidase-conjugated bovine anti-mouse IgG secondary antibody (Jackson ImmunoResearch, West Grove, PA, USA; 1:2000) for 1 h at room temperature. Anti-β-actin (Santa Cruz Biotechnology; 1:1000 #sc47778) and anti-βTubulin (Santa Cruz Biotechnology; 1:1000 #sc5274) antibodies were used as controls. Protein expression was detected by chemiluminescence (ECL) system (Thermo, Waltham, MA, USA), visualized with the ChemiDoc XRS (Bio-Rad, USA), and analyzed by using Image Lab 3.0 software (Bio-Rad, Hercules, CA, USA) as previously reported [27].

2.9. ELISA Kit Assay for eNOS and IL-8, Pro-Collagen 1 Alpha, and TGFβ1

ELISA kit assay for endothelial nitric oxide synthase (eNOS) and interleukin 8 (IL-8) were performed respectively on saphene vein samples, as previously described [28]. In details, samples were thawed on ice and homogenized in 300 µL lysis buffer (750 µL, Pierce #87787, Thermo Fisher Scientific, Waltham, MA, USA) supplemented with a protease inhibitor cocktail (Sigma-Aldrich, Rehovot, Israel). Thereafter, the samples were homogenized and centrifuged at 14,000× g for 10 min at 4 °C; supernatants were collected, aliquoted, and stored at −20 °C. eNOS, IL-8, Pro-collagen 1 alpha, and TGFβ-1 were measured by ELISA kits according to the manufacturer's instructions. The following kits, for mouse protein identifications, were used: mouse eNOS ELISA Kit (ab230938; Abcam, Cambridge, UK), mouse IL-8 ELISA Kit (MBS7606860; MyBiosource, San Diego, CA, USA), mouse Pro-Collagen I alpha 1 ELISA Kit, (ab229425; Abcam), and mouse TGF beta 1 ELISA Kit (ab119557; Abcam).

2.10. Myeloperoxidase (MPO) Activity

Veins tissues were analyzed for myeloperoxidase (MPO) activity, an indicator of polymorphonuclear leukocyte accumulation, using a spectrophotometric assay with tetramethylbenzidine as substrate, according to a method previously described [29]. MPO activity, as the quantity of enzyme degrading 1 µmol of peroxide per min at 37 °C, was expressed in U/g wet tissue.

2.11. Primary Culture of Vascular Smooth Muscle Cells (VSMCs) from Murine Saphene Vein: Tissue Block Culture Study

To demonstrate the compatibility of the in vivo model experiment, a study from tissue block culture was performed, as previously described [30]. The procedure for the experimental method includes the following steps: replication of in vivo study, isolation of the saphene vein, removal of the fat tissue around the vein, separation of the media, cutting the media into small tissue blocks, transferring the tissue blocks to cell culture plates, and incubation until the cells reach confluence.

2.11.1. Experimental Groups for In Vivo Study

Mice were divided into three experimental groups:

Group 1: Sham + vehicle, control group to which the saphenous vein ligation was not performed, orally administered with saline for 7 days (n = 8);
Group 2: Saphene vein ligation (SVL), group subjected to ligation of the saphenous vein, orally administered with saline for 7 days (n = 8);
Group 3: SVL + KYP-2047, group subjected to ligation of the saphenous vein, administered with KYP-2047 (i.p., 10 mg/kg), 30 min after saphene vein ligation, for 7 days (n = 8).

After the sacrifice, saphene veins samples from each mouse, were picked up and longitudinally cut and placed in another cell culture dish containing DMEM. After removal of the fat tissue around the vein, the media tunica was extracted from the vein by pressing and pushing the vein with its blunt back side. Then, the media was cut into approximately 1-mm squares and transferred into cell culture. DMEM containing 20% FBS was carefully added and the tissue blocks were incubated in cell culture chamber for 5 days. The cells were identified as vascular smooth muscle cells (VSMCs). VSMCs obtained from vein samples of each experimental group were processed for immunofluorescence analysis and ELISA kit detection.

2.11.2. Immunofluorescence Analysis

Immunofluorescence analysis was performed as previously described [31]. VSMCs were plated (1×10^4/well) on glass coverslips in a culture dish of size 100 mm. After 24 h of adhesion at 37 °C and 5% CO_2, cells were fixed in 4% paraformaldehyde and rinsed briefly in phosphate-buffered saline (PBS: 0.15 M NaCl, 10 mM Na_2HPO_4, 3 mM NaN_3, pH 7.4), permeabilized in 0.2% Triton X-100/PBS and blocked with 10% goat serum. The cells were stained with an antibody against VEGF (1:200, Monoclonal Antibody JH121; Thermofisher, USA) O/N, followed by ITC-conjugated anti-mouse Alexa Fluor-488 antibody (1:2000 *v/v* Molecular Probes, UK) for 1 h at 37 °C. Sections were washed and for nuclear staining 4′,6′-diamidino-2-phenylindole (DAPI; Hoechst, Frankfurt; Germany) 2 µg/mL in PBS was added. Sections were observed and photographed at ×40 magnification using a Leica DM2000 microscope (Leica).

2.11.3. ELISA Kit for Endoglin

ELISA kits assay for and Endoglin (CD105) was performed on primary cells VSMCs obtained from vein samples, as previous described [28]. In details, cell culture supernatants collected from VSMCs after 24 h of adhesion at 37 °C and 5% CO_2, were thawed on ice, aliquoted, and stored at −20 °C. The following kit, for mouse protein identifications, was used: Endoglin/CD105 ELISA Kit (ab240677; Abcam, Cambridge, UK).

2.12. Materials

KYP-2047 (Sigma, cat#SML0208, Lot#032M4606V) was obtained by Sigma-Aldrich (Milan, Italy). Unless otherwise stated, all compounds were obtained from Sigma-Aldrich (St. Louis, MO, USA). All other chemicals were of the highest commercial grade available. All stock solutions were prepared in non-pyrogenic saline (0.9% NaCl, Baxter, Milan, Italy).

2.13. Statistical Analysis

All values are showed as mean ± standard error of the mean (SEM) of N observations. N denotes the number of animals employed. The experiment is representative of at least three experiments performed on different days on tissue sections collected from all animals in each group. Data were analyzed with GraphPad 5 software, by one-way ANOVA followed by a Bonferroni post-hoc test for multiple comparisons. A p-value of less than 0.05 was considered significant.

3. Results

3.1. The Expression of POP in CVI Mouse Model

To clearly demonstrate the role of POP in CVI insufficiency, we performed a Western Blot analysis to detect PREP in vein samples. A basal expression on PREP was observed in samples from control group (Figure 1A,B) compared to the significant increase observed in samples from CVI-injured group (Figure 1A,B), for the first time demonstrating an over expression of PREP in these venous pathology. Moreover, small peptides structure-activity studies have shown that POP covalent inhibitors, as KYP-2047, are more potent and effective than their non-covalent analogs [32] because of the transient covalent bond with the enzyme that is hydrolyzed after a short time. In this study we confirmed the inhibitory role of KYP-2047 on POP enzymatic activity, also demonstrating an important inhibition effect on POP expression (Figure 1A,B).

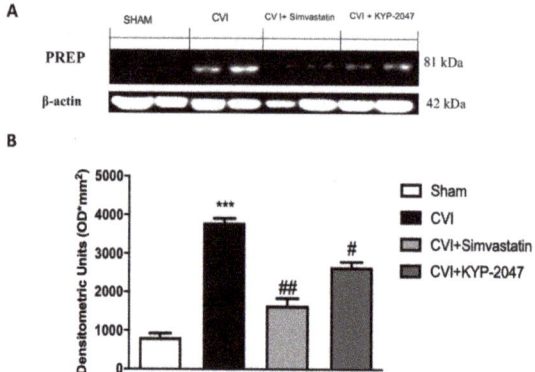

Figure 1. The inhibitor effects of KYP-2047 on prolyl oligopeptidase (POP) in chronic venous insufficiency (CVI) mouse model. Western Blot analysis to detect PREP in vein samples, clearly demonstrated the increase of POP in CVI insufficiency (**A**,**B**) compared to the control group (**A**,**B**). Moreover, this study confirmed the inhibitory role of KYP-2047 on POP enzymatic activity, also demonstrating an important inhibition effect on POP expression (**A**,**B**). Data represent the means of at least three independent experiments. One-way ANOVA followed by Bonferroni post-hoc. *** $p < 0.001$ versus Sham and ## $p < 0.01$ and # $p < 0.05$ versus CVI.

3.2. Role of KYP-2047 Treatment on TGF-β1 and IL-8, as Vascular Markers in CVI

An important mediator in CVI is represented by TGF-β1, a highly complex polypeptide involved in venous pathophysiology [33]; particularly, TGF-β1 contributes to specific pathological processes concerning the vessel wall [34], participating in vascular pathologies associated with matrix remodeling and fibrosis [35]. In this research, TGF-β1 expression levels were analyzed by Western Blot, suggesting an increment of this marker in SVL group, compared to control animals (Figure 2A, see the densitometric units score Figure 2B). Treatment with KYP-2047 significantly reduced TGF-β1 expression (Figure 2A, see the densitometric units score Figure 2B). GF-β1 is a pleiotropic factor that plays pivotal roles in angiogenesis and thus is indispensable for the development and homeostasis of the vascular system [36]. Moreover, varicose veins had a distinct chemokine expression pattern, since significant up-regulation of IL-8 [37], an angiogenic chemokine produced by a variety of cell types [38]. To understand the modulation of KYP-2047 treatment on IL-8 expression, we performed an ELISA kit for IL-8 on vein samples, observing a notable reduction on IL-8 expression in samples treated with KYP-2047 compared to CVI-damaged group (Figure 2C).

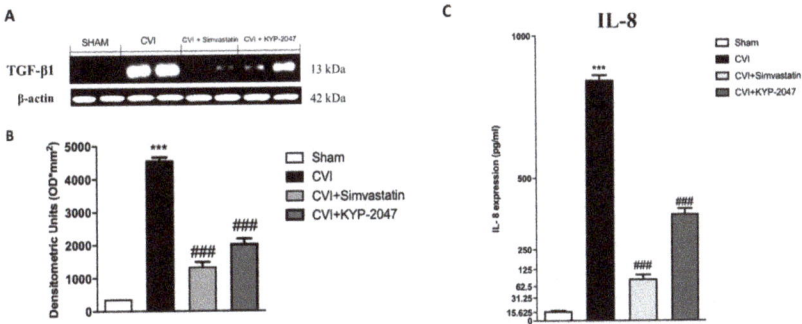

Figure 2. KYP-2047 treatment on TGF-β1 expression and IL-8 content. TGF-β1 expression was analyzed by Western Blot, suggesting an increment of this marker in CVI group, compared to control animals (**A**), see the densitometric units score (**B**); treatment with KYP-2047 significantly reduced TGF-β1 expression (**A**), see the densitometric units score (**B**). ELISA kit for IL-8 expression on saphene vein samples was performed; treatment with KYP-2047 (10 mg/kg, i.p.) significantly reduced IL-8 quantification (**C**), compared to the high amount of IL-8 released in the vein samples subjected to damage (**C**). Data represent the means of at least three independent experiments. One way ANOVA followed by Bonferroni post-hoc. *** $p < 0.001$ versus Sham; ### $p < 0.001$ versus CVI.

3.3. Role of KYP-2047 Treatment on Angiogenesis Modulation and Vasodilation

To highlight the in vivo modulatory action of KYP-2047 on angiogenesis, a Western blot analysis was performed to quantify VEGF and α-SMA expression on vein samples. In details, α-SMA, an isoform of the vascular smooth muscle actins, typically expressed in the vascular smooth muscle cells contributing to vascular motility and contraction, was found to be increased in varicose veins [39]; while VEGF plays an important role in maintaining the integrity of blood vessel walls [7]. VEGF/α-SMA ratio expression levels increased in CVI group, compared to control animals (Figure 3A, see the densitometric units score Figure 3A); treatment with KYP-2047 significantly reduced VEGF/α-SMA expression (Figure 3A, see the densitometric units score Figure 3B), like Simvastatin group (Figure 3A, see the densitometric units score Figure 3A). VEGF primarily exerts its effect through the production of vasodilatory mediators. VEGF signaling through VEGFR increases nitric oxide (NO) production, acutely bring to the eNOS activation [40]. The resultant increase in NO production promotes vascular permeability and endothelial cell survival, because NO also diffuses to adjacent vascular smooth muscle cells and mediates endothelium-dependent vasodilation [41]. To better understand the capacity of KYP-2047 treatment in vessel remodeling trough endothelium-derived nitric oxide, ELISA kit for eNOS expression on saphene vein samples was performed. Interestingly, treatment with KYP-2047 (10 mg/kg, i.p.) significantly reduced eNOS quantification (Figure 3C), compared to the high amount of eNOS released in the vein samples subjected to damage (Figure 3C).

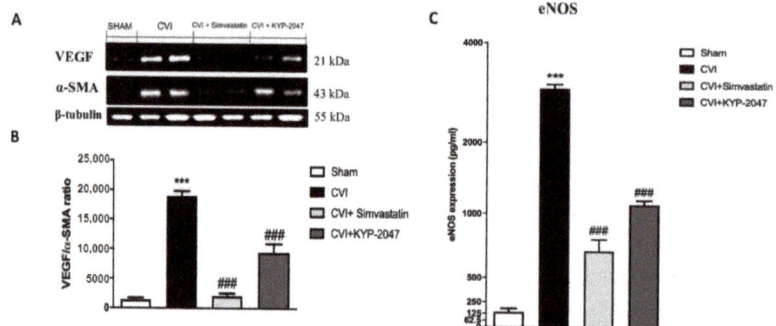

Figure 3. KYP-2047 treatment on VEGF/α-SMA expression and eNOS levels. VEGF/α-SMA ratio expression was analyzed by Western blot, suggesting an increment of this marker in CVI group, compared to control animals (**A**), see the densitometric units score (**B**); treatment with KYP-2047 significantly reduced VEGF/α-SMA expression (**A**), see the densitometric units score (**B**). ELISA kit for eNOS expression on saphene vein samples was performed; treatment with KYP-2047 (10 mg/kg, i.p.) significantly reduced IL-8 quantification (**C**), compared to the high amount of eNOS released in the CVI-damaged groups (**C**). Data represent the means of at least three independent experiments. One-way ANOVA followed by Bonferroni post-hoc. *** $p < 0.001$ versus Sham; ### $p < 0.001$ versus CVI.

3.4. Treatment with KYP-2047 on Cytokines Expression in SVL-Damaged Mice

Recent findings indicate that inflammatory processes are crucial for the development of incompetent valves and vein wall remodeling in CVI [42]. In this study, both IL-1β and TNF-α expression levels, assessed by Western blotting in saphene veins tissue, were increased in CVI group, compared to sham basal levels (respectively Figure 4A, see the densitometric units score Figure 4B,C, see the densitometric units score Figure 4D). KYP-2047 administration (10 mg/kg, i.p.) significantly reduced both cytokines proteins expression, similar to the election treatment represented by Simvastatin (20 mg/kg, oral) (respectively Figure 4A, see the densitometric units score Figure 4B,C, see the densitometric units score Figure 4D).

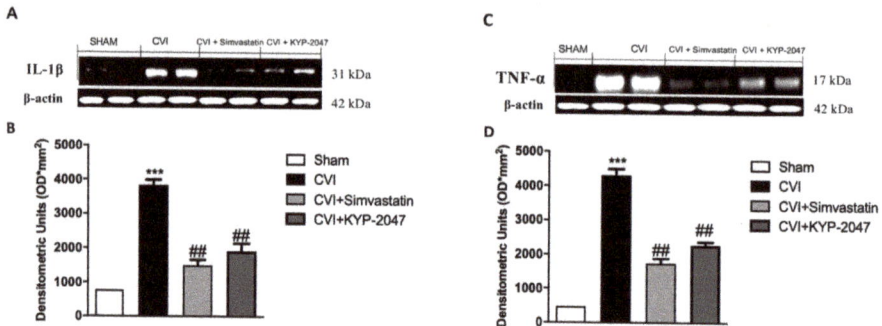

Figure 4. Treatment with KYP-2047 on cytokines expression. IL-1β and TNF-α expression levels, monitored by Western blotting in saphene veins tissue, were increased in CVI group, compared to sham basal levels respectively (**A**), see the densitometric units score (**B**,**C**), see the densitometric units score (**D**). KYP-2047 administration (10 mg/kg, i.p.) significantly reduced both cytokines proteins expression, similar to the election treatment represented by Simvastatin (20 mg/kg, oral) respectively (**A**), see the densitometric units score (**B**,**C**), see the densitometric units score (**D**). Data represent the means of at least three independent experiments. One-way ANOVA followed by Bonferroni post-hoc. *** $p < 0.001$ versus Sham; ## $p < 0.01$ versus CVI.

3.5. Effects of KYP-2047 on Histological Damage and Neutrophilic Activation Induced by SVL in Mice

Histopathologic examination of lateral saphene vein subjected to ligation for 7 days revealed inflammatory cell infiltration, structural alteration of the tunica adventitia, and formation of focal epithelial edema, with the collapse of the whole vessel wall (Figure 5D, see histological score Figure 5G) compared to sham group (Figure 5A, see histological score Figure 5G). In control mice, treated with Simvastatin or with KYP-2047, for 7 days, no major modification of the vessel lumen was found, as well as no signs of tissue inflammation (Figure 5B,C, see histological score Figure 5G); treatment with KYP-2047, at the dose of 10 mg/kg, for 7 days significantly reduced hypertrophy of the tunica media, holding the collapse of the wall caused by the ligation (Figure 5F, see the histological score Figure 5G). The effect of KYP-2047 treatment was similar to the elective Simvastatin treatment (20 mg/kg), that prevented vessel lumen modification and tissue inflammation (Figure 5E, see the histological score Figure 5G). Moreover, the venous hypertension contributes to leucocytes accumulation and neutrophilic activation [43], that was measured in this study by analyzing the Myeloperoxidase (MPO) activity on veins samples. Particularly, a significant increase in MPO activity was observed in CVI-injured group compared to control (Figure 5H), while treatment with KYP-2047, like Simvastatin, significantly reduced the neutrophilic activation detected in veins subjected to ligation (Figure 5H).

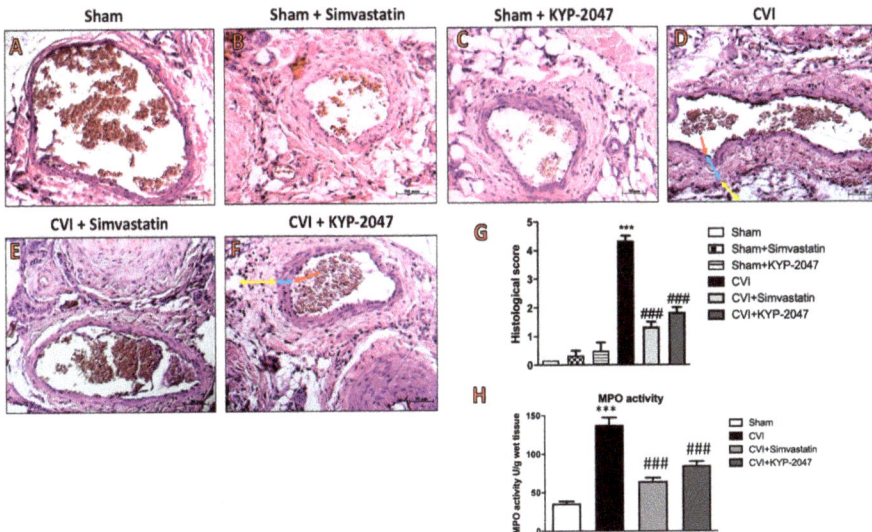

Figure 5. Treatment with KYP-2047 on histological damage and MPO activity induced by SVL in mice. Histopathologic examination, by hematoxylin-eosin staining, of lateral saphene vein subjected to ligation for 7 days (**D,G**) revealed an improvement in saphene vein treated with KYP-2047 (**F,G**) or Simvastatin (**E,G**). No damage or structural vein wall modification were observed in control groups (**A–C,G**). In particular, we detected tunica intima (red line), tunica media (blue line), tunica adventitia (yellow line), showing as treatment with KYP-2047 significantly reduced the hypertrophy of tunica media and preventing the collapse of the wall caused by the ligation (**F,G**). The MPO activity confirmed the protective effects of KYP-2047 to reduce inflammation in CVI mouse model (**H**). Data represent the means of at least three independent experiments. One-way ANOVA followed by Bonferroni post-hoc. *** $p < 0.001$ versus Sham and ### $p < 0.001$ versus CVI.

3.6. Role of KYP-2047 Treatment in the Collagen Content Reduction and Elastin Replacement

The nature and distribution of venous disease surrounding the development of varicose veins correlate with collagen and elastin as important components of the vein walls that affect their function [44]. The vein wall resistance to stretch depends on the collagen fibers and it is known that in CVI pathology, varicose veins show increased collagenosis and dilated distal varicosities [44]. This study confirmed that in CVI-provoked lesions, 7 days post-saphene ligation, the degree of fibrosis, assessed by Masson trichrome staining demonstrated a fibrotic area stained blue that was larger in the CVI group (Figure 6B, see collagen content score Figure 6E) compared to control group (Figure 6A, see collagen content score Figure 6E). Treatment with KYP-2047 significantly reduced the blue staining, which represents collagen depot, located in the tunica adventitia (Figure 6D, see collagen content score Figure 6E), almost similar to Simvastatin treatment (Figure 6C, see collagen content score Figure 6E). The collagen content in veins samples was also investigated by kit ELISA for pro-collagen I, highlighting a significant increase of collagen depot in CVI-injured groups compared to control mice (Figure 6K), also confirming the role of KYP-2047 and Simvastatin to reduce this collagen accumulation (Figure 6K). Contrarily, when the vessel wall is stretched, elastin generates a retractive force that opposes the lengthening force caused by pressure in the lumen vessel [45]. In this study, we observed a decreased amount of elastin in veins from SVL-damaged group, compared to control group (respectively, Figure 6F,G, see elastin content score Figure 6J); instead, treatment with KYP-2047, albeit to a lesser extent than Simvastatin treatment, notably increased elastin content thus preventing the collapse of lumen vessel (respectively, Figure 6H,I, see elastin content score Figure 6J).

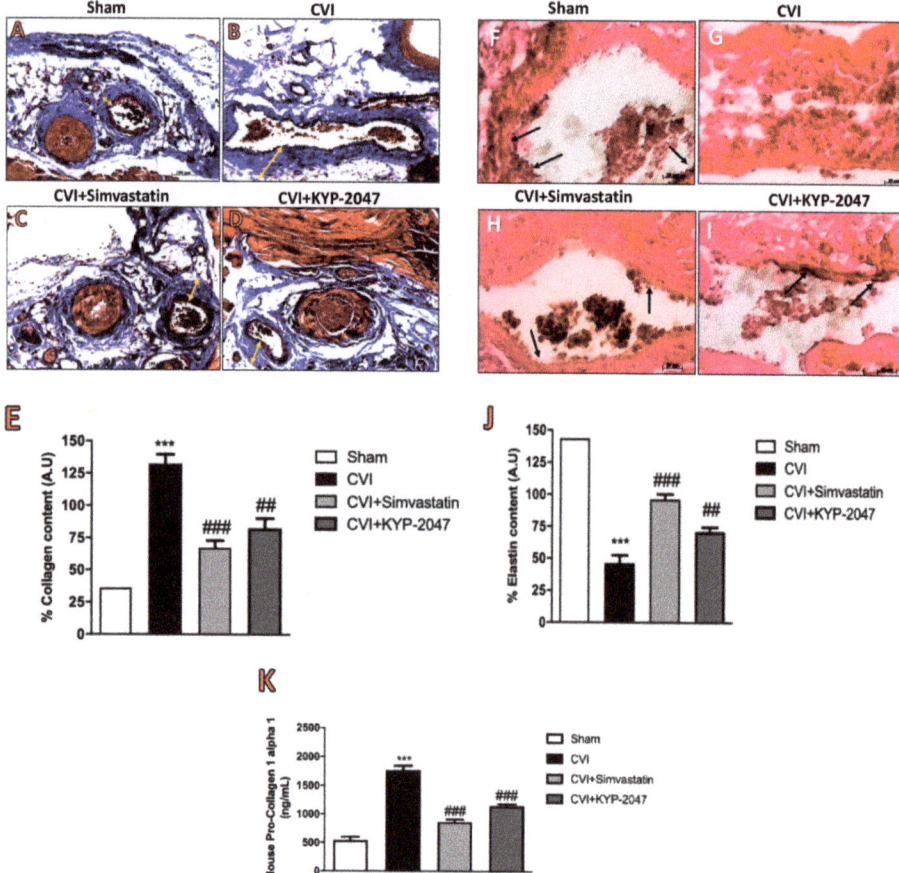

Figure 6. Treatment with KYP-2047 in the collagen content reduction and elastin increase. The degree of fibrosis, in CVI-provoked lesions, 7 days post-saphene vein ligation, was assessed by Masson trichrome

staining and Van Gieson staining, revealing a fibrotic area stained blue larger than in control group (**B**), see collagen content score (**E**); (**A**), see collagen content score (**E**) and a reduced elastin content highlighted in red purple (**G**), see elastin content score (**J**); (**F**), see elastin content score (**J**). Treatment with KYP-2047 significantly reduced collagen depot, located in the tunica adventitia (**D**), see collagen content score (**E**) and increased the content of elastin fibers (**I**), see collagen content score (**J**) almost similarly to Simvastatin treatment (respectively (**C**), see collagen content score (**E**), (**H**), see elastin content score (**J**). Pro-collagen 1 levels were detected by ELISA kit, confirm the results observed by Masson trichrome staining (**K**). Orange arrows identified collagen content, while black arrows identified elastin content. Scale bar for figures A–D: 100 µm, magnification 10 ×. Scale bar for figures F–I: 20 µm, magnification 40 ×. Data represent the means of at least three independent experiments. One-way ANOVA followed by Bonferroni post-hoc. *** $p < 0.001$ versus Sham; ## $p < 0.01$ and ### $p < 0.001$ versus CVI.

3.7. Role of KYP-2047 Treatment in Preventing Mast Cell Degranulation

Increased infiltration of activated mast cells has been recently implicated in the pathophysiology of varicose veins [46]; mast cells produce and release various kinds of vasoactive substance, including histamine, tryptase, prostaglandins, leukotrienes, and cytokines, that enhance local vasopermeability, leading to intimal thickening [47]. Mast cells were best identified in saphene veins tissues by their characteristic numerous metachromatic granules by using toluidine blue stain (Figure 7). Compared with the control group, the number of mast cells in CVI-damaged group was significantly increased (respectively, Figure 7A,B, see graph Figure 7E). Mast cells degranulation was appreciably reduced with KYP-2047 or Simvastatin treatment (respectively, Figure 7D,C, see graph Figure 7E).

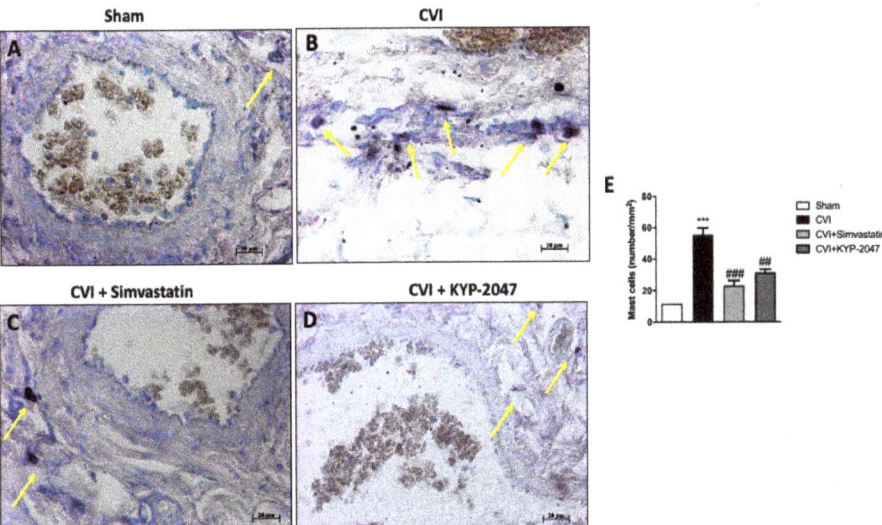

Figure 7. Treatment with KYP-2047 treatment on mast cells. Mast cells were evaluated in saphene veins tissues by staining with toluidine blue. Compared with sham animals, the mast cells number in CVI-damaged group was significantly increased (respectively, (**A**), (**B**), see graph (**E**). Neutrophilic accumulation was appreciably reduced by KYP-2047 or Simvastatin treatment (respectively, (**C**,**D**), see graph Figure 3E). Yellow arrows indicated the mast cells presented in the area. Scale bar for figures F-I: 20 µm, magnification 40 ×. Data represent the means of at least three independent experiments. One way ANOVA followed by Bonferroni post-hoc. *** $p < 0.001$ versus Sham; ## $p < 0.01$ and ### $p < 0.001$ versus CVI.

3.8. Evaluation of KYP-2047 in a Saphene Vein Block Culture Study

To underline the modulatory action of the KYP-2047 on angiogenesis, the in vivo study was repeated performing a new study from tissue block culture. The VSMCs obtained from tissue block culture study have been tested by immunofluorescence analysis for VEGF, considering that an up-regulation of this marker in the skin of patients with CVD has been demonstrated [48]. In this study, VEGF expression significantly increased in VSMCs from CVI-tissue block (Figure 8D–F, see VEGF positive score Figure 8J compared to control cells (Figure 8A–C, see VEGF positive score Figure 8J). Treatment with KYP-2047 on CVI-damaged veins reduced angiogenesis to bring it back to physiological conditions (Figure 8G–I, see VEGF positive score Figure 8J). Moreover, endoglin, plays an important role in vascular development, regulating angiogenesis through the interaction with VEGF receptor [49]. In this study, endoglin content was evaluated through an ELISA kit performed on supernatants of VSMCs cell extracted from tissue blocks; a notable increase in endoglin expression (pg/mL), was observed in CVI samples, compared to control group (Figure 9A), while VSMCs treated with KYP-2047 showed a reduced endoglin content, released in supernatants (Figure 9A). Furthermore, as Endoglin the co-receptor of the TGF-β, this marker was evaluated by ELISA kit also on VSMCs cell extracted from tissue blocks, observing for the first time a significant increase in TGF-β quantity (pg/mL) in CVI samples compared to control (Figure 9B), while treatment with KYP-2047 decreased this content (Figure 9B).

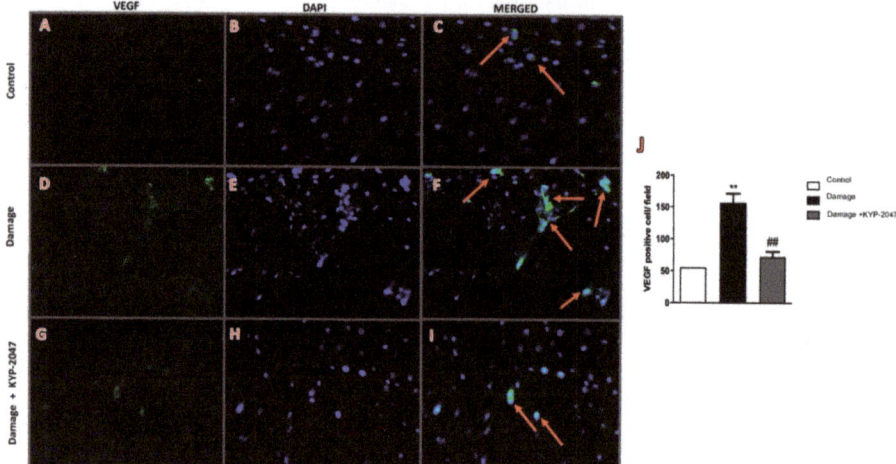

Figure 8. VEGF expression on saphene vein from tissue block culture study. Immunofluorescence analysis, on tissue block culture study, revealed an increase in VEGF expression in VSMCs from CVI-tissue block (**D–F**), see VEGF positive score (**J**) compared to control cells (**A–C**), see VEGF positive score (**J**). POP inhibition as treatment on CVI-damaged veins positively modulated angiogenesis as seen by VEGF expression in CVI + KYP-2047 tissue block group (**G–I**), see VEGF positive score (**J**). Red arrows refers to VEGF-positive cells merged with DAPI. Scale bar for figures F–I: 20 μm, magnification 40×. Data represent the means of at least three independent experiments. One-way ANOVA followed by Bonferroni post-hoc. ** $p < 0.01$ versus Control; ## $p < 0.01$ versus CVI.

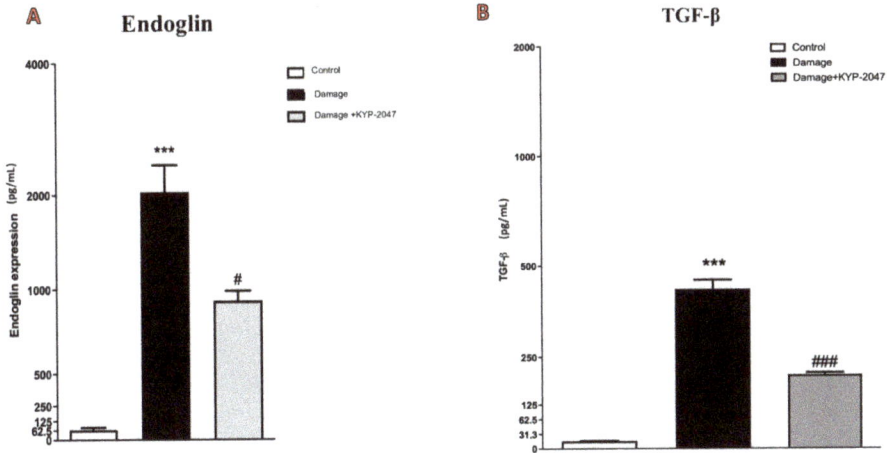

Figure 9. KYP-2047 treatment on TGF-β1 and Endoglin levels on saphene vein from tissue block culture study. The content of endoglin and TGF-β1 were evaluated by ELISA kits, revealing an increase in damaged group compared to control (**A,B**), while KYP-2047 treatment notably reduced this increase (**A,B**). Data represent the means of at least three independent experiments. One way ANOVA followed by Bonferroni post-hoc. *** $p < 0.001$ versus Sham; # $p < 0.05$ and ### $p < 0.001$ versus CVI.

4. Discussion

Chronic venous disease (CVD) is a very common problem [50], with an higher prevalence in Western countries where it already consumes up to 2% of the healthcare budgets [51]. However, because the most common manifestation of the pathology is represented by varicose veins, which begin as a result of incompetent valves and augmented venous pressure, the term CVI often symbolizes the full spectrum of manifestations of CVD, from simple telangiectases to skin fibrosis and venous ulceration [52]. Generally, when pressure is increased and return of blood is impaired, the onset of venous pathology is insured, resulting from valvular incompetence of axial deep or superficial veins, venous tributaries, venous obstruction, or a combination of these mechanisms [53]. The manifestations of CVI may be viewed in terms of a well-established clinical classification scheme; particularly, "The Clinical, Etiology, Anatomic, Pathophysiology" (CEAP) classification was developed by an international consensus conference to provide a basis of uniformity in reporting, diagnosing, and treating CVI [30]. The management of CVI starts with conservative measures to reduce the symptoms and prevent the development of secondary complications and progression of disease, then move on to further treatment if conservative measures fail or provide an unsatisfactory response [7]. Despite this, suitable pharmaceutical therapies to totally stem the pathology have not been explored to date and given the prevalence and socioeconomic impact of CVD, an understanding of new therapeutic options is warranted.

Recent findings indicate that inflammation and angiogenesis alterations greatly concur to the onset of CVI [31]; particularly, inflammatory tissue is often hypoxic and hypoxia can induce angiogenesis through upregulation of factors such as vascular endothelial growth factor (VEGF) and chemokines can both promote angiogenesis and stimulate the recruitment of inflammatory cells [54]. Recent findings indicate that peripheral POP may be involved in the inflammation [55] and in angiogenesis [56]. KYP-2047 is a very potent selective inhibitor of POP [57]. Thus, based on these evidences, the aim of this study was to evaluate the protective effect of KYP-2047 to counteract inflammatory and angiogenetic process involved in the pathophysiology of CVI through an in vivo mouse model of CVI.

First, the expression of POP was evaluated in this CVI mouse model, highlighting an increase expression of POP enzyme in vascular pathology, also confirming the efficiency of KYP-2047 to inhibit POP activity and, for the first time, to negatively modulate the expression of POP.

A critical inflammatory and angiogenetic mediator is represented by TGFβ-1, an important factor involved in regulating leucocyte and fibroblast recruitment and ECM remodeling, by both stimulating fibrogenesis and deposition of collagen [58]. TGF-β1 is able to act as a promoting and an inhibitory factor of angiogenesis and it is known to maintain a balance between apoptosis and cellular dysfunction, having a pivotal role in vessel remodeling during pathogenesis of vascular disorders [59]. In this study, an increased expression of TGFβ-1 was observed in damage conditions, while the treatment with KYP-2047 could greatly reduce the cascading events associated with the expression of TGFβ-1. TGF-β1 and IL-8 are important regulators of inflammation-induced angiogenesis [60], so the angiogenesis mechanisms involved in CVI disorders were also investigated evaluating IL-8, known as a potent factor, intricated in normal physiological processes and abnormal mechanisms, directly enhancing the endothelial cell survival and regulating vascular tone [61]; in this study, we confirmed an increased expression of IL-8 in vein samples compared to control, while treatment with Simvastatin and KYP-2047 reduced this increment.

Furthermore, the role of angiogenesis is correlated to the amount of α-SMA of primary varicose veins and various studies highlighted the relation between α-SMA and VEGF in vascular wall injury [59]. In this work, a significant reduction in VEGF/α-SMA ratio was observed in KYP-2047 and Simvastatin-treated groups, compared to CVI-damaged mice. VEGF primarily exerts its effect through the production of vasodilatory mediators, particularly causing an overproduction of NO [41]; NO release is associated with varicosity development [62] as well as a critical role for eNOS in controlling the magnitude of the acute inflammatory response for regulating microcirculatory endothelial barrier function [63]. In this study, we confirmed an increased augment of eNOS production in CVI-damaged group, closely associated with a significative decrement in Simvastatin and KYP-2047 groups.

The production/activity of vasodilatory mediators such as nitric oxide depends by cytokine-induced activation in vascular pathologies [64]. The state of inflammation in patients with venous disease highlighted that leucocytes sequestering leads to the activation of blood white cells, resulting in the release of free radicals, histamine, and neutrophil chemoattractants; the actions of these substances destroy the endothelial layer of the vessel and their basement membranes, increasing the vascular permeability and disturbing microcirculatory flow. Furthermore, the movement of white blood cells from the adventitia to the medial layer and the upper part of the vein wall, helped trigger the release of proinflammatory cytokines [65]. We demonstrated the capacity of KYP-2047, likewise Simvastatin treatment, to counteract the expression of pro-inflammatory markers, IL-1β and TNF-α, in SVL-damaged animals, highlighting the role of POP inhibitor to reduce the inflammation process in CVI.

The inflammatory response, the attraction of neutrophils, and the damage to veins are factors that perpetuate venous insufficiency and contribute to modify veins structure [66]. In this study, we observed by in vivo model of CVI that treatment with KYP-2047 significantly reduced the structural and morphological alterations provoked by ligation of saphene veins, decreasing the intimal and adventitia fibromuscular plaques, counteracting neutrophilic accumulation and preventing the whole wall vein collapse. Moreover, recent studies suggest that statins improve the microvascular function, effectively by inhibiting the development of varicose veins, thus representing a treatment option to prevent the growth of varicose veins and to limit the chances of recurrence after varicose vein surgery [21]. In this study, treatment with KYP-2047 showed to restore the histological alterations caused by vein ligation, similarly to the benefits observed by treatment with Simvastatin. Furthermore, as leucocytes become "trapped" in the circulation of the leg during periods of venous hypertension produced by sitting or standing, studies of the plasma levels of neutrophil granule enzymes show that these are increased during periods of venous hypertension, suggesting that this causes activation of the neutrophils. We confirmed neutrophilic activation in this CVI mouse model evaluating the MPO activity, that was significantly reduced by KYP-2047 or Simvastatin treatments.

Moreover, alterations on the connective tissue concentration and smooth muscle are visible in vascular disease [67]. In fact, the venous systems structure follows a pattern in concentric layers: intima, media, and adventitia, which undergo a modification of structural components of the vessel wall in conditions of altered venous flow, associated to loss of tone and the subsequent venous dilatation [68]. The Masson's trichrome stain revealed in red the smooth muscle fiber to smooth muscle fiber (SMF), while dyed bluish corresponded to the extracellular matrix (ECM), in which many elements such as collagen and elastin are arranged [69]. Some studies connected reflux with weakening of the venous walls, which may be due to an imbalance in the content of collagen and elastin in the vein [62]. In this study, the distribution and the relationships between SMF and ECM were impaired in CVI-damaged mice, thus reducing the tone and the progressive dilatation of the vein wall, compared to the control mice; while, treatment with KYP-2047 significantly reduced the disparity between the collagen and elastin content, similarly to Simvastatin treatment. The KYP-2047 modulation on collagen fibers in CVI was also confirmed by quantifying Pro-collagen 1, as the major vascular fibrillar collagen [63].

Recent findings suggest that inflammatory processes is crucial for the development of inept valves and vein wall remodeling; particularly, it is postulated that varicose lesions showed a greater extent of mast cell infiltration whereas baseline control veins had a smaller number of mast cells [70]. This research underlined the efficacy of KYP-2047 treatment to counteract mast cell degranulation phenomenon, compared to only CVI-damaged group, as noted by toluidine blue staining.

During the resolution phase of inflammation, the endothelial to mesenchymal (EndoMT)-mediated remodeling commonly occurs, bringing the endothelial cells from different vascular beds to respond differently to inflammatory stimuli [71]. Most of the signaling networks that are commonly utilized during epithelial-mesenchymal transition (EMT) are also responsible for EndoMT, seeing the active participation of TGF-β1 [72]. In this study, we confirmed for the first time, in a tissue block culture study, a significant increase in TGF-β1 quantity in VSMCs from CVI-tissue block, which might suggest a TGF-β-related induction of endothelial-to-mesenchymal transition. Furthermore, an important role in vascular development and tumor-associated angiogenesis is represent by Endoglin [73], a transmembrane auxillary receptor for TGF-β that is predominantly expressed on proliferating endothelial cells. It is known that endoglin promoted VEGF-induced tip cell formation, mechanistically, interacting with VEGF receptor (VEGFR) [49]. The demonstrated changes in mRNA expression, as well as in the contents of VEGF receptors, in the wall of varicose veins is accepted as one of the reasons for the clinical symptoms of the disease and can predispose to its progression [70]. In this research, we confirmed that the over-expression of both VEGF and endoglin in VSMCs cells from saphene veins subjected to binding, underlying as treatment with KYP-2047, inhibiting POP, significantly decreased these increases in angiogenesis markers.

CVI is a disease characterized by numerous risk factors, some of which suggest that disease can occur because of congenital valve or vessel abnormalities, but it most commonly occurs when the valves of the deep veins are damaged as a result of deep venous thrombosis, as well as, the expression of specific cardiovascular markers could contribute to the onset of the pathology. So, a limitation for this in vivo *study*, is represented by the difficulty in paying attention to all diagnostic markers that could contribute to the beginning of CVI.

Despite this, the results obtained in in vivo studies and confirmed in tissue block culture study, suggest a pivotal role of POP inhibition in the hypertrophy of the venous wall, although the exact mechanism leading to venous wall dilatations remains to be elucidated. Furthermore, the action of the KYP-2047, although it is similar to that of Simvastatin, today considered as the treatment most often used in CVD, has the extra gear because it is able to act in a targeted and effective manner.

5. Conclusions

Based on these results, POP inhibition due to KYP-2047, could represent a remarkable strategy to counteract the negative effects associated with vascular alterations. These data suggest a strong anti-inflammatory potential of KYP-2047 associated to its modulatory role on angiogenesis, that contribute to positively modulate CVI, offering new therapeutic target tools in the management of vascular pathologies.

Author Contributions: E.E. planned the experiments; G.C. and M.L. prepared the manuscript and analyzed the results; S.A.S., S.M. performed experiments; M.C. and I.P. supervised the research. All authors have read and agreed to the published version of the manuscript.

Funding: The authors have no other relevant affiliations or financial involvement with any organization or entity with a financial interest in or financial conflict with the subject matter or materials discussed in the manuscript apart from those disclosed. This research did not receive any specific grant from funding agencies in the public, commercial, or not-for-profit sectors.

Conflicts of Interest: The authors declare that they have no competing interests.

Availability of Data and Materials: The datasets used and/or analyzed during the current study are available from the corresponding author on reasonable request.

References

1. Onida, S.; Davies, A.H. Predicted burden of venous disease. *Phlebology* **2016**, *31*, 74–79. [CrossRef] [PubMed]
2. Raffetto, J.D.; Mannello, F. Pathophysiology of chronic venous disease. *Int. Angiol.* **2014**, *33*, 212–221. [PubMed]
3. Lichota, A.; Gwozdzinski, L.; Gwozdzinski, K. Therapeutic potential of natural compounds in inflammation and chronic venous insufficiency. *Eur. J. Med. Chem.* **2019**, *176*, 68–91. [CrossRef] [PubMed]
4. Santler, B.; Goerge, T. Chronic venous insufficiency-a review of pathophysiology, diagnosis, and treatment. *J. Dtsch. Derm. Ges.* **2017**, *15*, 538–556. [CrossRef]
5. Youn, Y.J.; Lee, J. Chronic venous insufficiency and varicose veins of the lower extremities. *Korean J. Int. Med.* **2019**, *34*, 269–283. [CrossRef]
6. Hyder, O.N.; Soukas, P.A. Chronic Venous Insufficiency: Novel Management Strategies for an Under-diagnosed Disease Process. *R. I. Med. J.* **2017**, *100*, 37–39.
7. Eberhardt, R.T.; Raffetto, J.D. Chronic venous insufficiency. *Circulation* **2014**, *130*, 333–346. [CrossRef]
8. Darvall, K.A.; Bate, G.R.; Adam, D.J.; Silverman, S.H.; Bradbury, A.W. Duplex ultrasound outcomes following ultrasound-guided foam sclerotherapy of symptomatic recurrent great saphenous varicose veins. *Eur. J. Vasc. Endovasc. Surg.* **2011**, *42*, 107–114. [CrossRef]
9. Verajankorva, E.; Rautio, R.; Giordano, S.; Koskivuo, I.; Savolainen, O. The Efficiency of Sclerotherapy in the Treatment of Vascular Malformations: A Retrospective Study of 63 Patients. *Plast. Surg. Int.* **2016**, *2016*, 2809152. [CrossRef]
10. Vuylsteke, M.E.; Martinelli, T.; Van Dorpe, J.; Roelens, J.; Mordon, S.; Fourneau, I. Endovenous laser ablation: The role of intraluminal blood. *Eur. J. Vasc. Endovasc. Surg.* **2011**, *42*, 120–126. [CrossRef]
11. Feldo, M.; Wojciak-Kosior, M.; Sowa, I.; Kocki, J.; Bogucki, J.; Zubilewicz, T.; Kesik, J.; Bogucka-Kocka, A. Effect of Diosmin Administration in Patients with Chronic Venous Disorders on Selected Factors Affecting Angiogenesis. *Molecules* **2019**, *24*, 3316. [CrossRef] [PubMed]
12. Potente, M.; Gerhardt, H.; Carmeliet, P. Basic and therapeutic aspects of angiogenesis. *Cell* **2011**, *146*, 873–887. [CrossRef]
13. Herouy, Y.; Kreis, S.; Mueller, T.; Duerk, T.; Martiny-Baron, G.; Reusch, P.; May, F.; Idzko, M.; Norgauer, Y. Inhibition of angiogenesis in lipodermatosclerosis: Implication for venous ulcer formation. *Int. J. Mol. Med.* **2009**, *24*, 645–651. [CrossRef] [PubMed]
14. Glowinski, J.; Glowinski, S. Generation of reactive oxygen metabolites by the varicose vein wall. *Eur. J. Vasc. Endovasc. Surg.* **2002**, *23*, 550–555. [CrossRef] [PubMed]
15. Saharay, M.; Shields, D.A.; Porter, J.B.; Scurr, J.H.; Coleridge Smith, P.D. Leukocyte activity in the microcirculation of the leg in patients with chronic venous disease. *J. Vasc. Surg.* **1997**, *26*, 265–273. [CrossRef]

16. Myohanen, T.T.; Tenorio-Laranga, J.; Jokinen, B.; Vazquez-Sanchez, R.; Moreno-Baylach, M.J.; Garcia-Horsman, J.A.; Mannisto, P.T. Prolyl oligopeptidase induces angiogenesis both in vitro and in vivo in a novel regulatory manner. *Br. J. Pharm.* **2011**, *163*, 1666–1678. [CrossRef] [PubMed]
17. Zhou, D.; Wang, J.; He, L.N.; Li, B.H.; Ding, Y.N.; Chen, Y.W.; Fan, J.G. Prolyl oligopeptidase attenuates hepatic stellate cell activation through induction of Smad7 and PPAR-gamma. *Exp. Med.* **2017**, *13*, 780–786. [CrossRef]
18. Brandt, I.; Scharpe, S.; Lambeir, A.M. Suggested functions for prolyl oligopeptidase: A puzzling paradox. *Clin. Chim. Acta* **2007**, *377*, 50–61. [CrossRef]
19. Boersma, D.; van Haelst, S.T.; van Eekeren, R.R.; Vink, A.; Reijnen, M.M.; de Vries, J.P.; de Borst, G.J. Macroscopic and Histologic Analysis of Vessel Wall Reaction After Mechanochemical Endovenous Ablation Using the ClariVein OC Device in an Animal Model. *Eur. J. Vasc. Endovasc. Surg.* **2017**, *53*, 290–298. [CrossRef]
20. Svarcbahs, R.; Julku, U.H.; Myohanen, T.T. Inhibition of Prolyl Oligopeptidase Restores Spontaneous Motor Behavior in the alpha-Synuclein Virus Vector-Based Parkinson's Disease Mouse Model by Decreasing alpha-Synuclein Oligomeric Species in Mouse Brain. *J. Neurosci.* **2016**, *36*, 12485–12497. [CrossRef]
21. Eschrich, J.; Meyer, R.; Kuk, H.; Wagner, A.H.; Noppeney, T.; Debus, S.; Hecker, M.; Korff, T. Varicose Remodeling of Veins Is Suppressed by 3-Hydroxy-3-Methylglutaryl Coenzyme A Reductase Inhibitors. *J. Am. Heart Assoc.* **2016**, *5*. [CrossRef] [PubMed]
22. Somers, P.; Knaapen, M. The histopathology of varicose vein disease. *Angiology* **2006**, *57*, 546–555. [CrossRef] [PubMed]
23. Lai, Y.J.; Chen, P.R.; Huang, Y.L.; Hsu, H.H. Unique wreath-like smooth muscle proliferation of the pulmonary vasculature in pulmonary veno-occlusive disease versus pulmonary arterial hypertension. *J. Med. Assoc.* **2020**, *119*, 300–309. [CrossRef]
24. Vass, D.G.; Ainsworth, R.; Anderson, J.H.; Murray, D.; Foulis, A.K. The value of an elastic tissue stain in detecting venous invasion in colorectal cancer. *J. Clin. Pathol.* **2004**, *57*, 769–772. [CrossRef] [PubMed]
25. Di Paola, R.; Fusco, R.; Gugliandolo, E.; Crupi, R.; Evangelista, M.; Granese, R.; Cuzzocrea, S. Co-micronized Palmitoylethanolamide/Polydatin Treatment Causes Endometriotic Lesion Regression in a Rodent Model of Surgically Induced Endometriosis. *Front Pharm.* **2016**, *7*, 382. [CrossRef] [PubMed]
26. Casili, G.; Lanza, M.; Filippone, A.; Campolo, M.; Paterniti, I.; Cuzzocrea, S.; Esposito, E. Dimethyl fumarate alleviates the nitroglycerin (NTG)-induced migraine in mice. *J. Neuroinflammation* **2020**, *17*, 59. [CrossRef] [PubMed]
27. Lanza, M.; Campolo, M.; Casili, G.; Filippone, A.; Paterniti, I.; Cuzzocrea, S.; Esposito, E. Sodium Butyrate Exerts Neuroprotective Effects in Spinal Cord Injury. *Mol. Neurobiol.* **2019**, *56*, 3937–3947. [CrossRef]
28. Campolo, M.; Casili, G.; Paterniti, I.; Filippone, A.; Lanza, M.; Ardizzone, A.; Scuderi, S.A.; Cuzzocrea, S.; Esposito, E. Effect of a Product Containing Xyloglucan and Pea Protein on a Murine Model of Atopic Dermatitis. *Int. J. Mol. Sci.* **2020**, *21*, 3596. [CrossRef]
29. Casili, G.; Campolo, M.; Lanza, M.; Filippone, A.; Scuderi, S.; Messina, S.; Ardizzone, A.; Esposito, E.; Paterniti, I. Role of ABT888, a Novel Poly(ADP-Ribose) Polymerase (PARP) Inhibitor in Countering Autophagy and Apoptotic Processes Associated to Spinal Cord Injury. *Mol. Neurobiol.* **2020**, *57*, 4394–4407. [CrossRef]
30. Porter, J.M.; Moneta, G.L. Reporting standards in venous disease: An update. International Consensus Committee on Chronic Venous Disease. *J. Vasc. Surg.* **1995**, *21*, 635–645. [CrossRef]
31. Castro-Ferreira, R.; Cardoso, R.; Leite-Moreira, A.; Mansilha, A. The Role of Endothelial Dysfunction and Inflammation in Chronic Venous Disease. *Ann. Vasc. Surg.* **2018**, *46*, 380–393. [CrossRef] [PubMed]
32. Guardiola, S.; Prades, R.; Mendieta, L.; Brouwer, A.J.; Streefkerk, J.; Nevola, L.; Tarrago, T.; Liskamp, R.M.J.; Giralt, E. Targeted Covalent Inhibition of Prolyl Oligopeptidase (POP): Discovery of Sulfonylfluoride Peptidomimetics. *Cell Chem. Biol.* **2018**, *25*, 1031–1037 e1034. [CrossRef] [PubMed]
33. Ligi, D.; Croce, L.; Mosti, G.; Raffetto, J.D.; Mannello, F. Chronic Venous Insufficiency: Transforming Growth Factor-beta Isoforms and Soluble Endoglin Concentration in Different States of Wound Healing. *Int. J. Mol. Sci.* **2017**, *18*, 2206. [CrossRef] [PubMed]
34. Sugden, W.W.; Meissner, R.; Aegerter-Wilmsen, T.; Tsaryk, R.; Leonard, E.V.; Bussmann, J.; Hamm, M.J.; Herzog, W.; Jin, Y.; Jakobsson, L.; et al. Endoglin controls blood vessel diameter through endothelial cell shape changes in response to haemodynamic cues. *Nat. Cell Biol.* **2017**, *19*, 653–665. [CrossRef]

35. Ruiz-Ortega, M.; Rodriguez-Vita, J.; Sanchez-Lopez, E.; Carvajal, G.; Egido, J. TGF-beta signaling in vascular fibrosis. *Cardiovasc. Res.* **2007**, *74*, 196–206. [CrossRef]
36. Gordon, K.J.; Blobe, G.C. Role of transforming growth factor-beta superfamily signaling pathways in human disease. *Biochim. Biophys. Acta* **2008**, *1782*, 197–228. [CrossRef]
37. Sola Ldel, R.; Aceves, M.; Duenas, A.I.; Gonzalez-Fajardo, J.A.; Vaquero, C.; Crespo, M.S.; Garcia-Rodriguez, C. Varicose veins show enhanced chemokine expression. *Eur. J. Vasc. Endovasc. Surg.* **2009**, *38*, 635–641. [CrossRef]
38. Heidemann, J.; Ogawa, H.; Dwinell, M.B.; Rafiee, P.; Maaser, C.; Gockel, H.R.; Otterson, M.F.; Ota, D.M.; Lugering, N.; Domschke, W.; et al. Angiogenic effects of interleukin 8 (CXCL8) in human intestinal microvascular endothelial cells are mediated by CXCR2. *J. Biol. Chem.* **2003**, *278*, 8508–8515. [CrossRef]
39. Surendran, S.; Ramegowda, K.S.; Suresh, A.; Binil Raj, S.S.; Lakkappa, R.K.B.; Kamalapurkar, G.; Radhakrishnan, N.; Chandrasekharan, C.K. Arterialization and anomalous vein wall remodeling in varicose veins is associated with upregulated FoxC2-Dll4 pathway. *Lab. Invest.* **2016**, *96*, 399–408. [CrossRef]
40. Papapetropoulos, A.; Garcia-Cardena, G.; Madri, J.A.; Sessa, W.C. Nitric oxide production contributes to the angiogenic properties of vascular endothelial growth factor in human endothelial cells. *J. Clin. Invest.* **1997**, *100*, 3131–3139. [CrossRef]
41. Marumo, T.; Schini-Kerth, V.B.; Busse, R. Vascular endothelial growth factor activates nuclear factor-kappaB and induces monocyte chemoattractant protein-1 in bovine retinal endothelial cells. *Diabetes* **1999**, *48*, 1131–1137. [CrossRef] [PubMed]
42. Grudzinska, E.; Czuba, Z.P. Immunological aspects of chronic venous disease pathogenesis. *Cent. Eur. J. Immunol.* **2014**, *39*, 525–531. [CrossRef] [PubMed]
43. Smith, P.D. Neutrophil activation and mediators of inflammation in chronic venous insufficiency. *J. Vasc. Res.* **1999**, *36*, 24–36. [CrossRef] [PubMed]
44. Wali, M.A.; Eid, R.A. Changes of elastic and collagen fibers in varicose veins. *Int. Angiol.* **2002**, *21*, 337–343.
45. Venturi, M.; Bonavina, L.; Annoni, F.; Colombo, L.; Butera, C.; Peracchia, A.; Mussini, E. Biochemical assay of collagen and elastin in the normal and varicose vein wall. *J. Surg. Res.* **1996**, *60*, 245–248. [CrossRef]
46. Kakkos, S.K.; Zolota, V.G.; Peristeropoulou, P.; Apostolopoulou, A.; Geroukalos, G.; Tsolakis, I.A. Increased mast cell infiltration in familial varicose veins: Pathogenetic implications? *Int. Angiol.* **2003**, *22*, 43–49.
47. Yamada, T.; Tomita, S.; Mori, M.; Sasatomi, E.; Suenaga, E.; Itoh, T. Increased mast cell infiltration in varicose veins of the lower limbs: A possible role in the development of varices. *Surgery* **1996**, *119*, 494–497. [CrossRef]
48. Shoab, S.S.; Scurr, J.H.; Coleridge-Smith, P.D. Increased plasma vascular endothelial growth factor among patients with chronic venous disease. *J. Vasc. Surg.* **1998**, *28*, 535–540. [CrossRef]
49. Tian, H.; Huang, J.J.; Golzio, C.; Gao, X.; Hector-Greene, M.; Katsanis, N.; Blobe, G.C. Endoglin interacts with VEGFR2 to promote angiogenesis. *Faseb J.* **2018**, *32*, 2934–2949. [CrossRef]
50. Beebe-Dimmer, J.L.; Pfeifer, J.R.; Engle, J.S.; Schottenfeld, D. The epidemiology of chronic venous insufficiency and varicose veins. *Ann. Epidemiol.* **2005**, *15*, 175–184. [CrossRef]
51. Davies, A.H. The Seriousness of Chronic Venous Disease: A Review of Real-World Evidence. *Adv. Ther.* **2019**, *36*, 5–12. [CrossRef] [PubMed]
52. McLafferty, R.B.; Passman, M.A.; Caprini, J.A.; Rooke, T.W.; Markwell, S.A.; Lohr, J.M.; Meissner, M.H.; Eklof, B.G.; Wakefield, T.W.; Dalsing, M.C. Increasing awareness about venous disease: The American Venous Forum expands the National Venous Screening Program. *J. Vasc. Surg.* **2008**, *48*, 394–399. [CrossRef] [PubMed]
53. McGuinness, C.L.; Humphries, J.; Waltham, M.; Burnand, K.G.; Collins, M.; Smith, A. Recruitment of labelled monocytes by experimental venous thrombi. *Thromb. Haemost.* **2001**, *85*, 1018–1024. [CrossRef] [PubMed]
54. Walsh, D.A.; Pearson, C.I. Angiogenesis in the pathogenesis of inflammatory joint and lung diseases. *Arthritis Res.* **2001**, *3*, 147–153. [CrossRef]
55. Gaggar, A.; Jackson, P.L.; Noerager, B.D.; O'Reilly, P.J.; McQuaid, D.B.; Rowe, S.M.; Clancy, J.P.; Blalock, J.E. A novel proteolytic cascade generates an extracellular matrix-derived chemoattractant in chronic neutrophilic inflammation. *J. Immunol.* **2008**, *180*, 5662–5669. [CrossRef]
56. Tenorio-Laranga, J.; Coret-Ferrer, F.; Casanova-Estruch, B.; Burgal, M.; Garcia-Horsman, J.A. Prolyl oligopeptidase is inhibited in relapsing-remitting multiple sclerosis. *J. Neuroinflammation* **2010**, *7*, 23. [CrossRef]

57. Venalainen, J.I.; Juvonen, R.O.; Garcia-Horsman, J.A.; Wallen, E.A.; Christiaans, J.A.; Jarho, E.M.; Gynther, J.; Mannisto, P.T. Slow-binding inhibitors of prolyl oligopeptidase with different functional groups at the P1 site. *Biochem. J.* **2004**, *382*, 1003–1008. [CrossRef]
58. Han, G.; Li, F.; Singh, T.P.; Wolf, P.; Wang, X.J. The pro-inflammatory role of TGFbeta1: A paradox? *Int. J. Biol. Sci.* **2012**, *8*, 228–235. [CrossRef]
59. Serralheiro, P.; Soares, A.; Costa Almeida, C.M.; Verde, I. TGF-beta1 in Vascular Wall Pathology: Unraveling Chronic Venous Insufficiency Pathophysiology. *Int. J. Mol. Sci.* **2017**, *18*, 2534. [CrossRef]
60. Kofler, S.; Nickel, T.; Weis, M. Role of cytokines in cardiovascular diseases: A focus on endothelial responses to inflammation. *Clin. Sci. (London)* **2005**, *108*, 205–213. [CrossRef]
61. Li, A.; Dubey, S.; Varney, M.L.; Dave, B.J.; Singh, R.K. IL-8 directly enhanced endothelial cell survival, proliferation, and matrix metalloproteinases production and regulated angiogenesis. *J. Immunol.* **2003**, *170*, 3369–3376. [CrossRef] [PubMed]
62. Sansilvestri-Morel, P.; Rupin, A.; Badier-Commander, C.; Fabiani, J.N.; Verbeuren, T.J. Chronic venous insufficiency: Dysregulation of collagen synthesis. *Angiology* **2003**, *54*, S13–S18. [CrossRef] [PubMed]
63. Barasch, E.; Gottdiener, J.S.; Aurigemma, G.; Kitzman, D.W.; Han, J.; Kop, W.J.; Tracy, R.P. The relationship between serum markers of collagen turnover and cardiovascular outcome in the elderly: The Cardiovascular Health Study. *Circ. Heart Fail* **2011**, *4*, 733–739. [CrossRef] [PubMed]
64. Sprague, A.H.; Khalil, R.A. Inflammatory cytokines in vascular dysfunction and vascular disease. *Biochem. Pharm.* **2009**, *78*, 539–552. [CrossRef]
65. Takase, S.; Bergan, J.J.; Schmid-Schonbein, G. Expression of adhesion molecules and cytokines on saphenous veins in chronic venous insufficiency. *Ann. Vasc. Surg.* **2000**, *14*, 427–435. [CrossRef]
66. Boada, J.N.; Nazco, G.J. Therapeutic effect of venotonics in chronic venous insufficiency—A meta-analysis. *Clin. Drug Invest.* **1999**, *18*, 413–432. [CrossRef]
67. Novotny, K.; Campr, V. Histopathological changes to the vascular wall after treatment of great saphenous veins using n-butyl-2-cyanoacrylate. *Vasa* **2019**, *48*, 399–404. [CrossRef]
68. Wali, M.A.; Dewan, M.; Eid, R.A. Histopathological changes in the wall of varicose veins. *Int. Angiol.* **2003**, *22*, 188–193.
69. Hernandez-Morera, P.; Castano-Gonzalez, I.; Travieso-Gonzalez, C.M.; Mompeo-Corredera, B.; Ortega-Santana, F. Quantification and Statistical Analysis Methods for Vessel Wall Components from Stained Images with Masson's Trichrome. *PLoS ONE* **2016**, *11*, e0146954. [CrossRef]
70. Satoh, T.; Sugama, K.; Matsuo, A.; Kato, S.; Ito, S.; Hatanaka, M.; Sasaguri, Y. Histamine as an activator of cell growth and extracellular matrix reconstruction for human vascular smooth muscle cells. *Atherosclerosis* **1994**, *110*, 53–61. [CrossRef]
71. Lu, X.; Gong, J.; Dennery, P.A.; Yao, H. Endothelial-to-mesenchymal transition: Pathogenesis and therapeutic targets for chronic pulmonary and vascular diseases. *Biochem. Pharm.* **2019**, *168*, 100–107. [CrossRef] [PubMed]
72. Kovacic, J.C.; Dimmeler, S.; Harvey, R.P.; Finkel, T.; Aikawa, E.; Krenning, G.; Baker, A.H. Endothelial to Mesenchymal Transition in Cardiovascular Disease: JACC State-of-the-Art Review. *J. Am. Coll. Cardiol.* **2019**, *73*, 190–209. [CrossRef] [PubMed]
73. ten Dijke, P.; Goumans, M.J.; Pardali, E. Endoglin in angiogenesis and vascular diseases. *Angiogenesis* **2008**, *11*, 79–89. [CrossRef] [PubMed]

Publisher's Note: MDPI stays neutral with regard to jurisdictional claims in published maps and institutional affiliations.

© 2020 by the authors. Licensee MDPI, Basel, Switzerland. This article is an open access article distributed under the terms and conditions of the Creative Commons Attribution (CC BY) license (http://creativecommons.org/licenses/by/4.0/).

Review

Molecular and Cellular Mechanisms of Electronegative Lipoproteins in Cardiovascular Diseases

Liang-Yin Ke [1,2,3,†], Shi Hui Law [1,†], Vineet Kumar Mishra [1], Farzana Parveen [1], Hua-Chen Chan [3], Ye-Hsu Lu [3,4] and Chih-Sheng Chu [3,4,5,*]

1. Department of Medical Laboratory Science and Biotechnology, College of Health Sciences, Kaohsiung Medical University, Kaohsiung 807378, Taiwan; kly@gap.kmu.edu.tw (L.-Y.K.); shlaw9994@gmail.com (S.H.L.); vineetkmishra.jh@gmail.com (V.K.M.); fparveen.jh@gmail.com (F.P.)
2. Graduate Institute of Medicine, College of Medicine and Drug Development and Value Creation Research Center, Kaohsiung Medical University, Kaohsiung 807378, Taiwan
3. Center for Lipid Biosciences, Kaohsiung Medical University Hospital, Kaohsiung Medical University, Kaohsiung 807377, Taiwan; huachen.chan@gmail.com (H.-C.C.); yehslu@cc.kmu.edu.tw (Y.-H.L.)
4. Division of Cardiology, Department of International Medicine, Kaohsiung Medical University Hospital, Kaohsiung 807377, Taiwan
5. Division of Cardiology, Department of Internal Medicine, Kaohsiung Municipal Ta-Tung Hospital, Kaohsiung 80145, Taiwan
* Correspondence: chucs@kmu.edu.tw; Tel.: +886-73121101 (ext. 2297); Fax: +886-73111996
† These authors contributed equally to this work.

Received: 27 October 2020; Accepted: 26 November 2020; Published: 29 November 2020

Abstract: Dysregulation of glucose and lipid metabolism increases plasma levels of lipoproteins and triglycerides, resulting in vascular endothelial damage. Remarkably, the oxidation of lipid and lipoprotein particles generates electronegative lipoproteins that mediate cellular deterioration of atherosclerosis. In this review, we examined the core of atherosclerotic plaque, which is enriched by byproducts of lipid metabolism and lipoproteins, such as oxidized low-density lipoproteins (oxLDL) and electronegative subfraction of LDL (LDL(−)). We also summarized the chemical properties, receptors, and molecular mechanisms of LDL(−). In combination with other well-known markers of inflammation, namely metabolic diseases, we concluded that LDL(−) can be used as a novel prognostic tool for these lipid disorders. In addition, through understanding the underlying pathophysiological molecular routes for endothelial dysfunction and inflammation, we may reassess current therapeutics and might gain a new direction to treat atherosclerotic cardiovascular diseases, mainly targeting LDL(−) clearance.

Keywords: electronegative LDL; LDL(−); L5 LDL; oxidized LDL; oxLDL; lectin-like oxLDL receptor-1; LOX-1; dyslipidemia; endothelial dysfunction; atherosclerosis; cardiovascular disease

1. Introduction

Approximately 1.9 billion people are obese or overweight worldwide [1]. Obesity is associated with excessive calorific intake and microvasculature damage, resulting in atherosclerosis, diabetes, and cardiovascular diseases (CVDs) [2]. The prevalence of CVDs has significantly increased in the past few decades [3]. Current strategies against CVDs mainly focus on lowering the level of low-density lipoprotein cholesterol (LDL-C) [4,5]. Intensive-dose statin therapy has been endorsed for clinical atherosclerotic vascular disease (ASCVD); however, it also increases statin-related side effects and intolerance [6,7]. To figure out this dilemma and find a balanced solution, here we address the mechanistic players behind these metabolic disturbances through the following disease progression steps: unhealthy lifestyle and unbalanced diet lead to obesity, chronic inflammation, and development of atherosclerosis and CVDs [8–10].

The onset of atherosclerosis initiates vascular lipid deposition, luminal narrowing, and plaque expansion. Unstable plaque deposits further lead to myocardial infarction and stroke [11]. Plaque consists of LDL-C variants, lipids, leukocytes, and inflammasomes in the vascular walls (Figure 1) [11,12]. In addition, several mediators of vasoconstriction, platelet aggregation, inflammatory chemokines, leukocyte adherence, and nitric oxide (NO) disturb the endothelial homeostasis [13]. LDL variants such as oxidized LDL (oxLDL) are essential constituents in the pathogenesis of atherosclerosis and CVDs [14–16]. Differing from the in vitro preparation of oxLDL, electronegative LDL (LDL(−)) is separated from human plasma using fast-protein liquid chromatography equipped with an anion exchange column [17]. According to the physical properties of LDL(−), it can be defined as the minimized oxLDL [18,19].

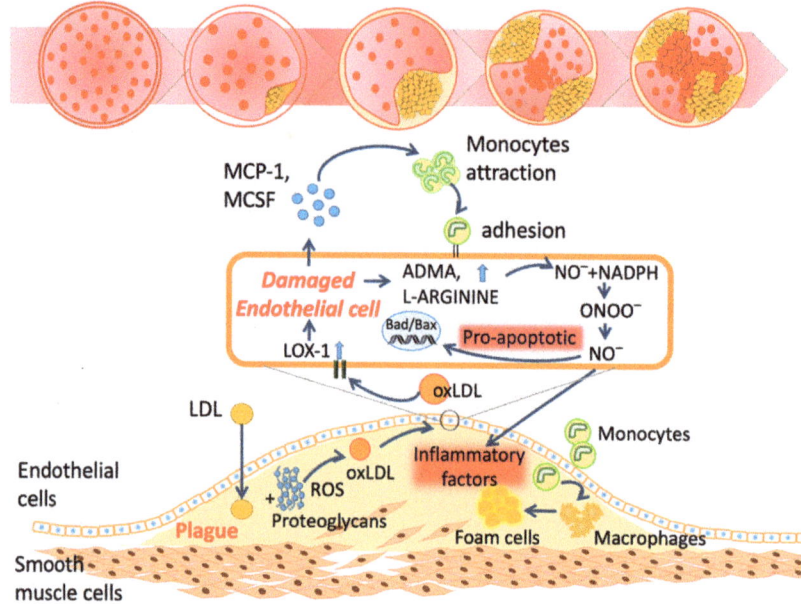

Figure 1. Schematic mechanism of atherosclerosis. LDL: low-density lipoprotein; ROS: reactive oxygen species; oxLDL: oxidized LDL; LOX-1: lectin-like oxidized LDL receptor-1; ADMA: asymmetric dimethylarginine; NO: nitric oxide; NADPH: nicotinamide adenine dinucleotide phosphate; ONOO: peroxynitrite; Bad: BCL2-associated agonist of cell death; Bax: Bcl-2-associated X protein; MCP-1: monocyte chemoattractant protein-1; MCSF: macrophage colony-stimulating factor.

Accumulating evidence shows that LDL(−) could be a novel marker for ASCVD, and levels of LDL(−) are positively correlated with the increasing severity of CVDs [20–22]. LDL(−) serves as a pivotal target for further studies and clinical development strategies beyond statins therapies. By targeting LDL(−), we summarize its pathophysiological links and highlight the molecular mechanisms of atherogenic lipids in the current review.

2. Properties of Electronegative Low-Density Lipoprotein (LDL(−))

2.1. Chemical Properties of LDL(−)

LDL(−) differs from LDL(+) in many aspects [23]. Regarding the lipid components, LDL(−) contains higher concentrations of triglycerides, non-esterified fatty acids (NEFA), lysophosphatidylcholine

(LPC), platelet-activating factor (PAF), and ceramide [24–27]. Notably, lipid extracts of LDL(−) contribute to the atherogenic effects on endothelial cells and immune cells [27,28]. Regarding its protein composition, LDL(−) shows additional proteins such as apolipoprotein AI (apoAI), apolipoprotein E (apoE), and apolipoprotein CIII (apoCIII) [29]. Furthermore, the conformation of apoB100 in LDL(−) is altered and has higher competency to bind with proteoglycans [30–32]. Based on the sodium chloride gradient, Chen et al. successfully divided LDL into five subfractions, L1–L5, with increasing electronegativity [29,33,34]. L1 LDL is unmodified; in contrast, L5 LDL is highly O-glycosylated on the apoB100 and apoE [28,35]. The terminal glycan of apoE glycosylation (94S, 194T, 289T) in L5 LDL is sialic acid. This sialic-acid-containing glycan increases the electronegativity and hydrophilicity [35]. However, by dividing human plasma LDL into either two subfractions ((+) and (−)) or five (L1–L5), the most electronegative subfractions show similar properties and apoptotic effects on endothelial cells. Thus, we will be using LDL(−) throughout this review.

2.2. Receptors of LDL(−)

LDL(−) is not recognized by the LDL receptor. Instead, it goes through lectin-like oxLDL receptor-1 (LOX-1), which is highly expressed in endothelial cells, immune cells, platelets, and adipocytes [36–39]. Transfection with LOX-1-specific small interfering RNAs (siLOX-1) to endothelial cells may attenuate LDL(−)-induced downstream signaling [36]. LOX-1-neutralizing antibodies such as TS20 (for bovine) [40], TS58 (for mouse) [41], and TS92 (for human) [42,43] can inhibit the internalization of LDL(−). Genetic knockout LOX-1 also protects against the harmful effects of LDL(−) [37,38]. Higher content of PAF on LDL(−) activates the PAF receptor (PAFR) and leads to endothelial cell apoptosis [33]. Incubating PAF acetylhydrolase (PAF-AH) with LDL(−) or pretreatment of WEB-2086 attenuates LDL(−)-induced apoptosis [33]. In addition, ceramide-rich LDL(−) activates toll-like receptor 4 (TLR4) and the cluster of differentiation 14 (CD14) on monocytes that results in cytokine release. Using the TLR4 inhibitor, the viral inhibitory peptide of TLR4 (VIPER), reduces these effects [44,45].

2.3. Structure Modifications and Enzymatic Functions of Electronegative LDL

Electronegativity and apolipoprotein misfolding are two independent features of LDL(−) [46]. The misfolded apoB100 of LDL(−) shows an increased binding affinity to proteoglycans, which may prolong LDL retention in the arterial wall and trigger inflammatory responses [31]. Stabilizing the LDL's structure through the use of 17-β-estradiol (E2) prevents aggregation; however, it cannot prevent the generation of LDL(−) [46,47]. The structural modifications of apoB100 are associated with phospholipolytic activities and exchange of lipid components [28,48,49]. The sphingomyelinase (SMase)-like activity of LDL(−) may hydrolyze sphingomyelin, which produces apoptotic factor, a ceramide [28,48]. The phospholipase D (PLD) activity of LDL(−) degrades phosphorylcholine, LPC, and sphingomyelin, which is associated with self-aggregation and atherogenic properties. Treatment with 400 µM of chlorpromazine may effectively inhibit both the SMase and PLD activities of LDL(−) [48].

2.4. Animal Models Showing Elevated Electronegative LDL

The overproduction of LDL(−) was demonstrated in animal models that consumed a high-fat diet. Lai et al. gave either a standard chow diet or high-fat & high-cholesterol (HFC) diet to each group of 8-week-old male golden Syrian hamsters for six weeks. Plasma LDL-C levels in HFC-diet-fed hamsters were significantly higher than for the control group. Additionally, LDL(−) accounted for 12.5% of all lipoproteins in control hamsters, whereas the value was drastically increased to 42% in HFC-diet-fed hamsters [50]. Recently, Chang et al. distributed an atherogenic diet to sixteen-week-old male New Zealand White rabbits. After six weeks, the LDL(−) from HFC-diet-fed rabbits accounted for about 17.2 ± 5.5% of the LDL fraction. On the other hand, it was almost undetectable in rabbits fed with a control chow diet [51]. Moreover, from the recent publication by Chan et al., LDL(+) and LDL(−) isolated from SLE patients' LDL samples were then injected into eight-week-old apoE knockout mice. Their results showed that only the LDL(−)-injected mice experienced a significant increase in the plasma CX3CL1 level. By observing histological staining results, LDL(−) can trigger endothelial dysfunction and the formation of atherosclerotic lesions in apoE knockout mice [27]. Taken together, we summarized that LDL(−) plays a vital role in atherosclerosis and plaque formation.

3. Mechanisms of Electronegative LDL on Endothelial Cells

The endothelium regulates fluid and molecule trafficking between the bloodstream and tissues for metabolism [52]. In addition, it inhibits platelet aggregation and adhesions by secreting prostacyclin, NO, and exosomes [53,54]. With LDL(−), the atherogenic components lead to endothelial activation and vascular inflammation. Chemokines such as monocyte chemotactic protein-1 (MCP-1) and interleukin-8 (IL-8) are released from the damaged endothelium. The vascular adhesion molecules are highly expressed to promote plaque formation [55]. The mechanisms behind this are listed below.

3.1. Phosphatidylinositol-3 Kinase (PI3K)-Serine/Threonine Kinase (Akt) Signaling

The phosphatidylinositol-3 kinase (PI3K)-serine/threonine kinase (Akt) signaling involves the proliferation and survival of endothelial cells through inhibiting pro-apoptotic proteins [56]. Both fibroblast growth factor 2 (FGF2) and vascular endothelial growth factor (VEGF) activate PI3K/Akt signaling [57,58]; in contrast, LDL(−) disrupts Akt phosphorylation, impairing the FGF2 mRNA expression, as well as induces endothelial cell apoptosis [40,59]. In their study, Lu et al. also demonstrated that the apoptotic effects of LDL(−) on endothelial cells could be attenuated by treatment with FGF2 or constitutively expressing active Akt [59]. LDL(−) inhibits B-cell lymphoma 2 (Bcl-2); in contrast, it triggers the expression of Bad/Bax (Bcl-2-associated agonist cell death) and inflammatory factor tumor necrosis factor-α (TNF-α). These actions result in the release of cytochrome c from mitochondria [36,59].

3.2. Lectin-Like oxLDL Receptor-1 (LOX-1) Signaling

Lectin-like oxLDL receptor-1 (LOX-1) reacts with multiple ligands in response to danger signals [60]. Patients with cerebral stroke and coronary artery diseases exhibited elevated levels of soluble-form LOX-1 (sLOX-1) [61,62]. Furthermore, patients with ST segment elevation myocardial infarction (STEMI) and rheumatoid arthritis (RA) showed increased sLOX-1 expression in the aspirated coronary thrombi [63,64]. Due to earlier release than biochemical markers of myocardial injury, sLOX-1 could be a novel biomarker for plaque instability [65]. In a hypercholesteremic mice model, the LOX-1 knockout reduced the plaque size and atherosclerotic lesions [66–68].

For the detailed mechanisms, LDL(−) leads to the overexpressed changes of LOX-1 on endothelial cells by inducing the expression changes of the pro-inflammatory molecules nuclear factor of kappa light polypeptide gene enhancer in B-cells (NF-κB), vascular cell adhesion molecule (1VCAM-1), and MCP-1 [69,70]. Recently, a similar cohort study was completed to show similar results of LOX-1-mediated inflammation in SLE patients [71]. In addition, the expression of LOX-1 dependents on vasoconstrictors (angiotensin II, endothelin-1) and inflammatory factors such as interferon-γ (IFN-γ), tumor necrosis factor-α (TNF-α), and IL-1β was observed [72]. In vitro, oxidized LDL may enhance the production of angiotensin-converting enzyme (ACE) and endothelin-1 [73,74].

Through LOX-1, LDL(−) downregulates the phosphorylation of Akt and endothelial nitric oxide synthase (eNOS) but increases C-reactive protein (CRP) [11,36,42]. LOX-1 activates Ras homolog family member A (RhoA) and the Ras-related C3 botulinum toxin substrate 1 (Rac1) pathway, leading to the inhibition of intracellular endothelial NO synthesis and overproduction of ROS [75]. Recently, NOS was reported to influence miR-122 expression in hypertension cases, leading to endothelial dysfunction; however, the expression changes of miR-122-mediating endothelial dysfunction remains unanswered. We, therefore, predict LOX-1 signaling of LDL(−) in such cases [76]. Similarly, ROS overproduction leads to p66shc protein phosphorylation, which further deteriorates mitochondrial DNA and contributes to plaque formation [77–79]. The phenomenon mentioned above can be attenuated by knocking out the LOX-1 gene [80,81].

3.3. Mitochondria Damage

The basal physiological mechanism of mitochondrial ROS formation is dependent on several factors such as NO, cytosolic Ca2+, and fatty acids [82]. NADPH oxidase 4 (NOX4) in vascular cells inhibits mitochondrial complex I and promotes ROS generation [83]. During the pro-apoptotic conditions, ROS formation is also boosted by growth factor adaptor protein p66Shc, which facilitates the cytochrome c oxidation. Moreover, ROS formation can be further increased by the expression and activation of p66Shc during hyperglycemic conditions [84,85]. LDL(−) inhibits endothelial nitric oxide synthase (eNOS) expression via the Akt signaling pathway, resulting in decreased NO production and leading to endothelial cell apoptosis [86]. Recently, Chen et al. demonstrated that apoE in LDL(−) is responsible for LDL-induced mitochondrial dysfunction. After LDL(−) internalization, apoE translocates from the lysosome to the mitochondria, leading to mitochondrial permeability transition pore (mPTP) opening, dynamin-related protein 1 (DRP1) phosphorylation, and mitochondrial fission [41].

3.4. Endoplasmic Reticulum Stress

The intraluminal oxidation in the endoplasmic reticulum (ER) plays a critical role in maintaining calcium concentration and proper folding of transmembrane proteins. The increased amount of lipoprotein promotes a condition known as ER stress, defined by the accumulation of unfolded protein in the ER lumen [87,88]. The molecular mechanism between LDL oxidation and UPR (unfolded protein response)-mediated expression of IL-8, IL-6, and MCP-1 in endothelial cells, which contributes to endothelial dysfunction, is poorly explained [89,90]. Apart from oxidation, glycation of LDL is also found to be a potent marker for dyslipidemia. Studies showed that glycated LDL could initiate nicotinamide adenine dinucleotide phosphate (NADPH) oxidation via ROS production and could induce apoptosis in endothelial cells [91,92]. Therefore, the LDL oxidation and glycation are involved in amplifying endothelial dysfunction and contributing to atherosclerosis.

4. Mechanisms of Electronegative LDL on Immune Cells

Alongside endothelial cells, immune cells play a significant role in the pathogenesis of atherosclerosis. Monocytes and T lymphocytes create an inflammatory milieu by releasing several cytokines and growth factors. As LDL(−) concentration is elevated in the blood plasma, it tend to interacts with these monocytes and lymphocytes via cytokines and growth factors [93,94]. LDL(−) impregnates the process of oxidation via the feedback loop mechanism shown in Figure 2 and enhances inflammation. The NEFA and ceramide in LDL(−) also show atherogenic properties [93,95–97]. The detailed mechanisms behind this are listed below.

4.1. Monocytes

Numerous studies have described the effects of LDL(−) on inducing cytokine release from monocytes, which may be important in atherosclerosis [25,98]. Remodeling of the vascular extracellular matrix (ECM) seemed to be an important landmark of atherosclerosis. LDL(−) induces the release of matrix metalloproteinase (MMP)-9 and tissue inhibitors of metalloproteinase (TIMP)-1 from monocytes through the TLR4/CD14 inflammatory pathway [45]. Additionally, the downstream signal cascade of TLR4/CD14 will then trigger PI3K/Akt signaling and promote p38 mitogen-activated protein kinase (p38 MAPK) phosphorylation, leading to LDL(−)-induced cytokine release from monocytes [99]. The elevated levels of those cytokines may regulate and contribute to vascular plaque formation.

4.2. Macrophages

Macrophages play a crucial role in the early stage pathogenesis of atherosclerosis [100]. Circulating monocytes undergo differentiation into macrophages and further polarization into classically activated (M1) or alternatively activated (M2) states in order to withstand environmental stimuli.

M1 macrophages are responsible for pro-inflammatory properties, whereas M2 macrophages exert opposing anti-inflammatory properties [101].

According to Yang et al., LDL(+) and LDL(−) isolated from patients with ST segment elevation myocardial infarction (STEMI) were treated with THP-1 macrophages. Their results indicated that only LDL(−) could induce the overproduction of interleukin (IL)-1β [102], granulocyte colony-stimulating factor (G-CSF), and granulocyte–macrophage colony-stimulating factor (GM-CSF) in macrophages through LOX-1-, extracellular signal-regulated kinase (ERK)1/2-, and NF-κB-dependent pathways. Inhibition of ERK1/2 and NF-κB activation can prevent G-CSF and GM-CSF production induced by LDL(−) [103].

In 2020, Chang et al. treated THP-1 with LDL(−), which resulted in increased pro-inflammatory cytokines such as IL-1β, IL-6, IL-8, and TNF-α, as well as M1 surface marker CD86; however, M2-related cytokines and surface marker CD206 were not changed by LDL(−) [39]. Additionally, the expression of CD11c, a marker of M1 macrophages, can also be induced by LDL(−) [104]. LDL(−) can induce M1 polarization of human macrophages responsible for secreting pro-inflammatory cytokines, resulting in foam cell formation and vascular plaque formation.

In addition to human macrophages, in treating LDL(+) and LDL(−) with RAW264.7 cell, the results showed that only LDL(−) can induce the expression of CD95 death receptor (Fas), its ligand CD95 L (FasL), and tumor necrosis factor ligand member 10 (Tnfsf10), which stimulate the activation of the caspases, resulting in cell apoptosis [105].

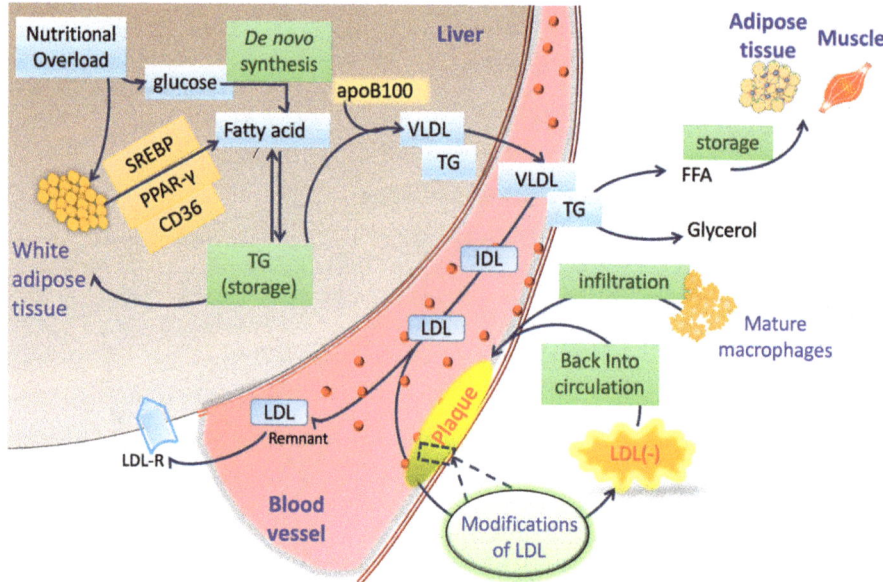

Figure 2. Schematic procedures of lipoprotein metabolism and LDL(−) formation. SREBP: sterol regulatory element-binding protein; PPAR-γ: peroxisome proliferator-activated receptor; CD36: cluster of differentiation 36; TG: triglycerides; apoB100: apolipoprotein B100; VLDL: very low-density lipoprotein; IDL: intermediate-density lipoprotein; LDL: low-density lipoprotein; LDLR: LDL receptor; FFA: free fatty acid.

4.3. Platelets

Apart from monocytes and macrophages, accumulating evidence has shown that LDL(−) may trigger platelet activation and aggregation. Platelet hyperreactivity is the most direct evidence contributing to thrombosis in the leading causes of cardiovascular diseases, such as STEMI [106] and stroke [43,107]. As above, Chan et al. separated LDL(+) and LDL(−) from patients with STEMI, with the results illustrating that only LDL(−) was augmented in patients compared to healthy controls. Treating LDL(−) to platelets enhanced their aggregation and adhesion to damaged human aortic endothelial cells (HAECs), which was through LOX-1 and PAFR activation [37]. Furthermore, LDL(−)-induced amyloid β (Aβ) release via IκB kinase 2 (IKK2) in human platelets was reported by Shen et al. in 2016. Besides, LDL(−) works synergistically with Aβ to induce glycoprotein IIb/IIIa receptor activation and phosphorylation of IKK2, IkBa, p65, and c-Jun N-terminal kinase 1 in order to enhance platelet aggregation. These results can be attenuated by inhibiting IKK2, LOX-1, or NF-kB with their inhibitors BMS-345541, TS92, and Bay 117-82, respectively [43]. To conclude, high levels of LDL(−) in patients can trigger platelet activation and aggregation through LOX-1 and PAFR receptors.

5. Electronegative LDL in Vascular Diseases

Figure 2 demonstrates the lipid and lipoprotein metabolism in the liver, blood, and peripheral tissues. Nutritional overload increases fatty acids via the overexpression of cluster of differentiation 36 (CD36) and peroxisome proliferator-activated receptor (PPAR-γ) [108–110]. This phenomenon is highly contrasted to the de novo synthesis pathway, although FFAs from either source in the liver are indistinguishable. The elevated level of free fatty acids ultimately increases triglyceride through esterification. Combined with apoB100 and triglyceride, the efflux of VLDL into circulation promotes the pro-atherogenic metabolic state. VLDL particles deliver lipids hydrolyzed by lipoprotein lipase (LPL) and release FFAs in plasma [111–113].

With the increasing incidence of LDL retention in endothelial cells [114–118], the LDL particles reportedly undergo oxidative modifications by macrophages and endothelial cells within arterial walls (Figure 2) [119–124]. The accumulation of oxLDL further boosts the electronegativity, ultimately generating LDL(−) in circulation [33]. LDL(−) is highly atherogenic and pro-apoptotic to the vascular system, including the endothelium of the blood–brain barrier (BBB). Wang et al. in 2017 explored the role of LDL(−) in pheochromocytoma-derived cell line (PC12) cells, where deliberate dosages of LDL(−) induced neurotoxic stress in a LOX-1-dependent manner [125].

The presence of LDL(−) in circulation correlates with atherosclerosis progression and endothelial dysfunction-mediated cardiovascular diseases. LDL(−) levels are significantly higher in frequent smokers, diabetic patients, and hypercholesterolemia patients [33,34,40,59]. In addition, LDL(−) levels were 10-times higher in STEMI and stroke patients, even though the LDL-C levels were similar to healthy controls [37,43].

6. Current Treatment Strategies Targeting Electronegative LDL

The diagnosis and treatment for endothelial damage are dependent on the ankle–brachial index, vascular imaging, surgery, and revascularization [126–128]. Currently, treatment for dyslipidemia and the prevention of microvasculature damage mainly revolve around reducing LDL-C levels [129–131]. A plethora of studies have demonstrated that excessive levels of lipids lead to endothelial damage; however, only a few studies have outlined strong mechanistic interactions between lipid alterations and endothelial dysfunction (Table 1).

Table 1. Primary dyslipidemia markers and pathways involved in different diseases.

Diseases	Dyslipidemia Markers	Drug Treatment	Effect on ED	Pathway/Phenomenon Involved	Studied on	References
Hypertension	NOS, ROS	α-Linolenic acid	Yes	SIRT-3	Mice	[132]
Hypertension	NOS, ROS	—	Yes	miR-122, CAT-1	Human	[76]
Hypertension, Angina	NOS, CRP, Hyperglycemia	Carvedilol	Yes	β-adrenergic mediate Vasodilation	Human	[133–136]
Heart failure	oxLDL, LDL	Rosuvastatin	Yes	Inflammatory markers	Human	[137,138]
ACS	oxLDL, LDL-C and cardiac fibrosis	perindopril	Yes	—	Human	[139,140]
CKD, CHF	Cardiac fibrosis	carvedilol	Yes	β-adrenergic mediate Vasodilation	Human	[141]
LVF, CKD	oxLDL, LDL and Cardiac fibrosis	Renal and heart transplant	—	—	Human	[142]
STEMI	—	Enoxaparin, Clopidogrel and β-blocker	No	Case study	Human	[143]
STEMI	Atherosclerotic Plaques	Statins, Aspirins, β-blocker, ACE-inhibitor	Yes	—	Human	[144,145]
STEMI	CRP and Atherosclerotic plaques	Vit B, B6, and B12	No	Homocysteine	Human	[146]
STEMI	LDL-C, Ox-LDL and L5	—	—	PKC/AKT pathway	Mice	[37]
CAD, Diabetes	NOS, Hyperinsulinemia, Hyperglycemia	Pioglitazone	Yes	Anti-inflammation, Vasodialation	Human	[147,148]
T1DM	Cardiac fibrosis	Fingolimod (FTY720)	Yes	Rag-1	Mice	[149]
T2DM	Hyperglycemia and Cardiac Fibrosis	H2/H3- RLX	Yes	α-SMA, MMP, TIMP and NLRP3	Rat	[150,151]
T2DM	NOS and Hyperglycemia	Berberine	Yes	AMPK and eNOS Phosphorylation	In-vitro, Ex-vivo	[152–154]
T2DM	Hyperglycemia, oxLDL, LDL, TG	Fenofibrate	Yes	PPAR-α/γ	Rat	[155,156]
RA	CRP, LDL, TG	MTX and Glucocorticoid	Yes	Hemodynamics	Human	[157]
RA	NOS, Myeloperoxidase, LDL	Tocilizimab	Yes	JAK/STAT and mTOR	Human	[158]
Stroke SLE	Atherosclerotic plaques	Glucocorticoids, Immunosuppressant	Yes	—	Human	[159,160]
SLE	Atherosclerotic plaques	Anifrolumab and tsDMARDs	Yes	JAK/BTK	Human Phase III	[161]

RP: C-reactive protein; LDL-C: Low-density lipoprotein cholesterol; TG: Triglyceride; MTX: Methotrexate; RA: Rheumatoid arthritis; NHC: Normal healthy control; T1DM: Type 1 diabetes mellitus; Rag-1: Recombination-activating gene 1; NOS: Nitric oxide synthase; JAK/STAT: Janus kinase/signal transducer activator of transcription protein; T2DM: Type 2 diabetes mellitus; H2/H3-RLX: Relaxin-1 and Relaxin3; mTOR: mammalian target of rapamycin; α-SMA: Alpha smooth muscle actin; MMP: Matrix metallopeptidase; TIMP: Tissue inhibitor of metalloproteinase; NLRP3: NOD-LRR and pyrin-domain-containing protein 3; Ox-LDL: oxidized low-density lipoprotein, CHF: Chronic heart failure; CVD: Cardiovascular disease; AMPK: AMP (Adenosine monophosphate)-activated protein kinase; eNOS: endothelial NOS; PPAR-γ/α: Peroxisome proliferator-activated receptor alpha/gamma; CKD: Chorionic kidney disease; LVF: Left ventricular failure; SLE: Systemic lupus erythematosus; tsDMARDs: Targeted synthetic disease-modifying antirheumatic drugs; JAK/BTK: JAK/Bruton's tyrosine kinase (inhibitor); STEMI: St-elevation myocardial infarction; ACE: Angiotensin-converting enzyme; ROS: Reactive oxygen species; SIRT-3: Nicotinamide adenosine diphosphate (NAD)-dependent deacetylase sirtuin-3; miR122: MicroRNA 122; CAT-1: Cationic amino acid transporter 1; PKC/AKT: Protein kinase C/protein kinase B.

Statins, the inhibitors of β-hydroxy β-methylglutaryl-CoA (HMG-CoA), are successful in lowering cholesterol loadings and expression of LOX-1; they also inhibit atherosclerotic progression and acute atherothrombosis [162–164]. Additionally, statins effectively reduce the proportion of LDL(−) [165–168]; discontinuation leads to LDL(−) approaching baseline levels [42]. However, the mechanisms of LDL(−) reduction are still not clear. Ezetimibe inhibits the Niemann–Pick C1-like 1 transporter (NPC1L1), which leads to decreased cholesterol absorption [169]. Proprotein convertase subtilisin kexin type 9 (PCSK9) is an enzyme for the degradation of LDL receptor (LDLR); blocking PCSK9 may increase LDLR, therefore lowering blood LDL-C concentrations. PCSK9 inhibitors such as alirocumab and evolocumab aggressively reduce the degradation of LDL receptors and increase the clearance of LDL cholesterol in hepatic cells [170]. They increase plaque stability but decrease the necrotic lipid core, as shown in Figure 1 [171–175]. However, other than statins, whether these drugs can decrease LDL(−) or not is currently unclear.

Several anti-inflammatory approaches were taken here to study the management of dyslipidemia, such as cell therapy using mesenchymal stem cells [176], leukotriene inhibitors [177], chemokine ligands (CC motif ligand), MCP-1, IL-1, and TNF-α blockers for the prevention of atherosclerotic plaque formation [178–184]. The currently used drugs significantly decrease LDL-C levels, stabilize vascular plaque, and slowdown atherosclerotic progression; however, new therapeutic strategies for LDL(−) and biomarkers are still needed.

7. Perspective

LDL(−) plays a critical role in the pathophysiology of atherogenesis. It triggers the dysfunction of endothelium by macrophage differentiation, monocyte migration, and platelet aggregation. Moreover, LDL(−) impairs endothelial cells by superoxide overproduction and platelet activation [185–187]. In combination with other well-known markers of inflammation, namely metabolic diseases, we concluded that LDL(−) can be a novel prognostic tool for these lipid disorders. Regarding treatment for the prevention of ASCVD, even though statins can partially reduce the concentration, finding a way to clear LDL(−) remains of utmost importance [22]. In particular, a method involving hydrolyzing atherogenic lipids in LDL(−) and producing harmless metabolites might be a novel therapeutic approach in the future.

Author Contributions: Conceptualization, C.-S.C.; validation, L.-Y.K.; resources, Y.-H.L.; data curation, H.-C.C.; writing—original draft preparation, S.H.L., V.K.M., F.P.; writing—review and editing, S.H.L., L.-Y.K.; visualization, F.P., H.-C.C.; supervision, C.-S.C., Y.-H.L.; project administration, L.-Y.K.; funding acquisition, C.-S.C., H.-C.C., L.-Y.K. All authors have read and agreed to the published version of the manuscript.

Funding: This work was supported in part by grants from the Kaohsiung Medical University (KMU-TC108A03-0), Kaohsiung Medical University Hospital (KMUH-M109017), Kaohsiung Municipal Ta-Tung Hospital (kmtth-101-001), Taiwan Ministry of Science and Technology (MOST109-2320-B-037-028-, 109-2628-B-037-010).

Conflicts of Interest: The authors declare no conflict of interest.

Abbreviations

Aβ	Amyloid β
ACE	Angiotensin-converting enzyme
ADP	Adenosine diphosphate
ADPase	Ecto-Adenosine diphosphate
ApoB100	Apolipoprotein B100
ApoCIII	Apolipoprotein CIII
ApoE	Apolipoprotein E
ASCVD	Atherosclerotic cardiovascular diseases
Bad/Bax	BCL2-associated agonist of cell death
Bcl-2	B-cell lymphoma 2
BBB	Blood–brain barrier
BP	Blood pressure
CAD	Coronary artery disease
CCL	Chemokine ligand
CD	Cluster of differentiation
CER	Ceramide
cIMTPWV	Carotid intermedia thickness and pulse wave velocity
COX	Cyclooxygenase
CD36	Cluster of differentiation 36
CRP	C-reactive protein
CVD	Cardiovascular disease
EC	Endothelial cell

ECM	Extracellular matrix
ED	Endothelial dysfunction
ERK	Extracellular signal-regulated kinase
eNOS	Endothelial nitric oxide synthase
ER	Endoplasmic reticulum
Fas	CD95 death receptor
FasL	Ligand CD95 L
FFA	Free fatty acids
FGF2	Fibroblast growth factor 2
FPLC	Fast-protein liquid chromatography
G-CSF	Granulocyte colony-stimulating factor
GDF	Growth differentiation factor
GM-CSF	Granulocyte–macrophage colony-stimulating factor
HDL	High-density lipoprotein
HFC	High-fat, high-cholesterol
HIF-1α	Hypoxia-inducible factor-1α
HMGCoA	β-hydroxy β-methylglutaryl-CoA
HUVECs	Human umbilical vein endothelial cells
ICAM	Intracellular adhesion molecule 1
IDL	Intermediate-density lipoprotein
IFN-γ	Interferon-γ
IKK2	IκB kinase 2
IL	Interleukin
iNOS	Inducible NO synthase
IR	Insulin resistance
IRAK2	Interleukin-1 receptor-associated kinase-2
IRE-1	Inositol requiring enzyme-1
Lp(a)	Lipoprotein (a)
LDL	Low-density lipoprotein
LDL(−)	Electronegative LDL
LDL-C	LDL cholesterol
LPC	Lysophosphatidylcholine
LPL	Lipoprotein lipase
LOX-1	Lectin-like oxidized low-density lipoprotein receptor-1
MAPK	Mitogen-activated protein kinase
MCP-1	Monocyte chemotactic protein-1
MetS	Metabolic syndrome
MMP	Metalloproteinase
MSC	Mesenchymal stem cell
NADPH	Nicotinamide adenine dinucleotide phosphate
NEFA	Non-esterified fatty acids
NF-κB	Nuclear factor kappa light-chain enhancer of activated B cells
NO	Nitric oxide
Nox	NADPH oxidase
NPC1L1	Niemann–pick C1-like
oxLDL	Oxidized LDL
PAFR	Platelet activating factor
PAFR	Platelet activating factor receptor
PC12	Pheochromocytoma cell-derived cell line
PGI2	Prostacyclin 2
PI3K	Phosphatidylinositol-3 kinase
PLD	Phospholipase D
PSCK9	Proprotein convertase subtilisin kexin type 9
RhoA	Ras homology family member A

Rac1	Ras-related C3 botulinum toxin substrate 1
Smase	Sphingomyelinase
STEMI	ST segment elevation myocardial infarction
TGF-β	Transforming growth factor- β
TIMP	Tissue inhibitors of metalloproteinase
TLR4	Toll-like receptor 4
TNF-α	Tumor necrosis factor-α
Tnfsf10	Tumor necrosis factor ligand, member 10
UCP 2	Uncoupling protein 2
UPR	Unfolded protein response
VCAM-1	Vascular cell adhesion molecule-1
VEGF	Vascular endothelial growth factor
VIPER	Viral inhibitory peptide of TLR4
VLDL	Very low-density lipoprotein
VSMCs	Vascular smooth muscle cells

References

1. Saltiel, A.R.; Olefsky, J.M. Inflammatory mechanisms linking obesity and metabolic disease. *J. Clin. Investig.* **2017**, *127*, 1–4. [CrossRef] [PubMed]
2. Benjamin, E.J.; Virani, S.S.; Callaway, C.W.; Chamberlain, A.M.; Chang, A.R.; Cheng, S.; Chiuve, S.E.; Cushman, M.; Delling, F.N.; Deo, R.; et al. Heart Disease and Stroke Statistics-2018 Update: A Report From the American Heart Association. *Circulation* **2018**, *137*, e67–e492. [CrossRef] [PubMed]
3. Hinton, W.; McGovern, A.; Coyle, R.; Han, T.S.; Sharma, P.; Correa, A.; Ferreira, F.; de Lusignan, S. Incidence and prevalence of cardiovascular disease in English primary care: A cross-sectional and follow-up study of the Royal College of General Practitioners (RCGP) Research and Surveillance Centre (RSC). *BMJ Open* **2018**, *8*, e020282. [CrossRef]
4. Goff, D.C., Jr.; Lloyd-Jones, D.M.; Bennett, G.; Coady, S.; D'Agostino, R.B.; Gibbons, R.; Greenland, P.; Lackland, D.T.; Levy, D.; O'Donnell, C.J.; et al. 2013 ACC/AHA guideline on the assessment of cardiovascular risk: A report of the American College of Cardiology/American Heart Association Task Force on Practice Guidelines. *Circulation* **2014**, *129*, S49–S73. [CrossRef] [PubMed]
5. Stone, N.J.; Robinson, J.G.; Lichtenstein, A.H.; Bairey Merz, C.N.; Blum, C.B.; Eckel, R.H.; Goldberg, A.C.; Gordon, D.; Levy, D.; Lloyd-Jones, D.M.; et al. 2013 ACC/AHA guideline on the treatment of blood cholesterol to reduce atherosclerotic cardiovascular risk in adults: A report of the American College of Cardiology/American Heart Association Task Force on Practice Guidelines. *Circulation* **2014**, *129*, S1–S45. [CrossRef] [PubMed]
6. Fernandez-Friera, L.; Fuster, V.; Lopez-Melgar, B.; Oliva, B.; Garcia-Ruiz, J.M.; Mendiguren, J.; Bueno, H.; Pocock, S.; Ibanez, B.; Fernandez-Ortiz, A.; et al. Normal LDL-Cholesterol Levels Are Associated With Subclinical Atherosclerosis in the Absence of Risk Factors. *J. Am. Coll. Cardiol.* **2017**, *70*, 2979–2991. [CrossRef] [PubMed]
7. Toth, P.P.; Patti, A.M.; Giglio, R.V.; Nikolic, D.; Castellino, G.; Rizzo, M.; Banach, M. Management of Statin Intolerance in 2018: Still More Questions Than Answers. *Am. J. Cardiovasc. Drugs* **2018**, *18*, 157–173. [CrossRef]
8. Schwandt, P.; Liepold, E.; Bertsch, T.; Haas, G.M. Lifestyle, Cardiovascular Drugs and Risk Factors in Younger and Elder Adults: The PEP Family Heart Study. *Int. J. Prev. Med.* **2010**, *1*, 56–61.
9. Yasue, H.; Hirai, N.; Mizuno, Y.; Harada, E.; Itoh, T.; Yoshimura, M.; Kugiyama, K.; Ogawa, H. Low-grade inflammation, thrombogenicity, and atherogenic lipid profile in cigarette smokers. *Circ. J.* **2006**, *70*, 8–13. [CrossRef]
10. Hansson, G.K. Inflammation, atherosclerosis, and coronary artery disease. *N. Engl. J. Med.* **2005**, *352*, 1685–1695. [CrossRef]
11. Stancel, N.; Chen, C.C.; Ke, L.Y.; Chu, C.S.; Lu, J.; Sawamura, T.; Chen, C.H. Interplay between CRP, Atherogenic LDL, and LOX-1 and Its Potential Role in the Pathogenesis of Atherosclerosis. *Clin. Chem.* **2016**, *62*, 320–327. [CrossRef] [PubMed]
12. Ross, R. Atherosclerosis—An inflammatory disease. *N. Engl. J. Med.* **1999**, *340*, 115–126. [CrossRef] [PubMed]

13. Mudau, M.; Genis, A.; Lochner, A.; Strijdom, H. Endothelial dysfunction: The early predictor of atherosclerosis. *Cardiovasc. J. Afr.* **2012**, *23*, 222–231. [CrossRef] [PubMed]
14. Carmena, R.; Duriez, P.; Fruchart, J.C. Atherogenic lipoprotein particles in atherosclerosis. *Circulation* **2004**, *109*, III-2–III-7. [CrossRef] [PubMed]
15. Libby, P. Inflammation in atherosclerosis. *Nature* **2002**, *420*, 868–874. [CrossRef]
16. Witztum, J.L.; Steinberg, D. The oxidative modification hypothesis of atherosclerosis: Does it hold for humans? *Trends Cardiovasc. Med.* **2001**, *11*, 93–102. [CrossRef]
17. Estruch, M.; Sanchez-Quesada, J.L.; Ordonez Llanos, J.; Benitez, S. Electronegative LDL: A circulating modified LDL with a role in inflammation. *Mediators Inflamm.* **2013**, *2013*, 181324. [CrossRef]
18. Nyyssonen, K.; Kaikkonen, J.; Salonen, J.T. Characterization and determinants of an electronegatively charged low-density lipoprotein in human plasma. *Scand. J. Clin. Lab. Investig.* **1996**, *56*, 681–689. [CrossRef]
19. Barros, M.R.; Bertolami, M.C.; Abdalla, D.S.; Ferreira, W.P. Identification of mildly oxidized low-density lipoprotein (electronegative LDL) and its auto-antibodies IgG in children and adolescents hypercholesterolemic offsprings. *Atherosclerosis* **2006**, *184*, 103–107. [CrossRef]
20. Ivanova, E.A.; Bobryshev, Y.V.; Orekhov, A.N. LDL electronegativity index: A potential novel index for predicting cardiovascular disease. *Vasc. Health Risk Manag.* **2015**, *11*, 525–532. [CrossRef]
21. Chu, C.S.; Chan, H.C.; Tsai, M.H.; Stancel, N.; Lee, H.C.; Cheng, K.H.; Tung, Y.C.; Chan, H.C.; Wang, C.Y.; Shin, S.J.; et al. Range of L5 LDL levels in healthy adults and L5's predictive power in patients with hyperlipidemia or coronary artery disease. *Sci. Rep.* **2018**, *8*, 11866. [CrossRef] [PubMed]
22. Chu, C.S.; Law, S.H.; Lenzen, D.; Tan, Y.H.; Weng, S.F.; Ito, E.; Wu, J.C.; Chen, C.H.; Chan, H.C.; Ke, L.Y. Clinical Significance of Electronegative Low-Density Lipoprotein Cholesterol in Atherothrombosis. *Biomedicines* **2020**, *8*, 254. [CrossRef]
23. Sanchez-Quesada, J.L.; Benitez, S.; Ordonez-Llanos, J. Electronegative low-density lipoprotein. *Curr. Opin. Lipidol.* **2004**, *15*, 329–335. [CrossRef] [PubMed]
24. Benitez, S.; Camacho, M.; Arcelus, R.; Vila, L.; Bancells, C.; Ordonez-Llanos, J.; Sanchez-Quesada, J.L. Increased lysophosphatidylcholine and non-esterified fatty acid content in LDL induces chemokine release in endothelial cells. Relationship with electronegative LDL. *Atherosclerosis* **2004**, *177*, 299–305. [CrossRef] [PubMed]
25. Estruch, M.; Sanchez-Quesada, J.L.; Beloki, L.; Ordonez-Llanos, J.; Benitez, S. The Induction of Cytokine Release in Monocytes by Electronegative Low-Density Lipoprotein (LDL) Is Related to Its Higher Ceramide Content than Native LDL. *Int. J. Mol. Sci.* **2013**, *14*, 2601–2616. [CrossRef] [PubMed]
26. Ke, L.Y.; Stancel, N.; Bair, H.; Chen, C.H. The underlying chemistry of electronegative LDL's atherogenicity. *Curr. Atheroscler. Rep.* **2014**, *16*, 428. [CrossRef]
27. Chan, H.C.; Chan, H.C.; Liang, C.J.; Lee, H.C.; Su, H.; Lee, A.S.; Shiea, J.; Tsai, W.C.; Ou, T.T.; Wu, C.C.; et al. Role of Low-Density Lipoprotein in Early Vascular Aging Associated With Systemic Lupus Erythematosus. *Arthritis Rheumatol.* **2020**, *72*, 972–984. [CrossRef]
28. Ke, L.Y.; Chan, H.C.; Chen, C.C.; Lu, J.; Marathe, G.K.; Chu, C.S.; Chan, H.C.; Wang, C.Y.; Tung, Y.C.; McIntyre, T.M.; et al. Enhanced Sphingomyelinase Activity Contributes to the Apoptotic Capacity of Electronegative Low-Density Lipoprotein. *J. Med. Chem.* **2016**, *59*, 1032–1040. [CrossRef]
29. Ke, L.Y.; Engler, D.A.; Lu, J.; Matsunami, R.K.; Chan, H.C.; Wang, G.J.; Yang, C.Y.; Chang, J.G.; Chen, C.H. Chemical composition-oriented receptor selectivity of L5, a naturally occurring atherogenic low-density lipoprotein. *Pure Appl. Chem.* **2011**, *83*. [CrossRef]
30. Jayaraman, S.; Chavez, O.R.; Perez, A.; Minambres, I.; Sanchez-Quesada, J.L.; Gursky, O. Binding to heparin triggers deleterious structural and biochemical changes in human low-density lipoprotein, which are amplified in hyperglycemia. *Biochim. Biophys. Acta Mol. Cell Biol. Lipids* **2020**, *1865*, 158712. [CrossRef]
31. Bancells, C.; Benitez, S.; Ordonez-Llanos, J.; Oorni, K.; Kovanen, P.T.; Milne, R.W.; Sanchez-Quesada, J.L. Immunochemical analysis of the electronegative LDL subfraction shows that abnormal N-terminal apolipoprotein B conformation is involved in increased binding to proteoglycans. *J. Biol. Chem.* **2011**, *286*, 1125–1133. [CrossRef] [PubMed]
32. Blanco, F.J.; Villegas, S.; Benitez, S.; Bancells, C.; Diercks, T.; Ordonez-Llanos, J.; Sanchez-Quesada, J.L. 2D-NMR reveals different populations of exposed lysine residues in the apoB-100 protein of electronegative and electropositive fractions of LDL particles. *J. Lipid Res.* **2010**, *51*, 1560–1565. [CrossRef] [PubMed]

33. Chen, C.H.; Jiang, T.; Yang, J.H.; Jiang, W.; Lu, J.; Marathe, G.K.; Pownall, H.J.; Ballantyne, C.M.; McIntyre, T.M.; Henry, P.D.; et al. Low-density lipoprotein in hypercholesterolemic human plasma induces vascular endothelial cell apoptosis by inhibiting fibroblast growth factor 2 transcription. *Circulation* **2003**, *107*, 2102–2108. [CrossRef] [PubMed]
34. Yang, C.Y.; Raya, J.L.; Chen, H.H.; Chen, C.H.; Abe, Y.; Pownall, H.J.; Taylor, A.A.; Smith, C.V. Isolation, characterization, and functional assessment of oxidatively modified subfractions of circulating low-density lipoproteins. *Arterioscler. Thromb. Vasc. Biol.* **2003**, *23*, 1083–1090. [CrossRef] [PubMed]
35. Ke, L.Y.; Chan, H.C.; Chen, C.C.; Chang, C.F.; Lu, P.L.; Chu, C.S.; Lai, W.T.; Shin, S.J.; Liu, F.T.; Chen, C.H. Increased APOE glycosylation plays a key role in the atherogenicity of L5 low-density lipoprotein. *FASEB J.* **2020**, *34*, 9802–9813. [CrossRef]
36. Lu, J.; Yang, J.H.; Burns, A.R.; Chen, H.H.; Tang, D.; Walterscheid, J.P.; Suzuki, S.; Yang, C.Y.; Sawamura, T.; Chen, C.H. Mediation of electronegative low-density lipoprotein signaling by LOX-1: A possible mechanism of endothelial apoptosis. *Circ. Res.* **2009**, *104*, 619–627. [CrossRef]
37. Chan, H.C.; Ke, L.Y.; Chu, C.S.; Lee, A.S.; Shen, M.Y.; Cruz, M.A.; Hsu, J.F.; Cheng, K.H.; Chan, H.C.; Lu, J.; et al. Highly electronegative LDL from patients with ST-elevation myocardial infarction triggers platelet activation and aggregation. *Blood* **2013**, *122*, 3632–3641. [CrossRef]
38. Ke, L.Y.; Chan, H.C.; Chan, H.C.; Kalu, F.C.U.; Lee, H.C.; Lin, I.L.; Jhuo, S.J.; Lai, W.T.; Tsao, C.R.; Sawamura, T.; et al. Electronegative Low-Density Lipoprotein L5 Induces Adipose Tissue Inflammation Associated With Metabolic Syndrome. *J. Clin. Endocrinol. Metab.* **2017**, *102*, 4615–4625. [CrossRef]
39. Chang, S.F.; Chang, P.Y.; Chou, Y.C.; Lu, S.C. Electronegative LDL Induces M1 Polarization of Human Macrophages Through a LOX-1-Dependent Pathway. *Inflammation* **2020**, *43*, 1524–1535. [CrossRef]
40. Tang, D.; Lu, J.; Walterscheid, J.P.; Chen, H.H.; Engler, D.A.; Sawamura, T.; Chang, P.Y.; Safi, H.J.; Yang, C.Y.; Chen, C.H. Electronegative LDL circulating in smokers impairs endothelial progenitor cell differentiation by inhibiting Akt phosphorylation via LOX-1. *J. Lipid Res.* **2008**, *49*, 33–47. [CrossRef]
41. Chen, W.Y.; Chen, Y.F.; Chan, H.C.; Chung, C.H.; Peng, H.Y.; Ho, Y.C.; Chen, C.H.; Chang, K.C.; Tang, C.H.; Lee, A.S. Role of apolipoprotein E in electronegative low-density lipoprotein-induced mitochondrial dysfunction in cardiomyocytes. *Metab. Clin. Exp.* **2020**, *107*, 154227. [CrossRef] [PubMed]
42. Chu, C.S.; Wang, Y.C.; Lu, L.S.; Walton, B.; Yilmaz, H.R.; Huang, R.Y.; Sawamura, T.; Dixon, R.A.; Lai, W.T.; Chen, C.H.; et al. Electronegative low-density lipoprotein increases C-reactive protein expression in vascular endothelial cells through the LOX-1 receptor. *PLoS ONE* **2013**, *8*, e70533. [CrossRef] [PubMed]
43. Shen, M.Y.; Chen, F.Y.; Hsu, J.F.; Fu, R.H.; Chang, C.M.; Chang, C.T.; Liu, C.H.; Wu, J.R.; Lee, A.S.; Chan, H.C.; et al. Plasma L5 levels are elevated in ischemic stroke patients and enhance platelet aggregation. *Blood* **2016**, *127*, 1336–1345. [CrossRef] [PubMed]
44. Estruch, M.; Sanchez-Quesada, J.L.; Ordonez-Llanos, J.; Benitez, S. Ceramide-enriched LDL induces cytokine release through TLR4 and CD14 in monocytes. Similarities with electronegative LDL. *Clin. Investig. Arterioscler.* **2014**, *26*, 131–137. [CrossRef] [PubMed]
45. Ligi, D.; Benitez, S.; Croce, L.; Rivas-Urbina, A.; Puig, N.; Ordonez-Llanos, J.; Mannello, F.; Sanchez-Quesada, J.L. Electronegative LDL induces MMP-9 and TIMP-1 release in monocytes through CD14 activation: Inhibitory effect of glycosaminoglycan sulodexide. *Biochim. Biophys. Acta Mol. Basis Dis.* **2018**, *1864*, 3559–3567. [CrossRef] [PubMed]
46. Brunelli, R.; Balogh, G.; Costa, G.; De Spirito, M.; Greco, G.; Mei, G.; Nicolai, E.; Vigh, L.; Ursini, F.; Parasassi, T. Estradiol binding prevents ApoB-100 misfolding in electronegative LDL(−). *Biochemistry* **2010**, *49*, 7297–7302. [CrossRef]
47. Brunelli, R.; De Spirito, M.; Mei, G.; Papi, M.; Perrone, G.; Stefanutti, C.; Parasassi, T. Misfolding of apoprotein B-100, LDL aggregation and 17-beta -estradiol in atherogenesis. *Curr. Med. Chem.* **2014**, *21*, 2276–2283. [CrossRef]
48. Bancells, C.; Benitez, S.; Villegas, S.; Jorba, O.; Ordonez-Llanos, J.; Sanchez-Quesada, J.L. Novel phospholipolytic activities associated with electronegative low-density lipoprotein are involved in increased self-aggregation. *Biochemistry* **2008**, *47*, 8186–8194. [CrossRef]
49. Sanchez-Quesada, J.L.; Villegas, S.; Ordonez-Llanos, J. Electronegative low-density lipoprotein. A link between apolipoprotein B misfolding, lipoprotein aggregation and proteoglycan binding. *Curr. Opin. Lipidol.* **2012**, *23*, 479–486. [CrossRef]

50. Lai, Y.S.; Yang, T.C.; Chang, P.Y.; Chang, S.F.; Ho, S.L.; Chen, H.L.; Lu, S.C. Electronegative LDL is linked to high-fat, high-cholesterol diet-induced nonalcoholic steatohepatitis in hamsters. *J. Nutr. Biochem.* **2016**, *30*, 44–52. [CrossRef]
51. Chang, P.Y.; Pai, J.H.; Lai, Y.S.; Lu, S.C. Electronegative LDL from Rabbits Fed with Atherogenic Diet Is Highly Proinflammatory. *Mediators Inflamm.* **2019**, *2019*, 6163130. [CrossRef] [PubMed]
52. Kruger-Genge, A.; Blocki, A.; Franke, R.P.; Jung, F. Vascular Endothelial Cell Biology: An Update. *Int. J. Mol. Sci.* **2019**, *20*, 4411. [CrossRef] [PubMed]
53. Yau, J.W.; Teoh, H.; Verma, S. Endothelial cell control of thrombosis. *BMC Cardiovasc. Disord.* **2015**, *15*, 130. [CrossRef] [PubMed]
54. Hamilos, M.; Petousis, S.; Parthenakis, F. Interaction between platelets and endothelium: From pathophysiology to new therapeutic options. *Cardiovasc. Diagn. Ther.* **2018**, *8*, 568–580. [CrossRef]
55. Ziouzenkova, O.; Asatryan, L.; Sahady, D.; Orasanu, G.; Perrey, S.; Cutak, B.; Hassell, T.; Akiyama, T.E.; Berger, J.P.; Sevanian, A.; et al. Dual roles for lipolysis and oxidation in peroxisome proliferation-activator receptor responses to electronegative low density lipoprotein. *J. Biol. Chem.* **2003**, *278*, 39874–39881. [CrossRef]
56. Rommel, C.; Clarke, B.A.; Zimmermann, S.; Nunez, L.; Rossman, R.; Reid, K.; Moelling, K.; Yancopoulos, G.D.; Glass, D.J. Differentiation stage-specific inhibition of the Raf-MEK-ERK pathway by Akt. *Science* **1999**, *286*, 1738–1741. [CrossRef]
57. Kanda, S.; Hodgkin, M.N.; Woodfield, R.J.; Wakelam, M.J.; Thomas, G.; Claesson-Welsh, L. Phosphatidylinositol 3′-kinase-independent p70 S6 kinase activation by fibroblast growth factor receptor-1 is important for proliferation but not differentiation of endothelial cells. *J. Biol. Chem.* **1997**, *272*, 23347–23353. [CrossRef]
58. Abid, M.R.; Guo, S.; Minami, T.; Spokes, K.C.; Ueki, K.; Skurk, C.; Walsh, K.; Aird, W.C. Vascular endothelial growth factor activates PI3K/Akt/forkhead signaling in endothelial cells. *Arterioscler. Thromb. Vasc. Biol.* **2004**, *24*, 294–300. [CrossRef]
59. Lu, J.; Jiang, W.; Yang, J.H.; Chang, P.Y.; Walterscheid, J.P.; Chen, H.H.; Marcelli, M.; Tang, D.; Lee, Y.T.; Liao, W.S.; et al. Electronegative LDL impairs vascular endothelial cell integrity in diabetes by disrupting fibroblast growth factor 2 (FGF2) autoregulation. *Diabetes* **2008**, *57*, 158–166. [CrossRef]
60. Sawamura, T.; Kakino, A.; Fujita, Y. LOX-1: A multiligand receptor at the crossroads of response to danger signals. *Curr. Opin. Lipidol.* **2012**, *23*, 439–445. [CrossRef]
61. Inoue, T.; Ishida, T.; Inoue, T.; Saito, A.; Ezura, M.; Uenohara, H.; Fujimura, M.; Sato, K.; Endo, T.; Omodaka, S.; et al. Lectin-Like Oxidized Low-Density Lipoprotein Receptor-1 Levels as a Biomarker of Acute Intracerebral Hemorrhage. *J. Stroke Cerebrovasc. Dis.* **2019**, *28*, 490–494. [CrossRef] [PubMed]
62. Lubrano, V.; Pingitore, A.; Traghella, I.; Storti, S.; Parri, S.; Berti, S.; Ndreu, R.; Andrenelli, A.; Palmieri, C.; Iervasi, G.; et al. Emerging Biomarkers of Oxidative Stress in Acute and Stable Coronary Artery Disease: Levels and Determinants. *Antioxidants* **2019**, *8*, 115. [CrossRef] [PubMed]
63. Lee, A.S.; Wang, Y.C.; Chang, S.S.; Lo, P.H.; Chang, C.M.; Lu, J.; Burns, A.R.; Chen, C.H.; Kakino, A.; Sawamura, T.; et al. Detection of a High Ratio of Soluble to Membrane-Bound LOX-1 in Aspirated Coronary Thrombi From Patients With ST-Segment-Elevation Myocardial Infarction. *J. Am. Heart Assoc.* **2020**, *9*, e014008. [CrossRef] [PubMed]
64. Ishikawa, M.; Ito, H.; Akiyoshi, M.; Kume, N.; Yoshitomi, H.; Mitsuoka, H.; Tanida, S.; Murata, K.; Shibuya, H.; Kasahara, T.; et al. Lectin-like oxidized low-density lipoprotein receptor 1 signal is a potent biomarker and therapeutic target for human rheumatoid arthritis. *Arthritis Rheum.* **2012**, *64*, 1024–1034. [CrossRef] [PubMed]
65. Hofmann, A.; Brunssen, C.; Wolk, S.; Reeps, C.; Morawietz, H. Soluble LOX-1: A Novel Biomarker in Patients With Coronary Artery Disease, Stroke, and Acute Aortic Dissection? *J. Am. Heart Assoc.* **2020**, *9*, e013803. [CrossRef] [PubMed]
66. Inoue, K.; Arai, Y.; Kurihara, H.; Kita, T.; Sawamura, T. Overexpression of lectin-like oxidized low-density lipoprotein receptor-1 induces intramyocardial vasculopathy in apolipoprotein E-null mice. *Circ. Res.* **2005**, *97*, 176–184. [CrossRef]
67. Mehta, J.L.; Sanada, N.; Hu, C.P.; Chen, J.; Dandapat, A.; Sugawara, F.; Satoh, H.; Inoue, K.; Kawase, Y.; Jishage, K.; et al. Deletion of LOX-1 reduces atherogenesis in LDLR knockout mice fed high cholesterol diet. *Circ. Res.* **2007**, *100*, 1634–1642. [CrossRef]

68. Kataoka, H.; Kume, N.; Miyamoto, S.; Minami, M.; Moriwaki, H.; Murase, T.; Sawamura, T.; Masaki, T.; Hashimoto, N.; Kita, T. Expression of lectinlike oxidized low-density lipoprotein receptor-1 in human atherosclerotic lesions. *Circulation* **1999**, *99*, 3110–3117. [CrossRef]
69. Mattaliano, M.D.; Huard, C.; Cao, W.; Hill, A.A.; Zhong, W.; Martinez, R.V.; Harnish, D.C.; Paulsen, J.E.; Shih, H.H. LOX-1-dependent transcriptional regulation in response to oxidized LDL treatment of human aortic endothelial cells. *Am. J. Physiol. Cell Physiol.* **2009**, *296*, C1329–C1337. [CrossRef]
70. Kattoor, A.J.; Goel, A.; Mehta, J.L. LOX-1: Regulation, Signaling and Its Role in Atherosclerosis. *Antioxidants* **2019**, *8*, 218. [CrossRef]
71. Sagar, D.; Gaddipati, R.; Ongstad, E.L.; Bhagroo, N.; An, L.L.; Wang, J.; Belkhodja, M.; Rahman, S.; Manna, Z.; Davis, M.A.; et al. LOX-1: A potential driver of cardiovascular risk in SLE patients. *PLoS ONE* **2020**, *15*, e0229184. [CrossRef] [PubMed]
72. Xu, S.; Ogura, S.; Chen, J.; Little, P.J.; Moss, J.; Liu, P. LOX-1 in atherosclerosis: Biological functions and pharmacological modifiers. *Cell Mol. Life Sci.* **2013**, *70*, 2859–2872. [CrossRef] [PubMed]
73. Li, D.; Singh, R.M.; Liu, L.; Chen, H.; Singh, B.M.; Kazzaz, N.; Mehta, J.L. Oxidized-LDL through LOX-1 increases the expression of angiotensin converting enzyme in human coronary artery endothelial cells. *Cardiovasc. Res.* **2003**, *57*, 238–243. [CrossRef]
74. Li, D.; Mehta, J.L. Upregulation of endothelial receptor for oxidized LDL (LOX-1) by oxidized LDL and implications in apoptosis of human coronary artery endothelial cells: Evidence from use of antisense LOX-1 mRNA and chemical inhibitors. *Arterioscler. Thromb. Vasc. Biol.* **2000**, *20*, 1116–1122. [CrossRef] [PubMed]
75. Sugimoto, K.; Ishibashi, T.; Sawamura, T.; Inoue, N.; Kamioka, M.; Uekita, H.; Ohkawara, H.; Sakamoto, T.; Sakamoto, N.; Okamoto, Y.; et al. LOX-1-MT1-MMP axis is crucial for RhoA and Rac1 activation induced by oxidized low-density lipoprotein in endothelial cells. *Cardiovasc. Res.* **2009**, *84*, 127–136. [CrossRef] [PubMed]
76. Zhang, H.G.; Zhang, Q.J.; Li, B.W.; Li, L.H.; Song, X.H.; Xiong, C.M.; Zou, Y.B.; Liu, B.Y.; Han, J.Q.; Xiu, R.J. The circulating level of miR-122 is a potential risk factor for endothelial dysfunction in young patients with essential hypertension. *Hypertens. Res.* **2020**. [CrossRef] [PubMed]
77. Shi, Y.; Cosentino, F.; Camici, G.G.; Akhmedov, A.; Vanhoutte, P.M.; Tanner, F.C.; Luscher, T.F. Oxidized low-density lipoprotein activates p66Shc via lectin-like oxidized low-density lipoprotein receptor-1, protein kinase C-beta, and c-Jun N-terminal kinase kinase in human endothelial cells. *Arterioscler. Thromb. Vasc. Biol.* **2011**, *31*, 2090–2097. [CrossRef]
78. Ma, S.C.; Hao, Y.J.; Jiao, Y.; Wang, Y.H.; Xu, L.B.; Mao, C.Y.; Yang, X.L.; Yang, A.N.; Tian, J.; Zhang, M.H.; et al. Homocysteineinduced oxidative stress through TLR4/NFkappaB/DNMT1mediated LOX1 DNA methylation in endothelial cells. *Mol. Med. Rep.* **2017**, *16*, 9181–9188. [CrossRef]
79. Yu, E.P.; Bennett, M.R. The role of mitochondrial DNA damage in the development of atherosclerosis. *Free Radic. Biol. Med.* **2016**, *100*, 223–230. [CrossRef]
80. Hu, C.; Dandapat, A.; Sun, L.; Chen, J.; Marwali, M.R.; Romeo, F.; Sawamura, T.; Mehta, J.L. LOX-1 deletion decreases collagen accumulation in atherosclerotic plaque in low-density lipoprotein receptor knockout mice fed a high-cholesterol diet. *Cardiovasc. Res.* **2008**, *79*, 287–293. [CrossRef]
81. Lu, J.; Mitra, S.; Wang, X.; Khaidakov, M.; Mehta, J.L. Oxidative stress and lectin-like ox-LDL-receptor LOX-1 in atherogenesis and tumorigenesis. *Antioxid. Redox. Signal.* **2011**, *15*, 2301–2333. [CrossRef]
82. Murphy, M.P. How mitochondria produce reactive oxygen species. *Biochem. J.* **2009**, *417*, 1–13. [CrossRef]
83. Koziel, R.; Pircher, H.; Kratochwil, M.; Lener, B.; Hermann, M.; Dencher, N.A.; Jansen-Durr, P. Mitochondrial respiratory chain complex I is inactivated by NADPH oxidase Nox4. *Biochem. J.* **2013**, *452*, 231–239. [CrossRef] [PubMed]
84. Giorgio, M.; Migliaccio, E.; Orsini, F.; Paolucci, D.; Moroni, M.; Contursi, C.; Pelliccia, G.; Luzi, L.; Minucci, S.; Marcaccio, M.; et al. Electron transfer between cytochrome c and p66Shc generates reactive oxygen species that trigger mitochondrial apoptosis. *Cell* **2005**, *122*, 221–233. [CrossRef] [PubMed]
85. Trinei, M.; Berniakovich, I.; Beltrami, E.; Migliaccio, E.; Fassina, A.; Pelicci, P.; Giorgio, M. P66Shc signals to age. *Aging (Albany N.Y.)* **2009**, *1*, 503–510. [CrossRef] [PubMed]
86. Chang, C.T.; Wang, G.J.; Kuo, C.C.; Hsieh, J.Y.; Lee, A.S.; Chang, C.M.; Wang, C.C.; Shen, M.Y.; Huang, C.C.; Sawamura, T.; et al. Electronegative Low-density Lipoprotein Increases Coronary Artery Disease Risk in Uremia Patients on Maintenance Hemodialysis. *Medicine* **2016**, *95*, e2265. [CrossRef] [PubMed]

87. Ron, D.; Walter, P. Signal integration in the endoplasmic reticulum unfolded protein response. *Nat. Rev. Mol. Cell Biol.* **2007**, *8*, 519–529. [CrossRef]
88. Sanson, M.; Auge, N.; Vindis, C.; Muller, C.; Bando, Y.; Thiers, J.C.; Marachet, M.A.; Zarkovic, K.; Sawa, Y.; Salvayre, R.; et al. Oxidized low-density lipoproteins trigger endoplasmic reticulum stress in vascular cells: Prevention by oxygen-regulated protein 150 expression. *Circ. Res.* **2009**, *104*, 328–336. [CrossRef]
89. Gora, S.; Maouche, S.; Atout, R.; Wanherdrick, K.; Lambeau, G.; Cambien, F.; Ninio, E.; Karabina, S.A. Phospholipolyzed LDL induces an inflammatory response in endothelial cells through endoplasmic reticulum stress signaling. *FASEB J.* **2010**, *24*, 3284–3297. [CrossRef]
90. Gargalovic, P.S.; Gharavi, N.M.; Clark, M.J.; Pagnon, J.; Yang, W.P.; He, A.; Truong, A.; Baruch-Oren, T.; Berliner, J.A.; Kirchgessner, T.G.; et al. The unfolded protein response is an important regulator of inflammatory genes in endothelial cells. *Arterioscler. Thromb. Vasc. Biol.* **2006**, *26*, 2490–2496. [CrossRef]
91. Xie, X.; Zhao, R.; Shen, G.X. Impact of cyanidin-3-glucoside on glycated LDL-induced NADPH oxidase activation, mitochondrial dysfunction and cell viability in cultured vascular endothelial cells. *Int. J. Mol. Sci.* **2012**, *13*, 15867–15880. [CrossRef] [PubMed]
92. Zhao, R.; Xie, X.; Le, K.; Li, W.; Moghadasian, M.H.; Beta, T.; Shen, G.X. Endoplasmic reticulum stress in diabetic mouse or glycated LDL-treated endothelial cells: Protective effect of Saskatoon berry powder and cyanidin glycans. *J. Nutr. Biochem.* **2015**, *26*, 1248–1253. [CrossRef]
93. Benitez, S.; Bancells, C.; Ordonez-Llanos, J.; Sanchez-Quesada, J.L. Pro-inflammatory action of LDL(−) on mononuclear cells is counteracted by increased IL10 production. *Biochim. Biophys. Acta* **2007**, *1771*, 613–622. [CrossRef]
94. Benitez, S.; Camacho, M.; Bancells, C.; Vila, L.; Sanchez-Quesada, J.L.; Ordonez-Llanos, J. Wide proinflammatory effect of electronegative low-density lipoprotein on human endothelial cells assayed by a protein array. *Biochim. Biophys. Acta* **2006**, *1761*, 1014–1021. [CrossRef] [PubMed]
95. De Castellarnau, C.; Sanchez-Quesada, J.L.; Benitez, S.; Rosa, R.; Caveda, L.; Vila, L.; Ordonez-Llanos, J. Electronegative LDL from normolipemic subjects induces IL-8 and monocyte chemotactic protein secretion by human endothelial cells. *Arterioscler. Thromb. Vasc. Biol.* **2000**, *20*, 2281–2287. [CrossRef] [PubMed]
96. Demuth, K.; Myara, I.; Chappey, B.; Vedie, B.; Pech-Amsellem, M.A.; Haberland, M.E.; Moatti, N. A cytotoxic electronegative LDL subfraction is present in human plasma. *Arterioscler. Thromb. Vasc. Biol.* **1996**, *16*, 773–783. [CrossRef] [PubMed]
97. Hodis, H.N.; Kramsch, D.M.; Avogaro, P.; Bittolo-Bon, G.; Cazzolato, G.; Hwang, J.; Peterson, H.; Sevanian, A. Biochemical and cytotoxic characteristics of an in vivo circulating oxidized low density lipoprotein (LDL-). *J. Lipid Res.* **1994**, *35*, 669–677.
98. Estruch, M.; Rajamaki, K.; Sanchez-Quesada, J.L.; Kovanen, P.T.; Oorni, K.; Benitez, S.; Ordonez-Llanos, J. Electronegative LDL induces priming and inflammasome activation leading to IL-1beta release in human monocytes and macrophages. *Biochim. Biophys. Acta* **2015**, *1851*, 1442–1449. [CrossRef]
99. Estruch, M.; Sanchez-Quesada, J.L.; Ordonez-Llanos, J.; Benitez, S. Inflammatory intracellular pathways activated by electronegative LDL in monocytes. *Biochim. Biophys. Acta* **2016**, *1861 Pt A*, 963–969. [CrossRef]
100. Glass, C.K.; Witztum, J.L. Atherosclerosis. the road ahead. *Cell* **2001**, *104*, 503–516. [CrossRef]
101. Gordon, S.; Martinez, F.O. Alternative activation of macrophages: Mechanism and functions. *Immunity* **2010**, *32*, 593–604. [CrossRef] [PubMed]
102. Yang, T.C.; Chang, P.Y.; Lu, S.C. L5-LDL from ST-elevation myocardial infarction patients induces IL-1beta production via LOX-1 and NLRP3 inflammasome activation in macrophages. *Am. J. Physiol. Heart Circ. Physiol.* **2017**, *312*, H265–H274. [CrossRef] [PubMed]
103. Yang, T.C.; Chang, P.Y.; Kuo, T.L.; Lu, S.C. Electronegative L5-LDL induces the production of G-CSF and GM-CSF in human macrophages through LOX-1 involving NF-kappaB and ERK2 activation. *Atherosclerosis* **2017**, *267*, 1–9. [CrossRef] [PubMed]
104. Chang, C.K.; Chen, P.K.; Lan, J.L.; Chang, S.H.; Hsieh, T.Y.; Liao, P.J.; Chen, C.H.; Chen, D.Y. Association of Electronegative LDL with Macrophage Foam Cell Formation and CD11c Expression in Rheumatoid Arthritis Patients. *Int. J. Mol. Sci.* **2020**, *21*, 5883. [CrossRef] [PubMed]
105. Pedrosa, A.M.; Faine, L.A.; Grosso, D.M.; de Las Heras, B.; Bosca, L.; Abdalla, D.S. Electronegative LDL induction of apoptosis in macrophages: Involvement of Nrf2. *Biochim. Biophys. Acta* **2010**, *1801*, 430–437. [CrossRef] [PubMed]

106. Parguina, A.F.; Grigorian-Shamagian, L.; Agra, R.M.; Lopez-Otero, D.; Rosa, I.; Alonso, J.; Teijeira-Fernandez, E.; Gonzalez-Juanatey, J.R.; Garcia, A. Variations in platelet proteins associated with ST-elevation myocardial infarction: Novel clues on pathways underlying platelet activation in acute coronary syndromes. *Arterioscler. Thromb. Vasc. Biol.* **2011**, *31*, 2957–2964. [CrossRef]
107. Podrez, E.A.; Byzova, T.V. Prothrombotic lipoprotein patterns in stroke. *Blood* **2016**, *127*, 1221–1222. [CrossRef]
108. Hou, X.; Summer, R.; Chen, Z.; Tian, Y.; Ma, J.; Cui, J.; Hao, X.; Guo, L.; Xu, H.; Wang, H.; et al. Lipid Uptake by Alveolar Macrophages Drives Fibrotic Responses to Silica Dust. *Sci. Rep.* **2019**, *9*, 399. [CrossRef]
109. Yoshida, H.; Quehenberger, O.; Kondratenko, N.; Green, S.; Steinberg, D. Minimally oxidized low-density lipoprotein increases expression of scavenger receptor A, CD36, and macrosialin in resident mouse peritoneal macrophages. *Arterioscler. Thromb. Vasc. Biol.* **1998**, *18*, 794–802. [CrossRef]
110. Nagy, L.; Tontonoz, P.; Alvarez, J.G.; Chen, H.; Evans, R.M. Oxidized LDL regulates macrophage gene expression through ligand activation of PPARgamma. *Cell* **1998**, *93*, 229–240. [CrossRef]
111. Rahalkar, A.R.; Hegele, R.A. Monogenic pediatric dyslipidemias: Classification, genetics and clinical spectrum. *Mol. Genet. Metab.* **2008**, *93*, 282–294. [CrossRef] [PubMed]
112. Tall, A.R. Protease variants, LDL, and coronary heart disease. *N. Engl. J. Med.* **2006**, *354*, 1310–1312. [CrossRef] [PubMed]
113. Sisman, G.; Erzin, Y.; Hatemi, I.; Caglar, E.; Boga, S.; Singh, V.; Senturk, H. Familial chylomicronemia syndrome related chronic pancreatitis: A single-center study. *Hepatobiliary Pancreat. Dis. Int.* **2014**, *13*, 209–214. [CrossRef]
114. Steinberg, D. Atherogenesis in perspective: Hypercholesterolemia and inflammation as partners in crime. *Nat. Med.* **2002**, *8*, 1211–1217. [CrossRef]
115. Brown, M.S.; Goldstein, J.L. A receptor-mediated pathway for cholesterol homeostasis. *Science* **1986**, *232*, 34–47. [CrossRef]
116. Stalenhoef, A.F.; van 't Laar, A. Clinical significance of current perspectives in cholesterol metabolism. *Ned. Tijdschr. Geneeskd.* **1986**, *130*, 951–955.
117. Goldstein, J.L.; Brown, M.S. Molecular medicine. The cholesterol quartet. *Science* **2001**, *292*, 1310–1312. [CrossRef]
118. Horton, J.D.; Goldstein, J.L.; Brown, M.S. SREBPs: Activators of the complete program of cholesterol and fatty acid synthesis in the liver. *J. Clin. Investig.* **2002**, *109*, 1125–1131. [CrossRef]
119. Steinberg, D. The LDL modification hypothesis of atherogenesis: An update. *J. Lipid Res.* **2009**, *50*, S376–S381. [CrossRef]
120. Parthasarathy, S.; Wieland, E.; Steinberg, D. A role for endothelial cell lipoxygenase in the oxidative modification of low density lipoprotein. *Proc. Natl. Acad. Sci. USA* **1989**, *86*, 1046–1050. [CrossRef]
121. Benz, D.J.; Mol, M.; Ezaki, M.; Mori-Ito, N.; Zelan, I.; Miyanohara, A.; Friedmann, T.; Parthasarathy, S.; Steinberg, D.; Witztum, J.L. Enhanced levels of lipoperoxides in low density lipoprotein incubated with murine fibroblast expressing high levels of human 15-lipoxygenase. *J. Biol. Chem.* **1995**, *270*, 5191–5197. [CrossRef] [PubMed]
122. Steinberg, D. Low density lipoprotein oxidation and its pathobiological significance. *J. Biol. Chem.* **1997**, *272*, 20963–20966. [CrossRef] [PubMed]
123. Yoshida, H.; Ishikawa, T.; Hosoai, H.; Suzukawa, M.; Ayaori, M.; Hisada, T.; Sawada, S.; Yonemura, A.; Higashi, K.; Ito, T.; et al. Inhibitory effect of tea flavonoids on the ability of cells to oxidize low density lipoprotein. *Biochem. Pharmacol.* **1999**, *58*, 1695–1703. [CrossRef]
124. Yoshida, H.; Sasaki, K.; Namiki, Y.; Sato, N.; Tada, N. Edaravone, a novel radical scavenger, inhibits oxidative modification of low-density lipoprotein (LDL) and reverses oxidized LDL-mediated reduction in the expression of endothelial nitric oxide synthase. *Atherosclerosis* **2005**, *179*, 97–102. [CrossRef] [PubMed]
125. Wang, J.Y.; Lai, C.L.; Lee, C.T.; Lin, C.Y. Electronegative Low-Density Lipoprotein L5 Impairs Viability and NGF-Induced Neuronal Differentiation of PC12 Cells via LOX-1. *Int. J. Mol. Sci.* **2017**, *18*, 1744. [CrossRef]
126. Gerhard-Herman, M.D.; Gornik, H.L.; Barrett, C.; Barshes, N.R.; Corriere, M.A.; Drachman, D.E.; Fleisher, L.A.; Fowkes, F.G.R.; Hamburg, N.M.; Kinlay, S.; et al. 2016 AHA/ACC Guideline on the Management of Patients With Lower Extremity Peripheral Artery Disease: A Report of the American College of Cardiology/American Heart Association Task Force on Clinical Practice Guidelines. *J. Am. Coll. Cardiol.* **2017**, *69*, e71–e126. [CrossRef]

127. Pollak, A.W.; Norton, P.T.; Kramer, C.M. Multimodality imaging of lower extremity peripheral arterial disease: Current role and future directions. *Circ. Cardiovasc. Imaging* **2012**, *5*, 797–807. [CrossRef]
128. Shishehbor, M.H.; White, C.J.; Gray, B.H.; Menard, M.T.; Lookstein, R.; Rosenfield, K.; Jaff, M.R. Critical Limb Ischemia: An Expert Statement. *J. Am. Coll. Cardiol.* **2016**, *68*, 2002–2015. [CrossRef]
129. Grundy, S.M.; Stone, N.J.; Bailey, A.L.; Beam, C.; Birtcher, K.K.; Blumenthal, R.S.; Braun, L.T.; de Ferranti, S.; Faiella-Tommasino, J.; Forman, D.E.; et al. 2018 AHA/ACC/AACVPR/AAPA/ABC/ACPM/ADA/AGS/APhA/ASPC/NLA/PCNA Guideline on the Management of Blood Cholesterol: A Report of the American College of Cardiology/American Heart Association Task Force on Clinical Practice Guidelines. *J. Am. Coll. Cardiol.* **2019**, *73*, e285–e350. [CrossRef]
130. Wadhera, R.K.; Steen, D.L.; Khan, I.; Giugliano, R.P.; Foody, J.M. A review of low-density lipoprotein cholesterol, treatment strategies, and its impact on cardiovascular disease morbidity and mortality. *J. Clin. Lipidol.* **2016**, *10*, 472–489. [CrossRef]
131. Koskinas, K.C.; Siontis, G.C.M.; Piccolo, R.; Mavridis, D.; Raber, L.; Mach, F.; Windecker, S. Effect of statins and non-statin LDL-lowering medications on cardiovascular outcomes in secondary prevention: A meta-analysis of randomized trials. *Eur. Heart J.* **2018**, *39*, 1172–1180. [CrossRef] [PubMed]
132. Li, G.; Wang, X.; Yang, H.; Zhang, P.; Wu, F.; Li, Y.; Zhou, Y.; Zhang, X.; Ma, H.; Zhang, W.; et al. alpha-Linolenic acid but not linolenic acid protects against hypertension: Critical role of SIRT3 and autophagic flux. *Cell Death Dis.* **2020**, *11*, 83. [CrossRef] [PubMed]
133. Stafylas, P.C.; Sarafidis, P.A. Carvedilol in hypertension treatment. *Vasc. Health Risk Manag.* **2008**, *4*, 23–30. [CrossRef] [PubMed]
134. Leonetti, G.; Egan, C.G. Use of carvedilol in hypertension: An update. *Vasc. Health Risk Manag.* **2012**, *8*, 307–322. [CrossRef] [PubMed]
135. Sy, R.G.; Nevado, J.B., Jr.; Llanes, E.J.B.; Magno, J.D.A.; Ona, D.I.D.; Punzalan, F.E.R.; Reganit, P.F.M.; Santos, L.E.G.; Tiongco, R.H.P., 2nd; Aherrera, J.A.M.; et al. The Klotho Variant rs36217263 Is Associated With Poor Response to Cardioselective Beta-Blocker Therapy Among Filipinos. *Clin. Pharmacol. Ther.* **2020**, *107*, 221–226. [CrossRef]
136. Silva, I.V.G.; de Figueiredo, R.C.; Rios, D.R.A. Effect of Different Classes of Antihypertensive Drugs on Endothelial Function and Inflammation. *Int. J. Mol. Sci.* **2019**, *20*, 3548. [CrossRef] [PubMed]
137. Erbs, S.; Beck, E.B.; Linke, A.; Adams, V.; Gielen, S.; Krankel, N.; Mobius-Winkler, S.; Hollriegel, R.; Thiele, H.; Hambrecht, R.; et al. High-dose rosuvastatin in chronic heart failure promotes vasculogenesis, corrects endothelial function, and improves cardiac remodeling—Results from a randomized, double-blind, and placebo-controlled study. *Int. J. Cardiol.* **2011**, *146*, 56–63. [CrossRef]
138. Szygula-Jurkiewicz, B.; Szczurek, W.; Krol, B.; Zembala, M. The role of statins in chronic heart failure. *Kardiochir Torakochirurgia Pol.* **2014**, *11*, 301–305. [CrossRef]
139. Cangiano, E.; Marchesini, J.; Campo, G.; Francolini, G.; Fortini, C.; Carra, G.; Miccoli, M.; Ceconi, C.; Tavazzi, L.; Ferrari, R. ACE inhibition modulates endothelial apoptosis and renewal via endothelial progenitor cells in patients with acute coronary syndromes. *Am. J. Cardiovasc. Drugs* **2011**, *11*, 189–198. [CrossRef]
140. Brugts, J.J.; Bertrand, M.; Remme, W.; Ferrari, R.; Fox, K.; MacMahon, S.; Chalmers, J.; Simoons, M.L.; Boersma, E. The Treatment Effect of an ACE-Inhibitor Based Regimen with Perindopril in Relation to Beta-Blocker use in 29,463 Patients with Vascular Disease: A Combined Analysis of Individual Data of ADVANCE, EUROPA and PROGRESS Trials. *Cardiovasc. Drugs Ther.* **2017**, *31*, 391–400. [CrossRef]
141. Rangaswami, J.; McCullough, P.A. Heart Failure in End-Stage Kidney Disease: Pathophysiology, Diagnosis, and Therapeutic Strategies. *Semin. Nephrol.* **2018**, *38*, 600–617. [CrossRef] [PubMed]
142. Weaver, D.J.; Mitsnefes, M. Cardiovascular Disease in Children and Adolescents With Chronic Kidney Disease. *Semin. Nephrol.* **2018**, *38*, 559–569. [CrossRef] [PubMed]
143. Kern, A.; Gil, R.; Gorny, J.; Sienkiewicz, E.; Bojko, K.; Wasilewski, G. Patient with ST-elevation myocardial infarction, coronary artery embolism and no signs of coronary atherosclerosis in angiography. *Postepy Kardiol. Interwencyjnej* **2015**, *11*, 334–336. [CrossRef] [PubMed]
144. Vernon, S.T.; Coffey, S.; D'Souza, M.; Chow, C.K.; Kilian, J.; Hyun, K.; Shaw, J.A.; Adams, M.; Roberts-Thomson, P.; Brieger, D.; et al. ST-Segment-Elevation Myocardial Infarction (STEMI) Patients Without Standard Modifiable Cardiovascular Risk Factors-How Common Are They, and What Are Their Outcomes? *J. Am. Heart Assoc.* **2019**, *8*, e013296. [CrossRef]

145. Corretti, M.C.; Anderson, T.J.; Benjamin, E.J.; Celermajer, D.; Charbonneau, F.; Creager, M.A.; Deanfield, J.; Drexler, H.; Gerhard-Herman, M.; Herrington, D.; et al. Guidelines for the ultrasound assessment of endothelial-dependent flow-mediated vasodilation of the brachial artery: A report of the International Brachial Artery Reactivity Task Force. *J. Am. Coll. Cardiol.* **2002**, *39*, 257–265. [CrossRef]
146. Chen, C.J.; Yang, T.C.; Chang, C.; Lu, S.C.; Chang, P.Y. Homocysteine is a bystander for ST-segment elevation myocardial infarction: A case-control study. *BMC Cardiovasc. Disord.* **2018**, *18*, 33. [CrossRef]
147. Zou, C.; Hu, H. Use of pioglitazone in the treatment of diabetes: Effect on cardiovascular risk. *Vasc. Health Risk Manag.* **2013**, *9*, 429–433. [CrossRef]
148. Yu, X.; Chen, P.; Wang, H.; Zhu, T. Pioglitazone ameliorates endothelial dysfunction in those with impaired glucose regulation among the first-degree relatives of type 2 diabetes mellitus patients. *Med. Princ. Pract.* **2013**, *22*, 156–160. [CrossRef]
149. Abdullah, C.S.; Li, Z.; Wang, X.; Jin, Z.Q. Depletion of T lymphocytes ameliorates cardiac fibrosis in streptozotocin-induced diabetic cardiomyopathy. *Int. Immunopharmacol.* **2016**, *39*, 251–264. [CrossRef]
150. Samuel, C.S.; Hewitson, T.D.; Zhang, Y.; Kelly, D.J. Relaxin ameliorates fibrosis in experimental diabetic cardiomyopathy. *Endocrinology* **2008**, *149*, 3286–3293. [CrossRef]
151. Zhang, X.; Pan, L.; Yang, K.; Fu, Y.; Liu, Y.; Chi, J.; Zhang, X.; Hong, S.; Ma, X.; Yin, X. H3 Relaxin Protects Against Myocardial Injury in Experimental Diabetic Cardiomyopathy by Inhibiting Myocardial Apoptosis, Fibrosis and Inflammation. *Cell Physiol. Biochem.* **2017**, *43*, 1311–1324. [CrossRef] [PubMed]
152. Zhang, M.; Wang, C.M.; Li, J.; Meng, Z.J.; Wei, S.N.; Li, J.; Bucala, R.; Li, Y.L.; Chen, L. Berberine protects against palmitate-induced endothelial dysfunction: Involvements of upregulation of AMPK and eNOS and downregulation of NOX4. *Mediators Inflamm.* **2013**, *2013*, 260464. [CrossRef] [PubMed]
153. Wang, Y.; Huang, Y.; Lam, K.S.; Li, Y.; Wong, W.T.; Ye, H.; Lau, C.W.; Vanhoutte, P.M.; Xu, A. Berberine prevents hyperglycemia-induced endothelial injury and enhances vasodilatation via adenosine monophosphate-activated protein kinase and endothelial nitric oxide synthase. *Cardiovasc. Res.* **2009**, *82*, 484–492. [CrossRef] [PubMed]
154. Suganya, N.; Bhakkiyalakshmi, E.; Sarada, D.V.; Ramkumar, K.M. Reversibility of endothelial dysfunction in diabetes: Role of polyphenols. *Br. J. Nutr.* **2016**, *116*, 223–246. [CrossRef]
155. Forcheron, F.; Basset, A.; Abdallah, P.; Del Carmine, P.; Gadot, N.; Beylot, M. Diabetic cardiomyopathy: Effects of fenofibrate and metformin in an experimental model—The Zucker diabetic rat. *Cardiovasc. Diabetol.* **2009**, *8*, 16. [CrossRef]
156. Baraka, A.; AbdelGawad, H. Targeting apoptosis in the heart of streptozotocin-induced diabetic rats. *J. Cardiovasc. Pharmacol. Ther.* **2010**, *15*, 175–181. [CrossRef]
157. Arosio, E.; De Marchi, S.; Rigoni, A.; Prior, M.; Delva, P.; Lechi, A. Forearm haemodynamics, arterial stiffness and microcirculatory reactivity in rheumatoid arthritis. *J. Hypertens.* **2007**, *25*, 1273–1278. [CrossRef] [PubMed]
158. Ruiz-Limon, P.; Ortega, R.; Arias de la Rosa, I.; Abalos-Aguilera, M.D.C.; Perez-Sanchez, C.; Jimenez-Gomez, Y.; Peralbo-Santaella, E.; Font, P.; Ruiz-Vilches, D.; Ferrin, G.; et al. Tocilizumab improves the proatherothrombotic profile of rheumatoid arthritis patients modulating endothelial dysfunction, NETosis, and inflammation. *Transl. Res.* **2017**, *183*, 87–103. [CrossRef] [PubMed]
159. Fanouriakis, A.; Pamfil, C.; Rednic, S.; Sidiropoulos, P.; Bertsias, G.; Boumpas, D.T. Is it primary neuropsychiatric systemic lupus erythematosus? Performance of existing attribution models using physician judgment as the gold standard. *Clin. Exp. Rheumatol.* **2016**, *34*, 910–917. [PubMed]
160. Nikolopoulos, D.; Fanouriakis, A.; Boumpas, D.T. Cerebrovascular Events in Systemic Lupus Erythematosus: Diagnosis and Management. *Mediterr. J. Rheumatol.* **2019**, *30*, 7–15. [CrossRef] [PubMed]
161. Aringer, M.; Leuchten, N.; Dorner, T. Biologicals and small molecules for systemic lupus erythematosus. *Z. Rheumatol.* **2020**, *79*, 232–240. [CrossRef] [PubMed]
162. Lee, S.E.; Chang, H.J.; Sung, J.M.; Park, H.B.; Heo, R.; Rizvi, A.; Lin, F.Y.; Kumar, A.; Hadamitzky, M.; Kim, Y.J.; et al. Effects of Statins on Coronary Atherosclerotic Plaques: The PARADIGM Study. *JACC Cardiovasc. Imaging* **2018**, *11*, 1475–1484. [CrossRef]
163. Nicholls, S.J.; Nelson, A.J. Monitoring the Response to Statin Therapy: One Scan at a Time. *JACC Cardiovasc. Imaging* **2018**, *11*, 1485–1486. [CrossRef] [PubMed]

164. Yla-Herttuala, S.; Bentzon, J.F.; Daemen, M.; Falk, E.; Garcia-Garcia, H.M.; Herrmann, J.; Hoefer, I.; Jauhiainen, S.; Jukema, J.W.; Krams, R.; et al. Stabilization of atherosclerotic plaques: An update. *Eur. Heart J.* **2013**, *34*, 3251–3258. [CrossRef] [PubMed]
165. Chu, C.S.; Ke, L.Y.; Chan, H.C.; Chan, H.C.; Chen, C.C.; Cheng, K.H.; Lee, H.C.; Kuo, H.F.; Chang, C.T.; Chang, K.C.; et al. Four Statin Benefit Groups Defined by The 2013 ACC/AHA New Cholesterol Guideline are Characterized by Increased Plasma Level of Electronegative Low-Density Lipoprotein. *Acta Cardiol. Sin.* **2016**, *32*, 667–675. [CrossRef] [PubMed]
166. Zhang, B.; Miura, S.; Yanagi, D.; Noda, K.; Nishikawa, H.; Matsunaga, A.; Shirai, K.; Iwata, A.; Yoshinaga, K.; Adachi, H.; et al. Reduction of charge-modified LDL by statin therapy in patients with CHD or CHD risk factors and elevated LDL-C levels: The SPECIAL Study. *Atherosclerosis* **2008**, *201*, 353–359. [CrossRef]
167. Zhang, B.; Matsunaga, A.; Rainwater, D.L.; Miura, S.; Noda, K.; Nishikawa, H.; Uehara, Y.; Shirai, K.; Ogawa, M.; Saku, K. Effects of rosuvastatin on electronegative LDL as characterized by capillary isotachophoresis: The ROSARY Study. *J. Lipid Res.* **2009**, *50*, 1832–1841. [CrossRef]
168. Sena-Evangelista, K.C.; Pedrosa, L.F.; Paiva, M.S.; Dias, P.C.; Ferreira, D.Q.; Cozzolino, S.M.; Faulin, T.E.; Abdalla, D.S. The hypolipidemic and pleiotropic effects of rosuvastatin are not enhanced by its association with zinc and selenium supplementation in coronary artery disease patients: A double blind randomized controlled study. *PLoS ONE* **2015**, *10*, e0119830. [CrossRef]
169. Vavlukis, M.; Vavlukis, A. Adding ezetimibe to statin therapy: Latest evidence and clinical implications. *Drugs Context* **2018**, *7*, 212534. [CrossRef]
170. Page, M.M.; Watts, G.F. PCSK9 inhibitors—Mechanisms of action. *Aust. Prescr.* **2016**, *39*, 164–167. [CrossRef]
171. Ikegami, Y.; Inoue, I.; Inoue, K.; Shinoda, Y.; Iida, S.; Goto, S.; Nakano, T.; Shimada, A.; Noda, M. The annual rate of coronary artery calcification with combination therapy with a PCSK9 inhibitor and a statin is lower than that with statin monotherapy. *NPJ Aging Mech. Dis.* **2018**, *4*, 7. [CrossRef] [PubMed]
172. Alonso, R.; Mata, P.; Muniz, O.; Fuentes-Jimenez, F.; Diaz, J.L.; Zambon, D.; Tomas, M.; Martin, C.; Moyon, T.; Croyal, M.; et al. PCSK9 and lipoprotein (a) levels are two predictors of coronary artery calcification in asymptomatic patients with familial hypercholesterolemia. *Atherosclerosis* **2016**, *254*, 249–253. [CrossRef] [PubMed]
173. Kuhnast, S.; van der Hoorn, J.W.; Pieterman, E.J.; van den Hoek, A.M.; Sasiela, W.J.; Gusarova, V.; Peyman, A.; Schafer, H.L.; Schwahn, U.; Jukema, J.W.; et al. Alirocumab inhibits atherosclerosis, improves the plaque morphology, and enhances the effects of a statin. *J. Lipid Res.* **2014**, *55*, 2103–2112. [CrossRef]
174. Pouwer, M.G.; Pieterman, E.J.; Worms, N.; Keijzer, N.; Jukema, J.W.; Gromada, J.; Gusarova, V.; Princen, H.M.G. Alirocumab, evinacumab, and atorvastatin triple therapy regresses plaque lesions and improves lesion composition in mice. *J. Lipid Res.* **2020**, *61*, 365–375. [CrossRef] [PubMed]
175. Berbee, J.F.; Wong, M.C.; Wang, Y.; van der Hoorn, J.W.; Khedoe, P.P.; van Klinken, J.B.; Mol, I.M.; Hiemstra, P.S.; Tsikas, D.; Romijn, J.A.; et al. Resveratrol protects against atherosclerosis, but does not add to the antiatherogenic effect of atorvastatin, in APOE*3-Leiden.CETP mice. *J. Nutr. Biochem.* **2013**, *24*, 1423–1430. [CrossRef] [PubMed]
176. Mishra, V.K.; Shih, H.H.; Parveen, F.; Lenzen, D.; Ito, E.; Chan, T.F.; Ke, L.Y. Identifying the Therapeutic Significance of Mesenchymal Stem Cells. *Cells* **2020**, *9*, 1145. [CrossRef]
177. Back, M. Leukotriene signaling in atherosclerosis and ischemia. *Cardiovasc. Drugs Ther.* **2009**, *23*, 41–48. [CrossRef]
178. Charo, I.F.; Taub, R. Anti-inflammatory therapeutics for the treatment of atherosclerosis. *Nat. Rev. Drug Discov.* **2011**, *10*, 365–376. [CrossRef]
179. Bertrand, M.J.; Tardif, J.C. Inflammation and beyond: New directions and emerging drugs for treating atherosclerosis. *Expert Opin. Emerg. Drugs* **2017**, *22*, 1–26. [CrossRef]
180. Tuttolomondo, A. Editorial: Treatment of atherosclerosis as an inflammatory disease. *Curr. Pharm. Des.* **2012**, *18*, 4265. [CrossRef]
181. Raggi, P.; Genest, J.; Giles, J.T.; Rayner, K.J.; Dwivedi, G.; Beanlands, R.S.; Gupta, M. Role of inflammation in the pathogenesis of atherosclerosis and therapeutic interventions. *Atherosclerosis* **2018**, *276*, 98–108. [CrossRef] [PubMed]
182. Riccioni, G.; Zanasi, A.; Vitulano, N.; Mancini, B.; D'Orazio, N. Leukotrienes in atherosclerosis: New target insights and future therapy perspectives. *Mediators Inflamm.* **2009**, *2009*, 737282. [CrossRef] [PubMed]

183. Lin, J.; Kakkar, V.; Lu, X. Impact of MCP-1 in atherosclerosis. *Curr. Pharm. Des.* **2014**, *20*, 4580–4588. [CrossRef] [PubMed]
184. Tousoulis, D.; Oikonomou, E.; Economou, E.K.; Crea, F.; Kaski, J.C. Inflammatory cytokines in atherosclerosis: Current therapeutic approaches. *Eur. Heart J.* **2016**, *37*, 1723–1732. [CrossRef] [PubMed]
185. Pieper, G.M. Enhanced, unaltered and impaired nitric oxide-mediated endothelium-dependent relaxation in experimental diabetes mellitus: Importance of disease duration. *Diabetologia* **1999**, *42*, 204–213. [CrossRef] [PubMed]
186. Li, H.; Forstermann, U. Nitric oxide in the pathogenesis of vascular disease. *J. Pathol.* **2000**, *190*, 244–254. [CrossRef]
187. McLeod, D.S.; Lefer, D.J.; Merges, C.; Lutty, G.A. Enhanced expression of intracellular adhesion molecule-1 and P-selectin in the diabetic human retina and choroid. *Am. J. Pathol.* **1995**, *147*, 642–653.

Publisher's Note: MDPI stays neutral with regard to jurisdictional claims in published maps and institutional affiliations.

© 2020 by the authors. Licensee MDPI, Basel, Switzerland. This article is an open access article distributed under the terms and conditions of the Creative Commons Attribution (CC BY) license (http://creativecommons.org/licenses/by/4.0/).

Article

Plasma APE1/Ref-1 Correlates with Atherosclerotic Inflammation in ApoE$^{-/-}$ Mice

Yu Ran Lee [1,2,†], Hee Kyoung Joo [1,2,†], Eun-Ok Lee [1,2], Myoung Soo Park [3], Hyun Sil Cho [4], Sungmin Kim [1,2], Hao Jin [1,2], Jin-Ok Jeong [1,5], Cuk-Seong Kim [1,2] and Byeong Hwa Jeon [1,2,*]

1. Research Institute for Medical Sciences, College of Medicine, Chungnam National University, 266 Munhwa-ro, Jung-gu, Daejeon 35015, Korea; lyr0913@cnu.ac.kr (Y.R.L.); hkjoo79@cnu.ac.kr (H.K.J.); y21c486@naver.com (E.-O.L.); s13845@naver.com (S.K.); jinhao0508@gmail.com (H.J.); jojeong@cnu.ac.kr (J.-O.J.); cskim@cnu.ac.kr (C.-S.K.)
2. Department of Physiology, College of Medicine, Chungnam National University, 266 Munhwa-ro, Jung-gu, Daejeon 35015, Korea
3. Preclinical Research Center, Chungnam National University Hospital, 266 Munhwa-ro, Jung-gu, Daejeon 35015, Korea; nova38@cnuh.co.kr
4. Department of Biomedical Sciences, College of Medicine, Chungnam National University, 266 Munhwa-ro, Jung-gu, Daejeon 35015, Korea; chsss@cnu.ac.kr
5. Division of Cardiology, Department of Internal Medicine, Chungnam National University Hospital, Chungnam National University School of Medicine, 282 Munhwa-ro, Jung-gu, Daejeon 35015, Korea
* Correspondence: bhjeon@cnu.ac.kr; Tel.: +82-42-580-8214
† These authors contributed equally to this work.

Received: 19 August 2020; Accepted: 17 September 2020; Published: 21 September 2020

Abstract: Apurinic/apyrimidinic endonuclease 1/redox factor-1 (APE1/Ref-1) is involved in DNA base repair and reducing activity. However, the role of APE1/Ref-1 in atherosclerosis is unclear. Herein, we investigated the role of APE1/Ref-1 in atherosclerotic apolipoprotein E (ApoE$^{-/-}$) mice fed with a Western-type diet. We found that serologic APE1/Ref-1 was strongly correlated with vascular inflammation in these mice. Neutrophil/lymphocyte ratio (NLR), endothelial cell/macrophage activation, and atherosclerotic plaque formation, reflected by atherosclerotic inflammation, were increased in the ApoE$^{-/-}$ mice fed with a Western-type diet. APE1/Ref-1 expression was upregulated in aortic tissues of these mice, and was co-localized with cells positive for cluster of differentiation 31 (CD31) and galectin-3, suggesting endothelial cell/macrophage expression of APE1/Ref-1. Interestingly, APE1/Ref-1 plasma levels of ApoE$^{-/-}$ mice fed with a Western-type diet were significantly increased compared with those of the mice fed with normal diet (15.76 ± 3.19 ng/mL vs. 3.51 ± 0.50 ng/mL, $p < 0.05$), and were suppressed by atorvastatin administration. Correlation analysis showed high correlation between plasma APE1/Ref-1 levels and NLR, a marker of systemic inflammation. The cut-off value for APE1/Ref-1 for predicting atherosclerotic inflammation at 4.903 ng/mL showed sensitivity of 100% and specificity of 91%. We conclude that APE1/Ref-1 expression is upregulated in aortic endothelial cells/macrophages of atherosclerotic mice, and that plasma APE1/Ref-1 levels could predict atherosclerotic inflammation.

Keywords: APE1/Ref-1; atherosclerosis; ApoE knockout mouse; atorvastatin; VCAM-1; galectin-3; neutrophil/lymphocyte ratio

---

1. Introduction

Atherosclerosis is chronic vascular inflammation characterized by excessive lipoprotein in macrophages and expression of proinflammatory molecules such as the vascular cell adhesion molecule [1]. Increased oxidative stress and proinflammatory gene induction initiates the formation of atherosclerotic lesions [2]. Under conditions of oxidative stress, reactive oxygen species produce these oxidative DNA lesions via mechanisms that involve oxidation and fragmentation of nucleobases [3]. Increasing evidence shows oxidative DNA damage in atherosclerotic plaques [4]. Atherosclerosis is characterized by lipid accumulation and inflammation within the arterial wall [1]. Chronic lipid accumulation promotes inflammation. Vascular inflammation is intimately involved in foam-cell formation and plaque stability, thereby contributing to all the stages of atherosclerosis [5]. Apolipoprotein E (ApoE) functions in the transport of lipids and plays a key role in the redistribution of lipids from local tissue [6].

Apurinic/apyrimidinic endonuclease 1/redox factor-1 (APE1/Ref-1) is an essential multifunctional protein involved in DNA base-excision repair and in redox regulation of several functional proteins including transcription factors. APE1/Ref-1 plays key roles in the maintenance of genomic stability and cellular homeostasis [7,8]. The sub-cellular localization of APE1/Ref-1 is regulated by post-translational modification such as acetylation [7]. Additionally, APE1/Ref-1 acts as an anti-inflammatory mediator by inhibiting reactive oxygen species and by increasing the levels of endothelial nitric oxide production [9,10]. Studies have shown extracellular secretion of APE1/Ref-1 in hyperacetylation and in the plasma of endotoxemic animals [11,12]. While biologically active APE1/Ref-1 can be secreted from cells, the biological function of extracellular APE1/Ref-1 remains unclear. Extracellular APE1/Ref-1 is believed to participate in circulating surveillance for oxidative damage [13]. Additionally, the biological utility of serologic APE1/Ref-1 in cardiovascular disorders has been reported in coronary arterial disorders [14] and murine myocarditis [15], suggesting that APE1/Ref-1 level in blood is correlated with angina or myocardial injury.

The specific cell types that secrete APE1/Ref-1 are not identified. Only limited information could be obtained from previous reports that preformed with in vitro experiments. In hyperacetylation condition, APE1/Ref-1 can be secreted from human embryonic kidney 293 (HEK293) cells [11] and vascular endothelial cells [16]. It also was proposed that APE1/Ref-1 is secreted from monocytes in response to lipopolysaccharide [17]. The ApoE knockout mouse (ApoE$^{-/-}$) model is widely used to investigate the pathogenesis of atherosclerosis. When challenged with a Western-type diet, ApoE$^{-/-}$ mice show increased cholesterol levels and develop atherosclerotic lesions [18]. High cholesterol levels and the ensuing inflammation increase cardiovascular events and mortality, this is considerably decreased by the use of statins [19]. In addition to their lipid-lowering effects, statins also show anti-inflammatory and antioxidative activity [19]. However, whether plasma APE1/Ref-1 levels are altered, and what hematological factors or which cells types are correlated with APE1/Ref-1 remains unclear in experimental models of atherosclerosis.

In this study, we investigated whether plasma APE1/Ref-1 levels were upregulated and correlated with vascular inflammation, and whether this process could be controlled by using statins. We also investigated the usefulness of serologic APE1/Ref-1 as a potential biomarker for vascular inflammation in ApoE$^{-/-}$ mice fed with a Western-type diet.

2. Experimental Section

2.1. Procedures Involving Animals

In this study, we used 8-week-old male apoprotein E-knockout mice (ApoE$^{-/-}$; Jackson Laboratory, Bar Harbor, ME, USA) and age- and sex-matched C57BL/6J mice (DooYeol Biotech, Seoul, Korea.) The mice were housed at 24 °C and with a 12-h day/12-h night cycle, with water and chow administered ad libitum. Mice were fed with either a normal diet (cat.# 2918, Envigo, Madison, MI, USA) or a Western-type diet containing 21% fat, 34% sucrose, 19.5% casein, and 0.2% cholesterol (cat.# TD 88137, Envigo, Madison, MI, USA) for 20 weeks. Our animal protocol was approved by the Ethics Committee of Animal Experimentation of the Chungnam National University (2019037-CNU-43, 26 Mar 2019). In addition to C57BL/6J wild-type (WT) control group, ApoE$^{-/-}$ mice ($n = 30$, male) were randomly subdivided into three groups ($n = 10$ per group): the normal diet (ND) group, the Western-type diet group (WD), and the atorvastatin-treated ApoE$^{-/-}$ mice fed with a Western-type diet (WD + statin). Atorvastatin (Pfizer Ltd., New York, NY, USA) (20 mg/kg/day) was administered orally.

2.2. Analysis of Blood Cells and Chemistry

Mice were sacrificed at 20 weeks after commencement of the diet. Blood samples were collected from the heart of the deeply anesthetized mice in the morning after an overnight starvation period. Whole blood was collected into ethylenediaminetetraacetic acid (EDTA) anticoagulated tubes. A ProCyte Dx® hematology analyzer (IDEXX Laboratories, Inc., Westbrook, ME, USA) was used to measure the hematological parameters of the collected blood. The following hematological parameters were assessed: red blood cell (RBC) count, hemoglobin, hematocrit, platelet count, and white blood cell (WBC) count. A differential WBC count was performed for neutrophils, lymphocytes, monocytes, eosinophils, and basophils. Whole blood was collected into heparin-coated tubes for plasma separation, and plasma was separated by centrifugation at 3000 rpm for 15 min. The plasma samples were used for blood chemistry analysis and measurements of cytokine levels. The levels of plasma cholesterol, triglycerides, lipoprotein, and glucose were measured using a BS-220 chemistry analyzer (Mindray, Shenzhen, China).

2.3. Quantification of Plasma APE1/Ref-1

Plasma levels of APE1/Ref-1 were determined using an APE1/Ref-1 sandwich enzyme-linked immunosorbent assay kit (MediRedox, Daejeon, Korea) according to the manufacturer's instructions. Briefly, plasma samples were added to the wells, the plates were incubated at 37 °C for 90 min, and then washed five times using phosphate-buffered saline with Tween® 20 (PBS-T). This was followed by the addition of 100 μL of the detection primary antibody at 1:200 dilution, and the plate was incubated at 37 °C for 2 h. The plate was then washed seven times with PBS-T, after which a horseradish peroxidase-conjugated secondary antibody (1:200) was added to the wells (100 μL per well), and the plate was incubated at room temperature for 30 min. After further washing, 100 μL freshly prepared tetramethyl benzidine substrate was added to each well. The color-development reaction was stopped by the addition of 100 μL stop solution, and absorbance was measured at 450 nm using a Glomax microplate reader (Promega, Madison, WI, USA). Each sample was assayed in duplicate, and mean values were determined. To establish a standard curve, purified recombinant human APE1/Ref-1 (MediRedox, Daejeon, Korea) was serially diluted (2-fold) and used in a concentration series from 0.312–20 ng/mL. Secreted APE1/Ref-1 levels (ng/mL) were calculated using a standard curve.

2.4. Immunoblotting

Aorta tissues, harvested from ApoE$^{-/-}$ and C57BL/6J mice, were chopped in radioimmunoprecipitation assay (RIPA) buffer (Cell Signaling Technology, Danvers, MA, USA) and homogenized using a sonicator (Hielscher, Teltow, Germany). Aorta tissue was centrifuged at 12,000 rpm for 15 min and the supernatant was collected. Aorta lysates were subjected to 10% sodium dodecyl sulfate polyacrylamide gel electrophoresis (SDS-PAGE), followed by immunoblotting using anti-vascular cell adhesion molecule-1 (VCAM-1) (1:1000, cat.# MAB6434) or anti-galectin-3 (1:000, cat.# AF1197) from R&D Systems (Minneapolis, MN, USA), anti-APE1/Ref-1 (1:1000, cat.# MR-MA14) from MediRedox (Daejeon, Korea), and anti-β-actin (1:5000, cat.# A5316) from Sigma-Aldrich (St. Louis, MI, USA).

2.5. Immunohistochemistry

Aortic tissues were fixed using 4% paraformaldehyde, paraffin-embedded, and sectioned at a thickness of 3 μm. Aortic sections were then incubated overnight at 4 °C with anti-VCAM-1 (1:200, cat.# MAB6434) or anti-galectin-3 (1:200, cat.# AF1197) from R&D Systems (Minneapolis, MN, USA), or anti-APE1/Ref-1 (1:300, cat.# MR-MA14) from MediRedox (Daejeon, Korea). Horseradish peroxidase-conjugated goat-anti-rabbit and anti-mouse secondary antibodies (1:1000) were then applied onto the sections. Specificity of immunostaining was assessed using nonimmune rabbit immunoglobulin G (IgG) and mouse IgG as negative controls. The sections were counterstained with hematoxylin, dehydrated, and mounted. Histological staining was digitalized using microscope (Motic, Richmond, BC, Canada) and analyzed using the TS view 7 software (Microscope.com, Roanoke, VA, USA).

2.6. Immunofluorescence

Aortic sections were dehydrated and quenched using 3% hydrogen peroxide. Sections were then blocked using 5% bovine serum albumin (BSA). The sections were incubated overnight at 4 °C with anti-cluster of differentiation 31 (CD31) (1:40, cat.# ab28364) or anti-smooth muscle protein 22α (SM22α, 1:400, cat.# ab14106) from Abcam (Cambridge, MA, USA), anti-galectin-3 (1:200, cat.# AF1197) from R&D Systems (Minneapolis, MN, USA), and anti-APE1/Ref-1 (1;300, Cat.# MR-MA14) from MediRedox (Daejeon, Korea). Then, Alexa Fluor® 488-conjugated anti-rabbit IgG, Alexa Fluor® 647-conjugated anti-mouse IgG, or Alexa Fluor® 647-conjugated anti-IgG was applied for 60 min at room temperature. Aortic sections were then counterstained with 4′,6-diamidino-2-phenylindole (Sigma-Aldrich, St. Louis, MI, USA). Digital microscopy was performed using a Leica confocal microscope (Leica-Microsystems, Wetzlar, Germany).

2.7. Oil Red O Staining

Next, mouse aortic tissues were stained with Oil red O to quantify advanced atherosclerotic lesions (20 weeks on Western-type diet). Oil Red O staining was performed using standard protocol [20]. Aortas were fixed in 10% formalin for 16 h at room temperature, washed three times in double-distilled water, and then exposed to 60% isopropanol for 3 min. The fixed aortas were subsequently incubated in Oil Red O solution (at 3:2 Oil Red O:double distilled water) for 30 min. Afterwards, the aortas were immersed into 60% isopropyl alcohol for 30 sec and then washed with double-distilled water. Images of Oil Red O-stained aortas were captured for quantification of areas containing atherosclerotic lesions. The digital images of the entire Oil Red O-stained aortas were evaluated using the ImageJ software [21].

2.8. Statistical Analysis

Values are expressed as mean ± standard error of the mean. Data were analyzed using one-way ANOVA and multiple comparison analysis with the post-hoc Bonferroni correction. $p < 0.05$ was considered statistically significant. All statistical analyses were performed using GraphPad Prism version 8 (GraphPad Software, La Jolla, CA, USA).

3. Results

3.1. Plasma Lipid Profile in ApoE$^{-/-}$ Mice Fed with Normal and Western-Type Diet

First, we evaluated how much the plasma lipid and glucose profile changed according to dietary conditions in all the experimental groups. As shown in Figure 1, total plasma cholesterol and low-density lipoprotein (LDL) in wild-type control mice (WT) fed with a normal diet was approximately 107 mg/dL and 10 mg/dL, respectively. ApoE$^{-/-}$ mice fed with a normal diet (ND) showed higher plasma cholesterol (335 mg/dL) and LDL levels (202 mg/dL) compared with those of wild-type control mice (WT). ApoE$^{-/-}$ mice fed with a Western-type diet (WD) for 20 weeks showed further increased levels of total cholesterol (1134 mg/dL) and LDL (723 mg/dL) compared with those of control mice or ApoE$^{-/-}$ mice fed with a normal diet (ND). However, blood glucose levels were unchanged by the Western-type diet. Interestingly, the groups treated with atorvastatin (20 mg/kg) did not show significantly improved lipid levels compared with those of ApoE$^{-/-}$ mice fed with a Western-type diet (WD).

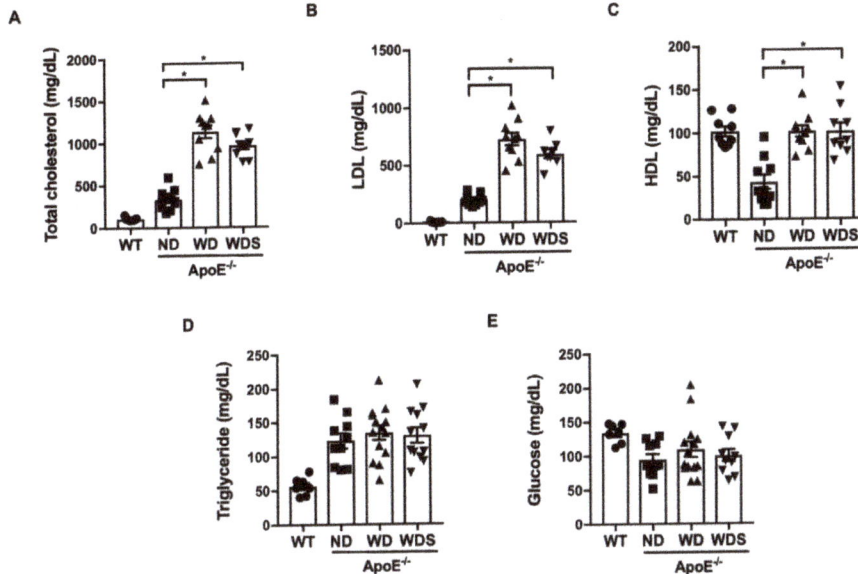

Figure 1. Plasma lipid and glucose level in apolipoprotein E (ApoE$^{-/-}$) mice fed with normal and Western-type diets. (**A**) Total cholesterol, (**B**) low density lipoprotein (LDL), (**C**) high density lipoprotein (HDL), (**D**) triglyceride, and (**E**) glucose were measured in experimental groups (WT; C57BL/6J wild-type control mice, ND; ApoE$^{-/-}$ mice fed with a normal diet, WD; ApoE$^{-/-}$ mice fed with a Western-type diet, WDS; atorvastatin-treated ApoE$^{-/-}$ mice fed with a Western-type diet) using a chemistry analyzer. All values represent mean ± SEM, n = 8–10 animals per group. * p < 0.05, vs. ND group was determined by one-way ANOVA followed by Bonferroni's multiple compare test.

3.2. Hematologic Parameters in ApoE$^{-/-}$ Mice Fed with Western-Type Diet

Using complete blood count and leukocyte/lymphocyte ratio (NLR), we examined which hematologic parameters were changed by a dietary condition for 20 weeks. RBC counts, hemoglobin hematocrit, platelet counts (Figure 2A–D) were unchanged in the experimental groups; however, the WBC differential counts were considerably altered. The percentage of neutrophils in ApoE$^{-/-}$ mice fed with a Western-type diet (WD) was significantly increased (33.8% for ND vs. 59.5% for WD, p < 0.05), while lymphocyte percentages were decreased (64.4% for ND vs. 33.9% for WD, p < 0.05) (Figure 2I,J). Interestingly, treatment with atorvastatin (20 mg/kg) significantly reduced changes in WBC differential count, suggesting that atorvastatin exerted anti-inflammatory effects. NLR is used as a marker for inflammation and is associated with atherosclerosis [22,23]. ApoE$^{-/-}$ mice fed with a Western-type diet showed an increased NLR; however, this effect was suppressed by atorvastatin (Figure 2K). Collectively, our results indicate that a Western-type diet administered for 20 weeks to ApoE$^{-/-}$ mice induced systemic inflammation and hypercholesterolemia in these animals.

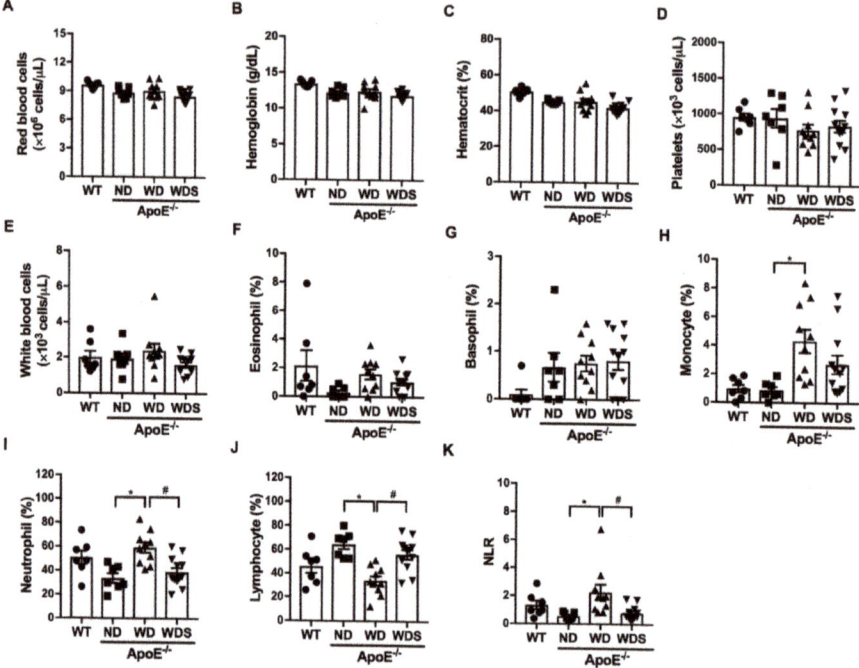

Figure 2. Hematologic parameters following 20 weeks of normal and Western-type diets. (**A**) Red blood cell count, (**B**) hemoglobin, (**C**) hematocrit, (**D**) platelets, (**E**) white blood cell count, (**F**) eosinophil, (**G**) basophil, (**H**) monocyte, (**I**) neutrophil, (**J**) lymphocyte were analyzed in experimental groups (WT; C57BL/6J wild-type control mice, ND; ApoE$^{-/-}$ mice fed with a normal diet, WD; ApoE$^{-/-}$ mice fed with a Western-type diet, WDS; atorvastatin-treated ApoE$^{-/-}$ mice fed with a Western-type diet) using a hematology analyzer. (**K**) Neutrophil-to-lymphocyte ratio (NLR) was calculated dividing the neutrophil by lymphocyte in experimental groups. All values represent mean ± SEM, n = 8–10 animals per group. * $p < 0.05$, vs. ND group, # $p < 0.05$ vs. WD group was determined by one-way ANOVA followed by Bonferroni's multiple compare test.

3.3. APE1/Ref-1 Expression Is Increased in Aorta of Atherosclerotic Mice

To explore whether APE1/Ref-1 expression was changed in atherosclerosis, we first evaluated the formation of atherosclerotic plaques. The aortas were excised and en face areas of the aortas were stained with Oil Red O. As shown in Figure 3A, ApoE$^{-/-}$ mice fed with a Western-type diet (WD) showed significantly increased plaque areas in the whole aortas and aortic arches compared with those of wild-type control mice (WT) and ApoE$^{-/-}$ mice fed with a normal diet (ND). However, the group treated with atorvastatin showed significantly reduced plaque areas compared with those of WD groups. As shown in Figure 3B, immunohistochemical labeling showed positive APE1/Ref-1 expression in the innermost endothelial layer of the aortas collected from WT and ND groups. In the aortic tissue of ApoE$^{-/-}$ mice fed with a Western-type diet (WD), APE1/Ref-1 expression was markedly increased in whole aortic wall, especially in the innermost layer of endothelial lining, in a fatty streak of plaque, in the smooth muscle layer, and in a thrombus in aortic lumen. Next, we used immunolabeling with vascular cell adhesion molecule-1 (VCAM-1), a vascular inflammation marker [24], and galectin-3, a macrophage marker [25–27], to investigate whether aortic tissue was activated or inflamed by a Western-type diet. Our results indicate low expression levels of VCAM-1 and galectin-3 in the aortas of WT and ND groups (Figure 3B). However, VCAM-1 and galectin-3 expression in the aortas of ApoE$^{-/-}$ mice fed with a Western-type diet was markedly increased. In the aortas of ApoE$^{-/-}$ mice fed with a Western-type diet, the expression of galectin-3 was particularly increased in fatty streaks or foam cells of plaques and areas of atherosclerotic microaneurysms. Next, we used Western blotting to quantitatively analyze whether APE1/Ref-1 expression was changed in aortic tissues (Figure 3C). As shown in Figure 3C, aortic APE1/Ref-1 expression was markedly increased in ApoE$^{-/-}$ mice fed with a Western-type diet compared with those of WT or ND groups. Additionally, the expression of VCAM-1 and galectin-3 were significantly increased in the aortas of ApoE$^{-/-}$ mice fed a Western-type diet (WD) compared with that of mice fed with a normal diet (ND). The upregulated expression of APE1/Ref-1, VCAM-1, and galectin-3 in ApoE$^{-/-}$ mice fed with a Western-type diet was robustly suppressed by treatment with atorvastatin. These results suggest that APE1/Ref-1 expression was increased in endothelial cell- and/or macrophage-activated aortic tissues of atherosclerotic mice.

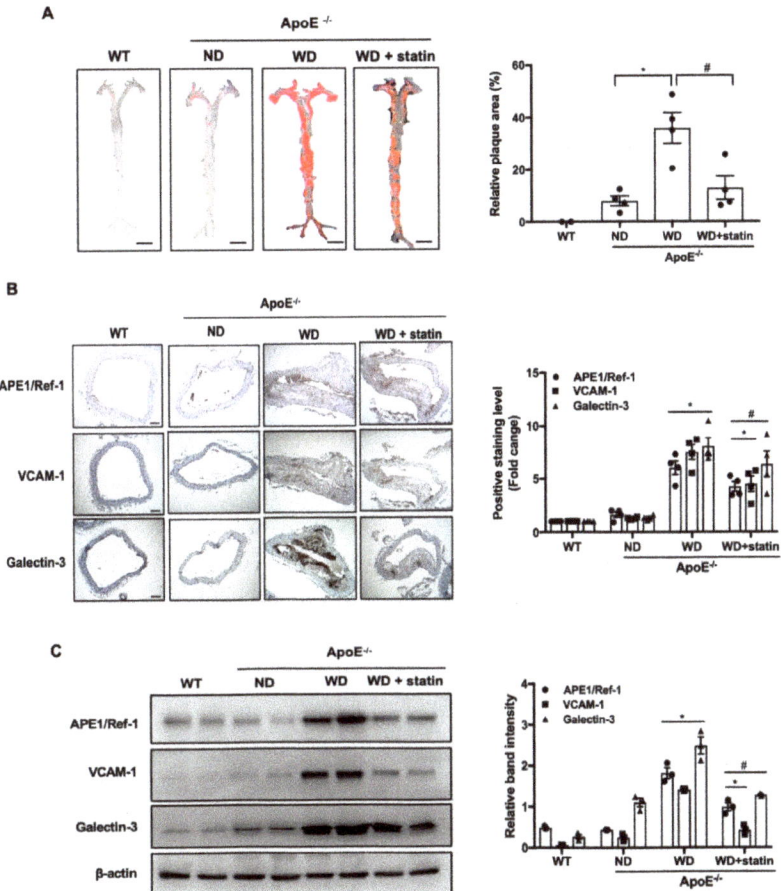

Figure 3. Apurinic/apyrimidinic endonuclease 1/redox factor-1 (APE1/Ref-1) expression is increased in the aortas of atherosclerotic mice. (A) Oil Red O staining and quantification of atherosclerotic plaques in the aorta. Representative images of whole aorta obtained from ApoE$^{-/-}$ mice (left). Relative percentage of Oil Red O positive areas in whole aorta of ApoE$^{-/-}$ mice analyzed using the ImageJ software (right). Each bar shows mean ± SEM (n = 3), * p < 0.05, vs. ND group, # p < 0.05, vs. WD group. (B) Immunohistochemistry for APE1/Ref-1, vascular cell adhesion molecule-1 (VCAM-1), and galectin-3 expression in the thoracic aortas of experimental groups (WT; C57BL/6J wild-type control mice, ND; ApoE$^{-/-}$ mice fed with a normal diet, WD; ApoE$^{-/-}$ mice fed with a Western-type diet, WD + statin;

atorvastatin-treated ApoE$^{-/-}$ mice fed with a Western-type diet). The aortic tissues were color developed using 3,3'-diaminobenzidine reagent and counter stained with hematoxylin (magnification 40×, scale bar 50 μm) (left). Fold changes in the levels of positive immunolabeling relative to those of the control group (WT) are shown for each experimental group (right). Each bar shows mean ± SEM ($n = 3$), * $p < 0.05$, vs. ND group, # $p < 0.05$, vs. WD group. (**C**) Immunoblotting for APE1/Ref-1, VCAM-1, and galectin-3 expression using aortic-tissue lysates obtained from each experimental group. Relative band intensities were normalized to that of β-actin. Each bar shows mean ± SEM ($n = 3$), * $p < 0.05$, vs. ND group, # $p < 0.05$, vs. WD group.

3.4. Upregulated APE1/Ref-1 Is Co-Localized in Macrophages and Endothelial Cells

Atherosclerotic plaques consist of a heterogeneous population of cells including endothelial and smooth muscle cells, macrophages, and/or transdifferentiated cells such as foam cells [28,29]. Therefore, we next examined cellular co-localization of APE1/Ref-1 in the aortas of ApoE$^{-/-}$ mice fed with a Western-type diet (WD). To identify macrophages, endothelial cells, and smooth muscle cells in atherosclerotic plaques, we utilized specific fluorescence imaging using cell-specific markers in tissue sections of the mouse aorta. As shown in Figure 4A, the expression of the macrophage marker [25] galectin-3 (red) was not detected in the aortas of ApoE$^{-/-}$ mice fed with a normal diet (ND), but was increased in ApoE$^{-/-}$ mice fed with a Western-type diet (WD). APE1/Ref-1 (green) was highly expressed in WD mice, and the signal for APE1/Ref-1 was merged with that for galectin-3 (red), thereby indicating the co-localization of APE1/Ref-1 in macrophages. As shown in Figure 4B, the signal for APE1/Ref-1 (green) was mainly merged with that for CD31 (red), which is a specific marker for endothelial cells [30] in the aortas of WD mice, this suggests that APE1/Ref-1 expression was upregulated in the endothelial layer. As shown in Figure 4C, the signal for SM22α (red), a specific marker for smooth muscle cells [31], was not merged with the signal for APE1/Ref-1 (green) in the aortic tissues of ApoE$^{-/-}$ mice fed with a normal (ND) or a Western-type diet (WD). These results suggest that upregulated APE1/Ref-1 expression in ApoE$^{-/-}$ mice fed with a Western-type diet may have been derived from macrophages and endothelial cells in atherosclerotic plaques.

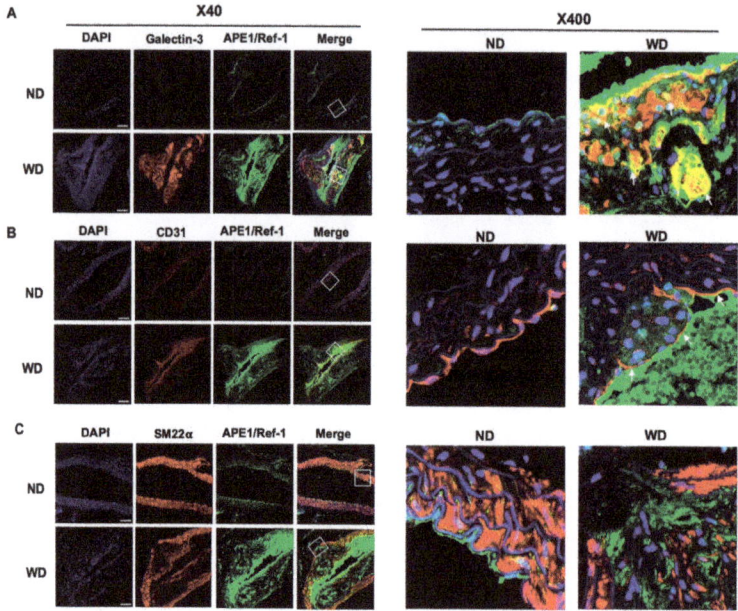

Figure 4. Upregulated APE1/Ref-1 expression is co-localized in macrophages and endothelial cells.

(**A**) Immunofluorescence labeling for APE1/Ref-1 and galectin-3 in the thoracic aorta of ApoE$^{-/-}$ mice fed with a normal diet (ND) or a Western-type diet (WD). Aortic tissues were immunolabeled using Alexa Fluor® 647-conjugated anti-galectin-3 antibody (red) and Alexa Fluor® 488-conjugated anti-APE1/Ref-1 (green). Red fluorescence signal for galectin-3 was used to detect macrophages. Right image (magnification 400×) shows a white square on the left image (magnification 40×, scale bar 50 μm). White arrows indicate co-localization of APE1/Ref-1 with galectin-3 in macrophages (orange-yellow signal) (right). Cell nuclei were stained with 4′,6-diamidino-2-phenylindole (DAPI) (blue). (**B**) Immunofluorescence labeling for APE1/Ref-1 and cluster of differentiation 31 (CD31) in the thoracic aorta of ApoE$^{-/-}$ mice fed with a normal diet (ND) or a Western-type diet (WD). Aortic tissues were immunolabeled using Alexa Fluor® 647-conjugated anti-CD31 antibody (red) and Alexa Fluor® 488-conjugated anti-APE1/Ref-1 (green). Red fluorescence signal for CD31 was used to detect endothelial cells. Right image (magnification 400×) shows a white square on the left image (magnification 40×, scale bar 50 μm). White arrows indicate co-localization of APE1/Ref-1 with CD31 in endothelial cells (orange-yellow signal) (right). Cell nuclei were stained with DAPI (blue). (**C**) Immunofluorescence analysis of APE1/Ref-1 and SM22α in the thoracic aorta of ApoE$^{-/-}$ mice fed with a normal diet (ND) or a Western-type diet (WD). Aortic tissues were immunolabeled with Alexa Fluor® 647-conjugated anti-SM22α antibody (red) and Alexa Fluor® 488-conjugated anti-APE1/Ref-1 (green). Red fluorescence signal for SM22α was used to detect vascular smooth muscle cells. Nuclei were stained with DAPI (blue fluorescence). Right image (magnification 400×) shows a white square on the left image (magnification 40×, scale bar 50 μm). Notably, APE1/Ref-1 signal did not merge with that for SM22α.

3.5. Plasma APE1/Ref-1 Levels Are Markedly Elevated in ApoE$^{-/-}$ Mice Fed with a Western-Type Diet

APE1/Ref-1 can be secreted into blood circulation in endotoxemia [12]. Having determined that the expression of APE1/Ref-1 was increased in atherosclerotic plaques of ApoE$^{-/-}$ mice fed with a Western-type diet, we investigated whether plasma APE1/Ref-1 levels were elevated in atherosclerotic inflammation. Plasma APE1/Ref-1 level in mice was evaluated with sandwich ELISA assay as described in the experimental section. Plasma APE1/Ref-1 in C57BL/6J wild-type control mice (WT) and ApoE$^{-/-}$ mice fed with a normal diet (ND) were 3.12 ± 0.57 ng/mL and 2.74 ± 0.82 ng/mL, respectively, thus showing no significant difference between the WT and ND groups. However, interestingly, the levels of plasma APE1/Ref-1 in ApoE$^{-/-}$ mice fed with a Western-type diet (WD) were significantly increased compared with those of the ND group (11.36 ± 2.17 ng/mL for WD vs. 2.74 ± 0.82 ng/mL as for ND). These increased plasma levels of APE1/Ref-1 detected in the WD group were suppressed by treatment with atorvastatin (11.36 ± 2.17 ng/mL for WD vs. 3.54 ± 0.52 ng/mL for WD + statin) (Figure 5A). This suggests that plasma levels of APE1/Ref-1 were increased under hypercholesterolemic conditions accompanied by inflammation. Analysis using receiver operating characteristic (ROC) was performed to evaluate the utility of plasma APE1/Ref-1 as a biomarker for atherosclerosis and to determine the optimal cut-off value. Based on the ROC curve, the cut-off value for plasma APE1/Ref-1 level for diagnosis of atherosclerosis in ApoE$^{-/-}$ mice fed with a Western-type diet (WD) as compared with wild-type control mice (WT) was set at 4.90 ng/mL, with an area under the ROC curve of 1.0, a sensitivity of 100%, and a specificity of 91% (Figure 5B). Similarly, the cut-off value for plasma APE1/Ref-1 level for diagnosis of atherosclerosis in ApoE$^{-/-}$ mice fed with a Western-type diet (WD) as compared with ApoE$^{-/-}$ mice fed with a normal diet (ND) was set at 5.64 ng/mL, with an area under ROC curve of 1.0, a sensitivity of 100%, and a specificity of 90%.

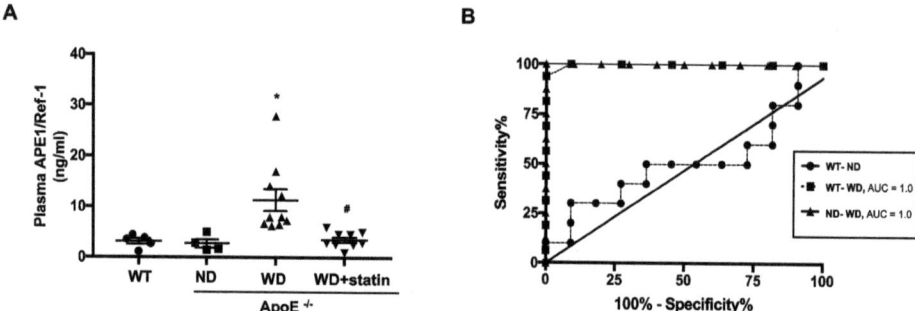

Figure 5. Plasma APE1/Ref-1 level was significantly elevated in ApoE$^{-/-}$ mice fed with a Western-type diet. (**A**) Quantitative analysis of plasma APE1/Ref-1 levels in each experimental group (WT; C57BL/6J wild-type control mice, ND; ApoE$^{-/-}$ mice fed with a normal diet, WD; ApoE$^{-/-}$ mice fed with a Western-type diet, WD + statin; atorvastatin-treated ApoE$^{-/-}$ mice fed with a Western-type diet) was performed using an enzyme-linked immunosorbent assay (ELISA). Results are represented as mean ± S.E.M., n = 5–10. * p < 0.05, vs. ND group, # p < 0.05, vs. WD group was determined using one-way ANOVA followed by the Bonferroni's multiple comparison test. (**B**) Receiver operating curves of plasma APE1/Ref-1 levels for the diagnosis of atherosclerosis in ApoE$^{-/-}$ mice fed with a Western-type diet (WD) compared with those of wildtype control mice (WT). Area under curve (AUC) (■) = 1.00; p < 0.0001. Receiver operating curves of plasma APE1/Ref-1 levels for the diagnosis of atherosclerosis in ApoE$^{-/-}$ mice fed with a Western-type diet (WD) compared with those of ApoE$^{-/-}$ mice fed with a normal diet (ND). Area under curve (AUC) (▲) = 1.00; p < 0.0001. Notably, there was no significant difference between WT and ND groups.

3.6. Correlation of Hematologic Parameters with Plasma APE1/Ref-1 Level

Having established that plasma APE1/Ref-1 levels were elevated in ApoE$^{-/-}$ mice fed with a Western-type diet, we next analyzed which hematologic parameters were correlated with plasma APE1/Ref-1 levels. Correlation analysis was first performed in three groups of wild-type control mice (WT) and ApoE$^{-/-}$ mice fed with a normal diet (ND) or a Western-type diet (WD). As shown in the left panel of Table 1, plasma APE1/Ref-1 levels were significantly correlated with the level of total cholesterol (r = 0.61), LDL (r = 0.62), and the neutrophil/lymphocyte ratio (NLR, r = 0.63). This suggests that high cholesterol, LDL, and NLR were important factors that correlated with increased APE1/Ref-1 levels. To evaluate the correlation between plasma APE1/Ref-1 levels and hematologic parameters during treatment with atorvastatin, correlation analysis was performed in ApoE$^{-/-}$ mice fed with a Western-type diet (WD) and atorvastatin-treated ApoE$^{-/-}$ mice fed with a Western-type diet (WD + statin). As shown in the right panel of Table 1, plasma APE1/Ref-1 level was significantly correlated with neutrophil counts (r = 0.61), lymphocyte counts (r = −0.62), and NLR (r = 0.79), but was not correlated with lipid profiles such as that for total cholesterol (r = 0.37). This indicates that plasma levels of APE1/Ref-1 were correlated mainly with the neutrophil/lymphocyte ratio, which is used as a marker of systemic inflammation.

Table 1. Correlation analysis of hematologic parameters with plasma APE1/Ref-1 levels.

Hematologic Parameters with Plasma APE1/Ref-1	Between Wild Type, ApoE$^{-/-}$ Mice Fed Normal Diet and Western Type Diet (WT, ND and WD)			Between ApoE$^{-/-}$ Mice Fed Western-Type Diet and Atorvastatin-Treated Group (WD and WD + statin)		
	r	95% CI [a]	p Value	r	95% CI [a]	p Value
Total cholesterol	0.609	0.213 to 0.833	0.006	0.366	−0.106 to 0.703	n.s
Low density lipoprotein	0.616	0.224 to 0.836	0.005	0.277	−0.203 to 0.649	n.s
Triglyceride	0.288	−0.192 to 0.656	n.s	−0.055	−0.497 to 0.409	n.s
monocyte	0.382	−0.088 to 0.712	n.s	0.059	−0.406 to 0.500	n.s
Neutrophil	0.313	−0.165 to 0.671	n.s	0.611	0.217 to 0.834	0.005
Lymphocyte	−0.347	−0.692 to 0.127	n.s	−0.616	−0.836 to −0.224	0.005
Neutrophil/lymphocyte ratio	0.633	0.251 to 0.845	0.004	0.786	0.515 to 0.914	<0.001

WT; C57BL/6J wild-type control mice, ND; ApoE$^{-/-}$ mice fed with normal diet, WD; ApoE$^{-/-}$ mice fed with a Western-type diet, WD + statin; atorvastatin-treated ApoE$^{-/-}$ mice fed with a Western-type diet, CI; confidence interval, n.s; not significant, [a] Pearson Correlation Coefficient between plasma APE1/Ref-1 and relevant parameters.

4. Discussion

In this study, we showed that APE1/Ref-1 expression is upregulated in the aortic tissues of atherosclerotic mice fed with a Western-type diet. In these atherosclerotic mice fed with a Western-type diet, elevated plasma levels of APE1/Ref-1 were correlated with the neutrophil/lymphocyte ratio, which were controlled with treatment with atorvastatin.

To evaluate the functions of APE1/Ref-1 in atherosclerosis, we selected ApoE$^{-/-}$ mice as our animal model of atherosclerosis. ApoE$^{-/-}$ mice demonstrate decreased cholesterol clearance of remnant lipoproteins, which results in hypercholesterolemia. In our present study, however, the level of plasma cholesterol (~335 mg/dL) did not lead to the formation of atherosclerotic plaques in ApoE$^{-/-}$ mice fed with a normal diet. However, ApoE$^{-/-}$ mice fed with a Western-type diet (high in fat and cholesterol) showed plasma cholesterol levels that were elevated to more than 1000 mg/dL (Figure 1), and atherosclerotic plaques were observed throughout the aortas of these mice. Collectively, ApoE$^{-/-}$ mice that chronically consumed a Western-type diet for 20 weeks developed atherosclerotic plaques with vascular inflammation. Inflammatory monocytes can secrete proinflammatory cytokines, thereby contributing to vascular dysfunction and accumulation of lipid-laden foam cells. The association of elevated monocyte and neutrophil levels with atherosclerotic progression has been reported in several studies [32,33]. The neutrophil/lymphocyte ratio (NLR), which is calculated by dividing the neutrophil count by the lymphocyte count, is also used as an indicator of systemic inflammation [34]. High NLR is particularly associated with atherosclerosis [22,23,35]. In our present study, ApoE$^{-/-}$ mice fed with a Western-type diet showed increased neutrophil or monocyte counts, decreased lymphocyte counts, and increased NLR, these effects were suppressed by treatment with atorvastatin (Figure 2).

Increasing evidence indicates that APE1/Ref-1 expression is upregulated in several types of cancer, and in cardiovascular and inflammatory disorders [13]. APE1/Ref-1 is highly expressed in hypertensive rats [36] and in human atherosclerotic plaques [37]. Oxidative stress and endothelial dysfunction are implicated in the pathogenesis of numerous cardiovascular diseases including hypercholesterolemia and atherosclerosis in humans [38] and in ApoE knockout mice [39]. In humans, endothelial dysfunction is thought to precede the development of atherosclerosis [40]. The incorporation of lipids within the endothelium, which is an early manifestation of atherosclerosis, and the associated oxidative processes, may contribute to the degradation of nitric oxide, resulting in vascular dysfunction [41]. Defective base excision repair of oxidative DNA damage promotes atherosclerosis [42]. Endogenous reactive oxygen species (ROS) may increase the level of DNA damage, which then leads to an increased APE1/Ref-1 level, thereby enhancing base excision repair capacity. Indeed, the upregulation of APE1/Ref-1 expression is an adaptive response to cytotoxicity of oxidative agents [43–45]. Therefore, increased expression of APE1/Ref-1 is necessary for the repair of damaged DNA and defense against oxidative stress in atherosclerotic lesions.

It is important to uncover the mechanisms behind APE1/Ref-1 secretion. APE/Ref-1 can be secreted by using non-classical secretion pathways because of the absence of a leading peptide sequence [13]. The secretion of APE1/Ref-1 is mediated by ATP-binding cassette A1 (ABCA1) transporter or vesicular formation [46,47]. The acetylation at the lysine residues of APE1/Ref-1 is a particularly important step for extracellular secretion [11]. Hyperacetylation induced the extracellular vesicular formation containing APE1/Ref-1, which was analyzed with gold particle-labelled APE1/Ref-1 in triple negative breast cancer cell lines [47]. However, research into the mechanisms behind how APE1/Ref-1 is secreted and by which cell types will be further needed.

The immunofluorescence images of co-localized APE1/Ref-1 show that APE1/Ref-1 is mainly overexpressed in macrophages and endothelial cells. APE1/Ref-1 overexpression was closely related to increased expression of galectin-3, a specific macrophage marker, and CD31, a specific endothelial cell marker. Galectin-3 is a carbohydrate-binding lectin implicated in the pathophysiology of cardiovascular diseases and highly expressed within atherosclerotic lesions of mice and humans [48–50]. Galectin-3 expression is related to oxidative stress in macrophages. Protein kinase C (PKC) activator, a nicotinamide adenine dinucleotide phosphate (NADPH) oxidase-dependent inducer of ROS, stimulates an increase in galectin-3 mRNA and protein expression; however, blocking with apocynin reverses these effects [51]. These previous findings show that upregulated APE1/Ref-1 expression in macrophages under hypercholesterolemic conditions is likely related to macrophage defense mechanisms against oxidative stress. CD31 is a protein that is encoded by platelet endothelial cell adhesion molecule-1. CD31 is fairly specific for endothelial differentiation [52]. CD31 is enriched at endothelial cell intercellular junctions. In these locales, it regulates leukocyte trafficking, mechanotransduction, and vascular permeability, and functions as an adhesive stress-response protein to maintain endothelial cell junctional integrity and to restore the vascular permeability barrier following inflammatory or thrombotic challenge [53]. APE1/Ref-1 expression in endothelial cell layers suggests that it plays an important role in the regulation of endothelial cell activation under hypercholesterolemic conditions. Previous studies have shown that APE1/Ref-1 overexpression inhibits tumor necrosis factor-α (TNF-α)-induced endothelial activation by inhibiting the generation of intracellular superoxide and phosphorylation of p38 mitogen-activated protein kinase [9]. Adenoviral APE1/Ref-1 gene transfer inhibits balloon injury-induced neointimal formation in the rat carotid artery and inhibits PKC-mediated p66shc phosphorylation [54,55]. Additionally, lipid-rich plaque formation within the arterial wall produces a hypoxic environment. Hypoxia leads to the formation of new blood vessels [56]. APE1/Ref-1 regulates several transcription factors involved in cell survival mechanisms and hypoxia signaling. APE1/Ref-1 redox signaling activity can regulate the transcriptional activation of hypoxia-inducible factor 1 alpha [57]. Therefore, increased APE1/Ref-1 expression in atherosclerotic plaques may be a defense mechanism to protect tissue or cells against hypoxic injury.

Several animal and human studies have been conducted in an attempt to find a correlation between plasma APE1/Ref-1 levels and cardiovascular diseases. Elevated APE1/Ref-1 levels are detected in cardiovascular disorders. The 37-kDa immunoreactive band, identified as rat APE1/Ref-1 using liquid chromatography/tandem mass spectrometry in lipopolysaccharide-induced endotoxemic rats [12], suggests that plasma APE1/Ref-1 level may serve as a serological biomarker for endotoxemia. The levels of serum APE1/Ref-1 are elevated in coronary artery disease, and these levels are higher in myocardial infarction than those in angina [14]. In our present study, we confirmed that serologic APE1/Ref-1 was increased in the plasma of ApoE$^{-/-}$ mice, and that its levels were decreased by treatment with atorvastatin. These results strongly suggest that APE1/Ref-1 could be used as a serologic biomarker to detect the progression of atherosclerosis. Previous reports have shown that atorvastatin decreases plasma levels of inflammatory markers, plasma levels of highly sensitive C-reactive protein, TNF-α, and monocyte chemoattractant protein-1 [58]. Based on results obtained in these studies, reduction in vascular inflammation using treatment with atorvastatin would decrease plasma or tissue levels of APE1/Ref-1, suggesting that APE1/Ref-1 can be used as a biomarker to assess the degree of vascular inflammation. Our results show that plasma levels of APE1/Ref-1 were correlated with vascular inflammation involved in atherosclerotic processes of ApoE$^{-/-}$ mice. Therefore, elevation of plasma APE1/Ref-1 can aid in the diagnosis or follow-up of patients with vascular inflammation and atherosclerosis. Furthermore, we evaluated the cut-off value for the plasma level of APE1/Ref-1 for use as a biomarker of atherosclerosis. Based on the ROC curve, the cut-off value was set at 4.903 ng/mL, with an area under the ROC curve of 1.0, a sensitivity of 100%, and a specificity of 91%. This suggests that the plasma level of APE1/Ref-1 could be a reliable serologic biomarker for the evaluation of atherosclerosis.

There are some limitations in the present work. Multiple strategies are necessary to find new biomarkers associated with atherosclerotic inflammation. However, this study was conducted in animals by using ApoE knockout animals as a model for arteriosclerosis. Experimental data such as cut-off values obtained from animal experiments are difficult to use directly in humans and need to be supplemented through human studies in the future. Research to determine the secreted cells has been conducted only in vascular tissues. In order to identify specific cells from which APE1/Ref-1 is released, it needs to be confirmed through the future development of tissue-specific knockout of APE1/Ref-1.

Author Contributions: Conceptualization, Y.R.L. and B.H.J.; data curation, H.K.J., E.-O.L., M.S.P., H.S.C., H.J.; formal analysis, Y.R.L., H.K.J., S.K.; writing—original draft writing, H.K.J.; supervision, C.-S.K., J.-O.J.; writing—review and editing, B.H.J. All authors have read and agreed to the published version of the manuscript.

Funding: This research was supported by grants from the Basic Science Research Program through the National Research Foundation of Korea (NRF) funded by the Ministry of Education (NRF-2014R1A6A1029617 to B.H.J. and 2017R1A6A3A11027834 to Y.R.L.) and Ministry of Science, ICT & Future Planning (2020R1C1C1014490 to H.K.J.).

Conflicts of Interest: The authors declare no conflict of interest. The funders played no role in the design of the study; in the collection, analyses, or interpretation of data; in the writing of the manuscript; or in the decision to publish the results.

References

1. Libby, P. Inflammation in atherosclerosis. *Nature* **2002**, *420*, 868–874. [CrossRef] [PubMed]
2. Godschalk, R.W.; Albrecht, C.; Curfs, D.M.; Schins, R.P.; Bartsch, H.; van Schooten, F.J.; Nair, J. Decreased levels of lipid peroxidation-induced DNA damage in the onset of atherogenesis in apolipoprotein E deficient mice. *Mutat. Res.* **2007**, *621*, 87–94. [CrossRef] [PubMed]
3. Whitaker, A.M.; Schaich, M.A.; Smith, M.R.; Flynn, T.S.; Freudenthal, B.D. Base excision repair of oxidative DNA damage: From mechanism to disease. *Front. Biosci. (Landmark Ed.)* **2017**, *22*, 1493–1522. [PubMed]
4. Cervelli, T.; Borghini, A.; Galli, A.; Andreassi, M.G. DNA damage and repair in atherosclerosis: Current insights and future perspectives. *Int. J. Mol. Sci.* **2012**, *13*, 16929–16944. [CrossRef]
5. Singh, R.B.; Mengi, S.A.; Xu, Y.J.; Arneja, A.S.; Dhalla, N.S. Pathogenesis of atherosclerosis: A multifactorial process. *Exp. Clin. Cardiol.* **2002**, *7*, 40–53.
6. Mahley, R.W. Apolipoprotein E: Cholesterol transport protein with expanding role in cell biology. *Science* **1988**, *240*, 622–630. [CrossRef]
7. Tell, G.; Quadrifoglio, F.; Tiribelli, C.; Kelley, M.R. The many functions of APE1/Ref-1: Not only a DNA repair enzyme. *Antioxid. Redox. Signal.* **2009**, *11*, 601–620. [CrossRef]
8. Whitaker, A.M.; Freudenthal, B.D. APE1: A skilled nucleic acid surgeon. *DNA Repair* **2018**, *71*, 93–100. [CrossRef]
9. Kim, C.S.; Son, S.J.; Kim, E.K.; Kim, S.N.; Yoo, D.G.; Kim, H.S.; Ryoo, S.W.; Lee, S.D.; Irani, K.; Jeon, B.H. Apurinic/apyrimidinic endonuclease1/redox factor-1 inhibits monocyte adhesion in endothelial cells. *Cardiovasc. Res.* **2006**, *69*, 520–526. [CrossRef]

10. Jeon, B.H.; Gupta, G.; Park, Y.C.; Qi, B.; Haile, A.; Khanday, F.A.; Liu, Y.X.; Kim, J.M.; Ozaki, M.; White, A.R.; et al. Apurinic/apyrimidinic endonuclease 1 regulates endothelial NO production and vascular tone. *Circ. Res.* **2004**, *95*, 902–910. [CrossRef]
11. Choi, S.; Lee, Y.R.; Park, M.S.; Joo, H.K.; Cho, E.J.; Kim, H.S.; Kim, C.S.; Park, J.B.; Irani, K.; Jeon, B.H. Histone deacetylases inhibitor trichostatin A modulates the extracellular release of APE1/Ref-1. *Biochem. Biophys. Res. Commun.* **2013**, *435*, 403–407. [CrossRef] [PubMed]
12. Park, M.S.; Lee, Y.R.; Choi, S.; Joo, H.K.; Cho, E.J.; Kim, C.S.; Park, J.B.; Jo, E.K.; Jeon, B.H. Identification of plasma APE1/Ref-1 in lipopolysaccharide-induced endotoxemic rats: Implication of serological biomarker for an endotoxemia. *Biochem. Biophys. Res. Commun.* **2013**, *435*, 621–626. [CrossRef] [PubMed]
13. Lee, Y.R.; Joo, H.K.; Jeon, B.H. The Biological Role of Apurinic/Apyrimidinic Endonuclease1/Redox Factor-1 as a Therapeutic Target for Vascular Inflammation and as a Serologic Biomarker. *Biomedicines* **2020**, *8*, 57. [CrossRef]
14. Jin, S.A.; Seo, H.J.; Kim, S.K.; Lee, Y.R.; Choi, S.; Ahn, K.T.; Kim, J.H.; Park, J.H.; Lee, J.H.; Choi, S.W.; et al. Elevation of the Serum Apurinic/Apyrimidinic Endonuclease 1/Redox Factor-1 in Coronary Artery Disease. *Korean Circ. J.* **2015**, *45*, 364–371. [CrossRef] [PubMed]
15. Jin, S.A.; Lim, B.K.; Seo, H.J.; Kim, S.K.; Ahn, K.T.; Jeon, B.H.; Jeong, J.O. Elevation of Serum APE1/Ref-1 in Experimental Murine Myocarditis. *Int. J. Mol. Sci.* **2017**, *18*, 2664. [CrossRef] [PubMed]
16. Park, M.S.; Choi, S.; Lee, Y.R.; Joo, H.K.; Kang, G.; Kim, C.S.; Kim, S.J.; Lee, S.D.; Jeon, B.H. Secreted APE1/Ref-1 inhibits TNF-alpha-stimulated endothelial inflammation via thiol-disulfide exchange in TNF receptor. *Sci. Rep.* **2016**, *6*, 23015.
17. Nath, S.; Roychoudhury, S.; Kling, M.J.; Song, H.; Biswas, P.; Shukla, A.; Band, H.; Joshi, S.; Bhakat, K.K. The extracellular role of DNA damage repair protein APE1 in regulation of IL-6 expression. *Cell. Signal.* **2017**, *39*, 18–31. [CrossRef]
18. Plump, A.S.; Smith, J.D.; Hayek, T.; Aalto-Setala, K.; Walsh, A.; Verstuyft, J.G.; Rubin, E.M.; Breslow, J.L. Severe hypercholesterolemia and atherosclerosis in apolipoprotein E-deficient mice created by homologous recombination in ES cells. *Cell* **1992**, *71*, 343–353. [CrossRef]
19. Zhou, Q.; Liao, J.K. Statins and cardiovascular diseases: From cholesterol lowering to pleiotropy. *Curr. Pharm. Des.* **2009**, *15*, 467–478. [CrossRef]
20. Mohanta, S.; Yin, C.; Weber, C.; Hu, D.; Habenicht, A.J. Aorta Atherosclerosis Lesion Analysis in Hyperlipidemic Mice. *Bio Protoc.* **2016**, *6*, e1833. [CrossRef]
21. Meng, Z.; Wang, M.; Xing, J.; Liu, Y.; Li, H. Myricetin ameliorates atherosclerosis in the low-density-lipoprotein receptor knockout mice by suppression of cholesterol accumulation in macrophage foam cells. *Nutr. Metab.* **2019**, *16*, 25. [CrossRef] [PubMed]
22. Meng, L.B.; Yu, Z.M.; Guo, P.; Wang, Q.Q.; Qi, R.M.; Shan, M.J.; Lv, J.; Gong, T. Neutrophils and neutrophil-lymphocyte ratio: Inflammatory markers associated with intimal-media thickness of atherosclerosis. *Thromb. Res.* **2018**, *170*, 45–52. [CrossRef] [PubMed]
23. Corriere, T.; Di Marca, S.; Cataudella, E.; Pulvirenti, A.; Alaimo, S.; Stancanelli, B.; Malatino, L. Neutrophil-to-Lymphocyte Ratio is a strong predictor of atherosclerotic carotid plaques in older adults. *Nutr. Metab. Cardiovasc. Dis.* **2018**, *28*, 23–27. [CrossRef] [PubMed]
24. Ley, K.; Huo, Y. VCAM-1 is critical in atherosclerosis. *J. Clin. Investig.* **2001**, *107*, 1209–1210. [CrossRef]
25. Papaspyridonos, M.; McNeill, E.; De Bono, J.P.; Smith, A.; Burnand, K.G.; Channon, K.M.; Greaves, D.R. Galectin-3 is an amplifier of inflammation in atherosclerotic plaque progression through macrophage activation and monocyte chemoattraction. *Arterioscler. Thromb. Vasc. Biol.* **2008**, *28*, 433–440. [CrossRef]
26. Ho, M.K.; Springer, T.A. Mac-2, a novel 32,000 Mr mouse macrophage subpopulation-specific antigen defined by monoclonal antibodies. *J. Immunol.* **1982**, *128*, 1221–1228.
27. Leenen, P.J.; De Bruijn, M.F.; Voerman, J.S.; Campbell, P.A.; Van Ewijk, W. Markers of mouse macrophage development detected by monoclonal antibodies. *J. Immunol. Methods* **1994**, *174*, 5–19. [CrossRef]
28. Reddick, R.L.; Zhang, S.H.; Maeda, N. Atherosclerosis in mice lacking apo E. Evaluation of lesional development and progression. *Arterioscler. Thromb.* **1994**, *14*, 141–147. [CrossRef]
29. Rong, J.X.; Li, J.; Reis, E.D.; Choudhury, R.P.; Dansky, H.M.; Elmalem, V.I.; Fallon, J.T.; Breslow, J.L.; Fisher, E.A. Elevating high-density lipoprotein cholesterol in apolipoprotein E-deficient mice remodels advanced atherosclerotic lesions by decreasing macrophage and increasing smooth muscle cell content. *Circulation* **2001**, *104*, 2447–2452. [CrossRef]

30. Muller, A.M.; Hermanns, M.I.; Skrzynski, C.; Nesslinger, M.; Muller, K.M.; Kirkpatrick, C.J. Expression of the endothelial markers PECAM-1, vWf, and CD34 in vivo and in vitro. *Exp. Mol. Pathol.* **2002**, *72*, 221–229. [CrossRef]
31. Winder, S.J.; Sutherland, C.; Walsh, M.P. Biochemical and functional characterization of smooth muscle calponin. *Adv. Exp. Med. Biol.* **1991**, *304*, 37–51. [PubMed]
32. Patino, R.; Ibarra, J.; Rodriguez, A.; Yague, M.R.; Pintor, E.; Fernandez-Cruz, A.; Figueredo, A. Circulating monocytes in patients with diabetes mellitus, arterial disease, and increased CD14 expression. *Am. J. Cardiol.* **2000**, *85*, 1288–1291. [CrossRef]
33. Guasti, L.; Dentali, F.; Castiglioni, L.; Maroni, L.; Marino, F.; Squizzato, A.; Ageno, W.; Gianni, M.; Gaudio, G.; Grandi, A.M.; et al. Neutrophils and clinical outcomes in patients with acute coronary syndromes and/or cardiac revascularisation. A systematic review on more than 34,000 subjects. *Thromb. Haemost.* **2011**, *106*, 591–599. [CrossRef] [PubMed]
34. Balta, S.; Demirkol, S.; Unlu, M.; Arslan, Z.; Celik, T. Neutrophil to lymphocyte ratio may be predict of mortality in all conditions. *Br. J. Cancer* **2013**, *109*, 3125–3126. [CrossRef]
35. Horne, B.D.; Anderson, J.L.; John, J.M.; Weaver, A.; Bair, T.L.; Jensen, K.R.; Renlund, D.G.; Muhlestein, J.B.; Intermountain Heart Collaborative Study Group. Which white blood cell subtypes predict increased cardiovascular risk? *J. Am. Coll. Cardiol.* **2005**, *45*, 1638–1643. [CrossRef]
36. Song, S.H.; Cho, E.J.; Park, M.S.; Lee, Y.R.; Joo, H.K.; Kang, G.; Kang, S.K.; Choi, S.; Jeon, B.H. Redox Regulating Protein APE1/Ref-1 Expression is Increased in Abdominal Aortic Coarctation-induced Hypertension Rats. *J. Korean Soc. Hypertens* **2012**, *18*, 126–135. [CrossRef]
37. Martinet, W.; Knaapen, M.W.; De Meyer, G.R.; Herman, A.G.; Kockx, M.M. Elevated levels of oxidative DNA damage and DNA repair enzymes in human atherosclerotic plaques. *Circulation* **2002**, *106*, 927–932. [CrossRef]
38. Stroes, E.; Kastelein, J.; Cosentino, F.; Erkelens, W.; Wever, R.; Koomans, H.; Luscher, T.; Rabelink, T. Tetrahydrobiopterin restores endothelial function in hypercholesterolemia. *J. Clin. Investig.* **1997**, *99*, 41–46. [CrossRef]
39. Laursen, J.B.; Somers, M.; Kurz, S.; McCann, L.; Warnholtz, A.; Freeman, B.A.; Tarpey, M.; Fukai, T.; Harrison, D.G. Endothelial regulation of vasomotion in apoE-deficient mice: Implications for interactions between peroxynitrite and tetrahydrobiopterin. *Circulation* **2001**, *103*, 1282–1288. [CrossRef]
40. Landmesser, U.; Hornig, B.; Drexler, H. Endothelial function: A critical determinant in atherosclerosis? *Circulation* **2004**, *109*, II27–II33. [CrossRef]
41. Bonthu, S.; Heistad, D.D.; Chappell, D.A.; Lamping, K.G.; Faraci, F.M. Atherosclerosis, vascular remodeling, and impairment of endothelium-dependent relaxation in genetically altered hyperlipidemic mice. *Arterioscler. Thromb. Vasc. Biol.* **1997**, *17*, 2333–2340. [CrossRef] [PubMed]
42. Shah, A.; Gray, K.; Figg, N.; Finigan, A.; Starks, L.; Bennett, M. Defective Base Excision Repair of Oxidative DNA Damage in Vascular Smooth Muscle Cells Promotes Atherosclerosis. *Circulation* **2018**, *138*, 1446–1462. [CrossRef] [PubMed]
43. Fritz, G.; Grosch, S.; Tomicic, M.; Kaina, B. APE/Ref-1 and the mammalian response to genotoxic stress. *Toxicology* **2003**, *193*, 67–78. [CrossRef]
44. Grosch, S.; Kaina, B. Transcriptional activation of apurinic/apyrimidinic endonuclease (Ape, Ref-1) by oxidative stress requires CREB. *Biochem. Biophys. Res. Commun.* **1999**, *261*, 859–863. [CrossRef] [PubMed]
45. Ramana, C.V.; Boldogh, I.; Izumi, T.; Mitra, S. Activation of apurinic/apyrimidinic endonuclease in human cells by reactive oxygen species and its correlation with their adaptive response to genotoxicity of free radicals. *Proc. Natl. Acad. Sci. USA* **1998**, *95*, 5061–5066. [CrossRef] [PubMed]
46. Lee, Y.R.; Joo, H.K.; Lee, E.O.; Cho, H.S.; Choi, S.; Kim, C.S.; Jeon, B.H. ATP Binding Cassette Transporter A1 is Involved in Extracellular Secretion of Acetylated APE1/Ref-1. *Int. J. Mol. Sci.* **2019**, *20*, 3178. [CrossRef] [PubMed]
47. Lee, Y.R.; Kim, K.M.; Jeon, B.H.; Choi, S. Extracellularly secreted APE1/Ref-1 triggers apoptosis in triple-negative breast cancer cells via RAGE binding, which is mediated through acetylation. *Oncotarget* **2015**, *6*, 23383–23398. [CrossRef]
48. MacKinnon, A.C.; Liu, X.; Hadoke, P.W.; Miller, M.R.; Newby, D.E.; Sethi, T. Inhibition of galectin-3 reduces atherosclerosis in apolipoprotein E-deficient mice. *Glycobiology* **2013**, *23*, 654–663. [CrossRef]

49. Nachtigal, M.; Al-Assaad, Z.; Mayer, E.P.; Kim, K.; Monsigny, M. Galectin-3 expression in human atherosclerotic lesions. *Am. J. Pathol.* **1998**, *152*, 1199–1208.
50. Arar, C.; Gaudin, J.C.; Capron, L.; Legrand, A. Galectin-3 gene (LGALS3) expression in experimental atherosclerosis and cultured smooth muscle cells. *FEBS Lett.* **1998**, *430*, 307–311. [CrossRef]
51. Madrigal-Matute, J.; Lindholt, J.S.; Fernandez-Garcia, C.E.; Benito-Martin, A.; Burillo, E.; Zalba, G.; Beloqui, O.; Llamas-Granda, P.; Ortiz, A.; Egido, J.; et al. Galectin-3, a biomarker linking oxidative stress and inflammation with the clinical outcomes of patients with atherothrombosis. *J. Am. Heart Assoc.* **2014**, *3*, e000785. [CrossRef]
52. Cabanas, C.; Sanchez-Madrid, F.; Bellon, T.; Figdor, C.G.; Te Velde, A.A.; Fernandez, J.M.; Acevedo, A.; Bernabeu, C. Characterization of a novel myeloid antigen regulated during differentiation of monocytic cells. *Eur. J. Immunol.* **1989**, *19*, 1373–1378. [CrossRef]
53. Lertkiatmongkol, P.; Liao, D.; Mei, H.; Hu, Y.; Newman, P.J. Endothelial functions of platelet/endothelial cell adhesion molecule-1 (CD31). *Curr. Opin. Hematol.* **2016**, *23*, 253–259. [CrossRef]
54. Lee, H.M.; Jeon, B.H.; Won, K.J.; Lee, C.K.; Park, T.K.; Choi, W.S.; Bae, Y.M.; Kim, H.S.; Lee, S.K.; Park, S.H.; et al. Gene transfer of redox factor-1 inhibits neointimal formation: Involvement of platelet-derived growth factor-beta receptor signaling via the inhibition of the reactive oxygen species-mediated Syk pathway. *Circ. Res.* **2009**, *104*, 219–227, 215p following 227. [CrossRef]
55. Lee, S.K.; Chung, J.I.; Park, M.S.; Joo, H.K.; Lee, E.J.; Cho, E.J.; Park, J.B.; Ryoo, S.; Irani, K.; Jeon, B.H. Apurinic/apyrimidinic endonuclease 1 inhibits protein kinase C-mediated p66shc phosphorylation and vasoconstriction. *Cardiovasc. Res.* **2011**, *91*, 502–509. [CrossRef]
56. Bjornheden, T.; Levin, M.; Evaldsson, M.; Wiklund, O. Evidence of hypoxic areas within the arterial wall in vivo. *Arterioscler. Thromb. Vasc. Biol.* **1999**, *19*, 870–876. [CrossRef]
57. Logsdon, D.P.; Grimard, M.; Luo, M.; Shahda, S.; Jiang, Y.; Tong, Y.; Yu, Z.; Zyromski, N.; Schipani, E.; Carta, F.; et al. Regulation of HIF1alpha under Hypoxia by APE1/Ref-1 Impacts CA9 Expression: Dual Targeting in Patient-Derived 3D Pancreatic Cancer Models. *Mol. Cancer Ther.* **2016**, *15*, 2722–2732. [CrossRef]
58. Nie, P.; Li, D.; Hu, L.; Jin, S.; Yu, Y.; Cai, Z.; Shao, Q.; Shen, J.; Yi, J.; Xiao, H.; et al. Atorvastatin improves plaque stability in ApoE-knockout mice by regulating chemokines and chemokine receptors. *PLoS ONE* **2014**, *9*, e97009. [CrossRef]

© 2020 by the authors. Licensee MDPI, Basel, Switzerland. This article is an open access article distributed under the terms and conditions of the Creative Commons Attribution (CC BY) license (http://creativecommons.org/licenses/by/4.0/).

Review

Arginine and Endothelial Function

Jessica Gambardella [1,2,3,4], Wafiq Khondkar [1], Marco Bruno Morelli [1,2], Xujun Wang [1], Gaetano Santulli [1,2,3,4,*] and Valentina Trimarco [5]

1. Department of Medicine (Division of Cardiology), Wilf Family Cardiovascular Research Institute, Albert Einstein College of Medicine—Montefiore University Hospital, New York City, NY 10461, USA; jessica.gambardella@einsteinmed.org (J.G.); wakhonda22@herricksk12.org (W.K.); marco.morelli@einsteinmed.org (M.B.M.); xujun.wang@einsteinmed.org (X.W.)
2. Department of Molecular Pharmacology, Fleischer Institute for Diabetes and Metabolism, Albert Einstein College of Medicine, New York City, NY 10461, USA
3. Department of Advanced Biomedical Sciences, "Federico II" University, 80131 Naples, Italy
4. International Translational Research and Medical Education (ITME), 80100 Naples, Italy
5. Department of Neuroscience, "Federico II" University, 80131 Naples, Italy; valentina.trimarco@unina.it
* Correspondence: gsantulli001@gmail.com

Received: 30 June 2020; Accepted: 5 August 2020; Published: 6 August 2020

Abstract: Arginine (L-arginine), is an amino acid involved in a number of biological processes, including the biosynthesis of proteins, host immune response, urea cycle, and nitric oxide production. In this systematic review, we focus on the functional role of arginine in the regulation of endothelial function and vascular tone. Both clinical and preclinical studies are examined, analyzing the effects of arginine supplementation in hypertension, ischemic heart disease, aging, peripheral artery disease, and diabetes mellitus.

Keywords: ADMA; arginine; arginine paradox; BH4; blood pressure; COVID-19; dietary supplements; endothelial dysfunction; endothelium; eNOS uncoupling; heart failure; hypertension; L-arginine; myocardial infarction; NADPH; nitric oxide; oxidative stress; peripheral artery disease

---

1. Pleiotropic Effects of Arginine

L-arginine, hereinafter referred to as arginine, is a semi-essential or conditionally essential amino acid, since it can be synthetized by healthy individuals but not by preterm infants [1]. From a chemical point of view, arginine is a 2-amino-5-guanidinopentanoic acid (Figure 1). Its name derives from the Greek word ἄργυρος (silver), indicating the color of arginine nitrate crystals.

Arginine is involved in a number of biological processes, it is the substrate for a series of reactions leading to the synthesis of other amino acids, and it is a substrate for two enzymes, namely nitric oxide (NO) synthase (NOS) and arginase, which are fundamental for the generation of NO and urea, respectively. Arginine is known to act as a substrate for NO production by endothelial cells, thus regulating vascular tone and, overall, cardiovascular homeostasis [2]. NO is synthesized from arginine by the enzyme NOS in a reaction that involves the transfer of electrons from nicotinamide adenine dinucleotide phosphate (NADPH)—via the flavin adenine dinucleotide (FAD) and flavin mononucleotide (FMN) in the C-terminal reductase domain [3,4]—to the heme in the N-terminal oxygenase domain, where the substrate arginine is oxidized to citrulline and NO [5,6], as shown in Figure 1. Arginine is also implicated in T-cell proliferation and host immune responses, as well as in creatine and collagen synthesis [7–11].

There are three isoforms of NOS, two of which—endothelial (eNOS) [12,13] and neuronal (nNOS) [14–16]—are constitutively expressed, while the third one, inducible NOS (iNOS) [17–19], is expressed in response to cytokines and is related to the inflammatory response [6,20]. NO generation occurs in two steps: first, NOS hydroxylates arginine to N^{ω}-hydroxy-arginine (which remains largely bound to the enzyme); in a second step, NOS oxidizes N^{ω}-hydroxy-arginine to citrulline and NO [21–29].

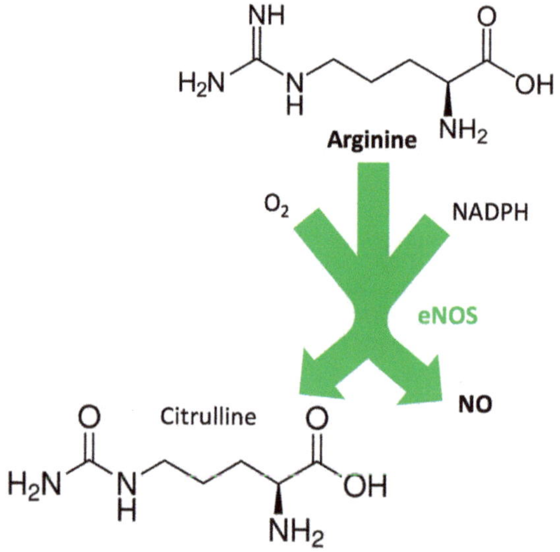

Figure 1. Functional role of arginine in the synthesis of nitric oxide (NO). NADPH: nicotinamide adenine dinucleotide phosphate; eNOS: endothelial NO synthase.

In normal conditions, NOS catalyzes the transformation of arginine, O_2, and NADPH-derived electrons to NO and citrulline (Figure 1). However, in the presence of pathologic conditions like atherosclerosis and diabetes, the NOS function is altered, and the enzyme catalyzes the reduction of O_2 to superoxide (O_2^-), a phenomenon that is generally referred to as "NOS uncoupling" [30–41], and has been linked to a limited bioavailability of tetrahydrobiopterin (BH4, also known as sapropterin) [42–47]. Indeed, the donation of an electron by BH4 to produce a transient BH4$^{\bullet+}$ radical is required for the oxidation of arginine to citrulline and the associated formation of a ferrous iron–NO complex at the NOS heme catalytic center [48–51]. BH4 is synthesized from guanosine triphosphate (GTP) by GTP cyclohydrolase I (GTPCH) and recycled from 7,8-dihydrobiopterin (BH2) by dihydrofolate reductase (Figure 2). Of note, NOS is inhibited by arginine analogs that are substituted at the guanidino nitrogen atom, like NG-monomethyl-arginine or NG-nitro-arginine [52–58].

As mentioned above, in the urea cycle arginine is converted by arginase, a manganese metalloenzyme, in ornithine and urea; this cycle is crucial not only for allowing urea excretion, but also for producing bicarbonate, which is critical for maintaining acid/base homeostasis [59–63]. Arginase exists in two distinct isoforms, arginase I and II, that share ~60% sequence homology; arginase I is a cytosolic enzyme mainly localized in the liver, whereas arginase II is a mitochondrial enzyme with a wide distribution and is expressed in the kidney, prostate, gastrointestinal tract, and the vasculature [64–67].

The enzyme arginase is a key modulator of NO production by competing for arginine: in other words, NO generation is dependent on the relative expression and activities of arginase and NOS. More specifically, increased arginase activity may lead to a decreased bioavailability of arginine for NOS, thereby diminishing NO production. This mechanism has emerged as an essential factor underlying impaired endothelial functions [68,69]. Specifically, an increased arginase activity has been associated with endothelial dysfunction in a number of experimental models of hypertension, atherosclerosis, diabetes, and aging [70–92].

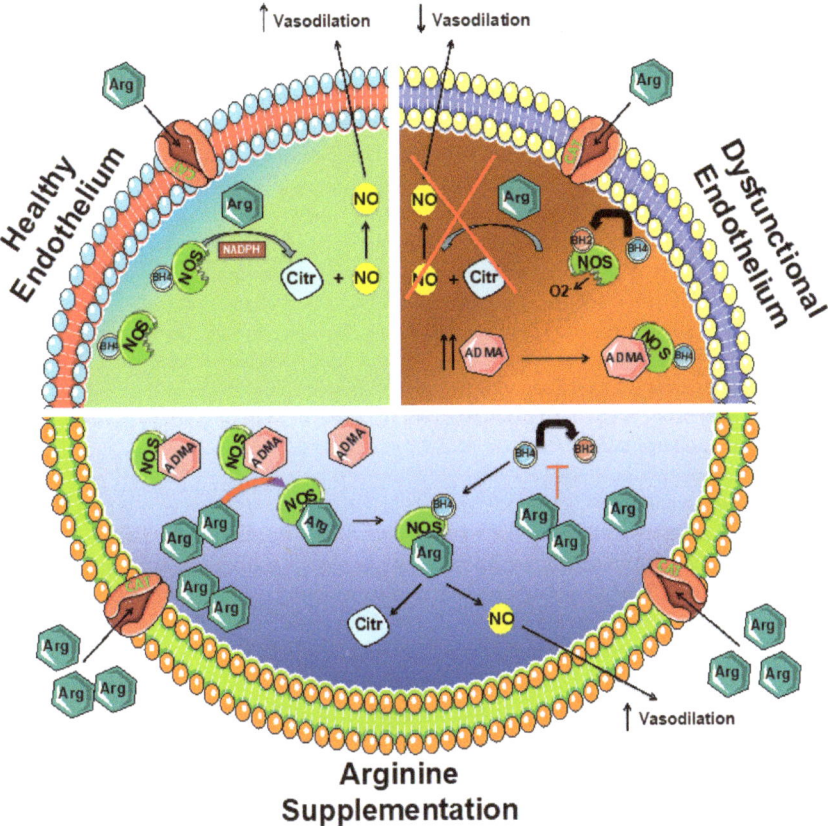

Figure 2. Functional role of arginine in endothelial (dys)function. ADMA: asymmetric dimethylarginine; Arg: arginine; BH2: 7,8-dihydrobiopterin; BH4: tetrahydrobiopterin; CAT: cationic amino acid transporter; Citr: citrulline; NADPH: nicotinamide adenine dinucleotide phosphate; NO: nitric oxide; NOS: NO synthase.

2. Arginine and NO Production in Physiological Conditions: The Arginine Paradox

Indeed, endothelial dysfunction is a leading cause of several pathological conditions affecting the cardiovascular system, including hypertension, atherosclerosis, diabetes, and atherothrombosis [46,93–119]. Moreover, in April 2020, we were the first group to show that the systemic manifestations observed in coronavirus disease (COVID-19), caused by the severe acute respiratory syndrome coronavirus 2 (SARS-CoV-2), could be explained by endothelial dysfunction [120]. Indeed, alterations in endothelial function have been linked to hypertension, diabetes, thromboembolism, and kidney failure, all featured, to different extents, in COVID-19 patients [121–123]. Other investigators have later confirmed our view [124–133]. On these grounds, based on the positive effects of arginine on endothelial function, we can also speculate that arginine supplementation could be helpful, while not being harmful, for contrasting endothelial dysfunction in COVID-19 patients.

An increasing interest in the potential therapeutic effects of arginine supplementation, especially in cardiovascular disorders, has recently emerged. An impaired NO synthesis is considered a main feature of a dysfunctional endothelium [107,134–136]; however, several studies suggest that arginine supplementation in healthy subjects does not lead to a significant increase in NO production [11,137–140]. For instance, the daily administration of arginine for 1 week did not affect the serum concentration of two established indicators of NO production, namely NO_2^- and NO_3^-, in twelve healthy subjects [138]. In another study, 20 healthy subjects received daily arginine supplementation in both sustained-release or immediate-release form; despite the significant increase in the plasma arginine concentration, which proved the effectiveness of the administration protocol, the authors did not observe significant differences in urinary extraction of nitrate [141].

One reason for the absence of significant results in normal conditions could be that the NO synthesis machinery seems to be saturated by the endogenous arginine. Indeed, the Michaelis–Menten constant (K_m) of NO synthase is in the micromolar range, specifically 2.9 µmol/L, as demonstrated by Bredt and colleagues [142]. Arginine plasma levels measured in healthy humans are 15–30-fold higher than this K_m, thereby making the levels of the substrate a non-limiting factor in the enzymatic reaction leading to NO production. Despite such a biochemical ratio, which in fact makes the enzyme physiologically saturated, various studies are also showing beneficial effects of arginine supplementation in healthy subjects. For instance, arginine supplementation has been tested in athletes, as vasodilation favors muscle perfusion and nutrient/oxygen delivery during exercise, enhancing muscle strength and recovery [143]. Controversial results come from these studies, sometimes yielding no effects of arginine supplementation on muscle performance, and sometimes demonstrating a significant improvement in exercise capability [137,144–148].

The phenomenon known as "arginine paradox" is born from this scenario, and indicates that we were losing part of the story concerning the alternative ways by which arginine can act on endothelial NO production. The arginine paradox refers to the fact that despite intracellular physiological concentrations of arginine being several hundred micromoles per liter, thereby exceeding the K_m of eNOS, the acute provision of exogenous arginine still increases NO production [149–151].

One of the mechanisms that may help explain the arginine paradox comes from the discovery of asymmetric dimethylarginine (ADMA), an endogenous inhibitor of NOS [152–155]. Given its own structure similar to arginine, ADMA is a direct competitor for NOS binding. Moreover, both ADMA and arginine are both transported into the cell via the cationic amino acid transporter (CAT, also known as "y+ system"), a high-affinity, Na^+-independent transporter of the basic amino acids [156,157], and therefore also compete with each other on this level (Figure 2). Since ADMA competes with arginine for NOS and for cell transport, the bioavailability of NO depends on the balance between the two [158]. Plasma levels of ADMA increase during hypertension, hypercholesterolemia, diabetes, and atherosclerosis [95,159–170]. Hence, despite the range of endogenous arginine levels, they could still be sufficient to guarantee eNOS saturation, and so the arginine/ADMA ratio would be reduced, resulting in a net inhibition of NO production [171–173].

The arginine/ADMA ratio is widely considered to be an important indicator of NO bioavailability as well as of the risk of formation of atherosclerotic plaques [174]. The ratio has been shown to be a better predictor for all-cause mortality compared to ADMA alone [174,175]. Similarly, although plasma ADMA levels were a significant predictor of all-cause mortality in an elderly population, the effect disappeared in subjects with higher arginine levels [176], and the arginine/ADMA ratio (but not ADMA alone) was a significant risk factor for microangiopathy-related cerebral damage in an elderly population [177].

Arginine supplementation can equilibrate the arginine/ADMA ratio, recovering the production of NO. In other terms, the increased arginine availability, resulting from supplementation, competes with ADMA in binding eNOS (Figure 2). This interesting mechanism sheds light on the effectiveness of the increased arginine availability, implicating further therapeutic options for arginine supplementation. Furthermore, this phenomenon can explain some conflicting results about arginine supplementation studies, as ADMA levels should be considered in the study populations. Specifically, cardiovascular patients with increased ADMA plasma levels could be the best target of arginine supplementation.

Another potential explanation of the arginine paradox may be that arginine could be compartmentalized in the cytoplasm, and local concentrations in the vicinity of NOS may be lower than expected from arginine levels in whole-cell homogenates [178].

3. Impaired NO Production as a Mechanism of Endothelial Dysfunction and Arginine Intervention

The major determinants of cardiovascular risk, including dyslipidemia, glucose intolerance, smoking, hypercholesterolemia, and aging, have a direct impact on the endothelium [179–181]. Exposing the vasculature to these conditions induces endothelial dysfunction and alterations as an early phenomenon, able to evolve and contribute to the progression towards clinically relevant disorders like hypertension, atherosclerosis, and diabetes mellitus. Hence, the endothelium plays a key role in cardiovascular physiology and pathophysiology [182–194]. Fervent research has been conducted in recent years in order to understand the underlying mechanisms and identify therapeutic strategies to prevent or counteract endothelial dysfunction.

The ability of the endothelium to regulate vascular homeostasis is largely dependent on NO production, making endothelial vasodilator failure the main sign of endothelial dysfunction and a hot point to be targeted. The impaired endothelial NO availability in perturbed vasculature can be attributable to a diminished synthesis of NO or, indirectly, to an increased ROS production, which inactivates the NO source [195,196]. In addition to counteracting oxidative stress, the stimulation of NO synthesis represents an alternative and a potentially effective approach [197,198], for instance, by providing further substrates to NO synthase. Theoretically, arginine supplementation meets these needs, and thus, it has been tested in many cardiovascular disorders as a potential therapeutic strategy [199]. However, human studies on arginine supplementation have often been a source of debate. Indeed, in healthy subjects as well as in patients suffering from cardiovascular disorders, levels of plasma arginine range from ~45 to ~100 μmol/L [137,200–202], significantly higher than the eNOS K_m of 2.9 μmol/L [203]. Endocrine mechanisms may also contribute to vasodilation induced by arginine. Indeed, arginine stimulates the release of both insulin [204–206] and glucagon [207] from pancreatic islets of Langerhans. Interestingly, an intravenous infusion of arginine has been shown to induce vasodilation and insulin release in healthy humans, but when insulin secretion was blocked by octreotide co-infusion, no vasodilation occurred, whereas vasodilation was restored by insulin co-administration [208]. Since high intravenous doses of arginine (30 g) have also been shown to induce growth hormones (GHs), and secretion [209], the vasodilation induced by arginine could also be mediated by GHs via a signaling pathway that includes insulin-like growth factor-1 [210,211].

Substantial data indicate that endothelial dysfunction is highly prevalent in elderly individuals [212,213]. Endothelial dysfunction has also been implicated in age-associated declines in cognitive function, physical function, as well as in the pathogenesis of stroke, erectile dysfunction, and renal dysfunction. Clinical trials testing the effects of arginine in aging-induced endothelial dysfunction have yielded controversial results. An acute intravenous infusion of arginine (1 g/min for 30 min) had no effect on endothelial-dependent vasodilation in healthy older individuals [214]. Similarly, the intravenous infusion of arginine induced a significant increase in the renal plasma flow, glomerular filtration rate, natriuresis, and kaliuresis, in young but not in aged hypertensives [215]. Another study conducted in healthy postmenopausal women taking 9 g of arginine per day for 1 month confirmed that plasma arginine increased without a concomitant significant change in flow-mediated dilation [216]. On the contrary, in a prospective, double-blind, randomized crossover trial in 12 healthy, old participants (age 73.8 ± 2.7 years), chronic arginine supplementation (16 g/day for 2 weeks) markedly increased their plasma levels of arginine (114.9 ± 11.6 vs. 57.4 ± 5.0 mM) and significantly improved endothelial-dependent vasodilation [217].

4. Arginine Supplementation in Hypertension

The majority of studies in animal models supports a beneficial effect of arginine supplementation in hypertension, especially in the presence of salt-sensitive hypertension. For instance, both oral [218–220] and intraperitoneal [221,222] arginine administration in Dahl salt-sensitive (DSS) rats was shown to prevent the increase in blood pressure induced by a high salt diet. However, arginine was not effective in DSS pretreated with high salt for three weeks [218], suggesting that arginine is able to prevent and counteract hypertension when it is in the early stages, but probably not when some changes and pathological remodeling have already occurred.

The outcome of arginine supplementation could also depend on the method of administration. For instance, renal medullary interstitial infusion of arginine prevents the increase in blood pressure in high salt-treated rats, while the intravenous dose necessary to obtain a similar increase in plasma arginine does not affect blood pressure [223]. A rat model of type 1 diabetes mellitus shows an important reduction in blood pressure after 4 weeks of oral arginine treatment [224]; oral arginine administration prevents fructose-induced hypertension [225]. Oral arginine administration does not correct hypertension in spontaneously hypertensive rats, although markedly reduces renal damage [226].

Although the beneficial effect of arginine supplementation in hypertension appears to be largely attributable to its impact on NO synthesis, arginine has also been shown to have antioxidant properties, thus affecting the activity of redox-sensitive proteins and lowering blood pressure [227–234]. Indeed, supplementation with 3 g/day arginine for two months increases the serum total antioxidant capacity in obese patients with prediabetes [235]; of note, in vitro experiments performed in endothelial cells have revealed that arginine reduces superoxide release and the cell-mediated breakdown of NO [236].

In the clinical scenario, the oral administration of arginine acutely improves endothelium-dependent, flow-mediated dilatation of the brachial artery in patients with essential hypertension [237]; however, the long-term effects of arginine were not investigated in this study [237]. In a Japanese population, the acute intravenous infusion of arginine (500 mg/kg for 30 min) is able to decrease arterial pressure of both salt-sensitive and salt-insensitive patients [238]. In a similar study, conducted on African-Americans, the same amount of arginine administration reduces blood pressure with a greater effect in the salt-sensitive population [239]. Interestingly, in hypertensive patients in which the control of blood pressure with angiotensin converting enzyme (ACE)-inhibitors and diuretics for three months was unsuccessful, the addition of oral arginine (6 g/day) was effective in reducing both systolic and diastolic blood pressure levels [240]. Unfortunately, many of the findings on the effects of arginine supplementation in hypertension derive from small clinical studies and, despite the promising efficacy, further investigations are needed, especially large, randomized, and controlled trials. The ability to modulate the renin-angiotensin-aldosterone system (RAAS) is another mechanism by which arginine can regulate blood pressure: specifically, arginine inhibits ACE activity, reducing angiotensin II production and its effects on vascular tone [241].

5. Arginine Supplementation in Ischemic Heart Disease and Peripheral Artery Disease

Alongside the preservation of endothelial-dependent vasodilation, the enhanced bioavailability of NO reduces the activation of pro-inflammatory genes and the expression of endothelial adhesion molecules [242]. These events strongly regulate the development and the fate of atherosclerosis [243–245]. For these reasons, it is not surprising that arginine has a powerful effect on atherogenesis and its evolution. In particular, preclinical investigations have shown that chronic arginine administration in LDL-receptor KO mice significantly reduces the extension of atherosclerotic plaques [246]. Similarly, arginine supplementation in humans reverses the increased monocyte–endothelial adhesion, mirrored by a normalization of platelet aggregation [247]. These effects make arginine a promising drug for disorders like coronary artery disease (CAD), heart failure, and peripheral artery disease (PAD).

In 1997, two important studies investigating the effects of arginine in CAD were published [248,249]. In a placebo-controlled study, Adams and collaborators showed that oral administration of arginine (21 g/day for 3 days) significantly improved the vasodilatory response of the brachial artery in premature CAD [248]. A double-blind placebo-controlled study conducted on 22 patients with stable angina pectoris revealed that the administration of arginine was able to improve their exercise capacity in just 3 days [249]. The following year, a clinical study confirmed the beneficial effects of long-term arginine supplementation (9 g for 6 months), showing significantly enhanced vascular responses to acetylcholine in patients with coronary atherosclerosis [250]. Preclinical studies were consistent with these findings. For instance, oral administration of arginine reduced the intimal hyperplasia in balloon-injured carotid arteries in spontaneously hypertensive rats [251]. This first encouraging evidence prompted further investigations about arginine's effects on CAD. Again, arginine treatment for 4 weeks preserved endothelial function in CAD patients, markedly reducing LDL oxidation [252]. Another study highlighted the method of administration as a major determinant of the efficacy of high dose arginine supplementation: intra-arterial infusion, but not oral administration, was able to improve endothelial-dependent vasodilation in patients with stable angina pectoris [253].

The therapeutic potential of arginine has been also investigated in heart failure [254–258] and ischemia-reperfusion injury [259–261], often yielding controversial results. Endothelium-dependent vasodilation in response to acetylcholine and ischemic vasodilation during reactive hyperemia is attenuated in the forearm of patients with heart failure [262]. In a seminal paper, Hirooka and collaborators demonstrated that the intra-arterial infusion of arginine was effective in reversing the blunted endothelium-dependent vasodilation observed in heart failure [263]. Moreover, oral arginine supplementation (6 g twice a day for 6 weeks) enhanced endurance exercise tolerance in heart failure patients, an important determinant of daily-life activity in patients with chronic stable heart failure [264]. In line with these results, a clinical study carried out in 21 patients with class II/III heart failure (New York Heart Association, NYHA) established that improved endothelial function following exercise training is associated with increased arginine transport [265]. However, another investigation in 20 patients with NYHA class III/IV heart failure demonstrated that responses to acetylcholine and sodium nitroprusside determined using forearm plethysmography were not affected by arginine (20 g/day every day for 28 days), although the actual levels of arginine in the blood were not measured [266]. Exogenous arginine (3 g three times a day for 6 months) administered to patients after an acute myocardial infarction did not improve vascular stiffness measurements or ejection fractions; this clinical trial had to be interrupted due to excess mortality in the treated patients [267].

The improvement in peripheral circulation is critical in patients with PAD, as in severe cases the extensive damage of leg tissues can result in gangrene and amputation [268–270]. Intravenous arginine administration to PAD patients is able to increase the calf blood flow and walking distance [271]. Similarly, an acute intravenous arginine infusion (30 g in 60 min) improves NO production and blood flow of the femoral artery in PAD patients [272]. The oral consumption of arginine for 2 weeks is able to increase the pain-free walking distance, improving the quality of life of patients with hypercholesterolemia [273]. Nevertheless, if the short-term arginine administration seems to be effective in treating PAD, the results on long-term administration are less consistent. A randomized clinical trial testing the long-term (6 months) effects of arginine supplementation was conducted on 133 subjects. Despite an increase in plasma levels of arginine, the study revealed no significant effect of arginine treatment on NO-dependent vasodilation, as well as on the relative functional phenotype of PAD patients [274].

6. Arginine Supplementation in Diabetes Mellitus

Given the fundamental pathogenic role of endothelial dysfunction in diabetes and its complications [275,276], the therapeutic use of arginine supplementation has been tested. In addition to the direct impact of arginine on endothelial vasodilator capacity, a crosstalk with the insulin pathway has been suggested [150,277]. In particular, as mentioned above, arginine can induce the release of insulin from pancreatic beta cells [204–206]. On the other hand, insulin is able to reduce ADMA concentrations [278] and to stimulate the secretion of arginine [279,280]. The stimulation of insulin receptors induces NO release, producing an insulin-dependent vasodilation [281–285]. Of note, such a protective effect of insulin on arginine mobility and endothelial NO production is compromised in diabetes [286]. Henceforth, diabetic patients could be an optimal target population for arginine supplementation.

Preclinical studies corroborate this theory: in diabetic rats, the oral administration of arginine reverses endothelial dysfunction [287], restoring endothelium-dependent relaxation and decreasing oxidative stress [224]. Arginine administration in tap water (free base, 50 mg/kg/day) for 4 months has been shown to reduce both cardiac [288] and renal [289] fibrosis in *db/db* mice, by the interaction of arginine with reactive carbonyl residues of glycosylation adducts of collagen, thereby inhibiting glucose-mediated abnormal cross-linking of collagenous structures. These results were later confirmed in a clinical setting, showing that 2 g of arginine free base administered orally as two daily doses of 1 g each reduced the lipid peroxidation product malondialdehyde in diabetic patients [290].

Clinical studies confirmed the reduction in blood pressure, platelet aggregation, and hemodynamic function in diabetic patients treated with intravenous arginine [291]. While in healthy subjects arginine treatment does not seem to affect insulin receptor sensitivity or density [292], in conditions of insulin resistance, arginine improves insulin sensitivity; indeed, the intravenous injection of arginine in obese or type 2 diabetic patients stimulates insulin responsiveness, restoring insulin-dependent vasodilation [151,293]. Similarly, the oral administration of arginine improves hepatic and peripheral insulin sensitivity in a cGMP dependent fashion [294]. A prospective, crossover clinical trial conducted in mildly hypertensive type 2 diabetic patients revealed a significant decrease in blood pressure in response to arginine, occurring two hours after the oral administration; the effect of lowering blood pressure was associated with increased plasma levels of citrulline, whereas no significant changes in insulin levels were detected, suggesting that the observed phenotype was dependent on arginine-induced NO synthesis [295].

Overall, the mentioned studies substantiate the use of arginine in the diabetic population, at least as a prophylactic treatment able to prevent cardiovascular complications of diabetes. One potential limitation for the use of arginine is the risk of reaction with precursors of advanced glycosylated products [296], which are particularly abundant in diabetes. Since the addition of methylglyoxal (abundant in diabetic patients [297]) to arginine has been shown in vitro to produce potent superoxide radicals in a dose-dependent manner [298], arginine supplementation has been suggested to be combined with antioxidants. A double-blind study on 24 diabetic patients verified this assumption evaluating the combination of *N*-acetylcysteine and arginine oral treatments: the combined treatment was able to reduce systolic and diastolic blood pressure, total cholesterol, C-reactive proteins, vascular adhesion molecules, and improved the intima-media thickness during endothelial post-ischemic vasodilation [299]. This last evidence indicates that the combination of arginine with an antioxidant agent should be potentially effective and well-tolerated.

7. Conclusions and Perspective: Arginine as a Therapeutic Tool

Overall, data available in the literature support and encourage the use of arginine supplementation in cardiovascular disorders, especially in preventing the evolution of hypertension and atherosclerosis. One limitation of using arginine supplementation remains the selection of the optimal target population. In this sense, we believe that ADMA levels could be very useful in selecting the target population, and patients with increased ADMA/arginine ratios are probably the most suitable population, in which arginine supplementation can actually be effective. Another limitation about arginine use concerns its dose. Indeed, available studies suggest a number of different doses, sometimes effective, sometimes not. For instance, the acute oral administration of arginine (9 g/day) has been shown to be not successful in inducing an effective NO production [216]. Instead, chronic administration of oral arginine (e.g., vials containing arginine salts-free 1.66 g/20 mL), has been shown to favor the utilization of arginine for NO synthesis [300], and we have data showing that oral arginine (3 g/day of Bioarginina®, Farmaceutici Damor, 2 vials/day) improves endothelial function in hypertensive patients via the regulation of non-coding RNAs (Gambardella et al., personal communication). Large, prospective randomized clinical trials are needed to better define the target population for arginine supplementation, alongside with correct dosage definitions. To date, a dose of ~3 g/day of arginine (e.g., Bioarginina®, 2 vials/day) seems to be effective in favoring the utilization of arginine for NO synthesis, without toxic effects.

Author Contributions: Conceptualization, G.S.; data curation, writing—original draft preparation, J.G., W.K., M.B.M., X.W. and G.S.; writing—review and editing, J.G., X.W.; G.S. and V.T.; supervision, G.S.; funding acquisition, G.S. and J.G. All authors have read and agreed to the published version of the manuscript.

Funding: The Santulli's lab is supported in part by the NIH (R01-DK123259, R01-HL146691, R01-DK033823, and R00-DK107895 to G.S.) and by the American Heart Association (AHA-20POST35211151 to J.G.).

Conflicts of Interest: The authors declare no conflict of interest. The funders had no role in the design of the study; in the collection, analyses, or interpretation of data; in the writing of the manuscript, or in the decision to publish the paper.

References

1. Lopez, M.J.; Mohiuddin, S.S. Biochemistry, Essential Amino Acids. In *StatPearls*; StatPearls Publishing: Treasure Island, FL, USA, 2020.
2. Luiking, Y.C.; Ten Have, G.A.; Wolfe, R.R.; Deutz, N.E. Arginine de novo and nitric oxide production in disease states. *Am. J. Physiol. Endocrinol. Metab.* **2012**, *303*, E1177–E1189. [CrossRef] [PubMed]
3. Agapie, T.; Suseno, S.; Woodward, J.J.; Stoll, S.; Britt, R.D.; Marletta, M.A. NO formation by a catalytically self-sufficient bacterial nitric oxide synthase from Sorangium cellulosum. *Proc. Natl. Acad. Sci. USA* **2009**, *106*, 16221–16226. [CrossRef] [PubMed]
4. Jachymova, M.; Martasek, P.; Panda, S.; Roman, L.J.; Panda, M.; Shea, T.M.; Ishimura, Y.; Kim, J.J. Masters, B.S. Recruitment of governing elements for electron transfer in the nitric oxide synthase family. *Proc. Natl. Acad. Sci. USA* **2005**, *102*, 15833–15838. [CrossRef] [PubMed]
5. Serpe, M.J.; Zhang, X. The Principles, Development and Application of Microelectrodes for the In Vivo Determination of Nitric Oxide. In *Electrochemical Methods for Neuroscience*; Michael, A.C., Borland, L.M., Eds.; CRC Press/Taylor & Francis: Boca Raton, FL, USA, 2007.
6. Andrew, P.J.; Mayer, B. Enzymatic function of nitric oxide synthases. *Cardiovasc. Res.* **1999**, *43*, 521–531. [CrossRef]
7. Sax, H.C. Arginine stimulates wound healing and immune function in elderly human beings. *JPEN J. Parenter. Enteral. Nutr.* **1994**, *18*, 559–560. [CrossRef]
8. Barbul, A.; Lazarou, S.A.; Efron, D.T.; Wasserkrug, H.L.; Efron, G. Arginine enhances wound healing and lymphocyte immune responses in humans. *Surgery* **1990**, *108*, 331–336.
9. Durante, W.; Liao, L.; Reyna, S.V.; Peyton, K.J.; Schafer, A.I. Physiological cyclic stretch directs L-arginine transport and metabolism to collagen synthesis in vascular smooth muscle. *FASEB J.* **2000**, *14*, 1775–1783. [CrossRef]
10. Da Silva, R.P.; Nissim, I.; Brosnan, M.E.; Brosnan, J.T. Creatine synthesis: Hepatic metabolism of guanidinoacetate and creatine in the rat in vitro and in vivo. *Am. J. Physiol. Endocrinol. Metab.* **2009**, *296*, E256–E261. [CrossRef]
11. Bai, Y.; Sun, L.; Yang, T.; Sun, K.; Chen, J.; Hui, R. Increase in fasting vascular endothelial function after short-term oral L-arginine is effective when baseline flow-mediated dilation is low: A meta-analysis of randomized controlled trials. *Am. J. Clin. Nutr.* **2009**, *89*, 77–84. [CrossRef]
12. Raman, C.S.; Li, H.; Martasek, P.; Kral, V.; Masters, B.S.; Poulos, T.L. Crystal structure of constitutive endothelial nitric oxide synthase: A paradigm for pterin function involving a novel metal center. *Cell* **1998**, *95*, 939–950. [CrossRef]
13. Garthwaite, J.; Charles, S.L.; Chess-Williams, R. Endothelium-derived relaxing factor release on activation of NMDA receptors suggests role as intercellular messenger in the brain. *Nature* **1988**, *336*, 385–388. [CrossRef] [PubMed]
14. Bec, N.; Gorren, A.C.; Voelker, C.; Mayer, B.; Lange, R. Reaction of neuronal nitric-oxide synthase with oxygen at low temperature. Evidence for reductive activation of the oxy-ferrous complex by tetrahydrobiopterin. *J. Biol. Chem.* **1998**, *273*, 13502–13508. [CrossRef]
15. Iwai, N.; Hanai, K.; Tooyama, I.; Kitamura, Y.; Kinoshita, M. Regulation of neuronal nitric oxide synthase in rat adrenal medulla. *Hypertension* **1995**, *25*, 431–436. [CrossRef] [PubMed]
16. O'Dell, T.J.; Huang, P.L.; Dawson, T.M.; Dinerman, J.L.; Snyder, S.H.; Kandel, E.R.; Fishman, M.C. Endothelial NOS and the blockade of LTP by NOS inhibitors in mice lacking neuronal NOS. *Science* **1994**, *265*, 542–546. [CrossRef] [PubMed]

17. Geller, D.A.; Lowenstein, C.J.; Shapiro, R.A.; Nussler, A.K.; Di Silvio, M.; Wang, S.C.; Nakayama, D.K.; Simmons, R.L.; Snyder, S.H.; Billiar, T.R. Molecular cloning and expression of inducible nitric oxide synthase from human hepatocytes. *Proc. Natl. Acad. Sci. USA* **1993**, *90*, 3491–3495. [CrossRef]
18. Geller, D.A.; Nussler, A.K.; Di Silvio, M.; Lowenstein, C.J.; Shapiro, R.A.; Wang, S.C.; Simmons, R.L.; Billiar, T.R. Cytokines, endotoxin, and glucocorticoids regulate the expression of inducible nitric oxide synthase in hepatocytes. *Proc. Natl. Acad. Sci. USA* **1993**, *90*, 522–526. [CrossRef]
19. Hokari, A.; Zeniya, M.; Esumi, H. Cloning and functional expression of human inducible nitric oxide synthase (NOS) cDNA from a glioblastoma cell line A-172. *J. Biochem.* **1994**, *116*, 575–581. [CrossRef]
20. Forstermann, U.; Sessa, W.C. Nitric oxide synthases: Regulation and function. *Eur. Heart J.* **2012**, *33*, 829–837, 837a–837d. [CrossRef]
21. Rafikov, R.; Fonseca, F.V.; Kumar, S.; Pardo, D.; Darragh, C.; Elms, S.; Fulton, D.; Black, S.M. eNOS activation and NO function: Structural motifs responsible for the posttranslational control of endothelial nitric oxide synthase activity. *J. Endocrinol.* **2011**, *210*, 271–284. [CrossRef]
22. Alderton, W.K.; Cooper, C.E.; Knowles, R.G. Nitric oxide synthases: Structure, function and inhibition. *Biochem. J.* **2001**, *357 Pt 3*, 593–615. [CrossRef]
23. Stuehr, D.J. Enzymes of the L-arginine to nitric oxide pathway. *J. Nutr.* **2004**, *134* (Suppl. 10), 2748S–2751S. [CrossRef] [PubMed]
24. Meulemans, A. Electrochemical detection of nitroso-arginine as an intermediate between N-hydroxy-arginine and citrulline. An in vitro versus in vivo study using microcarbon electrodes in neuronal nitric oxide synthase and mice brain. *Neurosci. Lett.* **2000**, *294*, 125–129. [CrossRef]
25. Tsuboi, T.; Maeda, M.; Hayashi, T. Administration of L-arginine plus L-citrulline or L-citrulline alone successfully retarded endothelial senescence. *PLoS ONE* **2018**, *13*, e0192252. [CrossRef] [PubMed]
26. De Betue, C.T.I.; Garcia Casal, X.C.; van Waardenburg, D.A.; Schexnayder, S.M.; Joosten, K.F.M.; Deutz, N.E.P.; Engelen, M. 24-Hour protein, arginine and citrulline metabolism in fed critically ill children—A stable isotope tracer study. *Clin. Nutr.* **2017**, *36*, 876–887. [CrossRef]
27. Arnett, D.C.; Persechini, A.; Tran, Q.K.; Black, D.J.; Johnson, C.K. Fluorescence quenching studies of structure and dynamics in calmodulin-eNOS complexes. *FEBS Lett.* **2015**, *589*, 1173–1178. [CrossRef]
28. Li, H.; Raman, C.S.; Glaser, C.B.; Blasko, E.; Young, T.A.; Parkinson, J.F.; Whitlow, M.; Poulos, T.L. Crystal structures of zinc-free and -bound heme domain of human inducible nitric-oxide synthase. Implications for dimer stability and comparison with endothelial nitric-oxide synthase. *J. Biol. Chem.* **1999**, *274*, 21276–21284. [CrossRef]
29. Crane, B.R.; Rosenfeld, R.J.; Arvai, A.S.; Ghosh, D.K.; Ghosh, S.; Tainer, J.A.; Stuehr, D.J.; Getzoff, E.D. N-terminal domain swapping and metal ion binding in nitric oxide synthase dimerization. *EMBO J.* **1999**, *18*, 6271–6281. [CrossRef]
30. Daiber, A.; Kroller-Schon, S.; Oelze, M.; Hadad, O.; Li, H.; Schulz, R.; Steven, S.; Munzel, T. Oxidative stress and inflammation contribute to traffic noise-induced vascular and cerebral dysfunction via uncoupling of nitric oxide synthases. *Redox Biol.* **2020**, *34*, 101506. [CrossRef]
31. Jayaram, R.; Goodfellow, N.; Zhang, M.H.; Reilly, S.; Crabtree, M.; De Silva, R.; Sayeed, R.; Casadei, B. Molecular mechanisms of myocardial nitroso-redox imbalance during on-pump cardiac surgery. *Lancet* **2015**, *385* (Suppl. 1), S49. [CrossRef]
32. Siu, K.L.; Lotz, C.; Ping, P.; Cai, H. Netrin-1 abrogates ischemia/reperfusion-induced cardiac mitochondrial dysfunction via nitric oxide-dependent attenuation of NOX4 activation and recoupling of NOS. *J. Mol. Cell Cardiol.* **2015**, *78*, 174–185. [CrossRef]
33. Gebhart, V.; Reiss, K.; Kollau, A.; Mayer, B.; Gorren, A.C.F. Site and mechanism of uncoupling of nitric-oxide synthase: Uncoupling by monomerization and other misconceptions. *Nitric Oxide* **2019**, *89*, 14–21. [CrossRef] [PubMed]
34. Yang, Y.M.; Huang, A.; Kaley, G.; Sun, D. eNOS uncoupling and endothelial dysfunction in aged vessels. *Am. J. Physiol. Heart Circ. Physiol.* **2009**, *297*, H1829–H1836. [CrossRef]
35. Musicki, B.; Burnett, A.L. Constitutive NOS uncoupling and NADPH oxidase upregulation in the penis of type 2 diabetic men with erectile dysfunction. *Andrology* **2017**, *5*, 294–298. [CrossRef] [PubMed]

36. Bohmer, A.; Gambaryan, S.; Flentje, M.; Jordan, J.; Tsikas, D. [Ureido-(1)(5)N]citrulline UPLC-MS/MS nitric oxide synthase (NOS) activity assay: Development, validation, and applications to assess NOS uncoupling and human platelets NOS activity. *J. Chromatogr. B Analyt. Technol. Biomed. Life Sci.* **2014**, *965*, 173–182. [CrossRef] [PubMed]
37. Kietadisorn, R.; Juni, R.P.; Moens, A.L. Tackling endothelial dysfunction by modulating NOS uncoupling: New insights into its pathogenesis and therapeutic possibilities. *Am. J. Physiol. Endocrinol. Metab.* **2012**, *302*, E481–E495. [CrossRef] [PubMed]
38. Xie, L.; Liu, Z.; Lu, H.; Zhang, W.; Mi, Q.; Li, X.; Tang, Y.; Chen, Q.; Ferro, A.; Ji, Y. Pyridoxine inhibits endothelial NOS uncoupling induced by oxidized low-density lipoprotein via the PKCalpha signalling pathway in human umbilical vein endothelial cells. *Br. J. Pharmacol.* **2012**, *165*, 754–764. [CrossRef]
39. Crijns, H.J.; Schotten, U.; Moens, A.L. Is NOS uncoupling the missing link between atrial fibrillation and chronic non-ischaemic cardiomyopathy? *Cardiovasc. Res.* **2011**, *91*, 556. [CrossRef]
40. Moens, A.L.; Leyton-Mange, J.S.; Niu, X.; Yang, R.; Cingolani, O.; Arkenbout, E.K.; Champion, H.C.; Bedja, D.; Gabrielson, K.L.; Chen, J.; et al. Adverse ventricular remodeling and exacerbated NOS uncoupling from pressure-overload in mice lacking the beta3-adrenoreceptor. *J. Mol. Cell Cardiol.* **2009**, *47*, 576–585. [CrossRef]
41. Mollnau, H.; Schulz, E.; Daiber, A.; Baldus, S.; Oelze, M.; August, M.; Wendt, M.; Walter, U.; Geiger, C.; Agrawal, R.; et al. Nebivolol prevents vascular NOS III uncoupling in experimental hyperlipidemia and inhibits NADPH oxidase activity in inflammatory cells. *Arterioscler. Thromb. Vasc. Biol.* **2003**, *23*, 615–621. [CrossRef]
42. Dikalova, A.; Aschner, J.L.; Kaplowitz, M.R.; Cunningham, G.; Summar, M.; Fike, C.D. Combined l-citrulline and tetrahydrobiopterin therapy improves NO signaling and ameliorates chronic hypoxia-induced pulmonary hypertension in newborn pigs. *Am. J. Physiol. Lung Cell Mol. Physiol.* **2020**, *318*, L762–L772. [CrossRef]
43. Picciano, A.L.; Crane, B.R. A nitric oxide synthase-like protein from Synechococcus produces NO/NO3 (-) from l-arginine and NADPH in a tetrahydrobiopterin- and Ca(2+)-dependent manner. *J. Biol. Chem.* **2019**, *294*, 10708–10719. [CrossRef]
44. Nagarkoti, S.; Sadaf, S.; Awasthi, D.; Chandra, T.; Jagavelu, K.; Kumar, S.; Dikshit, M. L-Arginine and tetrahydrobiopterin supported nitric oxide production is crucial for the microbicidal activity of neutrophils. *Free Radic. Res.* **2019**, *53*, 281–292. [CrossRef] [PubMed]
45. Latini, A.; de Bortoli da Silva, L.; da Luz Scheffer, D.; Pires, A.C.S.; de Matos, F.J.; Nesi, R.T.; Ghisoni, K.; de Paula Martins, R.; de Oliveira, P.A.; Prediger, R.D.; et al. Tetrahydrobiopterin improves hippocampal nitric oxide-linked long-term memory. *Mol. Genet. Metab.* **2018**, *125*, 104–111. [CrossRef] [PubMed]
46. Chuaiphichai, S.; Rashbrook, V.S.; Hale, A.B.; Trelfa, L.; Patel, J.; McNeill, E.; Lygate, C.A.; Channon, K.M.; Douglas, G. Endothelial Cell Tetrahydrobiopterin Modulates Sensitivity to Ang (Angiotensin) II-Induced Vascular Remodeling, Blood Pressure, and Abdominal Aortic Aneurysm. *Hypertension* **2018**, *72*, 128–138. [CrossRef] [PubMed]
47. Ramasamy, S.; Haque, M.M.; Gangoda, M.; Stuehr, D.J. Tetrahydrobiopterin redox cycling in nitric oxide synthase: Evidence supports a through-heme electron delivery. *FEBS J.* **2016**, *283*, 4491–4501. [CrossRef] [PubMed]
48. Stuehr, D.J.; Kwon, N.S.; Nathan, C.F.; Griffith, O.W.; Feldman, P.L.; Wiseman, J. N omega-hydroxy-L-arginine is an intermediate in the biosynthesis of nitric oxide from L-arginine. *J. Biol. Chem.* **1991**, *266*, 6259–6263.
49. Vasquez-Vivar, J.; Kalyanaraman, B.; Martasek, P.; Hogg, N.; Masters, B.S.; Karoui, H.; Tordo, P.; Pritchard, K.A., Jr. Superoxide generation by endothelial nitric oxide synthase: The influence of cofactors. *Proc. Natl. Acad. Sci. USA* **1998**, *95*, 9220–9225. [CrossRef] [PubMed]
50. Xia, Y.; Tsai, A.L.; Berka, V.; Zweier, J.L. Superoxide generation from endothelial nitric-oxide synthase. A Ca2+/calmodulin-dependent and tetrahydrobiopterin regulatory process. *J. Biol. Chem.* **1998**, *273*, 25804–25808. [CrossRef] [PubMed]
51. Landmesser, U.; Dikalov, S.; Price, S.R.; McCann, L.; Fukai, T.; Holland, S.M.; Mitch, W.E.; Harrison, D.G. Oxidation of tetrahydrobiopterin leads to uncoupling of endothelial cell nitric oxide synthase in hypertension. *J. Clin. Invest.* **2003**, *111*, 1201–1209. [CrossRef] [PubMed]
52. Moore, P.K.; al-Swayeh, O.A.; Chong, N.W.; Evans, R.A.; Gibson, A. L-NG-nitro arginine (L-NOARG), a novel, L-arginine-reversible inhibitor of endothelium-dependent vasodilatation in vitro. *Br. J. Pharmacol.* **1990**, *99*, 408–412. [CrossRef] [PubMed]

53. Gibson, A.; Mirzazadeh, S.; Hobbs, A.J.; Moore, P.K. L-NG-monomethyl arginine and L-NG-nitro arginine inhibit non-adrenergic, non-cholinergic relaxation of the mouse anococcygeus muscle. *Br. J. Pharmacol.* **1990**, *99*, 602–606. [CrossRef] [PubMed]
54. O'Kane, K.P.; Webb, D.J.; Collier, J.G.; Vallance, P.J. Local L-NG-monomethyl-arginine attenuates the vasodilator action of bradykinin in the human forearm. *Br. J. Clin. Pharmacol.* **1994**, *38*, 311–315. [CrossRef] [PubMed]
55. Fukuda, N.; Izumi, Y.; Soma, M.; Watanabe, Y.; Watanabe, M.; Hatano, M.; Sakuma, I.; Yasuda, H. L-NG-monomethyl arginine inhibits the vasodilating effects of low dose of endothelin-3 on rat mesenteric arteries. *Biochem. Biophys. Res. Commun.* **1990**, *167*, 739–745. [CrossRef]
56. Gaw, A.J.; Aberdeen, J.; Humphrey, P.P.; Wadsworth, R.M.; Burnstock, G. Relaxation of sheep cerebral arteries by vasoactive intestinal polypeptide and neurogenic stimulation: Inhibition by L-NG-monomethyl arginine in endothelium-denuded vessels. *Br. J. Pharmacol.* **1991**, *102*, 567–572. [CrossRef] [PubMed]
57. Toda, N.; Minami, Y.; Okamura, T. Inhibitory effects of L-NG-nitro-arginine on the synthesis of EDRF and the cerebroarterial response to vasodilator nerve stimulation. *Life Sci.* **1990**, *47*, 345–351. [CrossRef]
58. Cozzi, M.R.; Guglielmini, G.; Battiston, M.; Momi, S.; Lombardi, E.; Miller, E.C.; De Zanet, D.; Mazzucato, M.; Gresele, P.; De Marco, L. Visualization of nitric oxide production by individual platelets during adhesion in flowing blood. *Blood* **2015**, *125*, 697–705. [CrossRef]
59. Haussinger, D.; Gerok, W.; Sies, H. The effect of urea synthesis on extracellular pH in isolated perfused rat liver. *Biochem. J.* **1986**, *236*, 261–265. [CrossRef] [PubMed]
60. Mavri-Damelin, D.; Eaton, S.; Damelin, L.H.; Rees, M.; Hodgson, H.J.; Selden, C. Ornithine transcarbamylase and arginase I deficiency are responsible for diminished urea cycle function in the human hepatoblastoma cell line HepG2. *Int. J. Biochem. Cell Biol.* **2007**, *39*, 555–564. [CrossRef]
61. Callery, E.M.; Elinson, R.P. Developmental regulation of the urea-cycle enzyme arginase in the direct developing frog Eleutherodactylus coqui. *J. Exp. Zool.* **1996**, *275*, 61–66. [CrossRef]
62. Snellman, K.; Aperia, A.; Broberger, O. Studies of renal urea cycle enzymes. II. Human renal arginase activity and location of the adaptive changes of renal arginase in the protein deprived rat. *Scand. J. Clin. Lab. Invest.* **1979**, *39*, 337–342. [CrossRef]
63. Chan, P.Y.; Cossins, E.A. Regulation of arginase levels by urea and intermediates of the Krebs-Henseleit cycle in Saccharomyces cerevisiae. *FEBS Lett.* **1972**, *19*, 335–339. [CrossRef]
64. Pernow, J.; Jung, C. Arginase as a potential target in the treatment of cardiovascular disease: Reversal of arginine steal? *Cardiovasc. Res.* **2013**, *98*, 334–343. [CrossRef] [PubMed]
65. Pandey, D.; Romer, L.; Berkowitz, D.E. Arginase II: Atherogenesis beyond enzyme activity. *J. Am. Heart Assoc.* **2013**, *2*, e000392. [CrossRef] [PubMed]
66. Kuhn, N.J.; Ward, S.; Piponski, M.; Young, T.W. Purification of human hepatic arginase and its manganese (II)-dependent and pH-dependent interconversion between active and inactive forms: A possible pH-sensing function of the enzyme on the ornithine cycle. *Arch. Biochem. Biophys.* **1995**, *320*, 24–34. [CrossRef]
67. Sumitani, A. Immunological studies of liver arginase in man and various kinds of vertebrates. Part I: Microquantification of arginase enzyme in liver tissue by quantitative immunoelectrophoresis. Part II: Immunological studies of human liver arginase. *Hiroshima J. Med. Sci.* **1977**, *26*, 59.
68. Kim, J.H.; Bugaj, L.J.; Oh, Y.J.; Bivalacqua, T.J.; Ryoo, S.; Soucy, K.G.; Santhanam, L.; Webb, A.; Camara, A.; Sikka, G.; et al. Arginase inhibition restores NOS coupling and reverses endothelial dysfunction and vascular stiffness in old rats. *J. Appl. Physiol. (1985)* **2009**, *107*, 1249–1257. [CrossRef]
69. Romero, M.J.; Platt, D.H.; Tawfik, H.E.; Labazi, M.; El-Remessy, A.B.; Bartoli, M.; Caldwell, R.B.; Caldwell, R.W. Diabetes-induced coronary vascular dysfunction involves increased arginase activity. *Circ. Res.* **2008**, *102*, 95–102. [CrossRef]
70. Mahdi, A.; Kovamees, O.; Pernow, J. Improvement in endothelial function in cardiovascular disease—Is arginase the target? *Int. J. Cardiol.* **2020**, *301*, 207–214. [CrossRef]
71. Wernly, B.; Pernow, J.; Kelm, M.; Jung, C. The role of arginase in the microcirculation in cardiovascular disease. *Clin. Hemorheol. Microcirc.* **2020**, *74*, 79–92. [CrossRef]
72. Chandrasekharan, U.M.; Wang, Z.; Wu, Y.; Wilson Tang, W.H.; Hazen, S.L.; Wang, S.; Elaine Husni, M. Elevated levels of plasma symmetric dimethylarginine and increased arginase activity as potential indicators of cardiovascular comorbidity in rheumatoid arthritis. *Arthritis Res. Ther.* **2018**, *20*, 123. [CrossRef]

73. Yang, Z.; Ming, X.F. Functions of arginase isoforms in macrophage inflammatory responses: Impact on cardiovascular diseases and metabolic disorders. *Front. Immunol.* **2014**, *5*, 533. [CrossRef] [PubMed]
74. Bagnost, T.; Ma, L.; da Silva, R.F.; Rezakhaniha, R.; Houdayer, C.; Stergiopulos, N.; Andre, C.; Guillaume, Y.; Berthelot, A.; Demougeot, C. Cardiovascular effects of arginase inhibition in spontaneously hypertensive rats with fully developed hypertension. *Cardiovasc. Res.* **2010**, *87*, 569–577. [CrossRef] [PubMed]
75. Huang, J.; Liu, C.; Ming, X.F.; Yang, Z. Inhibition of p38mapk Reduces Adipose Tissue Inflammation in Aging Mediated by Arginase-II. *Pharmacology* **2020**, 1–14. [CrossRef] [PubMed]
76. Masi, S.; Colucci, R.; Duranti, E.; Nannipieri, M.; Anselmino, M.; Ippolito, C.; Tirotta, E.; Georgiopoulos, G.; Garelli, F.; Nericcio, A.; et al. Aging Modulates the Influence of Arginase on Endothelial Dysfunction in Obesity. *Arterioscler. Thromb. Vasc. Biol.* **2018**, *38*, 2474–2483. [CrossRef] [PubMed]
77. Cecilio, C.A.; Costa, E.H.; Simioni, P.U.; Gabriel, D.L.; Tamashiro, W.M. Aging alters the production of iNOS, arginase and cytokines in murine macrophages. *Braz. J. Med. Biol. Res.* **2011**, *44*, 671–681. [CrossRef] [PubMed]
78. Katusic, Z.S. Mechanisms of endothelial dysfunction induced by aging: Role of arginase I. *Circ. Res.* **2007**, *101*, 640–641. [CrossRef]
79. Sakai, Y.; Masuda, H.; Kihara, K.; Kurosaki, E.; Yamauchi, Y.; Azuma, H. Involvement of increased arginase activity in impaired cavernous relaxation with aging in the rabbit. *J. Urol.* **2004**, *172*, 369–373. [CrossRef]
80. Berkowitz, D.E.; White, R.; Li, D.; Minhas, K.M.; Cernetich, A.; Kim, S.; Burke, S.; Shoukas, A.A.; Nyhan, D.; Champion, H.C.; et al. Arginase reciprocally regulates nitric oxide synthase activity and contributes to endothelial dysfunction in aging blood vessels. *Circulation* **2003**, *108*, 2000–2006. [CrossRef]
81. Cheng, H.; Lu, T.; Wang, J.; Xia, Y.; Chai, X.; Zhang, M.; Yao, Y.; Zhou, N.; Zhou, S.; Chen, X.; et al. *HuangqiGuizhiWuwu* Decoction Prevents Vascular Dysfunction in Diabetes via Inhibition of Endothelial Arginase 1. *Front. Physiol* **2020**, *11*, 201. [CrossRef]
82. Folley, S.J.; Greenbaum, A.L. Effect of experimental diabetes on tissue arginase levels. *J. Endocrinol.* **1949**, *6*. [PubMed]
83. Yang, J.; Zheng, X.; Mahdi, A.; Zhou, Z.; Tratsiakovich, Y.; Jiao, T.; Kiss, A.; Kovamees, O.; Alvarsson, M.; Catrina, S.B.; et al. Red Blood Cells in Type 2 Diabetes Impair Cardiac Post-Ischemic Recovery Through an Arginase-Dependent Modulation of Nitric Oxide Synthase and Reactive Oxygen Species. *JACC Basic Transl. Sci.* **2018**, *3*, 450–463. [CrossRef] [PubMed]
84. Zhou, Z.; Mahdi, A.; Tratsiakovich, Y.; Zahoran, S.; Kovamees, O.; Nordin, F.; Uribe Gonzalez, A.E.; Alvarsson, M.; Ostenson, C.G.; Andersson, D.C.; et al. Erythrocytes From Patients With Type 2 Diabetes Induce Endothelial Dysfunction Via Arginase, I.J. *Am. Coll. Cardiol.* **2018**, *72*, 769–780. [CrossRef] [PubMed]
85. Zhang, H.; Liu, J.; Qu, D.; Wang, L.; Wong, C.M.; Lau, C.W.; Huang, Y.; Wang, Y.F.; Huang, H.; Xia, Y.; et al. Serum exosomes mediate delivery of arginase 1 as a novel mechanism for endothelial dysfunction in diabetes. *Proc. Natl. Acad. Sci. USA* **2018**, *115*, E6927–E6936. [CrossRef] [PubMed]
86. Shosha, E.; Xu, Z.; Narayanan, S.P.; Lemtalsi, T.; Fouda, A.Y.; Rojas, M.; Xing, J.; Fulton, D.; Caldwell, R.W.; Caldwell, R.B. Mechanisms of Diabetes-Induced Endothelial Cell Senescence: Role of Arginase 1. *Int. J. Mol. Sci.* **2018**, *19*, 1215. [CrossRef] [PubMed]
87. Kovamees, O.; Shemyakin, A.; Checa, A.; Wheelock, C.E.; Lundberg, J.O.; Ostenson, C.G.; Pernow, J. Arginase Inhibition Improves Microvascular Endothelial Function in Patients with Type 2 Diabetes Mellitus. *J. Clin. Endocrinol. Metab.* **2016**, *101*, 3952–3958. [CrossRef]
88. Yao, L.; Chandra, S.; Toque, H.A.; Bhatta, A.; Rojas, M.; Caldwell, R.B.; Caldwell, R.W. Prevention of diabetes-induced arginase activation and vascular dysfunction by Rho kinase (ROCK) knockout. *Cardiovasc. Res.* **2013**, *97*, 509–519. [CrossRef]
89. Elms, S.C.; Toque, H.A.; Rojas, M.; Xu, Z.; Caldwell, R.W.; Caldwell, R.B. The role of arginase I in diabetes-induced retinal vascular dysfunction in mouse and rat models of diabetes. *Diabetologia* **2013**, *56*, 654–662. [CrossRef]
90. Shemyakin, A.; Kovamees, O.; Rafnsson, A.; Bohm, F.; Svenarud, P.; Settergren, M.; Jung, C.; Pernow, J. Arginase inhibition improves endothelial function in patients with coronary artery disease and type 2 diabetes mellitus. *Circulation* **2012**, *126*, 2943–2950. [CrossRef]
91. Ren, B.; Van Kampen, E.; Van Berkel, T.J.; Cruickshank, S.M.; Van Eck, M. Hematopoietic arginase 1 deficiency results in decreased leukocytosis and increased foam cell formation but does not affect atherosclerosis. *Atherosclerosis* **2017**, *256*, 35–46. [CrossRef]

92. Teupser, D.; Burkhardt, R.; Wilfert, W.; Haffner, I.; Nebendahl, K.; Thiery, J. Identification of macrophage arginase I as a new candidate gene of atherosclerosis resistance. *Arterioscler. Thromb. Vasc. Biol.* **2006**, *26*, 365–371. [CrossRef]
93. Gimbrone, M.A., Jr.; Garcia-Cardena, G. Endothelial Cell Dysfunction and the Pathobiology of Atherosclerosis. *Circ. Res.* **2016**, *118*, 620–636. [CrossRef] [PubMed]
94. Rajendran, P.; Rengarajan, T.; Thangavel, J.; Nishigaki, Y.; Sakthisekaran, D.; Sethi, G.; Nishigaki, I. The vascular endothelium and human diseases. *Int. J. Biol. Sci.* **2013**, *9*, 1057–1069. [CrossRef] [PubMed]
95. Chirinos, J.A.; David, R.; Bralley, J.A.; Zea-Diaz, H.; Munoz-Atahualpa, E.; Corrales-Medina, F.; Cuba-Bustinza, C.; Chirinos-Pacheco, J.; Medina-Lezama, J. Endogenous nitric oxide synthase inhibitors, arterial hemodynamics, and subclinical vascular disease: The Prevencion Study. *Hypertension* **2008**, *52*, 1051–1059. [CrossRef] [PubMed]
96. Bermejo-Martin, J.F.; Almansa, R.; Torres, A.; Gonzalez-Rivera, M.; Kelvin, D.J. COVID-19 as a cardiovascular disease: The potential role of chronic endothelial dysfunction. *Cardiovasc. Res.* **2020**, *116*, e132–e133. [CrossRef]
97. Daiber, A.; Chlopicki, S. Revisiting pharmacology of oxidative stress and endothelial dysfunction in cardiovascular disease: Evidence for redox-based therapies. *Free Radic. Biol. Med.* **2020**. [CrossRef]
98. Lima, B.B.; Hammadah, M.; Kim, J.H.; Uphoff, I.; Shah, A.; Levantsevych, O.; Almuwaqqat, Z.; Moazzami, K.; Sullivan, S.; Ward, L.; et al. Association of Transient Endothelial Dysfunction Induced by Mental Stress With Major Adverse Cardiovascular Events in Men and Women With Coronary Artery Disease. *JAMA Cardiol.* **2019**, *4*, 988–996. [CrossRef]
99. Yepuri, G.; Ramasamy, R. Significance and Mechanistic Relevance of SIRT6-Mediated Endothelial Dysfunction in Cardiovascular Disease Progression. *Circ. Res.* **2019**, *124*, 1408–1410. [CrossRef]
100. Daiber, A.; Xia, N.; Steven, S.; Oelze, M.; Hanf, A.; Kroller-Schon, S.; Munzel, T.; Li, H. New Therapeutic Implications of Endothelial Nitric Oxide Synthase (eNOS) Function/Dysfunction in Cardiovascular Disease. *Int. J. Mol. Sci.* **2019**, *20*, 187. [CrossRef]
101. Maruhashi, T.; Soga, J.; Fujimura, N.; Idei, N.; Mikami, S.; Iwamoto, Y.; Iwamoto, A.; Kajikawa, M.; Matsumoto, T.; Oda, N.; et al. Endothelial Dysfunction, Increased Arterial Stiffness, and Cardiovascular Risk Prediction in Patients with Coronary Artery Disease: FMD-J (Flow-Mediated Dilation Japan) Study A. *J. Am. Heart Assoc.* **2018**, *7*, e008588. [CrossRef]
102. Akasaka, T.; Sueta, D.; Tabata, N.; Takashio, S.; Yamamoto, E.; Izumiya, Y.; Tsujita, K.; Kojima, S.; Kaikita, K.; Matsui, K.; et al. Effects of the Mean Amplitude of Glycemic Excursions and Vascular Endothelial Dysfunction on Cardiovascular Events in Nondiabetic Patients With Coronary Artery Disease. *J. Am. Heart Assoc.* **2017**, *6*, e004841. [CrossRef]
103. Chello, M.; Nenna, A. Ethnicity, ABO group, endothelial dysfunction and cardiovascular disease: Multiple connections, multiple implications. *Atherosclerosis* **2016**, *251*, 514–515. [CrossRef] [PubMed]
104. Erqou, S.; Kip, K.E.; Mulukutla, S.R.; Aiyer, A.N.; Reis, S.E. Endothelial Dysfunction and Racial Disparities in Mortality and Adverse Cardiovascular Disease Outcomes. *Clin. Cardiol.* **2016**, *39*, 338–344. [CrossRef] [PubMed]
105. Bodolay, E.; Prohaszka, Z.; Paragh, G.; Csipo, I.; Nagy, G.; Laczik, R.; Demeter, N.; Zold, E.; Nakken, B.; Szegedi, G.; et al. Increased levels of anti-heat-shock protein 60 (anti-Hsp60) indicate endothelial dysfunction, atherosclerosis and cardiovascular diseases in patients with mixed connective tissue disease. *Immunol. Res.* **2014**, *60*, 50–59. [CrossRef] [PubMed]
106. Moody, W.E.; Edwards, N.C.; Madhani, M.; Chue, C.D.; Steeds, R.P.; Ferro, C.J.; Townend, J.N. Endothelial dysfunction and cardiovascular disease in early-stage chronic kidney disease: Cause or association? *Atherosclerosis* **2012**, *223*, 86–94. [CrossRef] [PubMed]
107. Versari, D.; Daghini, E.; Virdis, A.; Ghiadoni, L.; Taddei, S. Endothelial dysfunction as a target for prevention of cardiovascular disease. *Diabetes Care* **2009**, *32* (Suppl. 2), S314–S321. [CrossRef] [PubMed]
108. Nozaki, T.; Sugiyama, S.; Koga, H.; Sugamura, K.; Ohba, K.; Matsuzawa, Y.; Sumida, H.; Matsui, K.; Jinnouchi, H.; Ogawa, H. Significance of a multiple biomarkers strategy including endothelial dysfunction to improve risk stratification for cardiovascular events in patients at high risk for coronary heart disease. *J. Am. Coll. Cardiol.* **2009**, *54*, 601–608. [CrossRef]

109. Martin, B.J.; Anderson, T.J. Risk prediction in cardiovascular disease: The prognostic significance of endothelial dysfunction. *Can. J. Cardiol.* **2009**, *25* (Suppl. A), 15A–20A. [CrossRef]
110. Nin, J.W.; Ferreira, I.; Schalkwijk, C.G.; Prins, M.H.; Chaturvedi, N.; Fuller, J.H.; Stehouwer, C.D.; EURODIAB Prospective Complications Study Group. Levels of soluble receptor for AGE are cross-sectionally associated with cardiovascular disease in type 1 diabetes, and this association is partially mediated by endothelial and renal dysfunction and by low-grade inflammation: The Eurodiab Prospective Complications Study. *Diabetologia* **2009**, *52*, 705–714.
111. Friedewald, V.E.; Giles, T.D.; Pool, J.L.; Yancy, C.W.; Roberts, W.C. The Editor's Roundtable: Endothelial dysfunction in cardiovascular disease. *Am. J. Cardiol.* **2008**, *102*, 418–423. [CrossRef]
112. Rodford, J.L.; Torrens, C.; Siow, R.C.; Mann, G.E.; Hanson, M.A.; Clough, G.F. Endothelial dysfunction and reduced antioxidant protection in an animal model of the developmental origins of cardiovascular disease. *J. Physiol.* **2008**, *586*, 4709–4720. [CrossRef]
113. Xu, Y.; Buikema, H.; van Gilst, W.H.; Henning, R.H. Caveolae and endothelial dysfunction: Filling the caves in cardiovascular disease. *Eur. J. Pharmacol.* **2008**, *585*, 256–260. [CrossRef] [PubMed]
114. Subah Packer, C. Estrogen protection, oxidized LDL, endothelial dysfunction and vasorelaxation in cardiovascular disease: New insights into a complex issue. *Cardiovasc. Res.* **2007**, *73*, 6–7. [CrossRef] [PubMed]
115. Ruef, J.; Marz, W.; Winkelmann, B.R. Markers for endothelial dysfunction, but not markers for oxidative stress correlate with classical risk factors and the severity of coronary artery disease. (A subgroup analysis from the Ludwigshafen Risk and Cardiovascular Health Study). *Scand. Cardiovasc. J.* **2006**, *40*, 274–279. [CrossRef] [PubMed]
116. Hanson, M.; Gluckman, P. Endothelial dysfunction and cardiovascular disease: The role of predictive adaptive responses. *Heart* **2005**, *91*, 864–866. [CrossRef]
117. Brevetti, G.; Silvestro, A.; Schiano, V.; Chiariello, M. Endothelial dysfunction and cardiovascular risk prediction in peripheral arterial disease: Additive value of flow-mediated dilation to ankle-brachial pressure index. *Circulation* **2003**, *108*, 2093–2098. [CrossRef]
118. Gokce, N.; Keaney, J.F., Jr.; Hunter, L.M.; Watkins, M.T.; Nedeljkovic, Z.S.; Menzoian, J.O.; Vita, J.A. Predictive value of noninvasively determined endothelial dysfunction for long-term cardiovascular events in patients with peripheral vascular disease. *J. Am. Coll. Cardiol.* **2003**, *41*, 1769–1775. [CrossRef]
119. Erhardt, L.R. Endothelial dysfunction and cardiovascular disease: The promise of blocking the renin-angiotensin system. *Int. J. Clin. Pract.* **2003**, *57*, 211–218.
120. Sardu, C.; Gambardella, J.; Morelli, M.B.; Wang, X.; Marfella, R.; Santulli, G. Is COVID-19 an Endothelial Disease? Clinical and Basic Evidence. *Preprints* **2020**. [CrossRef]
121. Sardu, C.; Gambardella, J.; Morelli, M.B.; Wang, X.; Marfella, R.; Santulli, G. Hypertension, Thrombosis, Kidney Failure, and Diabetes: Is COVID-19 an Endothelial Disease? A Comprehensive Evaluation of Clinical and Basic Evidence. *J. Clin. Med.* **2020**, *9*, 1417. [CrossRef]
122. Paules, C.I.; Marston, H.D.; Fauci, A.S. Coronavirus Infections-More Than Just the Common Cold. *JAMA* **2020**, *323*, 707–708. [CrossRef]
123. Hui, D.S.; Azhar, E.; Madani, T.A.; Ntoumi, F.; Kock, R.; Dar, O.; Ippolito, G.; McHugh, T.D.; Memish, Z.A.; Drosten, C.; et al. The continuing 2019-nCoV epidemic threat of novel coronaviruses to global health—The latest 2019 novel coronavirus outbreak in Wuhan, China. *Int. J. Infect. Dis.* **2020**, *91*, 264–266. [CrossRef] [PubMed]
124. Colmenero, I.; Santonja, C.; Alonso-Riano, M.; Noguera-Morel, L.; Hernandez-Martin, A.; Andina, D.; Wiesner, T.; Rodriguez-Peralto, J.L.; Requena, L.; Torrelo, A. SARS-CoV-2 endothelial infection causes COVID-19 chilblains: Histopathological, immunohistochemical and ultraestructural study of 7 paediatric cases. *Br. J. Dermatol.* **2020**. [CrossRef]
125. Hanafi, R.; Roger, P.A.; Perin, B.; Kuchcinski, G.; Deleval, N.; Dallery, F.; Michel, D.; Hacein-Bey, L.; Pruvo, J.P.; Outteryck, O.; et al. COVID-19 Neurologic Complication with CNS Vasculitis-Like Pattern. *AJNR Am. J. Neuroradiol.* **2020**. [CrossRef] [PubMed]
126. Mosleh, W.; Chen, K.; Pfau, S.E.; Vashist, A. Endotheliitis and Endothelial Dysfunction in Patients with COVID-19: Its Role in Thrombosis and Adverse Outcomes. *J. Clin. Med.* **2020**, *9*, 1862. [CrossRef] [PubMed]
127. Pons, S.; Fodil, S.; Azoulay, E.; Zafrani, L. The vascular endothelium: The cornerstone of organ dysfunction in severe SARS-CoV-2 infection. *Crit. Care* **2020**, *24*, 353. [CrossRef]

128. Konopka, K.E.; Nguyen, T.; Jentzen, J.M.; Rayes, O.; Schmidt, C.J.; Wilson, A.M.; Farver, C.F.; Myers, J.L. Diffuse Alveolar Damage (DAD) from Coronavirus Disease 2019 Infection is Morphologically Indistinguishable from Other Causes of DAD. *Histopathology* **2020**. [CrossRef]
129. Benger, M.; Williams, O.; Siddiqui, J.; Sztriha, L. Intracerebral haemorrhage and COVID-19: Clinical characteristics from a case series. *Brain Behav. Immun.* **2020**, *88*, 940–944. [CrossRef]
130. Wang, J.; Saguner, A.M.; An, J.; Ning, Y.; Yan, Y.; Li, G. Dysfunctional Coagulation in COVID-19: From Cell to Bedside. *Adv. Ther.* **2020**, *37*, 3033–3039. [CrossRef]
131. Ackermann, M.; Verleden, S.E.; Kuehnel, M.; Haverich, A.; Welte, T.; Laenger, F.; Vanstapel, A.; Werlein, C.; Stark, H.; Tzankov, A.; et al. Pulmonary Vascular Endothelialitis, Thrombosis, and Angiogenesis in Covid-19. *N. Engl. J. Med.* **2020**. [CrossRef]
132. Varga, Z.; Flammer, A.J.; Steiger, P.; Haberecker, M.; Andermatt, R.; Zinkernagel, A.S.; Mehra, M.R.; Schuepbach, R.A.; Ruschitzka, F.; Moch, H. Endothelial cell infection and endotheliitis in COVID-19. *Lancet* **2020**, *395*, 1417–1418. [CrossRef]
133. Jones, V.G.; Mills, M.; Suarez, D.; Hogan, C.A.; Yeh, D.; Bradley Segal, J.; Nguyen, E.L.; Barsh, G.R.; Maskatia, S.; Mathew, R. COVID-19 and Kawasaki Disease: Novel Virus and Novel Case. *Hosp. Pediatr.* **2020**, *10*, 537–540. [CrossRef] [PubMed]
134. Cyr, A.R.; Huckaby, L.V.; Shiva, S.S.; Zuckerbraun, B.S. Nitric Oxide and Endothelial Dysfunction. *Crit. Care Clin.* **2020**, *36*, 307–321. [CrossRef] [PubMed]
135. Chen, J.; Zhang, J.; Shaik, N.F.; Yi, B.; Wei, X.; Yang, X.F.; Naik, U.P.; Summer, R.; Yan, G.; Xu, X.; et al. The histone deacetylase inhibitor tubacin mitigates endothelial dysfunction by up-regulating the expression of endothelial nitric oxide synthase. *J. Biol. Chem.* **2019**, *294*, 19565–19576. [CrossRef] [PubMed]
136. Lomeli, O.; Perez-Torres, I.; Marquez, R.; Criales, S.; Mejia, A.M.; Chiney, C.; Hernandez-Lemus, E.; Soto, M.E. The Evaluation of Flow-Mediated Vasodilation in the Brachial Artery Correlates With Endothelial Dysfunction Evaluated by Nitric Oxide Synthase Metabolites in Marfan Syndrome Patients. *Front. Physiol.* **2018**, *9*, 965. [CrossRef]
137. Evans, R.W.; Fernstrom, J.D.; Thompson, J.; Morris, S.M., Jr.; Kuller, L.H. Biochemical responses of healthy subjects during dietary supplementation with L-arginine. *J. Nutr. Biochem.* **2004**, *15*, 534–539. [CrossRef]
138. Alvares, T.S.; Conte-Junior, C.A.; Silva, J.T.; Paschoalin, V.M. Acute L-Arginine supplementation does not increase nitric oxide production in healthy subjects. *Nutr. Metab.* **2012**, *9*, 54. [CrossRef]
139. Meirelles, C.M.; Matsuura, C.; Silva, R.S., Jr.; Guimaraes, F.F.; Gomes, P.S.C. Acute Effects of L-Arginine Supplementation on Oxygen Consumption Kinetics and Muscle Oxyhemoglobin and Deoxyhemoglobin during Treadmill Running in Male Adults. *Int. J. Exerc. Sci.* **2019**, *12*, 444–455.
140. Liu, T.H.; Wu, C.L.; Chiang, C.W.; Lo, Y.W.; Tseng, H.F.; Chang, C.K. No effect of short-term arginine supplementation on nitric oxide production, metabolism and performance in intermittent exercise in athletes. *J. Nutr. Biochem.* **2009**, *20*, 462–468. [CrossRef]
141. Schwedhelm, E.; Maas, R.; Freese, R.; Jung, D.; Lukacs, Z.; Jambrecina, A.; Spickler, W.; Schulze, F.; Boger, R.H. Pharmacokinetic and pharmacodynamic properties of oral L-citrulline and L-arginine: Impact on nitric oxide metabolism. *Br. J. Clin. Pharmacol.* **2008**, *65*, 51–59. [CrossRef]
142. Bredt, D.S.; Snyder, S.H. Isolation of nitric oxide synthetase, a calmodulin-requiring enzyme. *Proc. Natl. Acad. Sci. USA* **1990**, *87*, 682–685. [CrossRef]
143. Joyner, M.J.; Casey, D.P. Regulation of increased blood flow (hyperemia) to muscles during exercise: A hierarchy of competing physiological needs. *Physiol. Rev.* **2015**, *95*, 549–601. [CrossRef] [PubMed]
144. Alvares, T.S.; Meirelles, C.M.; Bhambhani, Y.N.; Paschoalin, V.M.; Gomes, P.S. L-Arginine as a potential ergogenic aid in healthy subjects. *Sports Med.* **2011**, *41*, 233–248. [CrossRef] [PubMed]
145. Doutreleau, S.; Rouyer, O.; Di Marco, P.; Lonsdorfer, E.; Richard, R.; Piquard, F.; Geny, B. L-arginine supplementation improves exercise capacity after a heart transplant. *Am. J. Clin. Nutr.* **2010**, *91*, 1261–1267. [CrossRef] [PubMed]
146. Maughan, R.J. Nutritional ergogenic aids and exercise performance. *Nutr. Res. Rev.* **1999**, *12*, 255–280. [CrossRef] [PubMed]
147. Suzuki, I.; Sakuraba, K.; Horiike, T.; Kishi, T.; Yabe, J.; Suzuki, T.; Morita, M.; Nishimura, A.; Suzuki, Y. A combination of oral L-citrulline and L-arginine improved 10-min full-power cycling test performance in male collegiate soccer players: A randomized crossover trial. *Eur. J. Appl. Physiol.* **2019**, *119*, 1075–1084. [CrossRef]

148. Andrade, W.B.; Jacinto, J.L.; da Silva, D.K.; Roveratti, M.C.; Estoche, J.M.; Oliveira, D.B.; Balvedi, M.C.W.; da Silva, R.A.; Aguiar, A.F. l-Arginine supplementation does not improve muscle function during recovery from resistance exercise. *Appl. Physiol. Nutr. Metab.* **2018**, *43*, 928–936. [CrossRef]
149. Dioguardi, F.S. To give or not to give? Lessons from the arginine paradox. *J. Nutrigenet. Nutrige.* **2011**, *4*, 90–98. [CrossRef]
150. Ueda, S.; Petrie, J.R.; Cleland, S.J.; Elliott, H.L.; Connell, J.M. Insulin vasodilatation and the "arginine paradox". *Lancet* **1998**, *351*, 959–960. [CrossRef]
151. Kurz, S.; Harrison, D.G. Insulin and the arginine paradox. *J. Clin. Investig.* **1997**, *99*, 369–370. [CrossRef]
152. Vallance, P.; Leone, A.; Calver, A.; Collier, J.; Moncada, S. Endogenous dimethylarginine as an inhibitor of nitric oxide synthesis. *J. Cardiovasc. Pharmacol.* **1992**, *20* (Suppl. 12), S60–S62. [CrossRef]
153. Bartnicki, P.; Kowalczyk, M.; Franczyk-Skora, B.; Baj, Z.; Rysz, J. Evaluation of Endothelial (dys)Function, Left Ventricular Structure and Function in Patients with Chronic Kidney Disease. *Curr. Vasc. Pharmacol.* **2016**, *14*, 360–367. [CrossRef] [PubMed]
154. Bouras, G.; Deftereos, S.; Tousoulis, D.; Giannopoulos, G.; Chatzis, G.; Tsounis, D.; Cleman, M.W.; Stefanadis, C. Asymmetric Dimethylarginine (ADMA): A promising biomarker for cardiovascular disease? *Curr. Top. Med. Chem.* **2013**, *13*, 180–200. [CrossRef] [PubMed]
155. Janes, F.; Cifu, A.; Pessa, M.E.; Domenis, R.; Gigli, G.L.; Sanvilli, N.; Nilo, A.; Garbo, R.; Curcio, F.; Giacomello, R.; et al. ADMA as a possible marker of endothelial damage. A study in young asymptomatic patients with cerebral small vessel disease. *Sci. Rep.* **2019**, *9*, 14207. [CrossRef] [PubMed]
156. Strobel, J.; Muller, F.; Zolk, O.; Endress, B.; Konig, J.; Fromm, M.F.; Maas, R. Transport of asymmetric dimethylarginine (ADMA) by cationic amino acid transporter 2 (CAT2), organic cation transporter 2 (OCT2) and multidrug and toxin extrusion protein 1 (MATE1). *Amino Acids* **2013**, *45*, 989–1002. [CrossRef] [PubMed]
157. Closs, E.I.; Basha, F.Z.; Habermeier, A.; Forstermann, U. Interference of L-arginine analogues with L-arginine transport mediated by the y+ carrier hCAT-2B. *Nitric Oxide* **1997**, *1*, 65–73. [CrossRef]
158. Wijnands, K.A.; Hoeksema, M.A.; Meesters, D.M.; van den Akker, N.M.; Molin, D.G.; Briede, J.J.; Ghosh, M.; Kohler, S.E.; van Zandvoort, M.A.; de Winther, M.P.; et al. Arginase-1 deficiency regulates arginine concentrations and NOS2-mediated NO production during endotoxemia. *PLoS ONE* **2014**, *9*, e86135. [CrossRef]
159. Boger, R.H.; Bode-Boger, S.M.; Szuba, A.; Tsao, P.S.; Chan, J.R.; Tangphao, O.; Blaschke, T.F.; Cooke, J.P. Asymmetric dimethylarginine (ADMA): A novel risk factor for endothelial dysfunction: Its role in hypercholesterolemia. *Circulation* **1998**, *98*, 1842–1847. [CrossRef]
160. Nijveldt, R.J.; Teerlink, T.; Van Der Hoven, B.; Siroen, M.P.; Kuik, D.J.; Rauwerda, J.A.; van Leeuwen, P.A. Asymmetrical dimethylarginine (ADMA) in critically ill patients: High plasma ADMA concentration is an independent risk factor of ICU mortality. *Clin. Nutr.* **2003**, *22*, 23–30. [CrossRef]
161. Abedini, S.; Meinitzer, A.; Holme, I.; Marz, W.; Weihrauch, G.; Fellstrom, B.; Jardine, A.; Holdaas, H. Asymmetrical dimethylarginine is associated with renal and cardiovascular outcomes and all-cause mortality in renal transplant recipients. *Kidney Int.* **2010**, *77*, 44–50. [CrossRef]
162. Krzyzanowska, K.; Mittermayer, F.; Wolzt, M.; Schernthaner, G. Asymmetric dimethylarginine predicts cardiovascular events in patients with type 2 diabetes. *Diabetes Care* **2007**, *30*, 1834–1839. [CrossRef]
163. Nanayakkara, P.W.; Teerlink, T.; Stehouwer, C.D.; Allajar, D.; Spijkerman, A.; Schalkwijk, C.; ter Wee, P.M.; van Guldener, C. Plasma asymmetric dimethylarginine (ADMA) concentration is independently associated with carotid intima-media thickness and plasma soluble vascular cell adhesion molecule-1 (sVCAM-1) concentration in patients with mild-to-moderate renal failure. *Kidney Int.* **2005**, *68*, 2230–2236. [CrossRef] [PubMed]
164. Zoccali, C.; Bode-Boger, S.; Mallamaci, F.; Benedetto, F.; Tripepi, G.; Malatino, L.; Cataliotti, A.; Bellanuova, I.; Fermo, I.; Frolich, J.; et al. Plasma concentration of asymmetrical dimethylarginine and mortality in patients with end-stage renal disease: A prospective study. *Lancet* **2001**, *358*, 2113–2117. [CrossRef]
165. Abbasi, F.; Asagmi, T.; Cooke, J.P.; Lamendola, C.; McLaughlin, T.; Reaven, G.M.; Stuehlinger, M.; Tsao, P.S. Plasma concentrations of asymmetric dimethylarginine are increased in patients with type 2 diabetes mellitus. *Am. J. Cardiol.* **2001**, *88*, 1201–1203. [CrossRef]
166. Furuki, K.; Adachi, H.; Matsuoka, H.; Enomoto, M.; Satoh, A.; Hino, A.; Hirai, Y.; Imaizumi, T. Plasma levels of asymmetric dimethylarginine (ADMA) are related to intima-media thickness of the carotid artery: An epidemiological study. *Atherosclerosis* **2007**, *191*, 206–210. [CrossRef]

167. Valkonen, V.P.; Paiva, H.; Salonen, J.T.; Lakka, T.A.; Lehtimaki, T.; Laakso, J.; Laaksonen, R. Risk of acute coronary events and serum concentration of asymmetrical dimethylarginine. *Lancet* **2001**, *358*, 2127–2128. [CrossRef]
168. Surdacki, A.; Nowicki, M.; Sandmann, J.; Tsikas, D.; Boeger, R.H.; Bode-Boeger, S.M.; Kruszelnicka-Kwiatkowska, O.; Kokot, F.; Dubiel, J.S.; Froelich, J.C. Reduced urinary excretion of nitric oxide metabolites and increased plasma levels of asymmetric dimethylarginine in men with essential hypertension. *J. Cardiovasc. Pharmacol.* **1999**, *33*, 652–658. [CrossRef]
169. Maas, R.; Xanthakis, V.; Polak, J.F.; Schwedhelm, E.; Sullivan, L.M.; Benndorf, R.; Schulze, F.; Vasan, R.S.; Wolf, P.A.; Boger, R.H.; et al. Association of the endogenous nitric oxide synthase inhibitor ADMA with carotid artery intimal media thickness in the Framingham Heart Study offspring cohort. *Stroke* **2009**, *40*, 2715–2719. [CrossRef]
170. Bai, Y.; Sun, L.; Du, L.; Zhang, T.; Xin, W.; Lan, X.; Du, G. Association of circulating levels of asymmetric dimethylarginine (ADMA) with carotid intima-media thickness: Evidence from 6168 participants. *Ageing Res. Rev.* **2013**, *12*, 699–707. [CrossRef]
171. Sundar, U.M.; Ugusman, A.; Chua, H.K.; Latip, J.; Aminuddin, A. Piper sarmentosum Promotes Endothelial Nitric Oxide Production by Reducing Asymmetric Dimethylarginine in Tumor Necrosis Factor-alpha-Induced Human Umbilical Vein Endothelial Cells. *Front. Pharmacol* **2019**, *10*, 1033. [CrossRef]
172. Arlouskaya, Y.; Sawicka, A.; Glowala, M.; Giebultowicz, J.; Korytowska, N.; Talalaj, M.; Nowicka, G.; Wrzosek, M. Asymmetric Dimethylarginine (ADMA) and Symmetric Dimethylarginine (SDMA) Concentrations in Patients with Obesity and the Risk of Obstructive Sleep Apnea (OSA). *J. Clin. Med.* **2019**, *8*, 897. [CrossRef]
173. Hsu, C.P.; Zhao, J.F.; Lin, S.J.; Shyue, S.K.; Guo, B.C.; Lu, T.M.; Lee, T.S. Asymmetric Dimethylarginine Limits the Efficacy of Simvastatin Activating Endothelial Nitric Oxide Synthase. *J. Am. Heart Assoc.* **2016**, *5*, e003327. [CrossRef] [PubMed]
174. Notsu, Y.; Yano, S.; Shibata, H.; Nagai, A.; Nabika, T. Plasma arginine/ADMA ratio as a sensitive risk marker for atherosclerosis: Shimane CoHRE study. *Atherosclerosis* **2015**, *239*, 61–66. [CrossRef] [PubMed]
175. Boger, R.H.; Endres, H.G.; Schwedhelm, E.; Darius, H.; Atzler, D.; Luneburg, N.; von Stritzky, B.; Maas, R.; Thiem, U.; Benndorf, R.A.; et al. Asymmetric dimethylarginine as an independent risk marker for mortality in ambulatory patients with peripheral arterial disease. *J. Intern. Med.* **2011**, *269*, 349–361. [CrossRef] [PubMed]
176. Pizzarelli, F.; Maas, R.; Dattolo, P.; Tripepi, G.; Michelassi, S.; D'Arrigo, G.; Mieth, M.; Bandinelli, S.; Ferrucci, L.; Zoccali, C. Asymmetric dimethylarginine predicts survival in the elderly. *Age* **2013**, *35*, 2465–2475. [CrossRef] [PubMed]
177. Notsu, Y.; Nabika, T.; Bokura, H.; Suyama, Y.; Kobayashi, S.; Yamaguchi, S.; Masuda, J. Evaluation of asymmetric dimethylarginine and homocysteine in microangiopathy-related cerebral damage. *Am. J. Hypertens.* **2009**, *22*, 257–262. [CrossRef] [PubMed]
178. McDonald, K.K.; Zharikov, S.; Block, E.R.; Kilberg, M.S. A caveolar complex between the cationic amino acid transporter 1 and endothelial nitric-oxide synthase may explain the "arginine paradox". *J. Biol. Chem.* **1997**, *272*, 31213–31216. [CrossRef]
179. Kim, J.A.; Montagnani, M.; Chandrasekran, S.; Quon, M.J. Role of lipotoxicity in endothelial dysfunction. *Heart Fail. Clin.* **2012**, *8*, 589–607. [CrossRef]
180. Benowitz, N.L. Cigarette smoking and cardiovascular disease: Pathophysiology and implications for treatment. *Prog. Cardiovasc. Dis.* **2003**, *46*, 91–111. [CrossRef]
181. Januzzi, J. *Cardiac Biomarkers in Clinical Practice*; Jones & Bartlett Learning: Burlington, MA, USA, 2009.
182. Rohde, D.; Busch, M.; Volkert, A.; Ritterhoff, J.; Katus, H.A.; Peppel, K.; Most, P. Cardiomyocytes, endothelial cells and cardiac fibroblasts: S100A1's triple action in cardiovascular pathophysiology. *Future Cardiol.* **2015**, *11*, 309–321. [CrossRef]
183. Davies, P.F. Hemodynamic shear stress and the endothelium in cardiovascular pathophysiology. *Nat. Clin. Pract. Cardiovasc. Med.* **2009**, *6*, 16–26. [CrossRef]
184. Santulli, G. microRNAs Distinctively Regulate Vascular Smooth Muscle and Endothelial Cells: Functional Implications in Angiogenesis, Atherosclerosis, and In-Stent Restenosis. *Adv. Exp. Med. Biol.* **2015**, *887*, 53–77. [PubMed]

185. Foster, W.; Carruthers, D.; Lip, G.Y.; Blann, A.D. Relationships between endothelial, inflammatory and angiogenesis markers in rheumatoid arthritis: Implications for cardiovascular pathophysiology. *Thromb. Res.* **2009**, *123*, 659–664. [CrossRef] [PubMed]
186. Ciccarelli, M.; Santulli, G.; Campanile, A.; Galasso, G.; Cervero, P.; Altobelli, G.G.; Cimini, V.; Pastore, L.; Piscione, F.; Trimarco, B.; et al. Endothelial alpha1-adrenoceptors regulate neo-angiogenesis. *Br. J. Pharmacol.* **2008**, *153*, 936–946. [CrossRef] [PubMed]
187. Luksha, L.; Agewall, S.; Kublickiene, K. Endothelium-derived hyperpolarizing factor in vascular physiology and cardiovascular disease. *Atherosclerosis* **2009**, *202*, 330–344. [CrossRef]
188. Santulli, G.; Cipolletta, E.; Sorriento, D.; Del Giudice, C.; Anastasio, A.; Monaco, S.; Maione, A.S.; Condorelli, G.; Puca, A.; Trimarco, B.; et al. CaMK4 Gene Deletion Induces Hypertension. *J. Am. Heart Assoc.* **2012**, *1*, e001081. [CrossRef]
189. Deedwania, P.C. Diabetes is a vascular disease: The role of endothelial dysfunction in pathophysiology of cardiovascular disease in diabetes. *Cardiol. Clin.* **2004**, *22*, 505–509. [CrossRef]
190. Sorriento, D.; Santulli, G.; Del Giudice, C.; Anastasio, A.; Trimarco, B.; Iaccarino, G. Endothelial cells are able to synthesize and release catecholamines both in vitro and in vivo. *Hypertension* **2012**, *60*, 129–136. [CrossRef]
191. Li, J.M.; Shah, A.M. Endothelial cell superoxide generation: Regulation and relevance for cardiovascular pathophysiology. *Am. J. Physiol. Regul. Integr. Comp. Physiol.* **2004**, *287*, R1014–R1030. [CrossRef]
192. Ciccarelli, M.; Cipolletta, E.; Santulli, G.; Campanile, A.; Pumiglia, K.; Cervero, P.; Pastore, L.; Astone, D.; Trimarco, B.; Iaccarino, G. Endothelial beta2 adrenergic signaling to AKT: Role of Gi and SRC. *Cell Signal.* **2007**, *19*, 1949–1955. [CrossRef]
193. Sellke, F.W.; Boyle, E.M., Jr.; Verrier, E.D. Endothelial cell injury in cardiovascular surgery: The pathophysiology of vasomotor dysfunction. *Ann. Thorac. Surg.* **1996**, *62*, 1222–1228. [CrossRef]
194. Iaccarino, G.; Ciccarelli, M.; Sorriento, D.; Galasso, G.; Campanile, A.; Santulli, G.; Cipolletta, E.; Cerullo, V.; Cimini, V.; Altobelli, G.G.; et al. Ischemic neoangiogenesis enhanced by beta2-adrenergic receptor overexpression: A novel role for the endothelial adrenergic system. *Circ. Res.* **2005**, *97*, 1182–1189. [CrossRef] [PubMed]
195. Ogita, H.; Liao, J. Endothelial function and oxidative stress. *Endothelium* **2004**, *11*, 123–132. [CrossRef] [PubMed]
196. Skrypnyk, I.; Maslova, G.; Lymanets, T.; Gusachenko, I. L-arginine is an effective medication for prevention of endothelial dysfunction, a predictor of anthracycline cardiotoxicity in patients with acute leukemia. *Exp. Oncol.* **2017**, *39*, 308–311. [CrossRef]
197. Ignarro, L.J.; Napoli, C. Novel features of nitric oxide, endothelial nitric oxide synthase, and atherosclerosis. *Curr. Diab. Rep.* **2005**, *5*, 17–23. [CrossRef]
198. Nguyen, M.C.; Park, J.T.; Jeon, Y.G.; Jeon, B.H.; Hoe, K.L.; Kim, Y.M.; Lim, H.K.; Ryoo, S. Arginase Inhibition Restores Peroxynitrite-Induced Endothelial Dysfunction via L-Arginine-Dependent Endothelial Nitric Oxide Synthase Phosphorylation. *Yonsei Med. J.* **2016**, *57*, 1329–1338. [CrossRef]
199. Wu, G.; Meininger, C.J. Arginine nutrition and cardiovascular function. *J. Nutr.* **2000**, *130*, 2626–2629. [CrossRef]
200. Scaglia, F.; Brunetti-Pierri, N.; Kleppe, S.; Marini, J.; Carter, S.; Garlick, P.; Jahoor, F.; O'Brien, W.; Lee, B. Clinical consequences of urea cycle enzyme deficiencies and potential links to arginine and nitric oxide metabolism. *J. Nutr.* **2004**, *134* (Suppl. 10), 2775S–2782S. [CrossRef]
201. Luneburg, N.; Xanthakis, V.; Schwedhelm, E.; Sullivan, L.M.; Maas, R.; Anderssohn, M.; Riederer, U.; Glazer, N.L.; Vasan, R.S.; Boger, R.H. Reference intervals for plasma L-arginine and the L-arginine:asymmetric dimethylarginine ratio in the Framingham Offspring Cohort. *J. Nutr.* **2011**, *141*, 2186–2190. [CrossRef]
202. Contreras, M.T.; Gallardo, M.J.; Betancourt, L.R.; Rada, P.V.; Ceballos, G.A.; Hernandez, L.E.; Hernandez, L.F. Correlation between plasma levels of arginine and citrulline in preterm and full-term neonates: Therapeutical implications. *J. Clin. Lab. Anal.* **2017**, *31*, e22134. [CrossRef]
203. Fernandez Diaz-Rullo, F.; Zamberlan, F.; Mewis, R.E.; Fekete, M.; Broche, L.; Cheyne, L.A.; Dall'Angelo, S.; Duckett, S.B.; Dawson, D.; Zanda, M. Synthesis and hyperpolarisation of eNOS substrates for quantification of NO production by (1)H NMR spectroscopy. *Bioorg. Med. Chem.* **2017**, *25*, 2730–2742. [CrossRef]
204. Schmidt, H.H.; Warner, T.D.; Ishii, K.; Sheng, H.; Murad, F. Insulin secretion from pancreatic B cells caused by L-arginine-derived nitrogen oxides. *Science* **1992**, *255*, 721–723. [CrossRef] [PubMed]

205. Fajans, S.S.; Floyd, J.C., Jr.; Knopf, R.F.; Conn, F.W. Effect of amino acids and proteins on insulin secretion in man. *Recent Prog. Horm. Res.* **1967**, *23*, 617–662. [PubMed]
206. Sener, A.; Best, L.C.; Yates, A.P.; Kadiata, M.M.; Olivares, E.; Louchami, K.; Jijakli, H.; Ladriere, L.; Malaisse, W.J. Stimulus-secretion coupling of arginine-induced insulin release: Comparison between the cationic amino acid and its methyl ester. *Endocrine* **2000**, *13*, 329–340. [CrossRef]
207. Gerich, J.E.; Lorenzi, M.; Schneider, V.; Kwan, C.W.; Karam, J.H.; Guillemin, R.; Forsham, P.H. Inhibition of pancreatic glucagon responses to arginine by somatostatin in normal man and in insulin-dependent diabetics. *Diabetes* **1974**, *23*, 876–880. [CrossRef]
208. Giugliano, D.; Marfella, R.; Verrazzo, G.; Acampora, R.; Coppola, L.; Cozzolino, D.; D'Onofrio, F. The vascular effects of L-Arginine in humans. The role of endogenous insulin. *J. Clin. Investig.* **1997**, *99*, 433–438. [CrossRef]
209. Merimee, T.J.; Rabinowitz, D.; Riggs, L.; Burgess, J.A.; Rimoin, D.L.; McKusick, V.A. Plasma growth hormone after arginine infusion. Clinical experiences. *N. Engl. J. Med.* **1967**, *276*, 434–439. [CrossRef]
210. Fryburg, D.A. NG-monomethyl-L-arginine inhibits the blood flow but not the insulin-like response of forearm muscle to IGF- I: Possible role of nitric oxide in muscle protein synthesis. *J. Clin. Investig.* **1996**, *97*, 1319–1328. [CrossRef]
211. Bode-Boger, S.M.; Boger, R.H.; Loffler, M.; Tsikas, D.; Brabant, G.; Frolich, J.C. L-arginine stimulates NO-dependent vasodilation in healthy humans–effect of somatostatin pretreatment. *J. Investig. Med.* **1999**, *47*, 43–50.
212. Taddei, S.; Virdis, A.; Mattei, P.; Ghiadoni, L.; Gennari, A.; Fasolo, C.B.; Sudano, I.; Salvetti, A. Aging and endothelial function in normotensive subjects and patients with essential hypertension. *Circulation* **1995**, *91*, 1981–1987. [CrossRef]
213. Celermajer, D.S.; Sorensen, K.E.; Gooch, V.M.; Spiegelhalter, D.J.; Miller, O.I.; Sullivan, I.D.; Lloyd, J.K.; Deanfield, J.E. Non-invasive detection of endothelial dysfunction in children and adults at risk of atherosclerosis. *Lancet* **1992**, *340*, 1111–1115. [CrossRef]
214. Gates, P.E.; Boucher, M.L.; Silver, A.E.; Monahan, K.D.; Seals, D.R. Impaired flow-mediated dilation with age is not explained by L-arginine bioavailability or endothelial asymmetric dimethylarginine protein expression. *J. Appl. Physiol.* **2007**, *102*, 63–71. [CrossRef] [PubMed]
215. Campo, C.; Lahera, V.; Garcia-Robles, R.; Cachofeiro, V.; Alcazar, J.M.; Andres, A.; Rodicio, J.L.; Ruilope, L.M. Aging abolishes the renal response to L-arginine infusion in essential hypertension. *Kidney Int. Suppl.* **1996**, *55*, S126–S128. [PubMed]
216. Blum, A.; Hathaway, L.; Mincemoyer, R.; Schenke, W.H.; Kirby, M.; Csako, G.; Waclawiw, M.A.; Panza, J.A.; Cannon, R.O., III. Effects of oral L-arginine on endothelium-dependent vasodilation and markers of inflammation in healthy postmenopausal women. *J. Am. Coll. Cardiol.* **2000**, *35*, 271–276. [CrossRef]
217. Bode-Boger, S.M.; Muke, J.; Surdacki, A.; Brabant, G.; Boger, R.H.; Frolich, J.C. Oral L-arginine improves endothelial function in healthy individuals older than 70 years. *Vasc. Med.* **2003**, *8*, 77–81. [CrossRef] [PubMed]
218. Chen, P.Y.; St John, P.L.; Kirk, K.A.; Abrahamson, D.R.; Sanders, P.W. Hypertensive nephrosclerosis in the Dahl/Rapp rat. Initial sites of injury and effect of dietary L-arginine supplementation. *Lab. Investig.* **1993**, *68*, 174–184. [PubMed]
219. Zhou, M.S.; Kosaka, H.; Tian, R.X.; Abe, Y.; Chen, Q.H.; Yoneyama, H.; Yamamoto, A.; Zhang, L. L-Arginine improves endothelial function in renal artery of hypertensive Dahl rats. *J. Hypertens.* **2001**, *19*, 421–429. [CrossRef]
220. Fujii, S.; Zhang, L.; Igarashi, J.; Kosaka, H. L-arginine reverses p47phox and gp91phox expression induced by high salt in Dahl rats. *Hypertension* **2003**, *42*, 1014–1020. [CrossRef]
221. Patel, A.; Layne, S.; Watts, D.; Kirchner, K.A. L-arginine administration normalizes pressure natriuresis in hypertensive Dahl rats. *Hypertension* **1993**, *22*, 863–869. [CrossRef]
222. Chen, P.Y.; Sanders, P.W. L-arginine abrogates salt-sensitive hypertension in Dahl/Rapp rats. *J. Clin. Investig.* **1991**, *88*, 1559–1567. [CrossRef]
223. Miyata, N.; Zou, A.P.; Mattson, D.L.; Cowley, A.W., Jr. Renal medullary interstitial infusion of L-arginine prevents hypertension in Dahl salt-sensitive rats. *Am. J. Physiol.* **1998**, *275*, R1667–R1673. [CrossRef]

224. Ozcelikay, A.T.; Tay, A.; Guner, S.; Tasyaran, V.; Yildizoglu-Ari, N.; Dincer, U.D.; Altan, V.M. Reversal effects of L-arginine treatment on blood pressure and vascular responsiveness of streptozotocin-diabetic rats. *Pharmacol. Res.* **2000**, *41*, 201–209. [CrossRef] [PubMed]
225. Tay, A.; Ozcelikay, A.T.; Altan, V.M. Effects of L-arginine on blood pressure and metabolic changes in fructose-hypertensive rats. *Am. J. Hypertens.* **2002**, *15 Pt 1*, 72–77. [CrossRef]
226. Ono, H.; Ono, Y.; Frohlich, E.D. L-arginine reverses severe nephrosclerosis in aged spontaneously hypertensive rats. *J. Hypertens.* **1999**, *17*, 121–128. [CrossRef] [PubMed]
227. Shan, L.; Wang, B.; Gao, G.; Cao, W.; Zhang, Y. L-Arginine supplementation improves antioxidant defenses through L-arginine/nitric oxide pathways in exercised rats. *J. Appl. Physiol.* **2013**, *115*, 1146–1155. [CrossRef] [PubMed]
228. Lin, W.T.; Yang, S.C.; Chen, K.T.; Huang, C.C.; Lee, N.Y. Protective effects of L-arginine on pulmonary oxidative stress and antioxidant defenses during exhaustive exercise in rats. *Acta Pharmacol. Sin.* **2005**, *26*, 992–999. [CrossRef] [PubMed]
229. Huang, C.C.; Tsai, S.C.; Lin, W.T. Potential ergogenic effects of L-arginine against oxidative and inflammatory stress induced by acute exercise in aging rats. *Exp. Gerontol.* **2008**, *43*, 571–577. [CrossRef] [PubMed]
230. de Nigris, F.; Lerman, L.O.; Ignarro, S.W.; Sica, G.; Lerman, A.; Palinski, W.; Ignarro, L.J.; Napoli, C. Beneficial effects of antioxidants and L-arginine on oxidation-sensitive gene expression and endothelial NO synthase activity at sites of disturbed shear stress. *Proc. Natl. Acad. Sci. USA* **2003**, *100*, 1420–1425. [CrossRef] [PubMed]
231. Ahmad, A.; Sattar, M.Z.; Rathore, H.A.; Hussain, A.I.; Khan, S.A.; Fatima, T.; Afzal, S.; Abdullah, N.A.; Johns, E.J. Antioxidant Activity and Free Radical Scavenging Capacity of L-Arginine and Nahs: A Comparative in Vitro Study. *Acta Pol. Pharm.* **2015**, *72*, 245–252.
232. Wallner, S.; Hermetter, A.; Mayer, B.; Wascher, T.C. The alpha-amino group of L-arginine mediates its antioxidant effect. *Eur. J. Clin. Investig.* **2001**, *31*, 98–102. [CrossRef]
233. Suliburska, J.; Bogdanski, P.; Krejpcio, Z.; Pupek-Musialik, D.; Jablecka, A. The effects of L-arginine, alone and combined with vitamin C, on mineral status in relation to its antidiabetic, anti-inflammatory, and antioxidant properties in male rats on a high-fat diet. *Biol. Trace Elem. Res.* **2014**, *157*, 67–74. [CrossRef]
234. Zheng, P.; Yu, B.; He, J.; Tian, G.; Luo, Y.; Mao, X.; Zhang, K.; Che, L.; Chen, D. Protective effects of dietary arginine supplementation against oxidative stress in weaned piglets. *Br. J. Nutr.* **2013**, *109*, 2253–2260. [CrossRef] [PubMed]
235. Fazelian, S.; Hoseini, M.; Namazi, N.; Heshmati, J.; Sepidar Kish, M.; Mirfatahi, M.; Some Olia, A.S. Effects of L-Arginine Supplementation on Antioxidant Status and Body Composition in Obese Patients with Pre-diabetes: A Randomized Controlled Clinical Trial. *Adv. Pharm. Bull.* **2014**, *4* (Suppl. 1), 449–454. [PubMed]
236. Wascher, T.C.; Posch, K.; Wallner, S.; Hermetter, A.; Kostner, G.M.; Graier, W.F. Vascular effects of L-arginine: Anything beyond a substrate for the NO-synthase? *Biochem Biophys Res. Commun* **1997**, *234*, 35–38. [CrossRef] [PubMed]
237. Lekakis, J.P.; Papathanassiou, S.; Papaioannou, T.G.; Papamichael, C.M.; Zakopoulos, N.; Kotsis, V.; Dagre, A.G.; Stamatelopoulos, K.; Protogerou, A.; Stamatelopoulos, S.F. Oral L-arginine improves endothelial dysfunction in patients with essential hypertension. *Int. J. Cardiol.* **2002**, *86*, 317–323. [CrossRef]
238. Higashi, Y.; Oshima, T.; Watanabe, M.; Matsuura, H.; Kajiyama, G. Renal response to L-arginine in salt-sensitive patients with essential hypertension. *Hypertension* **1996**, *27 Pt 2*, 643–648. [CrossRef]
239. Campese, V.M.; Amar, M.; Anjali, C.; Medhat, T.; Wurgaft, A. Effect of L-arginine on systemic and renal haemodynamics in salt-sensitive patients with essential hypertension. *J. Hum. Hypertens* **1997**, *11*, 527–532. [CrossRef]
240. Pezza, V.; Bernardini, F.; Pezza, E.; Pezza, B.; Curione, M. Study of supplemental oral l-arginine in hypertensives treated with enalapril + hydrochlorothiazide. *Am. J. Hypertens* **1998**, *11*, 1267–1270.
241. Higashi, Y.; Oshima, T.; Ono, N.; Hiraga, H.; Yoshimura, M.; Watanabe, M.; Matsuura, H.; Kambe, M.; Kajiyama, G. Intravenous administration of L-arginine inhibits angiotensin-converting enzyme in humans. *J. Clin. Endocrinol. Metab.* **1995**, *80*, 2198–2202.

242. De Caterina, R.; Libby, P.; Peng, H.B.; Thannickal, V.J.; Rajavashisth, T.B.; Gimbrone, M.A., Jr.; Shin, W.S.; Liao, J.K. Nitric oxide decreases cytokine-induced endothelial activation. Nitric oxide selectively reduces endothelial expression of adhesion molecules and proinflammatory cytokines. *J. Clin. Investig.* **1995**, *96*, 60–68. [CrossRef]

243. Sperone, A.; Dryden, N.H.; Birdsey, G.M.; Madden, L.; Johns, M.; Evans, P.C.; Mason, J.C.; Haskard, D.O.; Boyle, J.J.; Paleolog, E.M.; et al. The transcription factor Erg inhibits vascular inflammation by repressing NF-kappaB activation and proinflammatory gene expression in endothelial cells. *Arterioscler. Thromb. Vasc. Biol.* **2011**, *31*, 142–150. [CrossRef]

244. Gambardella, J.; Santulli, G. Integrating diet and inflammation to calculate cardiovascular risk. *Atherosclerosis* **2016**, *253*, 258–261. [CrossRef] [PubMed]

245. Huo, Y.; Ley, K. Adhesion molecules and atherogenesis. *Acta Physiol. Scand.* **2001**, *173*, 35–43. [CrossRef] [PubMed]

246. Blum, A.; Miller, H.; Blum, A.; Miller, H. The effects of L-arginine on atherosclerosis and heart disease. *Int. J. Cardiovasc. Intervent.* **1999**, *2*, 97–100. [CrossRef] [PubMed]

247. Adams, M.R.; Jessup, W.; Hailstones, D.; Celermajer, D.S. L-arginine reduces human monocyte adhesion to vascular endothelium and endothelial expression of cell adhesion molecules. *Circulation* **1997**, *95*, 662–668. [CrossRef]

248. Adams, M.R.; McCredie, R.; Jessup, W.; Robinson, J.; Sullivan, D.; Celermajer, D.S. Oral L-arginine improves endothelium-dependent dilatation and reduces monocyte adhesion to endothelial cells in young men with coronary artery disease. *Atherosclerosis* **1997**, *129*, 261–269. [CrossRef]

249. Ceremuzynski, L.; Chamiec, T.; Herbaczynska-Cedro, K. Effect of supplemental oral L-arginine on exercise capacity in patients with stable angina pectoris. *Am. J. Cardiol.* **1997**, *80*, 331–333. [CrossRef]

250. Lerman, A.; Burnett, J.C., Jr.; Higano, S.T.; McKinley, L.J.; Holmes, D.R., Jr. Long-term L-arginine supplementation improves small-vessel coronary endothelial function in humans. *Circulation* **1998**, *97*, 2123–2128. [CrossRef]

251. Chen, C.; Mattar, S.G.; Lumsden, A.B. Oral administration of L-arginine reduces intimal hyperplasia in balloon-injured rat carotid arteries. *J. Surg. Res.* **1999**, *82*, 17–23. [CrossRef]

252. Yin, W.H.; Chen, J.W.; Tsai, C.; Chiang, M.C.; Young, M.S.; Lin, S.J. L-arginine improves endothelial function and reduces LDL oxidation in patients with stable coronary artery disease. *Clin. Nutr.* **2005**, *24*, 988–997. [CrossRef]

253. Walker, H.A.; McGing, E.; Fisher, I.; Boger, R.H.; Bode-Boger, S.M.; Jackson, G.; Ritter, J.M.; Chowienczyk, P.J. Endothelium-dependent vasodilation is independent of the plasma L-arginine/ADMA ratio in men with stable angina: Lack of effect of oral L-arginine on endothelial function, oxidative stress and exercise performance. *J. Am. Coll. Cardiol.* **2001**, *38*, 499–505. [CrossRef]

254. Padilla, F.; Garcia-Dorado, D.; Agullo, L.; Inserte, J.; Paniagua, A.; Mirabet, S.; Barrabes, J.A.; Ruiz-Meana, M.; Soler-Soler, J. L-Arginine administration prevents reperfusion-induced cardiomyocyte hypercontracture and reduces infarct size in the pig. *Cardiovasc. Res.* **2000**, *46*, 412–420. [CrossRef]

255. Rector, T.S.; Bank, A.J.; Mullen, K.A.; Tschumperlin, L.K.; Sih, R.; Pillai, K.; Kubo, S.H. Randomized, double-blind, placebo-controlled study of supplemental oral L-arginine in patients with heart failure. *Circulation* **1996**, *93*, 2135–2141. [CrossRef] [PubMed]

256. Bednarz, B.; Jaxa-Chamiec, T.; Gebalska, J.; Herbaczynska-Cedro, K.; Ceremuzynski, L. L-arginine supplementation prolongs exercise capacity in congestive heart failure. *Kardiol. Pol.* **2004**, *60*, 348–353. [PubMed]

257. Rodrigues-Krause, J.; Krause, M.; Rocha, I.; Umpierre, D.; Fayh, A.P.T. Association of l-Arginine Supplementation with Markers of Endothelial Function in Patients with Cardiovascular or Metabolic Disorders: A Systematic Review and Meta-Analysis. *Nutrients* **2018**, *11*, 15. [CrossRef] [PubMed]

258. Hambrecht, R.; Hilbrich, L.; Erbs, S.; Gielen, S.; Fiehn, E.; Schoene, N.; Schuler, G. Correction of endothelial dysfunction in chronic heart failure: Additional effects of exercise training and oral L-arginine supplementation. *J. Am. Coll. Cardiol.* **2000**, *35*, 706–713. [CrossRef]

259. Weyrich, A.S.; Ma, X.L.; Lefer, A.M. The role of L-arginine in ameliorating reperfusion injury after myocardial ischemia in the cat. *Circulation* **1992**, *86*, 279–288. [CrossRef]

260. Huk, I.; Nanobashvili, J.; Orljanski, W.; Neumayer, C.; Punz, A.; Holzaepfel, A.; Fuegl, A.; Mittlboeck, M.; Polterauer, P.; Roth, E. L-arginine treatment in ischemia/reperfusion injury. *Cas. Lek. Cesk.* **1998**, *137*, 496–499.
261. Takeuchi, K.; McGowan, F.X.; Danh, H.C.; Glynn, P.; Simplaceanu, E.; del Nido, P.J. Direct detrimental effects of L-arginine upon ischemia–reperfusion injury to myocardium. *J. Mol. Cell Cardiol.* **1995**, *27*, 1405–1414. [CrossRef]
262. Kubo, S.H.; Rector, T.S.; Bank, A.J.; Williams, R.E.; Heifetz, S.M. Endothelium-dependent vasodilation is attenuated in patients with heart failure. *Circulation* **1991**, *84*, 1589–1596. [CrossRef]
263. Hirooka, Y.; Imaizumi, T.; Tagawa, T.; Shiramoto, M.; Endo, T.; Ando, S.; Takeshita, A. Effects of L-arginine on impaired acetylcholine-induced and ischemic vasodilation of the forearm in patients with heart failure. *Circulation* **1994**, *90*, 658–668. [CrossRef]
264. Doutreleau, S.; Mettauer, B.; Piquard, F.; Rouyer, O.; Schaefer, A.; Lonsdorfer, J.; Geny, B. Chronic L-arginine supplementation enhances endurance exercise tolerance in heart failure patients. *Int. J. Sports Med.* **2006**, *27*, 567–572. [CrossRef] [PubMed]
265. Parnell, M.M.; Holst, D.P.; Kaye, D.M. Augmentation of endothelial function following exercise training is associated with increased L-arginine transport in human heart failure. *Clin. Sci.* **2005**, *109*, 523–530. [CrossRef] [PubMed]
266. Chin-Dusting, J.P.; Kaye, D.M.; Lefkovits, J.; Wong, J.; Bergin, P.; Jennings, G.L. Dietary supplementation with L-arginine fails to restore endothelial function in forearm resistance arteries of patients with severe heart failure. *J. Am. Coll. Cardiol.* **1996**, *27*, 1207–1213. [CrossRef]
267. Schulman, S.P.; Becker, L.C.; Kass, D.A.; Champion, H.C.; Terrin, M.L.; Forman, S.; Ernst, K.V.; Kelemen, M.D.; Townsend, S.N.; Capriotti, A.; et al. L-arginine therapy in acute myocardial infarction: The Vascular Interaction With Age in Myocardial Infarction (VINTAGE MI) randomized clinical trial. *JAMA* **2006**, *295*, 58–64. [CrossRef]
268. Shu, J.; Santulli, G. Update on peripheral artery disease: Epidemiology and evidence-based facts. *Atherosclerosis* **2018**, *275*, 379–381. [CrossRef]
269. Azab, S.M.; Zamzam, A.; Syed, M.H.; Abdin, R.; Qadura, M.; Britz-McKibbin, P. Serum Metabolic Signatures of Chronic Limb-Threatening Ischemia in Patients with Peripheral Artery Disease. *J. Clin. Med.* **2020**, *9*, 1877. [CrossRef]
270. Misra, S.; Shishehbor, M.H.; Takahashi, E.A.; Aronow, H.D.; Brewster, L.P.; Bunte, M.C.; Kim, E.S.H.; Lindner, J.R.; Rich, K.; American Heart Association Council on Peripheral Vascular Disease; et al. Perfusion Assessment in Critical Limb Ischemia: Principles for Understanding and the Development of Evidence and Evaluation of Devices: A Scientific Statement From the American Heart Association. *Circulation* **2019**, *140*, e657–e672. [CrossRef]
271. Boger, R.H.; Bode-Boger, S.M.; Thiele, W.; Junker, W.; Alexander, K.; Frolich, J.C. Biochemical evidence for impaired nitric oxide synthesis in patients with peripheral arterial occlusive disease. *Circulation* **1997**, *95*, 2068–2074. [CrossRef]
272. Maxwell, A.J.; Cooke, J.P. Cardiovascular effects of L-arginine. *Curr. Opin. Nephrol. Hypertens.* **1998**, *7*, 63–70. [CrossRef]
273. Maxwell, A.J.; Anderson, B.E.; Cooke, J.P. Nutritional therapy for peripheral arterial disease: A double-blind, placebo-controlled, randomized trial of HeartBar. *Vasc. Med.* **2000**, *5*, 11–19.
274. Wilson, A.M.; Harada, R.; Nair, N.; Balasubramanian, N.; Cooke, J.P. L-arginine supplementation in peripheral arterial disease: No benefit and possible harm. *Circulation* **2007**, *116*, 188–195. [CrossRef] [PubMed]
275. Avogaro, A.; Albiero, M.; Menegazzo, L.; de Kreutzenberg, S.; Fadini, G.P. Endothelial dysfunction in diabetes: The role of reparatory mechanisms. *Diabetes Care* **2011**, *34* (Suppl. 2), S285–S290. [CrossRef] [PubMed]
276. Sena, C.M.; Pereira, A.M.; Seica, R. Endothelial dysfunction—A major mediator of diabetic vascular disease. *Biochim. Biophys. Acta* **2013**, *1832*, 2216–2231. [CrossRef] [PubMed]
277. Bogdanski, P.; Suliburska, J.; Grabanska, K.; Musialik, K.; Cieslewicz, A.; Skoluda, A.; Jablecka, A. Effect of 3-month L-arginine supplementation on insulin resistance and tumor necrosis factor activity in patients with visceral obesity. *Eur. Rev. Med. Pharmacol. Sci.* **2012**, *16*, 816–823.
278. Eid, H.M.; Reims, H.; Arnesen, H.; Kjeldsen, S.E.; Lyberg, T.; Seljeflot, I. Decreased levels of asymmetric dimethylarginine during acute hyperinsulinemia. *Metabolism* **2007**, *56*, 464–469. [CrossRef]

279. Sobrevia, L.; Nadal, A.; Yudilevich, D.L.; Mann, G.E. Activation of L-arginine transport (system y+) and nitric oxide synthase by elevated glucose and insulin in human endothelial cells. *J. Physiol.* **1996**, *490 Pt 3*, 775–781. [CrossRef]
280. Mann, G.E.; Yudilevich, D.L.; Sobrevia, L. Regulation of amino acid and glucose transporters in endothelial and smooth muscle cells. *Physiol. Rev.* **2003**, *83*, 183–252. [CrossRef]
281. Lembo, G.; Iaccarino, G.; Vecchione, C.; Barbato, E.; Morisco, C.; Monti, F.; Parrella, L.; Trimarco, B. Insulin enhances endothelial alpha2-adrenergic vasorelaxation by a pertussis toxin mechanism. *Hypertension* **1997**, *30*, 1128–1134. [CrossRef]
282. Zeng, G.; Nystrom, F.H.; Ravichandran, L.V.; Cong, L.N.; Kirby, M.; Mostowski, H.; Quon, M.J. Roles for insulin receptor, PI3-kinase, and Akt in insulin-signaling pathways related to production of nitric oxide in human vascular endothelial cells. *Circulation* **2000**, *101*, 1539–1545. [CrossRef]
283. Vecchione, C.; Aretini, A.; Maffei, A.; Marino, G.; Selvetella, G.; Poulet, R.; Trimarco, V.; Frati, G.; Lembo, G. Cooperation between insulin and leptin in the modulation of vascular tone. *Hypertension* **2003**, *42*, 166–170. [CrossRef]
284. Scherrer, U.; Randin, D.; Vollenweider, P.; Vollenweider, L.; Nicod, P. Nitric oxide release accounts for insulin's vascular effects in humans. *J. Clin. Investig.* **1994**, *94*, 2511–2515. [CrossRef] [PubMed]
285. Iaccarino, G.; Ciccarelli, M.; Sorriento, D.; Cipolletta, E.; Cerullo, V.; Iovino, G.L.; Paudice, A.; Elia, A.; Santulli, G.; Campanile, A.; et al. AKT participates in endothelial dysfunction in hypertension. *Circulation* **2004**, *109*, 2587–2593. [CrossRef] [PubMed]
286. Muniyappa, R.; Montagnani, M.; Koh, K.K.; Quon, M.J. Cardiovascular actions of insulin. *Endocr. Rev.* **2007**, *28*, 463–491. [CrossRef] [PubMed]
287. Pieper, G.M.; Siebeneich, W.; Dondlinger, L.A. Short-term oral administration of L-arginine reverses defective endothelium-dependent relaxation and cGMP generation in diabetes. *Eur. J. Pharmacol.* **1996**, *317*, 317–320. [CrossRef]
288. Khaidar, A.; Marx, M.; Lubec, B.; Lubec, G. L-arginine reduces heart collagen accumulation in the diabetic db/db mouse. *Circulation* **1994**, *90*, 479–483. [CrossRef]
289. Lubec, G.; Bartosch, B.; Mallinger, R.; Adamiker, D.; Graef, I.; Frisch, H.; Hoger, H. The effect of substance L on glucose-mediated cross-links of collagen in the diabetic db/db mouse. *Nephron* **1990**, *56*, 281–284. [CrossRef]
290. Lubec, B.; Hayn, M.; Kitzmuller, E.; Vierhapper, H.; Lubec, G. L-Arginine reduces lipid peroxidation in patients with diabetes mellitus. *Free Radic. Biol. Med.* **1997**, *22*, 355–357. [CrossRef]
291. Pieper, G.M. Review of alterations in endothelial nitric oxide production in diabetes: Protective role of arginine on endothelial dysfunction. *Hypertension* **1998**, *31*, 1047–1060. [CrossRef]
292. De Pirro, R.; Tamburrano, G.; Fusco, A.; Lauro, R. Arginine does not influence insulin binding on circulating monocytes. *Endokrinologie* **1980**, *75*, 243–246.
293. Wascher, T.C.; Graier, W.F.; Dittrich, P.; Hussain, M.A.; Bahadori, B.; Wallner, S.; Toplak, H. Effects of low-dose L-arginine on insulin-mediated vasodilatation and insulin sensitivity. *Eur. J. Clin. Investig.* **1997**, *27*, 690–695. [CrossRef]
294. Piatti, P.M.; Monti, L.D.; Valsecchi, G.; Magni, F.; Setola, E.; Marchesi, F.; Galli-Kienle, M.; Pozza, G.; Alberti, K.G. Long-term oral L-arginine administration improves peripheral and hepatic insulin sensitivity in type 2 diabetic patients. *Diabetes Care* **2001**, *24*, 875–880. [CrossRef] [PubMed]
295. Huynh, N.T.; Tayek, J.A. Oral arginine reduces systemic blood pressure in type 2 diabetes: Its potential role in nitric oxide generation. *J. Am. Coll. Nutr.* **2002**, *21*, 422–427. [CrossRef] [PubMed]
296. Lai, Y.L.; Aoyama, S.; Nagai, R.; Miyoshi, N.; Ohshima, H. Inhibition of L-arginine metabolizing enzymes by L-arginine-derived advanced glycation end products. *J. Clin. Biochem. Nutr.* **2010**, *46*, 177–185. [CrossRef] [PubMed]
297. Fiory, F.; Lombardi, A.; Miele, C.; Giudicelli, J.; Beguinot, F.; Van Obberghen, E. Methylglyoxal impairs insulin signalling and insulin action on glucose-induced insulin secretion in the pancreatic beta cell line INS-1E. *Diabetologia* **2011**, *54*, 2941–2952. [CrossRef]
298. Polykretis, P.; Luchinat, E.; Boscaro, F.; Banci, L. Methylglyoxal interaction with superoxide dismutase 1. *Redox Biol.* **2020**, *30*, 101421. [CrossRef]

299. Martina, V.; Masha, A.; Gigliardi, V.R.; Brocato, L.; Manzato, E.; Berchio, A.; Massarenti, P.; Settanni, F.; Della Casa, L.; Bergamini, S.; et al. Long-term N-acetylcysteine and L-arginine administration reduces endothelial activation and systolic blood pressure in hypertensive patients with type 2 diabetes. *Diabetes Care* **2008**, *31*, 940–944. [CrossRef]
300. Deveaux, A.; Fouillet, H.; Petzke, K.J.; Hermier, D.; Andre, E.; Bunouf, P.; Lantoine-Adam, F.; Benamouzig, R.; Mathe, V.; Huneau, J.F.; et al. A Slow-Compared with a Fast-Release Form of Oral Arginine Increases Its Utilization for Nitric Oxide Synthesis in Overweight Adults with Cardiometabolic Risk Factors in a Randomized Controlled Study. *J. Nutr.* **2016**, *146*, 1322–1329. [CrossRef]

© 2020 by the authors. Licensee MDPI, Basel, Switzerland. This article is an open access article distributed under the terms and conditions of the Creative Commons Attribution (CC BY) license (http://creativecommons.org/licenses/by/4.0/).

Review

Clinical Significance of Electronegative Low-Density Lipoprotein Cholesterol in Atherothrombosis

Chih-Sheng Chu [1,2,3], Shi Hui Law [4], David Lenzen [4], Yong-Hong Tan [4], Shih-Feng Weng [5], Etsuro Ito [4,6,7], Jung-Chou Wu [8], Chu-Huang Chen [9], Hua-Chen Chan [1,*] and Liang-Yin Ke [4,10,*]

1. Center for Lipid Biosciences, Kaohsiung Medical University Hospital, Kaohsiung Medical University, Kaohsiung 807377, Taiwan; chucs@kmu.edu.tw
2. Division of Cardiology, Department of International Medicine, Kaohsiung Medical University Hospital, Kaohsiung 807377, Taiwan
3. Division of Cardiology, Department of Internal Medicine, Kaohsiung Municipal Ta-Tung Hospital, Kaohsiung 80145, Taiwan
4. Department of Medical Laboratory Science and Biotechnology, College of Health Sciences, Kaohsiung Medical University, Kaohsiung 807378, Taiwan; shlaw_0909@hotmail.com (S.H.L.); dav.lenzen@gmail.com (D.L.); yonghongtan96@hotmail.com (Y.-H.T.); eito@waseda.jp (E.I.)
5. Department of Healthcare Administration and Medical Informatics, College of Health Sciences, Kaohsiung Medical University, Kaohsiung 807378, Taiwan; sfweng@kmu.edu.tw
6. Department of Biology, Waseda University, Tokyo 162-8480, Japan
7. Waseda Research Institute for Science and Engineering, Waseda University, Tokyo 162-8480, Japan
8. Division of Cardiology, Department of Internal Medicine, Pingtung Christian Hospital, Pingtung 90059, Taiwan; wujcdd@gmail.com
9. Vascular and Medicinal Research, Texas Heart Institute, Houston, TX 77030, USA; cchen@texasheart.org
10. Graduate Institute of Medicine, College of Medicine, & Drug Development and Value Creation Research Center, Kaohsiung Medical University, Kaohsiung 807378, Taiwan
* Correspondence: huachen.chan@gmail.com (H.-C.C.); kly@gap.kmu.edu.tw (L.-Y.K.); Tel.: +886-73121101 (ext. 2296); Fax: +886-73111996 (L.-Y.K.)

Received: 30 June 2020; Accepted: 28 July 2020; Published: 30 July 2020

Abstract: Despite the numerous risk factors for atherosclerotic cardiovascular diseases (ASCVD), cumulative evidence shows that electronegative low-density lipoprotein (L5 LDL) cholesterol is a promising biomarker. Its toxicity may contribute to atherothrombotic events. Notably, plasma L5 LDL levels positively correlate with the increasing severity of cardiovascular diseases. In contrast, traditional markers such as LDL-cholesterol and triglyceride are the therapeutic goals in secondary prevention for ASCVD, but that is controversial in primary prevention for patients with low risk. In this review, we point out the clinical significance and pathophysiological mechanisms of L5 LDL, and the clinical applications of L5 LDL levels in ASCVD can be confidently addressed. Based on the previously defined cut-off value by receiver operating characteristic curve, the acceptable physiological range of L5 concentration is proposed to be below 1.7 mg/dL. When L5 LDL level surpass this threshold, clinically relevant ASCVD might be present, and further exams such as carotid intima-media thickness, pulse wave velocity, exercise stress test, or multidetector computed tomography are required. Notably, the ultimate goal of L5 LDL concentration is lower than 1.7 mg/dL. Instead, with L5 LDL greater than 1.7 mg/dL, lipid-lowering treatment may be required, including statin, ezetimibe or PCSK9 inhibitor, regardless of the low-density lipoprotein cholesterol (LDL-C) level. Since L5 LDL could be a promising biomarker, we propose that a high throughput, clinically feasible methodology is urgently required not only for conducting a prospective, large population study but for developing therapeutics strategies to decrease L5 LDL in the blood.

Keywords: electronegative low-density lipoprotein; LDL(−); L5 LDL; oxidized LDL; oxLDL; cardiovascular disease; atherosclerosis

1. Introduction

Blood cholesterol remains the critical therapeutic target for primary and secondary prevention in clinical atherosclerotic cardiovascular disease (ASCVD), according to the international guidelines published by the American College of Cardiology (ACC) and the American Heart Association (AHA) [1,2]. Incorporating both low-density lipoprotein-cholesterol (LDL-C) and high-density lipoprotein-cholesterol (HDL-C) as the essential parameters, many proposed risks calculators, such as Pooled Cohort Equation or ASCVD Risk Estimator Plus, are currently available for assessing a person's overall risk and monitoring statin therapy [1]. Some reports advocate "the lower the LDL-C, the better" [3], yet other studies criticized that this cannot preclude the occurrence of ASCVD but instead brings more individuals to suffering the side effects after statin exposure, such as muscle pain, impaired liver function, and new-onset diabetes mellitus [4,5], adversely impacting the quality of life.

We previously reported that by using fast-protein liquid chromatography, LDL can be divided into five subfractions, L1-L5, based on increasing electronegativity. Of those, L5 LDL exhibits atherothrombogenic and proinflammatory properties in vitro and in vivo [6]. The concentration of L5 LDL is low in normal healthy subjects [7], but increased in patients with chronic cardiometabolic disorders (e.g., type 2 diabetes [8], metabolic syndrome [9]) or acute ischemic events (e.g., ST-elevation myocardial infarction [10], ischemic stroke [11]), regardless of their plasma LDL-C concentrations (Table 1). More recently, we demonstrated that L5 LDL also plays an atherogenic role in patients with systemic lupus erythematosus (SLE) as well as rheumatoid arthritis (RA), who often have severe atherosclerotic complications that, however, cannot be attributed to conventional risk factors [12,13].

Table 1. Clinical significance of electronegative low-density lipoprotein (L5 LDL)–data collected in Taiwan.

Publications	Sci Rep [7]	JCEM [9]	Blood [10]	Blood [11]	JCM [13]	AR [12]
Subjects	HLP	MetS	STEMI	stroke	RA	SLE
n	35	29	30	35	30	45
T-CHOL	235.9 ± 36.6	232.9 ± 31.6 [a]	179.1 ± 33.9	151.4 ± 34.3	219 (193–245) [c]	NA
TG	164.5 ± 90.6	259.6 ± 209.1 [a]	119.6 ± 65.6	123.8 ± 72.5	123 (87–170) [c]	NA
HDL-C	53.1 ± 16.4	45.4 ± 9.7 [a,**]	38.5 ± 8.6 [***]	32.7 ± 6.6	58.5 (48–66) [c]	48.9 ± 17.5 [**]
LDL-C	146.0 ± 34.9 [***]	142.2 ± 41.8 [a]	116.7 ± 32.4	105.4 ± 34.5	142 (111–168) [c]	105.1 ± 32.3 [*]
L5%	2.3 ± 1.3 [***]	5.3 ± 6.9 [a,***]	15.4 ± 14.5 [***]	19.1 ± 10.6 [***]	2.0 (1.3–4.5) [c,***]	2.4 ± 1.3 [***]
[L5]	3.2 ± 2.0 [***]	7.3 ± 9.8 [a,**]	18.9 ± 21.0 [***]	20.6 ± 13.5 [***]	2.9 (1.7–5.7) [c,***]	2.4 ± 1.3 [***]
Controls	NHC	None-MetS	NHC	NHC	NHC	NHC
n	35	29	30	25	12	37
T-CHOL	173.4 ± 32.8	215.3 ± 50.8 [b]	179.3 ± 32.9	150.8 ± 32.9	208 (201–231) [c]	NA
TG	79.7 ± 56.1	91.6 ± 47.5 [b]	78.6 ± 59.8	109 ± 38.5	90 (72.8–126) [c]	NA
HDL-C	54.4 ± 14.0	56.5 ± 17.4 [b]	55.6 ± 14.2	41.8 ± 12.1	59 (46–78) [c]	58 ± 16
LDL-C	103.3 ± 27.6	140.9 ± 44.5 [b]	108.1 ± 28.4	92.6 ± 33.5	131 (120–155) [c]	118.2 ± 23.3
L5%	1.3 ± 0.7	2.1 ± 1.4 [b]	1.5 ± 1.1	0.5 ± 0.3	0.9 (0.6–1.1) [c]	0.7 ± 0.3
[L5]	1.3 ± 0.7	3.0 ± 2.0 [b]	1.7 ± 1.5	0.5 ± 0.4	1.3 (0.8–1.5) [c]	0.8 ± 0.4
L5% [P't-NHC]	1.0 ± 0.2	3.2 ± 1.3	13.9 ± 2.7	18.6 ± 1.8	NA	1.7 ± 0.2
[L5] [P't-NHC]	1.9 ± 0.4	4.3 ± 1.9	17.2 ± 3.8	20.1 ± 2.3	NA	1.6 ± 0.2

Data are presented as the mean ± SD unless indicated otherwise. * $p < 0.05$; ** $p < 0.01$; *** $p < 0.001$; [a] Patient who met criteria of metabolic syndrome (MetS); [b] Individual who met two or fewer criteria; [c] Data are presented as the median (interquartile range). HLP: hyperlipidemia; STEMI: ST-segment elevation myocardial infarction; RA: rheumatoid arthritis with subclinical atherosclerosis; SLE: systemic lupus erythematosus; NHC: normal healthy control; T-CHOL: total cholesterol; TG: triglyceride; HDL-C: high-density lipoprotein cholesterol; LDL-C: low-density lipoprotein cholesterol; NA: not available; L5%: percentage of L5 subfraction in total LDL; [L5]: L5 concentration; L5% [P't – NHC]: the difference of L5% between two groups; [L5] [P't – NHC]: the difference of [L5] between two groups.

In 2018, Chu et al. determined the cut-off values of L5 levels for clinical ASCVD using receiver operating characteristic (ROC) curve analysis [7]. In their original study, individuals with a plasma L5 LDL level of less than 1.7 mg/dL showed no clinical evidence of ASCVD. In contrast, individuals with the plasma L5 LDL concentration exceeding the range of 2.3~2.6 mg/dL exhibited subclinical atherosclerosis or coronary artery disease (CAD) even when LDL-C or triglyceride (TG) levels were not elevated [7,13]. While statin can reduce L5 LDL in quantity by lowering the total LDL volume, their mechanism of action is not in quality targeting on the clearance of L5 LDL [14]. Discontinued statin therapy results in rebounds of both total LDL-C and L5 LDL to the pretreatment levels in three months [15]. These findings indicate that L5 LDL should be a potential clinical biomarker to be accurately measured for ASCVD risk stratification. We hereby propose a plasma L5 LDL concentration of greater than 1.7 mg/dL as a therapeutic threshold of initiating lipid-lowering treatments, on the basis that the odds ratio for CAD would reach 17.68 [7]. Furthermore, precision medicine targeting the removal of L5 LDL is of great importance in the future.

In this review, we summarize the latest research advances in the field of L5 LDL cytotoxicity and pathogenic significance, with the goals of (1) identifying L5 LDL as the primary biomarker for clinical ASCVD risk and the guide for statin or other lipid-lowering therapies; (2) establishing the recommended therapeutic threshold of plasma L5 LDL; (3) encouraging further research in the development of rapid quantification of L5 LDL for large scale epidemiological surveys among different cohorts; (4) providing insightful information for research on the new therapeutic strategies targeting L5 LDL or its atherogenic moieties. The most recent advances, notably the newly-established connection to autoimmune vascular diseases, actively support L5 LDL as a promising new domain of lipoprotein research.

2. Characteristics of Electronegative Low-Density Lipoprotein (L5 LDL)

2.1. Definition and Methodolgy

The concept of "electronegative LDL" was first proposed by Gotto and Hoff at Baylor College of Medicine in 1979. They purified the lipoprotein from human aortic plaques and normal intima by using differential ultracentrifugation. Their immunoelectrophoresis results showed that a group of LDL-like particle from aortic extracts was more electronegative than plasma LDL and associated with the atherosclerotic progression [16]. In 1987, the presence of "modified LDL" was reported by Dr. Avogaro and his colleagues in Italy. They proposed that the atherogenic properties and endothelial cells (ECs) cytotoxicity of modified LDL can be enhanced due to the occurrence of lipoprotein oxidation [17]. Later in 1988, they used fast-protein liquid chromatography (FPLC) equipped with an ion-exchange column to separate plasma LDL into electropositive LDL(+) and electronegative LDL(−) subfractions [18]. LDL(−) particles are heterogeneous in morphology and size and have a tendency to aggregate by electronic microscopic exam. Electronegative LDL has emerged as a naturally occurring, atherogenic entity irrespective of the concentration of plasma LDL-C [19–21]. In 2003, Yang and Chen modified the protocol of separation, LDL can be chromatographically divided into five subfractions with increasing electronegativity, L1–L5 [19,20] (Figure 1). L5 LDL is the most electronegative subfraction.

2.2. Glycosylation of Apolipoproteins in L5 LDL

With the technique developed by Yang and Chen, the least electronegative subfraction of LDL is termed as L1, whereas the most electronegative LDL is L5 [20]. It can be more accurate because L1 does not appear electropositive. Besides, the intermediary subfractions (i.e., L2-L4) can be useful while investigating the transitional changes of electronegativity from L1 to L5. Based on the definition that one particle of LDL contains one mole of apolipoprotein B100 (apoB100), L5 LDL isolated from plasma contains many other proteins such as apo(a), apolipoprotein CIII (apoCIII), apolipoprotein J (apoJ), platelet-activating factor acetylhydrolase (PAF-AH), and paraoxonase 1 (PON1) [22,23], which are not in L1 LDL. Besides, L5 LDL has significantly higher levels of apolipoprotein E (apoE) and apolipoprotein AI (apoAI).

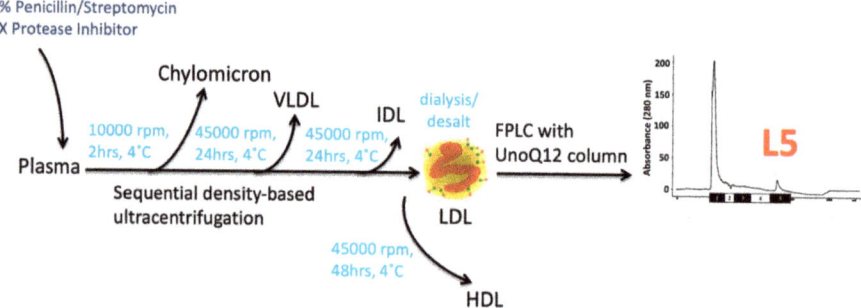

Figure 1. Schematic procedures of L5 LDL isolation. EDTA, antibiotics, and protease inhibitors are materials for the prevention of protein degradation. Samples undergo sequential density-based ultracentrifugation (10,000 rpm at 4 °C for 2 h; d = 1.004, 45,000 rpm at 4 °C for 24 h; d = 1.019, 45,000 rpm at 4 °C for 24 h; d = 1.063, 45,000 rpm at 4 °C for 48 h), and after that, LDL (d = 1.019~1.063) can be purified. Additional three times dialyzed against TRIS/EDTA buffer at pH 8.0 and later sterilized by 0.22 μm filter, the LDL sample can be further isolated into five subfractions by a fast-protein liquid chromatography (FPLC) system equipped with an UnoQ12 column. L5 LDL is the most electronegative subfraction. VLDL: very-low-density lipoprotein; IDL: intermediate-density lipoprotein; LDL: low-density lipoprotein; HDL: high-density lipoprotein; L5: electronegative LDL; FPLC: fast-protein liquid chromatography; rpm: revolutions per minute; 1× Protease Inhibitor: cOmplete™ (Roche Diagnostics, Basel, Switzerland).

In cardiomyocytes, apoE interacts with the voltage-dependent anion-selective channel (VDAC), leading to dynamin-related protein 1 (Drp1) phosphorylation and mitochondrial fission [24]. Additionally, apolipoproteins associated with L5 LDL from human plasma particles are found to be highly glycosylated. For instance, apoE that shows 94S, 194T, and 289T glycosylation with sialic acid terminal glycan, which alters the receptor selectivity and lipid-binding capability [25]. These findings may support that single nucleotide polymorphisms (SNPs) of apoE with changing electrical charges are associated with metabolic disorders [26,27]. ApoB100 glycosylation is also associated with the sphingomyelinase-like activity of electronegative LDL [28]. With sphingomyelinase activity, ceramide can be overproduced through the sphingomyelin hydrolysis pathway and therefore induces endothelial cell apoptosis.

2.3. Atherogenic Lipid Moieties of L5 LDL

By using colorimetric methods, L5 LDL shows triglyceride-rich but reduced cholesteryl ester in the lipid composition [19,29,30]. These findings match clinical observations regarding the higher plasma triglyceride content in patients with metabolic syndrome [31]. Other than that, L5 LDL from patients with familial hypercholesterolemia or diabetes has been shown to contain higher levels of lipoprotein-associated phospholipase A_2 (Lp-PLA2) [32–34]. The function of Lp-PLA2 is to hydrolyze phospholipids and generate lysophosphatidylcholine (LPC) and non-esterified fatty acids (NEFA) [35].

By using mass spectrometry, L5 LDL shows higher levels of ceramide, lysophosphatidylcholine (LPC) and platelet-activating factor (PAF) in comparison to L1 LDL [12,28]. Ceramide plays an essential role in stress-related cellular responses and apoptosis [36–39]. Alterations in ceramide levels have been recognized in pathological conditions such as Alzheimer's disease [40], type 2 diabetes [41], and cardiovascular diseases [42]. LPC stimulates inflammatory chemokine expression from endothelial cells [43–46], impairs arterial relaxation [47], increases oxidative stress [48,49], and inhibits endothelial cell migration and proliferation [50,51]. The level of LPC increases in cardiovascular diseases (CVDs), diabetes, and renal failure [52–54]. Besides, our recent studies also showed that LPC and PAF are inflammatory mediators that lead to the differentiation of monocytes into proinflammatory CD16[+] cells and contribute to endothelial dysfunction and vascular aging, thereby providing a novel explanation for the early onset of atherosclerosis-associated complications [12,55].

3. Cellular Signaling of L5 LDL

3.1. Signaling in Endothelial Cells

Accumulating evidence suggests that L5 LDL interacts with multiple cells such as endothelial cells [8,15,56–58], platelets [10,11], monocytes [9,59–63], and cardiomyocytes [24,64–67]. L5 LDL attracts both monocytes and lymphocytes to endothelial cells (ECs) [58], indicating the contribution in the early stage of atherosclerosis. L5 LDL is not recognized by the LDL receptor (LDLR) [68], but rather, it signals through the lectin-like oxidized LDL receptor-1 (LOX-1) and platelet-activating factor receptor (PAFR) [10,20,57]. LOX-1, initially identified as the major receptor for oxLDL in ECs, is expressed at high levels in pro-atherogenic settings and has been shown to have a critical role in atherogenesis [69,70].

Upon internalization through LOX-1, L5 LDL induces TNF-α expression, which subsequently triggers the expression of new LOX-1, making surrounding vascular epithelial cells increasingly susceptible to damage and apoptosis [8]. In cardiomyocyte, L5 LDL can enhance the ECs' activities by secreting Glu-Leu-Arg (ELR)$^+$, lipopolysaccharide-induced CXC chemokine (LIX) and interleukin-8 (IL-8), which further initiated CXCR2/PI3K/NF-κB signaling. These signals will then contribute and induce cardiomyocyte apoptosis through the release of the proinflammatory cytokines TNF-α and IL-1β [64].

On the other hand, L5 LDL possesses the ability to impair vascular ECs integrity and induce ECs apoptosis by suppressing the fibroblast growth factor 2 (FGF2) transcription and disrupting its autoregulation repairing system [8]. Cellular exogenous FGF2 plays a pivotal role in promoting cell metabolism, proliferation, cell survival, growth, and preventing apoptosis through the PI3K-Akt pathway [71]. These findings indicate that ECs dysfunction can be augmented by disrupting the FGF2 formation.

3.2. Signaling in Platelets

In the field of platelets, adenosine diphosphate (ADP), one of the major soluble agonists can mainly regulate the P2Y12/phosphatidylinositol-3 kinase (PI3K) pathway for platelet aggregation [72]. Apart from that, ADP increases LOX-1 expression and glycoprotein (GP)IIb/IIIa activation [73]. Through LOX-1 and PAFR, L5 LDL enhances ADP signaling of platelets. Besides, L5 LDL increases P-selectin and tissue factor expression on ECs. Particularly, P-selectin shows to be capable of interaction with the PAFR and induces platelet adherence and activation. These platelet-EC interactions triggered by L5 LDL may promote thrombosis formation, leading to STEMI [10].

In another study, L5 LDL induces amyloid β (Aβ) secretion through LOX-1 and IkB kinase 2 (IKK2) activation. Synergistically, L5 LDL and Aβ promote the platelet aggregation and activation [11]. These findings suggest that L5 is the thrombogenic fraction of LDL and may contribute to platelet hyper-reactivity, STEMI, and stroke complications [74,75].

3.3. Signaling in Immune Cells

During the past three decades, the autoimmune hypothesis of atherosclerosis is prospering under the evidence of LDL-containing circulating immune complexes (LDL-CIC) accumulation in atherogenesis [76,77]. Different from native LDL, the LDL-CIC is more electronegative and may alter lipid and lipoprotein levels in ECs and macrophages [78]. Besides, LDL(−) can induce inflammatory cytokines to release from monocytes, such as monocyte chemoattractant protein 1 (MCP1), interleukin-6 (IL-6), IL-8, growth-related oncogene (GRO), granulocyte-monocyte-colony stimulating factor (GM-CSF) [79], matrix metalloproteinase-9 (MMP-9) and its inhibitor tissue inhibitors of metalloproteinase-1 (TIMP-1) [63]. The release of these cytokines by LDL(−) might be mediated through CD14/toll-like receptor 4 (TLR4) signaling pathways [80].

LDL(−) also involves apoptosis and cytokine induction by upregulation of proapoptotic factor Fas on mononuclear leukocytes [81]. Apart from that, Klimov et al. have reported that mouse macrophages cultured with the presence of LDL-CIC can increased uptake of LDL [82]. Additionally, incubation of human peritoneal macrophages with the same condition causes the transformation of macrophages into foam cells [83].

In our recent studies, L5 LDL triggers the differentiation of $CD16^+$ monocytes. Through the CX3CR1 and CD16 expressing monocytes interact with CX3CL1-positive activated ECs, L5 LDL induces monocyte-endothelial cell adhesion [12]. L5 LDL also enhances the polarization of M1 macrophages that infiltrate to adipose tissue and lead to dysfunction and inflammation [9]. In macrophages, L5 LDL induces granulocyte colony-stimulating factor (G-CSF) and GM-CSF overproduction [84,85]. These biomarkers are associated with inflammation, increased risk of cardiovascular complications and STEMI [85,86]. Besides, L5 LDL enhances the overexpression of interleukin (IL)-1β (IL-1β) through the activation of the nucleotide-binding oligomerization domain (NOD)-like receptor pyrin domain containing 3 (NLRP3) inflammasomes [87].

4. Clinical Significance of L5 LDL

4.1. Cardiometabolic Disorders

In healthy subjects, the LDL constitution comprises mostly L1 LDL and the least of L5 LDL. The majority of their LDL, L1 LDL, is endocytosed by the low-density lipoprotein receptor (LDL-R) and processed by endolysosomes to provide nutrients to the cells [57]. In contrast, L5 LDL is the atherogenic component and its levels are elevated in the plasma of patients with increased cardiovascular risk [7,9,10]. The elevated levels of plasma L5 LDL can be found in patients with CAD, hyperlipidemia (HLP) [7], metabolic syndrome (MetS) [9], familial hypercholesterolemia (FH) [88], diabetes mellitus (DM) [8], and in smokers [89]. The reference range of L5% in total LDL and absolute L5 concentration ([L5] = L5% × LDL-C) for healthy adults were determined to be less than 1.6% and less than 1.7 mg/dL [7]. The increasing levels of L5% and absolute L5 concentration (mg/dL) can be found in patients mentioned above, respectively (Table 1). The level of absolute L5 concentration of STEMI patients could be 11.1 ± 14.0 times higher than in the healthy controls [10]. According to previous studies, L5 LDL can disrupt the integrity of ECs through LOX-1 [8] and promote the cell apoptosis by activating the signaling cascade downstream of CD14/TLR4 [80]. Hence, lowering the levels of L5 LDL could further be the goal of treatment in those diseases.

4.2. Acute Ischemic Events

According to the American Stroke Association (ASA) statement, acute ischemic stroke stays as the fifth cause of death and a leading cause of disability in the United States. Previous studies showed that L5% and L5 concentration in healthy control are 0.5 ± 0.3 and 0.5 ± 0.4 mg/dL, respectively. Nevertheless, in stroke patients, these markers will be elevated as 19.1 ± 10.6 and 20.6 ± 13.5 mg/dL, respectively [11]. L5 LDL can further reduce the viability of cultured ECs [8,20,22] and promote EC dysfunction by triggering procoagulant activity. Even more, L5 LDL may increase EC-platelet interactions to induce platelet activation [10,75]. On the basic concept of ASAs, 80 percent of stroke events are preventable. Early diagnosis may be relevant to the prevention of stroke. Overall, L5 LDL plays a critical role in the development of stroke. The ablation of L5 LDL may be a compelling goal of treatment in stroke.

4.3. Autoimmune Diseases

Patients with systemic lupus erythematosus (SLE) frequently accompanied by early vascular aging (EVA) [12] and severe atherosclerosis complications [90–92], though their low-density lipoprotein (LDL) levels remain low. Comparing LDL-C and L5 LDL levels between controls and patients, results showed that LDL-C was lower in SLE patients than in controls (105.1 ± 32.3 mg/dL versus 118.2 ± 23.3 mg/dL), but the concentration of L5 LDL in SLE patients were three times higher than in controls (2.4 ± 1.3 mg/dL versus 0.8 ± 0.4 mg/dL) [12].

As per SLE, the plasma L5% and L5 levels were significantly higher in rheumatoid arthritis (RA) patients than in controls. The expression levels of LDLR in PBMCs were no significant difference in observed, but the levels of LOX-1 in PBMCs of RA patients were two to three times higher than healthy controls [13]. This phenomenon indicates that the higher expression levels of LOX-1 and L5 LDL, the more can be uptake through LOX-1 to cause ECs dysfunction and atherosclerosis.

5. Implication of L5 LDL

5.1. Diagnostic Value of L5 LDL

According to the 2020 report from the American Heart Association (AHA), cardiovascular diseases remain the leading cause of death worldwide [93]. Although several new drugs and more aggressive approaches are proposed [2,94], the laboratory identification of atherogenic factors is still not specific. Currently, the clinical evaluation of ASCVD risk is based on (1) HDL-C in men: [<40 mg/dL]; in women: [<50 mg/dL]; (2) LDL-C [>160 mg/dL]; (3) TG [≥200 mg/dL]; (4) cholesterol ratio, i.e., total cholesterol/HDL. Other important issues include who should be treated, dosage, and goal of treatment.

Due to the lack of a quick and feasible measurement, now estimating the 10-year ASCVD risk by the Estimator App (developed by the American College of Cardiology Foundation) is essential. Several factors are calculated, including age, gender, blood pressure, total cholesterol, HDL, LDL, histories such as diabetes, smoking, statin, and aspirin usage. Recent ACC/AHA 2018 cholesterol guidelines for the prevention of cardiovascular diseases adopted the risk-enhancers concept, focusing on more detailed patients' history, including metabolic syndrome, chronic kidney disease, inflammatory condition, premature menopause, high-risk race, and some other new lipid parameters and biomarkers such as high sensitivity C-reactive protein (hs-CRP), Lipoprotein (a) (Lp(a)), apoB100, and ankle-brachial index (ABI) [95] in sharing decision-making.

Our recent clinical studies show that the four statin benefit groups are characterized by higher levels of L5 LDL [14]. L5 LDL markers, either percentage or the absolute plasma concentration, can be more reliable than those listed above (Table 1) [7–13,15,67]. Besides, the molecular mechanisms and the clinical relevance of L5 LDL provide strong evidence-based supports to demonstrate that L5 LDL can be potential biomarkers in the early diagnosis of vascular aging, atherosclerosis and cardiovascular diseases [6,24,25,28,55,56,60,63,70,84].

Based on the reference range of L5 LDL and the fact that higher concentration increases the risks of ASCVD, we propose that individuals with absolute plasma L5 LDL level of less than 1.7 mg/dL are within the acceptable safe range. These individuals should avoid lipid-lowering therapy despite that their LDL-C levels might exceed 190 mg/dL or above. For patients with increasing plasma L5 LDL levels, the screening tests for CVD such as carotid intima-media thickness, pulse wave velocity, and ankle-brachial index for peripheral artery disease, exercise stress test, and multidetector computerized tomography (MDCT) for CAD are highly recommended.

For those patients with established clinical ASCVD, acute coronary syndrome, CAD with percutaneous coronary intervention or bypass surgery, the therapeutic goal of absolute L5 LDL should still less than 1.7 mg/dL, even if their LDL-C has been lowered to less than 70 mg/dL that recommended by current guideline. In fact, the LDL-C goal has been revised to less than 55 mg/dL for patients with very-high risk and to 40 mg/dL for patients with recurrent myocardial infarction within two years by recent ESC/EAS 2019 dyslipidemia guideline. It reflects that these patients are of high toxic lipid components and these patients carry poor cardiovascular outcomes.

By keeping L5 LDL goal <1.7 mg/dL, more aggressive lipid-lowering agents, e.g., ezetimibe or PSCK9 inhibitors, on top of maximally tolerated statin use, should be considered regardless of the LDL-C level. That is why the 2013 ACC/AHA cholesterol guideline emphasized the intensity instead of the LDL-C target for four statin-benefit groups. Notably, current lipid-lowering agents available have not been shown to specifically reduce the toxic L5 LDL, but lower the overall LDL-C. Targeting specifically on toxic L5 LDL therapeutic strategies has been developing and hopefully can be beneficial for better CV outcomes in the near future.

5.2. Clinical Implication of L5 LDL

"No treatment goal", "the lower the LDL-C, the better", and "beyond absolute goals toward personalized risk" are the current strategies of primary and secondary prevention of CVDs [96,97]. The incidence of subclinical atherosclerosis remains as high as 64% in the LDL-C range of 150~160 mg/dL and 11% in the range of 60~70 mg/dL [4,98]. However, patients receiving statin may complain about the side effects of muscle pain or weakness; and some of them stopped taking statins because of the intolerance of side effects [99,100].

Based on ROC curve analysis, a L5 LDL level of more than 1.7 mg/dL is highly associated with subclinical atherosclerosis or increased odds ratio of CVDs [7,13]. We propose that individuals with a plasma L5 LDL level of more than 1.7 mg/dL are required for intervention. For example, statin therapies may be beneficial in reducing the electronegative subfraction of LDL [14,15,101,102]; while three months after discontinuation, the concentration of L5 LDL may return to the untreated-baseline levels [15]. Besides, the treating goal of LDL-lowering therapies should be the L5 LDL concentration of less than 1.7 mg/dL (Figure 2).

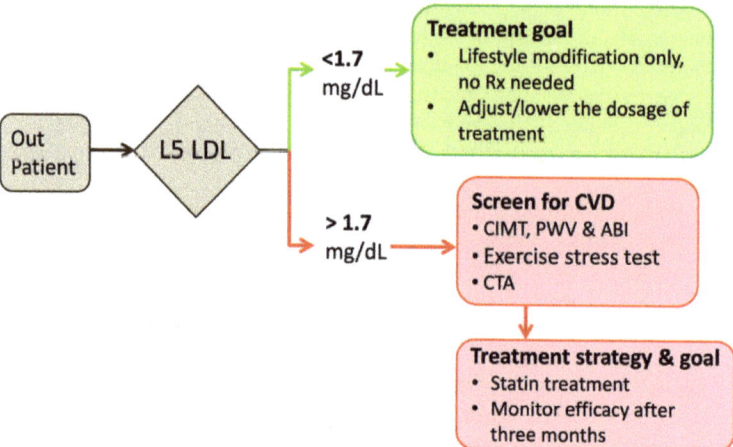

Figure 2. Reference ranges of L5 LDL and the strategy of statin treatment. Examining the plasma levels of L5 LDL can be useful to determine (1) whether to receive statin or not, (2) more detailed physical exams for cardiovascular functions and lesions, (3) the treatment strategy and goal. Rx: medical prescription; CIMT: carotid intima-media thickness; PWV: pulse wave velocity; ankle-brachial index; CTA: computed tomography angiography.

5.3. Drawback and Limitation of L5 LDL Quantification

Even though the clinical significance of L5 LDL can be high, some drawbacks and limitations must be mentioned. First of all, a large scale of clinical study must be done to validate the significance of being a novel clinical biomarker. Dr. Sanchez-Quesada and many other study groups published a variety of papers mentioning the clinical prevalence and the pathogenic mechanisms of electronegative LDL [103–108]. The level difference between patients and controls are comparable, and the atherogenic properties are noticeable. However, since there are algorithm differences in both experimental materials and isolation protocols, it would be necessary to adjust the reference range of controls. Moreover, due to the laborious work and long turn-around time (Figure 1), the diagnostic method's efficiency must be significantly improved. A rapid and clinically feasible diagnostic approach must be invented. Moreover, a precision medicine targeting the removal of L5 LDL is of great clinical importance.

6. Conclusions

In summary, the recognition of L5 LDL keeps rising. Until now, it has been a promising clinical biomarker for cardiovascular diseases despite the levels of LDL-C. We propose that a large-scale population survey and a high throughput methodology are required. Undisputedly, new strategies for directly eliminating L5 LDL from the bloodstream are essential works in the future.

Author Contributions: Conceptualization, C.-S.C. and L.-Y.K.; methodology, H.-C.C.; software, H.-C.C., S.-F.W.; validation and formal analysis, data curation, H.-C.C., S.-F.W.; writing—original draft preparation, S.H.L., D.L., Y.-H.T.; writing—review and editing, C.-S.C., E.I. and L.-Y.K.; visualization, H.-C.C.; supervision, C.-H.C. and J.-C.W.; funding acquisition, H.-C.C. and L.-Y.K. All authors have read and agreed to the published version of the manuscript.

Funding: This work was supported in part by grants from the Kaohsiung Medical University (KMU-DK108004, KMU-TC108A03-0), Kaohsiung Medical University Hospital (KMUH107-7R92, kmtth-103-023), Taiwan Ministry of Science and Technology (MOST109-2320-B-037-028-, MOST109-2628-B-037-010-) and Taiwan Ministry of Health and Welfare (MOHW106-TDU-B-212-113006).

Conflicts of Interest: The authors declare no conflict of interest.

References

1. Stone, N.J.; Robinson, J.G.; Lichtenstein, A.H.; Bairey Merz, C.N.; Blum, C.B.; Eckel, R.H.; Goldberg, A.C.; Gordon, D.; Levy, D.; Lloyd-Jones, D.M.; et al. 2013 ACC/AHA guideline on the treatment of blood cholesterol to reduce atherosclerotic cardiovascular risk in adults: A report of the American College of Cardiology/American Heart Association Task Force on Practice Guidelines. *J. Am. Coll. Cardiol.* **2014**, *63*, 2889–2934. [PubMed]
2. Grundy, S.M.; Stone, N.J.; Bailey, A.L.; Beam, C.; Birtcher, K.K.; Blumenthal, R.S.; Braun, L.T.; de Ferranti, S.; Faiella-Tommasino, J.; Forman, D.E.; et al. 2018 AHA/ACC/AACVPR/AAPA/ABC/ACPM/ADA/AGS/APhA/ASPC/NLA/PCNA Guideline on the Management of Blood Cholesterol: Executive Summary: A Report of the American College of Cardiology/American Heart Association Task Force on Clinical Practice Guidelines. *J. Am. Coll. Cardiol.* **2019**, *73*, 3168–3209. [CrossRef] [PubMed]
3. Mach, F.; Baigent, C.; Catapano, A.L.; Koskinas, K.C.; Casula, M.; Badimon, L.; Chapman, M.J.; De Backer, G.G.; Delgado, V.; Ference, B.A.; et al. 2019 ESC/EAS Guidelines for the management of dyslipidaemias: Lipid modification to reduce cardiovascular risk. *Eur. Heart J.* **2020**, *41*, 111–188. [CrossRef] [PubMed]
4. Fernandez-Friera, L.; Fuster, V.; Lopez-Melgar, B.; Oliva, B.; Garcia-Ruiz, J.M.; Mendiguren, J.; Bueno, H.; Pocock, S.; Ibanez, B.; Fernandez-Ortiz, A.; et al. Normal LDL-Cholesterol Levels Are Associated With Subclinical Atherosclerosis in the Absence of Risk Factors. *J. Am. Coll. Cardiol.* **2017**, *70*, 2979–2991. [CrossRef]
5. Toth, P.P.; Patti, A.M.; Giglio, R.V.; Nikolic, D.; Castellino, G.; Rizzo, M.; Banach, M. Management of Statin Intolerance in 2018: Still More Questions Than Answers. *Am. J. Cardiovasc. Drugs* **2018**, *18*, 157–173. [CrossRef]
6. Ke, L.Y.; Stancel, N.; Bair, H.; Chen, C.H. The underlying chemistry of electronegative LDL's atherogenicity. *Curr. Atheroscler. Rep.* **2014**, *16*, 428. [CrossRef]
7. Chu, C.S.; Chan, H.C.; Tsai, M.H.; Stancel, N.; Lee, H.C.; Cheng, K.H.; Tung, Y.C.; Chan, H.C.; Wang, C.Y.; Shin, S.J.; et al. Range of L5 LDL levels in healthy adults and L5's predictive power in patients with hyperlipidemia or coronary artery disease. *Sci. Rep.* **2018**, *8*, 11866. [CrossRef]
8. Lu, J.; Jiang, W.; Yang, J.H.; Chang, P.Y.; Walterscheid, J.P.; Chen, H.H.; Marcelli, M.; Tang, D.; Lee, Y.T.; Liao, W.S.; et al. Electronegative LDL impairs vascular endothelial cell integrity in diabetes by disrupting fibroblast growth factor 2 (FGF2) autoregulation. *Diabetes* **2008**, *57*, 158–166. [CrossRef]
9. Ke, L.Y.; Chan, H.C.; Chan, H.C.; Kalu, F.C.U.; Lee, H.C.; Lin, I.L.; Jhuo, S.J.; Lai, W.T.; Tsao, C.R.; Sawamura, T.; et al. Electronegative Low-Density Lipoprotein L5 Induces Adipose Tissue Inflammation Associated With Metabolic Syndrome. *J. Clin. Endocrinol. Metab.* **2017**, *102*, 4615–4625. [CrossRef]

10. Chan, H.C.; Ke, L.Y.; Chu, C.S.; Lee, A.S.; Shen, M.Y.; Cruz, M.A.; Hsu, J.F.; Cheng, K.H.; Chan, H.C.; Lu, J.; et al. Highly electronegative LDL from patients with ST-elevation myocardial infarction triggers platelet activation and aggregation. *Blood* **2013**, *122*, 3632–3641. [CrossRef]
11. Shen, M.Y.; Chen, F.Y.; Hsu, J.F.; Fu, R.H.; Chang, C.M.; Chang, C.T.; Liu, C.H.; Wu, J.R.; Lee, A.S.; Chan, H.C.; et al. Plasma L5 levels are elevated in ischemic stroke patients and enhance platelet aggregation. *Blood* **2016**, *127*, 1336–1345. [CrossRef]
12. Chan, H.C.; Bonnie Chan, H.C.; Liang, C.J.; Lee, H.C.; Su, H.; Lee, A.S.; Shiea, J.; Tsai, W.C.; Ou, T.T.; Wu, C.C.; et al. Role of Low-density Lipoprotein in Early Vascular Aging Associated With Systemic Lupus Erythematosus. *Arthritis Rheumatol.* **2020**, *72*, 972–984. [CrossRef]
13. Chang, C.Y.; Chen, C.H.; Chen, Y.M.; Hsieh, T.Y.; Li, J.P.; Shen, M.Y.; Lan, J.L.; Chen, D.Y. Association between Negatively Charged Low-Density Lipoprotein L5 and Subclinical Atherosclerosis in Rheumatoid Arthritis Patients. *J. Clin. Med.* **2019**, *8*, 177. [CrossRef]
14. Chu, C.S.; Ke, L.Y.; Chan, H.C.; Chan, H.C.; Chen, C.C.; Cheng, K.H.; Lee, H.C.; Kuo, H.F.; Chang, C.T.; Chang, K.C.; et al. Four Statin Benefit Groups Defined by the 2013 ACC/AHA New Cholesterol Guideline are Characterized by Increased Plasma Level of Electronegative Low-Density Lipoprotein. *Acta Cardiol. Sin.* **2016**, *32*, 667–675.
15. Chu, C.S.; Wang, Y.C.; Lu, L.S.; Walton, B.; Yilmaz, H.R.; Huang, R.Y.; Sawamura, T.; Dixon, R.A.; Lai, W.T.; Chen, C.H.; et al. Electronegative low-density lipoprotein increases C-reactive protein expression in vascular endothelial cells through the LOX-1 receptor. *PLoS ONE* **2013**, *8*, e70533. [CrossRef]
16. Hoff, H.F.; Bradley, W.A.; Heideman, C.L.; Gaubatz, J.W.; Karagas, M.D.; Gotto, A.M., Jr. Characterization of low density lipoprotein-like particle in the human aorta from grossly normal and atherosclerotic regions. *Biochim. Biophys. Acta* **1979**, *573*, 361–374. [CrossRef]
17. Avogaro, P.; Bittolo Bon, G.; Cazzolato, G. Meaning of a modified LDL in humans. *Adv. Exp. Med. Biol.* **1987**, *210*, 209–212. [PubMed]
18. Avogaro, P.; Bon, G.B.; Cazzolato, G. Presence of a modified low density lipoprotein in humans. *Arteriosclerosis* **1988**, *8*, 79–87. [CrossRef]
19. Yang, C.Y.; Raya, J.L.; Chen, H.H.; Chen, C.H.; Abe, Y.; Pownall, H.J.; Taylor, A.A.; Smith, C.V. Isolation, characterization, and functional assessment of oxidatively modified subfractions of circulating low-density lipoproteins. *Arter. Thromb Vasc. Biol.* **2003**, *23*, 1083–1090. [CrossRef]
20. Chen, C.H.; Jiang, T.; Yang, J.H.; Jiang, W.; Lu, J.; Marathe, G.K.; Pownall, H.J.; Ballantyne, C.M.; McIntyre, T.M.; Henry, P.D.; et al. Low-density lipoprotein in hypercholesterolemic human plasma induces vascular endothelial cell apoptosis by inhibiting fibroblast growth factor 2 transcription. *Circulation* **2003**, *107*, 2102–2108. [CrossRef]
21. Sanchez-Quesada, J.L.; Vinagre, I.; de Juan-Franco, E.; Sanchez-Hernandez, J.; Blanco-Vaca, F.; Ordonez-Llanos, J.; Perez, A. Effect of improving glycemic control in patients with type 2 diabetes mellitus on low-density lipoprotein size, electronegative low-density lipoprotein and lipoprotein-associated phospholipase A2 distribution. *Am. J. Cardiol.* **2012**, *110*, 67–71. [CrossRef] [PubMed]
22. Ke, L.Y.; Engler, D.A.; Lu, J.; Matsunami, R.K.; Chan, H.C.; Wang, G.J.; Yang, C.Y.; Chang, J.G.; Chen, C.H. Chemical composition-oriented receptor selectivity of L5, a naturally occurring atherogenic low-density lipoprotein. *Pure Appl. Chem.* **2011**, *83*, 1731–1740. [CrossRef]
23. Bancells, C.; Canals, F.; Benitez, S.; Colome, N.; Julve, J.; Ordonez-Llanos, J.; Sanchez-Quesada, J.L. Proteomic analysis of electronegative low-density lipoprotein. *J. Lipid Res.* **2010**, *51*, 3508–3515. [CrossRef] [PubMed]
24. Chen, W.Y.; Chen, Y.F.; Chan, H.C.; Chung, C.H.; Peng, H.Y.; Ho, Y.C.; Chen, C.H.; Chang, K.C.; Tang, C.H.; Lee, A.S. Role of apolipoprotein E in electronegative low-density lipoprotein-induced mitochondrial dysfunction in cardiomyocytes. *Metabolism* **2020**, *107*, 154227. [CrossRef] [PubMed]
25. Ke, L.Y.; Chan, H.C.; Chen, C.C.; Chang, C.F.; Lu, P.L.; Chu, C.S.; Lai, W.T.; Shin, S.J.; Liu, F.T.; Chen, C.H. Increased APOE glycosylation plays a key role in the atherogenicity of L5 low-density lipoprotein. *FASEB J.* **2020**. [CrossRef]
26. Santos-Ferreira, C.; Baptista, R.; Oliveira-Santos, M.; Costa, R.; Pereira Moura, J.; Goncalves, L. Apolipoprotein E2 Genotype Is Associated with a 2-Fold Increase in the Incidence of Type 2 Diabetes Mellitus: Results from a Long-Term Observational Study. *J. Lipids* **2019**, *2019*, 1698610. [CrossRef]
27. Safieh, M.; Korczyn, A.D.; Michaelson, D.M. ApoE4: An emerging therapeutic target for Alzheimer's disease. *BMC Med.* **2019**, *17*, 64. [CrossRef]

28. Ke, L.Y.; Chan, H.C.; Chen, C.C.; Lu, J.; Marathe, G.K.; Chu, C.S.; Chan, H.C.; Wang, C.Y.; Tung, Y.C.; McIntyre, T.M.; et al. Enhanced Sphingomyelinase Activity Contributes to the Apoptotic Capacity of Electronegative Low-Density Lipoprotein. *J. Med. Chem.* **2016**, *59*, 1032–1040.
29. Sanchez-Quesada, J.L.; Camacho, M.; Anton, R.; Benitez, S.; Vila, L.; Ordonez-Llanos, J. Electronegative LDL of FH subjects: Chemical characterization and induction of chemokine release from human endothelial cells. *Atherosclerosis* **2003**, *166*, 261–270. [CrossRef]
30. Benitez, S.; Perez, A.; Sanchez-Quesada, J.L.; Wagner, A.M.; Rigla, M.; Arcelus, R.; Jorba, O.; Ordonez-Llanos, J. Electronegative low-density lipoprotein subfraction from type 2 diabetic subjects is proatherogenic and unrelated to glycemic control. *Diabetes Metab. Res. Rev.* **2007**, *23*, 26–34. [CrossRef]
31. Ho, C.I.; Chen, J.Y.; Chen, S.Y.; Tsai, Y.W.; Weng, Y.M.; Tsao, Y.C.; Li, W.C. Relationship between TG/HDL-C ratio and metabolic syndrome risk factors with chronic kidney disease in healthy adult population. *Clin. Nutr.* **2015**, *34*, 874–880. [CrossRef]
32. Gaubatz, J.W.; Gillard, B.K.; Massey, J.B.; Hoogeveen, R.C.; Huang, M.; Lloyd, E.E.; Raya, J.L.; Yang, C.Y.; Pownall, H.J. Dynamics of dense electronegative low density lipoproteins and their preferential association with lipoprotein phospholipase A(2). *J. Lipid Res.* **2007**, *48*, 348–357. [CrossRef] [PubMed]
33. Sanchez-Quesada, J.L.; Vinagre, I.; De Juan-Franco, E.; Sanchez-Hernandez, J.; Bonet-Marques, R.; Blanco-Vaca, F.; Ordonez-Llanos, J.; Perez, A. Impact of the LDL subfraction phenotype on Lp-PLA2 distribution, LDL modification and HDL composition in type 2 diabetes. *Cardiovasc. Diabetol.* **2013**, *12*, 112. [CrossRef] [PubMed]
34. Yang, C.Y.; Chen, H.H.; Huang, M.T.; Raya, J.L.; Yang, J.H.; Chen, C.H.; Gaubatz, J.W.; Pownall, H.J.; Taylor, A.A.; Ballantyne, C.M.; et al. Pro-apoptotic low-density lipoprotein subfractions in type II diabetes. *Atherosclerosis* **2007**, *193*, 283–291. [CrossRef] [PubMed]
35. Zalewski, A.; Macphee, C. Role of lipoprotein-associated phospholipase A2 in atherosclerosis: Biology, epidemiology, and possible therapeutic target. *Arter. Thromb. Vasc. Biol.* **2005**, *25*, 923–931. [CrossRef]
36. Simons, K.; Ikonen, E. Functional rafts in cell membranes. *Nature* **1997**, *387*, 569–572. [CrossRef]
37. Perry, D.K.; Obeid, L.M.; Hannun, Y.A. Ceramide and the regulation of apoptosis and the stress response. *Trends Cardiovasc. Med.* **1996**, *6*, 158–162. [CrossRef]
38. Spiegel, S.; Cuvillier, O.; Edsall, L.; Kohama, T.; Menzeleev, R.; Olivera, A.; Thomas, D.; Tu, Z.; Van Brocklyn, J.; Wang, F. Roles of sphingosine-1-phosphate in cell growth, differentiation, and death. *Biochemistry* **1998**, *63*, 69–73.
39. Sassoli, C.; Pierucci, F.; Zecchi-Orlandini, S.; Meacci, E. Sphingosine 1-Phosphate (S1P)/ S1P Receptor Signaling and Mechanotransduction: Implications for Intrinsic Tissue Repair/Regeneration. *Int. J. Mol. Sci.* **2019**, *20*, 5545. [CrossRef]
40. Filippov, V.; Song, M.A.; Zhang, K.; Vinters, H.V.; Tung, S.; Kirsch, W.M.; Yang, J.; Duerksen-Hughes, P.J. Increased ceramide in brains with Alzheimer's and other neurodegenerative diseases. *J. Alzheimer's Dis.* **2012**, *29*, 537–547. [CrossRef]
41. Das, U.N. Is There a Role for Bioactive Lipids in the Pathobiology of Diabetes Mellitus? *Front. Endocrinol.* **2017**, *8*, 182. [CrossRef] [PubMed]
42. Laaksonen, R.; Ekroos, K.; Sysi-Aho, M.; Hilvo, M.; Vihervaara, T.; Kauhanen, D.; Suoniemi, M.; Hurme, R.; Marz, W.; Scharnagl, H.; et al. Plasma ceramides predict cardiovascular death in patients with stable coronary artery disease and acute coronary syndromes beyond LDL-cholesterol. *Eur. Heart J.* **2016**, *37*, 1967–1976. [CrossRef]
43. Huang, Y.H.; Schafer-Elinder, L.; Wu, R.; Claesson, H.E.; Frostegard, J. Lysophosphatidylcholine (LPC) induces proinflammatory cytokines by a platelet-activating factor (PAF) receptor-dependent mechanism. *Clin. Exp. Immunol.* **1999**, *116*, 326–331. [CrossRef]
44. Takahara, N.; Kashiwagi, A.; Maegawa, H.; Shigeta, Y. Lysophosphatidylcholine stimulates the expression and production of MCP-1 by human vascular endothelial cells. *Metabolism* **1996**, *45*, 559–564. [CrossRef]
45. Murugesan, G.; Sandhya Rani, M.R.; Gerber, C.E.; Mukhopadhyay, C.; Ransohoff, R.M.; Chisolm, G.M.; Kottke-Marchant, K. Lysophosphatidylcholine regulates human microvascular endothelial cell expression of chemokines. *J. Mol. Cell Cardiol.* **2003**, *35*, 1375–1384. [CrossRef] [PubMed]
46. Chang, M.C.; Lee, J.J.; Chen, Y.J.; Lin, S.I.; Lin, L.D.; Jein-Wen Liou, E.; Huang, W.L.; Chan, C.P.; Huang, C.C.; Jeng, J.H. Lysophosphatidylcholine induces cytotoxicity/apoptosis and IL-8 production of human endothelial cells: Related mechanisms. *Oncotarget* **2017**, *8*, 106177–106189. [CrossRef] [PubMed]

47. Kugiyama, K.; Kerns, S.A.; Morrisett, J.D.; Roberts, R.; Henry, P.D. Impairment of endothelium-dependent arterial relaxation by lysolecithin in modified low-density lipoproteins. *Nature* **1990**, *344*, 160–162. [CrossRef]
48. Kim, E.A.; Kim, J.A.; Park, M.H.; Jung, S.C.; Suh, S.H.; Pang, M.G.; Kim, Y.J. Lysophosphatidylcholine induces endothelial cell injury by nitric oxide production through oxidative stress. *J. Matern. Fetal Neonatal Med.* **2009**, *22*, 325–331. [CrossRef]
49. Li, B.; Tian, S.; Liu, X.; He, C.; Ding, Z.; Shan, Y. Sulforaphane protected the injury of human vascular endothelial cell induced by LPC through up-regulating endogenous antioxidants and phase II enzymes. *Food Funct.* **2015**, *6*, 1984–1991. [CrossRef]
50. Chaudhuri, P.; Colles, S.M.; Damron, D.S.; Graham, L.M. Lysophosphatidylcholine inhibits endothelial cell migration by increasing intracellular calcium and activating calpain. *Arter. Thromb. Vasc. Biol.* **2003**, *23*, 218–223. [CrossRef]
51. Rikitake, Y.; Kawashima, S.; Yamashita, T.; Ueyama, T.; Ishido, S.; Hotta, H.; Hirata, K.; Yokoyama, M. Lysophosphatidylcholine inhibits endothelial cell migration and proliferation via inhibition of the extracellular signal-regulated kinase pathway. *Arter. Thromb. Vasc. Biol.* **2000**, *20*, 1006–1012. [CrossRef] [PubMed]
52. Rabini, R.A.; Galassi, R.; Fumelli, P.; Dousset, N.; Solera, M.L.; Valdiguie, P.; Curatola, G.; Ferretti, G.; Taus, M.; Mazzanti, L. Reduced Na(+)-K(+)-ATPase activity and plasma lysophosphatidylcholine concentrations in diabetic patients. *Diabetes* **1994**, *43*, 915–919. [CrossRef] [PubMed]
53. Okita, M.; Gaudette, D.C.; Mills, G.B.; Holub, B.J. Elevated levels and altered fatty acid composition of plasma lysophosphatidylcholine(lysoPC) in ovarian cancer patients. *Int. J. Cancer* **1997**, *71*, 31–34. [CrossRef]
54. Sasagawa, T.; Suzuki, K.; Shiota, T.; Kondo, T.; Okita, M. The significance of plasma lysophospholipids in patients with renal failure on hemodialysis. *J. Nutr. Sci. Vitam.* **1998**, *44*, 809–818. [CrossRef]
55. Law, S.H.; Chan, M.L.; Marathe, G.K.; Parveen, F.; Chen, C.H.; Ke, L.Y. An Updated Review of Lysophosphatidylcholine Metabolism in Human Diseases. *Int. J. Mol. Sci.* **2019**, *20*, 1149. [CrossRef]
56. Wang, Y.C.; Lee, A.S.; Lu, L.S.; Ke, L.Y.; Chen, W.Y.; Dong, J.W.; Lu, J.; Chen, Z.; Chu, C.S.; Chan, H.C.; et al. Human electronegative LDL induces mitochondrial dysfunction and premature senescence of vascular cells in vivo. *Aging Cell* **2018**, *17*, e12792. [CrossRef]
57. Lu, J.; Yang, J.H.; Burns, A.R.; Chen, H.H.; Tang, D.; Walterscheid, J.P.; Suzuki, S.; Yang, C.Y.; Sawamura, T.; Chen, C.H. Mediation of electronegative low-density lipoprotein signaling by LOX-1: A possible mechanism of endothelial apoptosis. *Circ. Res.* **2009**, *104*, 619–627. [CrossRef]
58. Abe, Y.; Fornage, M.; Yang, C.Y.; Bui-Thanh, N.A.; Wise, V.; Chen, H.H.; Rangaraj, G.; Ballantyne, C.M. L5, the most electronegative subfraction of plasma LDL, induces endothelial vascular cell adhesion molecule 1 and CXC chemokines, which mediate mononuclear leukocyte adhesion. *Atherosclerosis* **2007**, *192*, 56–66. [CrossRef]
59. Estruch, M.; Rajamaki, K.; Sanchez-Quesada, J.L.; Kovanen, P.T.; Oorni, K.; Benitez, S.; Ordonez-Llanos, J. Electronegative LDL induces priming and inflammasome activation leading to IL-1beta release in human monocytes and macrophages. *Biochim. Biophys. Acta* **2015**, *1851*, 1442–1449. [CrossRef]
60. Chang, S.F.; Chang, P.Y.; Chou, Y.C.; Lu, S.C. Electronegative LDL Induces M1 Polarization of Human Macrophages Through a LOX-1-Dependent Pathway. *Inflammation* **2020**, *43*, 1524–1535. [CrossRef]
61. Puig, N.; Montolio, L.; Camps-Renom, P.; Navarra, L.; Jimenez-Altayo, F.; Jimenez-Xarrie, E.; Sanchez-Quesada, J.L.; Benitez, S. Electronegative LDL Promotes Inflammation and Triglyceride Accumulation in Macrophages. *Cells* **2020**, *9*, 583. [CrossRef] [PubMed]
62. Puig, N.; Estruch, M.; Jin, L.; Sanchez-Quesada, J.L.; Benitez, S. The Role of Distinctive Sphingolipids in the Inflammatory and Apoptotic Effects of Electronegative LDL on Monocytes. *Biomolecules* **2019**, *9*, 300. [CrossRef]
63. Ligi, D.; Benitez, S.; Croce, L.; Rivas-Urbina, A.; Puig, N.; Ordonez-Llanos, J.; Mannello, F.; Sanchez-Quesada, J.L. Electronegative LDL induces MMP-9 and TIMP-1 release in monocytes through CD14 activation: Inhibitory effect of glycosaminoglycan sulodexide. *Biochim. Biophys. Acta Mol. Basis Dis.* **2018**, *1864*, 3559–3567. [CrossRef] [PubMed]
64. Lee, A.S.; Wang, G.J.; Chan, H.C.; Chen, F.Y.; Chang, C.M.; Yang, C.Y.; Lee, Y.T.; Chang, K.C.; Chen, C.H. Electronegative low-density lipoprotein induces cardiomyocyte apoptosis indirectly through endothelial cell-released chemokines. *Apoptosis* **2012**, *17*, 1009–1018. [CrossRef] [PubMed]

65. Lee, A.S.; Xi, Y.; Lai, C.H.; Chen, W.Y.; Peng, H.Y.; Chan, H.C.; Chen, C.H.; Chang, K.C. Human electronegative low-density lipoprotein modulates cardiac repolarization via LOX-1-mediated alteration of sarcolemmal ion channels. *Sci. Rep.* **2017**, *7*, 10889. [CrossRef]
66. Revuelta-Lopez, E.; Cal, R.; Julve, J.; Rull, A.; Martinez-Bujidos, M.; Perez-Cuellar, M.; Ordonez-Llanos, J.; Badimon, L.; Sanchez-Quesada, J.L.; Llorente-Cortes, V. Hypoxia worsens the impact of intracellular triglyceride accumulation promoted by electronegative low-density lipoprotein in cardiomyocytes by impairing perilipin 5 upregulation. *Int. J. Biochem. Cell Biol.* **2015**, *65*, 257–267. [CrossRef]
67. Chang, K.C.; Lee, A.S.; Chen, W.Y.; Lin, Y.N.; Hsu, J.F.; Chan, H.C.; Chang, C.M.; Chang, S.S.; Pan, C.C.; Sawamura, T.; et al. Increased LDL electronegativity in chronic kidney disease disrupts calcium homeostasis resulting in cardiac dysfunction. *J. Mol. Cell Cardiol.* **2015**, *84*, 36–44. [CrossRef]
68. Urata, J.; Ikeda, S.; Koga, S.; Nakata, T.; Yasunaga, T.; Sonoda, K.; Koide, Y.; Ashizawa, N.; Kohno, S.; Maemura, K. Negatively charged low-density lipoprotein is associated with atherogenic risk in hypertensive patients. *Heart Vessel.* **2012**, *27*, 235–242. [CrossRef]
69. Sawamura, T.; Kakino, A.; Fujita, Y. LOX-1: A multiligand receptor at the crossroads of response to danger signals. *Curr. Opin. Lipidol.* **2012**, *23*, 439–445. [CrossRef]
70. Stancel, N.; Chen, C.C.; Ke, L.Y.; Chu, C.S.; Lu, J.; Sawamura, T.; Chen, C.H. Interplay between CRP, Atherogenic LDL, and LOX-1 and Its Potential Role in the Pathogenesis of Atherosclerosis. *Clin. Chem.* **2016**, *62*, 320–327. [CrossRef]
71. Raffioni, S.; Bradshaw, R.A. Activation of phosphatidylinositol 3-kinase by epidermal growth factor, basic fibroblast growth factor, and nerve growth factor in PC12 pheochromocytoma cells. *Proc. Natl. Acad. Sci. USA* **1992**, *89*, 9121–9125. [CrossRef]
72. Wu, C.C.; Wu, S.Y.; Liao, C.Y.; Teng, C.M.; Wu, Y.C.; Kuo, S.C. The roles and mechanisms of PAR4 and P2Y12/phosphatidylinositol 3-kinase pathway in maintaining thrombin-induced platelet aggregation. *Br. J. Pharm.* **2010**, *161*, 643–658. [CrossRef] [PubMed]
73. Marwali, M.R.; Hu, C.P.; Mohandas, B.; Dandapat, A.; Deonikar, P.; Chen, J.; Cawich, I.; Sawamura, T.; Kavdia, M.; Mehta, J.L. Modulation of ADP-induced platelet activation by aspirin and pravastatin: Role of lectin-like oxidized low-density lipoprotein receptor-1, nitric oxide, oxidative stress, and inside-out integrin signaling. *J. Pharm. Exp.* **2007**, *322*, 1324–1332. [CrossRef] [PubMed]
74. Podrez, E.A.; Byzova, T.V. Prothrombotic lipoprotein patterns in stroke. *Blood* **2016**, *127*, 1221–1222. [CrossRef] [PubMed]
75. Nichols, T.C. Bad cholesterol breaking really bad. *Blood* **2013**, *122*, 3551–3553. [CrossRef] [PubMed]
76. Grundtman, C.; Wick, G. The autoimmune concept of atherosclerosis. *Curr. Opin. Lipidol.* **2011**, *22*, 327–334. [CrossRef]
77. Cinoku, I.I.; Mavragani, C.P.; Moutsopoulos, H.M. Atherosclerosis: Beyond the lipid storage hypothesis. The role of autoimmunity. *Eur. J. Clin. Investig.* **2020**, *50*, e13195.
78. Sobenin, I.A.; Salonen, J.T.; Zhelankin, A.V.; Melnichenko, A.A.; Kaikkonen, J.; Bobryshev, Y.V.; Orekhov, A.N. Low density lipoprotein-containing circulating immune complexes: Role in atherosclerosis and diagnostic value. *Biomed. Res. Int.* **2014**, *2014*, 205697. [CrossRef]
79. Benitez, S.; Camacho, M.; Bancells, C.; Vila, L.; Sanchez-Quesada, J.L.; Ordonez-Llanos, J. Wide proinflammatory effect of electronegative low-density lipoprotein on human endothelial cells assayed by a protein array. *Biochim. Biophys. Acta* **2006**, *1761*, 1014–1021. [CrossRef]
80. Estruch, M.; Bancells, C.; Beloki, L.; Sanchez-Quesada, J.L.; Ordonez-Llanos, J.; Benitez, S. CD14 and TLR4 mediate cytokine release promoted by electronegative LDL in monocytes. *Atherosclerosis* **2013**, *229*, 356–362. [CrossRef]
81. Itoh, N.; Yonehara, S.; Ishii, A.; Yonehara, M.; Mizushima, S.; Sameshima, M.; Hase, A.; Seto, Y.; Nagata, S. The polypeptide encoded by the cDNA for human cell surface antigen Fas can mediate apoptosis. *Cell* **1991**, *66*, 233–243. [CrossRef]
82. Klimov, A.N.; Denisenko, A.D.; Popov, A.V.; Nagornev, V.A.; Pleskov, V.M.; Vinogradov, A.G.; Denisenko, T.V.; Magracheva, E.; Kheifes, G.M.; Kuznetzov, A.S. Lipoprotein-antibody immune complexes. Their catabolism and role in foam cell formation. *Atherosclerosis* **1985**, *58*, 1–15. [CrossRef]
83. Tertov, V.V.; Orekhov, A.N.; Sobenin, I.A.; Morriset, J.D.; Gotto, A.M., Jr.; Guevara, J.G., Jr. Carbohydrate composition of protein and lipid components in sialic acid-rich and -poor low density lipoproteins from subjects with and without coronary artery disease. *J. Lipid Res.* **1993**, *34*, 365–375. [PubMed]

84. Yang, T.C.; Chang, P.Y.; Kuo, T.L.; Lu, S.C. Electronegative L5-LDL induces the production of G-CSF and GM-CSF in human macrophages through LOX-1 involving NF-kappaB and ERK2 activation. *Atherosclerosis* **2017**, *267*, 1–9. [CrossRef]
85. Leone, A.M.; Rutella, S.; Bonanno, G.; Contemi, A.M.; de Ritis, D.G.; Giannico, M.B.; Rebuzzi, A.G.; Leone, G.; Crea, F. Endogenous G-CSF and CD34+ cell mobilization after acute myocardial infarction. *Int. J. Cardiol.* **2006**, *111*, 202–208. [CrossRef]
86. Cornish, A.L.; Campbell, I.K.; McKenzie, B.S.; Chatfield, S.; Wicks, I.P. G-CSF and GM-CSF as therapeutic targets in rheumatoid arthritis. *Nat. Rev. Rheumatol.* **2009**, *5*, 554–559. [CrossRef]
87. Yang, T.C.; Chang, P.Y.; Lu, S.C. L5-LDL from ST-elevation myocardial infarction patients induces IL-1beta production via LOX-1 and NLRP3 inflammasome activation in macrophages. *Am. J. Physiol. Heart Circ. Physiol.* **2017**, *312*, H265–H274. [CrossRef]
88. Tai, M.H.; Kuo, S.M.; Liang, H.T.; Chiou, K.R.; Lam, H.C.; Hsu, C.M.; Pownall, H.J.; Chen, H.H.; Huang, M.T.; Yang, C.Y. Modulation of angiogenic processes in cultured endothelial cells by low density lipoproteins subfractions from patients with familial hypercholesterolemia. *Atherosclerosis* **2006**, *186*, 448–457. [CrossRef]
89. Tang, D.; Lu, J.; Walterscheid, J.P.; Chen, H.H.; Engler, D.A.; Sawamura, T.; Chang, P.Y.; Safi, H.J.; Yang, C.Y.; Chen, C.H. Electronegative LDL circulating in smokers impairs endothelial progenitor cell differentiation by inhibiting Akt phosphorylation via LOX-1. *J. Lipid Res.* **2008**, *49*, 33–47. [CrossRef]
90. Lewandowski, L.B.; Kaplan, M.J. Update on cardiovascular disease in lupus. *Curr. Opin. Rheumatol.* **2016**, *28*, 468–476. [CrossRef]
91. Kahlenberg, J.M.; Kaplan, M.J. Mechanisms of premature atherosclerosis in rheumatoid arthritis and lupus. *Annu. Rev. Med.* **2013**, *64*, 249–263. [CrossRef] [PubMed]
92. Asanuma, Y.; Oeser, A.; Shintani, A.K.; Turner, E.; Olsen, N.; Fazio, S.; Linton, M.F.; Raggi, P.; Stein, C.M. Premature coronary-artery atherosclerosis in systemic lupus erythematosus. *N. Engl. J. Med.* **2003**, *349*, 2407–2415. [CrossRef] [PubMed]
93. Virani, S.S.; Alonso, A.; Benjamin, E.J.; Bittencourt, M.S.; Callaway, C.W.; Carson, A.P.; Chamberlain, A.M.; Chang, A.R.; Cheng, S.; Delling, F.N.; et al. Heart Disease and Stroke Statistics-2020 Update: A Report From the American Heart Association. *Circulation* **2020**, *141*, e139–e596. [CrossRef] [PubMed]
94. Hegele, R.A.; Tsimikas, S. Lipid-Lowering Agents. *Circ. Res.* **2019**, *124*, 386–404. [CrossRef]
95. Arnett, D.K.; Blumenthal, R.S.; Albert, M.A.; Buroker, A.B.; Goldberger, Z.D.; Hahn, E.J.; Himmelfarb, C.D.; Khera, A.; Lloyd-Jones, D.; McEvoy, J.W.; et al. 2019 ACC/AHA Guideline on the Primary Prevention of Cardiovascular Disease: A Report of the American College of Cardiology/American Heart Association Task Force on Clinical Practice Guidelines. *Circulation* **2019**, *140*, e596–e646. [CrossRef]
96. Leibowitz, M.; Cohen-Stavi, C.; Basu, S.; Balicer, R.D. Targeting LDL Cholesterol: Beyond Absolute Goals Toward Personalized Risk. *Curr. Cardiol. Rep.* **2017**, *19*, 52. [CrossRef]
97. Su, X.; Kong, Y.; Peng, D. Evidence for changing lipid management strategy to focus on non-high density lipoprotein cholesterol. *Lipids Health Dis.* **2019**, *18*, 134. [CrossRef]
98. Nambi, V.; Bhatt, D.L. Primary Prevention of Atherosclerosis: Time to Take a Selfie? *J. Am. Coll. Cardiol.* **2017**, *70*, 2992–2994. [CrossRef]
99. Ward, N.C.; Watts, G.F.; Eckel, R.H. Statin Toxicity. *Circ. Res.* **2019**, *124*, 328–350. [CrossRef]
100. Jacobson, T.A.; Khan, A.; Maki, K.C.; Brinton, E.A.; Cohen, J.D. Provider recommendations for patient-reported muscle symptoms on statin therapy: Insights from the Understanding Statin Use in America and Gaps in Patient Education survey. *J. Clin. Lipidol.* **2018**, *12*, 78–88. [CrossRef]
101. Zhang, B.; Matsunaga, A.; Rainwater, D.L.; Miura, S.; Noda, K.; Nishikawa, H.; Uehara, Y.; Shirai, K.; Ogawa, M.; Saku, K. Effects of rosuvastatin on electronegative LDL as characterized by capillary isotachophoresis: The ROSARY Study. *J. Lipid Res.* **2009**, *50*, 1832–1841. [CrossRef] [PubMed]
102. Zhang, B.; Miura, S.; Yanagi, D.; Noda, K.; Nishikawa, H.; Matsunaga, A.; Shirai, K.; Iwata, A.; Yoshinaga, K.; Adachi, H.; et al. Reduction of charge-modified LDL by statin therapy in patients with CHD or CHD risk factors and elevated LDL-C levels: The SPECIAL Study. *Atherosclerosis* **2008**, *201*, 353–359. [CrossRef] [PubMed]
103. Rivas-Urbina, A.; Rull, A.; Ordonez-Llanos, J.; Sanchez-Quesada, J.L. Electronegative LDL: An Active Player in Atherogenesis or a By-Product of Atherosclerosis? *Curr. Med. Chem.* **2019**, *26*, 1665–1679. [CrossRef] [PubMed]

104. Sawada, N.; Obama, T.; Koba, S.; Takaki, T.; Iwamoto, S.; Aiuchi, T.; Kato, R.; Kikuchi, M.; Hamazaki, Y.; Itabe, H. Circulating oxidized LDL, increased in patients with acute myocardial infarction, is accompanied by heavily modified HDL. *J. Lipid Res.* **2020**, *61*, 816–829. [CrossRef] [PubMed]
105. Faulin, T.; Kazuma, S.M.; Tripodi, G.L.; Cavalcante, M.F.; Wakasuqui, F.; Oliveira, C.L.P.; Degenhardt, M.F.S.; Michaloski, J.; Giordano, R.J.; Ketelhuth, D.F.J.; et al. Proinflammatory Action of a New Electronegative Low-Density Lipoprotein Epitope. *Biomolecules* **2019**, *9*, 386. [CrossRef]
106. Freitas, M.C.P.; Fernandez, D.G.E.; Cohen, D.; Figueiredo-Neto, A.M.; Maranhao, R.C.; Damasceno, N.R.T. Oxidized and electronegative low-density lipoprotein as potential biomarkers of cardiovascular risk in obese adolescents. *Clinics* **2018**, *73*, e189. [CrossRef]
107. Lobo, J.; Santos, F.; Grosso, D.; Lima, R.; Barreira, A.L.; Leite, M., Jr.; Mafra, D.; Abdalla, D.S. Electronegative LDL and lipid abnormalities in patients undergoing hemodialysis and peritoneal dialysis. *Nephron Clin. Pract.* **2008**, *108*, c298–c304. [CrossRef]
108. Oliveira, J.A.; Sevanian, A.; Rodrigues, R.J.; Apolinario, E.; Abdalla, D.S. Minimally modified electronegative LDL and its autoantibodies in acute and chronic coronary syndromes. *Clin. Biochem.* **2006**, *39*, 708–714. [CrossRef]

© 2020 by the authors. Licensee MDPI, Basel, Switzerland. This article is an open access article distributed under the terms and conditions of the Creative Commons Attribution (CC BY) license (http://creativecommons.org/licenses/by/4.0/).

Article

Novel Nargenicin A1 Analog Inhibits Angiogenesis by Downregulating the Endothelial VEGF/VEGFR2 Signaling and Tumoral HIF-1α/VEGF Pathway

Jang Mi Han [1,†], Ye Seul Choi [1,†], Dipesh Dhakal [1], Jae Kyung Sohng [1,2] and Hye Jin Jung [1,2,*]

1. Department of Life Science and Biochemical Engineering, Sun Moon University, Asan 31460, Korea; gkswkdal200@naver.com (J.M.H.); yesll96@naver.com (Y.S.C.); medipesh@gmail.com (D.D.); sohng@sunmoon.ac.kr (J.K.S.)
2. Department of Pharmaceutical Engineering and Biotechnology, Sun Moon University, Asan 31460, Korea
* Correspondence: poka96@sunmoon.ac.kr; Tel.: +82-41-530-2354; Fax: +82-41-530-2939
† These authors contributed equally to this work.

Received: 25 July 2020; Accepted: 28 July 2020; Published: 29 July 2020

Abstract: Targeting angiogenesis is an attractive strategy for the treatment of angiogenesis-related diseases, including cancer. We previously identified 23-demethyl 8,13-deoxynargenicin (compound **9**) as a novel nargenicin A1 analog with potential anticancer activity. In this study, we investigated the antiangiogenic activity and mode of action of compound **9**. This compound was found to effectively inhibit in vitro angiogenic characteristics, including the proliferation, invasion, capillary tube formation, and adhesion of human umbilical vein endothelial cells (HUVECs) stimulated by vascular endothelial growth factor (VEGF). Furthermore, compound **9** suppressed the neovascularization of the chorioallantoic membrane of growing chick embryos in vivo. Notably, the antiangiogenic properties of compound **9** were related to the downregulation of VEGF/VEGFR2-mediated downstream signaling pathways, as well as matrix metalloproteinase (MMP)-2 and MMP-9 expression in HUVECs. In addition, compound **9** was found to decrease the in vitro AGS gastric cancer cell-induced angiogenesis of HUVECs by blocking hypoxia-inducible factor-1α (HIF-1α) and VEGF expression in AGS cells. Collectively, our findings demonstrate for the first time that compound **9** is a promising antiangiogenic agent targeting both VEGF/VEGFR2 signaling in ECs and HIF-1α/VEGF pathway in tumor cells.

Keywords: angiogenesis; nargenicin A1; compound 9; VEGF; VEGFR2; HIF-1α

---

1. Introduction

Angiogenesis leads to the formation of new blood vessels by endothelial cells (ECs) from existing vessels. This process plays a crucial role in the growth, metastasis, and progression of solid tumors by providing oxygen, nutrients, and growth factors [1]. Therefore, targeting angiogenesis is regarded as a key strategy for the effective treatment of cancer [2,3]. Angiogenesis involves the enzymatic degradation of the vascular endothelial matrix and EC migration, adhesion, proliferation, and tube formation [4]. Diverse bioactive compounds with the antiangiogenic activities are involved in the downregulation of angiogenesis [5,6].

Tumor angiogenesis is generally initiated by various proangiogenic factors secreted from tumor cells. Vascular endothelial growth factor (VEGF) is the most potent inducer of angiogenesis and interacts with the tyrosine kinase VEGF receptors (VEGFRs), including VEGFR1, VEGFR2, and VEGFR3 [7]. Among these, VEGFR2-mediated signaling prominently induces cellular responses involved in angiogenesis. The binding of VEGF to VEGFR2 results in autophosphorylation of specific tyrosine residues in the cytoplasmic domain of VEGFR2 and sequentially promotes the activation of its downstream signaling effectors, including signal transducer and activator of transcription 3 (STAT3), serine/threonine protein kinase B (AKT), and extracellular signal-regulated kinase 1/2 (ERK1/2) [8–11]. Accordingly, blocking the VEGF/VEGFR2 signaling pathways is a powerful technique to disrupt tumor angiogenesis and tumor growth.

Nargenicin A1 is the major secondary metabolite produced by *Nocardia* species and a macrolide compound with effective antibacterial activity against various Gram-positive pathogenic bacteria [12]. In previous studies, several metabolic engineering and enzymatic modification approaches have been applied to boost production and generate novel glycosylated derivatives of nargenicin A1 [13,14]. More recently, we characterized the tailoring steps for the biosynthesis of nargenicin A1 in *Nocardia* sp. CS682, which resulted in the generation of several new analogs of the natural product [15]. An analysis of the bioactivity of the analogs revealed that 23-demethyl 8,13-deoxynargenicin (compound 9) possesses potential antitumor activity, unlike nargenicin A1 and the other analogs (Figure 1A). Compound 9 suppressed the growth of various cancer cell lines, including gastric, lung, skin, liver, colon, brain, breast, and cervical cancer, within a range of concentrations that did not affect normal cell growth [15]. Further analysis of the underlying molecular mechanisms of its anticancer effect demonstrated that compound 9 exhibits a growth inhibitory effect against AGS gastric cancer cells by inducing G2/M cell cycle arrest, reactive oxygen species (ROS)- and caspase-mediated apoptosis, and autophagy by downregulating the phosphatidylinositol 3-kinase (PI3K)/AKT/mammalian target of rapamycin (mTOR) pathway [15]. Although we identified the anticancer activity of compound 9, its effect on angiogenesis and VEGF/VEGFR2 signaling has not yet been studied.

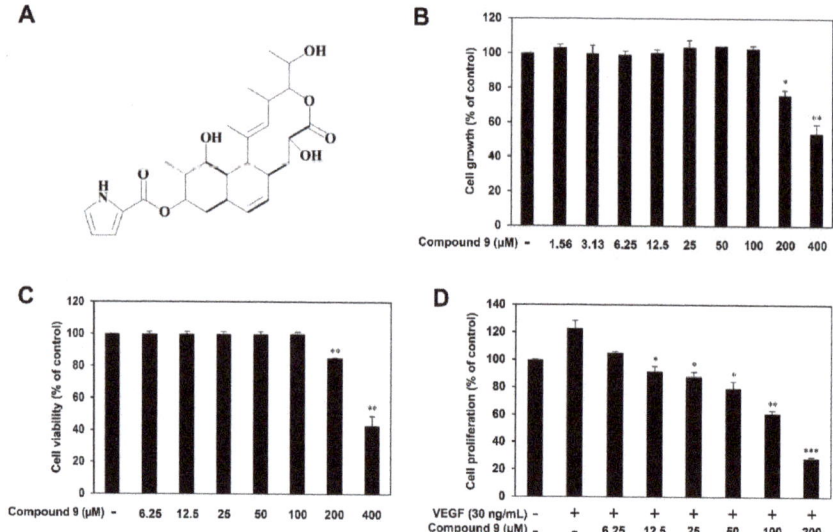

Figure 1. (**A**) The chemical structure of 23-demethyl 8,13-deoxynargenicin (compound 9). (**B,C**) The cytotoxic effect of compound 9 on HUVECs. Cells were treated with the indicated concentrations of compound 9 for 24 h. (**B**) Cell growth was measured by the MTT assay and the IC$_{50}$ value from obtained data was analyzed using the curve-fitting program GraphPad Prism 5. (**C**) Cell viability was measured by the trypan blue exclusion method. * $p < 0.05$, ** $p < 0.01$ vs. the control. (**D**) The antiproliferative effect of compound 9 in HUVECs. Cells were treated with the indicated concentrations (6.25–200 μM) of compound 9 in the presence of VEGF (30 ng/mL) for 72 h. * $p < 0.05$, ** $p < 0.01$, *** $p < 0.001$ vs. the VEGF control. Each value represents the mean ± SD from three independent experiments.

In the present study, we assessed the antiangiogenic activity of compound 9 using human umbilical vein endothelial cells (HUVECs). Compound 9 inhibited VEGF-induced angiogenesis by suppressing the proliferation, invasion, tube formation, and adhesion of HUVECs. The antiangiogenic activity of compound **9** was verified using the chorioallantoic membrane (CAM) assay in vivo. Compound **9** was found to downregulate the VEGFR2-mediated signaling pathways, as well as matrix metalloproteinase (MMP) expression. Furthermore, compound **9** inhibited the AGS gastric cancer cell-induced angiogenesis of HUVECs by decreasing the expression of hypoxia-inducible factor-1α (HIF-1α) and VEGF. Therefore, we propose that compound **9** has the potential to regress tumor progression by targeting angiogenesis.

2. Experimental Section

2.1. Materials

Compound **9** was isolated from the culture extract of the constructed *Nocardia* sp. CS682 mutant, as shown in our previous report [15] and prepared at a concentration of 100 mM using dimethyl sulfoxide (DMSO). Endothelial growth medium-2 (EGM-2) and antibiotics were purchased from Lonza (Walkersville, MD, USA), and fetal bovine serum (FBS) and RPMI-1640 medium were purchased from Invitrogen (Grand Island, NY, USA). Recombinant human vascular endothelial growth factor 165 (VEGF165), Transwell chamber system, and Matrigel were purchased from Koma Biotech (Seoul, Korea, cat. no. K0921148), Corning Costar (Acton, MA, USA), and BD Biosciences (San Jose, CA, USA), respectively. Gelatin, trypan blue, and 3-[4,5-dimethylthiazol-2-yl]2,5-diphenyl tetrazolium bromide (MTT) were purchased from Sigma-Aldrich (St. Louis, MO, USA). Antibodies against VEGFR2 (cat. no. 2479), phospho-VEGFR2 (cat. no. 2478), STAT3 (cat. no. 9139), phospho-STAT3 (cat. no. 9131), AKT (cat. no. 9272), phospho-AKT (cat. no. 4060), ERK1/2 (cat. no. 9102), phospho-ERK1/2 (cat. no. 9101), MMP-2 (cat. no. 4022), MMP-9 (cat. no. 3852), HIF-1α (cat. no. 3716), and β-actin (cat. no. 4967) were purchased from Cell Signaling Technology (Danvers, MA, USA).

2.2. Cell Culture and Hypoxic Conditions

Human umbilical vein endothelial cells (HUVECs) and AGS human gastric cancer cells were obtained from American Type Culture Collection (Manassas, VA, USA) and Korean Cell Line Bank (Seoul, Korea), respectively. HUVECs and AGS cells were grown in EGM-2 and RPMI supplemented with 10% fetal bovine serum (FBS). The cells were maintained at 37 °C in a humidified 5% CO_2 incubator (Thermo Scientific, Vantaa, Finland). For hypoxic conditions, the cells were incubated in a hypoxic chamber (SANYO, Chuou-ku, Osaka, Japan) under 5% CO_2 and 1% O_2 balanced with N_2.

2.3. Cell Viability Assay

HUVECs (1×10^4 cells/well) were seeded in 12-well culture plates and then treated with various concentrations of compound **9** (6.25–400 μM) for 24 h. The cells were stained with trypan blue and counted by a hemocytometer using an optical microscope (Olympus, Tokyo, Japan) at 200× magnification.

2.4. Cell Proliferation Assay

HUVECs (3×10^3 cells/well) were seeded in 96-well culture plates and then treated with various concentrations of compound **9** (6.25–200 μM) in the presence of VEGF (30 ng/mL) for 72 h. Cell proliferation was measured using a 3-(4,5-dimethylthiazol-2-yl)-2,5-diphenyltetrazolium bromide (MTT) colorimetric assay.

2.5. Chemoinvasion Assay

The invasiveness of HUVECs was investigated using a Transwell chamber system with polycarbonate filter inserts with a pore size of 8.0 μm. The lower surface of the filter was coated with 10 μL of gelatin (1 mg/mL) for 1 h, and the upper surface was coated with 10 μL of Matrigel (3 mg/mL) for 1 h. Serum-starved HUVECs (8×10^4 cells) were seeded in the upper chamber of the filter, and compound **9** (25–200 µM) were added to the lower chamber in the presence of VEGF (30 ng/mL). The chamber was incubated at 37 °C for 18 h, and then the cells were fixed with 70% methanol and stained with hematoxylin and eosin (H&E) at room temperature for 5 min. The total invaded cells were photographed and counted in randomly selected fields using an optical microscope (Olympus) at 200× magnification.

2.6. Capillary Tube Formation Assay

Serum-starved HUVECs (2×10^4 cells) were placed on a surface containing Matrigel (10 mg/mL) using an angiogenesis kit (Ibidi GmbH, Munich, Germany) and incubated with compound **9** (25–200 µM) for 6 h in the presence of VEGF (30 ng/mL). The morphological changes and the tube formation of the cells were visualized under an optical microscope (Olympus) and photographed at 100× magnification. The number of tubes formed in the cells was counted in randomly selected fields at 100× magnification.

2.7. Adhesion Assay

The cell-matrix adhesion assay was performed in a 24-well culture plate coated with gelatin overnight at 4 °C. HUVECs (6×10^4 cells/well) were seeded in each well and treated with compound **9** (25–200 µM) in the presence of VEGF (30 ng/mL). After 3 h, the unbound cells were carefully removed, and the attached cells were visualized and counted in randomly selected fields under a microscope at 200× magnification.

2.8. Chorioallantoic Membrane (CAM) Assay

Fertilized chick eggs were incubated in a humidified egg incubator at 37 °C and 50% humidity for 3 days. After incubation, approximately 6–9 mL of egg albumin was removed with a 10 mL hypodermic needle, allowing the CAM and yolk sac to drop away from the shell membrane. After 2 days, a small hole was punched on the broad end of the egg, and a window was carefully peeled away on the eggshell. Thermanox coverslips (Nalge Nunc International, Rochester, NY, USA) saturated with or without compound **9** (10 µg/egg) were air-dried and placed on the CAM surface. The windows were then sealed with cellophane tape. Two days later, 2 mL of 10% fat emulsion (Sigma-Aldrich) was injected into the chorioallantois and the vascular images were photographed.

2.9. Western Blot Analysis

After treatment, the cells were collected and lysed using RIPA buffer (Sigma-Aldrich) supplemented with a protease inhibitor cocktail (Roche Diagnostics, Indianapolis, IN, USA) on ice. Equal amounts of lysates were separated using 10% sodium dodecyl sulfate-polyacrylamide gel electrophoresis (SDS-PAGE). The separated proteins were then transferred to polyvinylidene difluoride (PVDF) membranes (EMD Millipore, Hayward, CA, USA) and blocked using Tris-buffered saline with Tween-20 (TBST) containing 5% skim milk at room temperature for 1 h. The membranes were then incubated with primary antibodies against phospho-VEGFR2 (dilution 1:2000), VEGFR2 (dilution 1:2000), phospho-STAT3 (dilution 1:2000), STAT3 (dilution 1:2000), phospho-AKT (dilution 1:2000), AKT (dilution 1:2000), phospho-ERK1/2 (dilution 1:2000), ERK1/2 (dilution 1:2000), MMP-2 (dilution 1:2000), MMP-9 (dilution 1:2000), HIF-1α (dilution 1:2000), and β-actin (dilution 1:2000) overnight at 4 °C. After washing with TBST three times, the membranes were incubated with horseradish peroxidase-conjugated anti-rabbit (dilution 1:3000) or anti-mouse (dilution 1:3000) secondary antibody for 1 h at room temperature. Immunolabeling was detected using an enhanced chemiluminescence (ECL) kit (Bio-Rad Laboratories, Hercules, CA, USA) according to the manufacturer's instructions.

2.10. Tumor Cell-Induced Chemoinvasion Assay

A tumor cell-induced chemoinvasion assay was performed using an in vitro co-culture system based on the chemoinvasion assay. AGS cells (1×10^5 cells/well) were seeded in the lower chamber and treated with compound **9** (50–200 µM) for 24 h. Then, the medium in each lower chamber was replaced with fresh medium without compound **9**, and serum-starved HUVECs (8×10^4 cells) were placed in the upper chamber. The chamber was incubated at 37 °C for 18 h. The HUVECs that invaded the lower chamber of the filter were analyzed using the same procedure as in the chemoinvasion assay.

2.11. Tumor Cell-Induced Capillary Tube Formation Assay

To perform the tumor cell-induced capillary tube formation assay, conditioned medium was obtained from AGS cells and used as the angiogenic stimulus for tube formation in HUVECs. Briefly, AGS cells were treated with compound **9** (50–200 µM) for 24 h. The medium was then replaced with fresh medium without compound **9**.

2.12. Measurement of VEGF by Enzyme-Linked Immunosorbent Assay

The VEGF concentration in the AGS cells was measured using a VEGF immunoassay kit (R&D Systems, Minneapolis, MN, USA). Cells were incubated with or without compound **9** (50–200 µM) for 11 h under the indicated conditions. The supernatants were collected and the VEGF protein levels were measured according to the manufacturer's instructions.

2.13. Statistical Analysis

The data are presented as the mean ± standard deviation (SD) of three independent experiments. Differences among groups were analyzed using analysis of variance (ANOVA) with SPSS statistics package (SPSS 9.0; SPSS Inc., Chicago, IL, USA). Post-hoc analysis was carried out using Tukey's test. A p-value of <0.05 was considered to indicate a statistically significant difference.

3. Results

3.1. Effect of Compound **9** on In Vitro Angiogenesis of HUVECs

Prior to assessing the antiangiogenic properties of compound **9**, the cytotoxic effect of compound **9** on HUVECs was first investigated using the MTT assay and trypan blue exclusion method. Compound **9** inhibited the growth of HUVECs, with IC_{50} value of 412.3 µM (Figure 1B). Treatment with 6.25–100 µM of compound **9** did not affect the viability of HUVECs, whereas the cell viability was 85 and 43% after treatment with 200 and 400 µM of compound **9**, respectively (Figure 1C). Thus, the effects of compound **9** on in vitro angiogenesis were evaluated at concentrations up to 200 µM with low cytotoxicity.

We next investigated the effects of compound **9** on key steps in VEGF-induced angiogenesis, including EC proliferation, invasion, tube formation, and adhesion. As shown in Figure 1D, the MTT assay revealed that compound **9** dose-dependently inhibited the VEGF-stimulated proliferation of HUVECs. We also examined whether compound **9** affects the invasive migration of HUVECs using the Matrigel chemoinvasion assay. Treatment with 50–200 µM of compound **9** significantly suppressed the invasion of HUVECs stimulated by VEGF (Figure 2A). Furthermore, the Matrigel-based tube formation assay revealed that 50–200 µM of compound **9** markedly disrupted the tubular structures of HUVECs induced by VEGF (Figure 2B). Its effect on EC-matrix adhesion was assessed, and 25–200 µM of compound **9** was found to effectively reduce the VEGF-stimulated adhesion of HUVECs to gelatin (Figure 2C). When compared to compound **9** treatment alone in the absence of VEGF, 50 µM of compound **9** inhibited the VEGF-induced angiogenic phenotypes to the basal levels as shown in Figure 2A–C. In addition, cytotoxicity was not observed at 25–100 µM of concentrations by trypan blue staining performed in parallel to the in vitro angiogenesis assays. However, treatment with 200 µM of compound **9** exhibited the cytotoxic effect on HUVECs. These findings indicate that compound **9** inhibits VEGF-induced angiogenesis in the concentration range of 25–100 µM where cytotoxicity was not observed.

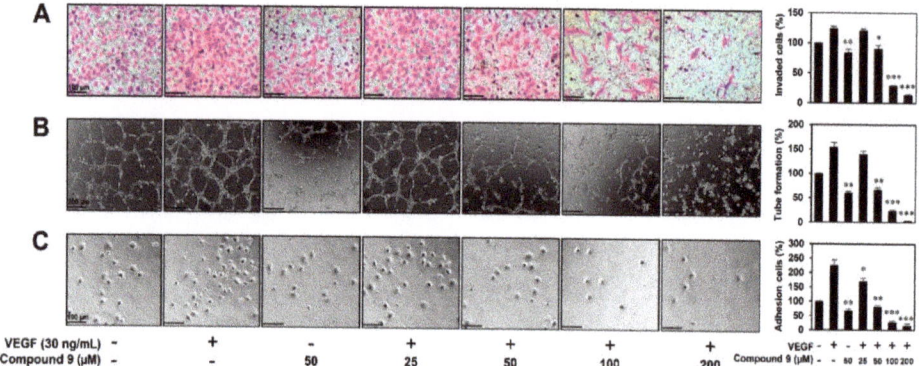

Figure 2. The in vitro antiangiogenic activity of compound **9** in HUVECs. (**A–C**) The inhibitory effects of compound **9** on the (**A**) invasion, (**B**) tube formation, and (**C**) adhesion of HUVECs induced by VEGF. Serum-starved HUVECs were stimulated with VEGF (30 ng/mL) in the presence or absence of compound **9** (25–200 µM). The basal levels of invasion, tube formation, and adhesion of HUVECs that were incubated in serum-free medium without VEGF were normalized to 100%. * $p < 0.05$, ** $p < 0.01$, *** $p < 0.001$ vs. the VEGF control. Each value represents the mean ± SD from three independent experiments.

3.2. Effect of Compound 9 on In Vivo Angiogenesis

We further validated the potential of compound **9** to suppress angiogenesis using the chick embryo CAM assay. As shown in Figure 3, compound **9** showed a significantly stronger inhibitory activity on CAM microvessel formation compared to the control group, with inhibition ratios of 81% and 20%, respectively. Furthermore, the blood vessel density was quantified and markedly decreased by compound **9** treatment without toxicity against pre-existing vessels. These results demonstrate that compound **9** exhibits promising antiangiogenic activity both in vitro and in vivo.

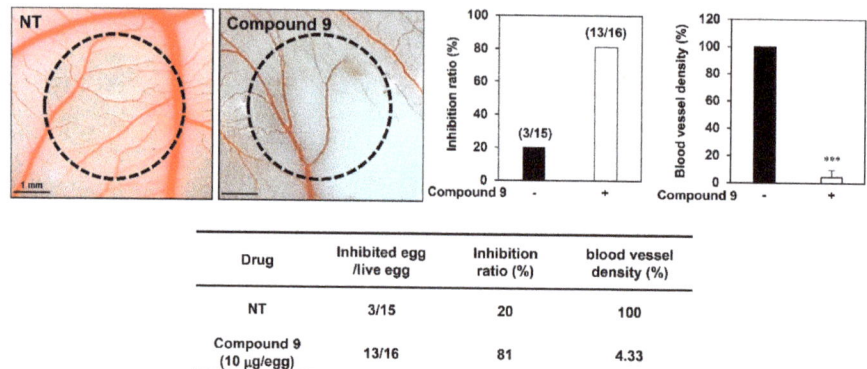

Drug	Inhibited egg /live egg	Inhibition ratio (%)	blood vessel density (%)
NT	3/15	20	100
Compound 9 (10 µg/egg)	13/16	81	4.33

Figure 3. The in vivo antiangiogenic activity of compound **9** in CAMs. Fertilized chick eggs were maintained in a humidified egg incubator at 37 °C. At embryonic day 5, coverslips filled with vehicle alone or compound **9** (10 µg/egg) were placed to the CAM surface. Two days later, the vascular images were photographed. Calculations were based on the ratio of inhibited eggs relative to the total number of live eggs. Microvessel density was counted and the basal levels of control were normalized to 100%. *** $p < 0.001$ vs. the control.

3.3. Effect of Compound 9 on VEGFR2-Mediated Downstream Signaling Pathways

The phosphorylation of VEGFR2 and its downstream protein kinases mediates VEGF-induced angiogenesis [16]. Thus, we examined the effect of compound 9 on VEGFR2-mediated signaling in HUVECs. As shown in Figure 4A, compound 9 clearly reduced the VEGF-stimulated phosphorylation of VEGFR2 and its downstream effectors, including STAT3, AKT, and ERK1/2. In addition, the total protein level of VEGFR2 was decreased after treatment with compound 9, whereas those of STAT3, AKT, and ERK1/2 were not significantly affected.

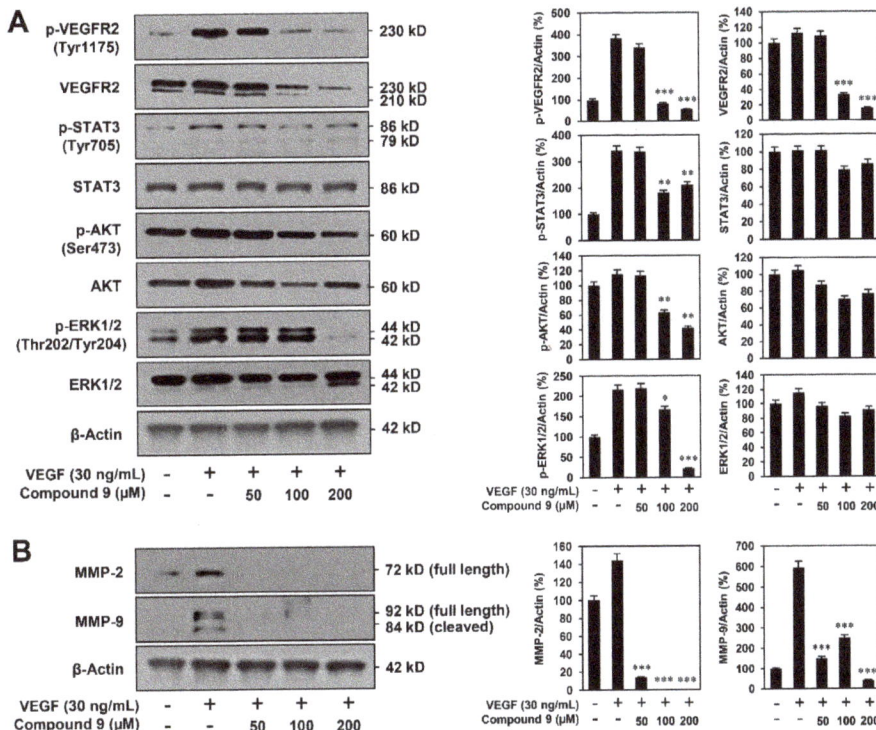

Figure 4. The downregulation of VEGF/VEGFR2-mediated signaling by compound 9. (**A**) VEGFR2 downstream signaling inhibitory activities of compound 9. Serum-starved HUVECs were pretreated with compound 9 (50–200 µM) for 1 h and then stimulated with VEGF (30 ng/mL) for 10 min. (**B**) MMP-2 and MMP-9 inhibitory activities of compound 9. Serum-starved HUVECs were treated with compound 9 (50–200 µM) in the presence of VEGF (30 ng/mL) for 24 h. (**A**,**B**) Protein levels were detected by Western blot analysis and further quantified by densitometry. The level of β-actin was used as an internal control. * $p < 0.05$, ** $p < 0.01$, *** $p < 0.001$ vs. the VEGF control. Each value represents the mean ± SD from three independent experiments.

Matrix metalloproteinases (MMPs) participate in the degradation of the vascular basement membrane, which is required for the invasive migration of ECs [17]. As such, we assessed the effect of compound 9 on the expression of MMP-2 and MMP-9 stimulated by VEGF. As shown in Figure 4B, compound 9 markedly reduced the expression levels of MMP-2 and MMP-9 in HUVECs. Taken together, these results suggest that compound 9 may inhibit the VEGF-induced angiogenesis of HUVECs by downregulating the VEGFR2-mediated downstream signaling pathways and MMP-2/MMP-9 expression.

3.4. Effect of Compound 9 on Tumor Cell-Induced Angiogenesis in Vitro

Our previous study demonstrated that compound **9** exerts effective anticancer activity against AGS gastric cancer cells by inducing G2/M cell cycle arrest, apoptosis, and autophagy [15]. To further evaluate whether compound **9** affects tumor cell-induced angiogenesis, we investigated the effect of compound **9** on the invasion and tube formation of HUVECs stimulated by AGS cells. As shown in Figure 5A, HUVEC tube formation was significantly induced by culture in conditioned medium (CM) from AGS cells compared to control (medium only). However, CM from AGS cells treated with compound **9** significantly inhibited the stimulated tube formation of HUVECs. In addition, the invasion of HUVECs co-cultured with AGS cells was increased compared to that of HUVECs alone, whereas the treatment of AGS cells with compound **9** prevented the increased invasion of HUVECs (Figure 5B). These data indicate that compound **9** inhibits the in vitro angiogenesis induced by AGS gastric cancer cells.

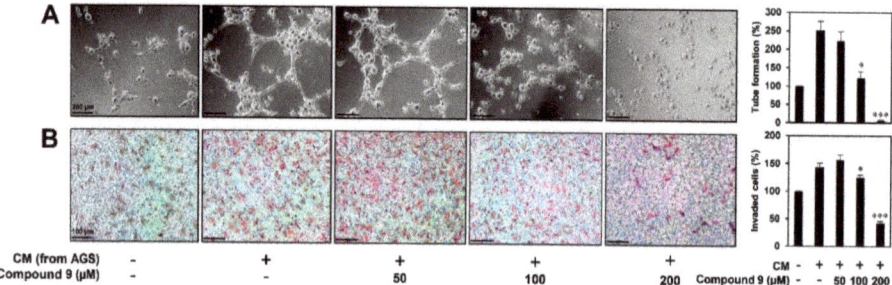

Figure 5. The inhibitory activity of compound **9** on tumor cell-induced angiogenesis. AGS gastric cancer cell-induced angiogenesis was assessed using (**A**) a conditioned medium from tumor cells for in vitro tube formation assay and (**B**) an in vitro co-culture system based on the chemoinvasion assay. (**A**) The basal level of the tube formation of HUVECs treated with non-conditioned medium without AGS cells was normalized to 100%. * $p < 0.05$, *** $p < 0.001$ vs. the conditioned medium from untreated AGS cells. (**B**) The basal level of the invasiveness of HUVECs that were incubated in serum-free medium without AGS cells was normalized to 100%. * $p < 0.05$, *** $p < 0.001$ vs. the control with untreated AGS cells. Each value represents the mean ± SD from three independent experiments.

3.5. Effect of Compound 9 on HIF-1α and VEGF Expression in Tumor Cells

Hypoxia-inducible factor-1α (HIF-1α) plays a critical role in promoting tumor angiogenesis by activating the transcription of major proangiogenic factors, including VEGF [18]. To confirm the role of HIF-1α in mediating the suppressive effect of compound **9** on tumor cell-induced angiogenesis, we investigated the effect of compound **9** on HIF-1α expression in AGS gastric cancer cells. As shown in Figure 6A, compound **9** dose-dependently decreased the accumulation of HIF-1α protein induced by hypoxia in AGS cells. Next, we examined the effect of compound **9** on VEGF expression in AGS cells. Compound **9** inhibited VEGF secretion in AGS cells stimulated by hypoxia in a dose-dependent manner (Figure 6B). Therefore, the inhibitory activity of compound **9** on the AGS cell-induced angiogenesis may be associated with the downregulation of HIF-1α and VEGF expression.

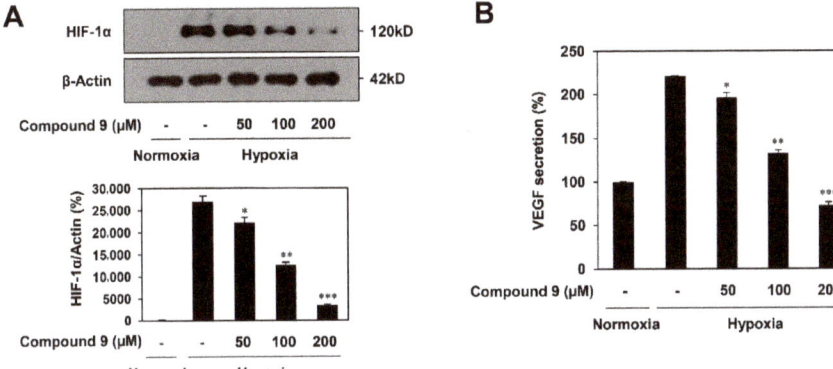

Figure 6. The HIF-1α inhibitory effect of compound **9**. (**A**,**B**) AGS gastric cancer cells were pretreated with compound **9** (50–200 μM) for 1 h and then exposed to 1% O_2 for 11 h. (**A**) The effect of compound **9** on HIF-1α protein accumulation. Protein levels were detected by Western blot analysis and further quantified by densitometry. The level of β-actin was used as an internal control. (**B**) The effect of compound **9** on VEGF expression. The concentration of VEGF protein in the supernatant was determined by a VEGF ELISA. * $p < 0.05$, ** $p < 0.01$, *** $p < 0.001$ vs. the hypoxic control. Each value represents the mean ± SD from three independent experiments.

4. Discussion

Tumor angiogenesis is a complicated process in which new blood vessels are formed in response to the interplay between tumor cells and ECs [19,20]. Tumor cells induce angiogenesis by secreting proangiogenic factors, such as VEGF [21]. Notably, VEGF binds to VEGFR2 on ECs with a high affinity, subsequently activating several key angiogenic signaling pathways, including STAT3, AKT, and ERK1/2, which stimulate EC growth, migration, and differentiation [22–25]. Therefore, strategies that inhibit VEGF-induced angiogenesis by targeting both ECs and tumor cells can be used to effectively block tumor angiogenesis. To the best of our knowledge, the present study is the first to demonstrate that 23-demethyl 8,13-deoxynargenicin, termed compound **9**, exhibits potential antiangiogenic activity through its inhibitory effects on VEGF/VEGFR2 signaling in ECs and the HIF-1α/VEGF pathway in tumor cells.

Nargenicin A1 has been previously found to act as an antibacterial macrolide against various Gram-positive bacteria isolated from *Nocardia* species, inhibiting DnaE involved in bacterial DNA replication [12,15,26]. In addition, nargenicin A1 has shown several other activities, including the activation of acute myeloid leukemia (AML) cell differentiation, anti-inflammation, and protection against oxidative stress [27–29]. Due to its valuable biological activities, various technical approaches to elucidate and modify its biosynthetic pathway have been employed to improve its production and generate novel derivatives [13,14]. In a recent study, we isolated a novel nargenicin A1 analog, termed compound **9**, in an attempt to characterize the key biosynthetic genes involved in post-polyketide synthase (PKS) tailoring of nargenicin A1 in *Nocardia* sp. CS682 [15]. Although compound **9** did not show any effective antibacterial activity as observed nargenicin A1, it exhibited potential anticancer properties, unlike the parent compound. Among the tested cancer cell lines, compound **9** most sensitively suppressed the growth of AGS gastric cancer cells without cytotoxic effects against normal cell lines. Notably, compound **9** inhibited the growth of AGS cells by inducing G2/M cell cycle arrest, ROS- and caspase-mediated apoptosis, and autophagy via the downregulation of the PI3K/AKT/mTOR pathway [15]. Furthermore, while compound **9** significantly decreased the migration and invasion of AGS cells, nargenicin A1 did not. The anti-metastatic effect of compound **9** was associated with the inhibition of MMP-2 and MMP-9, two zinc-dependent endopeptidases associated with tumor invasion and metastasis, as well as the induction of angiogenesis [15]. Although our previous study identified the anticancer activity of compound **9**, its antiangiogenic properties and underlying molecular mechanisms were not investigated.

In the present study, compound **9** was found to effectively inhibit the VEGF-stimulated angiogenic phenotypes of HUVECs, including their proliferation, invasion, capillary tube formation, and adhesion. In addition, compound **9** significantly reduced the neovascularization of the chick embryo CAM model without exhibiting toxicity against pre-existing vessels. As demonstrated by both the in vitro and in vivo results, compound **9** exhibits promising antiangiogenic activity. To elucidate the antiangiogenic mechanisms of compound **9**, we investigated its effect on endothelial VEGF/VEGFR2 signaling. As a result, compound **9** was found to inhibit the VEGF-stimulated phosphorylation of VEGFR2 and its downstream effectors, namely STAT3, AKT, and ERK1/2, in HUVECs. However, compound **9** did not significantly affect the total protein levels of the downstream effectors, but markedly reduced that of VEGFR2, implying that compound **9** may obstruct the dimerization of VEGFR2. MMPs have been previously found to degrade the vascular basement membrane to aid endothelial sprouting during angiogenesis [17]. In the present study, compound **9** markedly suppressed the expression of MMP-2 and MMP-9 stimulated by VEGF in HUVECs, indicating that it reduced MMP expression in both ECs and tumor cells.

Tumor cells express a variety of proangiogenic factors, including VEGF, through HIF-1, a heterodimeric transcription factor composed of two subunits, HIF-1α and HIF-1β [30–32]. HIF-1β is a constitutive nuclear protein, while HIF-1α is strongly induced in response to hypoxia, growth factor stimulation, and the activation of oncogenes. Therefore, HIF-1α is considered a crucial target for suppressing tumor angiogenesis. In the present study, compound **9** was found to effectively inhibit in vitro AGS tumor cell-induced angiogenesis. Moreover, compound **9** significantly decreased the expression of HIF-1α and VEGF induced by hypoxia in AGS cells. Taken together, our results suggest that compound **9** exerts an antiangiogenic activity via the dual downregulation of VEGF/VEGFR2-mediated signaling in ECs and HIF-1α/VEGF expression in tumor cells.

In conclusion, this study reveals for the first time the novel bioactivity and action mechanism of compound **9** and demonstrates its potential as an angiogenesis inhibitor. Further studies will be needed to identify the primary cellular target of compound **9** to improve our understanding of the molecular mechanisms responsible for its anticancer and antiangiogenic activities.

Author Contributions: Conceptualization, methodology, software, and validation: J.M.H. and H.J.J.; formal analysis and investigation: J.M.H. and Y.S.C.; resources: D.D., J.K.S., and H.J.J.; data curation: J.M.H. and H.J.J.; writing—original draft preparation: J.M.H.; writing—review and editing: H.J.J.; visualization: J.M.H.; supervision: H.J.J.; project administration and funding acquisition: H.J.J. All authors have read and agreed to the published version of the manuscript.

Funding: This research was supported by the Basic Science Research Program through the National Research Foundation of Korea (NRF) funded by the Ministry of Education (NRF-2016R1D1A1B03932956) and the NRF grant funded by the Ministry of Science and ICT (NRF-2019R1A2C1009033). This work was also supported by the Brain Korea 21 Plus Project, Republic of Korea.

Conflicts of Interest: The authors declare no conflict of interest.

References

1. Folkman, J. Seminars in Medicine of the Beth Israel Hospital, Boston. *Clin. Appl. Res. Angiogenesis N. Engl. J. Med.* **1995**, *333*, 1757–1763.
2. Andre, T.; Chastre, E.; Kotelevets, L.; Vaillant, J.C.; Louvet, C.; Balosso, J.; Le Gall, E.; Prévot, S.; Gespach, C. Tumoral angiogenesis: Physiopathology, prognostic value and therapeutic perspectives. *Rev. Med. Interne* **1998**, *19*, 904–913. [CrossRef]
3. Granci, V.; Dupertuis, Y.M.; Pichard, C. Angiogenesis as a potential target of pharmaconutrients in cancer therapy. *Curr. Opin. Clin. Nutr. Metab. Care* **2010**, *13*, 417–422. [CrossRef]
4. Risau, W. Mechanisms of angiogenesis. *Nature* **1997**, *386*, 671–674. [CrossRef]
5. Lim, H.N.; Jang, J.P.; Han, J.M.; Jang, J.H.; Ahn, J.S.; Jung, H.J. Antiangiogenic potential of microbial metabolite elaiophylin for targeting tumor angiogenesis. *Molecules* **2018**, *23*, 563. [CrossRef]

6. Shanmugam, M.K.; Warrier, S.; Kumar, A.P.; Sethi, G.; Arfuso, F. Potential role of natural compounds as anti-angiogenic agents in cancer. *Curr. Vasc. Pharmacol.* **2017**, *15*, 503–519. [CrossRef]
7. Roskoski, R., Jr. Vascular endothelial growth factor (VEGF) signaling in tumor progression. *Crit. Rev. Oncol. Hematol.* **2007**, *62*, 179–213. [CrossRef]
8. Ferrara, N.; Kerbel, R.S. Angiogenesis as a therapeutic target. *Nature* **2005**, *438*, 967–974. [CrossRef]
9. Chen, H.M.; Tsai, C.H.; Hung, W.C. Foretinib inhibits angiogenesis, lymphangiogenesis and tumor growth of pancreatic cancer in vivo by decreasing VEGFR-2/3 and TIE-2 signaling. *Oncotarget* **2015**, *6*, 14940–14952. [CrossRef]
10. Koch, S.; Tugues, S.; Li, X.; Gualandi, L.; Claesson-Welsh, L. Signal transduction by vascular endothelial growth factor receptors. *Biochem. J.* **2011**, *437*, 169–183. [CrossRef]
11. Karar, J.; Maity, A. PI3K/AKT/mTOR pathway in angiogenesis. *Front. Mol. Neurosci.* **2011**, *4*, 51. [CrossRef]
12. Sohng, J.K.; Yamaguchi, T.; Seong, C.N.; Baik, K.S.; Park, S.C.; Lee, H.J.; Jang, S.Y.; Simkhada, J.R.; Yoo, J.C. Production, isolation and biological activity of nargenicin from Nocardia sp. CS682. *Arch. Pharm. Res.* **2008**, *31*, 1339–1345. [CrossRef]
13. Dhakal, D.; Chaudhary, A.K.; Yi, J.S.; Pokhrel, A.R.; Shrestha, B.; Parajuli, P.; Shrestha, A.; Yamaguchi, T.; Jung, H.J.; Kim, S.Y.; et al. Enhanced production of nargenicin A1 and creation of a novel derivative using a synthetic biology platform. *Appl. Microbiol. Biotechnol.* **2016**, *100*, 9917–9931. [CrossRef]
14. Dhakal, D.; Le, T.T.; Pandey, R.P.; Jha, A.K.; Gurung, R.; Parajuli, P.; Pokhrel, A.R.; Yoo, J.C.; Sohng, J.K. Enhanced production of nargenicin A(1) and generation of novel glycosylated derivatives. *Appl. Biochem. Biotechnol.* **2015**, *175*, 2934–2949. [CrossRef] [PubMed]
15. Dhakal, D.; Han, J.M.; Mishra, R.; Pandey, R.P.; Kim, T.S.; Rayamajhi, V.; Jung, H.J.; Yamaguchi, T.; Sohng, J.K. Characterization of tailoring steps of nargenicin A1 biosynthesis reveals a novel analogue with anticancer activities. *ACS Chem. Biol.* **2020**, *15*, 1370–1380. [CrossRef]
16. Holmes, K.; Roberts, O.L.; Thomas, A.M.; Cross, M.J. Vascular endothelial growth factor receptor-2: Structure, function, intracellular signalling and therapeutic inhibition. *Cell. Signal.* **2007**, *19*, 2003–2012. [CrossRef]
17. Stetler-Stevenson, W.G. Matrix metalloproteinases in angiogenesis: A moving target for therapeutic intervention. *J. Clin. Investig.* **1999**, *103*, 1237–1241. [CrossRef]
18. Forsythe, J.A.; Jiang, B.H.; Iyer, N.V.; Agani, F.; Leung, S.W.; Koos, R.D.; Semenza, G.L. Activation of vascular endothelial growth factor gene transcription by hypoxia-inducible factor 1. *Mol. Cell. Biol.* **1996**, *16*, 4604–4613. [CrossRef]
19. Carmeliet, P. Angiogenesis in health and disease. *Nat. Med.* **2003**, *9*, 653–660. [CrossRef]
20. Bergers, G.; Benjamin, L.E. Tumorigenesis and the angiogenic switch. *Nat. Rev. Cancer* **2003**, *3*, 401–410. [CrossRef]
21. Chen, C.; Sun, M.Z.; Liu, S.; Yeh, D.; Yu, L.; Song, Y.; Gong, L.; Hao, L.; Hu, J.; Shao, S. Smad4 mediates malignant behaviors of human ovarian carcinoma cell through the effect on expressions of E-cadherin, plasminogen activator inhibitor-1 and VEGF. *Bmb Rep.* **2010**, *43*, 554–560. [CrossRef]
22. Shibuya, M. Differential roles of vascular endothelial growth factor receptor-1 and receptor-2 in angiogenesis. *J. Biochem. Mol. Biol.* **2006**, *39*, 469–478. [CrossRef] [PubMed]
23. Kim, S.H.; Schmitt, C.E.; Woolls, M.J.; Holland, M.B.; Kim, J.D.; Jin, S.W. Vascular endothelial growth factor signaling regulates the segregation of artery and vein via ERK activity during vascular development. *Biochem. Biophys. Res. Commun.* **2013**, *430*, 1212–1216. [CrossRef] [PubMed]
24. Niu, G.; Wright, K.L.; Huang, M.; Song, L.; Haura, E.; Turkson, J.; Zhang, S.; Wang, T.; Sinibaldi, D.; Coppola, D.; et al. Constitutive Stat3 activity up-regulates VEGF expression and tumor angiogenesis. *Oncogene* **2002**, *21*, 2000–2008. [CrossRef]
25. Pan, B.; Shen, J.; Cao, J.; Zhou, Y.; Shang, L.; Jin, S.; Cao, S.; Che, D.; Liu, F.; Yu, Y. Interleukin-17 promotes angiogenesis by stimulating VEGF production of cancer cells via the STAT3/GIV signaling pathway in non-small-cell lung cancer. *Sci. Rep.* **2015**, *5*, 16053. [CrossRef]
26. Painter, R.E.; Adam, G.C.; Arocho, M.; DiNunzio, E.; Donald, R.G.; Dorso, K.; Genilloud, O.; Gill, C.; Goetz, M.; Hairston, N.N.; et al. Elucidation of DnaE as the antibacterial target of the natural product, nargenicin. *Chem. Biol.* **2015**, *22*, 1362–1373. [CrossRef] [PubMed]
27. Kim, S.H.; Yoo, J.C.; Kim, T.S. Nargenicin enhances 1, 25-dihydroxyvitamin D3-and all-trans retinoic acid-induced leukemia cell differentiation via PKCβI/MAPK pathways. *Biochem. Pharmacol.* **2009**, *77*, 1694–1701. [CrossRef]

28. Yoo, J.C.; Cho, H.S.; Park, E.; Park, J.A.; Kim, S.; Kim, D.K.; Kim, C.S.; Kim, S.J.; Chun, H.S. Nargenicin attenuates lipopolysaccharide-induced inflammatory responses in BV-2 cells. *Neuroreport* **2009**, *20*, 1007–1012. [CrossRef]
29. Park, C.; Kwon, D.H.; Hwang, S.J.; Han, M.H.; Jeong, J.W.; Hong, S.H.; Cha, H.J.; Hong, S.H.; Kim, G.Y.; Lee, H.J.; et al. Protective effects of nargenicin A1 against tacrolimus-induced oxidative stress in Hirame Natural Embryo cells. *Int. J. Environ. Res. Public Health* **2019**, *16*, 1044. [CrossRef]
30. Vaupel, P.; Harrison, L. Tumor hypoxia: Causative factors, compensatory mechanisms and cellular response. *Oncologist* **2004**, *9* (Suppl. 5), 4–9. [CrossRef]
31. Semenza, G.L. HIF-1 and tumor progression: Pathophysiology and therapeutics. *Trends Mol. Med.* **2002**, *8*, S62–S67. [CrossRef]
32. Carmeliet, P.; Dor, Y.; Herbert, J.M.; Fukumura, D.; Brusselmans, K.; Dewerchin, M.; Neeman, M.; Bono, F.; Abramovitch, R.; Maxwell, P.; et al. Role of HIF-1alpha in hypoxia-mediated apoptosis, cell proliferation and tumour angiogenesis. *Nature* **1998**, *394*, 485–490. [CrossRef] [PubMed]

© 2020 by the authors. Licensee MDPI, Basel, Switzerland. This article is an open access article distributed under the terms and conditions of the Creative Commons Attribution (CC BY) license (http://creativecommons.org/licenses/by/4.0/).

Article

Serum γ-Glutamyltransferase Concentration Predicts Endothelial Dysfunction in Naïve Hypertensive Patients

Maria Perticone [1,*,†], Raffaele Maio [2,†], Benedetto Caroleo [2], Angela Sciacqua [3], Edoardo Suraci [3], Simona Gigliotti [3], Francesco Martino [4], Francesco Andreozzi [3], Giorgio Sesti [5] and Francesco Perticone [3]

1. Department of Experimental and Clinical Medicine, Magna Graecia University, 88100 Catanzaro, Italy
2. Geriatrics Division, University Hospital Mater Domini, 88100 Catanzaro, Italy; raf_maio@yahoo.it (R.M.); benedettocaroleo@libero.it (B.C.)
3. Department of Medical and Surgical Sciences, Magna Graecia University, 88100 Catanzaro, Italy; sciacqua@unicz.it (A.S.); edoardosuraci88@gmail.com (E.S.); simona_gigliotti@yahoo.it (S.G.); andreozzif@unicz.it (F.A.); perticone@unicz.it (F.P.)
4. Department of Pediatrics Gynecology and Obstetrics, Sapienza University of Rome, 00185 Rome, Italy; francesco.martino30@tin.it
5. Department of Clinical and Molecular Medicine, Sant' Andrea University Hospital, Sapienza University of Rome, 00185 Rome, Italy; giorgio.sesti@uniroma1.it
* Correspondence: mariaperticone@unicz.it; Tel.: +39-0961-364-7149
† These Authors contributed equally to this work.

Received: 4 June 2020; Accepted: 9 July 2020; Published: 11 July 2020

Abstract: Background: Serum gamma-glutamyltransferase (γ-GT) is recognized as a risk factor for cardiovascular diseases (CV). Traditional cardiovascular risk factors mediate endothelial dysfunction. Aim: to evaluate a possible correlation between serum γ-GT and endothelium-dependent vasodilation in naïve hypertensives. Methods: We enrolled 500 hypertensives. Endothelial function was studied by strain-gauge plethysmography. Receiver operating characteristic (ROC) analysis was used to assess the predictive value of γ-GT and to identify the optimal cut-off value of the same variable for endothelial dysfunction. Results: At univariate linear analysis peak percent increase in acetylcholine (ACh)-stimulated vasodilation was inversely related to γ-GT ($r = -0.587$), alanine aminotransferase (ALT) ($r = -0.559$), aspartate aminotransferase (AST) ($r = -0.464$), age ($r = -0.171$), body mass index (BMI) ($r = -0.152$), and fasting glucose ($r = -101$). In the stepwise multivariate regression model, endothelium-dependent vasodilation was significantly related to γ-GT ($\beta = -0.362$), ALT ($\beta = -0.297$), AST ($\beta = -0.217$), estimated glomerular filtration rate (e-GFR) ($\beta = 0.199$), gender ($\beta = 0.166$), and smoking ($\beta = -0.061$). The ROC analysis demonstrated that the accuracy of γ-GT for identifying patients with endothelial dysfunction was 82.1%; the optimal γ-GT cut-off value for discriminating patients with this alteration was 27 UI/L. Conclusions: Serum γ-GT values, within the normal range, are significantly associated with endothelial dysfunction in hypertensives, and may be considered a biomarker of early vascular damage.

Keywords: serum γ-glutamyltransferase; endothelial dysfunction; essential hypertension; atherosclerosis; cardiovascular risk factors

1. Introduction

Gamma-glutamyltransferase (γ-GT) is a glycoprotein located on the plasma membranes of most cells and organ tissues, especially of hepatocytes. It is involved in the extracellular catabolism of glutathione, recognized as the major thiol antioxidant in humans and other mammals. An increase in serum γ-GT levels >50 IU/mL is seen as a consequence of liver injury or bile ducts blockage. In clinical practice, serum γ-GT levels are routinely measured when hepatic/biliary disease and/or alcohol abuse are suspected [1], and it has recently been recognized as a risk factor for metabolic alterations [2], and chronic renal [3] and cardiovascular diseases (CV) [4–7]. These effects seem to be mediated by the capacity of γ-GT to increase the production of reactive oxygen species (ROS) in the presence of some transition metal such as iron [4,8]. Thus, on the basis of these findings and of other epidemiological studies, it was suggested to consider γ-GT, within its normal range, as an early and sensitive biomarker of oxidative stress [9]. In fact, subclinical inflammation related to oxidative stress is considered as the main pathogenetic mechanism involved in several cardio-metabolic diseases [10].

It is well established that endothelial dysfunction, an early event in the atherogenic process, is associated with some metabolic and hemodynamic [11–14] risk factors (i.e., arterial hypertension, obesity, etc.) sharing the same pathogenetic mechanisms, represented by an increased oxidative stress and subclinical inflammation. The activation of these pro-oxidant and pro-inflammatory pathways leads to the activation and progression of atherosclerotic disease by reducing nitric oxide (NO) bioavailability and its protective effect on vascular function [15,16]. In addition, it is important to remark that endothelial dysfunction has been demonstrated to predict the progression of subclinical target organ damage [17–19]; on this basis, it is possible to affirm that endothelial dysfunction has a key role in the pathogenetic mechanisms of CV diseases and associated outcomes [14]. Even if the association between essential hypertension and endothelial dysfunction is well established, to our knowledge there are no data testing a possible relationship between serum γ-GT concentrations and endothelial function in this setting of patients. Thus, we designed the present study to evaluate a possible association between serum γ-GT concentrations, within its normal range, and endothelium-dependent vasodilation, evaluated by strain-gauge plethysmography, in a large population of newly diagnosed, never-treated hypertensive patients.

2. Experimental Section

For this study we enrolled 500 Caucasian hypertensive outpatients (256 men and 244 women; mean age, 47 ± 11 years) referred to the Hypertension Clinic of the University Hospital of Catanzaro, Italy. Exclusion criteria were secondary forms of hypertension, clinical evidence or previous history of coronary artery disease, valvular heart disease, peripheral artery disease, diabetes mellitus, hypercholesterolemia, liver diseases, impaired renal function (defined as an estimated glomerular filtration rate [e-GFR] <60 mL/min per 1.73 m^2), coagulopathies, vasculitis and/or Raynaud's phenomenon, history of alcohol and/or drug abuse, and the use of drugs interfering with liver enzyme concentrations. To enter the study protocol, all subjects had to have serum values of alanine aminotransferase (ALT), aspartate aminotransferase (AST), and γ-GT in the normal range; in particular, we considered 50 IU/mL as the upper normal limit for all these three variables, as established by our laboratory. At the first evaluation, all subjects underwent routine blood tests, assessment of risk factors for atherosclerosis, and evaluation of vascular function through strain-gauge plethysmography. ALT and AST were measured by pyridoxal phosphate activated (liquid reagent) (COBAS Integra 800—Roche Diagnostics GmbH, Mannheim, Germany; normal values 0–50 UI/L), γ-GT was evaluated by standardized against Szasz (COBAS Integra 800—Roche Diagnostics GmbH, Mannheim, Germany; normal values 8–50 UI/L).

High-sensitivity C-reactive protein (hs-CRP) was measured by a high-sensitivity turbidimetric immunoassay (Cardio-Phase hs-CRP, Siemens Healthcare Diagnostics GmbH, Marburg, Germany) in a subgroup of 400 patients representative of the whole study population with regard to the variables listed in Table 1.

The local ethics committee approved the study (approval number 2012.63, 23 October 2012—Comitato Etico Azienda Ospedaliero-Universitaria Mater Domini of Catanzaro, Italy), and all participants gave written informed consent for all procedures. All the study procedures were conducted according to the Declaration of Helsinki.

Table 1. Baseline characteristics of the whole study population and of the subgroup of patients with hs-CRP values.

Variables	All $n = 500$	hs-CRP Group $n = 400$	p
Age, years	47 ± 11	47 ± 11	0.998
Gender, M (%)	256 (51)	183 (46)	0.0001
Body mass index, kg/m^2	27.3 ± 3.6	27.3 ± 3.6	0.997
SBP, mm Hg	149 ± 17	149 ± 17	0.999
DBP, mm Hg	91 + 12	91 ± 12	0.998
Heart rate, bpm	72 ± 9	73 ± 9	0.098
Total cholesterol, mg/dL	205 ± 31	205 ± 31	0.997
Smokers, No (%)	78 (16)	75 (19)	0.246
Fasting glucose, mg/dL	95 ± 11	95 ± 11	0.998
LDL cholesterol, mg/dL	130 ± 31	129 ± 31	0.631
HDL cholesterol, mg/dL	52 ± 12	52 ± 12	0.997
Triglycerides, mg/dL	116 ± 40	116 ± 41	0.999
e-GFR, ml/min/1.73/m^2	85 ± 20	88 ± 18	0.020
ALT, UI/L	21±11	21 ± 11	0.999
AST, UI/L	22 ± 9	22 ± 9	0.998
γ-GT, UI/L	33 ± 14	32 ± 14	0.287
hs-CRP, mg/L		4.1 ± 2.2	
FBF baseline, mL·0.100 tissue^{-1}·min^{-1}	3.4 ± 0.7	3.3 ± 0.6	0.024
FBF maximal response to acetylcholine, % of increase	303 ± 180	318 ± 183	0.218
Response to sodium nitroprusside, % of increase	317 ± 110	315 ± 107	0.784
Vascular resistance, U	34 + 8	34 + 7	0.998

ALT = alanine transaminase; AST = aspartate transaminase; SBP = systolic blood pressure; DBP = diastolic blood pressure; FBF = forearm blood flow; γ-GT = gamma glutamyltransferase; LDL= low-density lipoproteins; HDL = high-density lipoproteins; hs-CRP = high-sensitivity C-reactive protein.

2.1. Vascular Function Evaluation

Vascular function evaluation was made at 09:00 a.m. in a quiet and air-conditioned room (22–24 °C) with the fasting subjects lying supine. Forearm volume was determined by water displacement. A 20-gauge polyethylene catheter (Vasculon 2; Baxter Healthcare, Deerfield, IL, USA), introduced into the brachial artery of the non-dominant arm, was used for evaluation of blood pressure (BP) and for drug administration. Measurement of percent change in forearm volume was obtained by a mercury-filled silastic strain gauge placed on the widest part of the forearm, connected to a plethysmograph (model EC-4; DE Hokanson, Issaquah, WA, USA) that was connected to a chart recorder for detection of forearm blood flow (FBF) measurements. Exclusion of peripheral venous outflow was obtained by inflating to 40 mmHg a cuff placed on the upper arm with a rapid cuff inflator (model E-10; DE Hokanson, Issaquah, WA, USA). FBF was calculated as the slope of the change in forearm volume; the mean of 3 measurements was obtained at each time point.

For the evaluation of endothelial function, we used the protocol initially described by Panza et al. [12] and subsequently used by us [13,14,17,18]. Endothelium-dependent and endothelium-independent vasodilation was assessed by a dose–response curve during intra-arterial infusions of acetylcholine (ACh) (7.5, 15, and 30 µg/mL per minute, each for 5 min) and sodium nitroprusside (SNP) (0.8, 1.6, and 3.2 µg/mL per minute, each for 5 min), respectively. Prior to the administration, ACh (Sigma, Milan, Italy) was diluted with saline and SNP (Malesci, Florence, Italy) in 5% glucose solution, and protected from light with aluminum foil.

2.2. Statistical Analysis

Data are reported as mean ± SD or as percent frequency; we used *t*-test or the χ^2 test, as appropriate, for comparisons between groups. Relationships between paired parameters were tested by correlation coefficient of Pearson. Multivariate models (linear or logistic regression) were constructed using, as independent covariates, several traditional CV risk factors—age, gender, body mass index, glucose, LDL and HDL cholesterol, triglyceride, BP, smoking, and e-GFR—to test the independent relationship between γ-GT and the response to ACh. In an additional analysis, we tested the potential confounding effect of hs-CRP in a subgroup of 400 patients.

In the logistic regression analysis, endothelial dysfunction, as dichotomic variable, was expressed as a maximal response to ACh < 400% as previously reported [17,18].

In multiple linear regression models, data were expressed as standardized regression coefficient (beta) and *p* value. In multiple logistic regression analyses, data were expressed as odds ratio (OR), 95% confidence interval (CI), and *p* value.

Receiver operating characteristic (ROC) analysis was used to assess the predictive value of γ-GT (area under the curve) and to identify the optimal cut-off value of the same variable for endothelial dysfunction, i.e., the value which maximizes the difference between true positive and false positive rates of endothelial dysfunction.

To assess the internal consistency of study results, a sensitivity analysis was performed by randomly dividing the whole study population into two equally sized subgroups.

All calculations were made with a standard statistical package (SPSS for Windows version 20.0; SPSS, Inc., Chicago, IL, USA).

3. Results

Baseline demographic, clinical, and hemodynamic characteristics of the whole study population and of the subgroup of 400 patients, with hs-CRP dosage, are summarized in Table 1.

3.1. Correlational Analysis

The results of univariate linear analysis between ACh-stimulated vasodilation and different covariates in the study population are reported in Table 2. An inverse relationship was found between peak percent increase in ACh-stimulated vasodilation and the following: γ-GT (−0.587), accounting for 34.4% of its variation; ALT ($r = -0.559$), accounting for 31.2% of its variation; AST ($r = -0.464$), accounting for 21.5% of its variation; age ($r = -0.171$), accounting for 2.9% of its variation; BMI ($r = -0.152$), accounting for 2.3% of its variation; and fasting glucose ($r = -101$), accounting for 1% of its variation. On the contrary, a direct relationship was observed with the following covariates: e-GFR ($r = 0.257$), accounting for 6.6% of its variation and HDL cholesterol ($r = 0.108$), accounting for 1.2% of its variation. No significant relationships were detected between ACh-stimulated FBF and systolic and diastolic BP, triglyceride, and LDL cholesterol.

Table 2. Univariate relationships between FBF maximal response to acetylcholine and different covariates.

	R	p
γ-GT, UI/L	−0.587	0.0001
ALT, UI/L	−0.559	0.0001
AST, UI/L	−0.464	0.0001
e-GFR, mL/min/1.73/m^2	0257	0.0001
Gender, male vs. female	0.191	0.0001
Age, years	−0.171	0.0001
BMI, kg/m^2	−0.152	0.0001
Fasting glucose, mg/dL	−0.101	0.012
Systolic blood pressure, mmHg	−0.055	0.0108
Diastolic blood pressure, mmHg	−0.031	0.248
Smoking, yes vs. no	0.028	0.269
LDL cholesterol, mg/dL	−0.024	0.300
Triglycerides, mg/dL	−0.007	0.435
HDL cholesterol, mg/dL	0.001	0.494

Data are expressed as Pearson product-moment correlation coefficient (r) and p values. ALT = alanine transaminase; AST = aspartate transaminase; BMI = body mass index; γ-GT = gamma glutamyltransferase; HDL = high-density lipoproteins; LDL = low-density lipoproteins.

3.2. Multivariate Analysis

To evaluate the independent predictors of ACh-stimulated maximal FBF, covariates reaching statistical significance, with the addition of smoking and gender as dichotomic variables, were inserted into a stepwise multivariate regression model (Table 3). Results of this analysis demonstrated that the variables significantly associated with endothelium-dependent vasodilation were γ-GT (β = −0.362; p = 0.0001), ALT (β = −0.297; p = 0.0001), AST (β = −0.217; p = 0.0001), e-GFR (β = 0.199; p = 0.0001), gender (β = 0.166; p = 0.0001), and smoking (β = −0.061; p = 0.044). γ-GT and ALT account for 35.4% and 12.3% of the FBF variation, respectively, while other covariates retained in the final model explain another 9.8% of its variation. Overall, the final model accounts for 56.5% of FBF variation.

Table 3. Multiple linear regression models for FBF maximal response to acetylcholine.

	Partial r^2	Total r^2	b Coefficient	P
γ-GT	35.4	35.4	−0.362	0.0001
ALT	12.3	46.7	−0.297	0.0001
AST	5.4	52.1	−0.217	0.0001
e-GFR	1.9	54.0	0.199	0.0001
Gender	2.1	56.1	0.166	0.0001
Smoking	0.4	56.5	−0.061	0.044

ALT = alanine transaminase; AST = aspartate transaminase; γ-GT = gamma glutamyltransferase.

In the additional analysis including serum hs-CRP, conducted in a subgroup of 400 patients, the association between serum γ-GT and the peak percent of increase of ACh-stimulated FBF did not change (β = −0.362; p = 0.0001).

Finally, we performed multiple logistic regression analyses to estimate the odds of endothelial dysfunction, adopting the cut-off value of 400% of increase in FBF as the dependent variable, associated with serum γ-GT levels (Table 4). The probability of endothelial dysfunction was significantly increased by γ-GT (OR = 1.927 for 10 UI/L), ALT (OR = 2.175 for 10 UI/L), AST (OR = 1.973 for 10 UI/L), and gender (OR = 2.695 male vs. female); on the contrary, the risk of endothelial function impairment was significantly reduced by the preservation of e-GFR (OR = 0.699 for 10 mL/min/1.73 m^2). The additional analysis, including serum hs-CRP in the same model, demonstrated that the link between serum γ-GT and endothelial dysfunction did not change (OR = 1.953; 95% CI = 1.527–2.499; p = 0.0001).

Table 4. Multiple logistic analysis of endothelial dysfunction.

	OR	95% CI	p
γ-GT, 10 UI/L	1.927	1.548–2.399	0.0001
ALT, 10 UI/L	2.175	1.608–2.941	0.0001
AST, 10 UI/L	1.973	1.369–2.788	0.0001
Gender, male vs. female	2.695	1.413–5.141	0.003
e-GFR, 10 mL/min/1.73/m^2	0.699	0.583–0.837	0.0001

ALT = alanine transaminase; AST = aspartate transaminase; γ-GT = gamma glutamyltransferase.

The ROC analysis demonstrated that the accuracy of γ-GT for identifying patients with endothelial dysfunction was 82.1% (AUC = 0.821, p < 0.001) (Figure 1) and that the optimal γ-GT cut-off value for discriminating patients with this alteration from those without was 27 UI/L, a threshold providing a 81% sensitivity and a 74% specificity.

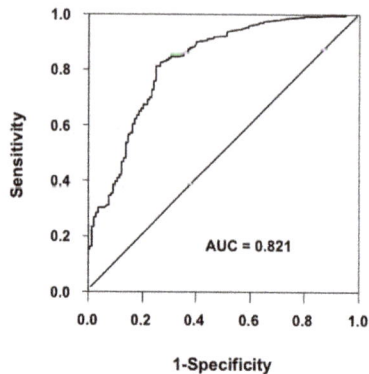

Figure 1. Receiver operating characteristic (ROC) curve for the accuracy of γ-GT for identifying patients with endothelial dysfunction. The accuracy of γ-GT for identifying patients with endothelial dysfunction is 82.1% (AUC = 0.821, p < 0.001).

3.3. Cross-Validation

To assess the robustness of study results, we performed a sensitivity analysis by randomly dividing the whole study population into two equally sized subgroups. This additional analysis showed that the strength of the relationships between the key risk factors (namely, γ-GT, ALT, AST, gender, and e-GFR) with the outcome variable (endothelial dysfunction) we found in the whole study population (see Table 4) was of similar magnitude to that found in subgroup A (see Table 5) and subgroup B (see Table 6) of the sensitivity analysis. The AUC in the two subgroups (Subgroup A: 76.9%; Subgroup B: 82.1%) provided by the above-mentioned risk factors was almost identical to that found in the whole study population (82.1%), indicating an adequate internal consistency of study results.

Table 5. Multiple logistic analysis of endothelial dysfunction in Subgroup A (n = 247).

	OR	95% CI	p
γ-GT, 10 UI/L	2.089	1.545–2.823	0.0001
ALT, 10 UI/L	1.556	1.045–2.316	0.029
AST, 10 UI/L	1.865	1.146–3.035	0.019
Gender, male vs. female	2.742	1.147–6.558	0.023
e-GFR, 10 mL/min/1.73/m^2	0.656	0.522–0.825	0.0001

Table 6. Multiple logistic analysis of endothelial dysfunction in Subgroup B (n = 253).

	OR	95% CI	p
γ-GT, 10 UI/L	1.819	1.335–2.480	0.0001
ALT, 10 UI/L	2.733	1.747–4.273	0.0001
AST, 10 UI/L	1.831	1.141–2.937	0.012
Gender, male vs. female	2.014	0.827–4.904	0.123
e-GFR, 10 mL/min/1.73/m^2	0.710	0.558–0.903	0.005

4. Discussion

To our knowledge, findings obtained in this study demonstrate, for the first time, the association between serum γ-GT within the normal range and endothelial dysfunction, evaluated by strain-gauge plethysmography, in a very large and well-characterized population of never-treated hypertensive patients. Particularly, the risk of endothelial dysfunction increases by 93% for each 10 IU/L elevation of this enzyme (Table 4).

It is well established that the primary role of γ-GT is to contribute to the maintenance of intracellular homeostasis of glutathione (GSH), one of the major intracellular antioxidant components [20], even if some experimental findings demonstrated that, in the presence of iron or other transition metals, it might also be involved in the generation of ROS [9]. Thus, on the basis of this evidence, γ-GT emerged over time as an early and sensitive enzyme related to oxidative stress. In fact, its circulating levels, within normal range, resulted in the increase of F2-isoprostanes, fibrinogen, and CRP, all markers of systemic inflammation [9]. Interestingly, this association was observed independently of the presence of metabolic alterations, typically related to γ-GT elevation [5,21]. Consistent with these findings, we also observed a significant relationship between γ-GT and hs-CRP, confirming previously published data suggesting that elevation of γ-GT is involved in the subclinical inflammatory response and oxidative stress [22], both conditions associated with endothelial dysfunction [13,23].

Another important finding of this study is that ALT levels, within the normal range, are also significantly associated with endothelial function. Particularly, the increase of 10 IU/L of ALT almost doubles the risk of endothelial dysfunction, as reported in Table 4. Although a significant relationship between γ-GT values within the normal range and the incidence of chronic elevation of ALT was previously reported [24], no data are available to demonstrate a possible pathogenetic role of ALT in the activation of an oxidative stress process. Although γ-GT can be considered as an early biomarker of systemic and hepatic oxidative stress, ALT elevation might reflect possible inflammatory liver damage as a consequence of this increased oxidative stress. In accordance with this, we previously reported that hypertensive patients with both metabolic syndrome and non-alcoholic fatty liver disease (NAFLD) had a reduced endothelium-dependent vasodilation in comparison with hypertensives with metabolic syndrome without NAFLD [25]. Serum ALT values were significantly higher in hypertensives with NAFLD than in those without (42.0 + 10.8 vs. 24.6 + 5.2 UI/L), leading to the hypothesis that NAFLD—and the associated elevation of ALT levels—could be considered as an early biomarker of endothelial dysfunction.

5. Conclusions

The results obtained in the present study showed a strong and inverse relationship between γ-GT and endothelium-dependent vasodilation; in particular, the novelty of this paper is the identification of a cut-off value of γ-GT (27 IU/L) to discriminate between patients with and without (γ-GT> and <27 IU/L, respectively) endothelial dysfunction. In addition, present data confirm previously published literature demonstrating the strong associations between serum γ-GT and many CV risk factors and/or events; in this context endothelial dysfunction could be considered as an established consequence of increased oxidative stress. Furthermore, it is important to remark that strain-gauge plethysmography represents the gold-standard technique for endothelial function testing, thus conferring robustness to the results obtained in a wide population. Thus, serum γ-GT may be considered an additional biomarker of early vascular damage; its usefulness is also supported by its wide availability and low cost.

Author Contributions: Conceptualization, R.M., M.P., F.P.; methodology, M.P., F.P.; formal analysis and investigation, A.S., E.S., F.M., B.C., S.G., F.M. and F.A.; writing—original draft preparation, R.M., M.P.; writing—review and editing: M.P., F.P.; supervision, F.P. and G.S. All authors have read and agreed to the published version of the manuscript.

Funding: This research received no external funding.

Conflicts of Interest: The authors declare no conflict of interest.

References

1. Kunutsor, S.K. Gamma-glutamyltransferase-friend or foe within? *Liv. Int.* **2016**, *36*, 1723–1734. [CrossRef] [PubMed]
2. Lee, D.; Silventoinen, K.; Jacobs, D.R.; Jousilahti, P.; Tuomileto, J. Glutamyltransferase, Obesity, and the Risk of Type 2 Diabetes: Observational Cohort Study among 20,158 Middle-Aged Men and Women. *J. Clin. Endocrinol. Metab.* **2004**, *89*, 5410–5414. [CrossRef] [PubMed]
3. Ko, S.H.; Baeg, M.K.; Han, K.D.; Ko, S.Y.; Shin, S.B.; Ko, S.H.; Ahn, Y.B. Association between gamma-glutamyltransferase and albuminuria in non diabetic adults with normal renal function. *Clin. Exp. Nephrol.* **2017**, *21*, 835–841. [CrossRef] [PubMed]
4. Mason, J.E.; Starke, R.D.; Van Kirk, J.E. Gamma-Glutamyl Transferase: A Novel Cardiovascular Risk BioMarker. *Prev. Cardiol.* **2010**, *13*, 36–41. [CrossRef]
5. Turgut, O.; Tandogan, I.; Tandoğan, I. Gamma-glutamyltransferase to Determine Cardiovascular Risk: Shifting the Paradigm Forward. *J. Atheroscler. Thromb.* **2011**, *18*, 177–181. [CrossRef]
6. Aksakal, E.; Tanboga, I.H.; Kurt, M.; Kaygın, M.A.; Kaya, A.; Isik, T.; Ekinci, M.; Sevimli, S.; Acikel, M. The relation of serum gamma-glutamyl transferase levels with coronary lesion complexity and long-term outcome in patients with stable coronary artery disease. *Atherosclerosis* **2012**, *221*, 596–601. [CrossRef]
7. Li, D.D.; Xu, T.; Cheng, X.Q.; Wu, W.; Ye, Y.C.; Guo, X.Z.; Cheng, Q.; Liu, Q.; Liu, L.; Zhu, G.J.; et al. Serum Gamma-Glutamyltransferase Levels are Associated with Cardiovascular Risk Factors in China: A Nationwide Population-Based Study. *Sci. Rep.* **2018**, *8*, 16533. [CrossRef]
8. Paolicchi, A.; Minotti, G.; Tonarelli, P.; Tongiani, R.; De Cesare, D.; Mezzetti, A.; Dominici, S.; Comporti, M.; Pompella, A. Gamma-glutamyl transpeptidase-dependent iron reduction and LDL oxidation—A potential mechanism in atherosclerosis. *J. Investig. Med.* **1999**, *47*, 151–160.
9. Lee, D.; Blomhoff, R.; Jacobs, D.R. ReviewIs Serum Gamma Glutamyltransferase a Marker of Oxidative Stress? *Free. Radic. Res.* **2004**, *38*, 535–539. [CrossRef]
10. Ross, R. Atherosclerosis—An inflammatory disease. *N. Engl. J. Med.* **1999**, *340*, 115–126. [CrossRef]
11. Vane, J.R.; Anggard, E.E.; Botting, R.M. Regulatory functions on the vascular endothelium. *N. Engl. J. Med.* **1990**, *323*, 27–36.
12. Panza, J.A.; Quyyumi, A.A.; Brush, J.E.; Epstein, S.E. Abnormal Endothelium-Dependent Vascular Relaxation in Patients with Essential Hypertension. *New. Engl. J. Med.* **1990**, *323*, 22–27. [CrossRef] [PubMed]
13. Perticone, F.; Ceravolo, R.; Candigliota, M.; Ventura, G.; Iacopino, S.; Sinopoli, F.; Mattioli, P.L. Obesity and body fat distribution induce endothelial dysfunction by oxidative stress: Protective effect of vitamin C. *Diabetes* **2001**, *50*, 159–165. [CrossRef] [PubMed]

14. Perticone, F.; Ceravolo, R.; Pujia, A.; Ventura, G.; Iacopino, S.; Scozzafava, A.; Ferraro, A.; Chello, M.; Mastroroberto, P.; Verdecchia, P.; et al. Prognostic significance of endothelial dysfunction in hypertensive patients. *Circulation* **2001**, *104*, 191–196. [CrossRef] [PubMed]
15. Quyyumi, A.A. Endothelial function in health and disease: New insights into the genesis of cardiovascular disease. *Am. J. Med.* **1998**, *105*, 32S–39S. [CrossRef]
16. Bonetti, O.; Lerman, P.; Lerman, L.O. Endothelial dysfunction: A marker of atherosclerotic risk. *Arter. Thromb. Vasc. Boil.* **2003**, *23*, 168–175. [CrossRef]
17. Perticone, F.; Maio, R.; Perticone, M.; Sciacqua, A.; Shehaj, E.; Naccarato, P.; Sesti, G. Endothelial Dysfunction and Subsequent Decline in Glomerular Filtration Rate in Hypertensive Patients. *Circulation* **2010**, *122*, 379–384. [CrossRef]
18. Perticone, F.; Maio, R.; Perticone, M.; Miceli, S.; Sciacqua, A.; Tassone, E.J.; Shehaj, E.; Tripepi, G.; Sesti, G. Endothelial dysfunction predicts regression of hypertensive cardiac mass. *Int. J. Cardiol.* **2013**, *167*, 1188–1192. [CrossRef]
19. Halcox, J.; Donald, A.E.; Ellins, E.; Witte, D.; Shipley, M.; Brunner, E.J.; Marmot, M.; Deanfield, J. Endothelial Function Predicts Progression of Carotid Intima-Media Thickness. *Circulation* **2009**, *119*, 1005–1012. [CrossRef]
20. Franco, R.; Schoneveld, O.J.; Pappa, A.; Panayiotidis, M.I. The central role of glutathione in the pathophysiology of human diseases. *Arch. Physiol. Biochem.* **2007**, *113*, 234–258. [CrossRef]
21. Yamada, J.; Tomiyama, H.; Yambe, M.; Koji, Y.; Motobe, K.; Shiina, K.; Yamamoto, Y.; Yamashina, A. Elevated serum levels of alanine aminotransferase and gamma glutamyltransferase are markers of inflammation and oxidative stress independent of the metabolic syndrome. *Atherosclerosis* **2006**, *189*, 198–205. [CrossRef]
22. Lee, D.-H.; Jacobs, D.R. Association between serum gamma-glutamyltransferase and C-reactive protein. *Atherosclerosis* **2005**, *178*, 327–330. [CrossRef] [PubMed]
23. Andreozzi, F.; Laratta, E.; Procopio, C.; Hribal, M.L.; Sciacqua, A.; Perticone, M.; Miele, C.; Perticone, F.; Sesti, G. Interleukin-6 Impairs the Insulin Signaling Pathway, Promoting Production of Nitric Oxide in Human Umbilical Vein Endothelial Cells. *Mol. Cell. Boil.* **2007**, *27*, 2372–2383. [CrossRef] [PubMed]
24. Lee, D.; Lim, J.-S.; Yang, J.-H.; Ha, M.-H.; Jacobs, D.R. Serum gamma-glutamyltransferase within its normal range predicts a chronic elevation of alanine aminotransferase: A four year follow-up study. *Free Radic. Res.* **2005**, *39*, 589–593. [CrossRef] [PubMed]
25. Sciacqua, A.; Perticone, M.; Miceli, S.; Laino, I.; Tassone, E.; Grembiale, R.D.; Andreozzi, F.; Sesti, G.; Perticone, F. Endothelial dysfunction and non-alcoholic liver steatosis in hypertensive patients. *Nutr. Metab. Cardiovasc. Dis.* **2011**, *21*, 485–491. [CrossRef] [PubMed]

© 2020 by the authors. Licensee MDPI, Basel, Switzerland. This article is an open access article distributed under the terms and conditions of the Creative Commons Attribution (CC BY) license (http://creativecommons.org/licenses/by/4.0/).

Review
Endothelial Dysfunction in Diabetes

Yusuke Takeda [1], Keiichiro Matoba [1,*], Kensuke Sekiguchi [1], Yosuke Nagai [1], Tamotsu Yokota [1], Kazunori Utsunomiya [2] and Rimei Nishimura [1]

1. Division of Diabetes, Metabolism, and Endocrinology, Department of Internal Medicine, The Jikei University School of Medicine, Tokyo 105-8461, Japan; ms05-takeda@jikei.ac.jp (Y.T.); k.sekiguchi.0322@gmail.com (K.S.); y.nagai@jikei.ac.jp (Y.N.); yokotat@jikei.ac.jp (T.Y.); rimei@jikei.ac.jp (R.N.)
2. Center for Preventive Medicine, The Jikei University School of Medicine, Tokyo 105-8461, Japan; kazu-utsunomiya@jikei.ac.jp
* Correspondence: matoba@jikei.ac.jp

Received: 25 May 2020; Accepted: 26 June 2020; Published: 29 June 2020

Abstract: Diabetes is a worldwide health issue closely associated with cardiovascular events. Given the pandemic of obesity, the identification of the basic underpinnings of vascular disease is strongly needed. Emerging evidence has suggested that endothelial dysfunction is a critical step in the progression of atherosclerosis. However, how diabetes affects the endothelium is poorly understood. Experimental and clinical studies have illuminated the tight link between insulin resistance and endothelial dysfunction. In addition, macrophage polarization from M2 towards M1 contributes to the process of endothelial damage. The possibility that novel classes of anti-hyperglycemic agents exert beneficial effects on the endothelial function and macrophage polarization has been raised. In this review, we discuss the current status of knowledge regarding the pathological significance of insulin signaling in endothelium. Finally, we summarize recent therapeutic strategies against endothelial dysfunction with an emphasis on macrophage polarity.

Keywords: endothelial dysfunction; insulin resistance; macrophage polarity

1. Introduction

Diabetes is a global health problem, characterized by defective insulin secretion and resistance to insulin. According to the International Diabetes Federation (IDF), the number of people with diabetes is estimated to rise from 425 million at present to more than 600 million by 2045 [1]. Diabetes carries a significant risk of microvascular pathologies, such as retinopathy, nephropathy, neuropathy, and atherosclerotic diseases. Indeed, the relative risk of cardiovascular disease increases by two- to four-fold in patients with diabetes compared to non-diabetes patients [2]. Endothelial dysfunction is an early marker for atherosclerosis, preceding angiographic or ultrasonic evidence of atherosclerotic plaque [3,4]. In addition, accumulating evidence implicates endothelial dysfunction as an event seen even in patients with prediabetic conditions, such as impaired fasting glucose (IFG) and impaired glucose tolerance (IGT) [5].

The vascular endothelium functions as a structural barrier between the lumen and vessel wall. Studies over the past decade have also shown that the endothelium secretes numerous growth factors and cytokines that regulate multiple vascular functions (e.g., vascular tone, proliferation of vascular smooth muscle, platelet aggregation, coagulation, and fibrinolysis). Furthermore, the endothelium mediates vasoconstriction by secreting mediators, such as endothelin-1 and thromboxane A2. In contrast, substances such as nitric oxide (NO), prostacyclin, and endothelium-derived hyperpolarizing factor (EDHF) regulate vasodilation [2]. NO, the primary contributor synthesized from L-arginine by endothelial NO synthase (eNOS), regulates the endothelium-dependent relaxation of arteries [6].

Endothelial dysfunction is characterized by a loss of molecular functions in endothelial cells. Factors promoting this event include metabolic disorders (e.g., diabetes [7,8], obesity [9], dyslipidemia [10]), smoking [11], a high salt intake [12], lack of exercise [13], and menopause [14]. The release of reactive oxygen species (ROS) and the generation of oxidative stress are considered critical factors for the pathogenesis of diabetic vascular complications. While endothelial dysfunction is associated with various pathological aspects, including local inflammation [15,16] and oxidative stress [17,18], the pivotal mechanisms are the decrease of NO production and inactivation of NO [19]. The inactivation of NO results from oxidative stress caused by uncoupling of eNOS [20] and an increase in ROS-generating enzymes, including nicotinamide adenine dinucleotide phosphate-oxidase (NADPH) oxidase (NOX), cyclooxygenases (COX), and xanthine oxidase (XO) [21–23].

A variety of clinical methods for assessing the endothelial function are used. Previous studies to assess the endothelial function in humans have often evaluated NO-dependent vasodilation. Measuring the changes in the diameter and blood flow of the coronary artery in response to intra-coronary infusion of acetylcholine is considered the standard method [24]. Non-invasive methods for measuring the endothelial function have been evaluated in previous studies [25]. One of the most commonly applied techniques is flow-mediated dilation (FMD), which is evaluated by brachial artery ultrasound [26]. This method is well-trusted and relevant to cardiovascular risk factors [27] but is highly dependent on the experience level of the operators, who need special training [28]. The analysis of the pulse amplitude tonometry (PAT) in the index finger after reactive hyperemia has been considered as another non-invasive method for assessing the endothelial function [29]. Elevation of plasma concentrations of biomarkers of hemostasis, inflammation, and oxidative stress are also used as indices suggesting endothelial dysfunction [24]. Circulating levels of markers such as P- and E-selectin, ICAM-1, VCAM-1, plasminogen activator inhibitor-1 (PAI-1), oxidized low-density lipoprotein (oxLDL), and asymmetrical dimethylarginine (ADMA) have been used as markers of endothelial dysfunction [24].

We herein review the underlying mechanisms of endothelial dysfunction in diabetes and discuss how endothelial metabolism is targeted by the clinical agents.

2. Insulin Resistance and Endothelial Dysfunction

Insulin plays a vital role in the maintenance of vascular homeostasis. Insulin resistance is defined as an impaired biologic sensitivity and/or responsiveness to insulin stimulation in target tissues including the muscle, adipose tissue, and liver. Substantial evidence supports insulin resistance as the essential pathophysiologic impairment responsible for metabolic and cardiovascular disorders, collectively known as metabolic syndrome. Disturbance of insulin signaling eventually leads to glucose intolerance, diabetes, dyslipidemia, and coronary artery disease. Over the past two decades, many studies have focused on mechanisms provoking endothelial dysfunction, including ROS-mediated eNOS uncoupling, loss of NO bioavailability, and hyperglycemia-induced apoptosis of vascular endothelium, which ultimately leads to impaired vascular relaxation, a common biomarker of endothelial dysfunction. Understanding the endothelial control of metabolism in detail may aid in the development of novel approaches for intervention in obesity and obesity-related diseases.

2.1. Insulin Signaling in Endothelium

Insulin binds to the cell surface receptor known as the insulin receptor (IR). Activated IR phosphorylates intracellular substrates, such as insulin receptor substrate (IRS) family members, Shc proteins, and Gap-1 [30]. In humans, three isoforms of IRS—1, 2, and 4—have been shown to play important roles that vary depending on the cell type and metabolic conditions. For example, IRS-1 regulates insulin action in skeletal muscle as evidenced by findings that genetic ablation of IRS-1 results in insulin resistance and hypertriglyceridemia. IRS-2 functions as a regulator of insulin action in liver and pancreatic β cells. Intriguingly, IRS-2-deficient mice are more susceptible to diabetes than IRS-1 knockout mice because of the impairment of insulin secretion [31], indicating that IRS-2 contributes to the molecular basis for diabetes. The phosphorylated IRS tyrosine activates phosphoinositide-3 kinase (PI3-K) and then converts phosphatidylinositol (3,4)-bisphosphate (PIP2) to phosphatidylinositol (3,4,5)-trisphosphate (PIP3). PIP3 initiates a cascade of serine kinases, resulting in the recruitment of phosphoinositide-dependent kinase-1 (PDK-1) and Akt to the membrane, where they are activated [32]. Activation of Akt greatly influences cellular functions by regulating NO production, angiogenesis, and glucose metabolism [33].

Both IRS-1 and -2 are expressed in the endothelium [34]. Akt activation promotes the cell survival and proliferation of tumor vasculature [35]. Under pathophysiological conditions including obesity and insulin resistance, selective endothelial insulin resistance is promoted by proteasomal degradation of IRS-2 [34]. In the setting of insulin resistance, the reduction of endothelial proliferation results in atherosclerosis, diminished collateral angiogenesis in occluded coronary arteries, and reduced reendothelialization [2]. Furthermore, emerging evidence has shown that the proangiogenic role of Akt is induced by the generation of hypoxia-inducible factor α (HIFα). HIFα activation leads to the expression and subsequent production of angiogenic factors, such as vascular endothelial growth factor (VEGF). Akt's ability to enhance the rate of glycolysis is dependent on HIFα and the subsequent expression of glycolytic enzymes [36].

Another insulin signaling pathway proceeds from Shc, which causes activation of the small GTP binding protein Ras and then initiates a phosphorylation cascade involving mitogen-activated protein kinase (MAPK). The MAPK pathway is associated with endothelial cells, mediating the secretion of ET-1 [37]. Insulin signal pathways form an extremely complicated network and multiple feedback loops. In other words, while MAPK pathways are weakly associated with regulating metabolic functions, PI3-kinase-dependent pathways function as pivotal branches to mediate the metabolic actions of insulin.

2.2. Insulin Resistance in the Endothelium

Insulin resistance is characterized by the deficiency in metabolic actions of insulin. A disorder of the PI3-K/Akt pathway results in a lack of insulin sensitivity in peripheral tissues. The MAPK pathway is strongly activated with compensatory hyperinsulinemia to produce inflammatory mediators (i.e., ICAM-1, VCAM-1, and E-selectin) when the PI3-K/Akt axis is downregulated [38]. The imbalance between these two signals leads to endothelial dysfunction, characterized by a decreased production of NO and increased generation of ET-1 in endothelial cells [39,40].

NOX is a key molecule in the development of endothelial dysfunction and is a major source of ROS production in endothelial cells. Type 2 diabetes is characterized by impaired control of the redox environment with overproduction of ROS [41]. The main factors playing a protective role are eNOS and NO. The biological balance at the endothelium is maintained by vasodilatory substances (i.e., prostaglandins, NO) and vasoconstricting factors (i.e., ET-1, angiotensin II). The activated PI-3/Akt pathway induces the phosphorylation of eNOS, transformation of L-arginine to L-citrulline, and production of NO. NO exerts a vasoprotective role by inhibiting the proliferation of vascular smooth muscle cells, expression of inflammatory cytokines, and platelet aggregation. In contrast, the lack of NO generation leads to the enhanced production of inflammatory and thrombotic cytokines [42]. Taken together, these findings indicate that the involvement of endothelial dysfunction and insulin resistance in pathological disorders contributes to the impairment of the cellular glucose uptake, NO-dependent vasodilation, enrichment of oxidative stress, and inflammation.

Elevation of circulating cytokine levels is strongly associated with insulin resistance and contributes to endothelial dysfunction. Increased levels of cytokines, including C-reactive protein (CRP), TNF-α, and interleukin-6 (IL-6), inhibit insulin-stimulated NO production by decreasing the eNOS expression, leading to the inhibition of the PI3K/Akt/eNOS pathway [43,44]. Obesity and type 2 diabetes are associated with elevated levels of leptin and resistin, which induce increases in TNF-α and IL-6 [45]. In addition, leptin enhances the serine phosphorylation of IRS-1, thereby disturbing insulin signaling through the PI-3K/Akt pathway [46]. In contrast, resistin reduces the expression of eNOS [47]. Although adiponectin and ghrelin stimulate NO production through the PI-3K/Akt signaling pathways and enhance the NO bioavailability, both cytokines are known to be reduced in patients with obesity or type 2 diabetes [48,49].

3. Crosstalk between Macrophage Polarization and Endothelial Cells

Macrophages and endothelial cells are closely related to each other. Endothelial cells produce cytokines pivotal for the differentiation and growth of macrophages. Macrophages constitute an important line of defense against infection and are essential for tissue repairing as well as wound healing [50,51]. These broad actions are mediated through macrophage conversion induced by environmental signals, such as lower temperatures and the secretion of colony-stimulating factor 1 (CSF-1) and interleukin (IL)-4. There are two types of macrophages: the proinflammatory M1 phenotype (classic activation) and the anti-inflammatory M2 phenotype (alternative activation). Adipose tissue macrophages (ATMs) from obese mice and humans are polarized toward an M1 phenotype, with the upregulation of tumor-necrosis factor (TNF) and inducible NO synthase (iNOS). In contrast, "lean" ATMs express high levels of M2 genes, including IL-10, Ym1, and Arginase 1 [52]. Emerging evidence indicates that proinflammatory M1 polarization induces adipose inflammation [53,54]. Consistently, a lack of M1 macrophages improves insulin sensitivity in obese mice [55,56]. In contrast, deletion of M2 macrophages has been shown to contribute to insulin resistance in wild-type mice [57]. These findings imply that macrophage polarization is implicated in metabolic disturbance.

NO exerts anti-inflammatory and antithrombotic effects. These actions are mediated by the activation of soluble guanylate cyclase, which in turn activates cyclic guanosine monophosphate (cGMP)-dependent protein kinase (PKG) through increased levels of cytoplasmic cGMP [58]. Vasodilator-stimulated phosphoprotein (VASP), a downstream target of PKG, has been identified as a regulator controlling cytoskeletal remodeling and cell migration [59]. Previous studies focusing on insulin resistance have revealed the endothelial NO/VASP-mediated suppression of inflammation in adipose tissue and liver [60,61]. Of note, activation of NO/VASP signaling promotes a phenotypic change into an M2 macrophage state. Conversely, a high-fat diet (HFD) attenuates M2 polarization and induces M1 activation in Kuppfer cells, which leads to insulin resistance in the liver [58]. Taken together, these findings suggest that a therapeutic approach targeting the NO/VASP pathway would promote anti-inflammatory actions and may thus be effective for managing metabolic disorders, including obesity and diabetes (Figure 1).

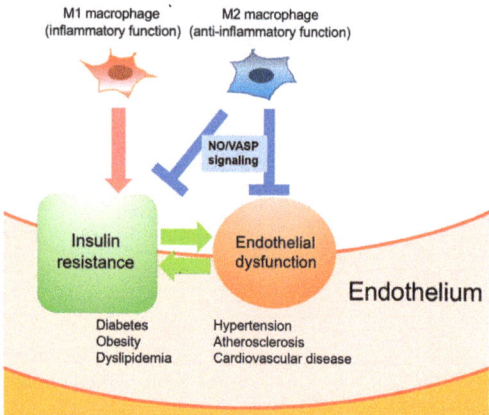

Figure 1. The metabolic network between macrophage polarization, insulin resistance, and endothelial cells. Macrophages play pleiotropic functions in the endothelium. Proinflammatory M1 macrophages are stimulated by LPS, IFN-γ, and TNF-α and promote the secretion of inflammatory cytokines, including IL-1, IL-6, and TNF-α. In contrast, anti-inflammatory M2 macrophages are stimulated by IL-4 and IL-10 and secrete anti-inflammatory cytokines, such as IL-10 and TGF-β. Through NO/VASP signaling, M2 macrophages contribute to the suppression of inflammation in the endothelium, leading to the improvement of insulin resistance and endothelial dysfunction. NO, Nitric oxide; VASP, Vasodilator-stimulated phosphoprotein.

4. Targeting Endothelial Dysfunction

The ultimate goal of the treatment of diabetes is to prevent microvascular and macrovascular complications. The endothelium lining the inner wall of the vasculature modulates basic hemostatic functions, including the circulation of blood cells, vascular tone, platelet activity, and inflammation. Endothelial dysfunction is considered an early predictor of future cardiovascular events and atherosclerosis. Growing knowledge concerning the diverse functions of the endothelium has focused attention on therapeutic strategies that may improve the endothelial function. From a clinical standpoint, a large amount of experimental evidence supports the notion that therapies targeting endothelial dysfunction reduce cardiovascular mortality and morbidity. It is important to consider whether or not drugs used in the clinical management of type 2 diabetes exert positive and pleiotropic effects on the endothelium independent of the glucose-lowering action. While statins have been reported to exert vascular protective effects that are independent of lowering the LDL-cholesterol level, some anti-diabetic agents have recently been suggested to exert beneficial effects against endothelial dysfunction. In addition to traditional drugs, clinical and experimental data support the possibility that novel classes of anti-hyperglycemic agents have beneficial effects on the endothelial function and macrophage polarization.

4.1. SGLT2 Inhibitors

SGLT2 inhibitors block the glucose uptake in the renal proximal tubule of the nephron, resulting in the induction of glycosuria and decreased blood glucose levels. Recent trials, such as the EMPA-REG-OUTCOME and the CANVAS Program have revealed that SGLT2 inhibitors, i.e., empagliflozin and canagliflozin, attenuate cardiovascular events and reduce the death rate compared to the patients treated with placebo [62,63]. SGLT2 inhibitors have been suggested to exert beneficial actions on the endothelial function. For example, Shigiyama et al. clearly demonstrated that dapagliflozin add-on therapy on metformin improved the endothelial function by improving the oxidative stress in patients with inadequate glycemic control [64]. Furthermore, dapagliflozin has been shown to improve systemic endothelial dysfunction and arterial stiffness, independent of the blood pressure and blood glucose levels [65]. Lee et al. also reported that dapagliflozin improves vascular smooth muscle dysfunction with alterations of gut microbiota in type 2 diabetic mouse [66]. Uthman et al. demonstrated the anti-inflammatory action of SGLT2 inhibitors by showing that empagliflozin rescued the TNF-α-induced reduction of the eNOS expression in human coronary arterial endothelial cells [67].

From the perspective of inflammation, empagliflozin is suggested to promote browning of white adipose tissue (WAT) by polarizing M2 ATMs [68]. Furthermore, dapagliflozin attenuates cardiac fibrosis by promoting M2 macrophage polarization in myocardial infarction in rodents [69]. As such, the inhibition of SGLT2 may shift the macrophage polarity to an M2 status, and thus, prevent metabolic disorders causing endothelial dysfunction.

4.2. GLP-1 Receptor Agonists

GLP-1 receptor (GLP-1R) is expressed not only in pancreatic β-cells but also in various tissues and organs, including endothelial cells, fat, brain, heart, liver, and muscle, and both GLP-1 and GLP-1R possess pleiotropic effects [70,71]. From the standpoint of vascular protection, the usefulness of GLP-1R agonists has been reported in basic research. For example, GLP-1R agonist reduces the production of inflammatory cytokines [72,73] and apoptosis of endothelial cells [74] and induces eNOS production [75]. The PI3K/Akt-eNOS activation pathway has been suggested as an underlying mechanism [76]. Cai et al. reported that GLP-1R agonists treatment induces a protective effect on endothelial cells through a GLP-1R-ERK1/2-dependent manner [77]. Furthermore, recent trials have shown that exenatide, a commonly used GLP-1R agonist, improves the endothelial function in patients with type 2 diabetes and pre-diabetes [78–80].

As is the case with SGLT2 inhibitors, GLP-1R agonist is suggested to modulate macrophage polarity. The reprogramming of the macrophage phenotype towards the M2 phenotype has been shown in mice with apoE and IRS2 deficiency treated with lixenatide. This was associated with a reduction in the atheroma plaque size [81] and the regression of the early stage of atherogenesis [82]. However, these studies were only performed in mouse models. Further mechanistic investigations will be required in order to elucidate the precise role of GLP-1R agonists in macrophage polarity.

4.3. DPP-4 Inhibitors

Dipeptidyl peptidase-4 (DPP-4) is released from adipose tissue and acts as a pro-inflammatory adipokine, mediating local inflammation, insulin resistance, and metabolic syndrome [83,84]. The expression of adhesion molecules and inflammatory cytokines is attenuated by DPP-4 inhibitor [85]. In addition, DPP-4 inhibitors exert anti-diabetic and myocardial protective effects through the activation of PI3/Akt signaling and eNOS [86].

Of note, DPP-4 has been shown to regulate inflammation and insulin resistance in the setting of obesity by modulating the macrophage polarity. For instance, linagliptin promotes the shift of polarity toward the anti-inflammatory M2 macrophage phenotype in liver and adipose tissue, thereby improving local inflammation and insulin resistance [87].

Furthermore, clinical data indicate that DPP-4 inhibitors, including sitagliptin [88], vildagliptin [89], linagliptin [90], and saxagliptin [91], improve endothelial dysfunction. Treatment with sitagliptin for 12 weeks significantly improved the change in FMD and increased the circulating levels of CD34, a marker of endothelial progenitor cells [92]. In addition, Kajikawa showed that saxagliptin markedly increased FMD and massively decreased stromal cell-derived factor-1α (SDF-1α), a DPP4 substrate participating in the recovery of vascular injury by recruiting endothelial progenitor cells [91]. Further clinical trials and mechanistic investigations will be required in order to validate the role of DPP4-inhibitors in the pathogenesis of vascular events.

4.4. Biguanides

Metformin, the most common anti-diabetic agent, upregulates the blood flow in adipose tissue and skeletal muscle [93]. The metformin-induced production of eNOS and inhibition of leukocyte adhesion, vascular aging, and endothelial cell apoptosis has also been reported. The activation of AMP-activated protein kinase (AMPK) is the underlying mechanism [94,95].

Metformin also exerts an anti-inflammatory function in endothelium and adipose tissue through multiple pathways. For example, a clinical trial using long-term metformin treatment in patients with type 2 diabetes reported its efficacy in reducing levels of plasma markers (i.e., VCAM-1 and ICAM-1) independent of changes in HbA1c [96]. In addition, metformin can mediate macrophage polarization to the M2 phenotype and subsequent inhibition of the Jun N-terminal Kinase (JNK) pathway [97].

From the viewpoint of clinical trials, many prospective studies targeted at patients with type 2 diabetes have shown that metformin treatment improves the cardiovascular prognosis independently from glycemic control. In a study dealing with type 2 diabetes patients, metformin improved both the insulin resistance and acetylcholine-stimulated flow, with a strong statistical relationship between these parameters [98]. In another study, the long-term treatment of metformin improved the plasma levels of markers of the endothelial function independent of other variables, including the weight, blood glucose level, and insulin dose [96].

4.5. Thiazolidinediones

Thiazolidinediones (TZDs) are antidiabetic agents that bind and activate peroxisome proliferator activated receptor γ (PPARγ), which is a nuclear receptor superfamily that improves insulin sensitivity. In addition, TZDs have attracted growing interest because of their biological activities, such as their anti-inflammatory, antitumor, and anti-atherosclerotic activities [99].

PPARγ is expressed in not only adipose tissue but also endothelial cells. Endothelial PPARγ decreases the production of chemokines and adhesion molecules, such as ICAM-1 and VCAM-1, and suppresses the production of components of NOX, NOX1, NOX2, and NOX4, leading to the inhibition of generation of ROS [100]. Furthermore, PPARγ promotes NO production in endothelium and abrogates endothelin expression [101].

4.6. Sulfonylureas

Sulfonylureas (SUs) have been widely used for treatment of type 2 diabetes. Effects of SUs on vascular and endothelial cells is inconsistent. Studies have showed that glibenclamide, a kind of second-generation SU, have a pro-arrhythmic effect on reperfusion after an ischemic event in vivo [102] and that SUs may be coupled to an enhanced risk of congestive heart failure [103]. Meanwhile, there are reports suggesting positive effects of SUs on endothelium. For instance, it has been reported that gliclazide, one of the second-generation SUs, improves endothelial function in diabetic rabbits [104] and decreases the progression of atherosclerosis in human [105]. From the standpoint of the molecular mechanism, gliclazide has been suggested to protect endothelial cells from apoptosis by decreasing oxidative stress [106], and glimepiride also has been shown to stimulate NO production via PI3-dependent pathways in endothelium and lead to reduction of NF-κβ activation [107].

4.7. Medical Nutrition Therapy and Physical Activity

The aim of treatment in diabetes is to maintain an optimal level of blood glucose, lipids, and blood pressure to delay or prevent chronic diabetic complications [108]. Patients with diabetes should achieve good control of their blood glucose by following a nutritious meal plan and exercise program, losing excess weight, implementing necessary self-care behaviors, and taking oral medications or insulin therapy. Weight loss through restriction of the daily diet and physical exercise is essential for managing diabetes and preventing vascular complications. When medications are used to control diabetes, they should primarily augment lifestyle improvements.

Calorie restriction and physical exercise are known to improve not only the insulin sensitivity but also endothelial dysfunction. Calorie restriction promotes NO-dependent vasodilation and coincidentally reduces circulating ET-1 levels in patients with insulin resistance [109,110]. In addition, regular physical exercise increases the expression of vascular eNOS via PI3K/Akt-dependent phosphorylation in humans [111]. Taken together, the favorable effects of these lifestyle modifications induce increased insulin signaling, enhanced eNOS activity, and reduced inflammatory and oxidative stress, leading to the right balance between the vasodilator and vasoconstrictor actions of insulin.

Indeed, a meta-analysis Montero performed pointed out that, in patients with type 2 diabetes, physical exercise greatly increased FMD [112]. Furthermore, another recent meta-analysis revealed that aerobic and combined aerobic and resistance exercise notably improved the endothelial function in patients with type 2 diabetes, as reflected by an elevated FMD. This observation was independent of changes in cardiometabolic markers, such as the blood pressure, body mass index, and glycemic control [113].

5. ROCK Inhibitors as Preclinical Agents

The small GTP-binding protein Rho and its downstream Rho-associated coiled-coil containing protein kinase (Rho-kinase, ROCK) mediate a variety of cellular processes such as cell contraction, proliferation, and migration. ROCK signaling is activated by many factors, including angiotensin II, glucose, and cytokines, all of which are upregulated under diabetic condition [114–116]. Previous studies have elucidated ROCK as a key molecule of endothelial dysfunction. For instance, statins have been reported to inhibit the RhoA/ROCK pathway indirectly, acting by reducing the synthesis of isoprenoids. The intravenous administration of pravastatin prevented impaired NO-dependent vasodilation by blocking the activation of Rho A and Rac and the inactivation of Akt/eNOS pathways in vivo [117]. Fasudil, the first ROCK inhibitor approved for clinical use, suppresses the migration of human pulmonary microvascular endothelial cells and the proliferation of pulmonary artery smooth muscle cells caused by ET-1 [118]. Moreover, fasudil has the potential to improve endothelial dysfunction via restoring NO bioavailability in humans with atherosclerosis [119].

ROCK initiates endothelial dysfunction via NF-κB activation. IκB kinase (IKK) phosphorylates IκB through activation signals, which induces the degradation of IκBα via the ubiquitin system, leads to NF-κB RelA/p65 translocation to the nucleus, and activates the transcription of target genes. ROCK mediates the NF-κB signaling through various pathways. Our laboratory showed that ROCK regulates thrombin-mediated p65 phosphorylation and IκBα phosphorylation in endothelial cells [120]. Moreover, we reported that ROCK regulates the nuclear translocation of RelA/p65 in mesangial cells [121]. Recent researches from Antoniellis et al. showed that RhoA, an upstream factor of ROCK, controls the translocation of NF-κB (p50) in neutrophils [122]. These studies suggest that ROCK controls the nuclear translocation of multiple NF-κB components and that the way of NF-κB regulation varies depending on the stimulus and type of cells. Taken together, ROCK is a principal determinant of endothelial dysfunction.

ROCK has two isoforms: ROCK1 and ROCK2. While these sequences share 65% sequence homology, each isoform plays different roles and has unique pathways of activation. ROCK1 and ROCK2 exert different roles in endothelial dysfunction. For example, endothelium without ROCK2 has shown the reduction of chemokines and adhesion molecules through NF-κB [116]. ROCK1 is required in oxidized LDL-induced cell adhesion, while ROCK2 is involved in both endothelial adhesion and apoptosis by regulating adhesion molecules [123]. ROCK2 has been shown to be a pivotal regulator of endothelial inflammation and functions as an essential factor in the development of atherosclerosis. Because ROCK1 and ROCK2 cannot completely compensate for each other's loss, distinctive roles of them have been pointed out [124]. ROCK2 is distributed in human vascular endothelial cells. Shimada et al. suggested that ROCK2 mediates the production of VCAM-1 and ICAM-1 and induces endothelial inflammation [125]. Furthermore, Shimokawa et al. reported that ROCK2 in a vascular smooth muscle cell (VSMC) leads to the progression of cardiovascular diseases including pulmonary arterial hypertension [126]. An elegant study from Sawada et al. suggested that loss of ROCK2 in bone marrow-derived cells decreased lipid accumulation and atherosclerotic lesions in the LDL receptor-null mice [127]. Though whether or not ROCK2 is engaged in modulating monocytic migration and adhesion toward endothelial cells is unclear, we recently demonstrated for the first time that ROCK2—but not ROCK1—is involved in the regulation of these processes [116]. These findings underscored the importance of ROCK2's involvement in endothelial dysfunction. Therefore, ROCK2 represents an attractive target for studying critical regulators of endothelial dysfunction.

With regard to macrophage polarity, the importance of Rho/ROCK signaling has been clarified gradually. Recent studies have shown that ROCK1 and ROCK2 have different roles in the regulation of macrophage polarization into classical pro-inflammatory macrophage type 1 (M1), producing IL-12, and alternative anti-inflammatory macrophage type 2 (M2), producing TGF-β and IL-10. Though ROCK2 inhibition is suggested to result in a decreased population of M2 macrophages with the upregulation of M1 markers in age-related macular degeneration (AMD) [128], the commitment of ROCK1 and ROCK2 in the conversion of the macrophage subtype in other organs and diseases remains to be elucidated. Further mechanistic analyses will be indispensable for clarifying the role of ROCK in regulating macrophage polarity.

6. Conclusions and Future Perspectives

Insulin signaling pathways and endothelial cells conduct crosstalk, thus, understanding the correlation between insulin resistance and endothelial dysfunction is essential for treating diabetes-related vascular complications. Insulin resistance and endothelial dysfunction lead to the failure of NO-dependent vasodilatation, the glucose uptake by cells, and the induction of inflammation in tissues, eventually leading to atherosclerosis. In addition, endothelial NO activates VASP signaling in macrophages by increasing the M2 macrophage polarization and exerting an anti-inflammatory function under conditions of insulin resistance. A novel class of anti-hyperglycemic agents is suggested to exert their beneficial effects through this mechanism. SGLT2 inhibitors and GLP-1 agonists may promote browning of WAT by polarizing M2 macrophages and protecting against endothelial dysfunction. A firm understanding of the mechanism underlying each drug's pleiotropic effect will be needed to establish new treatment approaches for endothelial dysfunction.

Author Contributions: Y.T. wrote the manuscript. K.M., K.S., Y.N., T.Y., K.U., and R.N. helped edit the manuscript and revised it for important intellectual content. All authors have read and agreed to the published version of the manuscript.

Funding: This work was supported by a Grant-in-Aid for Scientific Research from Japan Society for the Promotion of Science (to Keiichiro Matoba and Rimei Nishimura), the MSD Life Science Foundation (to Keiichiro Matoba), and the Takeda Science Foundation (to Keiichiro Matoba), along with the Suzuken Memorial Foundation (to Keiichiro Matoba).

Conflicts of Interest: Keiichiro Matoba has received research support from Sanofi KK, Tanabe Pharma, and Takeda Pharmaceutical. Kazunori Utsunomiya has received research support from Terumo, Novo Nordisk Pharma, Taisho Pharmaceutical, Böehringer Ingelheim, Kyowa Hakko Kirin, Sumitomo Dainippon Pharma, and Ono Pharmaceutical, as well as speaker honoraria from Tanabe Pharma, Sanofi KK, Sumitomo Dainippon Pharma, Eli Lilly, and Böehringer Ingelheim. Rimei Nishimura has received speaker honoraria from Astellas Pharma, Nippon Boehringer Ingelheim, Eli Lilly Japan KK, Kissei Pharmaceutical, Medtronic Japan, MSD, Novartis Pharma KK, Novo Nordisk Pharma, Sanofi KK, and Takeda Pharmaceutical and contract research fees for collaborative research with Taisho Pharmaceutical, Ono Pharmaceutical, Takeda Pharmaceutical, and Böehringer Ingelheim. All the sponsors have no role in paper processing.

References

1. International Diabetes Federation. IDF Diabetes Atlas 9th Edition 2019. Available online: https://www.diabetesatlas.org/en/resources/ (accessed on 20 May 2020).
2. Avogaro, A.; Albiero, M.; Menegazzo, L.; de Kreutzenberg, S.; Fadini, G.P. Endothelial dysfunction in diabetes: The role of reparatory mechanisms. *Diabetes Care* **2011**, *34* (Suppl. 2), S285–S290. [CrossRef] [PubMed]
3. Ross, R. Atherosclerosis—An Inflammatory Disease. *N. Engl. J. Med.* **1999**, *340*, 115–126. [CrossRef] [PubMed]
4. Davignon, J.; Ganz, P. Role of endothelial dysfunction in atherosclerosis. *Circulation* **2004**, *109*, III27–III32. [CrossRef] [PubMed]
5. Kirpichnikov, D.; Sowers, J.R. Diabetes mellitus and diabetes-associated vascular disease. *Trends Endocrinol. Metab. TEM* **2001**, *12*, 225–230. [CrossRef]
6. Siasos, G.; Tousoulis, D.; Antoniades, C.; Stefanadi, E.; Stefanadis, C. L-Arginine, the substrate for NO synthesis: An alternative treatment for premature atherosclerosis. *Int. J. Cardiol.* **2007**, *116*, 300–308. [CrossRef]
7. Kaur, R.; Kaur, M.; Singh, J. Endothelial dysfunction and platelet hyperactivity in type 2 diabetes mellitus: Molecular insights and therapeutic strategies. *Cardiovasc. Diabetol.* **2018**, *17*, 121. [CrossRef]
8. Xu, J.; Zou, M.H. Molecular insights and therapeutic targets for diabetic endothelial dysfunction. *Circulation* **2009**, *120*, 1266–1286. [CrossRef]
9. van Guilder, G.P.; Hoetzer, G.L.; Dengel, D.R.; Stauffer, B.L.; DeSouza, C.A. Impaired endothelium-dependent vasodilation in normotensive and normoglycemic obese adult humans. *J. Cardiovasc. Pharmacol.* **2006**, *47*, 310–313. [CrossRef]
10. Eelen, G.; de Zeeuw, P.; Treps, L.; Harjes, U.; Wong, B.W.; Carmeliet, P. Endothelial cell metabolism. *Physiol. Rev.* **2018**, *98*, 3–58. [CrossRef]
11. Fujii, N.; Reinke, M.C.; Brunt, V.E.; Minson, C.T. Impaired acetylcholine-induced cutaneous vasodilation in young smokers: Roles of nitric oxide and prostanoids. *Am. J. Physiol. Heart Circ. Physiol.* **2013**, *304*, H667–H673. [CrossRef]
12. Kusche-Vihrog, K.; Schmitz, B.; Brand, E. Salt controls endothelial and vascular phenotype. *Pflugers Arch.* **2015**, *467*, 499–512. [CrossRef]
13. Bender, S.B.; Laughlin, M.H. Modulation of endothelial cell phenotype by physical activity: Impact on obesity-related endothelial dysfunction. *Am. J. Physiol. Heart Circ. Physiol.* **2015**, *309*, H1–H8. [CrossRef]
14. Moreau, K.L.; Hildreth, K.L. Vascular aging across the menopause transition in healthy women. *Adv. Vasc. Med.* **2014**, *2014*, 204390. [CrossRef] [PubMed]

15. Antoniades, C.; Demosthenous, M.; Tousoulis, D.; Antonopoulos, A.S.; Vlachopoulos, C.; Toutouza, M.; Marinou, K.; Bakogiannis, C.; Mavragani, K.; Lazaros, G.; et al. Role of asymmetrical dimethylarginine in inflammation-induced endothelial dysfunction in human atherosclerosis. *Hypertension* **2011**, *58*, 93–98. [CrossRef]
16. Kharbanda, R.K.; Walton, B.; Allen, M.; Klein, N.; Hingorani, A.D.; MacAllister, R.J.; Vallance, P. Prevention of inflammation-induced endothelial dysfunction: A novel vasculo-protective action of aspirin. *Circulation* **2002**, *105*, 2600–2604. [CrossRef] [PubMed]
17. Channon, K.M.; Guzik, T.J. Mechanisms of superoxide production in human blood vessels: Relationship to endothelial dysfunction, clinical and genetic risk factors. *J. Physiol. Pharmacol.* **2002**, *53*, 515–524.
18. Matoba, K.; Takeda, Y.; Nagai, Y.; Yokota, T.; Utsunomiya, K.; Nishimura, R. Targeting redox imbalance as an approach for diabetic kidney disease. *Biomedicines* **2020**, *8*, 40. [CrossRef] [PubMed]
19. Cai, H.; Harrison, D.G. Endothelial dysfunction in cardiovascular diseases: The role of oxidant stress. *Circ. Res.* **2000**, *87*, 840–844. [CrossRef]
20. Wu, F.; Szczepaniak, W.S.; Shiva, S.; Liu, H.; Wang, Y.; Wang, L.; Wang, Y.; Kelley, E.E.; Chen, A.F.; Gladwin, M.T.; et al. Nox2-dependent glutathionylation of endothelial NOS leads to uncoupled superoxide production and endothelial barrier dysfunction in acute lung injury. *Am. J. Physiol. Lung Cell. Mol. Physiol.* **2014**, *307*, L987–L997. [CrossRef]
21. Zhang, Y.; Janssens, S.P.; Wingler, K.; Schmidt, H.H.H.W.; Moens, A.L. Modulating endothelial nitric oxide synthase: A new cardiovascular therapeutic strategy. *Am. J. Physiol. Heart Circ. Physiol.* **2011**, *301*, H634–H646. [CrossRef]
22. Virdis, A.; Bacca, A.; Colucci, R.; Duranti, E.; Fornai, M.; Materazzi, G.; Ippolito, C.; Bernardini, N.; Blandizzi, C.; Bernini, G.; et al. Endothelial dysfunction in small arteries of essential hypertensive patients: Role of cyclooxygenase-2 in oxidative stress generation. *Hypertension* **2013**, *62*, 337–344. [CrossRef]
23. Stuehr, D.; Pou, S.; Rosen, G.M. Oxygen reduction by nitric-oxide synthases. *J. Biol. Chem.* **2001**, *276*, 14533–14536. [CrossRef] [PubMed]
24. Barac, A.; Campia, U.; Panza, J.A. Methods for evaluating endothelial function in humans. *Hypertension* **2007**, *49*, 748–760. [CrossRef]
25. Deanfield, J.E.; Halcox, J.P.; Rabelink, J.T. Endothelial function and dysfunction: Testing and clinical relevance. *Circulation* **2007**, *115*, 1285–1295. [CrossRef] [PubMed]
26. Corretti, M.C.; Anderson, T.J.; Benjamin, E.J.; Celermajer, D.; Charbonneau, F.; Creager, M.A.; Deanfield, J.; Drexler, H.; Gerhard-Herman, M.; Herrington, D.; et al. Guidelines for the ultrasound assessment of endothelial-dependent flow-mediated vasodilation of the brachial artery: A report of the International Brachial Artery Reactivity Task Force. *J. Am. Coll. Cardiol.* **2002**, *39*, 257–265. [CrossRef]
27. Hamburg, N.M.; Palmisano, J.; Larson, M.G.; Sullivan, L.M.; Lehman, B.T.; Vasan, R.S.; Levy, D.; Mitchell, G.F.; Vita, J.A.; Benjamin, E.J. Relation of brachial and digital measures of vascular function in the community: The Framingham heart study. *Hypertension* **2011**, *57*, 390–396. [CrossRef] [PubMed]
28. Ghiadoni, L.; Versari, D.; Giannarelli, C.; Faita, F.; Taddei, S. Non-invasive diagnostic tools for investigating endothelial dysfunction. *Curr. Pharm. Des.* **2008**, *14*, 3715–3722. [CrossRef] [PubMed]
29. Moerland, M.; Kales, A.J.; Schrier, L.; van Dongen, M.G.; Bradnock, D.; Burggraaf, J. Evaluation of the EndoPAT as a tool to assess endothelial function. *Int. J. Vasc. Med.* **2012**, *2012*, 904141. [CrossRef]
30. Youngren, J.F. Regulation of insulin receptor function. *Cell. Mol. Life Sci. CMLS* **2007**, *64*, 873–891. [CrossRef]
31. Withers, D.J.; Gutierrez, J.S.; Towery, H.; Burks, D.J.; Ren, J.M.; Previs, S.; Zhang, Y.; Bernal, D.; Pons, S.; Shulman, G.I.; et al. Disruption of IRS-2 causes type 2 diabetes in mice. *Nature* **1998**, *391*, 900–904. [CrossRef]
32. Miao, B.; Skidan, I.; Yang, J.; Lugovskoy, A.; Reibarkh, M.; Long, K.; Brazell, T.; Durugkar, K.A.; Maki, J.; Ramana, C.V.; et al. Small molecule inhibition of phosphatidylinositol-3,4,5-triphosphate (PIP3) binding to pleckstrin homology domains. *Proc. Natl. Acad. Sci. USA* **2010**, *107*, 20126–20131. [CrossRef] [PubMed]
33. Lawlor, M.A.; Alessi, D.R. PKB/Akt: A key mediator of cell proliferation, survival and insulin responses? *J. Cell Sci.* **2001**, *114*, 2903–2910. [PubMed]
34. Park, K.; Li, Q.; Rask-Madsen, C.; Mima, A.; Mizutani, K.; Winnay, J.; Maeda, Y.; D'Aquino, K.; White, M.F.; Feener, E.P.; et al. Serine phosphorylation sites on IRS2 activated by angiotensin II and protein kinase C to induce selective insulin resistance in endothelial cells. *Mol. Cell. Biol.* **2013**, *33*, 3227–3241. [CrossRef] [PubMed]

35. Snyder, S.H.; Jaffrey, S.R. Vessels vivified by Akt acting on NO synthase. *Nat. Cell Biol.* **1999**, *1*, E95–E96. [CrossRef] [PubMed]
36. Manning, B.D.; Cantley, L.C. AKT/PKB signaling: Navigating downstream. *Cell* **2007**, *129*, 1261–1274. [CrossRef]
37. Ferri, C.; Pittoni, V.; Piccoli, A.; Laurenti, O.; Cassone, M.R.; Bellini, C.; Properzi, G.; Valesini, G.; de Mattia, G.; Santucci, A. Insulin stimulates endothelin-1 secretion from human endothelial cells and modulates its circulating levels in vivo. *J. Clin. Endocrinol. Metab.* **1995**, *80*, 829–835.
38. Cusi, K.; Maezono, K.; Osman, A.; Pendergrass, M.; Patti, M.E.; Pratipanawatr, T.; DeFronzo, R.A.; Kahn, C.R.; Mandarino, L.J. Insulin resistance differentially affects the PI 3-kinase- and MAP kinase-mediated signaling in human muscle. *J. Clin. Investig.* **2000**, *105*, 311–320. [CrossRef]
39. Brown, M.S.; Goldstein, J.L. Selective versus total insulin resistance: A pathogenic paradox. *Cell Metab.* **2008**, *7*, 95–96. [CrossRef]
40. Muniyappa, R.; Montagnani, M.; Koh, K.K.; Quon, M.J. Cardiovascular actions of insulin. *Endocr. Rev.* **2007**, *28*, 463–491. [CrossRef]
41. Meza, C.A.; la Favor, J.D.; Kim, D.-H.; Hickner, R.C. Endothelial dysfunction: Is there a hyperglycemia-induced imbalance of NOX and NOS? *Int. J. Mol. Sci.* **2019**, *20*, 3775. [CrossRef]
42. Janus, A.; Szahidewicz-Krupska, E.; Mazur, G.; Doroszko, A. Insulin resistance and endothelial dysfunction constitute a common therapeutic target in cardiometabolic disorders. *Mediat. Inflamm.* **2016**, *2016*, 3634948. [CrossRef] [PubMed]
43. Anderson, H.D.; Rahmutula, D.; Gardner, D.G. Tumor necrosis factor-alpha inhibits endothelial nitric-oxide synthase gene promoter activity in bovine aortic endothelial cells. *J. Biol. Chem.* **2004**, *279*, 963–969. [CrossRef] [PubMed]
44. Zhang, L.; Wheatley, C.M.; Richards, S.M.; Barrett, E.J.; Clark, M.G.; Rattigan, S. TNF-alpha acutely inhibits vascular effects of physiological but not high insulin or contraction. *Am. J. Physiol. Endocrinol. Metab.* **2003**, *285*, E654–E660. [CrossRef] [PubMed]
45. Kougias, P.; Chai, H.; Lin, P.H.; Lumsden, A.B.; Yao, Q.; Chen, C. Adipocyte-derived cytokine resistin causes endothelial dysfunction of porcine coronary arteries. *J. Vasc. Surg.* **2005**, *41*, 691–698. [CrossRef] [PubMed]
46. Hennige, A.M.; Stefan, N.; Kapp, K.; Lehmann, R.; Weigert, C.; Beck, A.; Moeschel, K.; Mushack, J.; Schleicher, E.; Häring, H.U. Leptin down-regulates insulin action through phosphorylation of serine-318 in insulin receptor substrate 1. *FASEB J.* **2006**, *20*, 1206–1208. [CrossRef]
47. Verma, S.; Li, S.H.; Wang, C.H.; Fedak, P.W.; Li, R.K.; Weisel, R.D.; Mickle, D.A. Resistin promotes endothelial cell activation: Further evidence of adipokine-endothelial interaction. *Circulation* **2003**, *108*, 736–740. [CrossRef]
48. Chen, H.; Montagnani, M.; Funahashi, T.; Shimomura, I.; Quon, M.J. Adiponectin stimulates production of nitric oxide in vascular endothelial cells. *J. Biol. Chem.* **2003**, *278*, 45021–45026. [CrossRef]
49. Tesauro, M.; Schinzari, F.; Iantorno, M.; Rizza, S.; Melina, D.; Lauro, D.; Cardillo, C. Ghrelin improves endothelial function in patients with metabolic syndrome. *Circulation* **2005**, *112*, 2986–2992. [CrossRef] [PubMed]
50. Murray, P.J.; Wynn, T.A. Protective and pathogenic functions of macrophage subsets. *Nat. Rev. Immunol.* **2011**, *11*, 723–737. [CrossRef] [PubMed]
51. Odegaard, J.I.; Chawla, A. Pleiotropic actions of insulin resistance and inflammation in metabolic homeostasis. *Science* **2013**, *339*, 172–177. [CrossRef] [PubMed]
52. Sica, A.; Mantovani, A. Macrophage plasticity and polarization: In vivo veritas. *J. Clin. Investig.* **2012**, *122*, 787–795. [CrossRef] [PubMed]
53. Gordon, S.; Taylor, P.R. Monocyte and macrophage heterogeneity. *Nat. Rev. Immunol.* **2005**, *5*, 953–964. [CrossRef] [PubMed]
54. Charo, I.F. Macrophage polarization and insulin resistance: PPARgamma in control. *Cell Metab.* **2007**, *6*, 96–98. [CrossRef] [PubMed]
55. Patsouris, D.; Li, P.P.; Thapar, D.; Chapman, J.; Olefsky, J.M.; Neels, J.G. Ablation of CD11c-positive cells normalizes insulin sensitivity in obese insulin resistant animals. *Cell Metab.* **2008**, *8*, 301–309. [CrossRef]
56. Lumeng, C.N.; Bodzin, J.L.; Saltiel, A.R. Obesity induces a phenotypic switch in adipose tissue macrophage polarization. *J. Clin. Investig.* **2007**, *117*, 175–184. [CrossRef]

57. Odegaard, J.I.; Ricardo-Gonzalez, R.R.; Goforth, M.H.; Morel, C.R.; Subramanian, V.; Mukundan, L.; Eagle, A.R.; Vats, D.; Brombacher, F.; Ferrante, A.W.; et al. Macrophage-specific PPARgamma controls alternative activation and improves insulin resistance. *Nature* **2007**, *447*, 1116–1120. [CrossRef]
58. Lee, W.J.; Tateya, S.; Cheng, A.M.; Rizzo-DeLeon, N.; Wang, N.F.; Handa, P.; Wilson, C.L.; Clowes, A.W.; Sweet, I.R.; Bomsztyk, K.; et al. M2 macrophage polarization mediates anti-inflammatory effects of endothelial nitric oxide signaling. *Diabetes* **2015**, *64*, 2836–2846. [CrossRef]
59. Trichet, L.; Sykes, C.; Plastino, J. Relaxing the actin cytoskeleton for adhesion and movement with Ena/VASP. *J. Cell Biol.* **2008**, *181*, 19–25. [CrossRef]
60. Handa, H.; Tateya, S.; Rizzo, N.O.; Cheng, A.M.; Morgan-Stevenson, V.; Han, C.Y.; Clowes, A.W.; Daum, G.; O'Brien, K.D.; Schwartz, M.W.; et al. Reduced vascular nitric oxide-cGMP signaling contributes to adipose tissue inflammation during high-fat feeding. *Arter. Thromb. Vasc. Biol.* **2011**, *31*, 2827–2835. [CrossRef] [PubMed]
61. Tateya, S.; Rizzo, N.O.; Handa, P.; Cheng, A.M.; Morgan-Stevenson, V.; Daum, G.; Clowes, A.W.; Morton, G.J.; Schwartz, M.W.; Kim, F. Endothelial NO/cGMP/VASP signaling attenuates Kupffer cell activation and hepatic insulin resistance induced by high-fat feeding. *Diabetes* **2011**, *60*, 2792–2801. [CrossRef] [PubMed]
62. Zinman, B.; Wanner, C.; Lachin, J.M.; Fitchett, D.; Bluhmki, E.; Hantel, S.; Mattheus, M.; Devins, T.; Johansen, O.E.; Woerle, H.J.; et al. Empagliflozin, Cardiovascular Outcomes, and Mortality in Type 2 Diabetes. *N. Engl. J. Med.* **2015**, *373*, 2117–2128. [CrossRef] [PubMed]
63. Neal, B.; Perkovic, V.; Mahaffey, K.W.; de Zeeuw, D.; Fulcher, G.; Erondu, N.; Shaw, W.; Law, G.; Desai, M.; Matthews, D.R. Canagliflozin and Cardiovascular and Renal Events in Type 2 Diabetes. *N. Engl. J. Med.* **2017**, *377*, 644–657. [CrossRef] [PubMed]
64. Shigiyama, F.; Kumashiro, N.; Miyagi, M.; Ikehara, K.; Kanda, E.; Uchino, H.; Hirose, T. Effectiveness of dapagliflozin on vascular endothelial function and glycemic control in patients with early-stage type 2 diabetes mellitus: DEFENCE study. *Cardiovasc. Diabetol.* **2017**, *16*, 84. [CrossRef] [PubMed]
65. Solini, A.; Giannini, L.; Seghieri, M.; Vitolo, E.; Taddei, S.; Ghiadoni, L.; Bruno, R.M. Dapagliflozin acutely improves endothelial dysfunction, reduces aortic stiffness and renal resistive index in type 2 diabetic patients: A pilot study. *Cardiovasc. Diabetol.* **2017**, *16*, 138. [CrossRef]
66. Lee, D.M.; Battson, M.L.; Jarrell, D.K.; Hou, S.; Ecton, K.E.; Weir, T.L.; Gentile, C.L. SGLT2 inhibition via dapagliflozin improves generalized vascular dysfunction and alters the gut microbiota in type 2 diabetic mice. *Cardiovasc. Diabetol.* **2018**, *17*, 62. [CrossRef]
67. Uthman, L.; Homayr, A.; Hollmann, M.W.; Zuurbier, C.J.; Weber, N.C. Administration of SGLT2 inhibitor empagliflozin against TNF-α induced endothelial dysfunction in human venous and arterial endothelial cells. *FASEB J.* **2018**, *32*, 25. [CrossRef]
68. Xu, L.; Ota, T. Emerging roles of SGLT2 inhibitors in obesity and insulin resistance: Focus on fat browning and macrophage polarization. *Adipocyte* **2018**, *7*, 121–128. [CrossRef]
69. Lee, T.M.; Chang, N.C.; Lin, S.Z. Dapagliflozin, a selective SGLT2 Inhibitor, attenuated cardiac fibrosis by regulating the macrophage polarization via STAT3 signaling in infarcted rat hearts. *Free Radic. Biol. Med.* **2017**, *104*, 298–310. [CrossRef]
70. Seino, Y.; Yabe, D. Glucose-dependent insulinotropic polypeptide and glucagon-like peptide-1: Incretin actions beyond the pancreas. *J. Diabetes Investig.* **2013**, *4*, 108–130. [CrossRef]
71. Drucker, D.J. The biology of incretin hormones. *Cell Metab.* **2006**, *3*, 153–165. [CrossRef]
72. Drucker, D.J. Incretin action in the pancreas: Potential promise, possible perils, and pathological pitfalls. *Diabetes* **2013**, *62*, 3316–3323. [CrossRef] [PubMed]
73. Shiraki, A.; Oyama, J.; Komoda, H.; Asaka, M.; Komatsu, A.; Sakuma, M.; Kodama, K.; Sakamoto, Y.; Kotooka, N.; Hirase, T.; et al. The glucagon-like peptide 1 analog liraglutide reduces TNF-α-induced oxidative stress and inflammation in endothelial cells. *Atherosclerosis* **2012**, *221*, 375–382. [CrossRef] [PubMed]
74. Hirano, T.; Mori, Y. Anti-atherogenic and anti-inflammatory properties of glucagon-like peptide-1, glucose-dependent insulinotropic polypepide, and dipeptidyl peptidase-4 inhibitors in experimental animals. *J. Diabetes Investig.* **2016**, *1* (Suppl. 7), 80–86. [CrossRef] [PubMed]
75. Helmstädter, J.; Frenis, K.; Filippou, K.; Grill, A.; Dib, M.; Kalinovic, S.; Pawelke, F.; Kus, K.; Kröller-Schön, S.; Oelze, M.; et al. Endothelial GLP-1 (Glucagon-Like Peptide-1) Receptor mediates cardiovascular protection by liraglutide in mice with experimental arterial hypertension. *Arter. Thromb. Vasc. Biol.* **2020**, *40*, 145–158. [CrossRef]

76. Erdogdu, O.; Nathanson, D.; Sjöholm, A.; Nyström, T.; Zhang, Q. Exendin-4 stimulates proliferation of human coronary artery endothelial cells through eNOS-, PKA- and PI3K/Akt-dependent pathways and requires GLP-1 receptor. *Mol. Cell. Endocrinol.* **2010**, *325*, 26–35. [CrossRef]
77. Cai, X.; She, M.; Xu, M.; Chen, H.; Li, J.; Chen, X.; Zheng, D.; Liu, J.; Chen, S.; Zhu, J.; et al. GLP-1 treatment protects endothelial cells from oxidative stress-induced autophagy and endothelial dysfunction. *Int. J. Biol. Sci.* **2018**, *14*, 1696–1708. [CrossRef]
78. Lovshin, J.; Cherney, D. GLP-1R agonists and endothelial dysfunction: More than just glucose lowering? *Diabetes* **2015**, *64*, 2319. [CrossRef]
79. Irace, C.; de Luca, S.; Shehaj, E.; Carallo, C.; Loprete, A.; Scavelli, F.; Gnasso, A. Exenatide improves endothelial function assessed by flow mediated dilation technique in subjects with type 2 diabetes: Results from an observational research. *Diabetes Vasc. Dis. Res.* **2013**, *10*, 72–77. [CrossRef]
80. Kelly, A.S.; Bergenstal, R.M.; Gonzalez-Campoy, J.M.; Katz, H.; Bank, A.J. Effects of exenatide vs. metformin on endothelial function in obese patients with pre-diabetes: A randomized trial. *Cardiovasc. Diabetol.* **2012**, *11*, 64. [CrossRef]
81. Vinué, Á.; Navarro, J.; Herrero-Cervera, A.; García-Cubas, M.; Andrés-Blasco, I.; Martínez-Hervás, S.; Real, J.T.; Ascaso, J.F.; González-Navarro, H. The GLP-1 analogue lixisenatide decreases atherosclerosis in insulin-resistant mice by modulating macrophage phenotype. *Diabetologia* **2017**, *60*, 1801–1812.
82. Bruen, R.; Curley, S.; Kajani, S.; Crean, D.; O'Reilly, M.E.; Lucitt, M.B.; Godson, C.G.; McGillicuddy, F.C.; Belton, O. Liraglutide dictates macrophage phenotype in apolipoprotein E null mice during early atherosclerosis. *Cardiovasc. Diabetol.* **2017**, *16*, 143. [CrossRef]
83. Chowdhury, H.H.; Velebit, J.; Radić, N.; Frančič, V.; Kreft, M.; Zorec, R. Hypoxia alters the expression of dipeptidyl peptidase 4 and induces developmental remodeling of human preadipocytes. *J. Diabetes Res.* **2016**, *2016*, 7481470. [CrossRef]
84. Varin, E.M.; Mulvihill, E.E.; Beaudry, J.L.; Pujadas, G.; Fuchs, S.; Tanti, J.-F.; Fazio, S.; Kaur, K.; Cao, X.; Baggio, L.L.; et al. Circulating levels of soluble dipeptidyl peptidase-4 are dissociated from inflammation and induced by enzymatic DPP4 inhibition. *Cell Metab.* **2019**, *29*, 320–334. [CrossRef] [PubMed]
85. Matsubara, J.; Sugiyama, S.; Sugamura, K.; Nakamura, T.; Fujiwara, Y.; Akiyama, E.; Kurokawa, H.; Nozaki, T.; Ohba, K.; Konishi, M.; et al. A dipeptidyl peptidase-4 inhibitor, des-fluoro-sitagliptin, improves endothelial function and reduces atherosclerotic lesion formation in apolipoprotein E-deficient mice. *J. Am. Coll. Cardiol.* **2012**, *59*, 265–276. [CrossRef] [PubMed]
86. Birnbaum, Y.; Castillo, A.C.; Qian, J.; Ling, S.; Ye, H.; Perez-Polo, J.R.; Bajaj, M.; Ye, Y. Phosphodiesterase III inhibition increases cAMP levels and augments the infarct size limiting effect of a DPP-4 inhibitor in mice with type-2 diabetes mellitus. *Cardiovasc. Drugs Ther.* **2012**, *26*, 445–456. [CrossRef]
87. Zhuge, F.; Ni, Y.; Nagashimada, M.; Nagata, N.; Xu, L.; Mukaida, N.; Kaneko, S.; Ota, T. DPP-4 Inhibition by linagliptin attenuates obesity-related inflammation and insulin resistance by regulating M1/M2 macrophage polarization. *Diabetes* **2016**, *65*, 2966–2979. [CrossRef] [PubMed]
88. Matsubara, J.; Sugiyama, S.; Akiyama, E.; Iwashita, S.; Kurokawa, H.; Ohba, K.; Maeda, H.; Fujisue, K.; Yamamoto, E.; Kaikita, K.; et al. Dipeptidyl peptidase-4 inhibitor, sitagliptin, improves endothelial dysfunction in association with its anti-inflammatory effects in patients with coronary artery disease and uncontrolled diabetes. *Circ. J.* **2013**, *77*, 1337–1344. [CrossRef]
89. Leung, M.; Leung, D.Y.; Wong, V.W. Effects of dipeptidyl peptidase-4 inhibitors on cardiac and endothelial function in type 2 diabetes mellitus: A pilot study. *Diabetes Vasc. Dis. Res.* **2016**, *13*, 236–243. [CrossRef]
90. Shigiyama, F.; Kumashiro, N.; Miyagi, M.; Iga, R.; Kobayashi, Y.; Kanda, E.; Uchino, H.; Hirose, T. Linagliptin improves endothelial function in patients with type 2 diabetes: A randomized study of linagliptin effectiveness on endothelial function. *J. Diabetes Investig.* **2017**, *8*, 330–340. [CrossRef]
91. Kajikawa, M.; Maruhashi, T.; Hidaka, T.; Matsui, S.; Hashimoto, H.; Takaeko, Y.; Nakano, Y.; Kurisu, S.; Kihara, Y.; Yusoff, F.M.; et al. Effect of Saxagliptin on Endothelial Function in Patients with Type 2 Diabetes: A Prospective Multicenter Study. *Sci. Rep.* **2019**, *9*, 10206. [CrossRef]
92. Nakamura, K.; Oe, H.; Kihara, H.; Shimada, K.; Fukuda, S.; Watanabe, K.; Takagi, T.; Yunoki, K.; Miyoshi, T.; Hirata, K.; et al. DPP-4 inhibitor and alpha-glucosidase inhibitor equally improve endothelial function in patients with type 2 diabetes: EDGE study. *Cardiovasc. Diabetol.* **2014**, *13*, 110. [CrossRef] [PubMed]

93. Jansson, P.A.; Gudbjörnsdóttir, H.S.; Andersson, O.K.; Lönnroth, P.N. The effect of metformin on adipose tissue metabolism and peripheral blood flow in subjects with NIDDM. *Diabetes Care* **1996**, *19*, 160–164. [CrossRef] [PubMed]
94. Davis, B.J.; Xie, Z.; Viollet, B.; Zou, M.H. Activation of the AMP-activated kinase by antidiabetes drug metformin stimulates nitric oxide synthesis in vivo by promoting the association of heat shock protein 90 and endothelial nitric oxide synthase. *Diabetes* **2006**, *55*, 496–505. [CrossRef] [PubMed]
95. Nafisa, A.; Gray, S.G.; Cao, Y.; Wang, T.; Xu, S.; Wattoo, F.H.; Barras, M.; Cohen, N.; Kamato, D.; Little, P.J. Endothelial function and dysfunction: Impact of metformin. *Pharmacol. Ther.* **2018**, *192*, 150–162. [CrossRef]
96. de Jager, J.; Kooy, A.; Schalkwijk, C.; van der Kolk, J.; Lehert, P.; Bets, D.; Wulffelé, M.G.; Donker, A.J.; Stehouwer, C.D. Long-term effects of metformin on endothelial function in type 2 diabetes: A randomized controlled trial. *J. Intern. Med.* **2014**, *275*, 59–70. [CrossRef]
97. Zhou, J.; Massey, S.; Story, D.; Li, L. Metformin: An old drug with new applications. *Int. J. Mol. Sci.* **2018**, *19*, 2863. [CrossRef]
98. Mather, K.J.; Verma, S.; Anderson, T.J. Improved endothelial function with metformin in type 2 diabetes mellitus. *J. Am. Coll. Cardiol.* **2001**, *37*, 1344–1350. [CrossRef]
99. Shih, M.H.; Xu, Y.Y.; Yang, Y.S.; Lin, G.L. A facile synthesis and antimicrobial activity evaluation of sydnonyl-substituted thiazolidine derivatives. *Molecules* **2015**, *20*, 6520–6532. [CrossRef]
100. Jiang, C.; Ting, A.T.; Seed, B. PPAR-gamma agonists inhibit production of monocyte inflammatory cytokines. *Nature* **1998**, *391*, 82–86. [CrossRef]
101. Knouff, C.; Auwerx, J. Peroxisome proliferator-activated receptor-gamma calls for activation in moderation: Lessons from genetics and pharmacology. *Endocr. Rev.* **2004**, *25*, 899–918. [CrossRef]
102. del Valle, H.F.; Lascano, E.C.; Negroni, J.A.; Crottogini, A.J. Glibenclamide effects on reperfusion-induced malignant arrhythmias and left ventricular mechanical recovery from stunning in conscious sheep. *Cardiovasc. Res.* **2001**, *50*, 474–485. [CrossRef]
103. McAlister, F.A.; Eurich, D.T.; Majumdar, S.R.; Johnson, J.A. The risk of heart failure in patients with type 2 diabetes treated with oral agent monotherapy. *Eur. J. Heart Fail.* **2008**, *10*, 703–708. [CrossRef] [PubMed]
104. Pagano, P.J.; Griswold, M.C.; Ravel, D.; Cohen, R.A. Vascular action of the hypoglycaemic agent gliclazide in diabetic rabbits. *Diabetologia* **1998**, *41*, 9–15. [CrossRef]
105. Katakami, N.; Yamasaki, Y.; Hayaishi-Okano, R.; Ohtoshi, K.; Kaneto, H.; Matsuhisa, M.; Kosugi, K.; Hori, M. Metformin or gliclazide, rather than glibenclamide, attenuate progression of carotid intima-media thickness in subjects with type 2 diabetes. *Diabetologia* **2004**, *47*, 1906–1913. [CrossRef] [PubMed]
106. Corgnali, M.; Piconi, L.; Ihnat, M.; Ceriello, A. Evaluation of gliclazide ability to attenuate the hyperglycaemic 'memory' induced by high glucose in isolated human endothelial cells. *Diabetes/Metab. Res. Rev.* **2008**, *24*, 301–309. [CrossRef]
107. Jojima, T.; Suzuki, K.; Hirama, N.; Uchida, K.; Hattori, Y. Glimepiride upregulates eNOS activity and inhibits cytokine-induced NF-kappaB activation through a phosphoinoside 3-kinase-Akt-dependent pathway. *Diabetes Obes. Metab.* **2009**, *11*, 143–149. [CrossRef]
108. American Diabetes Association. Standards of medical care in diabetes—2010. *Diabetes Care* **2010**, *1* (Suppl. 33), 11–61.
109. Hamdy, O.; Ledbury, S.; Mullooly, C.; Jarema, C.; Porter, S.; Ovalle, K.; Moussa, A.; Caselli, A.; Caballero, A.E.; Economides, P.A.; et al. Lifestyle modification improves endothelial function in obese subjects with the insulin resistance syndrome. *Diabetes Care* **2003**, *26*, 2119–2125. [CrossRef]
110. Hotta, K.; Funahashi, T.; Arita, Y.; Takahashi, M.; Matsuda, M.; Okamoto, Y.; Iwahashi, H.; Kuriyama, H.; Ouchi, N.; Maeda, K.; et al. Plasma concentrations of a novel, adipose-specific protein, adiponectin, in type 2 diabetic patients. *Arter. Thromb. Vasc. Biol.* **2000**, *20*, 1595–1599. [CrossRef]
111. Hambrecht, R.; Adams, V.; Erbs, S.; Linke, A.; Kränkel, N.; Shu, Y.; Baither, Y.; Gielen, S.; Thiele, H.; Gummert, J.F.; et al. Regular physical activity improves endothelial function in patients with coronary artery disease by increasing phosphorylation of endothelial nitric oxide synthase. *Circulation* **2003**, *107*, 3152–3158. [CrossRef]
112. Montero, D.; Walther, G.; Benamo, E.; Perez-Martin, A.; Vinet, A. Effects of exercise training on arterial function in type 2 diabetes mellitus: A systematic review and meta-analysis. *Sports Med.* **2013**, *43*, 1191–1199. [CrossRef] [PubMed]

113. Qiu, S.; Cai, X.; Yin, H.; Sun, Z.; Zügel, M.; Steinacker, J.M.; Schumann, U. Exercise training and endothelial function in patients with type 2 diabetes: A meta-analysis. *Cardiovasc. Diabetol.* **2018**, *17*, 64. [CrossRef]
114. Matoba, K.; Kawanami, D.; Nagai, Y.; Takeda, Y.; Akamine, T.; Ishizawa, S.; Kanazawa, Y.; Yokota, T.; Utsunomiya, K. Rho-Kinase blockade attenuates podocyte apoptosis by inhibiting the notch signaling pathway in diabetic nephropathy. *Int. J. Mol. Sci.* **2017**, *18*, 1795. [CrossRef] [PubMed]
115. Nagai, Y.; Matoba, K.; Kawanami, D.; Takeda, Y.; Akamine, T.; Ishizawa, S.; Kanazawa, Y.; Yokota, T.; Utsunomiya, K.; Nishimura, R. ROCK2 regulates TGF-β-induced expression of CTGF and profibrotic genes via NF-κB and cytoskeleton dynamics in mesangial cells. *Am. J. Physiol. Renal. Physiol.* **2019**, *317*, F839–F851. [CrossRef]
116. Takeda, Y.; Matoba, K.; Kawanami, D.; Nagai, Y.; Akamine, T.; Ishizawa, S.; Kanazawa, Y.; Yokota, T.; Utsunomiya, K. ROCK2 regulates monocyte migration and cell to cell adhesion in vascular endothelial Cells. *Int. J. Mol. Sci.* **2019**, *20*, 1331. [CrossRef] [PubMed]
117. Yao, L.; Romero, M.J.; Toque, H.A.; Yang, G.; Caldwell, R.B.; Caldwell, R.W. The role of RhoA/Rho kinase pathway in endothelial dysfunction. *J. Cardiovasc. Dis. Res.* **2010**, *1*, 165–170.
118. Zhuang, R.; Wu, J.; Lin, F.; Han, L.; Liang, X.; Meng, Q.; Jiang, Y.; Wang, Z.; Yue, A.; Gu, Y.; et al. Fasudil preserves lung endothelial function and reduces pulmonary vascular remodeling in a rat model of end-stage pulmonary hypertension with left heart disease. *Int. J. Mol. Med.* **2018**, *42*, 1341–1352. [CrossRef] [PubMed]
119. Nohria, A.; Grunert, M.E.; Rikitake, Y.; Noma, K.; Prsic, A.; Ganz, P.; Liao, J.K.; Creager, M.A. Rho kinase inhibition improves endothelial function in human subjects with coronary artery disease. *Circ. Res.* **2006**, *99*, 1426–1432. [CrossRef] [PubMed]
120. Kawanami, D.; Matoba, K.; Kanazawa, Y.; Ishizawa, S.; Yokota, T.; Utsunomiya, K. Thrombin induces MCP-1 expression through Rho-kinase and subsequent p38MAPK/NF-kappaB signaling pathway activation in vascular endothelial cells. *Biochem. Biophys Res. Commun.* **2011**, *411*, 798–803. [CrossRef]
121. Matoba, K.; Kawanami, D.; Tsukamoto, M.; Kinoshita, J.; Ito, T.; Ishizawa, S.; Kanazawa, Y.; Yokota, T.; Murai, N.; Matsufuji, S.; et al. Rho-kinase regulation of TNF-α-induced nuclear translocation of NF-κB RelA/p65 and M-CSF expression via p38 MAPK in mesangial cells. *Am. J. Physiol. Ren. Physiol.* **2014**, *307*, F571–F580. [CrossRef]
122. Silveira, A.A.A.; Dominical, V.M.; Vital, D.M.; Ferreira, W.A., Jr.; Costa, F.T.M.; Werneck, C.C.; Costa, F.F.; Conran, N. Attenuation of TNF-induced neutrophil adhesion by simvastatin is associated with the inhibition of Rho-GTPase activity, p50 activity and morphological changes. *Int. Immunopharmacol.* **2018**, *58*, 160–165. [CrossRef] [PubMed]
123. Huang, L.; Dai, F.; Tang, L.; Bao, X.; Liu, Z.; Huang, C.; Zhang, T.; Yao, W. Distinct roles for ROCK1 and ROCK2 in the regulation of oxldl-mediated endothelial dysfunction. *Cell Physiol. Biochem.* **2018**, *49*, 565–577. [CrossRef] [PubMed]
124. Pelosi, M.; Marampon, F.; Zani, B.M.; Prudente, S.; Perlas, E.; Caputo, V.; Cianetti, L.; Berno, V.; Narumiya, S.; Kang, S.W.; et al. ROCK2 and its alternatively spliced isoform ROCK2m positively control the maturation of the myogenic program. *Mol. Cell. Biol.* **2007**, *27*, 6163–6176. [CrossRef] [PubMed]
125. Shimada, H.; Rajagopalan, L.E. Rho-kinase mediates lysophosphatidic acid-induced IL-8 and MCP-1 production via p38 and JNK pathways in human endothelial cells. *FEBS Lett.* **2010**, *584*, 2827–2832. [CrossRef] [PubMed]
126. Shimizu, T.; Fukumoto, Y.; Tanaka, S.; Satoh, K.; Ikeda, S.; Shimokawa, H. Crucial role of ROCK2 in vascular smooth muscle cells for hypoxia-induced pulmonary hypertension in mice. *Arterioscler. Thromb. Vasc. Biol.* **2013**, *33*, 2780–2791. [CrossRef] [PubMed]
127. Sawada, N.; Liao, J.K. Rho/Rho-associated coiled-coil forming kinase pathway as therapeutic targets for statins in atherosclerosis. *Antioxid. Redox Signal.* **2014**, *20*, 1251–1267. [CrossRef]
128. Zandi, S.; Nakao, S.; Chun, K.H.; Fiorina, P.; Sun, D.; Arita, R.; Zhao, M.; Kim, E.; Schueller, O.; Campbell, S.; et al. ROCK-isoform-specific polarization of macrophages associated with age-related macular degeneration. *Cell Rep.* **2015**, *10*, 1173–1186. [CrossRef]

© 2020 by the authors. Licensee MDPI, Basel, Switzerland. This article is an open access article distributed under the terms and conditions of the Creative Commons Attribution (CC BY) license (http://creativecommons.org/licenses/by/4.0/).

Article

Aerobic Exercise Training Inhibits Neointimal Formation via Reduction of PCSK9 and LOX-1 in Atherosclerosis

Wei Li [1,2], Heegeun Park [1], Erling Guo [1], Wooyeon Jo [1], Kyu Min Sim [1] and Sang Ki Lee [1,*]

1. Department of Sport Science, College of Natural Science, Chungnam National University, 99 Daehak-ro, Yuseong-gu, Daejeon 34134, Korea; ty1986@zjnu.edu.cn (W.L.); exepre@cnu.ac.kr (H.P.); chuang042@gmail.com (E.G.); dndus7942@naver.com (W.J.); skm477@naver.com (K.M.S.)
2. Exercise and Metabolism Research Center, College of Physical Education and Health Sciences, Zhejiang Normal University, Jinhua 321004, China
* Correspondence: nicelsk@cnu.ac.kr; Tel.: +82-42-821-6456; Fax: +82-42-823-0387

Received: 30 March 2020; Accepted: 17 April 2020; Published: 19 April 2020

Abstract: The purpose of this study was to investigate whether aerobic exercise training inhibits atherosclerosis via the reduction of proprotein convertase subtilisin/kexin type 9 (PCSK9) expression in balloon-induced common carotid arteries of a high-fat-diet rats. Male SD (Sprague Dawley) rats fed an eight-weeks high-fat diet were randomly divided into three groups; these were the sham-operated control (SC), the balloon-induced control (BIC) and the balloon-induced exercise (BIE). The aerobic exercise training groups were performed on a treadmill. The major findings were as follows: first, body weight gain was significantly decreased by aerobic exercise training compared to the BIC without change of energy intake. Second, neointimal formation was significantly inhibited by aerobic exercise training in the balloon-induced common carotid arteries of high-fat-diet rats compared to the BIC. Third, low-density lipoprotein (LDL) receptor (LDLr) expression was significantly increased by aerobic exercise training in the livers of the high-fat diet group compared to the BIC, but not the proprotein convertase subtilisin/kexin type 9 (PCSK9) expression. Fourth, aerobic exercise training significantly decreased the expression of PCSK9, the lectin-like oxidized LDL receptor-1 (LOX-1), and vascular cell adhesion molecule-1 (VCAM-1) in balloon-induced common carotid arteries of high-fat-diet rats compared to the BIC. In conclusion, our results suggest that aerobic exercise training increases LDLr in the liver and inhibits neointimal formation via the reduction of PCSK9 and LOX-1 in balloon-induced common carotid arteries of high-fat-diet-induced rats.

Keywords: atherosclerosis; aerobic exercise; endothelial dysfunction; PCSK9; LOX-1

1. Introduction

Atherosclerosis is a condition in which an artery is occluded by a plaque that leads to cardiovascular disease, with high morbidity and mortality [1]. A high-fat diet (HFD) is the main factor that causes atherosclerosis, such as elevated serum levels of low-density lipoprotein cholesterol (LDL-C) and oxidative stress thereby promoting endothelial damage [2,3].

Previous studies have indicated that lectin-like oxidized LDL (low-density lipoprotein) receptor-1 (LOX-1) binds to oxidized LDL (ox-LDL) in vasculature, which plays an important role in endothelial damage [4]. High levels of LDL-C, particularly in the form of ox-LDL, were shown to increase intracellular reactive oxygen stress (ROS) generation via LOX-1 activation [5]. Numerous studies have shown that the reduction of the LOX-1 expression can significantly delay the development of atherosclerosis [6,7].

Proprotein convertase subtilisin/kexin type 9 (PCSK9) was initially proposed in 2003 [8]. It is mainly produced and secreted in the liver and is expressed in the small intestine, kidneys and brain [9,10].

PCSK9 overexpression increases intracellular LDL receptor (LDLr) degradation in hepatic cells [11] and the circulation of PCSK9 inhibits LDLr recycling to the cell surface [12,13]. By contrast, the inhibition of PCSK9 significantly reduces plasma LDL-C concentration [14], thus preventing lipid accumulation on the vessel wall and inhibiting atherogenic plaque formation [15]. Interestingly, Ferri et al. demonstrated that PCSK9 was detectable in human atherosclerotic plaques and PCSK9 activation reduced LDLr expression in macrophages [16]. These results indicated that PCSK9 could directly affect the formation of foam cells and atherosclerosis.

Current understanding suggests that exercising regularly as a therapeutic strategy can effectively inhibit atherosclerosis [17,18]. Exercise training improves damaged endothelial cells (ECs) and inhibits neointimal formation, which prompts the occurrence of atherosclerosis [19]. Exercising regularly can also markedly reduce plasma LDL-C and ox-LDL during oxidative stress [20,21]. It markedly decreases LOX-1 expression and mediates ox-LDL-induced apoptosis in vascular tissues [22]. However, studies on the effect of exercise on the prevention of atherosclerosis via the LOX-1-mediated pathway in the molecular aspect are rarely conducted.

Several studies have indicated that exercise could also reduce plasma PCSK9 concentration in high-fat-diet-induced mice [23] and decrease hepatic PCSK9 mRNA in ovariectomized rats [24]. This study mainly focused on the role of exercise on PCSK9 in the liver [24,25] and the intestine [26] or the application of drugs, such as antibodies, to PCSK9 to reduce the risk of atherosclerosis [27,28]. However, it did not establish the effect of exercise training on the relationships of PCSK9 and LDLr in the livers and PCSK9 and LOX-1 in atherosclerotic regions of high-fat-diet rats.

Therefore, this study was conducted to determine whether aerobic exercise training contributes to treat atherosclerosis via PCSK9 and LOX-1 in balloon-induced common carotid arteries of high-fat-diet rats.

2. Materials and Methods

2.1. Experimental Animals

Six-week-old male Sprague-Dawley (SD, $n = 30$) rats were used in this experiment. The rats were adapted to a new environment for one week, and then fed a 60% high-fat (D12492, Open Source Diets, Research Diets, New Brunswick, NJ, USA) and a chow-diet for 8 weeks [23]. After 8 weeks, they were randomly divided into SC (sham-operated control group, $n = 10$), BIC (balloon-injured control group, $n = 10$), and BIE (balloon-injured exercise group, $n = 10$), respectively. The rats were kept in animal cages (30 cm × 20 cm) by respective groups (temperature 20–25 °C, humidity 50–60% and contrast 12-h cycle). Diet-induced obesity was created by feeding a high-fat diet (60% fat, Raon Bio, Korea) ad libitum until the end of experiment. The food intake was measured every week. All experiments were approved by the Animal Care and Use Committee at the Chungnam National University (CNU-00818, 4 October 2016).

2.2. Balloon Injury Model

A balloon induced-atherosclerosis rat model was used in this experiment [29]. An anesthetic (a mixture of 80 mg/kg ketamine and 12 mg/kg xylazine) was injected intraperitoneally and then a 2 F Forgaty Catheter (Edwards Lifesciences, Irvine, CA, USA) was inserted into the left common carotid artery (CCA) through an external carotid artery (ECA). A pressure gauge was used to inflate the catheter balloon 1.5 times greater than the diameter of the carotid artery and then a 10 mm injury was induced by the withdrawal of inflated balloon catheter 5 times. The rats were allowed at least 3 days to recover from the surgery.

2.3. Exercise Protocol

The rats were exercised on a treadmill (motorized rodent treadmill) for 8 weeks in order to investigate the effects of aerobic exercise. The rats ran at 10 m/min with a 0% incline for 10 min on the first day. The speed and duration of the exercise were increased by 10 min and 2 m/min every day until the fourth day. From the fifth day to the end of the experiment, the rats ran at 16 m/min for 60 min [17]. This exercise intensity corresponded to 65–70% of the maximal oxygen uptake.

2.4. Experimental Animals

The rats were maintained on a feeding program and their individual body weights were recorded every week throughout the experimental period. In order to measure the food intakes, 120 g of diets per cage (3 rats) were supplied and the remaining amounts of supplied diets were measured two times per week after using an automatic electric balance (Cas, Seoul, Korea). These were regarded as the individual mean daily food consumption of rats (g/day/rats).

2.5. Hematoxylin and Eosin (H & E) Staining

In order to measure the neointimal formation, the balloon-injured rat carotid arteries were fixed in 4% formaldehyde and were paraffin-embedded. Serial cross sections (5 µm thick) of arteries were stained with hematoxylin and eosin (MHS-32, Sigma-Aldrich, MO, USA). A DP70 camera (Olympus, Tokyo) and a TSView version 7 (Fuzhou Tucsen Image Technology, Japan) were used to measure the size (μm^2) of the intima, media and lumen, to calculate the intima-media thickness and lumen diameter so as to compare the degree of the neointimal formation.

2.6. Western Blotting

The tissues of the carotid arteries and liver were homogenized in lysis buffer (20 mM Tris-HCl, 0.5% NP-40, 250 mM NaCl, 3 mM EDTA, 3 mM EGTA, 2 mM DTT, 0.5 mM phenylmethylsulfonylfluoride, 2 mM b-glycerophosphate, 1 mM sodium orthovanadate, 1 ug/mL leupeptin and pH 7.5, Sigma, USA), and then centrifuged for 30 min at 14,000 rpm at 4 °C so as to remove the supernatant and quantify the proteins by using a BCA assay kit (Bio-rad, Rockford, IL, USA). In addition, 50 ug of proteins were subjected to electrophoresis on a 9% SDS-PAGE and then transferred to a polyvinylidene difluoride (PVDF) membrane. The nonspecific reaction of the membrane was removed by blocking it for one h at room temperature in 5% nonfat dry milk in Tris buffered saline Tween-20 (TBST). PCSK9 (abcam, San Francisco, CA, USA), LDLr (Abcam, San Francisco, CA, USA), LOX-1 (Abcam, San Francisco, CA, USA), VCAM-1 (Cell signaling, USA) and beta-actin (Sigma, St. Louis, MO, USA) were incubated for 18 h at 4 °C in TBST with 5% nonfat dry milk. Horseradish peroxidase (HRP)-conjugated rabbit-antimouse IgG (Calbiochem, San Diego, CA, USA) was used for secondary antibodies. Blots were developed for visualization by using an enhanced chemiluminescence detection kit (Pierce Biotechnology, Rockford, IL, USA), and the Image Quant software (Molecular dynamics, Sunnyvale, CA, USA) was used to quantify the expression.

2.7. Statistical Analysis

SPSS 24.0 statistics were used to calculate the descriptive statistics quantity from the results of this study and a one-way ANOVA was used to verify each variable, while Duncan's test was used for post-hoc analysis. The differences were considered statistically significant at $p < 0.05$.

3. Results

3.1. Aerobic Exercise Training Inhibited Body Weight Gain

After the high-fat-diet induced weight gain was compared with the chow-diet groups for eight weeks ($p < 0.05$, 420.1 ± 19.2 vs. 463.9 ± 12.5) (Figure 1A), we investigated whether eight weeks of aerobic exercise training reduced the final body weight in the balloon-induced atherosclerotic rat model with a high-fat diet (Figure 1B,C). The high-fat diet significantly increased body weight gain compared with the chow-diet groups. Aerobic exercise training significantly decreased body weight gain compared with the SC and BIC groups ($p < 0.05$, 520 ± 24.4 and 543.9 ± 11.8 vs. 486.7 ± 14.3) without a change of energy intake ($p > 0.05$).

3.2. Aerobic Exercise Inhibited Neointimal Formation

In order to investigate the effect of aerobic exercise on the neointimal formation in the balloon-induced rat model with a high-fat diet, we measured the neointimal formation in the balloon-injured common carotid arteries (CCA) with a high-fat diet (Figure 2). The neointimal formation was significantly inhibited by the aerobic exercise training compared to the BIC group ($p < 0.05$, 1.31 ± 0.15 vs. 0.85 ± 0.12).

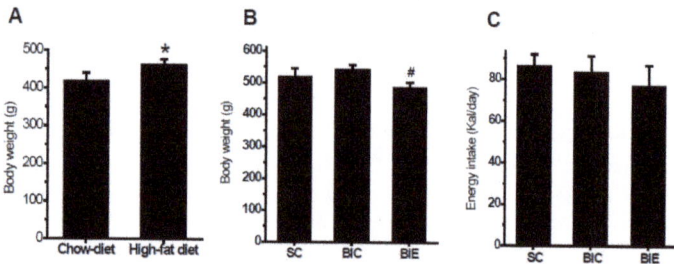

Figure 1. Aerobic exercise training significantly attenuated body weight without a change of energy intake in the balloon-induced rats with a high-fat diet. Body weight after 8 weeks of a high-fat diet (**A**), body weight after 8 weeks of exercise (**B**), and energy intake (**C**). Chow-diet ($n = 10$), high-fat diet ($n = 30$), SC ($n = 10$), sham control; BIC ($n = 10$), balloon-injured control; BIE ($n = 10$), balloon-injured exercise. Data showed a mean ±S.E. * $p < 0.05$ vs. Chow-diet; # $p < 0.05$ vs. BIC.

3.3. Aerobic Exercise Increased LDLr Expression in Liver of Rats, but Did not Affect PCSK9 Expression

We measured the hepatic PCSK9 and LDLr in the livers of rats (Figure 3). Aerobic exercise training did not significantly affect the hepatic PCSK9 protein expression. However, the LDLr expression in the liver was significantly increased by aerobic exercise compared with the BIC group ($p < 0.05$, 0.10 ± 0.02 vs. 0.05 ± 0.01).

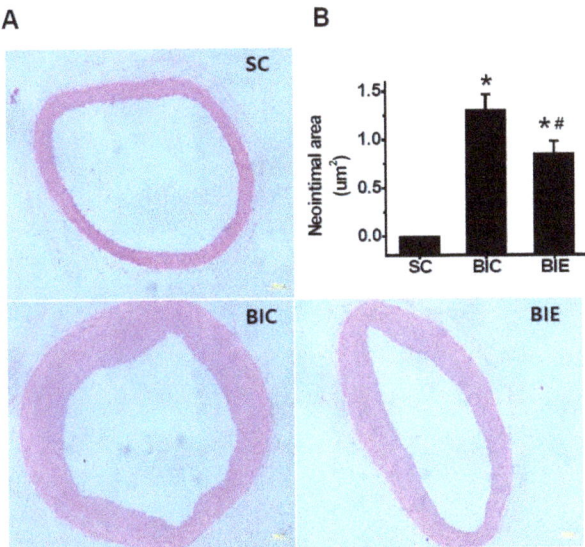

Figure 2. Aerobic exercise training inhibited the neointimal formation in the balloon-injured common carotid arteries (CCA) of the balloon-induced rats with high-fat diets. Representation of hematoxylin and eosin (H&E) staining in the injured CCA (**A**). Neointima of the cross-sectional area (μm^2) was quantified by planimetry (**B**). SC ($n = 7$), sham Control; BIC ($n = 8$), balloon-injured control; BIE ($n = 8$), balloon-injured exercise. Data showed a mean ±S.E. * $p < 0.05$ vs. SC; # $p < 0.05$ vs. BIC. Microscope were used by 100X and Scale bars are 100 μm.

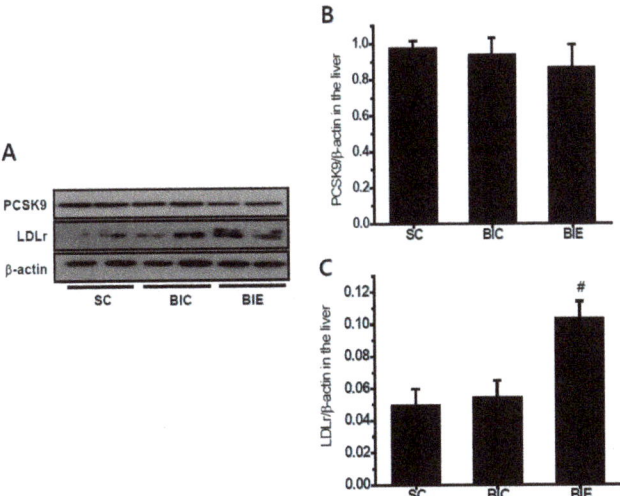

Figure 3. Aerobic exercise training did not affect the hepatic proprotein convertase subtilisin/kexin type 9 (PCSK9) expression but increased the low-density lipoprotein (LDL) receptor (LDLr) expression in the livers of balloon-induced rats with high-fat diets. Protein expression (**A**), and the densitometry analyses of PCSK9 (**B**) and LDLr (**C**). SC ($n = 6$), sham control; BIC ($n = 6$), balloon-injured control; BIE ($n = 6$), balloon-injured exercise. Data showed a mean ±S.E. # $p < 0.05$ vs. BIC.

3.4. Aerobic Exercise Suppressed Expression of PCSK9, LOX-1 and VCAM-1 in CCA

We investigated the expression of PCSK9, LOX-1 and VCAM-1 in the balloon-induced common carotid arteries of rats with a high-fat diet (Figure 4). Aerobic exercise training significantly inhibited the PCSK9 expression in the CCA compared to the SC and BIC group ($p < 0.05$, 0.38 ± 0.03 and 0.41 ± 0.04 vs. 0.28 ± 0.05). LOX-1 expression in the CCA was significantly increased in the BIC group compared to the SC ($p < 0.05$, 0.17 ± 0.01 vs. 0.20 ± 0.01). However, aerobic exercise training significantly recovered the increase of balloon-induced LOX-1 expression in the CCA compared to the BIC group ($p < 0.05$, 0.20 ± 0.01 vs. 0.15 ± 0.02). VCAM-1 expression in the CCA was significantly increased in the BIC compared to the SC ($p < 0.05$, 0.25 ± 0.02 vs. 0.39 ± 0.05). However, aerobic exercise training significantly recovered the increase of balloon-induced LOX-1 expression in the CCA compared to the BIC group ($p < 0.05$, 0.39 ± 0.05 vs. 0.26 ± 0.06).

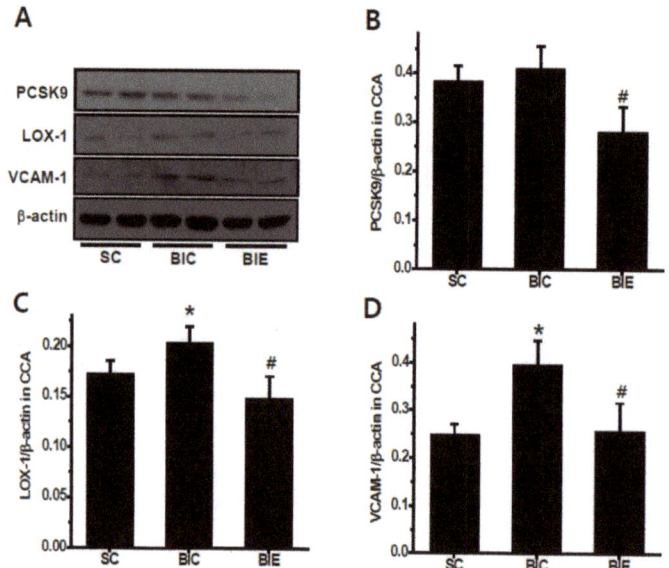

Figure 4. Aerobic exercise training reduced the expression of PCSK9, LOX-1 and VCAM-1 in the balloon-injured common carotid arteries (CCA) of the balloon-induced rats with high-fat diets. Protein expression (**A**), and the densitometry analyses of PCSK9 (**B**), LOX-1 (**C**) and VCAM-1 (**D**). SC ($n = 5$), sham control; BIC ($n = 5$), balloon-injured control; BIE ($n = 5$), balloon-injured exercise. Data showed a mean ±S.E. * $p < 0.05$ vs. BIC; # $p < 0.05$ vs. BIE.

4. Discussion

High-fat diets and physical inactivity are associated with obesity and lead to various cardiovascular system events, such as atherosclerosis. Physical activity as a therapeutic strategy to overcome obesity can effectively reduce body weight [30]. In our study, aerobic exercise training significantly inhibited weight gain in high-fat-diet rats.

Regular exercise suppresses atherosclerotic lesions in vascular walls and directly inhibits the neointimal formation as well [19,31]. Li et al. recently reported that exercise training can inhibit balloon-induced neointimal formation via overcoming endothelial dysfunction [18]. In this study, eight weeks of aerobic exercise training significantly suppressed the neointimal formation in the balloon-induced common carotid arteries of high-fat-diet rats. These findings provide direct evidence that aerobic exercise training can effectively inhibit atherosclerosis.

PCSK9 is a newly identified protein that shows potential in the treatment for dyslipidemia and atherosclerosis [32,33]. In fact, PCSK9 is mainly expressed in the liver and is secreted into circulation, and binds to LDLr on the hepatic surface, resulting in increased serum LDL-C concentrations. Experimental studies have indicated that the overexpression of PCSK9 increases plasma LDL-C level [33], but the inhibition of PCSK9 reduces LDL-C in serum [12,34].

One of the non-pharmacological strategies for the treatment of hypercholesterolemia is exercise training. Studies have recently shown controversial results that aerobic exercise training reduced the PCSK9 mRNA expression in the liver [23] of high-fat-diet obese animals. On the other hand, aerobic exercise training increased the PCSK9 mRNA expression in the liver [24] and intestines of ovariectomized rats. In the present study, exercise training did not affect the hepatic PCSK9. However, it has been consistently reported that aerobic exercise training increases the mRNA and protein expression of LDLr in the liver [23,35] and intestines [26].

In humans, LDL particles are the main carriers of cholesterol to the peripheral tissues, where they are internalized via the LDLr, a crucial mediator of plasma LDL concentrations [36]. Epidemiological evidence has consistently shown that increasing LDL concentrations were associated with the risk of vascular diseases [37]. Genetic studies have reported that early exposure to excessive LDL-C resulted in LDLr mutation and atherosclerotic lesion [38]. Exercise was shown to increase LDLr transcription in animal models [39,40]. Our results indicated that aerobic exercise significantly upregulated the hepatic LDLr protein level, which was consistent with the results of previous studies. However, considering that the main function of the hepatic LDLr is the removal of LDL-C in serum, our data showed that it effectively reduced hepatic LDLr but failed to show the positive effect of exercise training on hepatic PCSK9. It is possible that the effect of exercise training on the management of the LDL-C metabolism may take place more at the intestinal [24,25] and vasculature levels [16] than at the liver tissue, but the experimental evidence at the molecular level is unclear.

LDL-C was converted to ox-LDL by obesity-induced oxidative stress [5]. Ox-LDL was recognized by macrophages, which cause the formation of foam cells and lead to atherosclerotic plaque [41]. In addition, ox-LDL induces the apoptosis of ECs via LOX-1, whereas LOX-1 mRNA antisense inhibits ox-LDL in ECs [41]. Riahi et al. reported that chronic aerobic exercise could downregulate LOX-1 in the hearts of rats fed with a high-fat diet [42]. However, another study showed that the cessation of voluntary running significantly recovered LOX-1 in the vascular tissues [22]. A recent report indicated that LOX-1 expression was reduced in PCSK9 knockout animal models and that the overexpression of PCSK9 was reduced in LOX-1 knockout animal models [43–45]. These studies suggest crosstalk between PCSK9 and LOX-1 in the vascular tissues. Both LOX-1 and PCSK9 are also involved in inflammation-induced VCAM-1 [45]. VCAM-1 is expressed in endothelial cells (ECs) and smooth muscle cells (SMCs), and induces atherosclerosis [46]. Exercise training is widely known to decrease VCAM-1 expression in the vasculature [47–49].

Chronic endothelial injury causes atherosclerosis, but regular exercise training induces the fast recovery of damaged endothelial cells and inhibits the migration of smooth cells which is protective against balloon-induced atherosclerosis [17,19,31]. However, the role of exercise on PCSK9, LOX-1 and VCAM-1 in atherosclerosis has not been reported. Thus, we verified whether PCSK9 expression changed in atherosclerotic regions in response to exercise. Interestingly, our findings showed that aerobic exercise training significantly attenuated the expression levels of PCSK9, LOX-1 and VCAM-1 in balloon-induced sites of high-fat-diet rats.

Recently, Ding et al. suggested the possibility that an inflammatory stimulus activates the endothelial cell and smooth muscle cell to secret PCSK9, wherein induced LOX-1 activation and ox-LDL is up took, leading to atherosclerosis [43]. This study has shown a positive feedback between PCSK9 and the LOX-1 axis in the intra-arterial wall. Furthermore, it was reported that hemodynamic high shear stress reduced the expression of PCSK9 and LOX-1, but low shear stress increased the PCSK9 expression in human endothelial and smooth cells [45]. Importantly, both the PCSK9 and VCAM-1 expression are greater in aortic arch branch point and aorta–iliac bifurcation regions than in the thoracic aorta. To our knowledge, the carotid arteries used in the present study were also in bifurcation region and in low shear stress regions. Since lamina flow and high shear stress have atheroprotective effects on vascular endothelial and smooth muscle cells [45,50], it is possible that exercise-induced high shear stress may contribute to reduce the expression of PCSK9, LOX-1 and VCAM-1 in the atherosclerotic region.

Although circulating PCSK9 and LDL-C levels were not measured in the present study, our data showed at least the role of exercise training as a therapeutic option to suppress PCSK9 activation in atherosclerosis, but the evidence that it does affect PCSK9 in an isolated EC or SMC in the atherosclerotic region is lacking.

5. Conclusions

In conclusion, our results suggest that aerobic exercise training may increase LDLr expression in the liver and inhibits neointimal formation via the reduction of PCSK9, LOX-1 and VCAM-1 in atherosclerotic regions in the high-fat-diet-induced rat models.

Author Contributions: Conceptualization, S.K.L. and H.P.; methodology, W.L., H.P., K.M.S. and E.G.; software, W.L., K.M.S. and W.J.; validation, E.G.; formal analysis, H.P., W.J., K.M.S. and E.G.; investigation, W.J., E.G. and K.M.S.; data curation, W.J.; writing—original draft preparation, W.L.; writing—review and editing, S.K.L.; supervision, S.K.L.; project administration, H.P.; funding acquisition, S.K.L. All authors have read and agreed to the published version of the manuscript.

Funding: This work was supported by the Ministry of Education of the Republic of Korea and the National Research Foundation of Korea (NRF-2015S1A2A1A01026956, NRF-2017S1A5A2A01026677).

Conflicts of Interest: The authors declare no conflict of interest. The funders had no role in the design of the study; in the collection, analyses, or interpretation of data; in the writing of the manuscript, or in the decision to publish the results.

References

1. Ramli, J.; CalderonArtero, P.; Block, R.C.; Mousa, S.A. Novel therapeutic targets for preserving a healthy endothelium: Strategies for reducing the risk of vascular and cardiovascular disease. *Cardiol. J.* **2011**, *18*, 352–363.
2. Zhang, W.L.; Yan, W.J.; Sun, B.; Zou, Z.P. Synergistic effects of atorvastatin and rosiglitazone on endothelium protection in rats with dyslipidemia. *Lipids Health Dis.* **2014**, *13*, 168. [CrossRef] [PubMed]
3. Noeman, S.A.; Hamooda, H.E.; Baalash, A.A. Biochemical Study of Oxidative Stress Markers in the Liver, Kidney and Heart of High Fat Diet Induced Obesity in Rats. *Diabetol. Metab. Syndr.* **2011**, *3*, 17. [CrossRef] [PubMed]
4. Morawietz, H. LOX-1 and atherosclerosis: Proof of concept in LOX-1-knockout mice. *Circ. Res.* **2007**, *100*, 1534–1536. [CrossRef] [PubMed]
5. Colavitti, R.; Pani, G.; Bedogni, B.; Anzevino, R.; Borrello, S.; Waltenberger, J.; Galeotti, T. Reactive oxygen species as downstream mediators of angiogenic signaling by vascular endothelial growth factor receptor-2/KDR. *J. Biol. Chem.* **2002**, *277*, 3101–3108. [CrossRef] [PubMed]
6. Ding, Z.F.; Liu, S.J.; Wang, X.W.; Dai, Y.; Khaidakov, M.; Romeo, F.; Mehta, J.L. LOX-1, oxidant stress, mtDNA damage, autophagy, and immune response in atherosclerosis. *Can. J. Physiol. Pharmacol.* **2014**, *92*, 524–530. [CrossRef] [PubMed]
7. Lu, J.; Chen, X.N.; Xu, X.H.; Liu, J.Z.; Zhang, Z.P.; Wang, M.X.; Li, X.Z.; Chen, H.; Zhao, D.Q.; Wang, J.; et al. Active polypeptides from Hirudo inhibit endothelial cell inflammation and macrophage foam cell formation by regulating the LOX-1/LXR-alpha/ABCA1 pathway. *Biomed. Pharmacother.* **2019**, *115*, 108840. [CrossRef]
8. Abifadel, M.; Bernier, L.; Dubuc, G.; Nuel, G.; Rabes, J.P.; Bonneau, J.; Marques, A.; Marduel, M.; Devillers, M.; Munnich, A.; et al. A PCSK9 variant and familial combined hyperlipidaemia. *J. Med Genet.* **2008**, *45*, 780–786. [CrossRef]
9. Cohen, J.C.; Boerwinkle, E.; Mosley, T.H., Jr.; Hobbs, H.H. Sequence variations in PCSK9, low LDL, and protection against coronary heart disease. *N. Engl. J. Med.* **2006**, *354*, 1264–1272. [CrossRef]
10. Cunningham, D.; Danley, D.E.; Geoghegan, K.F.; Griffor, M.C.; Hawkins, J.L.; Subashi, T.A.; Varghese, A.H.; Ammirati, M.J.; Culp, J.S.; Hoth, L.R.; et al. Structural and biophysical studies of PCSK9 and its mutants linked to familial hypercholesterolemia. *Nat. Struct. Mol. Biol.* **2007**, *14*, 413–419. [CrossRef]

11. Poirier, S.; Mayer, G.; Benjannet, S.; Bergeron, E.; Marcinkiewicz, J.; Nassoury, N.; Mayer, H.; Nimpf, J.; Prat, A.; Seidah, N.G. The proprotein convertase PCSK9 induces the degradation of low density lipoprotein receptor (LDLR) and its closest family members VLDLR and ApoER2. *J. Biol. Chem.* **2008**, *283*, 2363–2372. [CrossRef] [PubMed]
12. Qian, Y.-W.; Schmidt, R.J.; Zhang, Y.; Chu, S.; Lin, A.; Wang, H.; Wang, X.; Beyer, T.P.; Bensch, W.R.; Li, W.; et al. Secreted PCSK9 downregulates low density lipoprotein receptor through receptor-mediated endocytosis. *J. Lipid Res.* **2007**, *48*, 1488–1498. [CrossRef] [PubMed]
13. Fisher, T.S.; Lo Surdo, P.; Pandit, S.; Mattu, M.; Santoro, J.C.; Wisniewski, D.; Cummings, R.T.; Calzetta, A.; Cubbon, R.M.; Fischer, P.A.; et al. Effects of pH and low density lipoprotein (LDL) on PCSK9-dependent LDL receptor regulation. *J. Biol. Chem.* **2007**, *282*, 20502–20512. [CrossRef] [PubMed]
14. Burke, A.C.; Dron, J.S.; Hegele, R.A.; Huff, M.W. PCSK9: Regulation and Target for Drug Development for Dyslipidemia. *Annu. Rev. Pharmacol. Toxicol.* **2017**, *57*, 223–244. [CrossRef]
15. Norata, G.D.; Ballantyne, C.M.; Catapano, A.L. New therapeutic principles in dyslipidaemia: Focus on LDL and Lp(a) lowering drugs. *Eur. Heart J.* **2013**, *34*, 1783–1789. [CrossRef]
16. Ferri, N.; Tibolla, G.; Pirillo, A.; Cipollone, F.; Mezzetti, A.; Pacia, S.; Corsini, A.; Catapano, A.L. Proprotein convertase subtilisin kexin type 9 (PCSK9) secreted by cultured smooth muscle cells reduces macrophages LDLR levels. *Atherosclerosis* **2012**, *220*, 381–386. [CrossRef]
17. Indolfi, C.; Torella, D.; Coppola, C.; Curcio, A.; Rodriguez, F.; Bilancio, A.; Leccia, A.; Arcucci, O.; Falco, M.; Leosco, D.; et al. Physical training increases eNOS vascular expression and activity and reduces restenosis after balloon angioplasty or arterial stenting in rats. *Circ. Res.* **2002**, *91*, 1190–1197. [CrossRef]
18. Li, W.; Jeong, J.H.; Park, H.G.; Lee, Y.R.; Li, M.; Lee, S.K. Endurance exercise training inhibits neointimal formation via enhancement of FOXOs expression in balloon-induced atherosclerosis rat model. *J. Exerc. Nutr. Biochem.* **2014**, *18*, 105–110. [CrossRef]
19. Laufs, U.; Werner, N.; Link, A.; Endres, M.; Wassmann, S.; Jurgens, K.; Miche, E.; Bohm, M.; Nickenig, G. Physical training increases endothelial progenitor cells, inhibits neointima formation, and enhances angiogenesis. *Circulation* **2004**, *109*, 220–226. [CrossRef]
20. Kannan, U.; Vasudevan, K.; Balasubramaniam, K.; Yerrabelli, D.; Shanmugavel, K.; John, N.A. Effect of exercise intensity on lipid profile in sedentary obese adults. *J. Clin. Diagn. Res.* **2014**, *8*, BC08–BC10. [CrossRef]
21. Tjonna, A.E.; Leinan, I.M.; Bartnes, A.T.; Jenssen, B.M.; Gibala, M.J.; Winett, R.A.; Wisloff, U. Low- and High-Volume of Intensive Endurance Training Significantly Improves Maximal Oxygen Uptake after 10-Weeks of Training in Healthy Men. *PLoS ONE* **2013**, *8*, e65382. [CrossRef] [PubMed]
22. Padilla, J.; Jenkins, N.T.; Roberts, M.D.; Arce-Esquivel, A.A.; Martin, J.S.; Laughlin, M.H.; Booth, F.W. Differential changes in vascular mRNA levels between rat iliac and renal arteries produced by cessation of voluntary running. *Exp. Physiol.* **2013**, *98*, 337–347. [CrossRef] [PubMed]
23. Wen, S.; Jadhav, K.S.; Williamson, D.L.; Rideout, T.C. Treadmill Exercise Training Modulates Hepatic Cholesterol Metabolism and Circulating PCSK9 Concentration in High-Fat-Fed Mice. *J. Lipids* **2013**, *2013*, 908048. [CrossRef] [PubMed]
24. Sock, E.T.N.; Chapados, N.A.; Lavoie, J.M. LDL Receptor and Pcsk9 Transcripts are Decreased in Liver of Ovariectomized Rats: Effects of Exercise Training. *Horm. Metab. Res.* **2014**, *46*, 550–555.
25. Sock, E.T.N.; Mayer, G.; Lavoie, J.M. Combined effects of rosuvastatin and exercise on gene expression of key molecules involved in cholesterol metabolism in ovariectomized rats. *PLoS ONE* **2016**, *11*, e0159550.
26. Farahnak, Z.; Chapados, N.; Lavoie, J.M. Exercise training increased gene expression of LDL-R and PCSK9 in intestine: Link to transintestinal cholesterol excretion. *Gen. Physiol. Biophys.* **2018**, *37*, 309–317. [CrossRef]
27. Careskey, H.E.; Davis, R.A.; Alborn, W.E.; Troutt, J.S.; Cao, G.Q.; Konrad, R.J. Atorvastatin increases human serum levels of proprotein convertase subtilisin/kexin type 9. *J. Lipid Res.* **2008**, *49*, 394–398. [CrossRef]
28. Welder, G.; Zineh, I.; Pacanowski, M.A.; Troutt, J.S.; Cao, G.Q.; Konrad, R.J. High-dose atorvastatin causes a rapid sustained increase in human serum PCSK9 and disrupts its correlation with LDL cholesterol. *J. Lipid Res.* **2010**, *51*, 2714–2721. [CrossRef]
29. Lee, H.M.; Jeon, B.H.; Won, K.J.; Lee, C.K.; Park, T.K.; Choi, W.S.; Bae, Y.M.; Kim, H.S.; Lee, S.K.; Park, S.H.; et al. Gene Transfer of Redox Factor-1 Inhibits Neointimal Formation Involvement of Platelet-Derived Growth Factor-beta Receptor Signaling via the Inhibition of the Reactive Oxygen Species-Mediated Syk Pathway. *Circ. Res.* **2009**, *104*, 219–227. [CrossRef]

30. Blair, S.N.; Archer, E.; Hand, G.A. Commentary: Luke and Cooper are wrong: Physical activity has a crucial role in weight management and determinants of obesity. *Int. J. Epidemiol.* **2013**, *42*, 1836–1838. [CrossRef]
31. Pynn, M.; Schafer, K.; Konstantinides, S.; Halle, M. Exercise training reduces neointimal growth and stabilizes vascular lesions developing after injury in apolipoprotein e-deficient mice. *Circulation* **2004**, *109*, 386–392. [CrossRef] [PubMed]
32. Wu, N.Q.; Li, J.J. PCSK9 gene mutations and low-density lipoprotein cholesterol. *Clin. Chim. Acta* **2014**, *431*, 148–153. [CrossRef] [PubMed]
33. Maxwell, K.N.; Breslow, J.L. Adenoviral-mediated expression of Pcsk9 in mice results in a low-density lipoprotein receptor knockout phenotype. *Proc. Natl. Acad. Sci. USA* **2004**, *101*, 7100–7105. [CrossRef] [PubMed]
34. Kosenko, T.; Golder, M.; Leblond, G.; Weng, W.; Lagace, T.A. Low density lipoprotein binds to proprotein convertase subtilisin/kexin type-9 (PCSK9) in human plasma and inhibits PCSK9-mediated low density lipoprotein receptor degradation. *J. Biol. Chem.* **2013**, *288*, 8279–8288. [CrossRef]
35. Pinto, P.R.; Rocco, D.D.; Okuda, L.S.; Machado-Lima, A.; Castilho, G.; da Silva, K.S.; Gomes, D.J.; Pinto, R.S.; Iborra, R.T.; Ferreira, G.S.; et al. Aerobic exercise training enhances the in vivo cholesterol trafficking from macrophages to the liver independently of changes in the expression of genes involved in lipid flux in macrophages and aorta. *Lipids Health Dis.* **2015**, *14*, 109. [CrossRef]
36. Goldstein, J.L.; Brown, M.S. The LDL Receptor. *Arterioscler. Thromb. Vasc. Biol.* **2009**, *29*, 431–438. [CrossRef]
37. Lambert, G.; Charlton, F.; Rye, K.A.; Piper, D.E. Molecular basis of PCSK9 function. *Atherosclerosis* **2009**, *203*, 1–7. [CrossRef]
38. Horton, J.D.; Cohen, J.C.; Hobbs, H.H. PCSK9: A convertase that coordinates LDL catabolism. *J. Lipid Res.* **2009**, *50*, S172–S177. [CrossRef]
39. Teodoro, B.G.; Natali, A.J.; Fernandes, S.A.T.; da Silva, L.A.; de Pinho, R.A.; da Matta, S.L.P.; Peluzio, M.D.G. Improvements of Atherosclerosis and Hepatic Oxidative Stress are Independent of Exercise Intensity in LDLr-/- Mice. *J. Atheroscler. Thromb.* **2012**, *19*, 904–911. [CrossRef]
40. Guizoni, D.M.; Dorighello, G.G.; Oliveira, H.C.F.; Delbin, M.A.; Krieger, M.H.; Davel, A.P. Aerobic exercise training protects against endothelial dysfunction by increasing nitric oxide and hydrogen peroxide production in LDL receptor-deficient mice. *J. Transl. Med.* **2016**, *14*, 213. [CrossRef]
41. Mehta, J.L.; Chen, J.; Hermonat, P.L.; Romeo, F.; Novelli, G. Lectin-like, oxidized low-density lipoprotein receptor-1 (LOX-1): A critical player in the development of atherosclerosis and related disorders. *Cardiovasc. Res.* **2006**, *69*, 36–45. [CrossRef] [PubMed]
42. Riahi, S.; Mohammadi, M.T.; Sobhani, V.; Soleimany, M. Chronic effects of aerobic exercise on gene expression of LOX-1 receptor in the heart of rats fed with high fat diet. *Iran. J. Basic Med Sci.* **2015**, *18*, 805–812. [PubMed]
43. Ding, Z.F.; Liu, S.J.; Wang, X.W.; Deng, X.Y.; Fan, Y.B.; Shahanawaz, J.; Reis, R.J.S.; Varughese, K.I.; Sawamura, T.; Mehta, J.L. Cross-talk between LOX-1 and PCSK9 in vascular tissues. *Cardiovasc. Res.* **2015**, *107*, 556–567. [CrossRef] [PubMed]
44. Wu, C.Y.; Tang, Z.H.; Jiang, L.; Li, X.F.; Jiang, Z.S.; Liu, L.S. PCSK9 siRNA inhibits HUVEC apoptosis induced by ox-LDL via Bcl/Bax-caspase9-caspase3 pathway. *Mol. Cell. Biochem.* **2012**, *359*, 347–358. [CrossRef]
45. Ding, Z.F.; Liu, S.J.; Wang, X.W.; Deng, X.Y.; Fan, Y.B.; Sun, C.Q.; Wang, Y.N.; Mehta, J.L. Hemodynamic Shear Stress via ROS Modulates PCSK9 Expression in Human Vascular Endothelial and Smooth Muscle Cells and Along the Mouse Aorta. *Antioxid. Redox Signal.* **2015**, *22*, 760–771. [CrossRef] [PubMed]
46. Balcells, M.; Martorell, J.; Olive, C.; Santacana, M.; Chitalia, V.; Cardoso, A.A.; Edelman, E.R. Smooth Muscle Cells Orchestrate the Endothelial Cell Response to Flow and Injury. *Circulation* **2010**, *121*, 2192–2199. [CrossRef]
47. Byrkjeland, R.; Njerve, I.U.; Arnesen, H.; Seljeflot, I.; Solheim, S. Reduced endothelial activation after exercise is associated with improved HbA(1c) in patients with type 2 diabetes and coronary artery disease. *Diabetes Vasc. Dis. Res.* **2017**, *14*, 94–103. [CrossRef]
48. Faulkner, J.; Lambrick, D.; Woolley, B.; Stoner, L.; Wong, L.K.; McGonigal, G. Effects of Early Exercise Engagement on Vascular Risk in Patients with Transient Ischemic Attack and Nondisabling Stroke. *J. Stroke Cerebrovasc. Dis.* **2013**, *22*, E388–E396. [CrossRef]
49. Chow, C.K.; Jolly, S.; Rao-Melacini, P.; Fox, K.A.A.; Anand, S.S.; Yusuf, S. Association of Diet, Exercise, and Smoking Modification With Risk of Early Cardiovascular Events After Acute Coronary Syndromes. *Circulation* **2010**, *121*, 750–758. [CrossRef]

50. Berk, B.C.; Min, W.; Yan, C.; Surapisitchat, J.; Liu, Y.; Hoefen, R. Atheroprotective mechanisms activated by fluid shear stress in endothelial cells. *Drug. News. Perspect.* **2002**, *15*, 133–139. [CrossRef]

© 2020 by the authors. Licensee MDPI, Basel, Switzerland. This article is an open access article distributed under the terms and conditions of the Creative Commons Attribution (CC BY) license (http://creativecommons.org/licenses/by/4.0/).

Review

The Biological Role of Apurinic/Apyrimidinic Endonuclease1/Redox Factor-1 as a Therapeutic Target for Vascular Inflammation and as a Serologic Biomarker

Yu Ran Lee [1,2,†], Hee Kyoung Joo [1,2,†] and Byeong Hwa Jeon [1,2,*]

1. Research Institute for Medical Sciences, College of Medicine, Chungnam National University, 266 Munhwa-ro, Jung-gu, Daejeon 35015, Korea; lyr0913@cnu.ac.kr (Y.R.L.); hkjoo79@cnu.ac.kr (H.K.J.)
2. Department of Physiology, College of Medicine, Chungnam National University, 266 Munhwa-ro, Jung-gu, Daejeon 35015, Korea
* Correspondence: bhjeon@cnu.ac.kr; Tel.: +82-42-580-8214; Fax: +82-42-585-8440
† These authors contributed equally to this work.

Received: 4 February 2020; Accepted: 8 March 2020; Published: 10 March 2020

Abstract: Endothelial dysfunction promotes vascular inflammation by inducing the production of reactive oxygen species and adhesion molecules. Vascular inflammation plays a key role in the pathogenesis of vascular diseases and atherosclerotic disorders. However, whether there is an endogenous system that can participate in circulating immune surveillance or managing a balance in homeostasis is unclear. Apurinic/apyrimidinic endonuclease 1/redox factor-1 (henceforth referred to as APE1/Ref-1) is a multifunctional protein that can be secreted from cells. It functions as an apurinic/apyrimidinic endonuclease in the DNA base repair pathway and modulates redox status and several types of transcriptional factors, in addition to its anti-inflammatory activity. Recently, it was reported that the secretion of APE1/Ref-1 into the extracellular medium of cultured cells or its presence in the plasma can act as a serological biomarker for certain disorders. In this review, we summarize the possible biological functions of APE1/Ref-1 according to its subcellular localization or its extracellular secretions, as therapeutic targets for vascular inflammation and as a serologic biomarker.

Keywords: endothelial dysfunction; vascular inflammation; APE1/Ref-1; cardiovascular diseases; subcellular localization; serological biomarkers

---

1. Endothelial Dysfunction and Vascular Inflammation

Endothelial cell activation or dysfunction is defined by the endothelial expression of cell-surface adhesion molecules. The expression of adhesion molecules and the subsequent monocyte adhesion are considered as early events in the development of atherosclerosis [1]. Vascular inflammation plays a key role in the pathogenesis of vascular diseases and atherosclerotic disorders [2]. The inflammatory reaction is a series of complex interactions between inflammatory cells or stimuli and defense cells, such as macrophages and endothelial cells [3]. This interactive reaction triggers an inflammatory response in vascular cells by the activating of increased proinflammatory mediators and/or molecules, and cytokines [4].

This type of interactive reaction helps to eliminate the initial cause of injury, clear out inflammatory foci or cells, and helps the host cells to survive. The adhesion of leukocytes to the vascular endothelium is a hallmark of the inflammatory process [5]. Several types of antiadhesion therapeutic molecules are being developed for inflammatory diseases [6]. Adhesion molecules such as intercellular adhesion molecule 1 (ICAM-1), vascular cell adhesion molecule 1 (VCAM-1), and platelet endothelial cell adhesion molecule, are involved in the recruitment of monocytes/macrophages to the inflamed sites in the vascular tissue [7]. The expression of cell adhesion molecules, such as VCAM-1, represents one of earliest pathological changes in vascular inflammation diseases such as atherosclerosis [2]. Atherosclerosis is a chronic inflammatory disease of the vascular tissue that is largely driven by an innate immune response from the macrophages [8]. Atherosclerosis is characterized by lipid accumulation and inflammatory infiltration of the arterial walls [9]. The accumulation of a lipid plaque and lipid-forming macrophage foam cells in the intima of the inflamed artery has been recognized as a hallmark of atherosclerosis [10]. Macrophages actively contribute in vascular inflammation by secreting proinflammatory cytokines, such as tumor necrosis factor (TNF)-alpha [11]. There is increasing evidence that TNF-blocking agents including TNF receptor blockade have successfully been used to treat systemic inflammatory disorders, such as rheumatoid arthritis [12]. A recent interesting study evaluated the inhibition of inflammatory cytokines for treating atherothrombosis [13], suggesting that cytokine inhibition can help resolve inflammation and maintain homeostasis, and is thus is crucial for atheroprotection. Because cholesterol is a key component of arterial plaques, a detailed understanding of the cholesterol transport system can lead to approaches that help to lower the risk of atherosclerosis. Intracellular cholesterol can be exported through cholesterol transporters. Macrophage cholesterol efflux depends on the ATP-binding cassette transporters ABCA1 or ABCG1 [14]. The combined efficiency of ABCA1 and ABCG1 promotes foam cell accumulation by inhibiting macrophage cholesterol efflux and accelerates atherosclerosis in mice [15,16] suggesting a target for atherosclerotic cardiovascular diseases. A new target molecule capable of efficiently monitoring vascular inflammation, extracellularly secreted as needed to act as a biomarker, and able to control vascular inflammation including sepsis or cytokine storms, is required. Here, we introduce APE1/Ref-1 as a potential new target capable of meeting these demands.

2. APE1/Ref-1 Protein Has Several Cellular Functions

Is there an endogenous system that can participate in circulating immune surveillance or managing the balance in homeostasis? The molecule that can act in circulatory surveillance is a functional protein, which can recognize the DNA damage, and is sensitive to their redox status and their existence in the biological fluids. To date, the cellular localization of APE1/Ref-1 exhibits three types—nuclear, cytoplasmic/mitochondrial, and secretory. Under basal conditions, APE1/Ref-1 is localized in the nucleus, and its localization is dynamically regulated, resulting in its cytoplasmic/mitochondrial translocation or extracellular secretion [17]. Overexpression of APE1/Ref-1 is inhibited by TNF-α-induced endothelial cell activation in cultured endothelial cells [18]. In contrast, heterozygous APE1/Ref-1 (+/−) mice showed endothelial dysfunction and hypertension [19], suggesting an important role for APE1/Ref-1 in endothelial functions. Conventional knockout of APE1/Ref-1 causes early embryonic lethality on embryonic day 5 to E9 [20,21]. Therefore, it is difficult to evaluate the biological function or phenotype changes in homozygous APE1/Ref-1-knockout mice. A recent study showed that secretory APE1/Ref-1 inhibited proinflammatory cytokines and inflammation in lipopolysaccharide-treated mice [22]. For approximately 20 decades, extranuclear functions in systemic inflammation and endothelial activation as well as basic nuclear functions in DNA basic repair and genomic stability have been revealed (Figure 1).

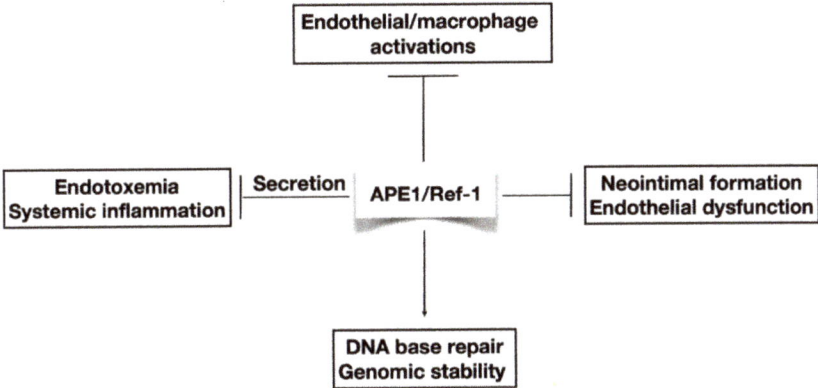

Figure 1. The role of apurinic/apyrimidinic endonuclease 1/redox factor-1 (APE1/Ref-1) in endothelial activation and systemic inflammation. Heterozygous APE1/Ref-1 mice showed endothelial dysfunction and hypertension [19]; gene transfer of APE1/Ref-1 inhibited neointimal formation of rat carotid arteries and inhibited endothelial activation in endothelial cells [18,23]. The secretory APE1/Ref-1 inhibited proinflammatory cytokines and inflammation in lipopolysaccharide-treated mice [22]. APE1/Ref-1 functions in DNA base repair and is essential for genomic stability. The arrow and T-bar represent activated and inhibitory interactions, respectively.

2.1. Nuclear Function of APE1/Ref-1

The primary subcellular localization of APE1/Ref-1 is in the nucleus in most cells or tissues [24]. This appears to be because of its fundamental activity in the base excision repair pathway of DNA lesions. APE1/Ref-1 hydrolyzes the DNA adjacent to the 5′-end of an apurinic/apyrimidinic site to produce a nick with a 3′-hydroxyl group and a 5′-deoxyribose phosphate group like a skilled nucleic acid surgeon [25]. The APE1/Ref-1-deficient cells show hypersensitivity to DNA damaging agents [26,27]. APE1/Ref-1 also regulates the redox activity of several transcription factors such as activator protein-1 (AP-1) and nuclear factor kappa B (NF-κB) [17]. The formation of disulfide bonds in APE1/Ref-1 is important in redox activity with cysteine residues C65 and C93 playing key roles in the thiol-mediated redox reactions [28,29]. The calcification of vascular smooth muscle cells is strongly correlated with intracellular ROS production and apoptosis [30]. Recently, Lee et al. showed that the redox function of APE1/Ref-1 prevents inorganic phosphate-induced calcification of vascular smooth muscle cells by inhibiting oxidative stress and osteoblastic differentiation [31]. As the overexpression of APE1/Ref-1 inhibits endothelial apoptosis, the redox-sensitive APE1/Ref-1 plays a critical role in endothelial cell survival in response to inflammatory cytokines including tumor necrosis factor-alpha [32].

2.2. Cytoplasmic Function of APE1/Ref-1

APE1/Ref-1 has also been detected in other areas in addition to the nucleus; cytoplasmic and mitochondrial APE1/Ref-1 have also been reported [33,34]. Cytoplasmic overexpression of APE1/Ref-1 is attenuated by the upregulation of high-mobility group box1 (HMGB-1)-mediated ROS generation, cytokine secretion, and cyclooxygenase-2 expression in macrophage cells [35]. S-Nitrosoglutathion (GSNO), a nitric oxide donor, induces the nuclear export of APE1/Ref-1 in a chromosome-region maintenance-1 (exportin-1)-independent manner [36]. This nuclear-cytoplasmic translocation of APE1/Ref-1 is dependent on the nitrosation at the target sites Cys93 and Cys310 in APE1/Ref-1. The N-terminal of 20 amino acids of APE1/Ref-1 includes the nuclear localization signal, as the cytoplasmic proportion of APE1/Ref-1 increased with the deletion of the N-terminal of 20–35 amino acids [19,29,36]. APE1/Ref-1 also contains a potential nuclear export sequences (NES) at amino acids

64-80. A deletion mutant of APE1/Ref-1 (60–80) showed a slight interference with cell viability, suggesting the important role of the cytoplasmic localization of APE1/Ref-1 in cell viability [36]. Additionally, the cytoplasmic expression of APE1/Ref-1 has antioxidant and anti-inflammatory functions in astrocyte or endothelial cells [29,37]. Hypoxia resulted in a significant decrease in APE1/Ref-1 expression in human umbilical vein endothelial cells [38]. A novel extranuclear function of APE1/Ref-1 in endothelial oxidative stress and apoptosis is that it protects against hypoxia-reoxygenation-induced apoptosis by modulating cytoplasmic rac-1-regulated ROS generation [39]. Recently, Hao et al. reported that APE1/Ref-1 overexpression inhibited hypoxia-reoxygenation, which induced an increase in ROS and NADPH oxidase expression and inhibited the mitochondrial dysfunction in H9c2 cardiomyocytes [38].

Endothelial mitochondria are a critical target of oxidative stress and DNA damage, and thus play a crucial role in the signaling during cellular responses [40]. Phorbol 12-myristate 13-acetate (PMA), an activator of protein kinase C, induces ROS generation and increases mitochondrial translocation of APE1/Ref-1 [41]. Moreover, the overexpression of APE1/Ref-1 suppresses PMA-induced mitochondrial dysfunction. In contrast, the gene silencing of APE1/Ref-1 increases the sensitivity of mitochondrial dysfunction, suggesting that the mitochondrial APE1/Ref-1 contributes to the protective role of protein kinase C-induced mitochondrial dysfunction in endothelial cells [41]. Mitochondrial APE1/Ref-1 is also involved in repairing mitochondrial DNA lesions caused by oxidative and alkylating agents [42]. APE1/Ref-1 interacts with the mitochondrial import and assembly protein Mia40, which is responsible for APE1/Ref-1 trafficking into the mitochondria [42]. A recent study using haploinsufficient APE1/Ref-1 mice revealed slower repair kinetics of azoxymethane-induced mitochondrial DNA damage, suggesting that APE1/Ref-1 is important for preventing changes in mitochondrial DNA integrity during azoxymethane-induced colorectal cancer [43].

2.3. Extracellular Function of APE1/Ref-1

Mammalian cells may secrete several types of cellular proteins. In 2013, the secretion of APE1/Ref-1 into the cultured medium in response to hyperacetylation [44] and the presence of plasma APE1/Ref-1 in lipopolysaccharide-induced endotoxemic mice were first reported [45]. Thus, secreted APE1/Ref-1 protein likely has a distinct function. It is thought that the fundamental function of an intracellular protein is performed even when the protein is secreted from the cells. The cysteine residues of APE1/Ref-1 have a reducing activity for the redox regulation of target proteins [46]. Nath et al. reported that the extracellular APE1/Ref-1 induces the production and secretion of the proinflammatory cytokine IL-6 and extracellular APE1/Ref-1 treatment activates the transcriptional factor NF-κB [47]. In contrast, the anti-inflammatory activities of secreted APE1/Ref-1 have been reported, which is thought to be exerted by the reducing activity of APE1/Ref-1 via thiol exchanges in the extracellular domain of cytokine receptors [48]. Recently, Joo et al. demonstrated the in vivo activity of extracellularly secreted APE1/Ref-1, which exerts inhibitory effects on lipopolysaccharide (LPS)-induced inflammation and has a potential for treating LPS-induced endotoxemia or systemic inflammation such as cytokine storms [22]. Under endotoxemic conditions, multiple organ failure is caused by uncontrolled inflammatory responses such as cytokine storms or cytokine overproduction [49]. Interestingly, the secreted APE1/Ref-1 inhibited the LPS-induced proinflammatory mediators such as TNF-α, IL-1β, and IL-6, and chemotactic cytokines such as monocyte chemoattractant protein-1 (MCP-1), suggesting that the secretory APE1/Ref-1 inhibits LPS-induced cytokine production [22]. Reports of the extracellular secretions of APE1/Ref-1 have shown consistent results but have not agreed on the extracellular functions. Taken together, the anti-inflammatory effects of secretory APE1/Ref-1 in vivo as well as the therapeutic potential of recombinant APE1/Ref-1 protein in endotoxemic or inflammatory conditions have been suggested (Figure 2). The diverse biological functions of APE1/Ref-1 according to its subcellular localization are summarized in Table 1.

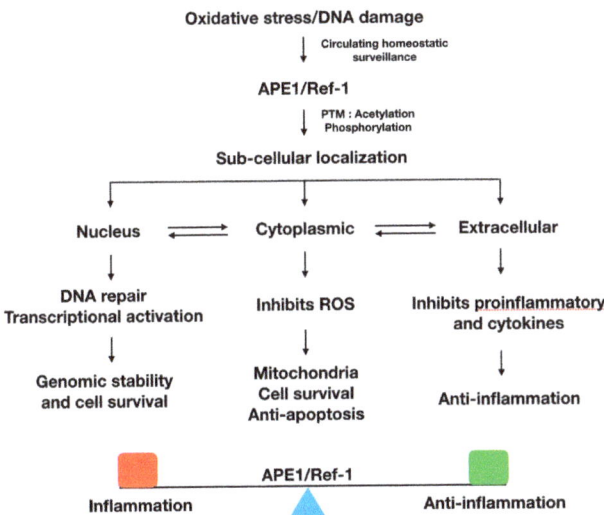

Figure 2. Flowchart model of APE1/Ref-1 and its subcellular localization and functions in response to oxidative stress and DNA damage. APE1/Ref-1 carries out circulating homeostatic surveillance in the human body by recognizing the cellular changes in response to oxidative stress or DNA damage. Subcellular localization of APE1/Ref-1 can be determined by post-translational modification including redox change, acetylation, phosphorylation, nitrosation, etc. Overall, APE1/Ref-1 is involved in DNA base repair and the modulating transcriptional factors, resulting in genomic stability or cell survival. Under basal conditions, APE1/Ref-1 is localized in the nucleus; its localization is dynamically regulated, which results in cytoplasmic/mitochondrial translocation or extracellular secretion.

Table 1. Summary of functions of APE1/Ref-1 according to subcellular localization.

APE1/Ref-1	Tissue/Cells	Functions	Reference
Intracellular	Endothelial cells	Inhibits endothelial dysfunction Inhibits cellular ROS and increases NO production Inhibits NF-kB and apoptosis Inhibits VCAM-1 expression	[18,19,23,29,50,51]
	Endothelial mitochondria	Inhibit mitochondrial dysfunctions Inhibits mitochondrial ROS	[41,52]
		Inhibits p66shc activation Mitochondrial DNA repair	[42]
	A549 cells	Inhibits COX-2 expression Inhibits p38 MAPK	[53]
	Vascular smooth muscle cells	Inhibits Pi-induced calcification Inhibits osteoblastic phenotype changes	[31]
Cytoplasmic	Endothelial cells	Inhibits rac1 or NADPH oxidase	[29,39]
	Glial cells	Inhibits neuroinflammatory response	[37]
	THP-1 cells	Inhibits HMGB1-mediated ROS and cytokines	[35]
Extracellular	HEK293 cells	Trichostatin A induced APE1/Ref-1 secretion	[44]
	MDA-MD-231 cells	Ac-APE1/Ref-1 induces apoptosis	[54]
	Endothelial cells	Inhibits VCAM-1 expression	[48]
		Inhibits COX-2 expression	[22]

3. Mechanism of APE1/Ref-1 Secretion

There are two possible mechanisms for the extracellular secretion of the APE1/Ref-1 protein—active secretion and passive release. APE1/Ref-1 is actively secreted by inflammatory cells such as macrophages or monocytes and endothelial cells in response to hyperacetylation signals [44]. However, different exogenous stimuli such as trichostatin A, LPS, testosterone, and coxsackievirus B3 can induce the secretion of APE1/Ref-1 [45,55–57]. Intracellular hyperacetylation conditions may be important intracellular signals for the secretion of APE1/Ref-1 in normal or tumor cells [48,54,58].

Until now, this active secretion of APE1/Ref-1 has been known to be initiated by transporter and vesicle formation; it is mediated by a nonclassical transport pathway (Figure 3). As evidence of this, brefeldin A, an inhibitor of the endoplasmic reticulum-to-Golgi classical transport pathway, did not affect APE1/Ref-1 secretion [57]. Active secretion of APE1/Ref-1 is not be involved in the classical endoplasmic reticulum-to-Golgi complex secretory pathway because of the absence of a leader peptide sequence. Trichostatin A-mediated acetylation was shown to cause post-translational modification of APE1/Ref-1 (including Lys 6 and Lys 7 of APE1/Ref-1) [59]. This acetylation reduces the net charge and increases the hydrophobicity of APE1/Ref-1, leading to cytoplasmic localization and secretion. Additionally, trichostatin A did not induce the secretion of lysine-mutated APE1/Ref-1 (K6R/K7R) [44]. Pharmacological inhibition by probenecid and glyburide on acetylation-induced APE1/Ref-1 secretion suggested the possible involvement of ABC transporters [57]. In a human monocyte cell line, APE1/Ref-1 was secreted from the monocytes upon inflammatory challenges via extracellular vesicle-mediated secretion pathways [47]. There is an interesting report describing vesicle formation in the release of APE1/Ref-1 in breast tumor cell lines. Hyperacetylated MDA-MD-231 cells, which were stimulated with aspirin, released vesicles containing APE1/Ref-1 according to analysis using gold particle-labelled APE1/Ref-1 [54]. Further research is required to determine the molecular mechanism of APE1/Ref-1 secretion and if this mechanism is dependent on the cell type or endogenous stimuli. Extracellular APE1/Ref-1 may be passively released following endogenous cell damage or from necrotic cells. In necrotic or apoptotic cells, APE1/Ref-1 may be released into the cultured medium from the cytoplasm or nucleus, like HMGB-1 [60]. Therefore, the secreted APE1/Ref-1 in the extracellular milieu may be considered as a cell death marker and/or a serologic biomarker of certain disorders.

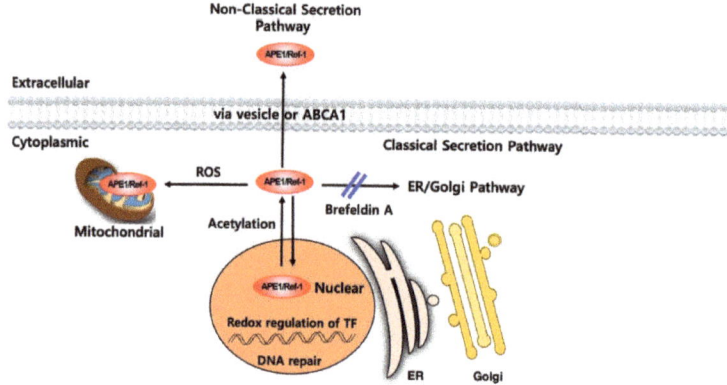

Figure 3. Proposed secretion mechanism of APE1/Ref-1. APE1/Ref-1 is mainly localized in the nucleus, which is dynamically regulated between the cytoplasm or mitochondria. Also, APE1/Ref-1 may be secreted in response to acetylation. Its secretion is not inhibited by brefeldin A, an inhibitor of the ER/Golgi pathway ('double slash' in blue), suggesting a nonclassical secretion pathway. Active secretion of APE1/Ref-1 may be mediated by the ABCA1 transporter or vesicle formation [54,57].

4. Extracellular APE1/Ref-1 as a Serological Biomarker

Since the concept of APE1/Ref-1 secretion was established in 2013, several studies have demonstrated the usefulness of APE1/Ref-1 as a serological biomarker for cardiovascular disorders and tumors (Table 2). Park et al. first reported APE1/Ref-1 in the plasma of endotoxemic rats as a 37 kDa immunoreactive band, suggesting that plasma APE1/Ref-1 is a useful biomarker for endotoxemia [45]. Jin et al. found that serum APE1/Ref-1 levels were elevated in the patients with coronary artery disease and were higher in myocadiac infarction than in angina in a study of clinical biomarkers [61]. Myocarditis is an inflammatory disease of the myocardium that causes cardiogenic shock, heart failure, and sudden death [62]. Myocarditis can only be diagnosed by endomyocardial biopsy [63]. Jin et al. reported that serum APE1/Ref-1 was elevated in experimental murine myocarditis; compared to N-terminal pro-B-type natriuretic peptide and troponin I, serum APE1/Ref-1 was more closely related to myocardial inflammation, reflecting the severity of myocardial injury in viral myocarditis without endomyocardial biopsy [55].

Vascular inflammation in the tumor microenvironment is associated with tumor angiogenesis or tumor metastasis [64]. In cancer research, the changes in the intracellular localization of APE1/Ref-1 in tissues have gained attention, as they are related to the prognosis of certain tumors. Overexpression of APE1/Ref-1 that is observed in tumor cells is associated with drug resistance of anticancer drugs and poor survival [65]. Moreover, gene silencing or the inhibition of redox activity of APE1/Ref-1 results in reduced drug resistance to anticancer drugs [66]. Therefore, APE1/Ref-1 is a target protein for tumor treatment. Recently, the usefulness of APE1/Ref-1 as a biomarker in various cancers has been demonstrated. Plasma or urine APE1/Ref-1 levels are significantly increased in patients with bladder cancer; area under the curve analysis revealed the diagnostic values of APE1/Ref-1 with high specificity and sensitivity [67,68]. There is increasing evidence for the role of serum APE1/Ref-1 as a new diagnostic biomarker for hepatocellular carcinoma [69], renal cell carcinoma and hepatobiliary carcinoma [70], cholangiocarcinoma [71], non-small cell lung cancer [72], and gastric cancer [73] as shown in Table 2.

Table 2. Summary of usefulness of APE1/Ref-1 as potential biomarker in vascular inflammatory disease or tumors.

Diseases	Clinical Significance	Patients (n)	Control (n)	Sensitivity (%)	Specificity (%)	AUC or 95% CI	Reference
Liposaccharide-induced endotoxemia (Preclinical study)	APE1/Ref-1 is elevated in plasma of lipopolysaccharide (LPS)-treated mice and reached a maximum at 12 h after injection of LPS.	-	-	NA	NA		[45]
Viral myocarditis (Preclinical study)	Serum APE1/Ref-1 is increased in coxsackievirus-induced myocarditis and is well-correlated with the degree of myocardial inflammation. Serum APE1/Ref-1 is useful for myocardial injury in viral myocarditis without endomyocardial biopsy	-	-	NA	NA		[55]
Coronary arterial diseases	Serum APE1/Ref-1 level was higher in coronary arterial diseases, which higher in myocardial infarction than angina	303	57	36	95	0.66	[61]
Bladder cancer	Urinary APE1/Ref-1 is increased in bladder cancer and it correlated with tumor grade and stage	169	108	82	80	0.83	[67]
Oral squamous cell carcinoma	Serum APE1/Ref-1 is a novel potential diagnostic biomarker of oral cancer and can reflect the treatment responses	58	40	67	87	0.80	[74]

Table 2. *Cont.*

Diseases	Clinical Significance	Patients (n)	Control (n)	Sensitivity (%)	Specificity (%)	AUC or 95% CI	Reference
Cholangiocarcinoma	Serum APE1/Ref-1 level is a potential diagnostic marker of cholangiocarcinoma and cytoplasmic expression in cancer cells predicts relapses	46	39	73.9	97.4	0.709–0.886	[71]
Hepatocellular carcinoma	Serum APE1/Ref-1 may be considered as a promising diagnostic biomarker for hepatocellular carcinoma	99	100	98	83	0.98	[69]
Renal cell carcinoma	Serum APE1/Ref-1 level may be a diagnostic markers of renal cell carcinoma	40	39	82.5	97.4	0.862–0.981	[70]
Non-small cell lung cancer	Serum APE1/Ref-1 is a biomarker for predicting prognosis and therapeutic efficacy in nonsmall cell lung cancer and post-treatment high serum APE1/Ref-1 level was associated with poor survival.	200	200	55.6	70.8	0.653	[72]
Gastric cancer (lymph node positive and negative)	Serum APE1/Ref-1 is a valuable marker for prediction of lymph node metastasis in patients with gastric cancer	52	35	49	85.7	0.666	[73]

5. Conclusions

In conclusion, APE1/Ref-1 has several cellular functions with an important role in DNA repair and redox regulation. In addition to the intracellular function of APE1/Ref-1, its extracellular function should be evaluated to develop therapeutic strategies. Recombinant APE1/Ref-1 protein including modifications may be used for circulating homeostatic surveillance, alone or in combination with treatment regimens, against endothelial dysfunction, inflammatory disorders or sepsis. Proteomic analysis of post-translational modifications of the APE1/Ref-1 protein in biological samples would improve the understanding of the diversity of APE1/Ref-1 function. Clinical studies of APE1/Ref-1 analysis as a biomarker in human samples will help in the diagnosis and follow-up of cardiovascular disorders, including coronary artery disease.

Author Contributions: Y.R.L. and H.K.J. wrote the manuscript and drew the figure and table. B.H.J. designed the contents, wrote and revised the manuscript, and approved the final version. All authors have read and agreed to the published version of the manuscript.

Funding: This research was supported by the Basic Science Research Program through the National Research Foundation of Korea (NRF) and funded by the Ministry of Education (NRF-2014R1A6A1029617 to B.H.J., 2016R1A6A3A11932015 to H.K.J., 2017R1A6A3A11027834 to Y.R.L.)

Conflicts of Interest: The authors declare no conflict of interest. The funders had no role in the design of the study; in the collection, analyses, or interpretation of data; in the writing of the manuscript, or in the decision to publish the results.

References

1. Liao, J.K. Linking endothelial dysfunction with endothelial cell activation. *J. Clin. Investig.* **2013**, *123*, 540–541. [CrossRef]
2. Ross, R. Atherosclerosis–an inflammatory disease. *N. Engl. J. Med.* **1999**, *340*, 115–126. [CrossRef]
3. Medzhitov, R. Inflammation 2010: New adventures of an old flame. *Cell* **2010**, *140*, 771–776. [CrossRef]
4. Irani, K. Oxidant signaling in vascular cell growth, death, and survival: A review of the roles of reactive oxygen species in smooth muscle and endothelial cell mitogenic and apoptotic signaling. *Circ. Res.* **2000**, *87*, 179–183. [CrossRef]
5. Gonzalez-Amaro, R.; Diaz-Gonzalez, F.; Sanchez-Madrid, F. Adhesion molecules in inflammatory diseases. *Drugs* **1998**, *56*, 977–988. [CrossRef]

6. Zundler, S.; Becker, E.; Weidinger, C.; Siegmund, B. Anti-Adhesion Therapies in Inflammatory Bowel Disease-Molecular and Clinical Aspects. *Front. Immunol.* **2017**, *8*, 891. [CrossRef]
7. Galkina, E.; Ley, K. Vascular adhesion molecules in atherosclerosis. *Arter. Thromb. Vasc. Biol.* **2007**, *27*, 2292–2301. [CrossRef] [PubMed]
8. Galkina, E.; Ley, K. Immune and inflammatory mechanisms of atherosclerosis. *Annu. Rev. Immunol.* **2009**, *27*, 165–197. [CrossRef] [PubMed]
9. Singh, R.B.; Mengi, S.A.; Xu, Y.J.; Arneja, A.S.; Dhalla, N.S. Pathogenesis of atherosclerosis: A multifactorial process. *Exp. Clin. Cardiol.* **2002**, *7*, 40–53. [PubMed]
10. Yu, X.H.; Fu, Y.C.; Zhang, D.W.; Yin, K.; Tang, C.K. Foam cells in atherosclerosis. *Clin. Chim. Acta* **2013**, *424*, 245–252. [CrossRef]
11. Ohta, H.; Wada, H.; Niwa, T.; Kirii, H.; Iwamoto, N.; Fujii, H.; Saito, K.; Sekikawa, K.; Seishima, M. Disruption of tumor necrosis factor-alpha gene diminishes the development of atherosclerosis in ApoE-deficient mice. *Atherosclerosis* **2005**, *180*, 11–17. [CrossRef] [PubMed]
12. Bluml, S.; Scheinecker, C.; Smolen, J.S.; Redlich, K. Targeting TNF receptors in rheumatoid arthritis. *Int. Immunol.* **2012**, *24*, 275–281. [CrossRef] [PubMed]
13. Ridker, P.M. Anticytokine Agents: Targeting Interleukin Signaling Pathways for the Treatment of Atherothrombosis. *Circ. Res.* **2019**, *124*, 437–450. [CrossRef] [PubMed]
14. Yvan-Charvet, L.; Wang, N.; Tall, A.R. Role of HDL, ABCA1, and ABCG1 transporters in cholesterol efflux and immune responses. *Arter. Thromb. Vasc. Biol.* **2010**, *30*, 139–143. [CrossRef] [PubMed]
15. Yvan-Charvet, L.; Ranalletta, M.; Wang, N.; Han, S.; Terasaka, N.; Li, R.; Welch, C.; Tall, A.R. Combined deficiency of ABCA1 and ABCG1 promotes foam cell accumulation and accelerates atherosclerosis in mice. *J. Clin. Investig.* **2007**, *117*, 3900–3908. [CrossRef] [PubMed]
16. Westerterp, M.; Tsuchiya, K.; Tattersall, I.W.; Fotakis, P.; Bochem, A.E.; Molusky, M.M.; Ntonga, V.; Abramowicz, S.; Parks, J.S.; Welch, C.L.; et al. Deficiency of ATP-Binding Cassette Transporters A1 and G1 in Endothelial Cells Accelerates Atherosclerosis in Mice. *Arter. Thromb. Vasc. Biol.* **2016**, *36*, 1328–1337. [CrossRef]
17. Choi, S.; Joo, H.K.; Jeon, B.H. Dynamic Regulation of APE1/Ref-1 as a Therapeutic Target Protein. *Chonnam Med. J.* **2016**, *52*, 75–80. [CrossRef]
18. Kim, C.S.; Son, S.J.; Kim, E.K.; Kim, S.N.; Yoo, D.G.; Kim, H.S.; Ryoo, S.W.; Lee, S.D.; Irani, K.; Jeon, B.H. Apurinic/apyrimidinic endonuclease1/redox factor-1 inhibits monocyte adhesion in endothelial cells. *Cardiovasc. Res.* **2006**, *69*, 520–526. [CrossRef]
19. Jeon, B.H.; Gupta, G.; Park, Y.C.; Qi, B.; Haile, A.; Khanday, F.A.; Liu, Y.X.; Kim, J.M.; Ozaki, M.; White, A.R.; et al. Apurinic/apyrimidinic endonuclease 1 regulates endothelial NO production and vascular tone. *Circ. Res.* **2004**, *95*, 902–910. [CrossRef]
20. Ludwig, D.L.; MacInnes, M.A.; Takiguchi, Y.; Purtymun, P.E.; Henrie, M.; Flannery, M.; Meneses, J.; Pedersen, R.A.; Chen, D.J. A murine AP-endonuclease gene-targeted deficiency with post-implantation embryonic progression and ionizing radiation sensitivity. *Mutat. Res.* **1998**, *409*, 17–29. [CrossRef]
21. Xanthoudakis, S.; Smeyne, R.J.; Wallace, J.D.; Curran, T. The redox/DNA repair protein, Ref-1, is essential for early embryonic development in mice. *Proc. Natl. Acad. Sci. USA* **1996**, *93*, 8919–8923. [CrossRef] [PubMed]
22. Joo, H.K.; Lee, Y.R.; Lee, E.O.; Park, M.S.; Choi, S.; Kim, C.S.; Park, J.B.; Jeon, B.H. The extracellular role of Ref-1 as anti-inflammatory function in lipopolysaccharide-induced septic mice. *Free Radic. Biol. Med.* **2019**, *139*, 16–23. [CrossRef] [PubMed]
23. Lee, H.M.; Jeon, B.H.; Won, K.J.; Lee, C.K.; Park, T.K.; Choi, W.S.; Bae, Y.M.; Kim, H.S.; Lee, S.K.; Park, S.H.; et al. Gene transfer of redox factor-1 inhibits neointimal formation: Involvement of platelet-derived growth factor-beta receptor signaling via the inhibition of the reactive oxygen species-mediated Syk pathway. *Circ. Res.* **2009**, *104*, 219–227. [CrossRef] [PubMed]
24. Kakolyris, S.; Kaklamanis, L.; Giatromanolaki, A.; Koukourakis, M.; Hickson, I.D.; Barzilay, G.; Turley, H.; Leek, R.D.; Kanavaros, P.; Georgoulias, V.; et al. Expression and subcellular localization of human AP endonuclease 1 (HAP1/Ref-1) protein: A basis for its role in human disease. *Histopathology* **1998**, *33*, 561–569. [CrossRef] [PubMed]
25. Whitaker, A.M.; Freudenthal, B.D. APE1: A skilled nucleic acid surgeon. *DNA Repair* **2018**, *71*, 93–100. [CrossRef] [PubMed]

26. Wang, D.; Luo, M.; Kelley, M.R. Human apurinic endonuclease 1 (APE1) expression and prognostic significance in osteosarcoma: Enhanced sensitivity of osteosarcoma to DNA damaging agents using silencing RNA APE1 expression inhibition. *Mol. Cancer* **2004**, *3*, 679–686.
27. Luo, M.; Kelley, M.R. Inhibition of the human apurinic/apyrimidinic endonuclease (APE1) repair activity and sensitization of breast cancer cells to DNA alkylating agents with lucanthone. *Anticancer Res.* **2004**, *24*, 2127–2134.
28. Luo, M.; Zhang, J.; He, H.; Su, D.; Chen, Q.; Gross, M.L.; Kelley, M.R.; Georgiadis, M.M. Characterization of the redox activity and disulfide bond formation in apurinic/apyrimidinic endonuclease. *Biochemistry* **2012**, *51*, 695–705. [CrossRef]
29. Park, M.S.; Kim, C.S.; Joo, H.K.; Lee, Y.R.; Kang, G.; Kim, S.J.; Choi, S.; Lee, S.D.; Park, J.B.; Jeon, B.H. Cytoplasmic localization and redox cysteine residue of APE1/Ref-1 are associated with its anti-inflammatory activity in cultured endothelial cells. *Mol. Cells* **2013**, *36*, 439–445. [CrossRef]
30. Proudfoot, D.; Skepper, J.N.; Hegyi, L.; Bennett, M.R.; Shanahan, C.M.; Weissberg, P.L. Apoptosis regulates human vascular calcification in vitro: Evidence for initiation of vascular calcification by apoptotic bodies. *Circ. Res.* **2000**, *87*, 1055–1062. [CrossRef]
31. Lee, K.M.; Lee, E.O.; Lee, Y.R.; Joo, H.K.; Park, M.S.; Kim, C.S.; Choi, S.; Jeong, J.O.; Jeon, B.H. APE1/Ref-1 Inhibits Phosphate-Induced Calcification and Osteoblastic Phenotype Changes in Vascular Smooth Muscle Cells. *Int. J. Mol. Sci.* **2017**, *18*, 2053. [CrossRef] [PubMed]
32. Hall, J.L.; Wang, X.; Van, A.; Zhao, Y.; Gibbons, G.H. Overexpression of Ref-1 Inhibits Hypoxia and Tumor Necrosis Factor–Induced Endothelial Cell Apoptosis Through Nuclear Factor-kappab-independent and -dependent pathway. *Circ. Res.* **2001**, *88*, 1247–1253. [CrossRef] [PubMed]
33. Tell, G.; Damante, G.; Caldwell, D.; Kelley, M.R. The intracellular localization of APE1/Ref-1: More than a passive phenomenon? *Antioxid. Redox Signal.* **2005**, *7*, 367–384. [CrossRef] [PubMed]
34. Tell, G.; Crivellato, E.; Pines, A.; Paron, I.; Pucillo, C.; Manzini, G.; Bandiera, A.; Kelley, M.R.; Di Loreto, C.; Damante, G. Mitochondrial localization of APE/Ref-1 in thyroid cells. *Mutat. Res.* **2001**, *485*, 143–152. [CrossRef]
35. Yuk, J.M.; Yang, C.S.; Shin, D.M.; Kim, K.K.; Lee, S.K.; Song, Y.J.; Lee, H.M.; Cho, C.H.; Jeon, B.H.; Jo, E.K. A dual regulatory role of apurinic/apyrimidinic endonuclease 1/redox factor-1 in HMGB1-induced inflammatory responses. *Antioxid. Redox Signal.* **2009**, *11*, 575–588. [CrossRef]
36. Qu, J.; Liu, G.H.; Huang, B.; Chen, C. Nitric oxide controls nuclear export of APE/Ref-1 through S-nitrosation of cysteines 93 and 310. *Nucleic Acids Res.* **2007**, *35*, 2522–2532. [CrossRef]
37. Baek, H.; Lim, C.S.; Byun, H.S.; Cho, H.S.; Lee, Y.R.; Shin, Y.S.; Kim, H.W.; Jeon, B.H.; Kim, D.W.; Hong, J.; et al. The anti-inflammatory role of extranuclear apurinic/apyrimidinic endonuclease 1/redox effector factor-1 in reactive astrocytes. *Mol. Brain* **2016**, *9*, 99. [CrossRef]
38. Hao, J.; Du, H.; Liu, F.; Lu, J.C.; Yang, X.C.; Cui, W. Apurinic/apyrimidinic endonuclease/redox factor 1 (APE1) alleviates myocardial hypoxia-reoxygenation injury by inhibiting oxidative stress and ameliorating mitochondrial dysfunction. *Exp. Med.* **2019**, *17*, 2143–2151. [CrossRef]
39. Angkeow, P.; Deshpande, S.S.; Qi, B.; Liu, Y.X.; Park, Y.C.; Jeon, B.H.; Ozaki, M.; Irani, K. Redox factor-1: An extra-nuclear role in the regulation of endothelial oxidative stress and apoptosis. *Cell Death Differ.* **2002**, *9*, 717–725. [CrossRef]
40. Davidson, S.M.; Duchen, M.R. Endothelial mitochondria: Contributing to vascular function and disease. *Circ. Res.* **2007**, *100*, 1128–1141. [CrossRef]
41. Joo, H.K.; Lee, Y.R.; Park, M.S.; Choi, S.; Park, K.; Lee, S.K.; Kim, C.S.; Park, J.B.; Jeon, B.H. Mitochondrial APE1/Ref-1 suppressed protein kinase C-induced mitochondrial dysfunction in mouse endothelial cells. *Mitochondrion* **2014**, *17*, 42–49. [CrossRef] [PubMed]
42. Barchiesi, A.; Wasilewski, M.; Chacinska, A.; Tell, G.; Vascotto, C. Mitochondrial translocation of APE1 relies on the MIA pathway. *Nucleic Acids Res.* **2015**, *43*, 5451–5464. [CrossRef] [PubMed]
43. Ballista-Hernandez, J.; Martinez-Ferrer, M.; Velez, R.; Climent, C.; Sanchez-Vazquez, M.M.; Torres, C.; Rodriguez-Munoz, A.; Ayala-Pena, S.; Torres-Ramos, C.A. Mitochondrial DNA Integrity Is Maintained by APE1 in Carcinogen-Induced Colorectal Cancer. *Mol. Cancer Res.* **2017**, *15*, 831–841. [CrossRef] [PubMed]
44. Choi, S.; Lee, Y.R.; Park, M.S.; Joo, H.K.; Cho, E.J.; Kim, H.S.; Kim, C.S.; Park, J.B.; Irani, K.; Jeon, B.H. Histone deacetylases inhibitor trichostatin A modulates the extracellular release of APE1/Ref-1. *Biochem. Biophys. Res. Commun.* **2013**, *435*, 403–407. [CrossRef]

45. Park, M.S.; Lee, Y.R.; Choi, S.; Joo, H.K.; Cho, E.J.; Kim, C.S.; Park, J.B.; Jo, E.K.; Jeon, B.H. Identification of plasma APE1/Ref-1 in lipopolysaccharide-induced endotoxemic rats: Implication of serological biomarker for an endotoxemia. *Biochem. Biophys. Res. Commun.* **2013**, *435*, 621–626. [CrossRef]
46. Ordway, J.M.; Eberhart, D.; Curran, T. Cysteine 64 of Ref-1 is not essential for redox regulation of AP-1 DNA binding. *Mol. Cell Biol.* **2003**, *23*, 4257–4266. [CrossRef]
47. Nath, S.; Roychoudhury, S.; Kling, M.J.; Song, H.; Biswas, P.; Shukla, A.; Band, H.; Joshi, S.; Bhakat, K.K. The extracellular role of DNA damage repair protein APE1 in regulation of IL-6 expression. *Cell. Signal.* **2017**, *39*, 18–31. [CrossRef]
48. Park, M.S.; Choi, S.; Lee, Y.R.; Joo, H.K.; Kang, G.; Kim, C.S.; Kim, S.J.; Lee, S.D.; Jeon, B.H. Secreted APE1/Ref-1 inhibits TNF-alpha-stimulated endothelial inflammation via thiol-disulfide exchange in TNF receptor. *Sci. Rep.* **2016**, *6*, 23015. [CrossRef]
49. Cinel, I.; Opal, S.M. Molecular biology of inflammation and sepsis: A primer. *Crit. Care Med.* **2009**, *37*, 291–304. [CrossRef]
50. Lee, K.H.; Lee, S.K.; Kim, H.S.; Cho, E.J.; Joo, H.K.; Lee, E.J.; Lee, J.Y.; Park, M.S.; Chang, S.J.; Cho, C.H.; et al. Overexpression of Ref-1 Inhibits Lead-induced Endothelial Cell Death via the Upregulation of Catalase. *Korean J. Physiol. Pharm.* **2009**, *13*, 431–436. [CrossRef]
51. Jung, S.B.; Kim, C.S.; Kim, Y.R.; Naqvi, A.; Yamamori, T.; Kumar, A.; Irani, K. Redox factor-1 anctivates endothelial SIRTUIN1 though reduction of conserved cystein sulfhydryls in its deacetylase domain. *PLoS ONE* **2013**, *8*, e65415. [CrossRef] [PubMed]
52. Lee, S.K.; Chung, J.I.; Park, M.S.; Joo, H.K.; Lee, E.J.; Cho, E.J.; Park, J.B.; Ryoo, S.; Irani, K.; Jeon, B.H. Apurinic/apyrimidinic endonuclease 1 inhibits protein kinase C-mediated p66shc phosphorylation and vasoconstriction. *Cardiovasc. Res.* **2011**, *91*, 502–509. [CrossRef] [PubMed]
53. Yoo, D.G.; Kim, C.S.; Lee, S.K.; Kim, H.S.; Cho, E.J.; Park, M.S.; Lee, S.D.; Park, J.B.; Jeon, B.H. Redox Factor-1 Inhibits Cyclooxygenase-2 Expression via Inhibiting of p38 MAPK in the A549 Cells. *Korean J. Physiol. Pharm.* **2010**, *14*, 139–144. [CrossRef] [PubMed]
54. Lee, Y.R.; Kim, K.M.; Jeon, B.H.; Choi, S. Extracellularly secreted APE1/Ref-1 triggers apoptosis in triple-negative breast cancer cells via RAGE binding, which is mediated through acetylation. *Oncotarget* **2015**, *6*, 23383–23398. [CrossRef] [PubMed]
55. Jin, S.A.; Lim, B.K.; Seo, H.J.; Kim, S.K.; Ahn, K.T.; Jeon, B.H.; Jeong, J.O. Elevation of Serum APE1/Ref-1 in Experimental Murine Myocarditis. *Int. J. Mol. Sci.* **2017**, *18*, 2664. [CrossRef] [PubMed]
56. Lee, Y.R.; Lim, J.S.; Shin, J.H.; Choi, S.; Joo, H.K.; Jeon, B.H. Altered Secretory Activity of APE1/Ref-1 D148E Variants Identified in Human Patients With Bladder Cancer. *Int. Neurourol. J.* **2016**, *20*, S30–S37. [CrossRef]
57. Lee, Y.R.; Joo, H.K.; Lee, E.O.; Cho, H.S.; Choi, S.; Kim, C.S.; Jeon, B.H. ATP Binding Cassette Transporter A1 is Involved in Extracellular Secretion of Acetylated APE1/Ref-1. *Int. J. Mol. Sci.* **2019**, *20*, 3178. [CrossRef]
58. Lee, Y.R.; Park, M.S.; Joo, H.K.; Kim, K.M.; Kim, J.; Jeon, B.H.; Choi, S. Therapeutic positioning of secretory acetylated APE1/Ref-1 requirement for suppression of tumor growth in triple-negative breast cancer in vivo. *Sci. Rep.* **2018**, *8*, 8701. [CrossRef]
59. Bhakat, K.K.; Izumi, T.; Yang, S.H.; Hazra, T.K.; Mitra, S. Role of acetylated human AP-endonuclease (APE1/Ref-1) in regulation of the parathyroid hormone gene. *EMBO J.* **2003**, *22*, 6299–6309. [CrossRef]
60. Scaffidi, P.; Misteli, T.; Bianchi, M.E. Release of chromatin protein HMGB1 by necrotic cells triggers inflammation. *Nature* **2002**, *418*, 191–195. [CrossRef]
61. Jin, S.A.; Seo, H.J.; Kim, S.K.; Lee, Y.R.; Choi, S.; Ahn, K.T.; Kim, J.H.; Park, J.H.; Lee, J.H.; Choi, S.W.; et al. Elevation of the Serum Apurinic/Apyrimidinic Endonuclease 1/Redox Factor-1 in Coronary Artery Disease. *Korean Circ. J.* **2015**, *45*, 364–371. [CrossRef] [PubMed]
62. Cooper, L.T., Jr. Myocarditis. *N. Engl. J. Med.* **2009**, *360*, 1526–1538. [CrossRef] [PubMed]
63. Magnani, J.W.; Dec, G.W. Myocarditis: Current trends in diagnosis and treatment. *Circulation* **2006**, *113*, 876–890. [CrossRef] [PubMed]
64. McDowell, S.A.C.; Quail, D.F. Immunological Regulation of Vascular Inflammation During Cancer Metastasis. *Front. Immunol.* **2019**, *10*, 1984. [CrossRef]
65. Yuan, C.L.; He, F.; Ye, J.Z.; Wu, H.N.; Zhang, J.Y.; Liu, Z.H.; Li, Y.-Q.; Luo, X.-L.; Lin, Y.; Liang, R. APE1 overexpression is associated with poor survival in patients with solid tumors: A meta-analysis. *Oncotarget* **2017**, *8*, 59720–59728. [CrossRef]

66. Fishel, M.L.; Kelley, M.R. The DNA base excision repair protein Ape1/Ref-1 as a therapeutic and chemopreventive target. *Mol. Asp. Med.* **2007**, *28*, 375–395. [CrossRef]
67. Choi, S.; Shin, J.H.; Lee, Y.R.; Joo, H.K.; Song, K.H.; Na, Y.G.; Chang, S.J.; Lim, J.S.; Jeon, B.H. Urinary APE1/Ref-1: A Potential Bladder Cancer Biomarker. *Dis. Markers* **2016**, *2016*, 7276502. [CrossRef]
68. Shin, J.H.; Choi, S.; Lee, Y.R.; Park, M.S.; Na, Y.G.; Irani, K.; Lee, S.D.; Park, J.B.; Kim, J.M.; Lim, J.S.; et al. APE1/Ref-1 as a Serological Biomarker for the Detection of Bladder Cancer. *Cancer Res. Treat.* **2015**, *47*, 823–833. [CrossRef]
69. Pascut, D.; Sukowati, C.H.C.; Antoniali, G.; Mangiapane, G.; Burra, S.; Mascaretti, L.G.; Buonocore, M.R.; Crocè, L.S.; Tiribelli, C.; Tell, G. Serum AP-endonuclease 1 (sAPE1) as novel biomarker for hepatocellular carcinoma. *Oncotarget* **2019**, *10*, 383–394. [CrossRef]
70. Kim, J.M.; Yeo, M.K.; Lim, J.S.; Song, I.S.; Chun, K.; Kim, K.H. APEX1 Expression as a Potential Diagnostic Biomarker of Clear Cell Renal Cell Carcinoma and Hepatobiliary Carcinomas. *J. Clin. Med.* **2019**, *8*, 1151. [CrossRef]
71. Tummanatsakun, D.; Proungvitaya, T.; Roytrakul, S.; Limpaiboon, T.; Wongkham, S.; Wongkham, C.; Silsirivanit, A.; Somintara, O.; Sangkhamanon, S.; Proungvitaya, S. Serum Apurinic/Apyrimidinic Endodeoxyribonuclease 1 (APEX1) Level as a Potential Biomarker of Cholangiocarcinoma. *Biomolecules* **2019**, *9*, 413. [CrossRef] [PubMed]
72. Zhang, S.; He, L.; Dai, N.; Guan, W.; Shan, J.; Yang, X.; Zhong, Z.; Qing, Y.; Jin, F.; Chen, C.; et al. Serum APE1 as a predictive marker for platinum-based chemotherapy of non-small cell lung cancer patients. *Oncotarget* **2016**, *7*, 77482–77494. [CrossRef] [PubMed]
73. Wei, X.; Li, Y.B.; Li, Y.; Lin, B.C.; Shen, X.M.; Cui, R.L.; Gu, Y.J.; Gao, M.; Li, Y.G.; Zhang, S. Prediction of Lymph Node Metastases in Gastric Cancer by Serum APE1 Expression. *J. Cancer* **2017**, *8*, 1492–1497. [CrossRef] [PubMed]
74. Xie, J.; Li, Y.; Kong, J.; Li, C. Apurinic/Apyrimidinic Endonuclease 1/Redox Factor-1 Could Serve as a Potential Serological Biomarker for the Diagnosis and Prognosis of Oral Squamous Cell Carcinoma. *J. Oral Maxillofac. Surg.* **2019**, *77*, 859–866. [CrossRef]

© 2020 by the authors. Licensee MDPI, Basel, Switzerland. This article is an open access article distributed under the terms and conditions of the Creative Commons Attribution (CC BY) license (http://creativecommons.org/licenses/by/4.0/).

Review

Endothelial Aldehyde Dehydrogenase 2 as a Target to Maintain Vascular Wellness and Function in Ageing

Ginevra Nannelli [1], Marina Ziche [2], Sandra Donnini [1] and Lucia Morbidelli [1,*]

1. Department of Life Sciences, University of Siena, 53100 Siena, Italy; gine_nannelli@hotmail.it (G.N.); sandra.donnini@unisi.it (S.D.)
2. Department of Medicine, Surgery and Neurosciences, University of Siena, 53100 Siena, Italy; marina.ziche@unisi.it
* Correspondence: lucia.morbidelli@unisi.it; Tel.: +39-0577-235-381

Received: 9 December 2019; Accepted: 31 December 2019; Published: 3 January 2020

Abstract: Endothelial cells are the main determinants of vascular function, since their dysfunction in response to a series of cardiovascular risk factors is responsible for disease progression and further consequences. Endothelial dysfunction, if not resolved, further aggravates the oxidative status and vessel wall inflammation, thus igniting a vicious cycle. We have furthermore to consider the physiological manifestation of vascular dysfunction and chronic low-grade inflammation during ageing, also known as inflammageing. Based on these considerations, knowledge of the molecular mechanism(s) responsible for endothelial loss-of-function can be pivotal to identify novel targets of intervention with the aim of maintaining endothelial wellness and vessel trophism and function. In this review we have examined the role of the detoxifying enzyme aldehyde dehydrogenase 2 (ALDH2) in the maintenance of endothelial function. Its impairment indeed is associated with oxidative stress and ageing, and in the development of atherosclerosis and neurodegenerative diseases. Strategies to improve its expression and activity may be beneficial in these largely diffused disorders.

Keywords: endothelial cells; oxidative stress; inflammageing; endothelial dysfunction; aldehyde dehydrogenase-2; cardiovascular disease; neurovascular disease

1. Introduction

We searched PubMed from its inception up to December, 2019, using the terms "ALDH2, endothelial cells, endothelial dysfunction, endothelial senescence, ageing, oxidative stress, inflammageing, cardiovascular diseases, neurovascular diseases" to identify publications in English that described the mechanism of action of ALDH2 activity in vascular function, preclinical evidence of beneficial effects of ALDH2 expression/activity for endothelial function, or clinical evidence of the benefit of ALDH2 activity modulation in cardio- or neuro-vascular diseases. We mostly selected publications from the past 10 years that we judged were relevant, but we did not exclude widely referenced and highly regarded older publications.

1.1. Endothelial Function and Dysfunction

Vascular function such as the heartbeat is an essential system for the good functioning of our body. Under normal conditions, endothelial cells (ECs) in blood vessels, through the release of vasoactive and anti-aggregatory mediators and the functioning of antioxidant systems, regulate blood pressure, protecting from hypertension and atherosclerosis. They also constitute a blood barrier, preventing leukocyte infiltration and inflammation into the vascular wall and surrounding tissues [1,2].

Conversely, endothelial dysfunction has been identified as a hallmark of most cardiovascular diseases. Dysfunction of ECs is correlated with an imbalance in the production of key regulators of the vascular homeostasis such as nitric oxide (NO) and growth factors, and/or impaired activity (uncoupling) of endothelial NO synthase (eNOS), associated with increased reactive oxygen species (ROS) levels and vascular oxidative stress [3,4]. Inflammatory factors such as inteleukin-6 (IL-6), tumor necrosis factor-α (TNF-α), intercellular adhesion molecule 1 (ICAM-1) and loss of the antioxidant mechanism are among the most important causes of vascular dysfunction [5].

When endothelial dysfunction occurs, the ability of endothelium to perform one or all of these functions is decreased. In this condition, ECs switch to a pro-inflammatory profile, characterized by loss of barrier integrity, and reduced release of vasoactive molecules associated with increased release of pro-thrombotic mediators [2,5]. Our studies and those of others showed that endothelial dysfunction is linked to impaired endothelial cell survival and physiological angiogenic outcomes with a later rearrangement of the microcirculation that contributes to the onset of various diseases [6–8]. Both anti- and pro-angiogenic therapeutic strategies have been developed for treating human diseases. While angiogenesis inhibitors have been shown to have success in many diseases as cancer, few treatment protocols with the aim to stimulate angiogenesis in ischemia-associated diseases have reached the clinic. Indeed, since endothelial function has been proposed as "barometer for cardiovascular risk", identification of the molecular determinants underlying endothelial integrity and functionality is a medical need [5,6,9–11]. In this scenario, risk factors for endothelial dysfunction are represented by pathological conditions as hypertension, diabetes and atherosclerosis, and lifestyles as high-fat diet, tobacco smoke, alcohol intake, and physical inactivity [12]. Since endothelium is a regulator of exchanges between the vascular wall and surrounding tissues, it is not surprising that dysfunctional ECs can lead to the impairment of other tissues [2,6,12]. Indeed, assessment of vascular function and structure, and in particular of endothelial dysfunction appears to play a crucial role in a broad array of human diseases as cardiovascular and neurodegenerative diseases, tumor growth and metastasis [2,5,8–12].

1.2. Endothelial Senescence

Cardiovascular diseases (CVDs) represent the major cause of disability and death in the elderly population of the Western World. As introduced above, a large number of risk factors plays a role in the development of CVDs, especially ageing which is associated with well-characterized phenotypes [8,13,14]. Over time, aged blood vessels become stiffer and thicker, and their ability to release vasoactive mediators, particularly NO, decreases, while vascular permeability increases, associated with the process of mild vessel inflammation, increased vessel thickness and compromised angiogenic response [14–18]. Of note, even if age remains the main determinant of vascular senescence, healthy vascular ageing can be achieved, and endothelial function is a key element of the heathy vasculature.

Senescent vascular endothelium and other tissues are frequently characterized by chronic mild inflammation [12,13]. In particular, the age-related low grade, chronic, and systemic inflammation is indicated by the term "inflammageing" [12,13]. The indicators of inflammageing include elevated levels of inflammatory mediators such as IL-1, IL-6 and TNF-α, which characterize many age-related pathological phenotypes. These indicators derive from an imbalance of the immune response (immunosenescence) [19].

A large population of senescent cells in organs has the potential to negatively impact organ renewal capabilities and functions, especially on ECs, accelerating the onset of several age-associated pathologies, as CVDs and cancer [15,20].

However, senescent cells may also exert various positive effects on individuals. As senescence is associated with irreversible growth arrest, it is considered to be a tumor suppressor process [21]. In particular, indirect evidence indicates that endothelial senescence plays a causal role in microvascular rarefaction and affects the ability of cells to proliferate and form capillary-like structures [20]. The timely and challenging issue for modern biomedical sciences focused on ageing is the identification of the biological targets and the pharmacological tools able to promote "healthy ageing".

1.3. Aldehyde Dehydrogenase-2 (ALDH2)

The aldehyde dehydrogenase (ALDH) gene superfamily encodes enzymes that are mainly responsible for the irreversible oxidation of various aldehydes. Three major classes of mammalian ALDHs (classes 1, 2 and 3) have been identified. Classes 1 and 3 contain both constitutive and inducible cytosolic isozymes. Class 2 consists of constitutive mitochondrial isozymes [22,23]. ALDH2 is primarily involved in the detoxification of acetaldehyde, the first intermediate of ethanol metabolism, into acetate, a less active byproduct [23]. In addition to acetaldehyde, other aldehydes are metabolized by ALDH2 [23,24]. The sources of aldehydic substrates of ALDH2 can be endogenous or exogenous. Endogenous reactive aldehydes are generated as byproducts of cell metabolism and include malondialdehyde, 4-hydroxynonenal (4-HNE), DOPAL, and acrolein [23]. Exogenous aldehydes are found in industrial and environmental pollutants and may be produced during metabolism of xenobiotics. Among these, acrolein, acetaldehyde and formaldehyde are the major reactive species found in tobacco smoke and car exhaust [23,24]. Generally, aldehydes are toxic molecules that react rapidly with amino acid residues, e.g., thiols to form the genotoxic DNA- and protein-adducts in cells [23,24].

ALDH proteins are found in all cellular compartments, including cytosol, mitochondria, endoplasmic reticulum and the nucleus. Some proteins have more than one cellular location. Nevertheless, ALDHs also catalyze some reactions involved in the formation of bioactive molecules that regulate important physiological functions. This is the case of some ALDHs as ALDH1A1, 1A2 and 1A3 that convert retinal aldehyde in retinoic acid. In turn, retinoic acid acts as a ligand for retinoic X receptor (RXR) and nuclear retinoic receptor (RAR), participating in a number of growth and developmental processes [23,25,26].

ALDH2 arises as an important gatekeeper of ROS overproduction that a cell is able to tolerate. In fact, the main function of mitochondrial ALDH2 is to protect mitochondria and cells from the damaging effect of aldehydes, by oxidizing the substrates into their corresponding non-toxic carboxylic acids. Some studies distinguish between direct and indirect anti-oxidative properties of ALDH2. The direct anti-oxidative properties are assumed to depend on its potent reductase function with its highly activated sulfhydryl groups [27]. In addition to oxidative capabilities, ALDH2 possesses nitrate reductase activity responsible for the bioconversion of nitroglycerin to 1,2-glyceryl dinitrate (GDN), thus inducing NO release [28–30]. ALDH2 protein can be the substrate of various post-translational modifications, including oxidation, S-nitrosylation, phosphorylation, nitration, acetylation, glycosylation, and adduct formation, most of which reduce its activity [31].

In the present review, we describe the effects of oxidative stress-linked accumulation of aldehydes in the cells, focusing on ECs, and discuss the contribution of ALDH2 in several endothelium-dependent disorders, including senescence.

2. ALDH2 and Endothelial-Related Diseases

Different diseases have in common EC impairment and dysfunction and more precisely an alteration in ALDH expression and enzymatic activity, in turn responsible for changes in oxidative status and inflammation (Figure 1).

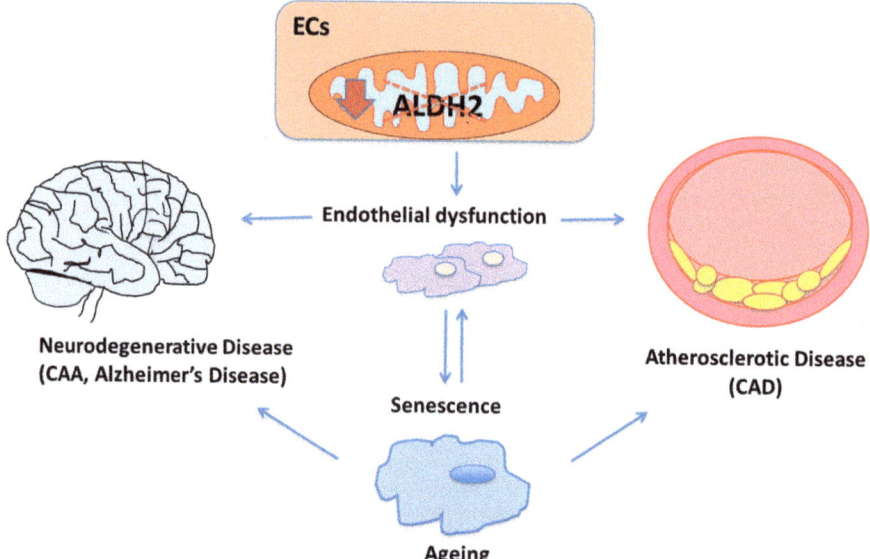

Figure 1. The role of aldehyde dehydrogenase-2 (ALDH2) in endothelial function and endothelial-related diseases. ALDH2 downregulation (down arrow) and/or catalytic inactivation (dashed red lines) into mitochondria in endothelial cells (ECs), leading to inefficient aldehyde metabolism, is recognized to affect endothelial functions and make senescence occur faster. Senescence itself might further aggravate endothelial dysfunction. In turn, ALDH2 is believed to be involved in the ageing process, in particular ageing related-cardiovascular diseases, such as atherosclerosis and coronary artery disease (CAD), and affect the cerebrovascular unit, contributing to the etiology of neurovascular degenerative diseases, including Cerebral Amyloid Angiopathy (CAA).

2.1. ALDH2, Oxidative Stress and Ageing

Advanced age is an independent risk factor for life-threatening diseases, including coronary artery disease, stroke, and neurodegenerative diseases, which are directly related to ageing-associated EC dysfunction [32]. Despite senescence being studied deeply for many centuries and many steps in our understanding of it accomplished, the process of ageing remains largely difficult to remedy and remains, unfortunately, inevitable [33]. In addition to the "inflammageing" hypothesis, many theories have been proposed to explain the process of ageing [33]. In particular, the general free radical theory of ageing indicates that ageing is caused by ROS-dependent accumulation of damage [34]. In other words, the traditional view in the field of free radical biology is that free radicals and ROS are toxic and able to directly damage a large plethora of biological targets. Furthermore, the accumulation of damage leads to the process of ageing resulting in various diseases [35]. Recent findings revealed that the accumulation of toxic aldehydes plays a key role in ischemia-reperfusion injuries and in ageing as much as ROS from which they derive [36–39].

Accordingly, some studies support the close link between ALDH2 dysfunction and ageing, especially in the heart [40,41]. In particular, ALDH2 has been reported to protect ECs against oxidative stress events due to aldehydes, and cell senescence [41,42].

One of our recent works showed that inhibition of ALDH2 activity negatively impacted EC function. We demonstrated that ALDH2 silencing or inhibition significantly affected cell proliferation, migration and altered cellular permeability, in terms of reduced VE–cadherin and ZO-1 expression at cell–cell contacts and increased cellular permeability in human umbilical vein endothelial cells (HUVECs) [42]. Although the mechanism of action has not been fully elucidated, our results indicate that ROS production and 4-hydroxy-2-nonenal (4-HNE) accumulation, a secondary end-product of lipid peroxidation, contribute to mediate endothelial dysfunction and the onset of senescence in siALDH2 and daidzin-treated cells [42]. These findings were supported by the observation that exposure to the ROS scavenger, N-acetylcysteine, affected the pattern of 4-HNE adducts, reduced ROS and recovered cell survival in siALDH2 cells [42]. The aldehyde 4-HNE is a reactive molecule which forms adducts with proteins, lipids and DNA. Excessive 4-HNE adduct formation has been reported in ischemic cardiovascular tissue isolated from rodents and humans [43,44]. From a bioenergetic standpoint, we observed that ALDH2 attenuation reduced basal and maximal respiration and abolished mitochondrial reserve capacity in ECs [42]. Intriguingly, this latter finding suggested that ALDH2 supports mitochondrial bioenergetics by increasing basal, maximal and consequently reserve capacity. Further, the changes in cell morphology, from polygonal to an enlarged and irregular shape, evoked the senescent phenotype and this led us to explore further details. The analysis of typical senescent markers, such as p21/p53 expression and SA-β-gal, further suggested that siALDH2 or daidzin-treated ECs present early signs of senescence [42]. This scenario suggests a protective role of ALDH2 on endothelium. The inhibition of ALDH2 affects endothelial functions, bioenergetics and metabolism, and makes the acquisition of a senescent phenotype faster. The acquisition of a senescent phenotype might have a defensive role, but the expansion of senescent cells might itself further aggravate the damage and accelerate ageing.

From a mechanistic point of view, the accumulation of endogenous reactive aldehydes, including 4-HNE, and inhibition of ALDH2 by these reactive species increased cell vulnerability to aldehyde-induced damage [45]. Indeed, susceptibility to lipid peroxidation byproducts of ALDH2 enzymatic activity was reported [46]. It was observed that ALDH2 is inactivated by aldehydes at low concentrations (<10 μM) and ROS. ALDH inactivation may interfere with the detoxification of lipid aldehydes, promoting their accumulation, which, in turn, is known to trigger ROS production, amplifying oxidative damage in the cell. Specifically, in ECs inactivation of ALDH2 correlates with vascular damage, vasoconstriction, and thrombosis [47].

Moreover, evidence showed the role of acetaldehyde in tissue and cell damage. In particular, a study reported that ALDH2 prevented acetaldehyde-related toxic effects by alleviating oxidative stress and apoptosis in HUVECs [47]. For instance, it has been observed that in HUVEC, ALDH2 over-expression protects cells from the toxic effects of exposure to different concentrations of acetaldehyde. As a result, the generation of ROS is decreased, as well as apoptosis and activation of stress signaling molecules, such as signal-regulated extracellular kinases (ERK1/2), and the p38 mitogenic activated MAP kinase protein [47]. Of note, the acetaldehyde-induced ROS generation, apoptosis and activation of stress molecules were prevented by the ALDH2 transgene in a manner similar to antioxidant alpha-tocopherol, indicating that facilitation of acetaldehyde detoxification by ALDH2 transgene overexpression is able to counteract acetaldehyde-induced EC injury and activation of stress signals. These data consecutively highlight the therapeutic potential of ALDH2 in the prevention of cell damage induced by ethanol consumption [47], features of which are very similar to those of accelerated ageing.

Consistently, mutations in ALDHs genes as well as their downregulation or ALDHs catalytic inactivation, leading to inefficient aldehyde metabolism, may contribute in the etiology of various diseases including cardiopathies and neurovascular degenerative diseases, including the cerebral amyloid angiopathy (CAA) and Alzheimer's disease (AD) [24,48].

2.2. ALDH2 and Atherosclerosis

Heart failure is a clinical manifestation characterized by alterations in cardiac morphology and functions that lead to reduced cardiac output and/or elevated intra-cardiac pressure at rest or during stress. Coronary artery disease (CAD), diabetes mellitus and alcohol abuse are common determinants of myocardial diseases, which result in ischemic injury, metabolic disorders or toxic damage [49]. Nevertheless, advanced ageing leads to alteration in morphology and function of the heart in the absence of other accompanying cardiovascular risk factors. In particular, ALDH2 is believed to be involved in the ageing process and ageing related-cardiovascular diseases. In fact, although ALDH2 is better known for its involvement in ethanol metabolism, it is also crucial for cardioprotection through the detoxification of reactive aldehydes such as 4-HNE and the bioconversion of nitrates into NO [27,29]. ALDH2 is largely seen as a critical enzyme involved in protecting the heart from ischemic injury. Much evidence, including meta-analysis, have assessed the associations between ALDH2 rs671 (ALDH2*2) polymorphism inactivating the enzymatic activity, and CAD [50]. Additionally, results from various studies conducted on Asian patients have highlighted the strong association between myocardial infarction and ALDH2 rs671 polymorphism [51]. In this context, the majority of studies focused on myocardium [52], with limited attention on ECs or vessels. Nonetheless, ALDH2 activity is impaired by oxidized low-density lipoproteins (ox-LDLs), possibly by post-translational modifications. In particular, ox-LDLs are found to exert an inhibitory effect on ALDH2 activity by preventing mitochondrial sirtuin 3 (SIRT3) expression [31]. SIRT3 is one of the members of the sirtuin family, which have a common core domain and an important role in ageing, stress resistance and metabolic regulation. In particular, SIRT3 has received much attention for its role in mitochondrial metabolism and ageing [53].

A study corroborates the association between ALDH2 and SIRT3 [54]. This study strengthens the concept that moderate ethanol consumption is associated with a positive effect on eNOS activation that results in antiatherogenic actions. In particular, Xue and colleagues showed that in human aortic ECs (HAECs), low-dose ethanol resulted in SIRT3 inactivation, leading to rapid activation of ALDH2. ALDH2 activity mediates ethanol-induced eNOS activation and prevents ROS accumulation [54]. Thus, ALDH2 activation promoted by SIRT3 inactivation improves HAEC function, resulting in an anti-atherogenic effect. Consistent with this finding, another recent study showed that ALDH2 silencing or inhibition aggravated the atherosclerotic process. In ApoE$^{-/-}$ mice the silencing of the ALDH2 gene was associated with a severe inflammation of the vascular wall and the formation of larger and more unstable plaques [55]. In-vitro experiments with HUVECs further illustrated that inhibition of ALDH2 activity resulted in elevated inflammatory molecules, an enhanced nuclear translocation or phosphorylation of pro-inflammatory transcription factor such as NF-κB or AP-1 involved in vessel permeability and plaque development [55]. Of note, in ECs, we and other reported that several natural agents such as polyphenols of extra virgin olive oil modulate, beyond ROS production, the expression and activity of enzymes and transcription factors involved in atherogenesis [56,57]. These data support the assumption that metabolic cell redox state has a marked impact on the cell transcriptome and proteome, and, in turn, on cell functions. In this context, it would be of interest to know the effects of nutraceuticals on ALDH2 activity in controlling vessel integrity.

Overall, these data convincingly indicate that ALDH2 silencing or inhibition exacerbates the atherosclerosis process, increasing plaque development and vulnerability with aggravated inflammation. Nevertheless, further work is needed to fully understand the clinical value of ALDH2 and its activation in the prevention or treatment of atherosclerotic diseases, such as CAD.

2.3. ALDH2 and Neurodegenerative Diseases

Alzheimer's disease (AD) is classified as a progressive neurodegenerative disorder, characterized by neuropathological changes in particular brain regions and in a variety of neuro-transmitter systems. The progressive neuronal degeneration that occurs in AD generally leads to dementia, the most common consequence, affecting cognitive functions in patients such as memory, thinking and reasoning [58]. Unfortunately, AD is becoming more prevalent as human lifespan increases. AD is a multifactorial disease driven by a combination of genetic and environmental factors and can be divided in two forms: familial and sporadic cases [59]. A characteristic feature of AD is the deposit of amyloid β peptides (mainly Aβ1-40 and Aβ1-42) in the brain that form extracellular amyloid plaques. These plaques lead to neuronal dysfunction, cell death, and loss of synaptic connections, notably due to the ensuing inflammation and oxidative stress [60]. Increased oxidative stress is reported to play a critical role in the pathogenesis of AD prior to the onset of Aβ deposition and cognitive impairment. Cerebrovascular dysfunction has emerged as a critical feature of neurodegenerative diseases [61].

Cerebral amyloid angiopathy (CAA), a cerebrovascular disease that is frequently associated with AD, is characterized by the accumulation of Aβ in cerebral microvessels. In CAA, the endothelial dysfunction is thought to alter Aβ homeostasis and to promote infiltration of the brain parenchyma with circulating toxic molecules [61]. At the same time, Aβ peptides, especially the vasculotropic isoform Aβ1–40, affect brain blood vessels, altering their fundamental functions, including impairment of vasoactive tone and barrier functions, vascular remodelling as well as suppression of the intrinsic angiogenic properties of endothelium [48,62–64]. Although CAA remains clinically distinct from AD, their common features have the potential to link cerebrovascular and neurodegenerative pathways in the ageing brain. The product 4-HNE is formed during oxidative stress, and is able to react with many molecules such as proteins, and to accumulate in the brain. Many studies have shown that 4-HNE covalently modifies Aβ via 1,4 conjugates, altering the function of Aβ and impairing cellular features such as metabolism, cell signaling and structural integrity [65]. Since ALDH2 protects mitochondrial functions through the detoxification of 4-HNE that accumulates in this organelle, it is not surprising that recent epidemiological studies showed a correlation between ALDH2*2 loss-of-function mutations in Asian patients and a higher incidence of AD in these people [24,45].

One potential intervention for the treatment of AD can be to reduce the toxicity caused by the Aβ peptides, particularly Aβ1-40. Previous studies from our lab elucidated the role of ALDH2 in the protection of endothelium against Aβ1-40 damage. In particular, the chronic exposure of cultured EC to Aβ results in profound modifications of cell pro-angiogenic functions, shifting their phenotype toward the senescence program and reducing its pro-angiogenic capability [66]. Aβ peptides also increase intracellular 4-HNE in ECs by impairment of mitochondrial ALDH2 activity [48]. Consistently, in an ALDH2$^{-/-}$ knockout null mouse model, mice exhibited both neuronal and vascular pathological changes associated with AD. In fact, mice exhibited progressive, age-related cognitive deficits in non-spatial and spatial working memory and other features, together with a multitude of AD-associated signs, including 4-HNE adducts as well as age-related increase in Aβ [67].

In addition to the observed AD-like changes in the brain, significant vascular alterations were found in cerebral microvessels (CMVs) of ALDH2$^{-/-}$ mice. In comparison to the wild type, CMVs of these mice displayed marked increases in HNE adducts and age-related increases in monomeric Aβ. Moreover, ALDH2$^{-/-}$ mice exhibited endothelial dysfunction, and increased Aβ deposits in microvessels [67].

From these studies it seems that ALDH2 activity may play a critical role for preserving endothelial function in cerebrovascular units and preventing age-related dysfunctions.

3. Conclusions

Global average lifespan is increasing as a result of many factors, such as lifestyle. But, as longevity continues to increase, ageing-related diseases and, consequentially, the need for new therapeutic interventions arise. Today, regenerative medicine plays a significant role relative to other therapies, including those for the treatment of CVD.

In this review we showed that ALDH2 asserts a protective role in the endothelium against different types of stressors including age-associated dysfunctions. ALDH2 has emerged in recent years as a crucial guardian against several stressor or toxic insults, supporting its potential role in many diseases, including cardiovascular and cerebrovascular diseases that are linked to ageing (Table 1).

Since senescence-related endothelial dysfunction plays a crucial role in the pathogenesis of several diseases, further studies are necessary to fully elucidate the role and protective mechanisms of ALDH2 in the endothelium.

Table 1. Summary of main findings of ALDH2 and ageing, atherosclerosis or neurodegenerative diseases.

ALDH2 Status	Tissue/Cells	Molecular Mechanism	Function	Ref.
ALDH2 and vascular ageing				
ALDH2 activation	Heart (rat)	(−) aldehydic adducts (carbonylation)	(−) cardiac dysfunction	[28]
ALDH2 activation	Liver (rat)	(−) ROS (−) pro-inflammatory cytokines (−) 4-HNE and MDA	(−) tissue apoptosis (+) mitochondrial membrane potential	[30]
ALDH2 transgenic mice	Heart	(+) ROS (−) AMPK phosphorylation (−) SIRT1	(+) ageing-induced cardiac hypertrophy (+) apoptosis (+) mitochondrial injury	[40]
ALDH2 activation	Heart (mouse)	(−) 4-HNE-protein adducts (−) protein carbonyls (+) autophagy flux (+) SIRT1	(+) cardiac function	[41]
ALDH2 gene silencing	Endothelial cells	(+) ROS (+) 4-HNE	(−) respiration (+) senescence	[42]
ALDH2 gene transfection	Endothelial cells	(−) ROS	(−) apoptosis (−) ERK/p38-MAPK	[47]
ALDH2 and atherosclerosis				
ALDH2*2 loss-of-function	—	(+) ox-LDLs	(+) coronary artery disease	[50–52]
ALDH2 gene silencing	ApoE$^{-/-}$ mice	(+) ROS	(+) vessel wall inflammation (+) plaque instability	[55]
ALDH2 activation	Endothelial cells	(−) ROS; (+) eNOS (−) SIRT3 activation	—	[54]
ALDH2 inhibition	Endothelial cells	(+) NF-κB (+) AP-1	(+) permeability (+) plaque formation	[55]
ALDH2 and neurodegenerative diseases				
ALDH2*2 loss-of-function	Brain	—	(+) incidence of Alzheimer's disease	[45]
ALDH2$^{-/-}$ mouse model	Brain	(+) 4-HNE adducts (+) Aβ	(+) cognitive deficits (+) endothelial dysfunction (+) Aβ in microvessels	[47]
ALDH2 inactivation	Endothelial cells	(+) 4-HNE	(−) angiogenesis (+) senescence	[48]

Author Contributions: G.N. wrote the manuscript and drew the figure; S.D., M.Z. and L.M. made substantial contribution in the revision process; L.M. gave approval of the submitted version. All authors have read and agreed to the published version of the manuscript All authors have read and agreed to the published version of the manuscript.

Funding: This research was funded by the Italian Ministry of Education, University and Research (MIUR- PRIN projects), grant number 2017XP72RF (to L.M.) and 20152HKF3Z (to M.Z.).

Conflicts of Interest: The authors declare no conflict of interest. The funders had no role in the design of the study; in the collection, analyses, or interpretation of data; in the writing of the manuscript, or in the decision to publish the results.

References

1. Wagner, D.D.; Frenette, P.S. The vessel wall and its interactions. *Blood* **2008**, *111*, 5271–5281. [CrossRef] [PubMed]

2. Rajendran, P.; Rengarajan, T.; Thangavel, J.; Nishigaki, Y.; Sakthisekaran, D.; Sethi, G.; Nishigaki, I. The vascular endothelium and human diseases. *Int. J. Biol. Sci.* **2013**, *9*, 1057–1069. [CrossRef] [PubMed]
3. Daiber, A.; Oelze, M.; Wenzel, P.; Wickramanayake, J.M.; Schuhmacher, S.; Jansen, T.; Lackner, K.J.; Torzewski, M.; Münzel, T. Nitrate tolerance as a model of vascular dysfunction: Roles for mitochondrial aldehyde dehydrogenase and mitochondrial oxidative stress. *Pharmacol. Rep.* **2009**, *61*, 33–48. [CrossRef]
4. Monti, M.; Donnini, S.; Giachetti, A.; Mochly-Rosen, D.; Ziche, M. deltaPKC inhibition or varepsilonPKC activation repairs endothelial vascular dysfunction by regulating eNOS post-translational modification. *J. Mol. Cell. Cardiol.* **2010**, *48*, 746–756. [CrossRef]
5. Haybar, H.; Shahrabi, S.; Rezaeeyan, H.; Shirzad, R.; Saki, N. Endothelial Cells: From Dysfunction Mechanism to Pharmacological Effect in Cardiovascular Disease. *Cardiovasc. Toxicol.* **2019**, *19*, 13–22. [CrossRef]
6. Donnini, S.; Cantara, S.; Morbidelli, L.; Giachetti, A.; Ziche, M. FGF-2 overexpression opposes the beta amyloid toxic injuries to the vascular endothelium. *Cell Death Differ.* **2006**, *13*, 1088–1096. [CrossRef]
7. Morbidelli, L.; Donnini, S.; Ziche, M. Targeting endothelial cell metabolism for cardio-protection from the toxicity of antitumor agents. *Cardio-Oncology* **2016**, *2*, 3. [CrossRef]
8. Ungvari, Z.; Tarantini, S.; Kiss, T.; Wren, J.; Giles, C.B.; Griffin, C.T.; Murfee, W.L.; Pacher, P.; Csiszar, A. Endothelial dysfunction and angiogenesis impairment in the ageing vasculature. *Nat. Rev. Cardiol.* **2018**, *15*, 555–565. [CrossRef]
9. Caruso, P.; Signori, R.; Moretti, R. Small vessel disease to subcortical dementia: A dynamic model, which interfaces ageing, cholinergic dysregulation and the neurovascular unit. *Vasc. Health Risk Manag.* **2019**, *15*, 259–281. [CrossRef]
10. Trott, D.W.; Fadel, P.J. Inflammation as a mediator of arterial ageing. *Exp. Physiol.* **2019**, *104*, 1455–1471. [CrossRef]
11. Hao, Y.M.; Yuan, H.Q.; Ren, Z.; Qu, S.L.; Liu, L.S.; Wei, D.-H.; Yin, K.; Fu, M.; Jiang, Z.S. Endothelial to mesenchymal transition in atherosclerotic vascular remodeling. *Clin. Chim. Acta* **2019**, *490*, 34–38. [CrossRef] [PubMed]
12. Donato, A.J.; Machin, D.R.; Lesniewski, L.A. Mechanisms of Dysfunction in the Ageing Vasculature and Role in Age-Related Disease. *Circ. Res.* **2018**, *123*, 825–848. [CrossRef]
13. Dantas, A.P.; Jiménez-Altayó, F.; Vila, E. Vascular ageing: Facts and factors. *Front. Physiol.* **2012**, *3*, 325. [CrossRef] [PubMed]
14. Ghebre, Y.T.; Yakubov, E.; Wong, W.T.; Krishnamurthy, P.; Sayed, N.; Sikora, A.G.; Bonnen, M.D. Vascular Ageing: Implications for Cardiovascular Disease and Therapy. *Transl. Med.* **2016**, *6*. [CrossRef] [PubMed]
15. North, B.J.; Sinclair, D.A. The intersection between ageing and cardiovascular disease. *Circ. Res.* **2012**, *110*, 1097–1108. [CrossRef] [PubMed]
16. Masi, S.; Colucci, R.; Duranti, E.; Nannipieri, M.; Anselmino, M.; Ippolito, C.; Tirotta, E.; Georgiopoulos, G.; Garelli, F.; Nericcio, A.; et al. Ageing Modulates the Influence of Arginase on Endothelial Dysfunction in Obesity. *Arterioscler. Thromb. Vasc. Biol.* **2018**, *38*, 2474–2483. [CrossRef]
17. Foote, K.; Reinhold, J.; Yu, E.P.K.; Figg, N.L.; Finigan, A.; Murphy, M.P.; Bennett, M.R. Restoring mitochondrial DNA copy number preserves mitochondrial function and delays vascular ageing in mice. *Ageing Cell* **2018**, *17*, e12773. [CrossRef]
18. De Silva, T.M.; Li, Y.; Kinzenbaw, D.A.; Sigmund, C.D.; Faraci, F.M. Endothelial PPARγ (Peroxisome Proliferator-Activated Receptor-γ) Is Essential for Preventing Endothelial Dysfunction with Ageing. *Hypertension* **2018**, *72*, 227–234. [CrossRef]
19. Cannizzo, E.S.; Clement, C.C.; Sahu, R.; Follo, C.; Santambrogio, L. Oxidative stress, inflamm-ageing and immunosenescence. *J. Proteom.* **2011**, *74*, 2313–2323. [CrossRef]
20. Campisi, J.; Andersen, J.K.; Kapahi, P.; Melov, S. Cellular senescence: A link between cancer and age-related degenerative disease? *Semin. Cancer Biol.* **2011**, *21*, 354–359. [CrossRef]
21. Li, Y.; Lui, K.O.; Zhou, B. Reassessing endothelial-to-mesenchymal transition in cardiovascular diseases. *Nat. Rev. Cardiol.* **2018**, *15*, 445–456. [CrossRef] [PubMed]
22. Vasiliou, V.; Thompson, D.C.; Smith, C.; Fujita, M.; Chen, Y. Aldehyde dehydrogenases: From eye crystallins to metabolic disease and cancer stem cells. *Chem. Biol. Interact.* **2013**, *202*, 2–10. [CrossRef] [PubMed]
23. Rodríguez-Zavala, J.S.; Calleja, L.F.; Moreno-Sánchez, R.; Yoval-Sánchez, B. Role of Aldehyde Dehydrogenases in Physiopathological Processes. *Chem. Res. Toxicol.* **2019**, *32*, 405–420. [CrossRef] [PubMed]

24. Chen, C.H.; Ferreira, J.C.; Gross, E.R.; Mochly-Rosen, D. Targeting aldehyde dehydrogenase 2: New therapeutic opportunities. *Physiol. Rev.* **2014**, *94*, 1–34. [CrossRef]
25. Rhinn, M.; Dollé, P. Retinoic acid signalling during development. *Development* **2012**, *139*, 843–858. [CrossRef]
26. Ciccone, V.; Terzuoli, E.; Donnini, S.; Giachetti, A.; Morbidelli, L.; Ziche, M. Stemness marker ALDH1A1 promotes tumor angiogenesis via retinoic acid/HIF-1α/VEGF signalling in MCF-7 breast cancer cells. *J. Exp. Clin. Cancer Res.* **2018**, *37*, 311, Correction in **2019**, *38*, 45. [CrossRef]
27. Wenzel, P.; Hink, U.; Oelze, M.; Seeling, A.; Isse, T.; Bruns, K.; Steinhoff, L.; Brandt, M.; Kleschyov, A.L.; Schulz, E.; et al. Number of nitrate groups determines reactivity and potency of organic nitrates: A proof of concept study in ALDH-2$^{-/-}$ mice. *Br. J. Pharmacol.* **2007**, *150*, 526–533. [CrossRef]
28. Sun, L.; Batista Ferreira, J.C.; Mochly-Rosen, D. ALDH2 activator inhibits increased myocardial infarction injury by nitroglycerin tolerance. *Sci. Transl. Med.* **2011**, *3*, 1–7. [CrossRef]
29. Opelt, M.; Eroglu, E.; Waldeck-Weiermair, M.; Russwurm, M.; Koesling, D.; Malli, R.; Graier, W.F.; Fassett, J.T.; Schrammel, A.; Mayer, B. Formation of Nitric Oxide by Aldehyde Dehydrogenase-2 Is Necessary and Sufficient for Vascular Bioactivation of Nitroglycerin. *J. Biol. Chem.* **2016**, *291*, 24076–24084. [CrossRef]
30. Zhang, T.; Zhao, Q.; Ye, F.; Huang, C.Y.; Chen, W.M.; Huang, W.Q. Alda-1 an ALDH2 activator, protects against hepatic ischemia/reperfusion injury in rats via inhibition of oxidative stress. *Free Radic. Res.* **2018**, *13*, 1–10. [CrossRef]
31. Wei, S.J.; Xing, J.H.; Wang, B.L.; Xue, L.; Wang, J.L.; Li, R.; Qin, W.D.; Wang, J.; Wang, X.P.; Zhang, M.X.; et al. Poly(ADPribose) polymerase inhibition prevents reactive oxygen species induced inhibition of aldehyde dehydrogenase2 activity. *Biochem. Biophys. Acta* **2013**, *1833*, 479–486. [CrossRef] [PubMed]
32. Tian, X.L.; Li, Y. Endothelial cell senescence and age-related vascular diseases. *J. Genet. Genom.* **2014**, *41*, 485–495. [CrossRef] [PubMed]
33. Sergiev, P.V.; Dontsova, O.A.; Berezkin, G.V. Theories of ageing: An ever-evolving field. *Acta Nat.* **2015**, *7*, 9–18. [CrossRef]
34. Gladyshev, V.N. The free radical theory of ageing is dead. Long live the damage theory! *Antioxid. Redox. Signal.* **2014**, *20*, 727–731. [CrossRef] [PubMed]
35. Liochev, S.I. Reactive oxygen species and the free radical theory of ageing. *Free Radic. Biol. Med.* **2013**, *60*, 1–4. [CrossRef]
36. Gomes, K.M.; Campos, J.C.; Bechara, L.R.; Queliconi, B.; Lima, V.M.; Disatnik, M.H.; Magno, P.; Chen, C.H.; Brum, P.C.; Kowaltowski, A.J.; et al. Aldehyde dehydrogenase 2 activation in heart failure restores mitochondrial function and improves ventricular function and remodelling. *Cardiovasc. Res.* **2014**, *103*, 498–508. [CrossRef]
37. Gomes, K.M.; Bechara, L.R.; Lima, V.M.; Ribeiro, M.A.; Campos, J.C.; Dourado, P.M.; Kowaltowski, A.J.; Mochly-Rosen, D.; Ferreira, J.C. Aldehydic load and aldehyde dehydrogenase 2 profile during the progression of post-myocardial infarction cardiomyopathy: Benefits of Alda-1. *Int. J. Cardiol.* **2015**, *179*, 129–138. [CrossRef]
38. Panisello-Roselló, A.; Lopez, A.; Folch-Puy, E.; Carbonell, T.; Rolo, A.; Palmeira, C.; Adam, R.; Net, M.; Roselló-Catafau, J. Role of aldehyde dehydrogenase 2 in ischemia reperfusion injury: An update. *World J. Gastroenterol.* **2018**, *24*, 2984–2994. [CrossRef]
39. Kimura, M.; Yokoyama, A.; Higuchi, S. Aldehyde dehydrogenase-2 as a therapeutic target. *Expert Opin. Ther. Targets* **2019**, *23*, 955–966. [CrossRef]
40. Zhang, Y.; Mi, S.L.; Hu, N.; Doser, T.A.; Sun, A.; Ge, J.; Ren, J. Mitochondrial aldehyde dehydrogenase 2 accentuates ageing-induced cardiac remodeling and contractile dysfunction: Role of AMPK, Sirt1, and mitochondrial function. *Free Radic. Biol. Med.* **2014**, *71*, 208–220. [CrossRef]
41. Wu, B.; Yu, L.; Wang, Y.; Wang, H.; Li, C.; Yin, Y.; Yang, J.; Wang, Z.; Zheng, Q.; Ma, H. Aldehyde dehydrogenase 2 activation in aged heart improves the autophagy by reducing the carbonyl modification on SIRT1. *Oncotarget* **2016**, *7*, 2175–2188. [CrossRef] [PubMed]
42. Nannelli, G.; Terzuoli, E.; Giorgio, V.; Donnini, S.; Lupetti, P.; Giachetti, A.; Bernardi, P.; Ziche, M. ALDH2 Activity Reduces Mitochondrial Oxygen Reserve Capacity in Endothelial Cells and Induces Senescence Properties. *Oxid. Med. Cell Longev.* **2018**, *2018*, 9765027. [CrossRef] [PubMed]
43. Uchidaand, K.; Stadtman, E.R. Modification of histidine residues in proteins by reaction with 4-hydroxynonenal. *Proc. Natl. Acad. Sci. USA* **1992**, *89*, 4544–4548. [CrossRef] [PubMed]

44. Chung, F.-L.; Nath, R.G.; Ocando, J.; Nishikawa, A.; Zhang, L. Deoxyguanosine Adducts of t-4-Hydroxy-2-nonenal Are Endogenous DNA Lesions in Rodents and Humans: Detection and Potential Source. *Cancer Res.* **2000**, *60*, 1507–1511. [PubMed]
45. Chen, C.H.; Joshi, A.U.; Mochly-Rosen, D. The Role of Mitochondrial Aldehyde Dehydrogenase 2 (ALDH2) in Neuropathology and Neurodegeneration. *Acta Neurol. Taiwan* **2016**, *25*, 111–123.
46. Yoval-Sánchez, B.; Rodríguez-Zavala, J.S. Differences in susceptibility to inactivation of human aldehyde dehydrogenases by lipid peroxidation byproducts. *Chem. Res. Toxicol.* **2012**, *25*, 722–729. [CrossRef]
47. Li, S.Y.; Gomelsky, M.; Duan, J.; Zhang, Z.; Gomelsky, L.; Zhang, X.; Epstein, P.N.; Ren, J. Overexpression of aldehyde dehydrogenase-2 (ALDH2) transgene prevents acetaldehyde-induced cell injury in human umbilical vein endothelial cells: Role of ERK and p38 mitogen-activated protein kinase. *J. Biol. Chem.* **2004**, *279*, 11244–11252. [CrossRef]
48. Solito, R.; Corti, F.; Chen, C.H.; Mochly-Rosen, D.; Giachetti, A.; Ziche, M.; Donnini, S. Mitochondrial aldehyde dehydrogenase-2 activation prevents β-amyloid-induced endothelial cell dysfunction and restores angiogenesis. *J. Cell Sci.* **2013**, *126 Pt 9*, 1952–1961. [CrossRef]
49. Pang, J.; Wang, J.; Zhang, Y.; Xu, F.; Chen, Y. Targeting acetaldehyde dehydrogenase 2 (ALDH2) in heart failure-Recent insights and perspectives. *Biochem. Biophys. Acta Mol. Basis Dis.* **2017**, *1863*, 1933–1941. [CrossRef]
50. Zhang, L.L.; Wang, Y.Q.; Fu, B.; Zhao, S.L.; Kui, Y. Aldehyde dehydrogenase 2 (ALDH2) polymorphism gene and coronary artery disease risk: A meta-analysis. *Genet. Mol. Res.* **2015**, *14*, 18503–18514. [CrossRef]
51. Wang, Q.; Zhou, S.; Wang, L.; Lei, M.; Wang, Y.; Miao, C.; Jin, Y. ALDH2 rs671 Polymorphism and coronary heart disease risk among Asian populations: A meta-analysis and meta-regression. *DNA Cell Biol.* **2013**, *32*, 393–399. [CrossRef] [PubMed]
52. Yasue, H.; Mizuno, Y.; Harada, E. Association of East Asian Variant Aldehyde Dehydrogenase 2 Genotype (ALDH2*2*) with Coronary Spasm and Acute Myocardial Infarction. *Adv. Exp. Med. Biol.* **2019**, *1193*, 121–134. [CrossRef] [PubMed]
53. Kincaid, B.; Bossy-Wetzel, E. Forever young: SIRT3 a shield against mitochondrial meltdown, ageing, and neurodegeneration. *Front. Ageing Neurosci.* **2013**, *5*, 48. [CrossRef] [PubMed]
54. Xue, L.; Xu, F.; Meng, L.; Wei, S.; Wang, J.; Hao, P.; Bian, Y.; Zhang, Y.; Chen, Y. Acetylation-dependent regulation of mitochondrial ALDH2 activation by SIRT3 mediates acute ethanol-induced eNOS activation. *FEBS Lett.* **2012**, *586*, 137–142. [CrossRef] [PubMed]
55. Pan, C.; Xing, J.H.; Zhang, C.; Zhang, Y.M.; Zhang, L.T.; Wei, S.J.; Zhang, M.X.; Wang, X.P.; Yuan, Q.H.; Xue, L.; et al. Aldehyde dehydrogenase 2 inhibits inflammatory response and regulates atherosclerotic plaque. *Oncotarget* **2016**, *7*, 35562–35576. [CrossRef] [PubMed]
56. Terzuoli, E.; Nannelli, G.; Giachetti, A.; Morbidelli, L.; Ziche, M.; Donnini, S. Targeting endothelial-to-mesenchymal transition: The protective role of hydroxytyrosol sulfate metabolite. *Eur. J. Nutr.* **2019**. [CrossRef]
57. Tressera-Rimbau, A.; Arranz, S.; Eder, M.; Vallverdú-Queralt, A. Dietary Polyphenols in the Prevention of Stroke. *Oxid. Med. Cell. Longev.* **2017**, *2017*, 7467962. [CrossRef]
58. Wenk, G.L. Neuropathologic changes in Alzheimer's disease. *J. Clin. Psychiatry* **2003**, *64*, 7–10.
59. Piaceri, I.; Nacmias, B.; Sorbi, S. Genetics of familial and sporadic Alzheimer's disease. *Front. Biosci.* **2015**, *5*, 167–177.
60. Zhao, Y.; Zhao, B. Oxidative stress and the pathogenesis of Alzheimer's disease. *Oxid. Med. Cell Longev.* **2013**, *2013*, 316523. [CrossRef]
61. Lyros, E.; Bakogiannis, C.; Liu, Y.; Fassbender, K. Molecular links between endothelial dysfunction and neurodegeneration in Alzheimer's disease. *Curr. Alzheimer Res.* **2014**, *11*, 18–26. [CrossRef] [PubMed]
62. Revesz, T.; Ghiso, J.; Lashley, T.; Plant, G.; Rostagno, A.; Frangione, B.; Holton, J.L. Cerebral amyloid angiopathies: A pathologic, biochemical, and genetic view. *J. Neuropathol. Exp. Neurol.* **2003**, *62*, 885–898. [CrossRef] [PubMed]
63. Patel, N.S.; Mathura, V.S.; Bachmeier, C.; Beaulieu-Abdelahad, D.; Laporte, V.; Weeks, O.; Mullan, M.; Paris, D. Alzheimer's beta-amyloid peptide blocks vascular endothelial growth factor mediated signaling via direct interaction with VEGFR-2. *J. Neurochem.* **2010**, *112*, 66–76. [CrossRef] [PubMed]

64. Solito, R.; Corti, F.; Fossati, S.; Mezhericher, E.; Donnini, S.; Ghiso, J.; Giachetti, A.; Rostagno, A.; Ziche, M. Dutch and Arctic mutant peptides of beta amyloid(1-40) differentially affect the FGF-2 pathway in brain endothelium. *Exp. Cell Res.* **2009**, *315*, 385–395. [CrossRef]
65. Siegel, S.J.; Bieschke, J.; Powers, E.T.; Kelly, J.W. The oxidative stress metabolite 4-hydroxynonenal promotes Alzheimer protofibril formation. *Biochemistry* **2007**, *46*, 1503–1510. [CrossRef]
66. Donnini, S.; Solito, R.; Cetti, E.; Corti, F.; Giachetti, A.; Carra, S.; Beltrame, M.; Cotelli, F.; Ziche, M. Abeta peptides accelerate the senescence of endothelial cells in vitro and in vivo, impairing angiogenesis. *FASEB J.* **2010**, *24*, 2385–2395. [CrossRef]
67. D'souza, Y.; Elharram, A.; Soon-Shiong, R.; Andrew, R.D.; Bennett, B.M. Characterization of Aldh2$^{-/-}$ mice as an age-related model of cognitive impairment and Alzheimer's disease. *Mol. Brain* **2015**, *8*, 27. [CrossRef]

© 2020 by the authors. Licensee MDPI, Basel, Switzerland. This article is an open access article distributed under the terms and conditions of the Creative Commons Attribution (CC BY) license (http://creativecommons.org/licenses/by/4.0/).

MDPI\
St. Alban-Anlage 66\
4052 Basel\
Switzerland\
Tel. +41 61 683 77 34\
Fax +41 61 302 89 18\
www.mdpi.com

Biomedicines Editorial Office\
E-mail: biomedicines@mdpi.com\
www.mdpi.com/journal/biomedicines